Understanding Biology
for Advanced Level

Second Edition

Glenn Toole
Vice Principal, Pendleton Sixth Form College, Salford

Susan Toole
Team Leader and Examiner for A-level Biology (London Board)
Head of Biology, The Hulme Grammar School for Girls, Oldham

Stanley Thornes (Publishers) Ltd

Originally published in 1987 by Hutchinson Education
ISBN 0 09 170051 5
Reprinted 1988, 1989 and 1990 by Stanley Thornes (Publishers) Ltd
ISBN 0 7487 0288 1

Second edition 1991 by
Stanley Thornes (Publishers) Ltd
Old Station Drive
Leckhampton
CHELTENHAM GL53 0DN

Reprinted 1992

British Library Cataloguing in Publication Data

Toole, Glenn
    Understanding biology for advanced level.
    I. Title   II. Toole, Susan   Second edition
    574

    ISBN 0 7487 0539 2

Printed and bound in Great Britain by
Butler & Tanner Ltd, Frome and London.

Contents

PART II THE CONTINUITY OF LIFE

PART III ENERGETICS

Chapter 21   Energy and organisms

Chapter 22   Enzymes

PART IV TRANSPORT AND EXCHANGE MECHANISMS

PART V COORDINATION, RESPONSE AND CONTROL

Preface to the Second Edition

Such is the pace of new scientific development that some aspects of a textbook are out-of-date even before it reaches the bookshelves. Regular revision of a text such as *Understanding Biology for Advanced Level* is therefore essential if it is to be of continued value to students and is the main reason for this new edition.

While an overall attempt has been made to keep the layout the same the need to incorporate new developments at the relevant point in the text was felt to be more important than maintaining the original page numbering. We appreciate the problems this creates for groups using the book alongside the first edition but hope that this disadvantage is outweighed by the greater coherence of the text which this arrangement allows.

The area of biological development with the greatest momentum has been that of Biotechnology. This is reflected in the addition of a new chapter devoted solely to this topic, incorporating such aspects as the growth of microorganisms, asepsis, fermentation techniques and their applications, cell, tissue and organ culture and monoclonal antibodies. Throughout the book there is an increased emphasis on the social aspects of the subject with new sections on heart disease, AIDS, the causes of infertility, food poisoning and food additives. More technological applications of biology have been included with sections on recombinant DNA technology, gene tracking, gene fingerprinting and kidney dialysis. The section on pollution has been expanded to include more detail on ozone depletion, the greenhouse effect and conservation. The chapters on classification have been fully revised and include new sections on keys and retroviruses. All the examination questions at the end of each chapter have been updated with older ones being replaced by more recent examples.

All nomenclature has been revised in accordance with the recent recommendations made by the Institute of Biology and Association for Science Education and a full review of all AS and A-level core examination syllabuses has been undertaken to ensure that the topics and examples are consistent with their requirements. Such has been the scale of the revision that few pages have escaped at least minor revision. With little from the first edition being excluded the book is now expanded in both its size and scope. Apart from being an up-to-date text for the AS or A-level biologist it is hoped it will be equally valuable to those taking Human and Social Biology syllabuses at these levels.

Compiling this second edition has been a complicated undertaking and we owe a particular debt of gratitude to Stephanie Brown of Pendleton College for her helpful comments on the biotechnology chapter and to Rita Chester of Hulme Grammar School for her useful observations. We are especially grateful for the encouragement of our editor, Adrian Wheaton and all others at Stanley Thornes who helped make its production as painless as possible.

Acknowledgements

We owe a considerable debt to many colleagues, students and friends without whose constructive criticisms this book would never have been completed in its present form.

Our particular thanks must go to Mr Martin Davis from the biology department of Woodhouse Sixth Form College who took on the marathon task of reading the entire manuscript. His comments and helpful advice were invaluable to us. Others at Woodhouse to whom we owe a debt of gratitude include Mr John Oakes, Mr David Hyde and Dr Tony Wilson, all of whom gave advice on the chemical aspects of the book and Dr Yvonne Bernstein who assisted with the mathematical sections.

Many students played an important role in reading and commenting on the manuscript, giving us much valued advice about its usefulness from their point of view. They are too numerous to name individually but Gaye Summers and Adwoa Oduro-Yeboah deserve particular mention for the time and effort they gave the task. Special thanks to Philippe Versluysen for the excellent photographs he produced for the book.

The following Examination Boards have kindly granted permission for the use of questions from recent examination papers: Associated Examining Board, Joint Matriculation Board, University of London Schools Examination Council, Northern Ireland Examination Board, Oxford Local Examinations, Oxford and Cambridge Joint Examinations Board, Southern Universities Joint Board, University of Cambridge Local Examinations Syndicate, and the Welsh Joint Education Committee.

Acknowledgements are due to the following for permission to reproduce photographs:

Heather Angel p. 120 (top), p. 135. p. 232 (top), p. 363; CEGB p. 299; Camera Press p. 595; Bruce Coleman p. 377, p. 484, p. 589, p. 608; Mary Evans Picture Library p. 180, p. 228; Express Foods p. 636 (bottom); Bob Gibbons p. 260 (top right), p. 355; Philip Harris Biological Ltd p. 223, p. 450; H. D. Hudson p. 68; Nestlé Company Ltd p. 638; Oxford Scientific Films p. 120 (bottom 3), p. 234, p. 361, p. 396 (bottom), p. 407, p. 539, p. 589; Panos Pictures p. 641; Picturepoint p. 402; RSPB p. 425; Science Photo Library p. 163, p. 343, p. 392, p. 396 (top), p. 397, p. 426, p. 437, p. 488, p. 504, p. 576; Scottish and Newcastle Beer Production Ltd p. 636 (top); SmithKline Beecham p. 640; Dr J. M. Squire, Imperial College, London p. 610; Thames Water Utilities p. 642; Phillipe Versluysen p. 65 (bottom right), p. 261, p. 348, p. 350, p. 352; Sylvan H. Wittwer p. 617.

Biophoto Associates provided all other photographs not listed above, and their help is gratefully acknowledged.

1 *Introduction to biology*

1.1 Biology in context

Biology is the study of life, and has in the past often been considered as a separate and independent subject. As our knowledge and understanding of science has improved, it has become clear that biology, physics and chemistry are so closely related that the study of one subject inevitably involves many aspects of the others. The overlap between the sciences is so great that whole new disciplines such as biophysics and biochemistry have arisen.

The emphasis in biology was at one time largely descriptive. This was reflected in branches of the subject such as **anatomy** and **morphology**, where the structure rather than the function was considered important. The more modern approach is to deal with the function, not in isolation but in relation to the structure. Disciplines such as **physiology** have become increasingly important. Biology could, at one time, be neatly divided into **botany**, the study of plants, and **zoology**, the study of animals. Clear differences could be drawn between them. The discovery of bacteria, which could not be easily placed in either group, was overcome by the creation of a new category – **microbiology**. The discovery of **viruses** was more perplexing and even questioned the fundamental differences between living and non-living things.

With more technical advances, biology moved into the study of the structure of cells – **cytology** – and the chemicals from which they were made – **molecular biology**. **Genetics** and **heredity** expanded as the purely descriptive work of Mendel was given a biochemical explanation with the elucidation of the structure of DNA by Watson and Crick. Much of this work indicated that, far from being easily distinguished, plants and animals were fundamentally the same. Although these two distinct groups are still recognized, some rearrangement of the traditional classification groups has become necessary (Chapter 6) to take account of new discoveries.

At the same time as the electron microscope and other technical discoveries were expanding knowledge at one end of the size scale, so a new branch of biology, called **ecology**, was developed in order to try to understand the biosphere as a whole. Ecology is the study of the inter-relationships of organisms with each other, and with their physical and climatic environment. Thus biology was expanded to incorporate aspects of geography and geology. The advent of space travel has even taken biology beyond earthly limits. Not only does the study of biology now involve the complete size range, it also spans an immense time scale. The reaction times involved in some biophysical processes take as little as 10^{-13} s while the study of fossils, **palaeontology**, and biological evolution cover the history of the world from its origins around 4500 million years ago. Almost all aspects of biology involve the measurement and the collection of data. An understanding of mathematics, especially statistics, has therefore become a useful skill.

With such a broad field of study, not to mention the variety produced by millions of different living species, biology presents a daunting prospect to the potential A-level candidate. To add to his or her apprehensions, the subject is changing at an alarming rate. New discoveries expand our knowledge of some topics and cause traditional ideas associated with others to be questioned. No one book could adequately cover all aspects of biology. It is essential that the A-level candidate reads as widely and as variedly as possible from all manner of sources. This book attempts to cover all aspects of the A-level biology syllabus as required by the ten A-level examination boards in the United Kingdom. As far as possible an integrated approach has been adopted to conform with modern thinking in biology and reflect the trend in present examination syllabuses. It does not claim to, nor could it, be exhaustive, but it attempts to give a sound foundation, good insight and adequate detail for A-level study. It is hoped at the same time to engender an interest and appreciation of the subject which will provoke the reader to further research.

1.2   The definition of life

It is not possible to give an absolute definition of life. It may seem strange to say that biology is the study of life, and then admit we do not know exactly what life is. Part of the problem is that the diversity of organisms is so great that it is difficult to find aspects which each has in common. While almost all organisms readily conform to one or other definition of life, one group, the viruses, exhibit features normally associated with both living and non-living things. On the other hand, reproduction and other processes cannot be performed by viruses independently; they require the assistance of living cells to carry them out. Their structure is not cellular and they can be crystallized – not features associated with living organisms.

What then is life? One description is that living organisms carry out seven characteristic processes.

Respiration – The release of energy from the breakdown of substances within the body. It often, although not always, involves oxygen which must therefore be obtained by the organism.

Nutrition – The acquisition of nutritive substances from the environment. They are used to build up the organism and provide energy for its various activities.

Excretion – The removal of unwanted, and often toxic, substances from the organism.

Movement – The ability to displace in space all or part of an organism. Most animals can move themselves from place to place (**locomotion**). Plants, however, move only parts of themselves in response to stimuli.

Sensitivity – The ability of organisms to respond to changes in their environment and within themselves.

Growth (and repair) – A quantitative increase in size during development. It involves not only an increase in size, but in numbers of cells and their differentiation into various forms.

Reproduction – The production of new individuals similar to the parents.

While this remains the usual means by which living organisms are recognized, there is a second method. This concerns the way living and non-living material uses energy. All material tries to attain its lowest energy state, i.e. it tends to lose energy. For example, an object thrown in the air falls to earth, hot bodies cool etc. A highly structured and organized system inevitably possesses much energy. A house with its bricks carefully organized one on top of the other, its tiles specifically positioned to form the roof, is highly ordered and therefore possesses much energy. Such a system, left untended, will lose its energy and become random and disordered in its structure. The wind may remove a tile and water will seep in and rot the wood supporting the roof which will, in turn, collapse. The weather and climbing plants may erode the mortar between the bricks until the house is reduced to a random, disordered pile of rubble. Such random disorder is termed **entropy**. A system which is disordered is said to have a high entropy, an ordered one has a low entropy. An ordered system has more useful energy than a disordered one. This useful energy of a system is termed **free energy**. Free energy and entropy are inversely proportional; high entropy means low free energy and vice versa. Non-living systems all tend to high entropy. Living organisms however are highly ordered and therefore have low entropy and much free energy. This they achieve by continually taking in energy, e.g. as food, which they use to maintain an orderly structure and counteract the natural tendency to become disordered. In a similar way the owner of the house can only maintain its ordered state by the input of energy. He must replace the loose tiles, treat and paint exposed wood, cut down the climbing plants and repoint the brickwork if the building is not to revert to a state of high entropy and collapse. In short, non-living systems have high entropy and little free energy whereas living organisms maintain a state of low entropy with much free energy.

Transverse section through liver tubules to show cellular organization (opposite)

Part I
Levels of organization

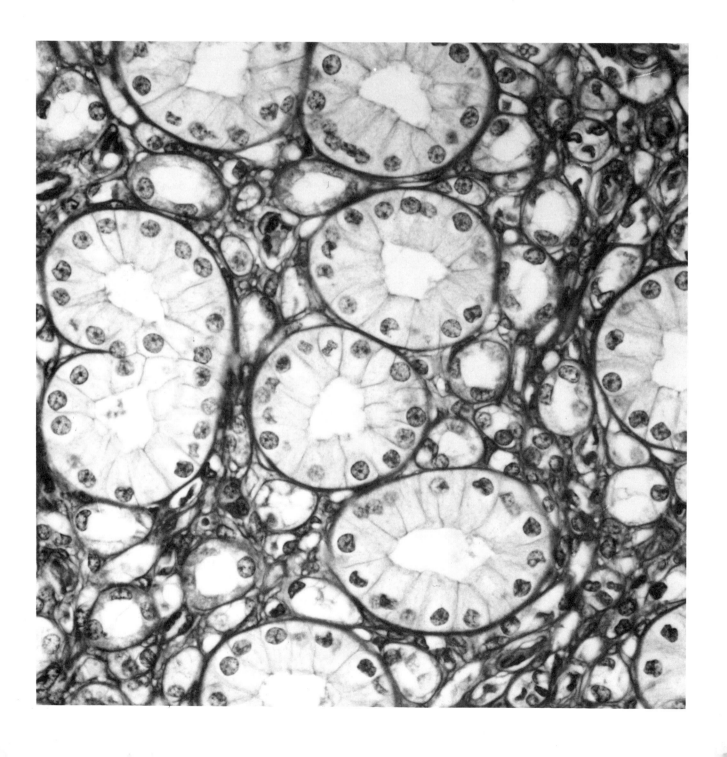

2 Size and complexity

TABLE 2.1 Metric units

Units of size

1 kilometre	(km) =	1000 (10^3) metres
1 metre	(m)	
1 centimetre	(cm) =	1/100 (10^{-2}) metre
1 millimetre	(mm) =	1/1000 (10^{-3}) metre
1 micrometre (micron)	(μm) =	1/1 000 000 (10^{-6}) metre
1 nanometre	(nm) =	1/1 000 000 000 (10^{-9}) metre
1 picometre	(pm) =	1/1 000 000 000 000 (10^{-12}) metre

Biology covers a wide field of information over a considerable size range. On the one hand it involves the movement of electrons in photosynthesis and on the other the migrations of individuals around the earth. Within this range it is possible to recognize seven levels of organization, each of which forms the basis of the next. The most fundamental unit is the **atom**; atoms group to form **molecules**, which, in turn, may be organized into **cells**. Cells are grouped into **tissues** which collectively form **organs**, which form **organisms**. A group of organisms of a single species may form a **population**.

2.1 Atomic organization

Atoms are the smallest unit of a chemical element which can exist independently. They comprise a nucleus which contains positively charged particles called **protons**, the number of which is referred to as the **atomic number**. For each proton there is a particle of equal negative charge called an **electron**, so the atom has no overall charge. The electrons are not within the nucleus, but orbit in fixed quantum shells around it. There is a fixed limit to the number of electrons in any one shell. There may be up to seven such shells each with its

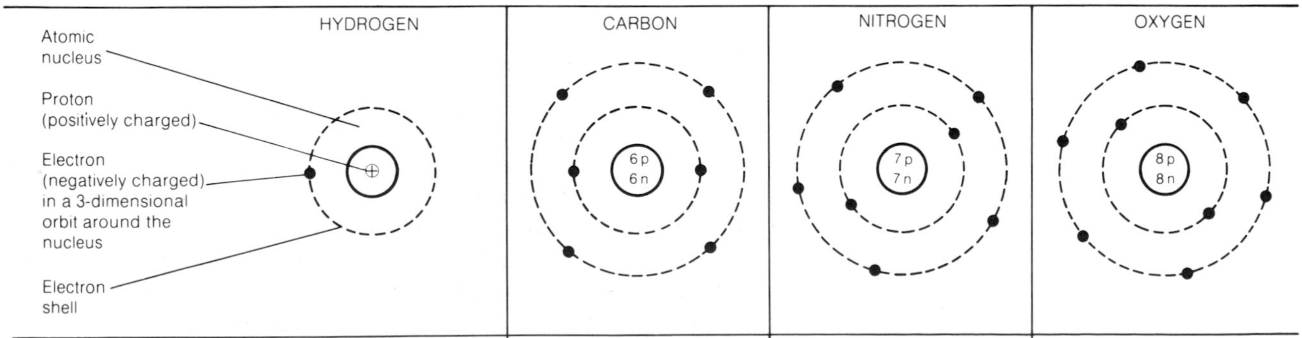

	HYDROGEN	CARBON	NITROGEN	OXYGEN
Atomic number	1	6	7	8
Number of protons	1	6	7	8
Number of neutrons	0	6	7	8
Relative atomic mass	1	12	14	16
Number of electrons				
First quantum shell	1	2	2	2
Second quantum shell	—	4	5	6
Total	1	6	7	8

Fig. 2.1 Atomic structure of four commonly occurring biological elements

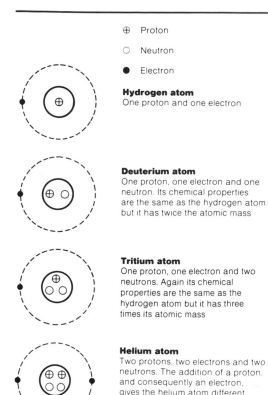

⊕   Proton

○   Neutron

●   Electron

Hydrogen atom
One proton and one electron

Deuterium atom
One proton, one electron and one neutron. Its chemical properties are the same as the hydrogen atom but it has twice the atomic mass

Tritium atom
One proton, one electron and two neutrons. Again its chemical properties are the same as the hydrogen atom but it has three times its atomic mass

Helium atom
Two protons, two electrons and two neutrons. The addition of a proton, and consequently an electron, gives the helium atom different chemical properties from those of hydrogen

Fig. 2.2 Atomic structure of the atoms of hydrogen, deuterium, tritium and helium

own energy level; electrons in the shells nearest the nucleus have the least energy. The addition of energy, e.g. in the form of heat or light, may promote an electron to a higher energy level within a shell. Such an electron almost immediately returns to its original level, releasing its newly absorbed energy as it does so. This electron movement is important biologically in processes such as photosynthesis (Chapter 23).

The nucleus of the atom also contains particles called **neutrons** which have no charge. Protons and neutrons contribute to the mass of an atom, but electrons have such a comparatively small mass that their contribution is negligible. However, the number of electrons determines the chemical properties of an atom. (See Fig. 2.1.)

2.1.1 Ions

As we have seen, atoms do not have any overall charge because the number of protons is always the same as the number of electrons and both have equal, but opposite, charges. If an atom loses or gains electrons it becomes an **ion**. The addition of electrons produces a negative ion while the loss of electrons gives rise to a positive ion. The loss of an electron is called **oxidation**, while the gain of an electron is called **reduction**. The atom losing an electron is said to be oxidized, while that gaining an electron is said to be reduced. The loss of an electron from a hydrogen atom, for instance, would leave a hydrogen ion, comprising just a single proton. Having an overall positive charge it is written as H^+. Where an atom, e.g. calcium, loses two electrons its overall charge is more positive and it is written Ca^{2+}. The process is similar where atoms gain electrons, except that the overall charge is negative, e.g. Cl^-. Ions may comprise more than one type of atom, e.g. the sulphate ion is formed from one sulphur and four oxygen atoms, with the addition of two electrons − $SO_4{}^{2-}$.

2.1.2 Isotopes

The properties of an element are determined by the number of protons and hence electrons it possesses. If protons (positively charged) are added to an element, then an equivalent number of electrons (negatively charged) must be added to maintain an overall neutral charge. The properties of the element would then change − indeed it now becomes a new element. For example, it can be seen from Fig. 2.1 that the addition of one proton, one electron and one neutron to the carbon atom, transforms it into a nitrogen atom.

If, however, a neutron (not charged) is added, there is no need for an additional electron and so its properties remain the same. As neutrons have mass, the element is heavier. Elements, which have the same chemical properties as the normal element, but have a different mass, are called **isotopes**. Hydrogen normally comprises one proton and one electron and consequently has an atomic mass of one. The addition of a neutron doubles the atomic mass to two, without altering the element's chemical properties. This isotope is called **deuterium**. Similarly, the addition of a further neutron forms the isotope **tritium**, which has an atomic mass of three.

Isotopes can be traced by various means, even when incorporated in living matter. This makes them exceedingly useful in tracing the route of certain elements in a variety of biological processes.

⊕ Proton

● Electron

Hydrogen atom – H
One proton and one electron. No overall charge. The electron shell is not full and the atom is therefore unstable

Hydrogen ion – H⁺
One proton only, leaving an overall positive charge

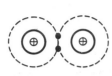

Hydrogen molecule – H₂
Two protons and two electrons. No overall charge. The electron from each atom is shared and so both atoms effectively have a full shell containing two electrons. The molecule is therefore relatively stable

Fig. 2.3 Atomic structure of a hydrogen atom, a hydrogen ion and a hydrogen molecule

2.2 Molecular organization

We have seen that the electron shells around an atom may each contain a maximum number of electrons. The shell nearest the nucleus may possess a maximum of two electrons and the next shell a maximum of eight. An atom is most stable, i.e. least reactive, when its outer electron shell contains the maximum possible number of electrons. For example, helium, with a full complement of two electrons in its outer shell, is inert. In a hydrogen atom, the electron shell has a single electron and so the atom is unstable. If two hydrogen atoms share their electrons they form a hydrogen **molecule**, which is more stable. The two atoms are effectively combined and the molecule is written as H_2. The sharing of electrons in order to produce stable molecules is called **covalent bonding**.

The oxygen atom contains eight protons and eight neutrons in the nucleus with eight orbiting electrons. The inner quantum shell contains its maximum of two electrons, leaving six electrons in the second shell (Fig. 2.1). As this second shell may contain up to eight electrons, it requires two electrons to complete the shell and become stable. It may therefore combine with two hydrogen atoms by sharing electrons to form a water molecule (Fig. 2.4). In this way the outer shells of the oxygen atom and both hydrogen atoms are completed and a relatively stable molecule is formed.

Carbon with its six electrons (Fig. 2.1) has an inner shell containing two, leaving four in the outer shell. It requires four more electrons to fill this shell. It may therefore combine with four hydrogen atoms each of which shares its single electron. This molecule is called methane CH_4 (Fig. 2.4). It may also combine with two oxygen atoms, each of which shares two electrons. This molecule is carbon dioxide (Fig. 2.4).

When an atom, e.g. hydrogen, requires one electron to complete its outer shell it is said to have a **combining power (valency)** of one. Oxygen, which requires two electrons to complete its outer shell, has a combining power of two. Likewise nitrogen has a combining power of three and carbon of four.

When two atoms share a single electron, the bond is referred to as a **single bond** and is written with a single line, e.g. the hydrogen molecule is H—H and water may be represented as H—O—H. If two atoms share two electrons a **double bond** is formed. It is represented by a double line, e.g. carbon dioxide may be written as $O=C=O$. To form stable molecules, hydrogen must therefore have a single bond; oxygen two bonds (either two singles or one double); nitrogen must have three bonds (either three singles, or one single and one double); and carbon must have four bonds. It should now be apparent that these four atoms can combine in a number of different ways to form a variety of molecules. This partly explains the abundance of these elements in living organisms although some are relatively rare in the earth's crust (Table 2.2).

Carbon in particular can be seen to be almost 200 times more abundant in living organisms than in the earth's crust. Why should this be so? In the first place, carbon with its combining power of four can form molecules with a wide variety of other elements such as hydrogen, oxygen, nitrogen, sulphur, phosphorus and chlorine. This versatility allows great diversity in living organisms. More importantly, carbon can form long chains linked by single, double and triple bonds. These chains may be thousands of carbon atoms long. Such large molecules are essential to living organisms, not least

TABLE 2.2 **Relative abundance by weight of elements in humans compared to the earth's crust**

Element	Human	Earth's crust
Oxygen	63.0	46.5
Carbon	19.5	0.1
Hydrogen	9.5	0.2
Nitrogen	5.0	0.0001
Phosphorus	0.5	1.5

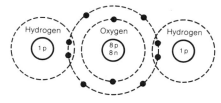

Water molecule (H₂O)
The oxygen atom shares 2 electrons with each hydrogen atom. Both molecules thereby complete their outer shell – the hydrogen atom with 2 electrons, the oxygen atom with 8

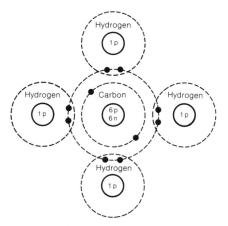

Methane molecule (CH₄)
The carbon atom shares 2 electrons with each hydrogen atom. Each hydrogen atom thus completes its outer shell with 2 electrons, while the carbon atom completes its outer shell with 8

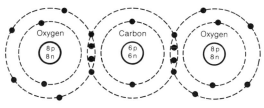

Carbon dioxide molecule (CO₂)
The carbon atom shares 4 electrons with each oxygen atom. All three atoms thereby complete their outer shells with 8 electrons

Fig. 2.4 Atomic models of the molecules of water, methane and carbon dioxide

as structural components. Furthermore, these chains have great stability – another essential feature. Carbon compounds may also form rings. These rings and chains may be combined with each other to give giant molecules of almost infinite variety. Examples of the size, diversity and complexity of carbon molecules can be found among the three major groups of biological compounds: carbohydrates, fats and proteins.

These are discussed in more detail in Chapter 3.

2.2.1 Ionic bonding

In addition to forming covalent bonds through the sharing of electrons, atoms may stabilize themselves by losing or gaining electrons to form ions. The loss of an electron (oxidation) leaves the atom positively charged (oxidized). The gain of an electron (reduction) leaves the atom negatively charged (reduced). Oppositely charged atoms attract one another forming **ionic bonds**. Sodium, for example, tends to lose an electron forming a Na^+ ion; chlorine tends to gain an electron forming a Cl^- ion. These two oppositely charged ions form ionic bonds and form sodium chloride (common salt).

2.2.2 Hydrogen bonds

The electrons in a molecule do not distribute themselves evenly but tend to group at one position. This region will consequently be more negative than the rest of the molecule. The molecule is said to be **polarized**. The negative region of such a molecule will be attracted to the positive region of a similarly polarized molecule. A weak electrostatic bond between the two is formed. In biological systems this type of bond is frequently a hydrogen bond. These bonds are weak individually, but collectively form important forces which alter the physical properties of molecules. Water forms hydrogen bonds which, as we shall see in Chapter 33, significantly affect its properties and hence its biological importance.

2.3 Cellular organization

In 1665, Robert Hooke, using a compound microscope, discovered that cork was composed of numerous small units. He called these units, **cells**. In the years which followed, Hooke and other researchers discovered that many other types of material were similarly composed of cells. By 1838, the amount of plant material shown to be composed of cells convinced Matthias Schleiden, a German botanist, that all plants were made up of cells. The following year, Theodor Schwann reached the same conclusion about the organization of animals. Their joint findings became known as the **cell theory**. It was of considerable biological significance as it suggested a common denominator for all living matter and so unified the nature of organisms. The theory makes the cell the fundamental unit of structure and function in living organisms. Hooke had originally thought the cell to be hollow, and that the wall represented the living portion. It soon became clear that cells were far from hollow. With the development of better light microscopes, first the nucleus and then organelles such as the chloroplasts became visible. One hundred years after Schleiden and Schwann put forward the cell theory, the development of the **electron microscope** revolutionized our understanding of cell structure. With its ability to magnify up to 500 times more than the light microscope, the electron microscope revealed the fine structure of cells including many new organelles. This detail is called the **ultrastructure** of the cell. The complexity of cellular structure so revealed led to the emergence of a new field of biology, **cytology** – the study of cell ultrastructure. This shows that while organisms are very diverse in their structures and cells vary considerably in size and shape, there is nevertheless a remarkable similarity in their basic structure and organization. This structure and organization is studied in Chapter 4.

2.4 Colonial organization

The first colonies may have arisen when individual unicells failed to separate after cell division. Within colonies each cell is capable of carrying out all the essential life processes. Indeed, if separated from the colony, any cell is capable of surviving independently. The only advantage of a colonial grouping is that the size of the unit probably deters some predators and thus increases the group's survival prospects.

If one cell in a colony should lose the ability to carry out a vital process, it could only survive by relying on other cells in the colony to perform the process on its behalf. The loss of one function, however, might permit the cell to perform one or other of its functions more efficiently, because the energy and resources required by the missing function could be directed towards the remaining ones. In this way, the individual cells within a colony could have become different from one another in both structure and function, a process known as **differentiation**. Further changes of this type would finally result in cells performing a single function. This is known as **specialization**. Clearly specialization must be organized in such a way that all essential functions are still performed by the colony as a whole. With increasing specialization, and the consequent loss of more and more functions, any cell becomes increasingly dependent on others in the colony for its survival. This **interdependence** of cells must be highly organized. Groups of cells must be coordinated so that the colony carries out its activities efficiently. Such coordination between the different cells is called **integration**. Once the cells become so dependent on each other that they are no longer capable of surviving independently, then the structure is no longer a colony but a **multicellular organism**.

2.5 Tissue organization

A tissue is a group of similar cells, along with any intercellular substance, which performs a particular function. Some cells, e.g. unicellular protozoans and algae, perform all functions which are essential to life. It is impossible for such cells to be efficient at all functions, because each function requires a different type of cellular organization. Whereas one function might require the cell to be long and thin, another might require it to be spherical. One function might require many mitochondria, another, very few. Acid conditions might suit one activity but not another. No one cell can possibly provide the optimum conditions for all activities. For this reason, cells are specialized to perform one, or at most a few, functions. To increase efficiency, cells performing the same functions are grouped together into a tissue. The study of tissues is called **histology** and the variety of plant and animal tissues is discussed in Chapter 5. Some organisms, e.g. cnidarians (coelenterates), are at the tissue level of organization. Their physiological activities are performed by tissues rather than organs.

2.6 Organ level of organization

An organ is a structural and functional unit of a plant or animal. It comprises a number of tissues which are coordinated to perform a variety of functions, although one major function often predominates. The majority of plants and animals are composed of organs. Most organs do not function independently but in groups called **organ systems**. A typical organ system is the digestive system which comprises organs such as the stomach, duodenum, ileum, liver and pancreas. Certain organs may belong to more than one system. The pancreas, for example, forms part of the **endocrine (hormone)** system as well as the digestive system, because it produces the hormones insulin and glucagon, as well as the digestive enzymes amylase and trypsinogen.

2.7 Social level of organization

A **population** is a number of individuals of the same species which occupy a particular area at the same time. In itself, a population is not a level of organization as no organization exists between the individual members. In some species, however, the individuals do exhibit some organization in which they cooperate for their mutual benefit. Such a population is more accurately termed a **society**. It differs from a colony (although the term is often used) in that the individuals are not physically connected to one another, but totally separate. As with a colony, the individuals can survive independently of others in the society, although usually somewhat less successfully. Unlike most colonies, there is considerable coordination between the society members and communication forms an integral part of their organization. Societies may exist simply because there is safety in numbers, e.g. schools of fish. They may enable more successful hunting, as in wolves, or aid the successful rearing of young, as in baboons. In insects, however, the degree of organization is considerable. There is **division of labour** which leads to differentiation of individuals in order to perform specialized functions. In a bee society for instance, the queen is the only fertile female and has a purely reproductive role. The drones (males) also function reproductively while the workers (sterile females) perform a variety of tasks such as collecting food, feeding the larvae and guarding and cleaning the hive. Complex societies can readily be compared to an organism with its organs each specialized for a major function. Some account of the organization of a bee colony is given in Section 38.7.5.

2.8 The advantages and disadvantages of large size

The earliest forms of life were unicellular organisms. These unicellular organisms alone occupied the earth for most of its history, multicellular organisms having evolved only in relatively recent times. Unicellular organisms are limited in the size they can attain because there is a limit to the volume of cytoplasm over which a nucleus can exert its influence. Any increase in the size of organisms therefore necessitated the development of the multicellular condition. When one considers the number and range of multicellular organisms which presently exist, it is obvious that increased size confers some selective advantage. What then are these advantages?

1. The larger an organism is, the less likely it is to be eaten by another.

2. A larger animal is generally better able to travel at speed. This means it is more likely to capture its prey or escape from a predator.

3. Large size gives a plant a competitive advantage. Being larger, it is better able to obtain more light for photosynthesis than its smaller neighbours.

4. A large organism has a relatively small surface area to volume ratio. This reduces the rate at which water and heat are lost and makes homeostatic control easier. It is therefore a particular advantage to terrestrial endotherms. While unicellular organisms are limited to moist environments, multicellular ones are able to survive in a wide range of habitats.

5. Being multicellular, large organisms are able to have cells specialized to particular functions. This division of labour makes the organisms more efficient.

An increase in size has its disadvantages too.

1. Support is a problem, more especially for terrestrial organisms, where the medium of air provides little assistance.

2. A large organism will usually have a greater amount of respiring tissue and therefore require more food to maintain itself in a proper condition.

3. By the same reasoning, larger organisms have more wastes to dispose of.

4. The smaller surface area to volume ratio of larger organisms does not allow the normal body surface to provide adequate supplies of nutrients and respiratory gases, nor can it adequately remove wastes. The development of alimentary canals, respiratory surfaces and excretory organs in order to provide the necessary surface area for each of these functions is only effective if an efficient internal transport system exists. A blood and circulatory system is therefore essential in these large organisms.

5. A large organism is more conspicuous to a predator and has more difficulty in hiding from it.

6. In plants especially, extreme large size makes them more vulnerable to damage from wind.

3 _Molecular organization_

3.1 Inorganic ions

Water is the most important inorganic molecule in biology and its chemical structure and properties are described in Chapter 33. Dissolved in the water within living organisms are a large number of inorganic ions. Typically they constitute about 1% of an organism by weight, but they are nonetheless essential. They are divided into two groups: the **macronutrients** or **major elements** which are needed in very small quantities, and the **micronutrients** or **trace elements** which are needed in minute amounts (a few parts per million). Although the elements mostly fall into the same category for plants and animals, there are a few exceptions. Chlorine, for example, is a major element in animals but a trace element in plants. In addition to the essential elements listed in Table 3.1 (on this and the next page), some organisms also have specific requirements such as vanadium, chromium and silicon.

TABLE 3.1 **Inorganic ions and their functions in plants and animals**

Macronutrients/ main elements	Functions	Notes
Nitrate NO_3^- Ammonium NH_4^+	Nitrogen is a component of amino acids, proteins, vitamins, coenzymes, nucleotides and chlorophyll. Some hormones contain nitrogen, e.g. auxins in plants and insulin in animals	A deficiency of nitrogen in plants causes chlorosis (yellowing of leaves) and stunted growth
Phosphate PO_4^{3-} Ortho-phosphate $H_2PO_4^-$	A component of nucleotides, ATP and some proteins. Used in the phosphorylation of sugars in respiration. A major constituent of bone and teeth. A component of cell membranes in the form of phospholipids	Deficiency of phosphates in plants leads to stunted growth, especially of roots, and the formation of dull, dark green leaves. In animals, deficiency can result in a form of bone malformation called rickets
Sulphate SO_4^{2-}	Sulphur is a component of some proteins and certain coenzymes, e.g. acetyl coenzyme A	Sulphur forms important bridges between the polypeptide chains of some proteins, giving them their tertiary structure. A deficiency in plants causes chlorosis and poor root development
Potassium K^+	Helps to maintain the electrical, osmotic and anion/cation balance across cell membranes. Assists active transport of certain materials across the cell membrane. Necessary for protein synthesis and is a co-factor in photosynthesis and respiration. A constituent of sap vacuoles in plants and so helps to maintain turgidity	Potassium plays an important role in the transmission of nerve impulses. A deficiency in plants leads to yellow-edged leaves and premature death
Calcium Ca^{2+}	In plants, calcium pectate is a major component of the middle lamella of cell walls and is therefore necessary for their proper development. It also aids the translocation of carbohydrates and amino acids. In animals, it is the main constituent of bones, teeth and shells. Needed for the clotting of blood and the contraction of muscle	In plants, deficiency causes the death of growing points and hence stunted growth. In animals, deficiency leads to rickets and delay in the clotting of blood

cont.

Table 3.1 *cont.*

Macronutrients/ main elements	Functions	Notes
Sodium Na$^+$	Helps to maintain the electrical, osmotic and anion/cation balance across cell membranes. Assists active transport of certain materials across the cell membrane. A constituent of the sap vacuole in plants and so helps maintain turgidity	In animals, it is necessary for the functioning of the kidney, nerves and muscles; deficiency may cause muscular cramps. Sodium is so common in soils that deficiency in plants is rare. Sodium ions have much the same function as potassium ions and may be exchanged for them
Chlorine Cl$^-$	Helps to maintain the electrical, osmotic and anion/cation balance across cell membranes. Needed for the formation of hydrochloric acid in gastric juice. Assists in the transport of carbon dioxide by blood (chloride shift)	In animals, deficiency may cause muscular cramps. Its widespread availability in soils makes deficiency in plants practically unknown
Magnesium Mg^{2+}	A constituent of chlorophyll. An activator for some enzymes, e.g. ATPase. A component of bone and teeth	Deficiency in plants leads to chlorosis
Iron Fe^{2+} or Fe^{3+}	A constituent of electron carriers, e.g. cytochromes, needed in respiration and photosynthesis. A constituent of certain enzymes, e.g. dehydrogenases, decarboxylases, peroxidases and catalase. Required in the synthesis of chlorophyll. Forms part of the haem group in respiratory pigments such as haemoglobin, haemoerythrin, myoglobin and chlorocruorin	Deficiency in plants leads to chlorosis and in animals to anaemia
Micro-nutrients/ trace elements		
Manganese Mn^{2+}	An activator of certain enzymes e.g. phosphatases. A growth factor in bone development	Deficiency in plants produces leaves mottled with grey and in animals, bone deformations
Copper Cu^{2+}	A constituent of some enzymes, e.g. cytochrome oxidase and tyrosinase. A component of the respiratory pigment haemocyanin	Deficiency in plants causes young shoots to die back at an early stage
Iodine I$^-$	A constituent of the hormone thyroxine, which controls metabolism in animals	Iodine is not required by higher plants. Deficiency in humans causes cretinism in children and goitre in adults; in some other vertebrates it is essential for metamorphic changes
Cobalt Co^{2+}	Constituent of vitamin B$_{12}$, which is important in the synthesis of RNA, nucleoprotein and red blood cells	Deficiency in animals causes pernicious anaemia
Zinc Zn^{2+}	An activator of certain enzymes, e.g. carbonic anhydrase. Required in plants for leaf formation, the synthesis of indole acetic acid (auxin) and anaerobic respiration (alcoholic fermentation)	Carbonic anhydrase is important in the transport of carbon dioxide in vertebrate blood. Deficiency in plants produces malformed, and sometimes mottled, leaves
Molybdenum Mo^{4+} or Mo^{5+}	Required by plants for the reduction of nitrate to nitrite in the formation of amino acids. Essential for nitrogen fixation by prokaryotes	Deficiency produces a reduction in crop yield. Not vital in most animals
Boron BO$_3^{3+}$ or B$_4$O^{2+}	Required for the uptake of Ca^{2+} by roots. Aids the germination of pollen grains and mitotic division in meristems	Boron is not required by animals. Deficiency in plants causes death of young shoots and abnormal growth. May cause specific diseases such as 'internal cork' of apples and 'heart rot' of beet and celery
Fluorine F$^-$	A component of teeth and bones	Not required by most plants. Associates with calcium to form calcium fluoride which strengthens teeth and helps prevent decay

3.2 Carbohydrates

Carbohydrates comprise a large group of organic compounds which contain carbon, hydrogen and oxygen and which are either aldehydes or ketones. The word carbohydrate suggests that these organic compounds are hydrates of carbon. Their general formula is $C_x(H_2O)_y$. The word carbohydrate is convenient rather than exact, because while most examples do conform to the formula, e.g. glucose – $C_6H_{12}O_6$, sucrose $C_{12}H_{22}O_{11}$, a few do not, e.g. deoxyribose – $C_5H_{10}O_4$. Carbohydrates are divided into three groups: the **monosaccharides** ('single-sugars'), the **disaccharides** ('double-sugars') and the **polysaccharides** ('many-sugars').

The functions of carbohydrates, although variable, are in the main concerned with storage and liberation of energy. A few, such as cellulose, have structural roles. A full list of individual carbohydrates and their functions is given in Table 3.3 on page 20.

3.3 Monosaccharides

Monosaccharides are a group of sweet, soluble crystalline molecules of relatively low molecular mass. They are named with the suffix -ose. Monosaccharides contain either an aldehyde group (—CHO), in which case they are called **aldoses** or **aldo-sugars**, or they contain a ketone group (C = O), in which case they are termed **ketoses** or **keto-sugars**. The general formula for a monosaccharide is $(CH_2O)_n$. Where $n = 3$, the sugar is called a **triose** sugar, $n = 5$, a **pentose** sugar, and $n = 6$, a **hexose** sugar. Table 3.2 classifies some of the more important monosaccharides.

TABLE 3.2   **Classification of monosaccharides**

	Trioses $(C_3H_6O_3)$	**Pentoses** $(C_5H_{10}O_5)$	**Hexoses** $(C_6H_{12}O_6)$
Aldoses (—CHO) (Aldo-sugars)	Glyceraldehyde	Ribose Arabinose Xylose	Glucose Galactose Mannose
Ketoses (C = O) (Keto-sugars)	Dihydroxyacetone	Ribulose Xylulose	Fructose Sorbose

3.3.1 Structure of monosaccharides

Probably the best known monosaccharide, glucose, has the formula $C_6H_{12}O_6$. All but one of the six carbon atoms possesses an hydroxyl group (—OH). The remaining carbon atom forms part of the aldehyde group. Glucose may be represented by a straight chain of six carbon atoms. These are numbered beginning at the carbon of the aldehyde group. Glucose in common with other hexoses and pentoses easily forms stable ring structures. At any one time most molecules exist as rings rather than a chain. In the case of glucose, carbon atom number 1 may combine with the oxygen atom on carbon 5. This forms a six-sided structure known as a **pyranose** ring. In the case of fructose, it is carbon atom number 2 which links with the oxygen on carbon atom 5. This forms a five-sided structure called a **furanose** ring (Fig. 3.1). Both glucose and fructose can exist in both pyranose and furanose forms.

Glucose, in common with most carbohydrates, can exist as a number of **isomers** (they possess the same molecular formula but differ in the arrangement of their atoms). One type of isomerism, called **sterioisomerism**, occurs when the same atoms, or groups, are joined together but differ in their arrangement in space. One form of sterioisomerism, called **optical isomerism**, results in isomers which can rotate the plane of polarized light (light which is vibrating in one plane only). The isomer which rotates the plane of polarized light to the right is called the **dextro(D or +) form**; the isomer rotating it to the left is called the **laevo(L or −) form**. (By present convention, however, the D and L forms are named by different criteria, regardless of the direction in which they rotate polarized light.) While the chemical and physical properties of the two forms are the same, many enzymes will only act on one type. There would seem to be no reason why one form should be preferred to another, and yet almost all naturally occurring carbohydrates are of the D(+) form. It must be assumed that at an early stage in evolution the D(+) form was arbitrarily adopted and the consequent development of enzymes specific to this type ensured that all subsequent development was based on this form. Both D(+) and L(−) forms of glucose are shown in Fig. 3.1. The D(+) and L(−) forms of glucose arise because the relevant carbon atom has four different groups attached to it. This is called an **asymmetric carbon atom**. Another asymmetric carbon atom arises when glucose forms a ring structure. This gives rise to two further isomers, the α-**form** and the β-**form**. Both types occur naturally and, as we shall see later, result in considerable biological differences when they form polymers. Fig. 3.1 again illustrates both types.

The ring structures are three dimensional. The ring itself lies in a plane at right angles to the paper with the attached groups lying above and below this plane. The black lines indicate that this part of the ring lies towards the reader in front of the part of the ring with thinner lines.

Straight chain arrangements

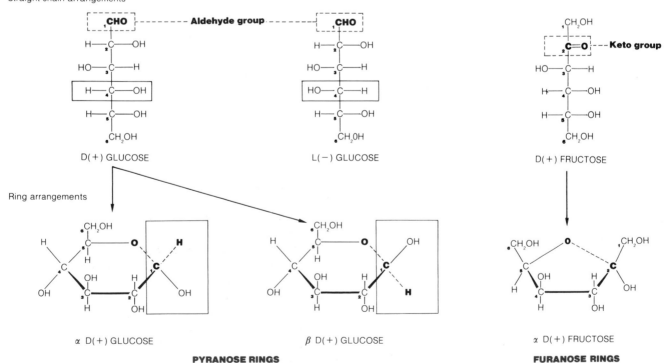

Ring arrangements

Fig. 3.1 Structure of various isomers of glucose and fructose

3.4 Disaccharides

Monosaccharides may combine together in pairs to give a **disaccharide** (double-sugar). The union involves the loss of a single water molecule and is therefore a **condensation reaction**. The addition of water, under suitable conditions, is necessary if the disaccharide is to be split into its constituent monosaccharides. This is called **hydrolysis** 'water-breakdown' or, more accurately, 'breakdown *by* water'. The bond which is formed is called a **glycosidic bond**. It is usually formed between carbon atom 1 of one monosaccharide and carbon atom 4 of the other, hence it is called a 1–4 glycosidic bond (see Fig. 3.2). Any two monosaccharides may be linked in this way to form a disaccharide of which maltose, sucrose and lactose are the most common.

Disaccharides, like monosaccharides, are sweet, soluble and crystalline. Maltose and lactose are reducing sugars, whereas sucrose is a non-reducing sugar. The significance of this is considered in Section 3.6.2.

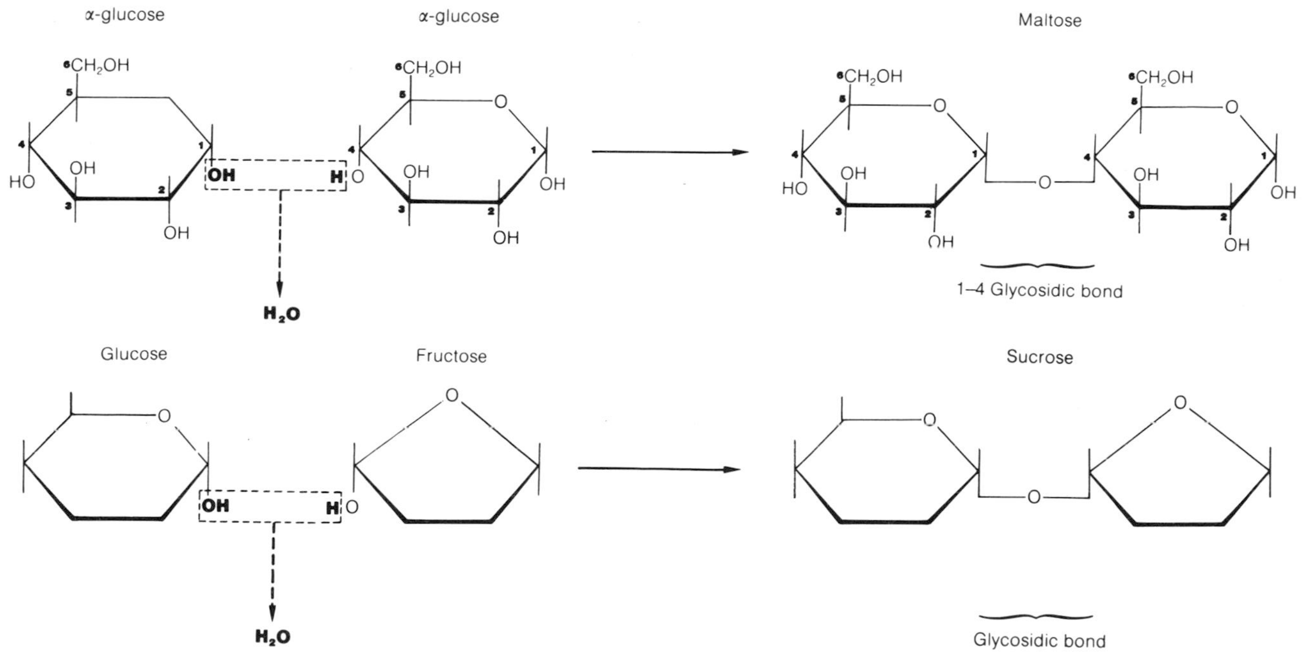

maltose (malt sugar) = glucose + glucose

sucrose (cane sugar) = glucose + fructose

lactose (milk sugar) = glucose + galactose

The removal of water (condensation) from the two hydroxyl groups (− OH) on carbons 1 and 4 of the respective glucose molecules, forms a maltose molecule. Some carbon and hydrogen atoms have been omitted for simplicity.

Sucrose is formed by a condensation reaction between one glucose and one fructose molecule. The process shown is much simplified.

Fig. 3.2 Formation of maltose and sucrose

3.5 Polysaccharides

In the same way that two monosaccharides may combine in pairs to give a disaccharide, many monosaccharides may combine by condensation reactions to give a **polysaccharide**. The number of monosaccharides which combine is variable and the chain produced can be branched or unbranched. The chains may be folded, thus making them compact and therefore ideal for storage. The size of the molecule makes them insoluble – another feature which suits them for storage as they exert no osmotic influence and do not easily diffuse out of the cell. Upon hydrolysis, polysaccharides can be converted to their constituent monosaccharides ready for use as respiratory substrates. Starch and glycogen are examples of storage polysaccharides. Not all polysaccharides are used for storage; cellulose, for example, is a structural polysaccharide giving strength and support to cell walls.

3.5.1 Starch

Starch is a polysaccharide which is found in most parts of the plant in the form of small granules. It is a reserve food formed from any excess glucose produced during photosynthesis. It is common in the seeds of some plants, e.g. maize, where it forms the food supply for germination. Indirectly these starch stores form an important food supply for animals.

Starch is a mixture of two substances: amylose and amylopectin. Starches differ slightly from one plant species to the next, but on the whole they comprise 20% amylose, 79% amylopectin, and 1% of other substances such as phosphates and fatty acids. A comparison of amylose and amylopectin is given in Fig. 3.3.

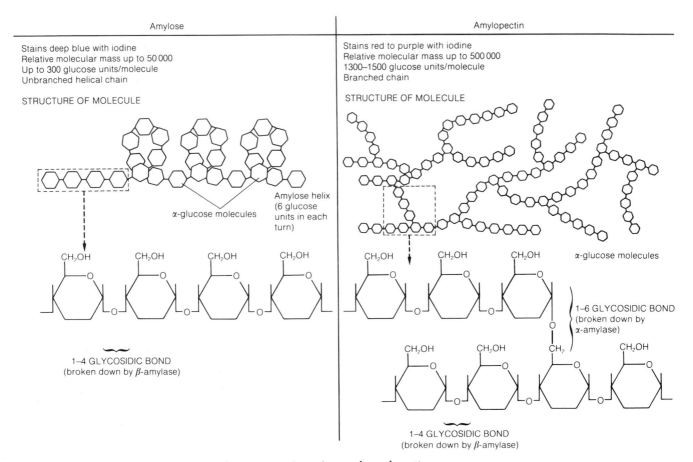

Fig. 3.3 Comparison of the properties and structures of amylose and amylopectin

3.5.2 Glycogen

Glycogen is the major polysaccharide storage material in animals and fungi and is often called 'animal starch'. It is stored mainly in the liver and muscles. Like starch it is made up of α-glucose molecules and exists as granules. It is similar to amylopectin in structure but it has shorter chains (10–20 glucose units) and is more highly branched.

3.5.3 Cellulose

Cellulose typically comprises up to 50% of a plant cell wall, and in cotton it makes up 90%. It is a polymer of around 10 000 β-glucose molecules forming a long unbranched chain. Many chains run parallel to each other and have cross linkages between them (Fig. 3.4). These help to give cellulose its considerable stability which makes it a valuable structural material. The stability of cellulose makes it difficult to digest and therefore not such a valuable food source to animals, which only rarely produce cellulose-digesting enzymes. Some, however, have formed symbiotic relationships with organisms which can digest cellulose (Section 24.6.2). To these organisms it is the major component of their diet. Cellulose's structural strength has long been recognized by man. Cotton is used in the manufacture of fabrics. Rayon is produced from cellulose extracted from wood and its remarkable tensile strength makes it especially useful in the manufacture of tyre cords. Cellophane, used in packaging, and celluloid, used in photographic film, are also cellulose derivatives. Paper is perhaps the best known cellulose product.

Being composed of β-glucose units, the chain, unlike that of starch, has adjacent glucose molecules rotated by 180°. This allows hydrogen bonds to be formed between the hydroxyl (—OH) groups on adjacent parallel chains which help to give cellulose its structural stability.

Simplified representation of the arrangement of glucose chains

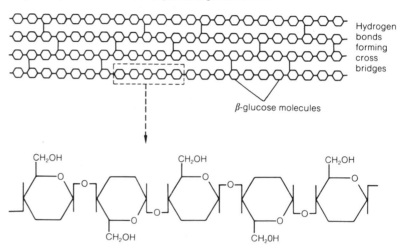

Hydrogen bonds forming cross bridges

β-glucose molecules

Fig. 3.4 Structure of the cellulose molecule

3.5.4 Other polysaccharides

Chitin – Chemically and structurally chitin resembles cellulose. It differs in possessing an acetyl-amino group ($NH.OCCH_3$) instead of one of the hydroxy (—OH) groups. Like cellulose it has a structural function and is a major component of the exoskeleton of insects and crustacea. It is also found in fungal cell walls.

Inulin – This is a polymer of fructose found as a storage carbohydrate in some plants, e.g. *Dahlia* root tubers.

Mucopolysaccharides – This group includes **hyaluronic acid**, which forms part of the matrix of vertebrate connective tissues. It is found in cartilage, bones, the vitreous humour of the eye and synovial fluid. The anticoagulant **heparin** is also a member of this group of polysaccharides.

3.6   Tests for carbohydrates

There are a number of relatively simple tests which can be used to identify groups of carbohydrates. Most are effective if carried out on a solution or suspension of the carbohydrate. If the material to be tested is in a solid form, it should be ground up with a little water in a mortar and pestle. The resultant liquid should be filtered or centrifuged and the tests carried out on the filtrate or supernatant liquid. Where heating is necessary, this should be carried out by placing test tubes in a water bath or beaker containing water near to boiling point. The tubes should not be heated directly.

TABLE 3.3 **Carbohydrates and their functions**

Group of carbohydrates	Name of carbohydrate	Type/composition	Function
Monosaccharides Trioses ($C_3H_6O_3$)	Glyceraldehyde	Aldose sugar	The phosphorylated form is the first formed sugar in photosynthesis, and as such may be used as a respiratory substrate or converted to starch for storage. It is an intermediate in Krebs cycle
	Dihydroxyacetone	Ketose sugar	Respiratory substrate. Intermediate in Krebs cycle
Pentoses ($C_5H_{10}O_5$)	Ribose/deoxyribose	Aldose sugars	Makes up part of nucleotides and as such gives structural support to the nucleic acids RNA and DNA. Constituent of hydrogen carriers such as NAD, NADP and FAD. Constituent of ATP
	Ribulose	Ketose sugar	Carbon dioxide acceptor in photosynthesis
Hexoses ($C_6H_{12}O_6$)	Glucose	Aldose sugar	Major respiratory substrate in plants and animals. Synthesis of disaccharides and polysaccharides. Constituent of nectar
	Galactose	Aldose sugar	Respiratory substrate. Synthesis of lactose
	Mannose	Aldose sugar	Respiratory substrate
	Fructose	Ketose sugar	Respiratory substrate. Synthesis of inulin. Constituent of nectar. Sweetens fruits to attract animals to aid seed dispersal
Disaccharides	Sucrose	Glucose + fructose	Respiratory substrate. Form in which most carbohydrate is transported in plants. Storage material in some plants, e.g. Allium (onion)
	Lactose	Glucose + galactose	Respiratory substrate. Mammalian milk contains 5% lactose, therefore major carbohydrate source for sucklings
	Maltose	Glucose + Glucose	Respiratory substrate
Polysaccharides	Amylose Amylopectin } starch	Unbranched chain of α-glucose with 1,4 glycosidic links + branched chain of α-glucose units with 1,4 and 1,6 glycosidic links	Major storage carbohydrate in plants
	Glycogen	Highly branched short chains of α-glucose units with 1,4 glycosidic links	Major storage carbohydrate in animals and fungi
	Cellulose	Unbranched chain of β-glucose units with 1,4 glycosidic links + cross bridges	Gives structural support to cell walls
	Inulin	Unbranched chain of fructose with 1,2 glycosidic links	Major storage carbohydrate in some plants, e.g. Jerusalem artichoke; Dahlia
	Chitin	Unbranched chain of β-acetylglucosamine units with 1,4 glycosidic links	Constituent of the exoskeleton of insects and crustacea

3.6.1 Controls

For each of the following food tests, the procedures should be repeated with an equal quantity of water being substituted for the carbohydrate solution under test. This is necessary to ensure that there is no contamination of apparatus or the reagents. The latter is quite common where these reagents are used communally. Provided there is no colour change in the control tube, all other results can be taken to be valid. If there is a colour change, the results must be disregarded and the experiment repeated with fresh, clean apparatus and new reagents.

TABLE 3.4 **Relationship between amount of reducing sugar and colour of precipitate on boiling with Benedict's reagent**

Amount of reducing sugar	Colour of solution and precipitate
No reducing sugar	Blue
Increasing quantity of reducing sugar	Green
	Yellow
	Brown
	Red

Apparatus
Test tubes
Test tube rack
Bunsen burner, tripod, gauze, beaker or thermostatically controlled water bath
1 cm³ syringe

Chemicals
Benedict's reagent
Carbohydrate solutions to be tested

Apparatus
Test tubes
Test tube rack
Bunsen burner, tripod, gauze, beaker or thermostatically controlled water bath
1 cm³ syringe

Chemicals
Benedict's reagent
Dilute hydrochloric acid
Sodium hydrogen carbonate solution
Carbohydrate solutions to be tested

3.6.2 Reducing and non-reducing sugars

All monosaccharides, whether aldo- or keto-sugars, are capable of reducing other chemicals such as copper (II) sulphate to copper (I) oxide. When monosaccharides combine to form disaccharides this reducing ability is often retained with the result that sugars such as lactose and maltose, although disaccharides, are still reducing sugars. In a few cases, however, the formation of a disaccharide results in the loss of this reducing ability. This is true of the formation of sucrose which is therefore a non-reducing sugar.

NB Throughout the following tests, Fehling's solution can be substituted for Benedict's reagent. It does, however, require the mixing of two Fehling's solutions (A and B) immediately prior to the tests.

3.6.3 Test for reducing sugars

Precautions
Once the water bath has reached boiling point, the source of heat should be turned down. Reactions will still take place even if the temperature is slightly below 100°C.

Method
1. Bring the water in the water bath up to boiling point and turn down the source of heat.
2. Take 2 cm³ of the solution to be tested and add 2 cm³ of Benedict's reagent. Mix the reagents thoroughly.
3. Place the test tube in the water bath and leave for 5 minutes, shaking occasionally.

Results
Where a reducing sugar is present it will reduce soluble copper (II) sulphate to insoluble copper (I) oxide which forms a **precipitate**. In addition, the blue copper (II) sulphate becomes the brick-red colour of the copper (I) oxide. The test is partially quantitative; the more reducing sugar the greater the amount and the darker the colour of the precipitate. The colour of the precipitate will range from green, through yellow, orange and brown to a deep red, as the quantity of reducing sugar increases.

3.6.4 Test for non-reducing sugars

Precautions
1. Dilute hydrochloric acid may be harmful on contact with the skin. If contact occurs, immediately wash the affected area with water and/or sodium hydrogen carbonate solution.
2. Care must be taken when neutralizing the acid as effervescence occurs which may cause liquid to splash out of the test tube. Add the sodium hydrogen carbonate solution *slowly*.
3. Turn down the source of heat once the water bath has reached boiling point.

Information
There is no specific test for a non-reducing sugar as such, but a non-reducing sugar can be detected by its inability to reduce Benedict's reagent *directly*. If it is then hydrolysed by boiling with dilute hydrochloric acid it will be broken down into its constituent

monosaccharides. These will then reduce Benedict's reagent in the normal way. A non-reducing sugar is thus identified by a negative reaction to Benedict's reagent *before* hydrolysis and a positive result *after* hydrolysis.

Method
1. Carry out the reducing sugar test (Section 3.6.3).
2. Add 1 cm³ of dilute hydrochloric acid to a fresh sample of 2 cm³ of the solution to be tested , mix the solution and **boil for 2–3 minutes**.
(*NB* It is a frequent error to carry out instruction 3 before the hydrochloric acid has had adequate time to effect hydrolysis.)
3. Add sodium hydrogen carbonate solution until the solution is neutral or preferably slightly alkaline. Use pH paper to test for this. (This procedure is necessary because Benedict's reagent is not effective in acid conditions.)
4. Carry out the reducing sugar test again (Section 3.6.3).

Results
A negative result (solution remains blue) after the first reducing sugar test, followed by a positive result (solution turns red/brown) after the second reducing sugar test is an indication of a non-reducing sugar.

3.6.5 Detecting the presence of a non-reducing sugar in a solution which also contains reducing sugar

As boiling with hydrochloric acid does not affect a reducing sugar, its presence means that a positive result will always be obtained when carrying out a non-reducing sugar test. It may, however, still be possible to detect a non-reducing sugar in the presence of a reducing sugar; much depends on the relative concentrations of each. If the amount of reducing sugar is small, detection of the non-reducing sugar should be possible. Proceed as follows:

Method
1. Add 2 cm³ of the solution under test to each of two test tubes.
2. Carry out the reducing sugar test (Section 3.6.3) on one solution and the non-reducing sugar test (Section 3.6.4) on the other.
(*NB* It is most important that the quantities of Benedict's reagent used in each case are *exactly* equal and that both are boiled for the *same length of time*.)
3. Compare the colour and amount of precipitate in each of the tubes.

Results
The amount of precipitate should be greater and the colour darker (more red) on carrying out the non-reducing sugar test. If one considers a solution containing equal concentrations of glucose (reducing sugar) and sucrose (non-reducing sugar), the glucose alone will reduce Benedict's reagent when boiled with it. Upon hydrolysis by hydrochloric acid, the sucrose will be converted to glucose and fructose, both reducing sugars. The amount of reducing sugar is now greater and so boiling this solution with Benedict's reagent produces more reduction. This second solution should therefore be brick red in colour and contain a greater quantity of precipitate.

TABLE 3.5 **Summary of typical results from reducing and non-reducing sugar tests.**

Solution content	Reducing sugar test (boil with Benedict's reagent)	Non-reducing sugar test (boil with hydrochloric acid, then boil with Benedict's reagent)
Reducing sugars, e.g. all monosaccharides + disaccharides such as maltose and lactose	Yellow/brown precipitate	Yellow/brown precipitate (the reducing sugar being unaffected by the acid)
Non-reducing sugars, e.g. disaccharides such as sucrose	No reaction – remains blue	Yellow/brown precipitate
Reducing and non-reducing sugars	Yellow/brown precipitate (small quantity)	Brick-red precipitate (larger quantity)

Apparatus
Spotting tile or test tube
Dropping pipette

Chemicals
Iodine in potassium iodide solution
Carbohydrate solutions to be tested

Boiling a starch solution causes temporary unwinding of the helix and the subsequent release of the iodine molecules. The blue-black colour therefore disappears. On cooling, the helix and the blue-black colour, reform.

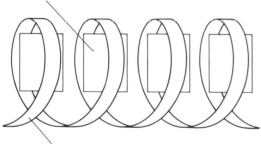

Iodine molecule in the centre of the starch helix. There is one iodine molecule for each turn of the helix

Starch helix formed by α-glucose molecules (6 per turn of helix). The dimensions of the centre are just sufficient to fit iodine molecules within it

Fig. 3.5 Starch–iodine staining reaction

The three triglycerides may all be the same, thereby forming a simple triglyceride, or they may be different in which case a mixed triglyceride is produced. In either case it is a condensation reaction.

3.6.6 Test for starch

Precautions
The test must be carried out at room temperature. Do **not** boil the solution.

Method
1. Place two drops of the solution to be tested in a depression in a spotting tile, or in a test tube.
2. Add a drop of iodine in potassium iodide solution.

Results
If starch is present, the yellow-orange iodine in potassium iodide solution becomes a blue-black colour. The iodine takes up a position in the centre of the starch helix (see Fig. 3.5), forming a starch-iodine complex and giving rise to the intense blue-black colouration.

3.7 Lipids

Lipids are a large and varied group of organic compounds. Like carbohydrates, they contain carbon, hydrogen and oxygen, although the proportion of oxygen is much smaller in lipids. They are insoluble in water but dissolve readily in organic solvents such as acetone, alcohols and others. They are of two types: fats and oils. There is no basic difference between these two; fats are simply solid at room temperatures (10–20°C) whereas oils are liquid. The chemistry of lipids is very varied but they are all esters of **fatty acids** and an alcohol, of which **glycerol** is by far the most abundant. Glycerol has three hydroxy (—OH) groups and each may combine with a separate fatty acid, forming a **triglyceride** (Fig. 3.6). It is a condensation reaction and thus hydrolysis of the triglyceride will again yield glycerol and three fatty acids.

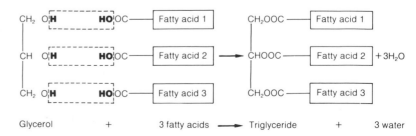

Fig. 3.6 Formation of a triglyceride

TABLE 3.6 **Nature and occurrence of some fatty acids**

Name of fatty acid	General formula	Saturated/ unsaturated	Occurrence
Butyric	C_3H_7COOH	Saturated	Butter fat
Linoleic	$C_{17}H_{31}COOH$	Unsaturated	Linseed oil
Oleic	$C_{17}H_{33}COOH$	Unsaturated	All fats
Palmitic	$C_{15}H_{31}COOH$	Saturated	Animal and vegetable fats
Stearic	$C_{17}H_{35}COOH$	Saturated	Animal and vegetable fats
Arachidic	$C_{19}H_{39}COOH$	Saturated	Peanut oil
Cerotic	$C_{25}H_{51}COOH$	Saturated	Wool oil

3.7.1 Fatty acids

As most naturally occurring lipids contain the same alcohol, namely glycerol, it is the nature of the fatty acids which determines the characteristics of any particular fat. All fatty acids contain a carboxyl group (—COOH). The remainder of the molecule is a hydrocarbon chain of varying length (examples are given in Table 3.6). This chain may possess one or more double bonds in which case it is said to be **unsaturated**. If, however, it possesses no double bonds it is said to be **saturated**.

It can be seen from Table 3.6 that the hydrocarbon chains may be very long. Within the fat they form long 'tails' which extend from the glycerol molecule. These 'tails' are **hydrophobic**, i.e. they repel water.

3.7.2 Phospholipids

Phospholipids are lipids in which one of the fatty acid groups is replaced by phosphoric acid (H_3PO_4) (Fig. 3.7). The phosphoric acid is **hydrophilic** (attracts water) in contrast to the remainder of the molecule which is hydrophobic (repels water). Having one end of the phospholipid attracting water while the other end repels it affects its role in the cell membrane.

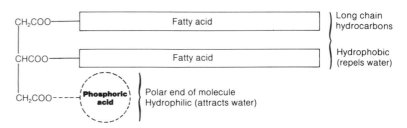

Fig. 3.7 *Structure of a phospholipid*

3.7.3 Waxes

Waxes are formed by combination with an alcohol other than glycerol. This alcohol is much larger than glycerol, and therefore waxes have a more complex chemical structure. Their main rôle is in waterproofing plants and animals, although they form storage compounds in a few organisms, e.g. castor oil and in fish.

3.7.4 Functions of lipids

1. An energy source – Upon breakdown they yield $38\,kJ\,g^{-1}$ of energy. This compares favourably with carbohydrates which yield $17\,kJ\,g^{-1}$.

2. Storage – On account of their high energy yield upon breakdown, they make excellent energy stores. For the equivalent amount of energy stored they possess less than half the mass of carbohydrate. This makes them especially useful for animals where locomotion requires mass to be kept to a minimum. In plants they are useful in seeds where dispersal by wind or insects makes small mass a necessity. This explains the abundance of oils extracted from seeds and fruits, e.g. olive, linseed, castor, peanut, coconut and sunflower. Their insolubility is another advantage, as they are not easily dissolved out of cells.

3. Insulation – Fats conduct heat only slowly and so are useful insulators. If fat is to be stored because of its concentrated energy supply, it may as well be put to a secondary use. In endothermic animals, such as mammals, it is stored beneath the skin (subcutaneous fat) where it helps to retain body heat. In aquatic mammals, such as whales, seals and manatees, hair is ineffective as an insulator because it cannot trap water in the same way as it can air. These animals therefore have extremely thick subcutaneous fat, called blubber, which forms an effective insulator.

4. Protection – Another secondary use to which stored fat is put is as a packing material around delicate organs. Fat surrounding the kidneys, for instance, helps to protect them from physical damage.

5. Waterproofing – Terrestrial plants and animals have a need to conserve water. Animal skins produce oily secretions, e.g. from the sebaceous glands in mammals, which waterproof the body. Oils also coat the fur, helping to repel water which would otherwise wet it and reduce its effectiveness as an insulator. Birds spread oil over their feathers, from a special gland near the cloaca, for the same purpose. Insects have a waxy cuticle to prevent evaporative loss in the same way that plant leaves have one to reduce transpiration.

6. Cell membranes – Phospholipids are major components of the cell membrane and contribute to many of its properties (see Section 4.3.3).

7. Other functions – Lipids perform a host of miscellaneous functions in different organisms. For example, plant scents are fatty acids (or their derivatives) and so aid the attraction of insects for pollination. Bees use wax in constructing their honeycombs.

3.7.5 Steroids

Steroids are related to lipids, and **cholesterol** is perhaps the best known. It is found in animals where it is important in the synthesis of steroid hormones, such as oestrogen and cortisone. Other important steroids include vitamin D and bile acids.

3.8 Tests for lipids

3.8.1 Emulsion test

Apparatus
Test tubes
Test tube rack
$1 \, cm^3$ syringe

Chemicals
Ethanol (absolute)
Test solutions (e.g. olive oil, castor oil)

Precautions
1. The test tubes must be completely dry and free from any grease.

Method
1. Place $2 \, cm^3$ of the oil in a test tube and add $5 \, cm^3$ of the ethanol.
2. Shake the tube thoroughly until all the oil is dissolved.
3. Add $5 \, cm^3$ of water and shake gently.
4. As a control, repeat procedures 1–3 using water instead of oil.

Results
A milky suspension indicates the presence of a lipid.

Explanation
Being insoluble in water and less dense, oil would simply float if added to water. In this test, the oil is first dissolved in alcohol, with

which water is miscible. When the water is added at stage 3, the alcohol, laden with oil, mixes with the water. The oil itself does not mix but is left as minute droplets dispersed throughout the water. Light rays entering the solution are scattered as they pass from water to oil, and vice versa. Since light must pass between the two substances many times, most never passes straight through the suspension but is scattered in all directions. This gives the suspension its milky white appearance.

3.8.2 Sudan III Test

Apparatus
Test tubes
Test tube rack

Chemicals
Sudan III
Test solutions (e.g. olive oil, castor oil)

Method
1. Place 2 cm³ of the oil in a test tube and add 2 cm³ of water.
2. Add 3 drops of Sudan III and shake gently.
3. Leave the tube to stand for 3 minutes.

Results
The oil layer, which settles out on top, is red in colour. The water beneath remains clear.

Explanation
Sudan III is a red dye which stains oils and fats. It is particularly useful for detecting fat in cells.

3.9 Proteins

Proteins are organic compounds of large molecular mass (up to 40 000 000 for some viral proteins but more typically several thousand, e.g. haemoglobin = 64 500). They are not truly soluble in water, but form colloidal suspensions (the nature of colloids is dealt with in Section 33.1.5). In addition to carbon, hydrogen and oxygen, they always contain nitrogen, usually sulphur and sometimes phosphorus. Whereas there are relatively few carbohydrates and fats, the number of proteins is almost limitless. A simple bacterium such as *Escherichia coli* has around 800, and man has over 10 000. They are specific to each species. Glucose is glucose in whatever organism it occurs, but proteins vary from one species to another. Indeed, it is the proteins rather than the fats or carbohydrates which determine the characteristics of a species. Proteins are rarely stored in organisms, except in eggs or seeds where they are used to form the new tissue. The word protein (from the Greek) means 'of first importance' and was coined by a Dutch chemist, Mulder, because he thought they played a fundamental rôle in cells. We now know that proteins form the structural basis of all living cells and that Mulder's judgement was sound.

3.9.1 Amino acids

Amino acids are a group of over a hundred chemicals of which around twenty commonly occur in proteins. They always contain a basic group, the amino group (—NH₂) and an acid group, the carboxyl group (—COOH). (See Fig. 3.8.) Most amino acids have one of each group and are therefore neutral, but a few have more amino groups than carboxyl ones (basic amino acids) while others have more carboxyl than amino groups (acidic amino acids). With

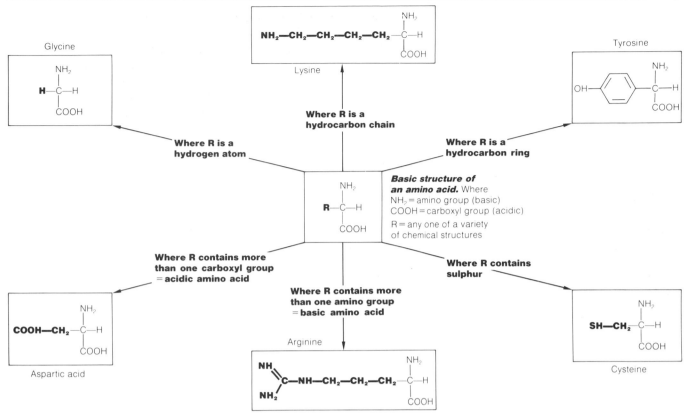

Fig. 3.8. Structure of a range of amino acids

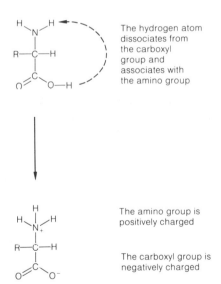

The hydrogen atom dissociates from the carboxyl group and associates with the amino group

The amino group is positively charged

The carboxyl group is negatively charged

Fig. 3.9. Zwitterion formation in amino acids

the exception of glycine, all amino acids have an asymmetric carbon atom and therefore exhibit optical isomerism, having both D(+) and L(−) forms. Whereas all naturally occurring carbohydrates are of the D(+) form, all naturally occurring amino acids are of the L(−) form. Amino acids are soluble in water where they form ions. These ions are formed by the loss of a hydrogen atom from the carboxyl group, making it negatively charged. This hydrogen atom associates with the amino group, making it positively charged. The ion is therefore **dipolar** — having a positive and a negative pole. Such ions are called **zwitterions** (see Fig. 3.9). Amino acids therefore have both acidic and basic properties, i.e. they are **amphoteric**. Being amphoteric means that amino acids act as **buffer solutions**. A buffer solution is one which resists the tendency to alter its pH even when small amounts of acid or alkali are added to it. Such a property is essential in biological systems where any sudden change in pH could adversely affect the performance of enzymes.

3.9.2 Formation of polypeptides

We have seen that monosaccharides may be linked to form disaccharides and polysaccharides by the loss of water (condensation reaction). Similarly, fats are formed from condensation reactions between fatty acids and glycerol. The formation of polypeptides follows the same pattern. A condensation reaction occurs between the amino group of one amino acid and the carboxyl group of another, to form a **dipeptide** (see Fig. 3.10, on the next page). Further combinations of this type extend the length of the chain to form a **polypeptide** (see Figs. 3.11 and 3.12).

A polypeptide usually contains many hundreds of amino acids. Polypeptides may be linked by forces such as disulphide bridges to give proteins comprising thousands of amino acids.

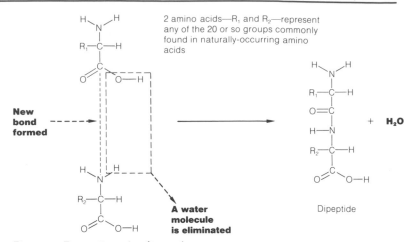

Fig. 3.10 *Formation of a dipeptide*

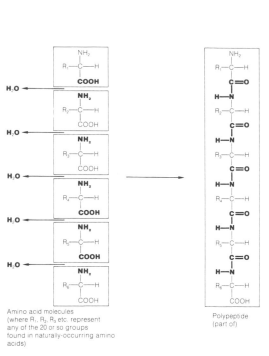

Fig. 3.11 *Formation of a polypeptide*

A simplified representation of a polypeptide chain to show three types of bonding responsible for shaping the chain. In practice the polypeptide chains are longer, contain more of these three types of bond and have a three dimensional shape.

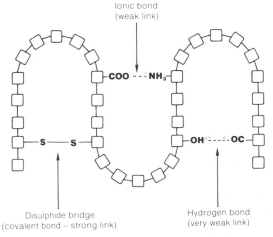

Fig. 3.12 *Types of bond in a polypeptide chain*

3.9.3 Structure of polypeptides

The chains of amino acids which make up a polypeptide have a specific three-dimensional shape (see Fig. 3.13). This shape is important in the functioning of proteins, especially enzymes. The shape of a polypeptide molecule is due to four types of bonding which occur between various amino acids in the chain.

The first type of bond is called a **disulphide bond**. It arises between sulphur-containing groups on any two cysteine molecules. These bonds may arise between cysteine molecules in the same amino acid chain (intrachain) or between molecules in different chains (interchain).

The second type of bond is the **ionic bond**. We have seen that amino acids form zwitterions (Section 3.9.1) which have NH_3^+ and COO^- groups. The formation of peptide bonds when making a polypeptide means that the COOH and NH_2 groups are not available to form ions. In the case of acidic amino acids, however, there are additional COOH groups which may ionise to give COO^- groups. In the same way, basic amino acids may still retain NH_3^+ groups even when combined into the structure of a polypeptide. In addition NH_3^+ and COO^- can occur at the ends of a polypeptide chain. Any of these available NH_3^+ and COO^- groups may form ionic bonds which help to give a polypeptide molecule its particular shape. These ionic bonds are weak and may be broken by alterations in the pH of the medium around the polypeptide.

The third type of bond is the **hydrogen bond**. This occurs between certain hydrogen atoms and certain oxygen atoms within the polypeptide chain. The hydrogen atoms have a small positive charge on them (electropositive) and the oxygen atoms a small negative charge (electronegative). The two charged atoms are attracted together and form a hydrogen bond. While each bond is very weak, the sheer number of bonds means that they play a considerable rôle in the shape and stability of a polypeptide molecule.

The fourth type is **hydrophobic interactions** which are interactions between non-polar R groups. These cause the protein to fold as hydrophobic side groups are shielded from water.

3.9.4 Fibrous proteins

The fibrous proteins have a primary structure of regular repetitive sequences. They form long chains which may run parallel to one another, being linked by cross bridges. They are very stable molecules and have structural rôles within organisms. Collagen is a good

(a) *The primary structure of a protein is the sequence of amino acids found in its polypeptide chains. This sequence determines its properties and shape. Following the elucidation of the amino acid sequence of the hormone insulin, by Frederick Sanger in 1954, the primary structure of many other proteins is now known.*

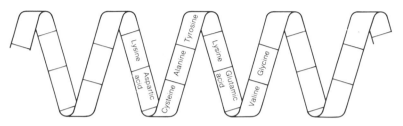

Lysine	Aspartic acid	Cysteine	Alanine	Tyrosine	Lysine	Glutamic acid	Valine	Glycine

(b) *The secondary structure is the shape which the polypeptide chain forms as a result of hydrogen bonding. This is most often a spiral known as the α-helix, although other configurations occur.*

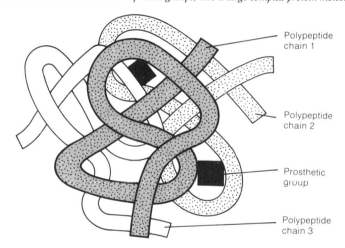

(c) *The tertiary structure is due to the bending and twisting of the polypeptide helix into a compact structure. All three types of bond, disulphide, ionic and hydrogen, contribute to the maintenance of the tertiary structure.*

(d) *The quarternary structure arises from the combination of a number of different polypeptide chains, and associated non-protein groups, into a large complex protein molecule.*

Fig. 3.13 *Structure of proteins*

example. It is a common constituent of animal connective tissue, especially in structures requiring physical strength, e.g. tendons. It has a primary structure which is largely a repeat of the tripeptide sequence, glycine – proline – alanine, and forms a long unbranched chain. Three such chains are wound into a triple helix, with cross bridges linking them to each other and providing additional structural support. (Compare the repeating glucose units, parallel chains and cross links of the structural carbohydrate cellulose.)

3.9.5 Globular proteins

In contrast to fibrous proteins, the globular proteins have highly irregular sequences of amino acids in their polypeptide chains. Their shape is also different, being compact globules. If a fibrous protein is likened to a series of strands of string twisted into a rope, then a globular protein can be thought of as the same string rolled into a ball. These molecules are far less stable and have metabolic rôles within organisms. All enzymes are globular proteins. Globular and fibrous proteins are compared in Table 3.7, at the top of the next page.

TABLE 3.7   Comparison of globular and fibrous proteins

Fibrous proteins	Globular proteins
Repetitive regular sequences of amino acids	Irregular amino acid sequences
Actual sequences may vary slightly between two examples of the same protein	Sequence highly specific and never varies between two examples of the same protein
Polypeptide chains form long parallel strands	Polypeptide chains folded into a spherical shape
Length of chain may vary in two examples of the same protein	Length always identical in two examples of the same protein
Stable structure	Relatively unstable structure
Insoluble	Soluble – forms colloidal suspensions
Support and structural functions	Metabolic functions
Examples include collagen and keratin	Examples include all enzymes, some hormones (e.g. insulin) and haemoglobin

TABLE 3.8   Examples of conjugated proteins

Name of protein	Where found	Prosthetic group
Haemoglobin	Blood	Haem (contains iron)
Mucin	Saliva	Carbohydrate
Casein	Milk	Phosphoric acid
Cytochrome oxidase	Electron carrier pathway of cells	Copper
Nucleoprotein	Ribosomes	Nucleic acid

3.9.6 Conjugated proteins

Many proteins incorporate other chemicals within their structure. These proteins are called **conjugated proteins** and the non-protein part is referred to as the **prosthetic group**. The prosthetic group plays a vital rôle in the functioning of the protein. Some examples are given in Table 3.8.

3.9.7 Denaturation of proteins

We have seen that the three-dimensional structure of a protein is, in part at least, due to fairly weak ionic and hydrogen bonds. Any agent which breaks these bonds will cause the three-dimensional shape to be changed. In many cases the globular proteins revert to a more fibrous form. This process is called **denaturation**. The actual sequence of amino acids is unaltered; only the overall shape of the

TABLE 3.9   Factors causing protein denaturation

Factor	Explanation	Example
Heat	Causes the atoms of the protein to vibrate more (increased kinetic energy), thus breaking hydrogen and ionic bonds	Coagulation of albumen (boiling eggs makes the white more fibrous and less soluble)
Acids	Additional H^+ ions in acids combine with COO^- groups on amino acids and form COOH. Ionic bonds are hence broken	The souring of milk by acid (e.g. *Lactobacillus* bacterium, produces lactic acid, lowering pH and causing it to denature the casein, making it insoluble and thus forming curds)
Alkalis	Reduced number of H^+ ions causes NH_3^+ groups to lose H^+ ions and form NH_2. Ionic bonds are hence broken	
Inorganic chemicals	The ions of heavy metals such as mercury and silver are highly electropositive. They combine with COO^- groups and disrupt ionic bonds. Similarly, highly electronegative ions, e.g. cyanide (CN^-), combine with NH_3^+ groups and disrupt ionic bonds	Many enzymes are inhibited by being denatured in the presence of certain ions, e.g. cytochrome oxidase (respiratory enzyme) is inhibited by cyanide
Organic chemicals	Organic solvents alter hydrogen bonding within a protein	Alcohol denatures certain bacterial proteins. This is what makes it useful for sterilization
Mechanical force	Physical movement may break hydrogen bonds	On stretching a hair, the hydrogen bonds in the keratin helix are broken. The helix is extended and the hair stretches. If released, the hair returns to its normal length. If, however, it is wetted and then dried under tension, it maintains its new length – the basis of hair styling

molecule is changed. This is still sufficient to prevent the molecule from carrying out its usual functions within an organism.

Denaturation may be temporary or permanent and is due to a variety of factors as shown in Table 3.9.

TABLE 3.10 **Functions of proteins**

Vital activity	Protein example	Function
Nutrition	Digestive enzymes, e.g. trypsin amylase lipase	Catalyses the hydrolysis of proteins to polypeptides Catalyses the hydrolysis of starch to maltose Catalyses the hydrolysis of fats to fatty acids and glycerol
	Fibrous proteins in granal lamellae	Help to arrange chlorophyll molecules in a position to receive maximum amount of light for photosynthesis
	Mucin	Assists trapping of food in filter feeders. Prevents autolysis. Lubricates gut wall
	Ovalbumin	Storage protein in egg white
	Casein	Storage protein in milk
Respiration and transport	Haemoglobin/haemoerythrin/ haemocyanin/chlorocruorin	Transport of oxygen
	Myoglobin	Stores oxygen in muscle
	Prothrombin/fibrinogen	Required for the clotting of blood
	Mucin	Keeps respiratory surface moist
	Antibodies	Essential to the defence of the body, e.g. against bacterial invasion
Growth	Hormones, e.g. thyroxine	Controls growth and metabolism
Excretion	Enzymes, e.g. urease; arginase	Catalyse reactions in ornithine cycle and therefore help in protein breakdown and urea formation
Support and movement	Actin/myosin	Needed for muscle contraction
	Ossein	Structural support in bone
	Collagen	Gives strength with flexibility in tendons and cartilage
	Elastin	Gives strength and elasticity to ligaments
	Keratin	Tough for protection, e.g. in scales, claws, nails, hooves, skin
	Sclerotin	Provides strength in insect exoskeleton
	Lipoproteins	Structural components of all cell membranes
Sensitivity and coordination	Hormones, e.g. insulin/glucagon ACTH vasopressin	Control blood sugar level Controls the activity of the adrenal cortex Controls blood pressure
	Rhodopsin/opsin	Visual pigments in the retina, sensitive to light
	Phytochromes	Plant pigments important in control of flowering, germination etc.
Reproduction	Hormones, e.g. prolactin	Induces milk production in mammals
	Chromatin	Gives structural support to chromosomes
	Gluten	Storage protein in seeds – nourishes the embryo
	Keratin	Forms horns and antlers which may be used for sexual display

3.9.8 Functions of proteins

Proteins perform a wide variety of functions in living organisms. They are involved in all living processes as shown by some of the examples given in Table 3.10.

3.10 Tests for proteins

There are a number of tests for proteins. The Biuret test detects peptide links and therefore all proteins give a positive result. Millon's test detects the amino acid tyrosine and will therefore only give a positive result with proteins possessing it. However, practically every commonly occurring protein, except gelatin, does contain tyrosine. As with the Biuret test, positive results may be obtained with non-protein material.

3.10.1 Biuret test for proteins

Apparatus
Test tubes
Test tube rack
1 cm^3 syringe

Chemicals
10% potassium hydroxide solution
0.5% copper sulphate solution
Protein solutions to be tested

Precautions
1. The test should be carried out at room temperature; the solutions must not be heated.
2. Excess copper sulphate will produce negative results – use it sparingly.
3. Potassium hydroxide is caustic. Wash affected area immediately if it comes into contact with the skin.

Method
1. To 2 cm^3 of the solution being tested add 2 cm^3 of 10% potassium hydroxide solution and shake the tube to mix the contents.
2. Add 0.5% copper sulphate solution *a drop at a time*, shaking the tube continuously. Do not exceed 10 drops.
3. Repeat procedures 1 and 2 using water in place of the test solution (the control).

Results
The presence of a protein is indicated by a purple/mauve colouration. The control remains clear, or very slightly blue (due to the copper sulphate solution).

3.10.2 Millon's test for proteins

Apparatus
Test tubes
Test tube rack
1 cm^3 syringe
Bunsen burner, tripod, gauze, beaker or thermostatically controlled water bath.

Chemicals
Millon's reagent
Protein solutions to be tested

Precautions
1. Millon's reagent contains concentrated acid. If it contacts the skin, wash affected area immediately with water or, better still, with sodium hydrogen carbonate solution.
2. In addition to the concentrated acid, Millon's reagent contains mercuric salts and is therefore poisonous. Under no circumstances should it be pipetted by mouth. If any is swallowed, medical advice should be sought.
3. In view of its contents the reagent should not be heated directly by a Bunsen flame because of the risk of its spitting from the test tube. Always use a water bath.
4. Gelatin is about the only commonly occurring protein which lacks tyrosine and will therefore give a negative result.

Method
1. To 2 cm³ of the solution being tested add 1 cm³ of Millon's reagent and shake the tube to mix the contents.
2. Place the tube in the water bath of boiling water for 5 minutes.
3. Repeat procedures 1 and 2 using water instead of the test solutions (the control).

Results
On boiling, the protein may coagulate, forming a white precipitate. This precipitate gradually turns pink or red. Where there is no coagulation, the solution goes pink or red. The control remains as a white precipitate or a clear solution.

3.11   Nucleic acids

Like proteins, nucleic acids are informational macromolecules. They are made up of chains of individual units called **nucleotides**. The structure of nucleic acids and their constituent nucleotides are closely related to their functions in heredity and protein synthesis. For this reason the details of their structure will be left until the nature of the genetic code is discussed in Chapter 12.

3.12   Questions

1. By means of symbols or diagrams, as appropriate, describe the structure of:

    (*a*) a typical amino acid;
    (*b*) the linkage of amino acids to form peptides;
    (*c*) the primary, secondary and tertiary structures of proteins.   (*8 marks*)

Southern Universities Joint Board June 1983, Paper I, No. 1

2. List three different classes of biochemical substance, used by plants for food storage. Give for each substance the organ or structure in a named plant in which the food is stored. To what use is such food put and how is it made available?   (*8 marks*)

Southern Universities Joint Board June 1985, Paper I, No. 1

3.   *Either* (*a*) Write an essay on the structure and functions of proteins in living organisms.
                         (*20 marks*)

  or   (*b*) Give an account of the structures of polysaccharides and lipids, and discuss their significance to living organisms.
                         (*20 marks*)

London Board January 1989, Paper II, No. 9

4. The two elements listed are essential to both plants and animals.
In the table below give **one** important use of each element in each kingdom.

Element	Important use in	
	Plants	Animals
Calcium	....	....
Iron	....	....

                         (*4 marks*)
Associated Examining Board June 1988, Paper I, No. 10

5. How would you show the distribution of lipid and starch in a large seed or cereal grain?   (*5 marks*)
Associated Examining Board June 1988, Paper III, No. 3

6. Proteins are macromolecules of between about 50 and about 500 amino acid units of a defined sequence.
    (*a*) What are the constituent chemical elements of amino acids (and hence, proteins)?   (*2 marks*)
    (*b*) How many naturally occurring amino acids are there?   (*1 mark*)
    (*c*) Proteins can be divided into two basic types,

globular and fibrous. Briefly describe the structure of the two types. (*4 marks*)

(*d*) Outline the experimental techniques by which
  (i) the composition, and
  (ii) the structure
  of a protein may be established. (*4 marks*)

(*e*) Enzymes have active sites.
  (i) What is the function of the active site?
  (ii) Why is the structure of the active site so important for its function, and hence the function of the enzyme? (*4 marks*)
  (*Total 15 marks*)

Northern Ireland Board June 1983, Paper I, No. 3

7.

CH$_2$OH

(*a*) What type of molecule is shown in the diagram above? (*1 mark*)

(*b*) If **two** of the molecules shown are linked together:
  (i) what **two** products will result? (*2 marks*)
  (ii) what biochemical term is used for the process by which the two molecules join? (*1 mark*)
  (iii) indicate by means of a diagram how two such molecules would join. (You need only draw the relevant parts of the molecule.) (*2 marks*)

(*c*) What biochemical term would be used to describe large numbers of such molecules joining together? (*1 mark*)

(*d*) If many of the molecules shown in the diagram are linked together, suggest a name for the product. (*1 mark*)

(*e*) Give the name of an enzyme that could reverse the reaction mentioned in (*b*) (*ii*) above. (*1 mark*)

(*f*) Give the name of an enzyme that could reverse the reaction mentioned in (*c*) above. (*1 mark*)
  (*Total 10 marks*)

Oxford Local June 1988, Paper 1, No. 3

4 Cellular organization

The cell is the fundamental unit of life. All organisms, whatever their type or size, are composed of cells. The modern theory of cellular organization states that:

1. All living organisms are composed of cells.

2. All new cells are derived from other cells.

3. Cells contain the hereditary material of an organism which is passed from parent to daughter cells.

4. All metabolic processes take place within cells.

Before looking at the detailed structure of cells it is necessary to review the means by which cellular structure is investigated.

4.1 Microscopy

4.1.1 Structure of the light microscope

The simplest form of microscope is a single lens, but it has very limited powers of magnification. In the compound microscope two lenses are used. Light from an object passes through the first lens and produces a magnified image. This magnified image acts as the object for the second lens which further magnifies it. The total magnification is therefore the product of the magnification of each lens. If one lens of a compound microscope magnifies $10 \times$ and the other $40 \times$, the total magnification is $10 \times 40 = 400 \times$.

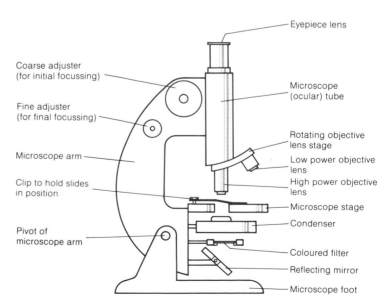

Coarse adjuster (for initial focussing)

Fine adjuster (for final focussing)

Microscope arm

Clip to hold slides in position

Pivot of microscope arm

Eyepiece lens

Microscope (ocular) tube

Rotating objective lens stage

Low power objective lens

High power objective lens

Microscope stage

Condenser

Coloured filter

Reflecting mirror

Microscope foot

Fig. 4.1 The compound light microscope

The degree of detail which can be seen with a microscope is called its **resolution** or **resolving power**. This measures its ability to distinguish two objects which are close together. A microscope with high resolution allows the user to distinguish two objects which are close together, whereas using a microscope with low resolution they will appear as a single object. The resolving power is inversely proportional to the wavelength of light being used; the shorter the wavelength the greater the resolution. This means that the resolving power of any light microscope is limited because the wavelength of light has a fixed range. While improvements may be made in preparing and staining slides, illuminating specimens and in the method of focussing, there is still a limit to the light microscope's resolving power. At best it can distinguish two points which are $0.2\mu m$ apart.

4.1.2 Staining and preparation of slides for the light microscope

In order to make transparent material visible or to enhance the colour of a specimen it is necessary to stain it before mounting it on a slide and viewing it under a microscope. There is a wide range of stains which colour different tissues and organelles in a variety of ways. As such, these stains are beyond the scope of this book, but it is important to realize that in most prepared slides the colours observed are not the natural colours of the material but the result of staining.

To make a temporary slide follow the procedures below:

1. Take a clean, dry microscope slide and coverslip.

2. Place a drop of the liquid (usually water or glycerine) in which the material is to be mounted, in the centre of the slide.

3. Put the material to be observed into the liquid without trapping any air-bubbles.

4. Place the coverslip at one end of the slide and at an angle of 45°.

5. Pull the coverslip along the slide using the finger and thumb, until it reaches the drop of liquid containing the material.

6. Place a seeker, or needle, under the upper edge of the coverslip to support it (Fig. 4.2).

7. Slowly lower the coverslip, using the seeker, until it is lying flat on the slide. Only by lowering it slowly and carefully will air have time to escape and bubbles be avoided.

8. Draw off any excess liquid from around the coverslip.

A well-prepared temporary mount should have:

1. The material central on the slide and more or less central under the coverslip.

2. The coverslip (if square) should have its sides parallel to the sides of the slide.

3. No air bubbles under the coverslip.

4. No excess liquid on the slide or coverslip.

4.1.3 Structure of the electron microscope

The limitations imposed upon the resolving power of the light microscope by the wavelength of light led scientists to consider the

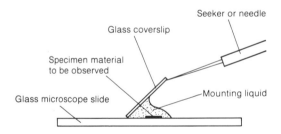

Fig. 4.2 *Making a temporary microscope mount*

Seeker or needle

Glass coverslip

Specimen material to be observed

Glass microscope slide

Mounting liquid

use of forms of radiation with shorter wavelengths. This led, in 1933, to the development of the first electron microscope. This instrument works on the same principles as the light microscope except that instead of light rays, with their wavelengths in the order 500 nm, a beam of electrons of wavelengths 0.005 nm is used. This means that in theory the electron microscope should be able to magnify objects up to 100 000 000 times. In practice it magnifies just over 500 000 times but it still compares favourably with the best light microscopes, which magnify only around 1500 times.

Whereas the light microscope uses glass lenses to focus the light rays, the electron beam of the electron microscope is focussed by means of powerful electromagnets. The image produced by the electron microscope cannot be detected directly by the naked eye. Instead, the electron beam is directed onto a screen from which black and white photographs, called **photoelectronmicrographs**, can be taken. A comparison of the radiation pathways in a light and electron microscope is given in Fig. 4.3.

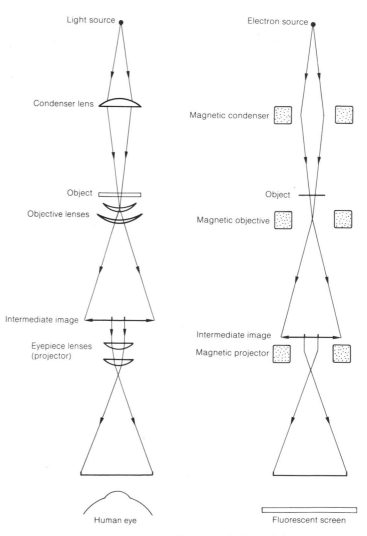

Light source

Condenser lens

Object
Objective lenses

Intermediate image

Eyepiece lenses
(projector)

Human eye

Electron source

Magnetic condenser

Object

Magnetic objective

Intermediate image

Magnetic projector

Fluorescent screen

Fig. 4.3 Comparison of radiation pathways in light and electron microscopes

Electron microscope

There are two types of electron microscope. In the **transmission electron microscope**, a beam of electrons is passed through thin, specially prepared slices of material (Section 4.1.4). Where electrons are absorbed by the material, and do not therefore reach the screen, the image is dark. Such areas are said to be **electron dense**. Where the electrons penetrate, the screen appears bright. These areas are termed **electron transparent**. As electrons have a very small mass, they do not easily penetrate materials and so sections need to be exceedingly thin. This sectioning creates a flat image and the natural contouring of a specimen cannot be seen. To overcome this problem, the **scanning electron microscope** was developed. In this instrument a fine beam of electrons is passed to and fro across the specimen, beginning at one end and working across to the other. The specimen scatters many electrons, whereas others are absorbed. Low energy, secondary electrons may be emitted by the specimen. The scattered electrons and the low energy secondary ones are amplified and transmitted to a screen. The resultant image shows holes and depressions as dark areas and ridges and extensions of the surface as bright areas. In this way the natural contouring of the material may be observed.

4.1.4 Preparation and staining of material for an electron microscope

The preparation of material for the electron microscope is a long and complicated process. In order that biological material, which is often fragile, can withstand the processes it must first be treated with a fixative to stabilize it. Commonly used fixatives are osmium tetroxide and glutaraldehyde.

Once fixed, the material often requires sectioning. A typical section is a mere 50 nm thick. To achieve this, without the section falling apart, it is embedded in a firm plastic or epoxy resin (the adhesive Araldite is an example). In this more rigid form, thin enough sections may be cut using a machine called a **microtome**.

Biological material may absorb or transmit light and so it may be viewed directly under a light microscope. At the thickness used in the electron microscope, biological material will transmit or scatter electrons equally and so different parts cannot be distinguished. The material must be stained. The use of dyes which absorb different light wavelengths, as used for a light microscope, is ineffectual. Instead, materials which scatter electrons must be used. The best materials are heavy metals such as uranium, tungsten and lead. The specimens may be soaked in a solution of certain salts of these metals, e.g. uranyl acetate or sodium phosphotungstate – **negative staining.** Another method, called **shadowing**, involves spraying evaporated heavy metals onto the specimen.

Material to be studied under an electron microscope cannot be supported on a glass slide because the glass would scatter the electron beam. It is therefore supported on a metal grid, usually of copper, which has a mesh of about 2 mm diameter, Across the mesh is a plastic film, less than 1 nm thick, which transmits electrons.

4.1.5 Comparison of light and electron microscopes

With its greater resolving power, the electron microscope clearly has a major advantage over the light microscope. Nevertheless, the light microscope is still much more widely used because the electron microscope is so expensive and cannot observe living material. The expertise needed to operate it and its sheer size prevent its use in most schools and colleges. A comparison of the advantages and disadvantages of each type of microscope is given in Table 4.1.

TABLE 4.1 **Comparison of advantages and disadvantages of the light and electron microscopes**

LIGHT MICROSCOPE	ELECTRON MICROSCOPE
Advantages	**Disadvantages**
Cheap to purchase (£100–500)	Expensive to purchase (over £1 000 000)
Cheap to operate – uses a little electricity where there is a built-in light source	Expensive to operate – requires up to 100 000 volts to produce the electron beam
Small and portable – can be used almost anywhere	Very large and must be operated in special rooms
Unaffected by magnetic fields	Affected by magnetic fields
Preparation of material is relatively quick and simple, requiring only a little expertise	Preparation of material is lengthy and requires considerable expertise and sometimes complex equipment
Material rarely distorted by preparation	Preparation of material may distort it
Living as well as dead material may be viewed	A high vacuum is required and living material cannot be observed
Natural colour of the material can be observed	All images are in black and white
Disadvantages	**Advantages**
Magnifies objects up to 1500 ×	Magnifies objects over 500 000 ×
Can resolve objects up to 200 nm apart	Has a resolving power for biological specimens of around 1 nm
The depth of field is restricted	It is possible to investigate a greater depth of field

4.2 Cytology – the study of cells

All cells are self-contained and more or less self-sufficient units. They are surrounded by a cell membrane and have a nucleus, or a nuclear area, at some stage of their existence. They show remarkable diversity, both in structure and function. They are basically spherical in shape, although they show some variation where they are modified to suit their function. In size they normally range from 10–30 μm.

4.2.1 The structure of prokaryotic cells

Prokaryotic cells (*pro* – 'before', *karyo* – 'nucleus') were probably the first forms of life on earth. Fossils of prokaryotic cells have been dated at 3500 million years old. Their hereditary material, DNA, is not enclosed within a nuclear membrane. This absence of a true nucleus only occurs in two groups, the bacteria and the blue-green bacteria (Section 7.2). There are no membrane-bounded organelles within a prokaryotic cell, the structure of which is shown below.

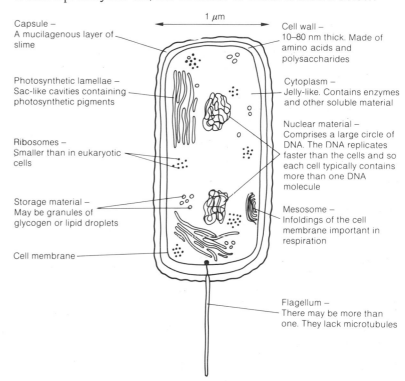

Fig. 4.4 Structure of the prokaryotic cell, e.g. a generalized bacterial cell

4.2.2 Structure of the eukaryotic cell

Eukaryotic cells (*Eu* – 'true', *karyo* – 'nucleus') probably arose a little over 1000 million years ago, nearly 2500 million years after their prokaryotic ancestors. The development of eukaryotic cells from prokaryotic ones involved considerable changes, as can be seen from Table 4.2. The essential change was the development of membrane-bounded organelles, such as mitochondria and chloroplasts, within the outer plasma membrane of the cell. The presence of membrane-bounded organelles confers four advantages:
1. Many metabolic processes involve enzymes being embedded in a membrane. As cells become larger, the proportion of membrane area to cell volume is reduced. This proportion is increased by the presence of organelle membranes.

TABLE 4.2 **Comparison of prokaryotic and eukaryotic cells**

PROKARYOTIC CELLS	EUKARYOTIC CELLS
No distinct nucleus; only diffuse area(s) of nucleoplasm with no nuclear membrane	A distinct, membrane-bounded nucleus
No chromosomes – circular strands of DNA	Chromosomes present on which DNA is located
No membrane-bounded organelles such as chloroplasts and mitochondria	Chloroplasts and mitochondria may be present
Ribosomes are smaller	Ribosomes are larger
Flagella (if present) lack internal 9 + 2 fibril arrangement	Flagella have 9 + 2 internal fibril arrangement
No mitosis or meiosis occurs	Mitosis and/or meiosis occurs

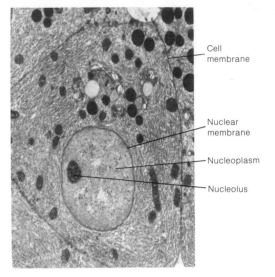

General animal cell (EM) (× 5000 approx.)

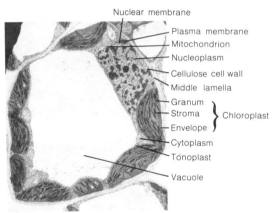

Plant cell (EM) (× 4500 approx.)

2. Containing enzymes for a particular metabolic pathway within organelles means that the products of one reaction will always be in close proximity to the next enzyme in the sequence. The rate of metabolic reactions will thereby be increased.

3. The rate of any metabolic pathway inside an organelle can be controlled by regulating the rate at which the membrane surrounding the organelle allows the first reactant to enter.

4. Potentially harmful reactants and/or enzymes can be isolated inside an organelle so they won't damage the rest of the cell.

It is possible that such organelles arose as separate prokaryotic cells which developed a symbiotic relationship with larger prokaryotic ones. This would explain the existence of one membraned structure inside another, the ability of mitochondria and chloroplasts to divide

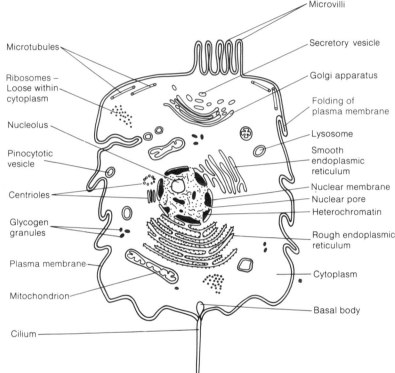

Fig. 4.5(a) A generalized animal cell

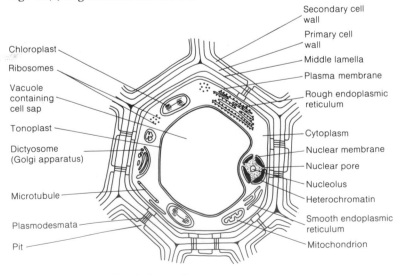

Fig. 4.5(b) A generalized plant cell

TABLE 4.3 **Differences between plant and animal cells**

PLANT CELLS	ANIMAL CELLS
Tough, slightly elastic cellulose cell wall present (in addition to the cell membrane)	Cell wall absent – only a membrane surrounds the cell
Pits and plasmodesmata present in the cell wall	No cell wall and therefore no pits or plasmadesmata
Middle lamella join cell walls of adjacent cells	Middle lamella absent – cells are joined by intercellular cement
Plastids, e.g. chloroplast and leucoplasts, present in large numbers	Plastids absent
Mature cells normally have a large single, central vacuole filled with cell sap	Vacuoles, e.g. contractile vacuoles, if present, are small and scattered throughout the cell
Tonoplast present around vacuole	Tonoplast absent
Cytoplasm normally confined to a thin layer at the edge of the cell	Cytoplasm present throughout the cell
Nucleus at edge of the cell	Nucleus anywhere in the cell but often central
Lysosomes not normally present	Lysosomes almost always present
Centrioles absent in higher plants	Centrioles present
Cilia and flagella absent in higher plants	Cilia or flagella often present
Starch grains used for storage	Glycogen granules used for storage
Only meristematic cells are capable of division	Almost all cells are capable of division
Few secretions are produced	A wide variety of secretions are produced

themselves (self-replication) and the presence of DNA within these two organelles. Alternatively, the organelles may have arisen by invaginations of the plasma membrane which became 'pinched off' to give a separate membrane-bounded structure within the main cell. Although many variations of the eukaryotic cell exist, there are two main types, the plant cell and the animal cell. (See Figs. 4.5(a) and (b) on the previous page.)

4.2.3 Differences between plant and animal cells

The major differences between plant and animal cells are given in Table 4.3.

4.3   Cell ultrastructure

4.3.1 Separation of cell organelles

The separation of different parts of the cell, or **cell fractionation**, is an important technique in the study of both cell structure and function. The tissue to be studied is taken and cut into small pieces or minced. These are then ground into small fragments in a homogenizer (rather like a sophisticated liquidizer). A suspension of these fragments is then placed in a **centrifuge**. A centrifuge is a machine which can spin tubes containing liquid suspensions at a very high speed. The effect is to exert a force on the contents of the tube similar to, but much greater than, that of gravity. The faster the speed at which the tubes are spun, the greater the force. At slower speeds (less force) the larger fragments collect at the bottom of the tube and the smaller ones remain in suspension in the liquid near the top of the tube – **supernatant liquid**. If the larger fragments are removed and the supernatant recentrifuged at a faster speed (more force), some of the smaller fragments will collect at the bottom. By continuing in this way, smaller and smaller fragments may be recovered. As the size of any organelle is relatively constant, each organelle will tend to separate from the supernatant at a specific speed of rotation. If the suspension of cell fragments is spun at a slower speed than that required to separate out a particular organelle, all larger fragments and organelles can be collected and discarded. Spinning the supernatant at the appropriate speed will now cause a new fraction to be collected. This fraction will be a relatively pure sample of the required organelles. Since the process involves centrifuging at different speeds, it is called **differential centrifugation**. Typically, spinning at slower speeds throws down large pieces of cell debris, e.g. cell wall fragments, where plant cells are used. At faster speeds nuclei are separated out, and as the speed is further increased smaller and smaller organelles can be isolated, e.g. chloroplasts, mitochondria, fragments of endoplasmic reticulum and finally ribosomes.

4.3.2 Cytoplasmic matrix

All the cell organelles are contained within a cytoplasmic matrix, sometimes called the **hyaloplasm** or **cytosol**. It is an aqueous material which is a solution or colloidal suspension of many vital cellular chemicals. These include simple ions such as sodium, phosphates and chlorides, organic molecules such as amino acids, ATP and nucleotides, and storage material such as oil droplets. Many important biochemical processes, including glycolysis, occur within the cytoplasm. It is not static but capable of mass flow, which is called **cytoplasmic streaming**.

4.3.3 Cell membranes

The cell membrane's main function is to serve as a boundary between the cell and its environment. It is not, however, inert but a functional organelle. It may permanently exclude certain substances from the cell while permanently retaining others. Some substances may pass freely in and out through the membrane. Yet others may be excluded at one moment only to pass freely across the membrane on another. occasion. On account of the membrane's ability to permit different substances to pass across it at different rates, it is said to be **partially permeable**.

There is little dispute that the cell membrane is made up almost entirely of two chemical groups – proteins and phospholipids. The precise arrangement of these chemicals is less certain. There are two main theories.

The protein-phospholipid sandwich (Davson-Danielli model)
We have seen in Section 3.7.2 that phospholipids comprise a hydrophilic (water-loving) head and a hydrophobic (water-repelling) tail. With this in mind, H. Davson and J. F. Danielli in 1935 proposed a model of the cell membrane in which the phospholipid molecules formed a bimolecular layer. The hydrophobic tails associated with each other at the centre of the membrane with the hydrophilic heads extending towards the surface. Each side of these phospholipids was a layer of protein molecules, rather like the bread on either side of a sandwich (Fig. 4.6).

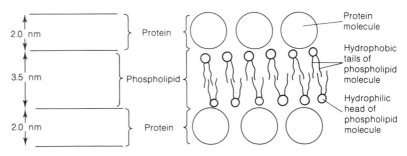

Fig. 4.6 The Davson–Danielli model of the structure of the cell membrane

The fluid-mosaic model (Singer-Nicholson model)
In 1972, J. J. Singer and G. L. Nicholson suggested a modified structure for the cell membrane. The bimolecular phospholipid layer with its inwardly directed hydrophobic tails remains unchanged. However, it is suggested that the protein molecules vary in size and have a much less regular arrangement (Fig. 4.7, on the next page). Some proteins occur on the surface of the phospholipid layer, while others extend into it; some even extend completely across. Viewed from the surface, the proteins are dotted throughout the phospholipid layer in a mosaic arrangement. Other research suggests that the phospholipid layer is capable of much movement, i.e. is fluid. It was these facts which gave rise to its name, the **fluid-mosaic model**.

The proteins in the membrane have a number of functions. Apart from giving structural support they are very specific, varying from cell to cell. It is this specificity which allows cells to be recognized

3-layered
structure of
cell membrane

Cell membrane (EM) (× 250 000)

by other agents in the body, e.g. enzymes, hormones and antibodies. In the fluid-mosaic model it is thought probable that the proteins also assist the active transport of materials across the membrane.

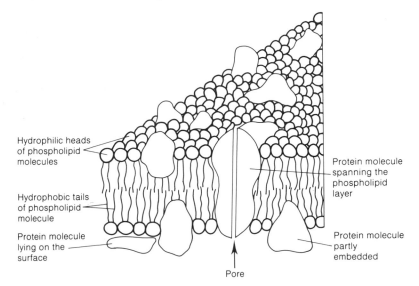

Fig. 4.7 *The fluid-mosaic model of the cell membrane*

Fig. 4.7 *The fluid-mosaic model of the cell membrane*

MEMBRANOUS ORGANELLES

4.3.4 The nucleus

When viewed under a microscope, the most prominent feature of a cell is the nucleus. While its shape, size, position and chemical composition vary from cell to cell, its functions are always the same, namely, to control the cell's activity and to retain the organism's hereditary material, the chromosomes. It is bounded by a double membrane, the **nuclear envelope**. It possesses many large pores, 40–100 nm in diameter, which permit the passage of large molecules, such as RNA, between it and the cytoplasm. The cytoplasm-like material within the nucleus is called **nucleoplasm**. It contains **chromatin** which is made up of coils of DNA bound to proteins. During division the chromatin condenses to form the chromosomes but these are rarely, if ever, visible in a non-dividing cell. The denser, more darkly staining areas of chromatin are called **heterochromatin**.

Within the nucleus are one or two small round bodies, each called a **nucleolus**. They are not distinct organelles as they are not bounded by a membrane. They manufacture ribosomal RNA, a substance in which they are especially rich, and assemble ribose.

The functions of a nucleus are:

1. To contain the genetic material of a cell in the form of chromosomes.
2. To act as a control centre for the activities of a cell.
3. The nuclear DNA carries the instructions for the synthesis of proteins.
4. It is involved in the production of ribosomes and RNA.
5. It is essential for cell division.

4.3.5 The chloroplast

Chloroplasts belong to a larger group of organelles known as

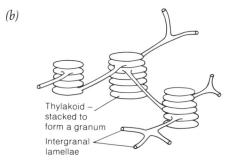

Nucleus (EM) (× 12 000 approx.)

(a)

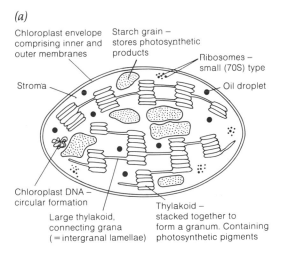

Chloroplast envelope comprising inner and outer membranes

Starch grain – stores photosynthetic products

Ribosomes – small (70S) type

Oil droplet

Stroma

Chloroplast DNA – circular formation

Large thylakoid, connecting grana (= intergranal lamellae)

Thylakoid – stacked together to form a granum. Containing photosynthetic pigments

(b)

Thylakoid – stacked to form a granum

Intergranal lamellae

Fig. 4.8 *Diagrams to show the structure of the chloroplast*

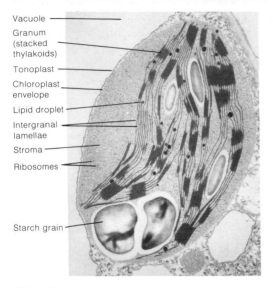

Vacuole
Granum (stacked thylakoids)
Tonoplast
Chloroplast envelope
Lipid droplet
Intergranal lamellae
Stroma
Ribosomes
Starch grain

Chloroplast (EM) (× 13 000 approx.)

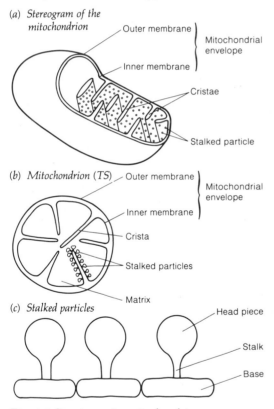

(a) *Stereogram of the mitochondrion*

Outer membrane
Inner membrane
} Mitochondrial envelope
Cristae
Stalked particle

(b) *Mitochondrion (TS)*

Outer membrane
Inner membrane
} Mitochondrial envelope
Crista
Stalked particles
Matrix

(c) *Stalked particles*

Head piece
Stalk
Base

Fig. 4.9 Structure of a mitochondrion

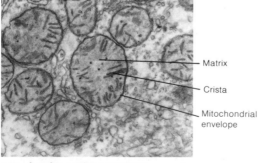

Matrix
Crista
Mitochondrial envelope

Mitochondrion (EM) (× 15 000 approx.)

plastids. In higher plants most chloroplasts are 4–10 μm long and are bounded by a double membrane, the **chloroplast envelope**, about 30 nm thick. While the outer membrane has a similar structure to the plasma membrane, the inner one is folded into a series of lamellae.

Within the chloroplast envelope are two distinct regions. The **stroma** is a colourless, structureless matrix in which are embedded structures rather like stacks of coins in appearance. These are the **grana**. Each granum, and there may be around fifty in a chloroplast, is made up of between two and a hundred closed flattened sacs called **thylakoids**. Within these are located the photosynthetic pigments such as chlorophyll, details of which are given in Section 23.2.1. Some thylakoids have tubular extensions which interconnect adjacent grana (Fig. 4.8).

Also present within the stroma are a series of starch grains which act as temporary stores for the products of photosynthesis. A number of smaller granules within the stroma readily take up osmium salts during the preparation of material for the electron microscope. They are called **osmiophilic granules** (*osmio* – 'osmium', *philo* – 'liking') but their function is not yet clear. A small amount of DNA is always present within the stroma, as are oil droplets.

4.3.6 The mitochondrion

Mitochondria are found within the cytoplasm of all eukaryotic cells, although in highly specialized cells such as mature red blood cells they may be absent. They range in shape from spherical to highly elongated. In length they vary from 1 to 10 μm and in width from 0.25 to 1.0 μm. They are bounded by a double membrane, the outer of which controls the entry and exit of chemicals. The inner membrane is folded inwards, giving rise to extensions called **cristae**, some of which extend across the entire organelle. They function to increase the surface area on which respiratory processes take place. The surface of these cristae has stalked granules along its length (Fig. 4.9).

The remainder of the mitochondrion is the **matrix**. It is a semi-rigid material containing protein, lipids and traces of DNA. Electron-dense granules of 25 nm diameter also occur.

Mitochondria function as sites for certain stages of respiration, details of which are given in Section 26.4.1. The number of mitochondria in a cell therefore varies with its metabolic activity. Highly active cells may possess up to 1000. Similarly the number of cristae increases in metabolically active cells, giving weight to the proposition that respiratory enzymes are located on them.

4.3.7 Endoplasmic reticulum

The endoplasmic reticulum (ER) is an elaborate system of membranes found throughout the cell, forming a cytoplasmic skeleton. It is an extension of the outer nuclear membrane with which it is continuous. The membranes form a series of sheets which enclose flattened sacs called **cisternae** (Fig. 4.10 on the next page). Its structure varies from cell to cell and can probably change its nature rapidly; the membranes of the ER may be loosely organized or tightly packed. Where the membranes are lined with ribosomes they are called **rough endoplasmic reticulum**. The rough ER is concerned with protein synthesis (Section 12.6) and is consequently most abundant in those cells which are rapidly growing or secrete enzymes. In the same way, damage to a cell often results in increased formation of ER in order to produce the proteins necessary for the cell's repair. Where the membranes lack ribosomes they are called **smooth endoplasmic reticulum**. The smooth ER is concerned with lipid synthesis and is consequently most abundant in those cells producing lipid-related secretions, e.g. the sebaceous glands of mammalian skin and cells secreting steroids.

The functions of the ER may thus be summarized as:

1. Providing a large surface area for chemical reactions.

2. Providing a pathway for the transport of materials through the cell.

3. Producing proteins, especially enzymes (rough ER).

4. Producing lipids and steroids (smooth ER).

5. Collecting and storing synthesized material.

6. Providing a structural skeleton to maintain cellular shape (e.g. the smooth ER of a rod cell from the retina of the eye).

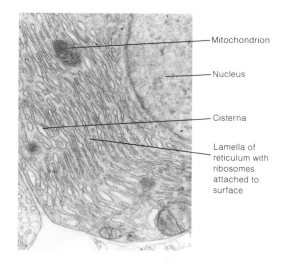

Endoplasmic reticulum (EM) (× 9000 approx.)

Labels: Mitochondrion; Nucleus; Cisterna; Lamella of reticulum with ribosomes attached to surface

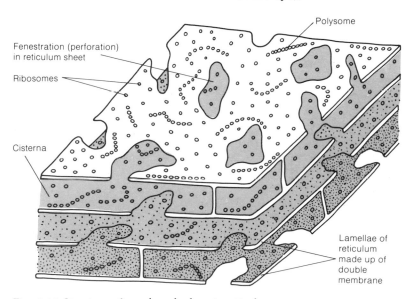

Labels: Polysome; Fenestration (perforation) in reticulum sheet; Ribosomes; Cisterna; Lamellae of reticulum made up of double membrane

Fig. 4.10 Structure of rough endoplasmic reticulum

4.3.8 Golgi apparatus (dictyosome)

The Golgi apparatus, named after its discoverer Camillo Golgi, has a similar structure to the smooth endoplasmic reticulum but is more compact. It is composed of stacks of flattened sacs made of membranes. The sacs are fluid-filled and pinch off smaller membranous sacs, called **vesicles**, at their ends. There is normally only one Golgi apparatus in each animal cell but in plant cells there may be a large number of stacks known as **dictyosomes**. Its position and size varies from cell to cell but it is well developed in secretory cells and neurones and is small in muscle cells. This suggests that the Golgi apparatus plays some role in the production of secretory material. In particular, it is thought to perform the following functions:

1. Produces glyco-proteins such as mucin required in secretions, by adding the carbohydrate part to the protein.

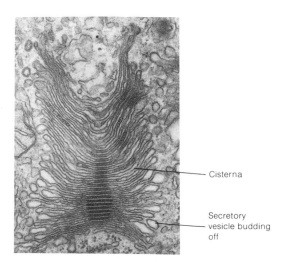

Golgi apparatus (EM) (× 30 000 approx.)

Labels: Cisterna; Secretory vesicle budding off

2. Produces secretory enzymes, e.g. the digestive enzymes of the pancreas.

3. Secretes carbohydrates such as those involved in the production of new cell walls.

4. Transports and stores lipids.

5. Forms lysosomes as described in Section 4.3.9.

4.3.9 Lysosomes

Lysosomes (*lysis* – 'splitting', *soma* – 'body') are similar in size to spherical mitochondria, being 0.2–0.5 μm in diameter. Separation of these two organelles by differential centrifugation is therefore difficult but can be achieved by altering the density of lysosomes by injecting a living organism with chemicals which are accumulated in the lysosomes. These chemicals make the lysosomes denser and so they are separated at slower centrifuge speeds than mitochondria during differential centrifugation.

Lysosomes are bounded by a single membrane but unlike mitochondria they have no cristae or other internal structure. They contain a large number of enzymes, mostly hydrolases, in acid solution. They isolate these enzymes from the remainder of the cell. By so doing they prevent them from acting upon other chemicals and organelles within the cell.

The functions of lysosomes are:

1. They digest material which the cell consumes from the environment. In the case of white blood cells, this may be bacteria or other harmful material. In Protozoa, it is the food which has been consumed by phagocytosis. In either case the material is broken down within the lysosome, useful chemicals are absorbed into the cytoplasm and any debris is egested by the cell (Fig. 4.11).

2. They digest parts of the cell, such as worn-out organelles, in a similar way to that described in **1**. After the death of the cell they are responsible for its complete breakdown, a process called **autolysis** (*auto* – 'self', *lysis* – 'splitting').

3. They release their enzymes outside the cell (**exocytosis**) in order to break down other cells, e.g. in the reabsorption of tadpole tails during metamorphosis.

In view of their function, it is hardly surprising that lysosomes are especially abundant in secretory cells and in phagocytic white blood cells.

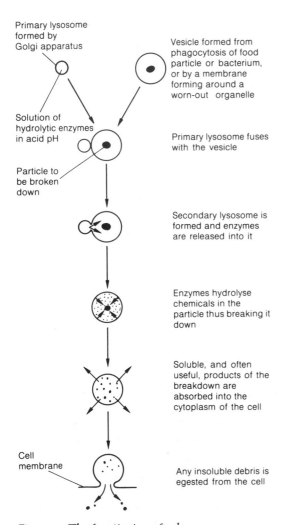

Primary lysosome formed by Golgi apparatus

Vesicle formed from phagocytosis of food particle or bacterium, or by a membrane forming around a worn-out organelle

Solution of hydrolytic enzymes in acid pH

Primary lysosome fuses with the vesicle

Particle to be broken down

Secondary lysosome is formed and enzymes are released into it

Enzymes hydrolyse chemicals in the particle thus breaking it down

Soluble, and often useful, products of the breakdown are absorbed into the cytoplasm of the cell

Cell membrane

Any insoluble debris is egested from the cell

Fig. 4.11 The functioning of a lysosome

4.3.10 Microbodies (peroxisomes)

Microbodies are small spherical membrane-bounded bodies which are between 0.5 and 1.5μm in diameter. Apart from being slightly granular, they have no internal structure. They contain a number of metabolically important enzymes, in particular the enzyme catalase. It catalyses the breakdown of hydrogen peroxide and hence these microbodies are sometimes called peroxisomes.

Hydrogen peroxide is a potentially toxic by-product of many biochemical reactions within organisms. Peroxisomes containing catalase are therefore particularly numerous in actively metabolizing cells like those of the liver.

$$2H_2O_2 \xrightarrow{\text{catalase}} 2H_2O + O_2$$

Hydrogen peroxide   Water   Oxygen

4.3.11 Vacuoles

A fluid-filled sac bounded by a single membrane may be termed a vacuole. Within mature plant cells there is usually one large central vacuole. The single membrane around it is called the **tonoplast**. It contains a solution of mineral salts, sugars, amino acids, wastes (e.g. tannins) and sometimes also pigments such as **anthocyanins**.

Plant vacuoles serve a variety of functions:

1. The sugars and amino acids may act as a temporary food store.

2. The anthocyanins are of various colours and so may colour petals to attract pollinating insects, or fruits to attract animals for dispersal.

3. They act as temporary stores for organic wastes, such as tannins. These may accumulate in the vacuoles of leaf cells and are removed when the leaves fall.

4. They occasionally contain hydrolytic enzymes and so perform functions similar to those of lysosomes (Section 4.3.9).

5. They support herbaceous plants, and herbaceous parts of woody plants by providing an osmotic system which creates a pressure potential (Section 33.2).

In animal cells, vacuoles are much smaller but may occur in larger numbers. Common types include food vacuoles, phagocytic vacuoles and contractile vacuoles. The latter are important in the osmoregulation of certain protozoans (Section 34.3.1).

Non-membranous structures

4.3.12 Ribosomes

Ribosomes are small cytoplasmic granules found in all cells. They are around 20 nm in diameter in eukaryotic cells (80S type) but slightly smaller in prokaryotic ones (70S type). They may occur in groups called **polysomes** and may be associated with endoplasmic reticulum or occur freely within the cytoplasm. Despite their small size, their enormous numbers mean that they can account for up to 20% of the mass of a cell.

Ribosomes are made up of relatively small RNA molecules and protein. They are important in protein synthesis (Section 12.6).

4.3.13 Storage granules

Every cell contains a limited store of food energy. This store may be in the form of soluble material such as the sugar found in the vacuoles of plant cells. It may also occur in insoluble form, as grains or granules, within cells or organelles.

Starch grains occur within chloroplasts and the cytoplasm of plant cells. Starch may also be stored in specialized leucoplasts called amyloplasts. **Glycogen granules** occur throughout the cytoplasm of animal cells. They store animal starch or glycogen. **Oil or lipid droplets** are found within the cytoplasm of both plant and animal cells.

(a) LS Basal region of a cilium

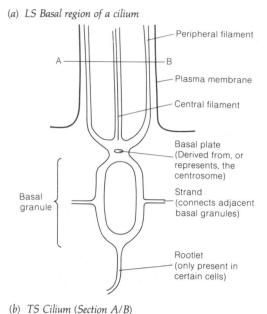

(b) TS Cilium (Section A/B)

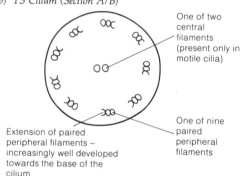

Fig. 4.12 Structure of a cilium

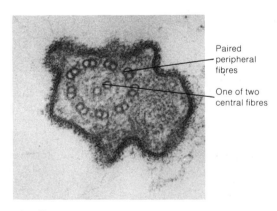

Flagellum (EM) (× 92 000 approx.)

4.3.14 Microtubules

Microtubules occur widely throughout living cells. They are slender, unbranched tubes about 20 nm in diameter and up to several microns in length.

Their functions are:

1. To provide an internal skeleton (**cytoskeleton**) for cells and so help determine their shape.

2. They may aid transport within cells by providing routes along which materials move.

3. They form a framework along which the cellulose cell wall of plants is laid down.

4. They are major components of cilia and flagella where they contribute to their movement.

5. They are found in the spindle during cell division and within the centrioles from which the spindle is formed.

4.3.15 Cilia and flagella

Cilia and flagella are almost identical, except that cilia are usually shorter and more numerous. Both are around 0.2 μm in diameter; cilia are about 10 μm long whereas flagella may be 100 μm long. They are found in a limited number of cells but are nevertheless of great importance. The structure of a cilium is shown in Fig. 4.12. They function to either move an entire organism, e.g. cilia on the protozoan *Paramecium*, or to move material within an organism, e.g. the cilia lining the respiratory tract move mucus towards the throat. The action of cilia is discussed in Section 39.2.

4.3.16 Centrioles

Centrioles have the same basic structure as the basal bodies of cilia. They are hollow cylinders about 0.2 μm in diameter. They arise in a distinct region of the cytoplasm known as the **centrosome**. It contains two centrioles. At cell division they migrate to opposite poles of the cell where they synthesize the microtubules of the spindle. Despite the absence of centrioles, the cells of higher plants do form spindles.

4.3.17 Microfilaments

Microfilaments are very thin strands about 6 nm in diameter. They are usually made up of the protein actin although a smaller proportion are of myosin. As these are the two proteins involved in muscle contraction, it seems probable that microfilaments play a role in movement within cells and possibly of the cells as a whole in some cases.

4.3.18 Microvilli

Microvilli are tiny finger-like projections about 0.6 μm in length on the membranes of certain cells, such as those of the intestinal epithelium and the kidney tubule. They should not be confused with the much larger villi which are multicellular structures. Microvilli massed together appear similar to the bristles of a brush, hence the term **brush border** given to the edge of cells bearing microvilli. Actin filaments within the microvilli allow them to contract, which, along with their large surface area, facilitates absorption.

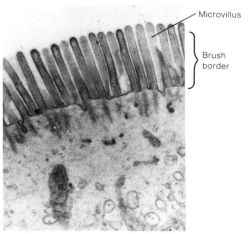

Microvilli (EM) (× 23 000 approx.)

4.3.19 Cellulose cell wall

A cell wall is a characteristic feature of plant cells. It consists of cellulose macrofibrils embedded in an amorphous polysaccharide matrix. The structure and properties of cellulose are discussed in Section 3.5.3 and the detailed structure of a cellulose macrofibril is given in Fig. 4.13. The matrix is usually composed of polysaccharides, e.g. pectin or lignin. The macrofibrils may be regular or irregular in arrangement.

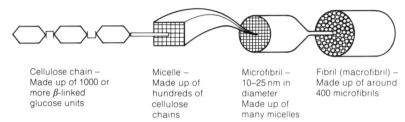

Cellulose chain – Made up of 1000 or more β-linked glucose units

Micelle – Made up of hundreds of cellulose chains

Microfibril – 10–25 nm in diameter Made up of many micelles

Fibril (macrofibril) – Made up of around 400 microfibrils

Fig. 4.13 Structure of a cellulose macrofibril

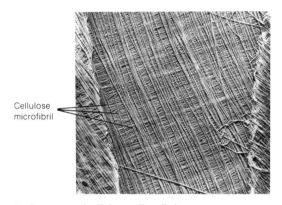

Surface view of cellulose cell wall showing arrangement of microfibrils (EM) (× 9500 approx.)

The main functions of the cell wall are:

1. To provide support in herbaceous plants. As water enters the cell osmotically the cell wall resists expansion and an internal pressure is created which provides turgidity for the plant.

2. To give direct support to the cell and the plant as a whole by providing mechanical strength. The strength may be increased by the presence of lignin in the matrix between the cellulose fibres.

3. To permit the movement of water through and along it and so contribute to the movement of water in the plant as a whole, in particular in the cortex of the root.

4. In some cell walls the presence of cutin, suberin or lignin in the matrix makes the cells less permeable to substances. Lignin helps to keep the water within the xylem; cutin in the epidermis of leaves prevents water being lost from the plant and suberin in root endodermal cells prevents movement of water across them, thus concentrating its movement through special passage cells.

5. The arrangement of the cellulose fibrils in the cell wall can determine the pattern of growth and hence the overall shape of a cell.

6. Occasionally cell walls act as food reserves.

4.4 Movement in and out of cells

The various organelles and structures within a cell require a variety of substances in order to carry out their functions. In turn they form products, some useful and some wastes. Most of these substances must pass in and out of the cell. This they do by **diffusion, osmosis, active transport, phagocytosis** and **pinocytosis.**

1. *If 10 particles occupying the left hand side of a closed vessel are in random motion, they will collide with each other and the sides of the vessel. Some particles from the left hand side move to the right, but initially there are no available particles to move in the opposite direction, so the movement is in one direction only. There is a large concentration gradient and diffusion is rapid.*

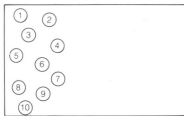

2. *After a short time the particles (still in random motion) have spread themselves more evenly. Particles can now move from right to left as well as left to right. However with a higher concentration of particles (7) on the left than on the right (3) there is a greater probability of a particle moving to the right than in the reverse direction. There is a smaller concentration gradient and diffusion is slower.*

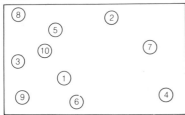

3. *Some time later, the particles will be evenly distributed throughout the vessel and the concentrations will be equal on each side. The system is in **equilibrium**. The particles are not however static but remain in random motion. With equal concentrations on each side, the probability of a particle moving from left to right is equal to the probability of one moving in the opposite direction. There is no concentration gradient and no net diffusion.*

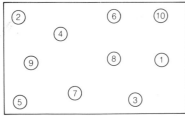

4. *At a later stage the particles remain evenly distributed and will continue to do so. Although the number of particles on each side remains the same, individual particles are continuously changing position. This situation is called a **dynamic equilibrium**.*

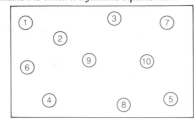

Fig. 4.14 Diffusion

4.4.1 Diffusion

Diffusion is the process by which a substance moves from a region of high concentration of that substance to a region of low concentration of the same substance. Diffusion occurs because the molecules of which substances are made are in random motion (kinetic theory). The process is explained in Fig. 4.14.

The rate of diffusion depends upon:

1. The concentration gradient – The greater the difference in concentration between two regions of a substance the greater the rate of diffusion. Organisms must therefore maintain a fresh supply of a substance to be absorbed by creating a stream over the diffusion surface. Equally, the substance, once absorbed, must be rapidly transported away.

2. The distance over which diffusion takes place – The shorter the distance between two regions of different concentration the greater the rate of diffusion. The rate is proportional to the reciprocal of the square of the distance (inverse square law). Any structure in an organism across which diffusion regularly takes place must therefore be thin. Cell membranes for example are only 7.5 nm thick and even epithelial layers such as those lining the alveoli of the lungs are as thin as 0.3 μm across.

3. The area over which diffusion takes place – The larger the surface area the greater the rate of diffusion. Diffusion surfaces frequently have structures for increasing their surface area and hence the rate at which they exchange materials. These structures include villi and microvilli.

4. The nature of any structure across which diffusion occurs – Diffusion frequently takes place across epithelial layers or cell membranes. Variations in their structure may affect diffusion. For example, the greater the number and size of pores in cell membranes the greater the rate of diffusion.

5. The size and nature of the diffusing molecule – Smaller molecules diffuse faster than large ones. Fat-soluble ones diffuse more rapidly through cell membranes than water-soluble ones.

Facilitated diffusion is a special form of diffusion which allows more rapid exchange. It appears to involve channels within a membrane which make diffusion of specific substances easier. Carrier molecules may also be involved. The process is passive, not involving any energy expenditure.

4.4.2 Osmosis

Osmosis is a special form of diffusion which involves the movement of solvent molecules. The solvent in biological systems is invariably water. Most cell membranes are permeable to water and certain solutes only. Such membranes are termed **partially permeable.** Osmosis in living organisms can therefore be defined as: **the passage of water from a region where it is highly concentrated to a region where its concentration is lower, through a partially permeable membrane.** The process is explained in Fig. 4.15.

If a solution is separated from its pure solvent, as in Fig. 4.15, the pressure which must be applied to stop water entering that solution, and so prevent osmosis, is called the **osmotic pressure**. The more

1. *Both solvent (water) and solute (glucose) molecules are in random motion, but only solvent (water) molecules are able to cross the partially permeable membrane. This they do until their concentration is equal on both sides of the membrane.*

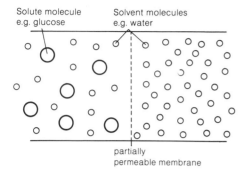

2. *Once the water molecules are evenly distributed, in theory a dynamic equilibrium should be established. However, the water molecules on the left of the membrane are impeded to some extent by the glucose molecules from crossing the membrane. With no glucose present on the right of the membrane, water molecules move more easily to the left than in the reverse direction.*

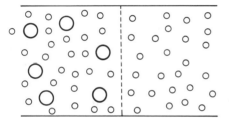

3. *A situation is reached whereby additional water molecules accumulate on the left of the membrane, until their greater concentration offsets the blocking effect of the glucose. The probability of water molecules moving in either direction is the same, and a dynamic equilibrium is established.*

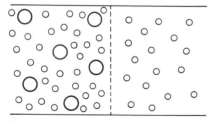

N.B. Solution = solute + solvent
e.g.   Glucose solution = glucose powder + water

Fig. 4.15 Osmosis

concentrated a solution the greater is its osmotic pressure. This is a hypothetical situation and, as a solution does not actually exert a pressure under normal circumstances, the term 'osmotic potential' is preferred. As the osmotic potential is in effect the potential of a solution to pull water into it, it always has a negative value. A more concentrated solution therefore has a more positive osmotic pressure but a more negative osmotic potential.

Osmosis not only occurs when a solution is separated from its pure solvent by a partially permeable membrane but also arises when such a membrane separates two solutions of different concentrations. In this case water moves from the more dilute, or **hypotonic**, solution, to the more concentrated, or **hypertonic**, solution. When a dynamic equilibrium is established and both solutions are of equal concentration they are said to be **isotonic**. The above terms should only be applied to animal cells. The osmotic relationships of plant cells should be described in terms of water potential (Section 33.2).

The **water potential** of a system is the difference between the chemical potential of water in the system and the chemical potential of pure water under standard conditions of temperature and pressure. It follows from this that, under standard conditions, the water potential of pure water is zero. In effect, water potential is a measure of the tendency of water to leave a solution. This tendency is increased by rises in pressure or temperature but decreased by the presence of solutes. As solute molecules in a solution tend to prevent water molecules leaving it, the solution will have a lower water potential than pure water, i.e. its value will be negative. The more concentrated the solution the more negative will be its water potential. Water will diffuse from a region of higher to a region of lower water potential.

Water potential values ('Ψ')

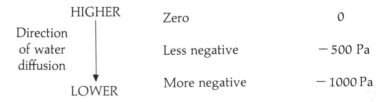

The relationship between water potential and other forces within the cell is described in Section 33.2.

4.4.3 Active transport

Diffusion and osmosis are passive processes, i.e. they occur without the expenditure of energy. Some molecules are transported in and out of cells by active means, i.e. energy is required to drive the process.

The energy is necessary because molecules are transported against a **concentration gradient**, i.e. from a region of low concentration to one of a high concentration. It is thought that the process occurs through the proteins that span the membrane (see Fig. 4.7). These accept the molecule to be transported on one side of the membrane and, by a change in the structure of the protein, convey it to the other side.

Due to the energy expenditure necessary to move molecules against a concentration gradient, cells and tissues carrying out active transport are characterized by:

1. The presence of numerous mitochondria.

2. A high concentration of ATP.

3. A high respiratory rate.

As a consequence of **3**, any factor which increases the rate of respiration, e.g. a higher temperature or increased concentration of oxygen, will increase the rate of active transport. Any factor reducing the rate of respiration or causing it to cease, e.g. the presence of cyanide, will cause active transport to slow or stop altogether.

4.4.4 Phagocytosis

Phagocytosis (*phago* – 'feeding', *cyto* - 'cell') is the process by which the cell can obtain particles which are too large to be absorbed by diffusion or active transport. The cell invaginates to form a cup-shaped depression in which the particle is contained. The depression is then pinched off to form a vacuole. Lysosomes fuse with the vacuole and their enzymes break down the particle, the useful contents of which may be absorbed. The process only occurs in a few specialized cells (called **phagocytes**), such as white blood cells where harmful bacteria can be ingested, or *Amoeba* where it is a means of feeding.

4.4.5 Pinocytosis

Pinocytosis or 'cell drinking' is very similar to phagocytosis except that the vesicles produced, called **pinocytic vesicles**, are smaller. The process is used for the intake of liquids rather than solids. Even smaller vesicles, called **micropinocytic vesicles**, may be pinched off in the same way.

Both pinocytosis and phagocytosis are methods by which materials are taken into the cell in bulk. This process is called **endocytosis**. By contrast, the reverse process, in which materials are removed from cells in bulk, is called **exocytosis**.

4.5 Questions

1. (a) What is meant by the term 'organelle'?
(3 marks)
(b) Describe the structure of each of the following as seen with an electron microscope, and indicate in each case how structure is related to function.
  (i) Golgi complex
  (ii) Nucleus
  (iii) Mitochondrion (12 marks)
(c) What are the functions of ribosomes in a cell?
(5 marks)
(Total 20 marks)

London Board June 1989, Paper II, No. 3

2. Give an account of
(a) the fluid-mosaic model of cell membrane structure and (6 marks)
(b) the different functions of the membranes of cells and their organelles. How do these functions relate to the structure of the membrane?
(14 marks)
(Total 20 marks)

Joint Matriculation Board (Nuffield) June 1989, Paper IIC, No. 4

3. How has the electron microscope improved our understanding of cell structure? (20 marks)

Associated Examining Board June 1984, Paper III, No 2

4. By stating differences in structure and function, distinguish between

(a) rough endoplasmic reticulum and Golgi apparatus;
(b) cell wall and cell membrane;
(c) cilia and flagella. (12 marks)

Southern Universities Joint Board June 1986, Paper I, No. 1

5. (a) List the differences between a eukaryotic cell and a prokaryotic cell. (4 marks)
(b) Describe the structure of each of the cytoplasmic inclusions seen in an animal cell with the aid of an electron microscope. (10 marks)
(c) Discuss the importance of membranes in dividing this animal cell into compartments.
(4 marks)
(Total 18 marks)

Cambridge Board November 1988, Paper I, No. 2

6. (a) The following schematic diagram represents the structure of the cell membrane as envisaged by Singer and Nicolson in their 'fluid mosaic' model.

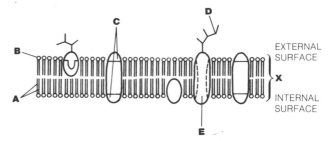

  (i) Where indicated by the letters **A–E**, give suitable names to the component parts.
(5 marks)
  (ii) Give a suitable average figure for the width **X**. (1 mark)
(b) The methods by which substances may pass across cell membranes are often described as 'active' or 'passive'. In the table below, list what you consider to be **four** important points of contrast between these two types of transport.

Active transport	Passive transport

(4 marks)
(Total 10 marks)

Oxford Local June 1989, Paper I, No. 12

7. The photograph below shows an electron micrograph of a part of a root cap cell of maize.

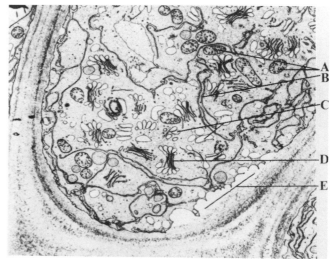

(a) Name the structures marked **A**, **B**, **C** and **D**.
(b) Why is the plasmalemma so convoluted at **E**?
(c) What is the function of **D** in this particular cell?

Cambridge Board November 1984, Paper II, No. 1

8. *The following passage refers to transport across plasma membranes. Read the passage carefully and then answer the questions that follow.*

Plasma membranes, although less than 10 nm wide, present barriers to the movement of ions and molecules, such as glucose and amino acids. Transport across membranes is essential for a number of reasons, for
5 example to generate the ionic gradients necessary for nervous and muscular activity. There are four basic methods of entry into, or exit from, cells, all of which involve membranes. These are diffusion, osmosis, active transport and endocytosis/exocytosis. The first two
10 processes are passive, the latter two active.

Two factors influence the direction in which ions move. One is their concentration and the other their electrical charge. An ion will diffuse from a region of high to a region of low concentration. It will also be attracted
15 towards a region of opposite charge, and move away from a region of like charge. Thus ions are said to move down electrochemical gradients. Active transport of ions is their movement against an electrochemical gradient. Cells maintain a potential difference, that is a charge,
20 across their plasma membranes. The inside of almost all cells is negative with respect to the outside medium.

The major ions of extracellular and intracellular fluids are sodium (Na^+), potassium (K^+) and chloride (Cl^-).

The table shows the concentration (m mol) of these
25 ions in a red blood cell and in the freshwater alga, *Nitella*, and in the fluid surrounding the cells.

Ion	Ion concentration (m mol)		Ion concentration (m mol)	
	Red blood cell		Nitella	
	Inside	Outside	Inside	Outside
Na^+	15	144	15	1.0
K^+	150	5	120	0.1
Cl^-	73	111	65	1.3

(a) Explain how plasma membranes act as barriers to molecules such as glucose and amino acids. (line 2)

(b) Give **three** reasons, excluding the generation of ionic gradients, why transport across membranes is essential for organisms. (line 3)

(c) (i) Why are diffusion and osmosis known as passive processes? (line 8)
   (ii) Distinguish between *diffusion* and *osmosis*.
   (iii) Explain why the uptake of most ions is an active process.

(d) What generalisations can be made from the data given in the table in comparing the concentration of ions inside and outside cells?

(e) Suggest an explanation for the following statements:

(i) When respiration of red blood cells is inhibited, the ionic composition of the cells gradually changes until it comes into equilibrium with the plasma.

(ii) When K^+ ions are removed from the blood plasma, the movement of sodium into the cells and potassium from the cells increases dramatically.

(12 marks)

Cambridge Board November 1989, Paper III, No. 2

9. The photograph below shows an electron micrograph of part of a meristematic cell.

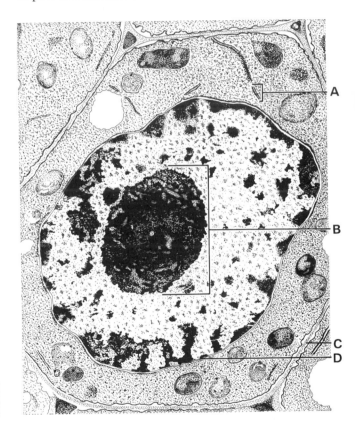

(a) Name the structures labelled A to D. *(4 marks)*

(b) State *two* functions of the structure B. *(2 marks)*

(c) State *two* features visible in this cell that are characteristic of meristematic cells. *(2 marks)*

(d) State *two* additional features that would be visible when the cell is dividing. *(2 marks)*
(Total 10 marks)

London Board June 1988, Paper I, No. 1

10. In order to separate ribosomes, nuclei and mitochondria, liver tissue was chopped, homogenized and centrifuged at different speeds for different lengths of time as indicated opposite.

'*g*' is a measure of gravitational force.

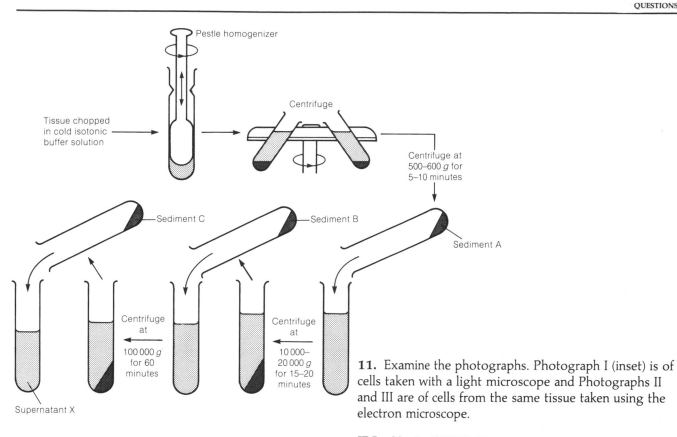

Pestle homogenizer

Tissue chopped in cold isotonic buffer solution

Centrifuge

Centrifuge at 500–600 g for 5–10 minutes

Sediment C

Sediment B

Sediment A

Centrifuge at 100 000 g for 60 minutes

Centrifuge at 10 000–20 000 g for 15–20 minutes

Supernatant X

11. Examine the photographs. Photograph I (inset) is of cells taken with a light microscope and Photographs II and III are of cells from the same tissue taken using the electron microscope.

(*a*) (i) In which sediment, A, B or C would you expect to find the following cell organelles?
Ribosomes
Nuclei
Mitochondria

(ii) Give a reason for your choice. (*4 marks*)

(*b*) In preparation of the tissue for homogenization, a buffer solution was used which was isotonic with the liver tissue. What would happen to the organelles if this solution were
(i) hypotonic,
(ii) hypertonic? (*2 marks*)

(*c*) This technique was used for the isolation of mitochondria in order to study their enzymic activity. However, liver tissue contains many lysosomes. Suggest why this feature makes the study of mitochondrial activity difficult. (*2 marks*)

(*d*) The biochemical activity of isolated organelles is rapidly lost. Suggest **two** features of this technique which will serve to reduce this loss. (*2 marks*)

(*e*) Supernatant X was found to contain several amino acids. Name a technique that would be used to separate and identify them. (*1 mark*)
(*Total 11 marks*)

Welsh Joint Education Committee June 1989, Paper A2, No. 10

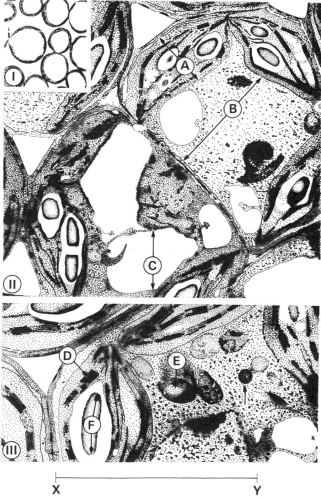

(*a*) Give **one** advantage and **one** disadvantage of examining biological material with the electron microscope rather than with the light microscope. (*2 marks*)

(*b*) Name the structures labelled **A** to **F** on the photographs. (*3 marks*)

(*c*) Name a type of organism and tissue from which the cells in the photographs were taken and give a reason in each case to support your identification. (*4 marks*)

(*d*) Photograph III has a magnification of × 10 000. What is the actual length between X and Y of the piece of tissue shown in this photograph? (*2 marks*)

(*Total 11 marks*)

Joint Matriculation Board June 1989, Paper IA, No. 1

5 Tissue organization

5.1 What is a tissue?

A tissue is a group of cells and their intercellular substance which are linked together and perform a particular function. Such tissues may comprise a single type of cell, as in the squamous epithelium of animals or parenchyma in plants. Alternatively, more than one cell type is included, as in xylem and phloem or blood. The study of tissues is undertaken using light and electron microscopes and is a branch of biology known as **histology.** Since cells are three-dimensional structures it is always necessary to study them from a number of different views, normally in longitudinal and transverse sections.

There are four main groups of animal tissues: epithelial, connective, muscular and nervous. Plant tissues may be simple, like parenchyma, collenchyma and sclerenchyma, or compound, like xylem and phloem.

5.2 Animal epithelial tissues

Most epithelial tissues are derived from an embryonic layer called the **ectoderm,** but parts of the alimentary canal and its associated organs are from another layer, the **endoderm**.

Epithelial tissue comprises single or compound sheets of cells held together by small amounts of intercellular substance based on hyaluronic acid. The bottom layer of cells is always attached to a **basement membrane** made mainly of collagen fibres. Nerves may penetrate the tissue but blood vessels do not and so oxygen and nutrients must diffuse into it from lymph vessels in the intercellular spaces. Epithelial tissues cover inner and outer surfaces where they often have a protective function. Cells may be thickened with keratin to resist abrasion and they divide rapidly to replace cells which are worn away. When diffusion must take place across a surface, the epithelial layer will be thin. Many epithelia are specialized for absorption, e.g. small intestine, or secretion, e.g. salivary glands.

TABLE 5.1 **Animal epithelial tissues**

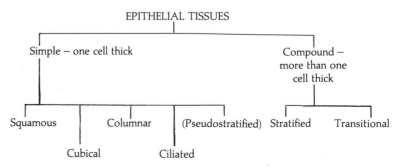

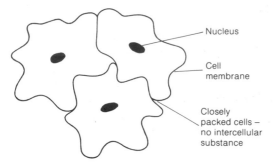

(a) Surface view

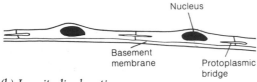

(b) Longitudinal section

Fig. 5.1 Squamous epithelium, e.g. from lining of mouth

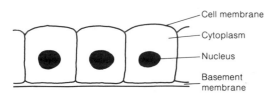

Fig. 5.2 Cubical epithelium (LS)

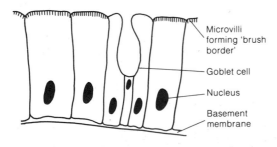

Fig. 5.3 Columnar epithelium (LS)

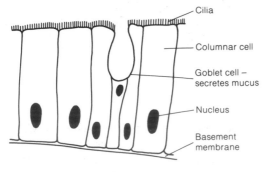

Fig. 5.4 Ciliated epithelium (LS)

5.2.1 Squamous epithelium

Squamous epithelial cells form a single layer attached to the basement membrane. In surface view the cell outlines are irregular and closely packed. The cells are shallow, the central nucleus often forming a bump in the surface. Adjacent cells may be joined by strands of cytoplasm. Such epithelia form ideal surfaces over which diffusion can occur and so are important in the alveoli of the lungs, in the Bowman's capsule and in capillary walls. Their smooth surface also provides a relatively friction-free lining for blood vessels (Fig 5.1).

5.2.2 Cubical epithelium

Cubical epithelium may have a secretory function in glands like the thyroid or a non-secretory function lining the kidney collecting ducts or ducts from the salivary glands. The cells are not very specialized and form a single layer of cuboid cells attached to a basement membrane. (See Fig. 5.2.)

5.2.3 Columnar epithelium

Columnar epithelial cells are tall and narrow, with a nucleus near the base. Their surface area is often increased by the presence of **microvilli** at the free end. This is especially true in areas which have an absorptive rôle. Secretory goblet cells are often found interspersed with the columnar epithelial cells. This tissue is especially important lining the digestive system where the secretion of mucus and the absorption of digested food take place. Columnar epithelium is also found lining various ducts and as a component of glands (Fig. 5.3).

5.2.4 Ciliated epithelium

Ciliated epithelium comprises columnar cells with cilia at their free edges. There are many mucus-secreting goblet cells present. The combination of mucus and cilia permits substances to be moved through ducts as in the oviducts, trachea and spinal cord. (See Fig. 5.4 and the photo at the top of the next page.)

5.2.5 Stratified epithelium

Stratified epithelium is made up of a number of layers of cells. Those attached to the basement membrane form the **germinative layer** and they undergo mitosis. As new cells form, older ones are pushed nearer the surface, changing shape and flattening to become **squames**. In the oesophagus these are not thickened but in the skin, which is subject to considerable abrasion, the squames are heavily thickened with **keratin**. These cells are also found lining the mouth and the vagina. (See Fig. 5.5 and the photo below it.)

5.2.6 Transitional epithelium

Transitional epithelium is found in structures which must be able to stretch, notably the urinary bladder and parts of the kidney. It comprises three or four layers of cells which may be flattened towards the surface, but which are not shed like those of stratified epithelium. (See Fig 5.6.)

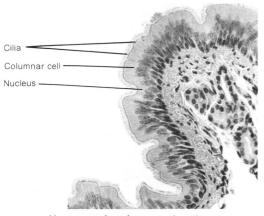

Cilia
Columnar cell
Nucleus

Section of human trachea showing ciliated epithelium (× 150 approx.)

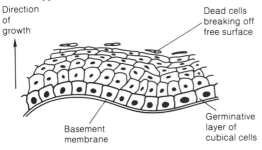

Direction of growth

Dead cells breaking off free surface

Basement membrane

Germinative layer of cubical cells

Fig. 5.5 Stratified epithelium, e.g. from vagina

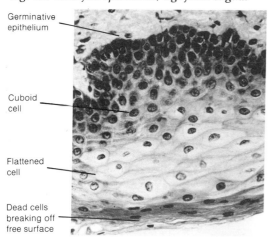

Germinative epithelium

Cuboid cell

Flattened cell

Dead cells breaking off free surface

Stratified epithelial tissue (× 200 approx.)

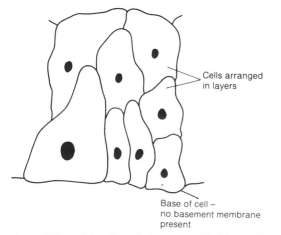

Cells arranged in layers

Base of cell – no basement membrane present

Fig. 5.6 Transitional epithelium, e.g. bladder wall

5.2.7 Glandular epithelium

A single glandular cell is called a **goblet cell**. Multicellular glands often have ducts through which their secretions are released onto the surface of the epithelium – these are the **exocrine glands.** Ductless glands are **endocrine glands** and their secretions directly enter the bloodstream. The cells comprising any gland may be **mucocytes**, producing viscous mucus, or **serocytes**, producing a clear fluid containing enzymes. A **mixed gland** has both types of glandular cell.

5.3   Connective tissue

Connective tissue is composed of a variety of cells embedded in a large amount of intercellular substance called the **matrix**. It develops from the embryonic mesoderm. Connective tissue provides the main supporting tissues in the form of **cartilage** and **bone**, and the main transport system – **blood**. Other connective tissues insulate the body (**adipose**), form the main packing tissue (**areolar**), form tendons (**white fibrous**) and ligaments (**yellow elastic**).

TABLE 5.2 **Connective tissues**

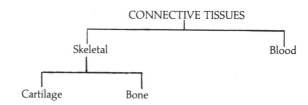

CONNECTIVE TISSUES

Skeletal   Blood

Cartilage   Bone

(The other forms of connective tissue are not covered in this book.)

5.3.1 Cartilage

The matrix of cartilage is made of chondrin and embedded in it are **chondrocytes** and fine fibres composed of **collagen**. The resulting tissue is hard but flexible and is found at the ends of bones, in the respiratory passages and parts of the ear. It comprises the skeleton of cartilaginous fish such as sharks. Cartilage may contain a high proportion of collagenous fibres, making it stronger and less flexible; this form makes up the intervertebral discs.

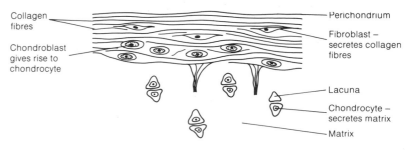

Collagen fibres

Chondroblast gives rise to chondrocyte

Perichondrium

Fibroblast – secretes collagen fibres

Lacuna

Chondrocyte – secretes matrix

Matrix

Fig. 5.7 Cartilage (TS)

5.3.2 Bone

The matrix of compact bone is made up of collagen together with inorganic substances such as calcium, magnesium and phosphorus. These components are arranged in concentric circles, called **lamellae**, around an **Haversian canal** containing an artery, a vein, lymph

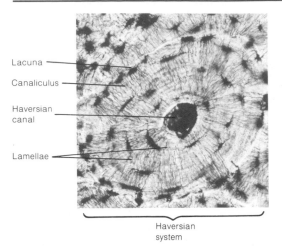

Lacuna

Canaliculus

Haversian canal

Lamellae

Haversian system

Compact bone (TS) (× 400 approx.)

vessels and nerve fibres. Bone cells, or **osteocytes**, are found in spaces in the lamellae known as **lacunae** and fine channels called **canaliculi** link lacunae. The system of lamellae around one Haversian canal is called an **Haversian system**. Bone is an extremely important and strong skeletal material. It is not static. Its various inorganic components may be deposited or absorbed at different times to meet new stresses put upon the tissue.

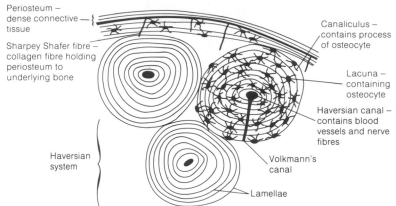

Periosteum – dense connective tissue

Sharpey Shafer fibre – collagen fibre holding periosteum to underlying bone

Haversian system

Canaliculus – contains process of osteocyte

Lacuna – containing osteocyte

Haversian canal – contains blood vessels and nerve fibres

Volkmann's canal

Lamellae

Fig. 5.8 Compact bone (TS)

5.3.3 Blood

In blood the matrix is the watery **plasma** which carries a variety of different cells. The majority of cells present are **erythrocytes** or red blood cells which are biconcave discs about 7 μm in diameter. They have no nucleus and are formed in the bone marrow. Their main function is the carriage of oxygen, which combines with the red pigment haemoglobin to form oxyhaemoglobin. The remaining cells are the larger, nucleated white blood cells or **leucocytes**. Most of these are also made in the bone marrow and their main rôle is one of defence. There are two basic types of leucocyte. **Granulocytes** have granular cytoplasm and a lobed nucleus; they can engulf bacteria by phagocytosis. Some of them are also thought to have antihistamine properties. **Agranulocytes** have a non-granular cytoplasm and a compact nucleus. Some of these also ingest bacteria but the **lymphocytes**, made mainly in the thymus gland and lymphoid tissues, produce **antibodies.** More sparsely distributed in the plasma are tiny cell fragments called **platelets.** These are important in the process of blood clotting.

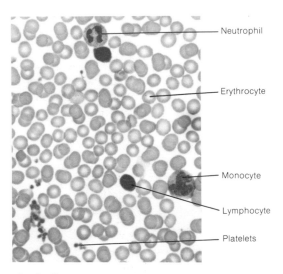

Neutrophil

Erythrocyte

Monocyte

Lymphocyte

Platelets

Blood cells (× 400 approx.)

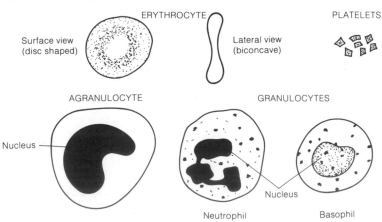

ERYTHROCYTE

Surface view (disc shaped)

Lateral view (biconcave)

PLATELETS

AGRANULOCYTE

GRANULOCYTES

Nucleus

Nucleus

Neutrophil

Basophil

Fig. 5.9 Blood cells

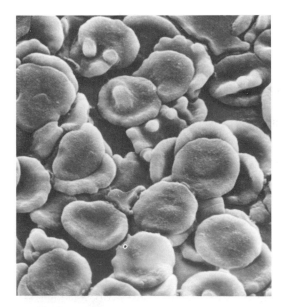

Blood cells (scanning EM) (× 2000 approx.)

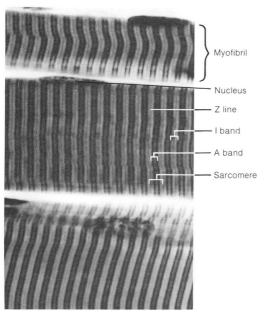

Striated muscle (LS) (× 1200 approx.)

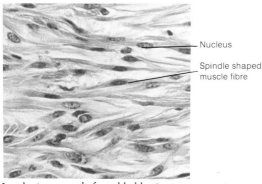

Involuntary muscle from bladder (× 500 approx.)

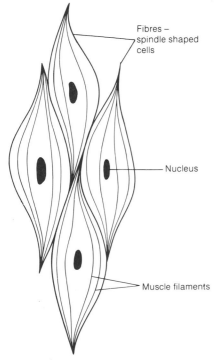

Fig. 5.11 Involuntary muscle (LS), e.g. from alimentary canal

Formation of blood

In the foetus red blood cells are formed in the liver, but in adults production moves to bones, such as the cranium, sternum, vertebrae and ribs, which have red bone marrow. White blood cells like lymphocytes are formed in the thymus gland and lymph nodes whereas other types are formed in bones, e.g. the long bones of the limbs, which have white bone marrow.

5.4 Muscular tissue

There are three main types of muscular tissue which together comprise about 40% of a mammal's body weight. They are all made up of cells or fibres which are capable of contraction and they are derived from the embryonic mesoderm.

5.4.1 Voluntary muscle

Voluntary muscle is also referred to as striped, striated or skeletal muscle. It is composed of long fibres held together by connective tissue. Within these muscle fibres are very fine **myofilaments** made of the proteins **actin** and **myosin**. The arrangement of these proteins gives the muscle its striped appearance. These muscles are attached to the skeleton and their contraction is under voluntary control. The detailed structure of voluntary muscle is given in Section 39.3.1.

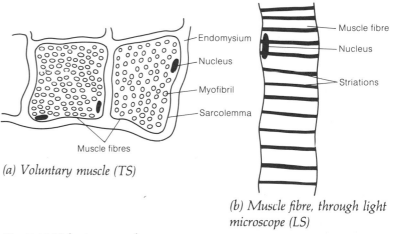

(a) Voluntary muscle (TS)

(b) Muscle fibre, through light microscope (LS)

Fig. 5.10 Voluntary muscle

5.4.2 Involuntary muscle

This muscle, also known as smooth, unstriped or unstriated muscle, is composed of spindle-shaped cells arranged in sheets or bundles. These muscles contract rhythmically and do not fatigue easily. Their action is controlled by nerves from the autonomic nervous system. They are important in the alimentary canal, the walls of the blood vessels and in tubes of the urino-genital system.

5.4.3 Cardiac muscle

This specialized type of muscle is found only in the heart. It is capable of rhythmical contraction and relaxation over a long period. Cardiac

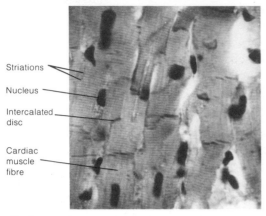

Striations

Nucleus

Intercalated disc

Cardiac muscle fibre

Cardiac muscle (× 600 approx.)

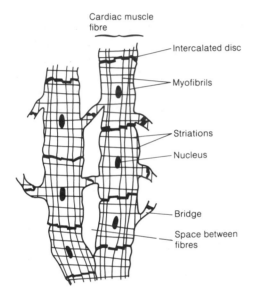

Cardiac muscle fibre

Intercalated disc

Myofibrils

Striations

Nucleus

Bridge

Space between fibres

Fig. 5.12 Cardiac muscle (LS), e.g. from ventricle

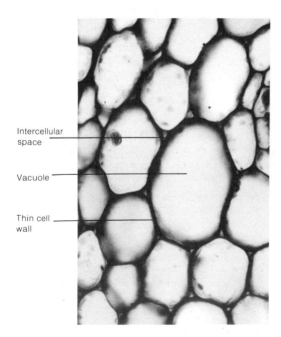

Intercellular space

Vacuole

Thin cell wall

Parenchyma cells (TS) (× 300 approx.)

muscle is myogenic, contraction being stimulated from within the heart itself. Each muscle cell or fibre has 1 or 2 nuclei and the net-like arrangement of the cells allows waves of contraction to spread rapidly over the heart. (See Fig. 5.12.)

5.5 Nervous tissue

Nervous tissue is derived from embryonic ectoderm. It comprises closely packed nerve cells or **neurones** with little intercellular space. The neurones are bound together by connective tissue.

5.5.1 Neurones

Neurones are capable of transmitting electrical impulses. These pass from **receptors** along **sensory neurones** to the **central nervous system**. Within the CNS **relay neurones** may be found. These connect with **effector (motor) neurones** along which impulses are transmitted to the **effector**. All neurones have a cell body containing the nucleus. This cell body has a number of processes called **dendrites** which transmit impulses to the cell body. Impulses leave via the **axon** which may be several metres in length. Some axons are covered by a fatty **myelin sheath** formed by **Schwann cells**.

Nerve fibres may be bundled together and wrapped in connective tissue to form **nerves.** Nerves may be **sensory**, comprising sensory neurones, **effector (motor)**, comprising effector (motor) neurones, or **mixed**, with both types present.

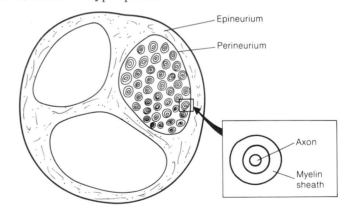

Epineurium

Perineurium

Axon

Myelin sheath

Fig. 5.13(a) Medullated nerve (TS)

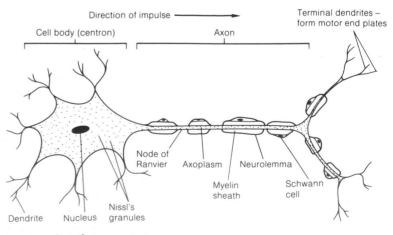

Direction of impulse

Terminal dendrites – form motor end plates

Cell body (centron)

Axon

Node of Ranvier

Axoplasm

Neurolemma

Myelin sheath

Schwann cell

Dendrite   Nucleus   Nissl's granules

Fig. 5.13(b) Effector (motor) neurone

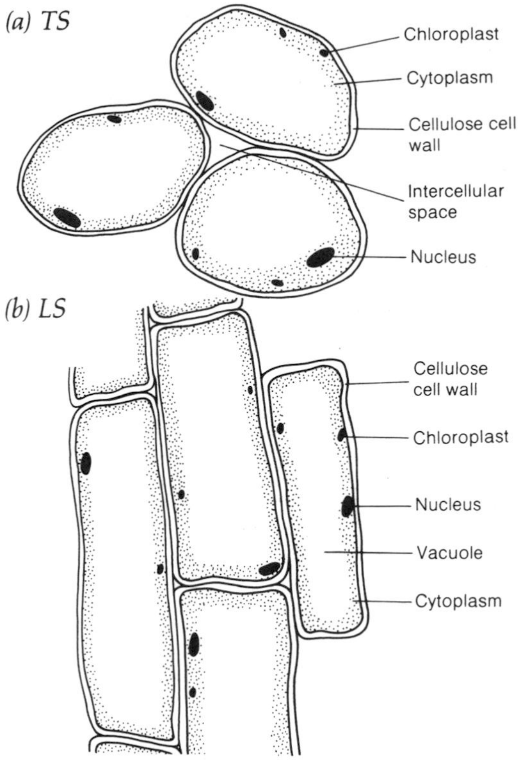

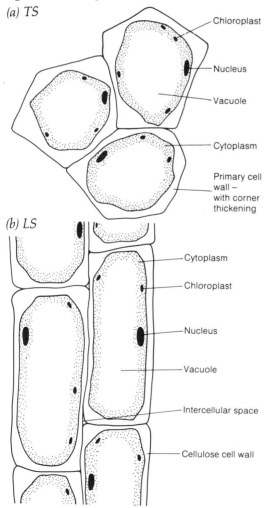

(a) TS
- Chloroplast
- Cytoplasm
- Cellulose cell wall
- Intercellular space
- Nucleus

(b) LS
- Cellulose cell wall
- Chloroplast
- Nucleus
- Vacuole
- Cytoplasm

Fig. 5.14 Parenchyma

(a) TS
- Chloroplast
- Nucleus
- Vacuole
- Cytoplasm
- Primary cell wall – with corner thickening

(b) LS
- Cytoplasm
- Chloroplast
- Nucleus
- Vacuole
- Intercellular space
- Cellulose cell wall

Fig. 5.16 Collenchyma

5.6 Simple plant tissues

Simple plant tissues each consist of only one type of cell. They are normally grouped according to the degree of thickening present in the cell wall.

5.6.1 Parenchyma

Parenchyma cells are usually spherical although their shape may be distorted by pressure from adjacent cells. These unspecialized living cells form the bulk of packing tissue within the plant. Parenchyma cells are metabolically active and may also store food. When tightly packed and turgid they provide support for herbaceous plants. Air spaces around parenchyma cells allow exchange of gases to take place. (See Fig. 5.14 and the photo at the bottom of the previous page.)

Photosynthetic parenchyma is called **chlorenchyma**. This tissue may be found in certain stems and also makes up the mesophyll of leaves. Epidermis may be considered as specialized parenchyma capable of producing a waxy cuticle of cutin to prevent desiccation. At intervals in the colourless epidermis are pairs of **guard cells** containing chloroplasts. The guard cells surround a pore or **stoma**.

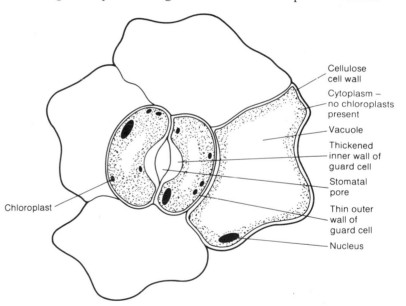

- Cellulose cell wall
- Cytoplasm – no chloroplasts present
- Vacuole
- Thickened inner wall of guard cell
- Stomatal pore
- Thin outer wall of guard cell
- Nucleus
- Chloroplast

Fig. 5.15 Epidermis and stoma

Unicellular or multicellular hairs may also be found on the epidermis. These reduce air movement and so cut down water loss by evaporation. In roots, unicellular hairs near the root tip assist with water uptake.

The endodermis of roots and certain stems is also a form of parenchyma.

5.6.2 Collenchyma

Collenchyma cells are living and have cell walls with additional cellulose deposited in the corners. This provides them with extra mechanical strength. They are elongated and important in growing stems since they are able to stretch. They are often found just under the epidermis of a stem or in the corners of angular stems such as *Lamium* (dead-nettle).

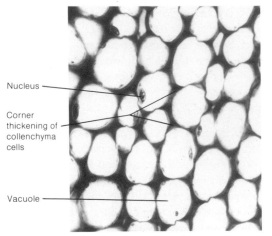

Nucleus

Corner thickening of collenchyma cells

Vacuole

Collenchyma cells (TS) (× 200 approx.)

(a) TS

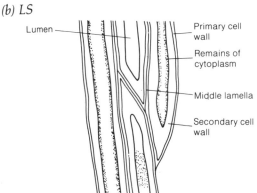

Cytoplasm

Nucleus

Primary cell wall

Secondary cell wall – lignified

Vacuole

Simple pit

(b) LS

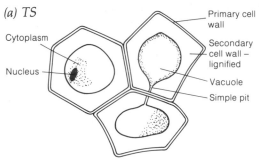

Lumen

Primary cell wall

Remains of cytoplasm

Middle lamella

Secondary cell wall

Fig. 5.17 Sclerenchyma

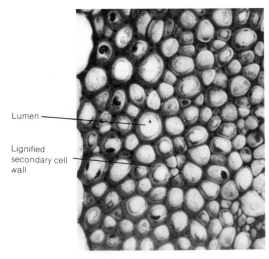

Lumen

Lignified secondary cell wall

Sclerenchyma cells (TS) (× 300 approx.)

5.6.3 Sclerenchyma

Mature sclerenchyma cells are dead and therefore incapable of growth. They develop fully when the growth of surrounding tissues is complete. Sclerenchyma cells have large deposits of lignin on the primary cell wall and the cell contents are lost. In places, lignin is not deposited due to the presence of **plasmodesmata** in the primary cell wall; such regions are called **pits**.

Some sclerenchyma cells are roughly spherical and are known as **sclereids**. These are usually found in small groups in fruits and seeds, cortex, pith and phloem. They toughen the structures in which they are found. (See Fig. 5.17 and the photo below it.)

Elongated sclerenchyma cells are called **fibres** and they provide the main supporting tissue of many mature stems. They may form a cylinder below the epidermis, are found in xylem and phloem and sometimes as masses associated with vascular bundles.

5.7 Compound plant tissues

The vascular tissues of plants, xylem and phloem, consist of more than one type of cell, some living and some dead.

5.7.1 Xylem

Xylem consists of parenchyma cells and fibres together with two specialized types of cells: vessels and tracheids. These tissues are both dead and serve the dual rôle of support and water transport. The types of **vessel** found depend upon the degree and nature of the cell wall thickening. In the **protoxylem** the lignin is deposited in rings or spirals so the cell is still capable of expansion.

(a) TS

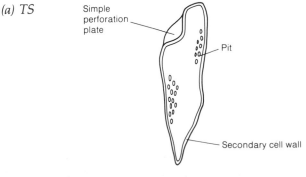

Simple perforation plate

Pit

Secondary cell wall

(b) LS   Thickenings of secondary cell walls in primary xylem

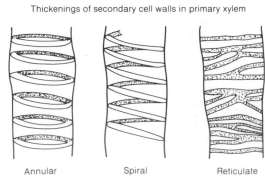

Annular   Spiral   Reticulate

Fig. 5.18 Vessels

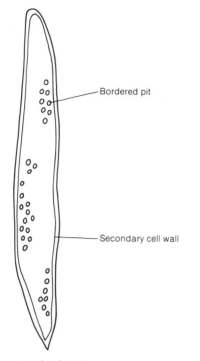

Fig. 5.19 Tracheid (LS)

In **metaxylem** there is more extensive lignification arranged in patterns known as **reticulate, scalariform** or **pitted**. All vessels are made up of cells whose cross walls have broken down, resulting in long tubes ideal for carrying water. (See Fig. 5.18.)

Tracheids are spindle-shaped cells arranged in rows with the ends of the cells overlapping. The cells have heavily lignified walls and so there are no cell contents. They provide mechanical strength and support to the plant. (See Fig. 5.19.)

5.7.2 Phloem

Phloem comprises parenchyma, sclereids and fibres together with specialized cells for translocation, called **sieve tube elements**, and **companion cells**. Long sieve tubes are formed by the fusion of sieve tube elements and the partial breakdown of the cross walls between them to form **sieve plates**. The cell walls are thickened with cellulose and pectin, the cell lacks a nucleus and the cytoplasm is confined to the edges of the cell. This cytoplasm is still living but is dependent on the nucleus and active cytoplasm of an adjacent companion cell.

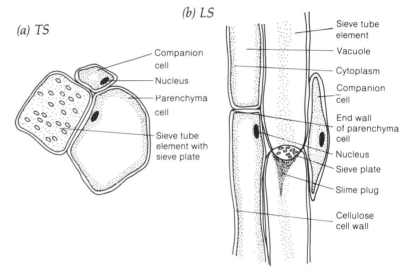

Fig. 5.20 Phloem

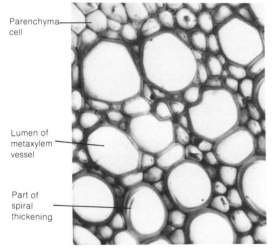

Xylem (TS) (× 300 approx.)

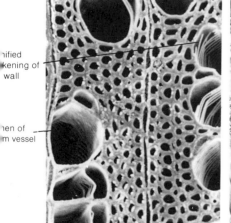

Wood of Alder showing vessels (scanning EM) (× 300 approx.)

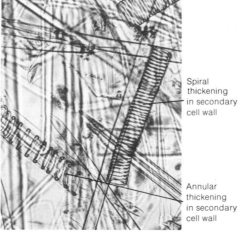

Xylem macerate (× 150 approx.)

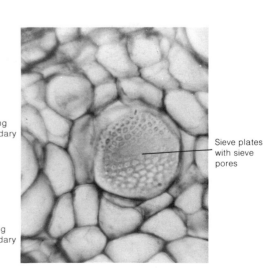

Phloem, with sieve plates (TS) (× 300 approx.)

5.8 Questions

1. (a) For **each** of the following, show how its structure is related to its function: (i) a red blood corpuscle; (ii) a motor neurone; (iii) a xylem vessel.
(5,5,5 marks)
   (b) What advantages do organisms gain from cell differentiation?
(3 marks)
(Total 18 marks)

Cambridge Board November 1984, Paper I, No. 1

2. By means of labelled diagrams only, illustrate the microscopic structure of mammalian bone and cartilage. Show how these tissues give support to the body.
(8 marks)

Southern Universities Joint Board June 1984, Paper I, No. 2

3. (a) Give an illustrated account of the structure of a meristematic cell of a plant
(6 marks)
   (b) How do the structure and function of each of the following differ from that of a meristematic cell?
      (i) a xylem fibre
      (ii) a sieve tube
      (iii) a guard cell
(14 marks)
(Total 20 marks)

London Board June 1986, Paper II, No. 6

4. (a) Collenchyma and sclerenchyma are two types of plant tissue involved in support of the plant. For each tissue state:
      (i) Whether it consists of living or dead cells, and
      (ii) the nature of the thickening of the cell wall.
(2 marks)
   (b) Make a fully labelled low-power diagram of a TS of a **named** herbaceous plant stem to show the distribution of collenchyma and sclerenchyma. (NB it is not necessary to show individual cells in your diagram.)
(4 marks)
   (c) Make labelled diagrams to show the appearance of collenchyma and sclerenchyma as seen in TS and LS under the high power of a light microscope.
(4 marks)
   (d) (i) What are the **two** main functions of xylem?
      (ii) What are the **four** types of cell in the xylem?
(3 marks)
   (e) (i) What is the main structural difference between xylem cells and sclerenchyma?
      (ii) How is this difference explained by the difference in function between the xylem and sclerenchyma?
(3 marks)
   (f) Processes in both the leaf and root contribute to the movement of water through the plant. Describe the process providing the main driving force for the uptake and transport of water through the plant.
(4 marks)
(Total 20 marks)

Northern Ireland Board June 1985, Paper II, No. 4

5. The drawings show cells from two different plant tissues as seen in transverse section.

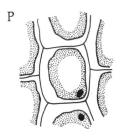
P

Q

   (a) (i) List **three** structural differences, visible in the drawings, between the two cells.
      (ii) State **one** further difference which could be shown by the use of a suitable stain.
(4 marks)
   (b) Make an accurate **scale** drawing of cell Q as it would appear in longitudinal section at a magnification of x**200**.
(3 marks)
   (c) Indicate clearly, by drawing on the outlines provided, where you would expect to find tissues P and Q in a typical dicotyledonous plant. (Do not draw individual cells.) Label the regions P and Q respectively.

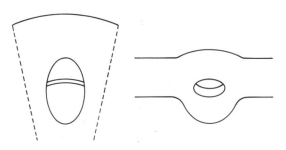

      (i) *TS Vascular bundle of young stem* (ii) *TS leaf midrib*
(4 marks)
   (d) (i) What is the function of the two tissues in plants?
      (ii) Name **two** tissues which perform a similar function in mammals.

(iii) Suggest how this function is performed in the flower stalk of a bluebell, which does not contain tissues P and Q. *(6 marks)*
(Total 17 marks)

Welsh Joint Education Committee June 1985, Paper AI, No. 5

6. The photomicrograph is of a transverse section of part of a plant organ.

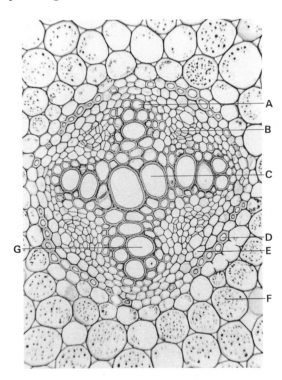

(a) Name the organ and list **four** features that enable you to recognise it as such.
(b) Name the cell types labelled **A, B, C, D, E** and **F**.
(c) State **one** function of the cell labelled **G**. Give **two** observable features of the cell which are important to this function and explain the importance of each feature to the functioning of the cell. *(10 marks)*

Cambridge Board November 1988, Paper III, No. 1

7. The diagram above right represents part of the PHLOEM of a dicotyledonous plant.

(a) Complete the diagram to show, if not already drawn, the following structures:
Sieve tube, sieve plate, sieve pore, nucleus, companion cell, plamodesmata, mitochondria, cytoplasm.
Label each of these structures. *(8 marks)*
(b) Name TWO different types of phloem cell not present in this diagram. *(2 marks)*

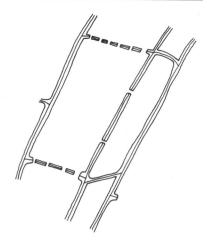

(c) State **four** points of comparison between the structure and function of xylem and phloem. *(4 marks)*
(d) Outline **two** pieces of experimental evidence to support the theory that phloem is the main route for the translocation of carbohydrates in plants. *(4 marks)*
(Total 18 marks)

Oxford and Cambridge Board June 1989, Paper I, No. 1

8. The diagrams below show cells from a tissue of a flowering plant.

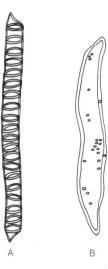

A   B

(a) (i) Name cells A and B. *(2 marks)*
(ii) State *one* way in which the structure of cell A differs from cell B. *(1 mark)*
(b) (i) In which tissue are these cells found? *(1 mark)*
(ii) State *two* functions of this tissue. *(2 marks)*
(c) Describe briefly *three* ways in which cell A is adapted for its functions in the plant. *(3 marks)*
(Total 9 marks)

London Board June 1989, Paper I, No. 7

9. The diagram shows part of a tissue from a plant stem as seen in transverse (**A**) and longitudinal (**B**) section.

(a) Name the tissue. What features enable you to recognize it as such?

(b) Name the structures labelled 1 and 2.

(c) (i) In what type of stem would you expect to find this tissue?

   (ii) Where would it be situated?

(d) (i) What is the main function of such a tissue?

   (ii) What features of its cell walls are important in adapting it to its particular function?

(e) Where else would you expect to find such a tissue? *(10 marks)*

Cambridge Board June 1984, Paper II, No. 3

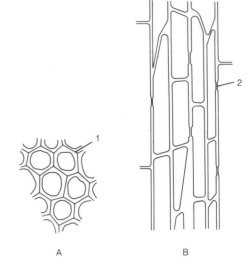

A   B

6 *Variety of organisms*

6.1 Principles of classification

Before any study can be made of living organisms it is necessary to devise a scheme whereby the enormous diversity of them can be organized into manageable groups. This grouping of organisms is known as **classification** and the study of biological classification is called **taxonomy**. A good universal system of classification aids communication between scientists and allows information about a particular organism to be found more readily. There is no 'correct' scheme of classification since organisms form a continuum and any division of them into groups has been devised by man solely for his own convenience.

6.1.1 Natural classification

During the eighteenth century, the Swedish botanist Linnaeus devised a scheme of classification which has become widely accepted. In this scheme organisms are grouped together according to their basic similarities. Relationships are based on homologous rather than analogous characteristics. **Homologous** characters are ones which have a fundamental similarity of origin, structure and position, regardless of their function in the adult. **Analogous** characters are ones which have a similar function in the adult but which are not homologous, i.e. they do not have the same origin. For example, wings of butterflies and birds are both used for flight but their origins are not similar. Classification based on homology is called **natural classification**. It now embraces biochemical and chromosome studies as well as the morphology and anatomy used by Linnaeus. A successful natural classification should reflect the true evolutionary relationships of organisms.

6.1.2 Artificial classification – Keys

Artificial classification is really classification according to differences rather than similarities. At each step organisms with a particular exclusive feature will be selected from the main group and placed in a subgroup. Each division is made on the basis of a single character and groups may contain many unrelated forms. However artificial this system, it is very useful for identifying organisms and is the basis of **dichotomous keys** used in the field. In order to use a key, such as the one given opposite, the item numbered **1** is read and a decision made as to the presence or absence of that feature in the organism being studied. For example, in the key opposite, if a shell is present you then read item **2**, and decide whether the shell is coiled or not. This allows an identification to be made. If there is no shell present then item **2** should be ignored and number **3** used instead.

---

1. Shell present – go to **2**
  Shell absent – go to **3**

2. Shell coiled   = WINKLE
  Shell not coiled = LIMPET

3. Fins present = FISH
  Fins absent – go to **4**

4. Tentacles present = SEA ANEMONE
  Tentacles absent = STARFISH

Construction of Keys

When constructing a simple key to separate a group of organisms, a useful starting point is to compile a comparative table of some obvious features shown by some of the organisms. For example, Table 6.1 compares four features of the specimens shown in Fig. 6.1.

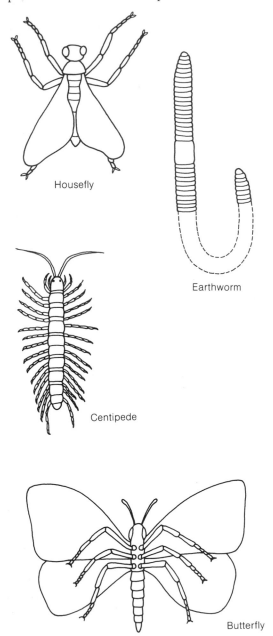

Fig. 6.1 Four animals to be used in constructing a key

TABLE 6.1 **Comparative features of four organisms useful in constructing a key.**

Feature	Earthworm	Housefly	Butterfly	Centipede
Legs	None	3 pairs	3 pairs	15 pairs
Segmentation	Yes	Yes	Yes	Yes
Wings	None	1 pair	2 pairs	None
Antennae	None	Short and stumpy	Long and club-shaped	Long and pointed

With practice, and provided the number of specimens is small, this technique can be performed mentally without the need to write anything down. The table shows that 'segmentation' is common to all four animals, and as such is useless as a means of separating one specimen from another. All other features can be used successfully — indeed, as the example below shows, just two are sufficient:

1. Wings present   — 2
    Wings absent   — 3

2. One pair of wings   — Housefly
    Two pairs of wings — Butterfly

3. 15 pairs of legs   — Centipede
    No legs   — Earthworm

There are many other equally suitable keys which can be made using only the features in the table. Try constructing some alternatives. Below are some basic guidelines to help in the construction of keys:

1. Use distinct, clearly recognizable features, e.g. number of legs, rather than their shape which might be difficult to describe accurately.

2. Use features which are consistent throughout all members of a species. For this reason it is advisable to avoid colour and size as these may vary between male and female, juvenile and adult or from season to season.

3. Be precise — avoid statements such as 'many' or 'few', but rather say 'more than 10' or 'less than 5'. Be objective rather than subjective.

4. Use features which clearly and unambiguously distinguish one or more specimens from the rest, i.e. features with clearly defined categories (number of wings or leaflets in a leaf) and not those with a range of intermediates (degree of hairiness or overall size).

5. Avoid grouping more than one feature at a time except where they refer to the same character and are exclusive to the specimen(s) being considered. E.g. it may be appropriate to say 'hairy leaves divided into 3 leaflets and arranged alternately on the stem' but it is not advisable to say 'hairy leaves, white flowers, stem with square cross-section and with smooth, grey-coloured bark'. In general, keep to one solitary aspect of a single feature if possible.

Artificial classifications may also be used by scientists studying a particular aspect of an organism's way of life. For example, in a study of locomotion, flying animals will be separated from burrowing or swimming forms, regardless of the evolutionary relationships between them.

6.1.3 Taxonomic ranks

It is convenient to distinguish large groups of organisms from smaller subgroups and a series of rank names has been devised to identify the different levels within this hierarchy. The rank names used today are largely derived from those used by Linnaeus over 200 years ago. The largest groups are known as **phyla** and the organisms in each phylum have a body plan radically different from organisms in any other phylum. Diversity within each phylum allows it to be divided into **classes**. Each class is divided into **orders** of organisms which have additional features in common. Each order is divided into

TABLE 6.2   **Classification of three organisms**

Rank	Cabbage white butterfly	Man	Sweet pea
Phylum	Arthropoda	Chordata	Spermatophyta
Class	Insecta	Mammalia	Angiospermae
Order	Lepidoptera	Primates	Rosales
Family	Pieridae	Hominidae	Leguminosae
Genus	*Pieris*	*Homo*	*Lathyrus*
Species	*brassica*	*sapiens*	*odoratus*

families and at this level differences are less obvious. Each family is divided into **genera** and each genus into **species**.

With the gradual acceptance that all species arose by adaptation of existing forms, the basis of this hierarchy became evolutionary. Species are groups that have diverged most recently, genera somewhat earlier and so on up the taxonomic ranks.

Every organism is given a scientific name according to an internationally agreed system of nomenclature, first devised by Linnaeus. The name is always in Latin and is in two parts. The first name indicates the genus and is written with an initial capital letter; the second name indicates the species and is written with a small initial letter. These names are always distinguished in text by italics or underlining. This system of naming organisms is known as **binomial nomenclature**.

Table 6.2 shows the use of rank names in classifying a cabbage white butterfly, man and a sweet pea. Only the obligate ranks of classification to which every organism must be assigned have been shown in the table. However, a taxonomist may use a large number of additional categories within this scheme as shown below:

kingdom, subkingdom, grade, **Phylum**, subphylum, superclass, **Class**, subclass, infraclass, superorder, **Order**, suborder, infraorder, superfamily, **Family**, subfamily, tribe, **Genus**, subgenus, **Species**, subspecies, variety.

6.2   Origin of multicellular organisms

6.2.1 Acellular organisms

As living organisms evolved from non-living material it is not surprising that the dividing line between some of them is thin. **Viruses**, for example, possess features of both living and non-living material. The other acellular organisms (see Chapter 7) are far from being the simple structures they are sometimes thought to be and show considerable specialization within the confines of a single cell. They are the oldest organisms in the world (3000 million years old). Indeed, more than three quarters of the time since life evolved on earth was occupied by acellular organisms alone.

Single-celled organisms can never become large since one nucleus can only control a limited volume of cytoplasm and because a large cell has a relatively small surface area to volume ratio. A small surface area to volume ratio will not allow diffusion of oxygen, metabolites and waste materials to occur fast enough to support life. Although a single cell develops specialized organelles it cannot develop a high degree of differentiation and so the number of habitats available to it are limited.

6.2.2 Differentiation

The division of a large mass of protoplasm into separate cells allows spaces to develop between them. This overcomes the reduction in surface area to volume ratio which occurs when single cells enlarge. The development of cell membranes, and cell walls in plants, also stiffens this large mass of protoplasm, enabling it to support itself to some extent. When an organism is made up of a number of different cells these cells may become specialized for particular purposes such as feeding, reproduction and locomotion. This division of labour depends on coordination between the different cells. Specialization produces functional efficiency and has enabled multicellular organisms to diversify into numerous habitats.

Sometimes it is difficult to decide whether a particular organism is acellular or multicellular. *Volvox* is generally considered to be a colony of acellular organisms which show a certain degree of coordination between cells. Similarly, sponges are not true multicellular animals (Metazoa) but are called Parazoa, meaning 'beside the animals'.

6.3 Metazoan organization

6.3.1 Symmetry

Symmetry refers to similarity of size and shape of the parts on either side of a median plane. **Spherical symmetry** is found in ball-shaped organisms such as a few of the Protozoa and some microscopic algae such as *Volvox*. **Radial symmetry** applies to most cnidarians and adult echinoderms which can be divided, like a pipe or wheel, into similar halves by any plane passing through the longitudinal axis. These forms normally have the mouth at one end, referred to as the oral end, and the opposite end is called aboral. Such organisms are well adapted to a sessile (fixed) mode of life since they react equally well on all sides. Most animals show **bilateral symmetry**, e.g. the mammals. In these only one plane divides the animal into equivalent right and left halves. This plane is referred to as **sagittal** and passes through the antero-posterior axis and the dorso-ventral axis. These animals are well adapted for forward movement. Associated with bilateral symmetry is the differentiation of a head end with a greater accumulation of sense organs than at the tail end. The differentiation of a definite head end is called **cephalization**. (See Chapter 35.)

6.3.2 Body layers and cavities

During the embryonic development of Metazoa definite layers of cells can be recognized. The simplest Metazoa, the cnidarians, are said to be **diploblastic** since they are made up of only two cell layers. The outer layer is the **ectoderm** and the inner one the **endoderm**. Between these lies a non-cellular material. The flatworms, or Platyhelminthes, are derived from three cell layers because the **mesoderm** develops between the ectoderm and the endoderm. The Platyhelminthes are therefore **triploblastic** as are all other Metazoa.

In the annelids the development of a true body cavity or **coelom** is seen for the first time. The coelom is a fluid-filled space surrounded by mesoderm. It is of great significance in animal evolution because it provides space in which organs develop and it allows an increase in size and complexity because more cell surfaces are exposed for diffusion. It also forms the spaces inside ducts of the excretory and reproductive systems and may act as a hydrostatic skeleton.

6.3.3 Metameric segmentation

Metameric segmentation or **metamerism** is seen in the annelids, arthropods and chordates. In its most primitive form the body is seen to comprise a linear series of similar segments or somites all of

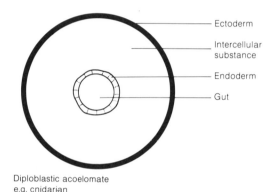

Diploblastic acoelomate
e.g. cnidarian

— Ectoderm
— Intercellular substance
— Endoderm
— Gut

— Ectoderm
— Mesoderm
— Endoderm
— Gut

Triploblastic acoelomate
e.g. platyhelminth

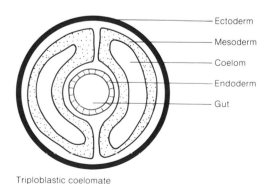

— Ectoderm
— Mesoderm
— Coelom
— Endoderm
— Gut

Triploblastic coelomate
e.g. annelid

Fig. 6.2 Body cavities

the same age. The segmental arrangement includes both internal and external structures with repetition of muscles, blood vessels and nerves in each segment. This repetition of structures gives great potential for specialization. Reproductive organs may only be repeated in a few segments. Segmentation is often much more apparent in the embryo than in the adult, e.g. segmentation of the muscles is obvious in the vertebrate embryo but not in the adult.

The segments may be grouped into functional units called **tagmata**. This is especially noticeable in the arthropods where insects have three tagmata: the head, thorax and abdomen.

6.4 Adaptations to an aquatic existence

Large bodies of seawater provide a relatively stable environment in terms of oxygen content, temperature and salt balance. Life is thought to have evolved in the sea and so many marine organisms are ideally adapted to it. Freshwater habitats have a smaller volume, less stable temperature, lower mineral content, greater light penetration and a greater content of suspended material. All aquatic habitats have a lower oxygen content, greater density and greater viscosity than those of air.

6.4.1 Gaseous exchange

Aquatic plants generally lack a cuticle and therefore gaseous exchange can occur over the whole surface. Protozoa and other animals with a large surface area to volume ratio can also exchange gases by diffusion over the whole surface. With increase in animal size, this process is not fast enough, and so special exchange surfaces develop with a large surface area to volume ratio and an efficient transport system to carry gases to and fro. These surfaces are called **gills**.

6.4.2 Reproduction

Aquatic plants and animals release their gametes into the water and fertilization occurs outside the body. External fertilization is wasteful of resources since large numbers of gametes do not fuse but drift away from each other and die. Eggs that are fertilized are also in great danger since they may be eaten or carried far from a suitable habitat. For this reason aquatic organisms produce large numbers of offspring. Larval stages are common in aquatic animals and may provide the only method of dispersal for sessile animals, like barnacles or mussels.

6.4.3 Excretion and osmoregulation

Marine invertebrates are isotonic with sea water, so there is no net loss or gain of water. Ammonia is the nitrogenous excretory product and may diffuse over the general body surface or from specialized structures such as **nephridia**, as in annelids. Marine vertebrates are hypotonic to sea water and lose water from the body. For this reason they cannot afford to excrete ammonia which is very toxic and must be dissolved in water before it is excreted. Marine teleosts excrete trimethylamine oxide, and cartilagenous fish excrete urea. Freshwater organisms normally have body fluids which are more concentrated than the water which surrounds them. For this reason freshwater *Amoeba* has **contractile vacuoles** to get rid of excess water. Ammonia is the nitrogenous excretory product.

6.4.4 Nutrition

Plants need sunlight for photosynthesis and so they are only found in the upper regions of deep water masses. While chlorophyll is the commonest photosynthetic pigment of surface-dwelling plants, brown and red pigments become important in deeper waters because these absorb light in the blue region of the spectrum – the only light to penetrate to any depth. Plants may be attached to rocks in order to maintain a suitable position for photosynthesis or they may have various flotation devices such as the air bladders of *Fucus*. Aquatic animals can exploit all the feeding mechanisms open to terrestrial forms with the addition of filter feeding, e.g. barnacle, *Balanus* (see Section 24.3.1). Water often contains large amounts of suspended organic material which may be filtered from it using cilia.

6.4.5 Support and movement

The dense, viscous nature of water provides the organisms which live in it with a degree of support not offered by air. For this reason the plants do not contain much lignified tissue and yet may reach quite a large size. The largest mammal ever to have lived, the blue whale, would not be able to support its weight on land. A range of locomotory devices is found in animals but the best aquatic method of locomotion is swimming. Fish are ideally adapted with a streamlined body, scales, a mucus covering, fins and muscles arranged in blocks either side of the body. The jet propulsion shown by squids is a form of locomotion only found in water.

6.4.6 Sensitivity

Free-swimming algae have light-sensitive **eye spots** and swim towards light. Light does not penetrate far in deep bodies of water and so eyes become less important with depth. However, some deep-sea fish do produce fluorescence, providing light at depths not reached by sunlight. The perception of sound is less important in water but fish have a well developed **lateral line** system which is sensitive to changes in water pressure on the surface of the animal. Fish, particularly bottom dwellers, have **taste cells** scattered over the whole surface of their bodies.

6.5 Adaptations to a terrestrial existence

Although there is more oxygen available in air than in water, the terrestrial environment is a very varied and difficult habitat for plants and animals. The biggest problem is one of desiccation and this affects many aspects of an organism's life. The first organisms to invade land were probably derived from green algae which developed the ability to absorb water from the soil via roots while photosynthesizing with aerial parts. The problems associated with the move to land are seen in bryophytes through to the most successful terrestrial plants, the angiosperms.

The first truly terrestrial animals were probably the arthropods whose cuticle prevented desiccation.

6.5.1 Gaseous exchange

If organisms are covered with a waterproof coat to prevent desiccation then gaseous exchange can no longer occur over the whole body surface. Terrestrial plants either have to live in very moist habitats, as do mosses, or have a waxy cuticle with small holes in it for gaseous exchange. In flowering plants, diffusion of gases occurs through pores called **stomata** in the leaves and green stems and through **lenticels** in woody tissues (Section 33.3). The most successful terrestrial animals are the vertebrates and the insects. Land vertebrates have internal structures for gaseous exchange called lungs, associated with a well developed blood system (Sections 31.2.7 and 32.1). Insects have pores in the exoskeleton called **spiracles** and these open into a series of tubes called **tracheae** which carry gases directly to and from the muscles (Section 31.2.3).

6.5.2 Reproduction

Mosses and ferns have two distinct stages in their life cycle. The **sporophyte** generation is well adapted to terrestrial life since it needs dry conditions for the liberation of the **spores**. However, the **gametophyte** is very dependent on water because the male **gametes** need to swim to the female gametes (see Section 8.1). In the Coniferophyta, an increasing level of protection for the gametes is seen, culminating in the complete enclosure of the gametophyte within the sporophyte in the Angiospermophyta (see Section 8.4).

Protection of the gametes on land is also a major problem for terrestrial animals. Some forms, such as the Amphibia, have not succeeded in overcoming the problem and must return to water to breed. Truly terrestrial animals have internal fertilization, the male gamete being transferred to the body of the female and fertilization taking place inside her. The next problem to be solved is protection of the highly vulnerable embryo. Most terrestrial animals lay eggs whose shells show an increasing level of waterproofing. Some animals retain the eggs inside the body of the female until they hatch. The most advanced are the mammals where the embryo develops inside the mother attached to the **placenta** through which respiratory gases, nutrients and waste products are exchanged. With an increasing level of protection fewer offspring are produced and a greater level of parental care is shown. (See Chapter 19.)

6.5.3 Excretion and osmoregulation

Water loss is a major problem for terrestrial organisms. In plants, the problem is exaggerated by the fact that photosynthesis requires leaves with a large surface area to volume ratio, a form also subject to the greatest water loss. Some loss of water is necessary in order to maintain a flow of water and dissolved mineral salts through the plant. The balance is critical.

For animals, excretion of nitrogenous waste potentially involves water loss. Terrestrial animals cannot excrete ammonia but need a more concentrated excretory product. This is urea in mammals and uric acid in birds and insects. (See Chapter 34.)

6.5.4 Nutrition

Terrestrial plants photosynthesize. They absorb water and mineral salts from the soil through their roots or, in the case of bryophytes, over the whole plant surface. Terrestrial animals exhibit a wide range of feeding mechanisms. Bees feed on nectar and pollen and so bring about the transfer of pollen in many angiosperms. The only method of feeding not available to land animals is filter feeding.

6.5.5 Support and movement

Terrestrial plants must either remain small, like the mosses, or develop supporting tissues. The first plants to develop supporting tissues were the ferns but they are most evolved in the Coniferophyta and Angiospermophyta, where heavily lignified cells help to support tall trees. Land animals may have a **hydrostatic skeleton** but most important are the **exoskeletons** of arthropods and the bony **endoskeletons** of the vertebrates. Both are hinged to permit muscular movement. Some insects, birds and bats are also able to fly. Very often a compromise must be found between the rôle of a limb for support and its rôle in locomotion (Section 39.5).

6.5.6 Sensitivity

Terrestrial plants are particularly sensitive to light, water and gravity, showing responses which position parts of the plant ideally for water uptake and photosynthesis. Some plants also respond to touch, curling around structures which will help to support them. (See Section 40.1.) The relative importance of the five senses varies in different animals according to the degree of development of the nervous system and the habitat in which they live. For example, tree-dwelling and predatory mammals have good, stereoscopic vision whereas many nocturnal species rely more on their senses of hearing and smell. While smell and taste are important on land it must be remembered that the essential chemicals need to be dissolved before they can be detected. Hearing is a more important sense on land than in water. (See Section 38.6.)

1. (a) Why do biologists classify organisms? (*4 marks*)
   (b) Define a 'natural classification' and discuss the principles used to create one, giving appropriate examples. (*11 marks*)
   (c) Do you regard viruses as plants, animals or neither? Give your reasons. (*3 marks*)
   (*Total 18 marks*)

   Cambridge Board June 1983, Paper I, No. 8

2. (a) List the problems faced by an animal living on dry land. (*4 marks*)
   (b) Compare the adaptations to life on land shown by insects and mammals. (*14 marks*)
   (*Total 18 marks*)

   Cambridge Board November 1983, Paper I, No. 7

3. With reference to a named phylum, discuss the factors which are used to assign organisms to smaller groups within the phylum. What is the value of classifying organisms?

   Southern Universities Joint Board June 1984, Paper II, No. 10

4. Define and illustrate the terms phylum, class, genus and species by reference to any **one** plant or animal group with which you are familiar. (*12 marks*)

   For the same group, explain what light their classification throws upon evolutionary relationships within the group. (*8 marks*)
   (*Total 20 marks*)

   Oxford and Cambridge Board July 1984, Paper II, No. 12

5. (a) Place the following list of taxonomic categories in their correct sequence beginning with the largest unit of classification and ending with the smallest:
   genus; phylum; family; species; order; class
   (*5 marks*)
   (b) Name a genus of flowering plant or conifer. (*1 mark*)
   (*Total 6 marks*)

   Oxford Local 1987, Specimen Paper, No. 1

6. The leaves, labelled A to H, are typical for eight species in a particular genus. Construct a key which will identify the species using the reference letters. All the drawings are to the same scale.

   Joint Matriculation Board June 1984, Paper IA, No. 7

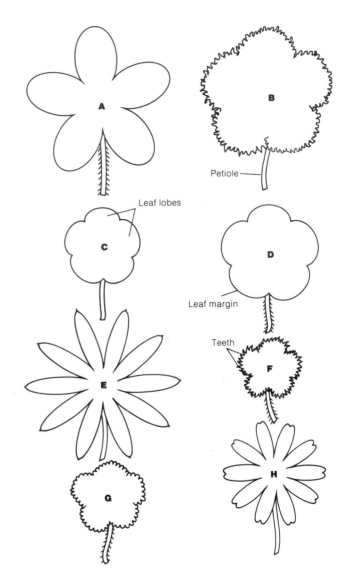

7. The following key distinguishes between the members of four different kingdoms:

1.	Cells prokaryotic	Kingdom A
	Cells eukaryotic	2
2.	Organisms with mycelia; undulipodia absent	Kingdom B
	Organisms without mycelia; undulipodia present	3
3.	Organisms multicellular; nutrition heterotrophic	Kingdom C
	Organisms multicellular; nutrition autotrophic	Kingdom D

Identify:
(a) Kingdom A
(b) Kingdom B

(c) Kingdom C

(d) Kingdom D

(4 marks)

Associated Examining Board June 1989, Paper I, No. 1

8. The drawings below illustrate seven common arthropods which may be found in or around the home.

(a) List *three* features which are visible on all seven of the organisms. (3 marks)

(b) Using only features shown in the drawings, construct a simple dichotomous key which could be used to separate the seven specimens.

(12 marks)

(Total 15 marks)

London Board June 1988, Paper IV, No. 2

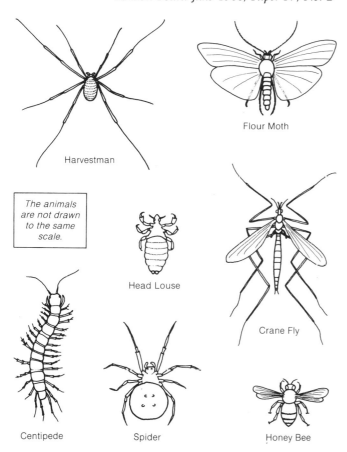

Harvestman

Flour Moth

The animals are not drawn to the same scale.

Head Louse

Crane Fly

Centipede

Spider

Honey Bee

9. Read through the following passage and answer the questions that follow.

Taxonomy is the branch of biology concerned with classification. The Linnaean system of classification is one based on hierarchy. The basic unit of classification is the species since species have an important reality in nature. Biologists, however, have always had difficulties in drawing up a species concept that will include all plants, animals and microorganisms. One complication is polymorphism.

The first task of the taxonomist is to determine the characteristics of a species; this requires deciding what criteria will be used to distinguish between species. A number of criteria may be used – these may include ecological, morphological and physiological factors.

(a) Explain what is meant by a *Linnaean system of classification* (line 2). (3 marks)

(b) Place the following in an ascending hierarchical order:

genus   kingdom   order   class   phylum

(2 marks)

(c) The scientific name for a certain type of dogwood is *Cornus alba sibirica variegata.* What is the generic name for dogwoods? (1 mark)

(d) Explain with reference to a named example, what is meant by *polymorphism* (line 8).

(4 marks)

(e) (i) Suggest a definition of *a species.*

(ii) Excluding polymorphism, why have biologists had difficulty in drawing up a 'species concept'? (line 6). (4 marks)

(f) Distinguish between the terms *phenotype* and *genotype.* (2 marks)

(g) What are the disadvantages of using a classification based on phenotypic characters?

(2 marks)

(Total 18 marks)

Oxford and Cambridge Board June 1988, Paper I, No. 7

10. Read through the following passage and then answer the questions set.

Classification is based on the assessment of similarities among organisms. Today we recognise that many varieties of organisms have evolved from a common ancestor by modification. The most useful system of classification is one which reflects a natural system based on evolutionary relationships. In general, anatomical structures seem to provide the best basis for a system of classification of this type. However, taxonomists have found that it is never valid to use a single criterion. For example, organisms which resemble each other in phenotype are not necessarily closely related. Distinguishing homologous from analogous structures is the most important procedure in taxonomy. However analogous structures are of interest and recognition of these is a major tool in the study of adaptation.

(a) Distinguish between the *genotype* and the *phenotype* of an organism. (3 marks)

(b) State **three** ways by which the genotype of an organism may become modified (line 4).

(3 marks)

(c) Outline **one** piece of evidence which suggests that present day organisms have evolved from a common ancestor (line 3). (4 marks)

(d) Do you agree that anatomical structures seem to provide the best basis of classification (line 6)? Give reasons for your answer. (3 marks)

(e) State **one** example of the danger of using a single criterion to classify a named group of animals. (2 marks)

(f) (i) State **one** pair of animals and **one** pair of plants to support the contention that 'organisms which resemble each other in external appearance are not necessarily closely related'. (line 10) (2 marks)

(ii) Referring to **one** of these examples, indicate why the organisms cited are classified in different groups. (2 marks)

(g) Quoting an example of each, explain what is meant by the terms
  (i) *homologous structure;*
  (ii) *analogous structure;*
  (iii) *adaptation.* (6 marks)
(*Total 25 marks*)

Cambridge, Oxford and Southern Board AS
Paper 2 (specimen), No. 1

7 *Lower organisms*

In this chapter, four groups of organisms will be studied. The first, the **viruses**, are obligate parasites often considered to be on the border between living and non-living. The **Prokaryotae** (bacteria) comprise cells which lack nuclei organized within membranes. Two kingdoms of eukaryotic organisms are also included in this chapter: **Fungi** and the **Protoctista**.

7.1 Viruses

Viruses are smaller than bacteria, ranging in size from about 20 nm to 300 nm. They cannot be seen through a light microscope and pass through filters which retain bacteria. Many can be crystallized and they can only multiply inside living cells. They do, however, contain nucleic acids such as DNA or RNA and must therefore be considered on the border between living and non-living. They are not classified with any other living organisms. They are made up of a nucleic acid core surrounded by a coat of protein; outside cells these inert particles are known as **virions**. Most viruses found in animal cells and those attacking bacteria (known as **bacteriophages**) have the nucleic acid DNA. Other animal viruses and plant viruses contain RNA. Electron microscopy and X-ray diffraction have shown viruses to be a variety of shapes such as spherical, e.g. poliomyelitis, straight rods, e.g. tobacco mosaic virus (TMV), or flexible rods, e.g. potato virus X. Bacteriophages have a distinct 'head' and 'tail'.

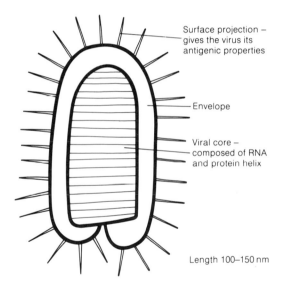

Fig. 7.1 *Simplified diagram of rabies virus*

Surface projection – gives the virus its antigenic properties

Envelope

Viral core – composed of RNA and protein helix

Length 100–150 nm

7.1.1 Transmission of viruses

Two viruses that have been widely studied are the **tobacco mosaic virus (TMV)**, which attacks tomato, blackcurrant, potato and orchid as well as tobacco itself, and the **T$_2$ phage**, a bacteriophage, which infects *Escherichia coli*. Tobacco mosaic virus is rod-shaped with a length of about 300 nm and a diameter of 15 nm. It comprises 94% protein and 6% RNA, the nucleic acid determining its characteristics. TMV is very infectious, being carried on seed coats, by grasshoppers and by mechanical means. The only effective way to limit its effect is to maintain virus-free stock.

T$_2$ phage is tadpole-shaped, the head having a diameter of about 70 nm and the tail a length of about 0.2 μm. The cycle of infection of this bacteriophage has been particularly well studied and is explained in Fig. 7.3.

T$_2$ phage immediately kills the bacterium it enters and is therefore known as a **virulent phage**. In **temperate phages** the process is much less rapid and the host and phage may exist together for many generations. Host DNA may become incorporated in the viral DNA,

Complete bacteriophage

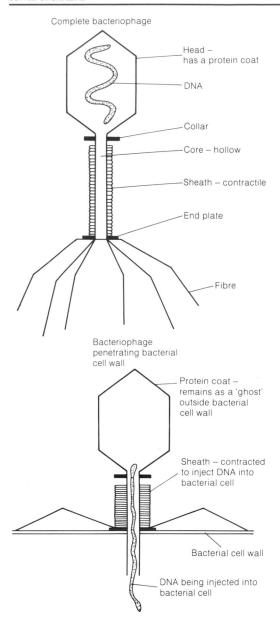

Head –
has a protein coat

DNA

Collar

Core – hollow

Sheath – contractile

End plate

Fibre

Bacteriophage
penetrating bacterial
cell wall

Protein coat –
remains as a 'ghost'
outside bacterial
cell wall

Sheath – contracted
to inject DNA into
bacterial cell

Bacterial cell wall

DNA being injected into
bacterial cell

Fig. 7.2 Structure of a bacteriophage

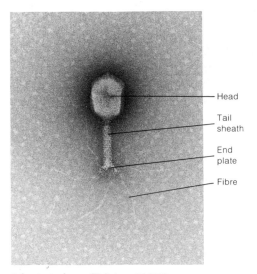

Head

Tail
sheath

End
plate

Fibre

A bacteriophage (EM) (× 108 000)

and this DNA is carried to the next host, thereby resulting in new characteristics. This process of **transduction** is an important method by which antibiotic resistance spreads throughout a population of bacteria.

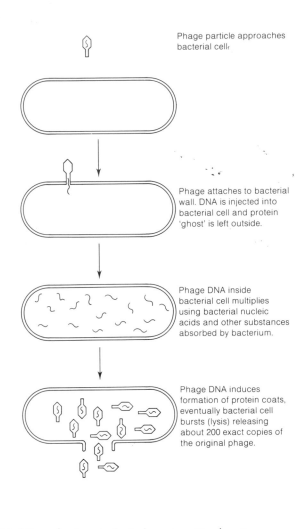

Phage particle approaches
bacterial cell.

Phage attaches to bacterial
wall. DNA is injected into
bacterial cell and protein
'ghost' is left outside.

Phage DNA inside
bacterial cell multiplies
using bacterial nucleic
acids and other substances
absorbed by bacterium.

Phage DNA induces
formation of protein coats,
eventually bacterial cell
bursts (lysis) releasing
about 200 exact copies of
the original phage.

Fig. 7.3 Life cycle of a virulent phage (e.g. T_2 phage)

7.1.2 Retroviruses

Probably the best known retrovirus is the Human Immunodeficiency (HIV) virus which causes AIDS (Acquired Immune Deficiency Syndrome), further details of which are given in Section 32.2.3.

The genetic information in a retrovirus is RNA. While many viruses possess RNA, retroviruses are different in that they can use it to synthesize DNA. This is a reversal of the usual genetic process in which RNA is made from DNA and the reason retroviruses are so called (retro = behind or backwards).

In 1970 the enzyme capable of synthesising DNA from RNA was discovered and given the name **reverse transcriptase** (as it catalyses the opposite process to transcriptase which synthesizes RNA from DNA). The discovery of this enzyme, more details of which are given in Section 12.8.2, has considerable importance for genetic engineering.

The DNA form of the retrovirus genes is called the **provirus** and

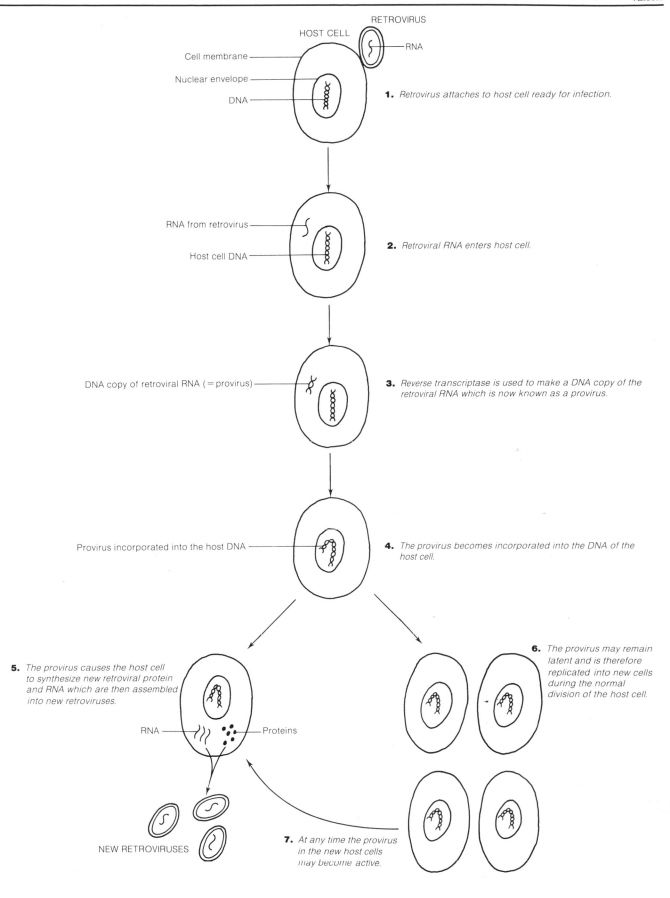

Fig. 7.4 Life cycle of a retrovirus

is significant in that it can be incorporated into the host's DNA. Here it may remain latent for long periods before the DNA of the provirus is again expressed and new viral RNA produced. During this time any division of the host cell results in the proviral DNA being duplicated as well. In this way the number of potential retroviruses can proliferate considerably. This explains why individuals infected with the HIV virus often display no symptoms for many years before suddenly developing full-blown AIDS.

When incorporated into the host DNA the provirus is capable of activating the host genes in its immediate vicinity. Where these genes are concerned with cell division or growth, and are 'switched off' at the time, their activation by the provirus can result in a malignant growth known as **cancer**. The RNA produced by these newly activated genes may become packaged inside new retrovirus particles being assembled inside the host cell. This RNA may then be delivered, along with the retroviral RNA, to the next cell the virus infects. This new cell will then become potentially cancerous.

Host genes which have been acquired by retroviruses in this way are called **oncogenes** (*oncos* = tumour). Very few human cancers are caused by retroviruses in this way but research into them has led to the discovery of similar genes found in human chromosomes. These genes can be activated by chemicals or forms of radiation rather than viruses, and their investigation has already helped to prevent some cancers and may, in time, provide a cure.

Retroviruses can cause diseases other than cancer, but most are harmless. Some proviral DNA has become such an integral part of the host-cell DNA that it is passed on from one generation to the next via the gametes and is, in effect, part of the host's genetic make-up. Such a virus is referred to as an **endogenous** virus.

7.1.3 Economic importance of viruses

Viruses cause a variety of infectious diseases in man, other animals and plants. The symptoms shown by plants may be localized or distributed throughout the plant. The same virus may have quite different effects in different hosts and these symptoms may be influenced by environmental conditions.

Viral diseases are often difficult to treat because antibiotics cannot be used. Vaccines may be produced but these are not always effective because one virus may exist in a variety of forms. Methods of control therefore depend primarily on prevention, such as breeding resistant species, removal of the source of infection and the protection of susceptible plants and animals.

Viral diseases of plants include those caused by TMV, potato virus X, barley yellow dwarf virus and turnip yellow mosaic virus. Some important viral diseases of animals, including man, are summarized in Table 7.1.

Retroviruses cause a number of diseases including a degenerative brain disease in sheep, anaemia in cattle and some cancers, but by far the most important one is AIDS caused by the HIV retrovirus. It may, however, prove possible to use retroviruses to cure diseases by utilizing them to insert useful genes into cells where particular genes are defective. Inherited diseases such as phenylketonuria and thalassaemia are the most likely to be cured by this means.

TABLE 7.1 **Important viral diseases of animals**

Disease	RNA or DNA virus	Host	Region affected/symptoms	Control
Foot and mouth	RNA	Mainly cattle; sheep; pigs	Lesions in mouth	Vaccine expensive and short lasting. Slaughter of diseased animals
Rabies	RNA	Mainly carnivorous mammals; man	Hyperexcitability; paralysis; death	Active and passive immunity after a bite. Quarantine regulations. Elimination of stray dogs
Poliomyelitis	RNA	Man	Inflammation of the spinal cord resulting in paralysis of various groups of muscles	Living attenuated virus; usually given orally
Mumps	RNA	Man	Inflammation of parotid gland	Living attenuated virus
Measles	RNA	Man	Coryza (the symptoms of a common cold); a blotchy rash	Living attenuated virus
Influenza	DNA	Man	Several different manifestations, e.g. common cold symptoms, aching limbs	Dead virus. Only effective if vaccinated with the right strain
Acquired Immune Deficiency Syndrome. AIDS	RNA	Man	T-lymphocytes. Absence of resistance to disease	No vaccine currently available

7.2 Prokaryotae

The cyanobacteria (blue-green bacteria) and bacteria which comprise the Prokaryotae are the only living prokaryotic organisms. As such they are the living organisms which most closely resemble the first forms of life. The differences between prokaryotic and eukaryotic cells are given in Section 4.2.2.

No Prokaryotae are truly multicellular although the blue-green bacteria are commonly found in filaments and clusters. This is either because their cell walls fail to separate completely at cell division or because they are held together by a mucilagenous sheath. Most blue-green bacteria can photosynthesize and many are capable of nitrogen fixation. They are important colonizers of bare land and were probably among the first organisms to evolve.

7.2.1 Bacteria

Bacteria are the smallest cellular organisms and are the most abundant. They may respire aerobically and exhibit a variety of feeding mechanisms. These features, together with their phenomenal rate of cell division, have made them a most successful group.

Fig. 7.5 shows the structure of a typical bacterial cell. Such cells may vary in the nature of the cell wall. In some forms the glycoprotein is supplemented by large molecules of lipopolysaccharide. Cells which lack the lipopolysaccharide combine with dyes like gentian violet and are said to be **gram positive**. Those with the lipopolysaccharide are not stained by gentian violet and are said to be **gram negative**. Gram positive bacteria are more susceptible to both antibiotics and lysozyme than are gram negative ones. Bacteria may be coated with a slime capsule which is thought to interfere

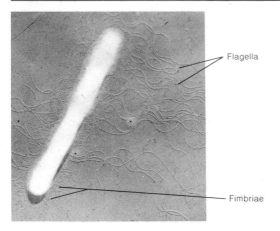

A rod bacterium showing flagella (EM) (× 1000 approx.)

Flagella

Fimbriae

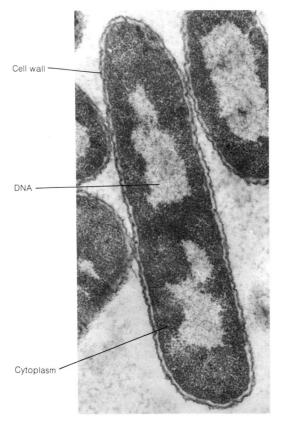

Cell wall

DNA

Cytoplasm

E. coli (EM) (× 38 000 approx.)

with phagocytosis by the white blood cells. Bacteria are generally distinguished from each other by their shape. Spherical ones are known as **cocci** (singular – coccus), rod-shaped as **bacilli** (singular – bacillus) and spiral ones as **spirilla** (singular – spirillum).

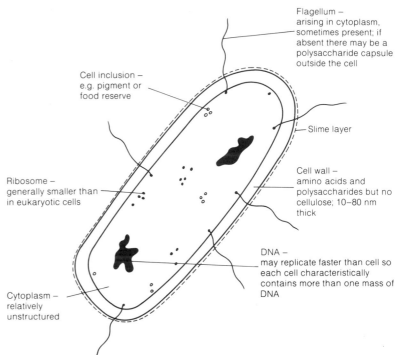

Flagellum –
arising in cytoplasm,
sometimes present; if
absent there may be a
polysaccharide capsule
outside the cell

Cell inclusion –
e.g. pigment or
food reserve

Slime layer

Ribosome –
generally smaller than
in eukaryotic cells

Cell wall –
amino acids and
polysaccharides but no
cellulose; 10–80 nm
thick

DNA –
may replicate faster than cell so
each cell characteristically
contains more than one mass of
DNA

Cytoplasm –
relatively
unstructured

Fig. 7.5 Generalized bacterial cell

Cocci may stick together in chains – **streptococcus**, or in clusters – **staphylococcus**. Bacteria show considerable diversity in their metabolism. The majority are heterotrophic and most of these are saprophytes. They are responsible, with the Fungi, for decaying and recycling organic material in the soil. Others are parasitic, some causing disease but many having little effect on their host. Numerous gut bacteria have a symbiotic relationship with their host, for example helping to digest the cellulose ingested by ruminants.

Other bacteria are autotrophic. Photosynthetic bacteria are anaerobic and often use sulphur compounds as electron donors rather than the water used by higher plants.

$$CO_2 + 2H_2S \xrightarrow{\text{light}} (CH_2O) + H_2O + 2S$$

Some bacteria derive their energy from inorganic molecules such as ammonia, nitrite, sulphur or hydrogen sulphide. These are the **chemosynthetic bacteria**, some of which are essential links in the nitrogen cycle. For example, one group oxidizes ammonia or ammonium compounds to nitrites and energy, and another oxidizes nitrites to nitrates and energy.

$$2NH_4^+ + 3O_2 \longrightarrow 2NO_2^- + 4H^+ + 2H_2O \ (\textit{Nitrosomonas})$$

$$2NO_2^- + O_2 \longrightarrow 2NO_3^- + \text{ energy } (\textit{Nitrobacter})$$

Bacteria reproduce by binary fission, one cell being capable of giving rise to over 4×10^{21} cells in 24 hours. Under certain circumstances conjugation occurs and new combinations of genetic material result. Bacteria may also produce thick-walled spores which are highly resistant, often surviving drought and extremes of temperature.

7.2.2 Economic importance of bacteria

It is easy to think of all bacteria as pathogens but it is important to remember that many are beneficial to man. These benefits include:

1. The breakdown of plant and animal remains and the recycling of nitrogen, carbon and phosphorus.

2. Symbiotic relationships with other organisms. For example supplying vitamin K and some of the vitamin B complex in man, breaking down cellulose in herbivores.

3. Food production, e.g. some cheeses, yoghurts, vinegar.

4. Manufacturing processes, e.g. making soap powders, tanning leather and retting flax to make linen.

5. They are easily cultured and may be used for research, particularly in genetics. They are also used for making antibiotics, amino acids, enzymes and SCP (single cell protein).

Further details of how man exploits these beneficial uses of bacteria are given in Chapter 41.

Detrimental effects of bacteria include deterioration of stored food and damage to buried metal pipes caused by sulphuric acid production by *Thiobacillus* and *Desulphovibrio*.

Bacterial diseases of man may be spread in a variety of ways including droplet infection, direct contact and contamination of food or water. A range of bacterial infections of plants and animals, including man, is given in Table 7.2.

TABLE 7.2 **Bacterial infections**

Disease	Bacterium	Host organism	Symptoms/region affected	Control/treatment
Soft rot	*Erwinia caratovora* (and *Pseudomonas* spp.)	Most vegetables, especially carrots, celery and potatoes	Pectin of cell walls broken down by enzymes	No treatment. Avoid contact with infected vegetables
Typhoid	*Salmonella typhi*	Man	Alimentary canal, blood, lungs, bone	Vaccination (TAB) with killed bacteria. Avoid contact with carriers
Whooping cough	*Bordetella pertussis*	Man	Upper respiratory tract; violent coughing	Vaccination with killed bacteria
Tuberculosis	*Mycobacterium tuberculosis*	Man	Mainly lungs	If not immune, use living attenuated vaccine (BCG). Treat with antibiotics
Gonorrhoea	*Neisseria gonorrhoea*	Man	Reproductive organs. Eyes of newborn babies affected if mother infected	Avoid contact with infection. Treat with antibiotics
Bacterial food poisoning	*Salmonella* spp. (not *S. typhi*)	Man	Alimentary canal	Avoid infected food. Thorough cooking needed. Antibiotic treatment not very effective

7.3 Fungi

The Fungi are a large group of organisms composed of about 80 000 named species. For many years they were classified with the plants but are now recognized as a separate kingdom. This separation is based on the presence of the polysaccharide chitin found in their cell walls, and never in those of plants. Their bodies are usually a

mycelium of thread-like **hyphae** without distinct cell boundaries. The Fungi lack chlorophyll and are therefore unable to photosynthesize. They feed heterotrophically, generally as saprophytes or parasites, and details of these methods of feeding will be found in Chapter 25.

Within this kingdom three phyla will be described in Table 7.3: Zygomycota, Ascomycota and Basidiomycota.

TABLE 7.3 **Classification of the fungi**

KINGDOM FUNGI	No chlorophyll; do not photosynthesize Heterotrophic Cell walls contain chitin Body usually a mycelium Carbohydrate stored as glycogen Reproduce by means of spores without flagella

Zygomycota	**Ascomycota**	**Basidiomycota**
No septa in hyphae; large branched mycelium formed	Septa in hyphae	Septa in hyphae; large 3-dimensional structures often formed
Asexual reproduction by sporangia producing spores or by conidia	Asexual reproduction by conidia	Asexual reproduction unusual but spores formed
Conjugation gives rise to a zygospore	Sexual reproduction by acospores forming in an ascus	Sexual reproduction by formation of basidiospores outside basidia
e.g. Mucor – pin mould *Rhizopus* – bread mould (See Fig. 7.6)	*e.g. Saccharomyces* – yeast *Erysiphe* – powdery mildew *Aspergillus* and *Penicillium* – saprophytic moulds (See Fig. 7.7)	*e.g. Agaricus campestris* – field mushroom *Coprinus* – ink cap toadstool (See Fig. 7.9)

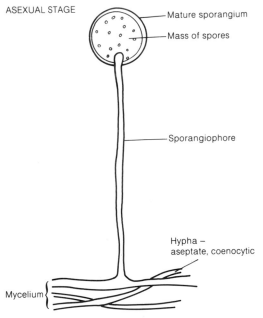

Fig. 7.6 Mucor

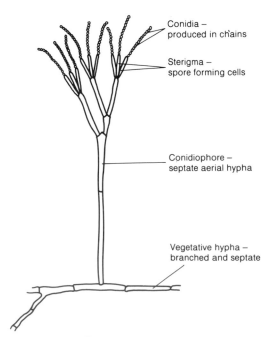

Fig. 7.7 Penicillium

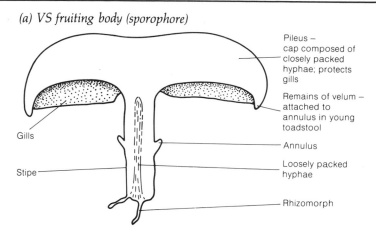

(a) VS fruiting body (sporophore)

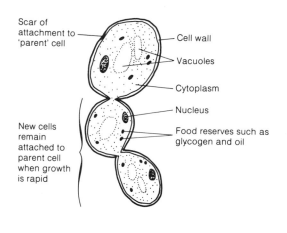

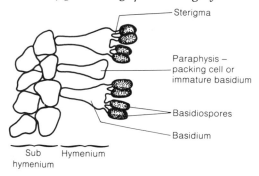

(b) Part of gill showing spore-bearing hymenium (VS)

Fig. 7.8 Saccharomyces

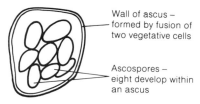

Fig. 7.9 Agaricus

TABLE 7.4 **Common fungal diseases of plants**

Disease	Fungus	Group	Host	Importance
Powdery mildew	*Erysiphe graminae*	Ascomycota	Cereals	Serious, rapidly spreading disease of cereal crops
Dutch elm disease	*Ceratocystis ulmi*	Ascomycota	Elm	Carried by beetles. Led to eradication of elms in much of Europe and North America
Brown rot	*Monilinia fructigena*	Ascomycota	Fruits such as plums and peaches	Causes serious economic damage throughout the world
Loose smut of oats	*Ustilago avenae*	Basidiomycota	Oats	Masses of powdery spores which may affect the grain itself, making it useless
Black stem rust of wheat	*Puccinia graminis*	Basidiomycota	Wheat	Airborne spores rapidly spread this economically serious disease

7.3.1 Economic importance of fungi

Many fungi are beneficial to man. Examples include:

1. Decomposition of sewage and organic material in the soil.

2. Production of antibiotics, notably from *Penicillium* and *Aspergillus*.

3. Production of alcohol for drinking and industry.

4. Production of other foods. Citric acid for lemonade is produced by the fermentation of glucose by *Aspergillus*. Yeasts are used in bread production and the food yeast *Candida utilis* has been investigated as a source of single cell protein (SCP).

5. Experimental use, especially for genetic investigations.

Many fungi are also harmful to man, causing decomposition of stored foods and deterioration of natural materials such as leather and wood. Fungi more commonly cause disease in plants than in animals but some of the plants infected are of great economic importance to man. Table 7.4 shows some common fungal diseases of plants.

7.4 Protoctista

The kingdom Protoctista is made up of single celled eukaryotic organisms. Apart from this one common feature the kingdom is very varied.

7.4.1 Oomycota

This group of organisms demonstrates the problems of trying to impose an artificial classification on organisms whose phyletic relationships are not well understood. Until recently they were considered to be Fungi, largely because they comprise hyphae, lack cellulose in their cell walls and have a non-photosynthetic absorptive mode of nutrition. However, the fact that their hyphae lack septa (cross-walls) and their spores have flagella, means they are now considered as protoctists.

The structure of one example of an oomycete, *Peronospora*, is shown in Fig. 7.10. It causes downy mildew of onions and crucifers although the damage it causes is rarely of economic importance. By contrast, another oomycete – *Phytophthora infestans* – created serious economic harm as the potato blight which caused the Irish potato famine of 1845.

7.4.2 Euglenophyta

For many years members of this phylum, such as *Euglena*, were classified by botanists as plants and by zoologists as animals. Their inclusion as a separate division of eukaryotic unicells avoids this dispute and recognizes their unique position on the boundary between plants and animals.

Euglena closely resembles the flagellate protozoa but its possession of numerous chloroplasts containing chlorophyll a and b means that it is able to photosynthesize. The rather animal-like possession of an eye-spot and flagella means that it is able to detect and swim towards light. It stores the products of photosynthesis as **paramylon**, a polysaccharide not found in any other group of organisms. *Euglena* does not have a cell wall but has strips of protein forming a pellicle inside the cell membrane. This pellicle is flexible and allows *Euglena* to change shape, providing an alternative means of locomotion in mud.

Osmoregulation is by means of a contractile vacuole which opens into the reservoir at the base of the flagella. The cells reproduce by longitudinal binary fission.

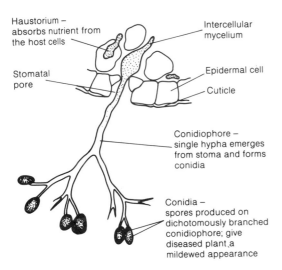

Haustorium – absorbs nutrient from the host cells

Intercellular mycelium

Stomatal pore

Epidermal cell

Cuticle

Conidiophore – single hypha emerges from stoma and forms conidia

Conidia – spores produced on dichotomously branched conidiophore; give diseased plant a mildewed appearance

Fig. 7.10 Peronospora

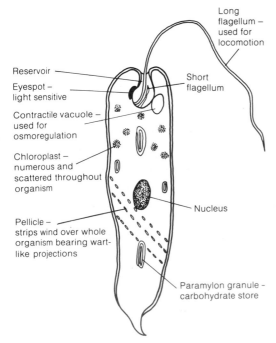

Long flagellum – used for locomotion

Reservoir

Eyespot – light sensitive

Short flagellum

Contractile vacuole – used for osmoregulation

Chloroplast – numerous and scattered throughout organism

Pellicle – strips wind over whole organism bearing wart-like projections

Nucleus

Paramylon granule - carbohydrate store

Fig. 7.11 Euglena

7.4.3 Protozoa

This is a collective name for four separate phyla of heterotrophic eukaryotic unicells. The protozoa are widely distributed organisms but are only active in aqueous media. They are highly specialized, with division of labour and specialization of organelles, which has resulted in a wide variation of form and physiology. One phylum of protozoa, the **Apicomplexa** (sporozoans), is entirely parasitic but members of the other three phyla may be free-living or parasitic. Division into phyla is made primarily according to the method of locomotion used.

Cilia and flagella are extremely important biologically and details of their structure are covered in Section 4.3.15. Similarly amoeboid, ciliary and flagellar locomotion are important and are considered, along with other methods of locomotion, in Sections 39.1 and 39.2.

TABLE 7.5 **Classification of the protozoa**

Features of protozoa — Unicellular (acellular)
No chlorophyll
Heterotrophic
Mostly microscopic ($2\,\mu m$–$10\,\mu m$ long)
Specialized organelles, but no tissues or organs
No cell wall

Phylum	Apicomplexa (sporozoans)	Rhizopoda	Zoomastigina (flagellates)	Ciliophora (ciliates)
	Mainly parasitic	Free-living or parasitic	Free-living or parasitic	Free-living
	One nucleus	One nucleus	One nucleus	Two nuclei, meganucleus and micronucleus
	Little, if any, self-induced movement	Move by pseudopodia	Move by one or more flagella	Move by cilia
	Definite shape; pellide present	Irregular shape; no pellicle; some secrete 'shells'	Definite shape; pellicle present	Definite shape; pellicle present
	Food absorbed directly; no feeding organelles	Food captured by pseudopodia	Feeding methods diverse	Food trapped by mucus and cilia
	Asexual reproduction by multiple fission	Asexual reproduction by binary fission	Asexual reproduction by longitudinal binary fission	Asexual reproduction by transverse binary fission
	Sexual process	Sometimes gametes may be formed	No sexual reproduction	Sexual reproduction by conjugation
Examples	*Plasmodium* *Monocystis*	*Arcella* *Amoeba* (See Fig. 7.12)	*Trichonympha* — symbiont in gut of termite *Trypanosoma* — parasite causing sleeping sickness (See Fig. 7.13)	*Stentor* *Vorticella* *Paramecium* (See Fig. 7.14)

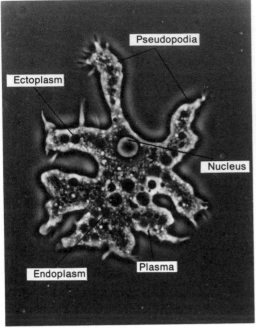

Amoeba

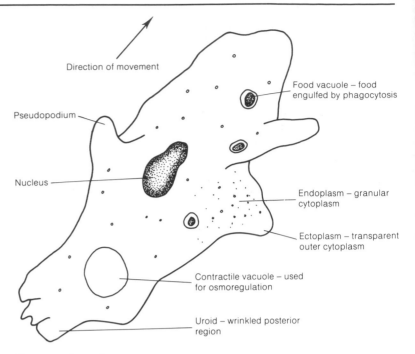

Direction of movement

Food vacuole – food engulfed by phagocytosis

Pseudopodium

Nucleus

Endoplasm – granular cytoplasm

Ectoplasm – transparent outer cytoplasm

Contractile vacuole – used for osmoregulation

Uroid – wrinkled posterior region

Fig. 7.12 Amoeba

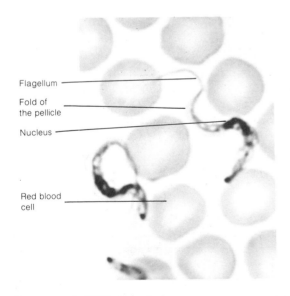

Flagellum

Fold of the pellicle

Nucleus

Red blood cell

Trypanosoma (× 2000 approx.)

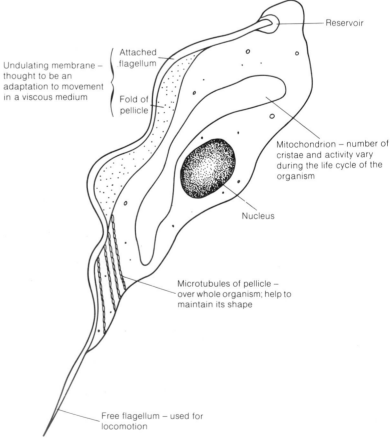

Reservoir

Undulating membrane – thought to be an adaptation to movement in a viscous medium

Attached flagellum

Fold of pellicle

Mitochondrion – number of cristae and activity vary during the life cycle of the organism

Nucleus

Microtubules of pellicle – over whole organism; help to maintain its shape

Free flagellum – used for locomotion

Fig. 7.13 Trypanosoma

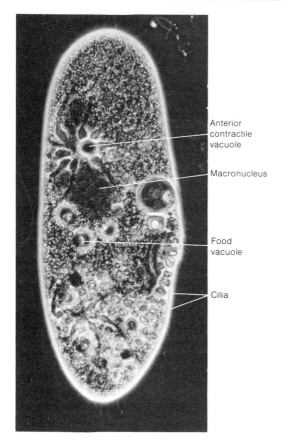

Paramecium (× 350 approx.)

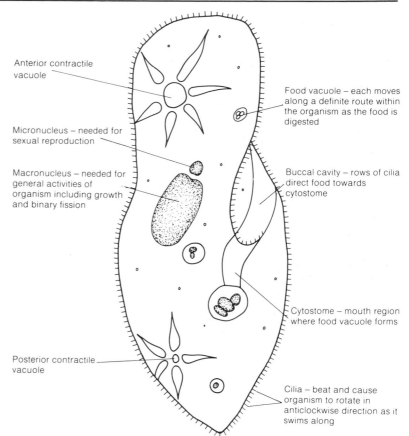

Fig. 7.14 Paramecium

TABLE 7.6   **Classification of the algae**

ALGAE	No stems, roots or leaves No sclerenchyma No vascular tissue No archegonia Other photosynthetic pigments, in addition to chlorophyll a	
Chlorophyta (green algae)	**Phaeophyta (brown algae)**
Chlorophyll a and b present	Chlorophyll a and c, xanthophylls (e.g. fucoxanthin)
Food reserve is starch	Food reserves include mannitol and laminarin
Cellulose cell walls	Cell wall includes alginic acid
2, 4 or 0 flagella present	Motile stages are pear-shaped with 2 flagella
Unicellular, filamentous or thalloid	No unicellular forms
Mostly fresh-water	Almost entirely marine

e.g. Chlamydomonas
  Chlorella
  Pleurococcus
  Spirogyra
	e.g. Ascophyllum
	  Fucus

7.4.4 Economic importance of protozoa

Protozoans form a significant part of **plankton** and as such are important food for surface dwelling fish and, when they die, provide food for animals at greater depths. Two orders of rhizopods, the **Radiolaria** and the **Foraminifera**, have hard 'shells' and over millions of years these have fallen to the bottom of oceans to form deposits thousands of feet thick. Changes in sea level have exposed some of these as chalk and limestone. The presence of fossil Foraminifera in rock strata can also be useful indicators of oil-bearing rock.

Most animals have one or more protozoan parasites, some of which cause disease in man including: *Trypanosoma*, causing sleeping sickness, *Plasmodium*, causing malaria and *Entamoeba histolytica*, which is responsible for amoebic dysentery. Protozoans also affect the taste and odour of drinking water, dinoflagellates such as *Dinobryon* giving it a definite fishy smell and *Synura* giving it a bitter taste.

7.4.5 Algae

This is a collective name for a varied group of phyla with no one diagnostic feature. They are normally aquatic or live in damp terrestrial habitats. Sub-divisions are mainly associated with biochemical differences related to photosynthesis.

The **chlorophyta** are green algae which range in form from unicells such as *Chlamydomonas* and *Chlorella* through colonies like *Volvox* and filaments like *Spirogyra* to delicate thalloid genera like *Ulva*. They contain the same photosynthetic pigments as higher plants but the

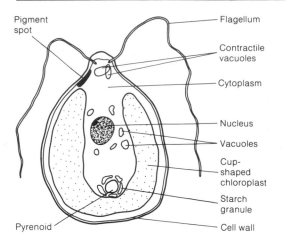

Fig. 7.15(a) Chlamydomonas

chloroplasts which contain them vary. *Chlamydomonas* has a single bowl-shaped chloroplast and that of *Spirogyra* is spiral. Both have starch deposits called **pyrenoids**. *Chlamydomonas* also has a light-sensitive spot and will swim, by means of flagella, towards the light. Both genera are capable of asexual and sexual reproduction.

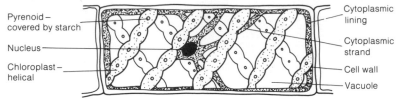

Fig. 7.15(b) Spirogyra

The phaeophyta are a phylum which shows great diversity in structure and method of reproduction. It includes all the larger seaweeds as well as small, branched filamentous ones such as *Ectocarpus*. Genera like *Fucus* are well adapted for life in the intertidal zone where they are frequently buffeted by waves and may be exposed at low tide. Both asexual and sexual reproduction are shown, although the latter is unusual in *Fucus*.

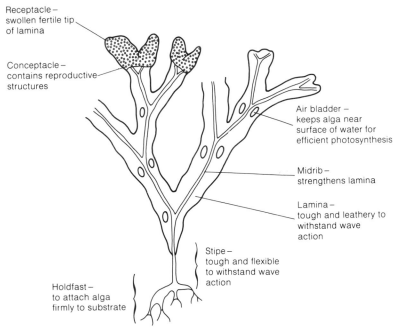

Fig. 7.16 Fucus

7.4.6 Economic importance of algae

At least half the carbon fixation of the earth is carried out by algae in the surface layers of oceans. This primary production is at the base of all aquatic food chains. These algae are also responsible for half the oxygen released by plants into the atmosphere.

Algae may be used in some parts of the world as a direct food source for man and they may be used as fertilizers on coastal farms. Unicellular green algae such as *Chlorella* are easy to cultivate and may be used as a source of single cell protein (SCP) for human and animal consumption.

Spirogyra (× 500 approx.)

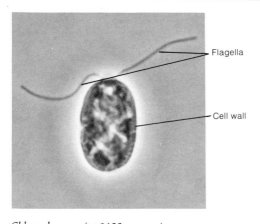

— Flagella

— Cell wall

Chlamydomonas (× 1600 approx.)

Green algae provide oxygen for the aerobic bacteria which break down sewage.

Derivatives of alginic acid found in the cell walls of many brown algae are non-toxic and readily form gels. These alginates are used as thickeners in many products including ice cream, hand cream, polish, medicine, paint, ceramic glazes and confectionery.

Excessive numbers of algae may develop in bodies of water following pollution by fertilizers or other chemicals. These 'blooms' cause the water to smell and taste unpleasant and may lead to oxygen depletion and the death of fish (Section 29.3.2).

7.5 Questions

1. (a) What size are viruses and how can they be studied?
   (b) Describe the structures of two different types of virus.
   (c) Describe how one of these viruses infects living cells and reproduces.
   (d) With the aid of named examples compare viruses and bacteria as agents of disease.

   Oxford Local June 1983, Paper II, No. 1

2. (a) List those features that distinguish viruses from bacteria. *(4 marks)*
   (b) Draw a typical virus and explain how it infects its host and reproduces. *(10 marks)*
   (c) Why is it that viruses frequently cause diseases in organisms? *(4 marks)*
   (Total 18 marks)

   Cambridge Board June 1984, Paper I, No. 8

3. By reference to specific examples, summarize the importance of bacteria to other living organisms.
   (8 marks)

   Southern Universities Joint Board June 1985, Paper I, No. 3

4. Briefly describe *Penicillium* and *Saccharomyces* and outline the different ways in which they have been utilized by man.

   Welsh Joint Education Committee June 1983, Paper AII, No.15

5. Opposite is a diagram illustrating the injection of genetic material through a bacterial cell wall by a T-even phage.

   (a) Name the structures labelled A, B, C and D. *(4 marks)*
   (b) The viral DNA controls the production of new viral RNA. Name the term which describes this process. *(1 mark)*
   (c) Give the name of the structure formed when the host ribosomes are linked into a chain by the new messenger RNA. *(1 mark)*
   (d) What is the term given to the process whereby viral polypeptides are produced in association with the host ribosomes and viral RNAs? *(1 mark)*
   (Total 7 marks)

   Oxford Local June 1984, Paper I, No. 3

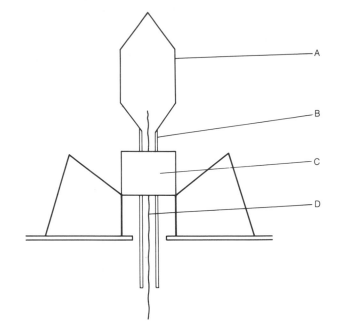

6. The following description was written after an acellular organism had been examined with an optical microscope:

   The body is slender and surrounded by a flexible membrane. It is approximately 50 μm in length and 10 μm across at its widest. One end of the body is tapered almost to a point. At the other end, which is more rounded, there is a short canal, opening by a pore to the exterior. The other end of the canal ends in the cytoplasm in a spherical reservoir. Growing from the base of the reservoir is a long flagellum, approximately the same length as the body. The flagellum passes through the canal into the surrounding water and is directed away from the body.
   Another short flagellum is present near to the base of the long flagellum but it is too short to project into the water. Approximately six small contractile vacuoles are attached to one side of the reservoir. In the canal, a small circular photoreceptor is attached to the side of the long flagellum. In the centre of the cell there is a darkly-staining circular nucleus with a diameter of about half that of the width of the body. The remainder of the cell is mostly occupied by a number of irregularly rounded chloroplasts about the same size as the nucleus.

   (a) State **two** structural features of this organism which would suggest that it is a protozoan.
   (2 marks)

(b) In having chloroplasts, the organism resembles a plant cell. State **two** ways in which it **differs** from a typical plant cell.

(2 marks)

(c) Draw a large labelled diagram to show the structure of this organism as you would expect it to appear from the description. Indicate the scale of your drawing.

(9 marks)

(Total 13 marks)

Welsh Joint Education Committee June 1988, Paper A2, No. 7

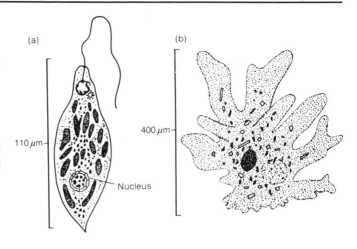

7. (a) The drawings show three examples of protozoans belonging to different groups. Name the group to which each belongs.

(3 marks)

(b) Describe briefly how locomotion is achieved by each of these organisms. *(6 marks)*

(c) State *two* ways in which the organism shown in A differs from a cell of a filamentous alga.

(2 marks)

(d) How do organisms like those shown assist in sewage treatment? *(3 marks)*

(Total 14 marks)

London Board June 1983, Paper 1, No. 12

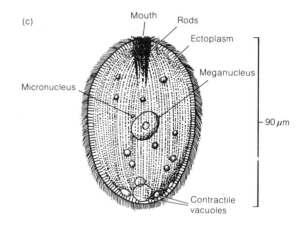

8 Plantae

The organisms included in this kingdom are made up of more than one eukaryotic cell, have cell walls containing cellulose and photosynthesize using chlorophyll as the main pigment (Sections 23.2.1 and 23.2.2).

The scheme of plant classification adopted here is one of many, but is the one being increasingly used and accepted. Under this system two terms, commonly used at present, disappear. The term 'Pteridophyta' is no longer used because each of the three classes of this group are promoted to phylum status. Thus the classes Lycopodiales, Equisitales and Filicales now become the phyla Lycopodophyta, Sphenophyta and Filicinophyta respectively. Similarly the term 'Spermatophyta' is not used as the two classes of this group, Gymnospermae and Angiospermae become the phyla Coniferophyta and Angiospermophyta respectively. The word 'Tracheophyta' is also dropped although the term 'tracheophytes' is still used as a collective noun for the five plant phyla with vascular tissue, namely Lycopodophyta, Sphenophyta, Filicinophyta, Coniferophyta and Angiospermophyta.

8.1 Bryophyta

The mosses and liverworts which make up the Bryophyta are small plants generally found in moist terrestrial habitats. They have no roots and no vascular tissue. They all show alternation of generations in which the sporophyte and gametophyte are almost equally conspicuous, although the sporophyte is attached to, and dependent on, the gametophyte throughout its life. Although it is thought that bryophytes arose from green algae and colonized land over 400 million years ago, they are still very dependent on water for their existence, the gametophyte antherozooids being particularly susceptible to desiccation. Spore dispersal, however, is well adapted to land, requiring dry conditions.

8.1.1 Musci

Fig. 8.1. gives an account of the life cycle of a moss.

8.1.2 Hepaticae

The internal structure of liverworts is less well differentiated than that of mosses. These small plate-like or leafy plants are still very dependent on water to prevent desiccation and for the liberation of their gametes.

Liverwort with capsules (*Marchantia*)

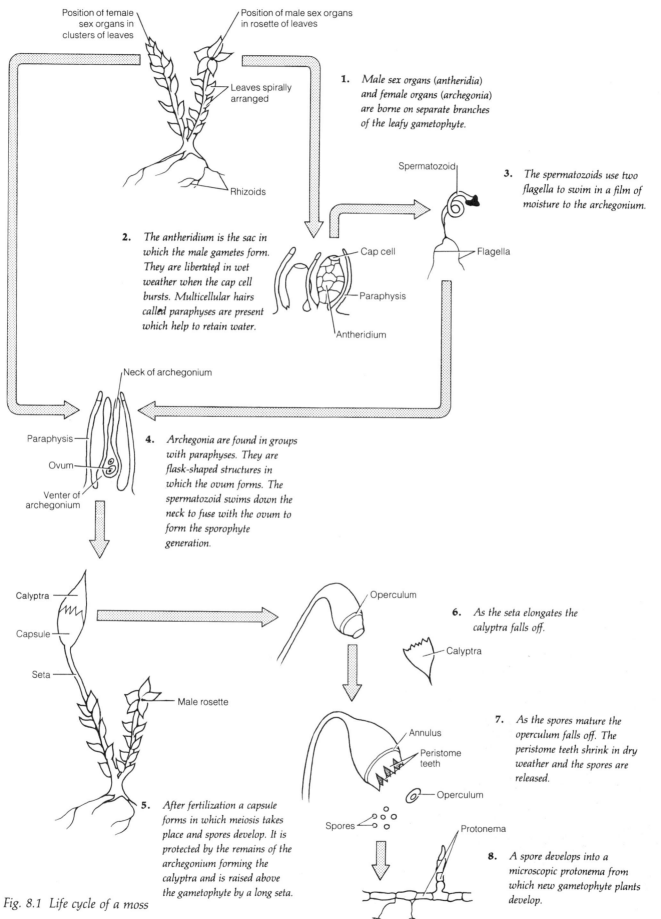

Position of female sex organs in clusters of leaves

Position of male sex organs in rosette of leaves

Leaves spirally arranged

Rhizoids

1. *Male sex organs (antheridia) and female organs (archegonia) are borne on separate branches of the leafy gametophyte.*

2. *The antheridium is the sac in which the male gametes form. They are liberated in wet weather when the cap cell bursts. Multicellular hairs called paraphyses are present which help to retain water.*

Cap cell

Paraphysis

Antheridium

Spermatozoid

3. *The spermatozoids use two flagella to swim in a film of moisture to the archegonium.*

Flagella

Neck of archegonium

Paraphysis

Ovum

Venter of archegonium

4. *Archegonia are found in groups with paraphyses. They are flask-shaped structures in which the ovum forms. The spermatozoid swims down the neck to fuse with the ovum to form the sporophyte generation.*

Calyptra

Capsule

Seta

Male rosette

Operculum

6. *As the seta elongates the calyptra falls off.*

Calyptra

Annulus

Peristome teeth

Operculum

Spores

7. *As the spores mature the operculum falls off. The peristome teeth shrink in dry weather and the spores are released.*

Protonema

5. *After fertilization a capsule forms in which meiosis takes place and spores develop. It is protected by the remains of the archegonium forming the calyptra and is raised above the gametophyte by a long seta.*

8. *A spore develops into a microscopic protonema from which new gametophyte plants develop.*

Fig. 8.1 Life cycle of a moss

TABLE 8.1   **Classification of the Bryophyta**

PHYLUM BRYOPHYTA	Alternation of generations with dominant gametophyte Sporophyte attached to and dependent on gametophyte No vascular tissue Simple stem and leaves or a thallus No true roots – plant anchored by rhizoids	
Class – Musci (mosses)	**Class – Hepaticae (liverworts)**	
Gametophyte with spirally arranged leaves	Gametophyte flattened or leafy (leaves in three rows)	
Multicellular rhizoids for attachment	Unicellular rhizoids	
Elaborate mechanism for dispersal of spores involving hygroscopic 'teeth'	Capsule splits in four to release spores	
Spores develop into protonema	Spores develop directly into photosynthetic gametophyte	

e.g. Funaria
   Sphagnum – bog moss
   Mnium
	e.g. Marchantia
	   Pellia	

8.1.3 Economic importance of the Bryophyta

Bryophytes have little importance except as early colonizers of bare land, helping to prevent erosion and enrich the ground for the growth of larger plants. Peat, used for fuel where alternatives are scarce, is formed from dead, compressed bog mosses.

8.2   Lycopodophyta (club-mosses)

Club mosses have small spirally arranged leaves and their sporangia are usually in cones. In species such as *Lycopodium* all spores are the same size – **homosporous**, whereas species like *Selaginella* produce small microspores and larger megaspores. These types are described as **heterosporous**. The microspores develop into a male prothallus and the megaspores into a female prothallus, both of which remain enclosed in the spore. A description of the life-cycle of *Selaginella* is shown in Fig. 8.2.

8.3   Sphenophyta (horsetails)

Horsetails have leaves arranged in whorls around the stem and their sporangia are in cones. *Equisetum* is the only surviving genus of this once abundant group. It produces spores which all look alike but on germination they produce two kinds of prothalli. The external features of *Equisetum* are shown in Fig. 8.3 on p. 102.

Strobilus

Vegetative shoot

Lycopodium

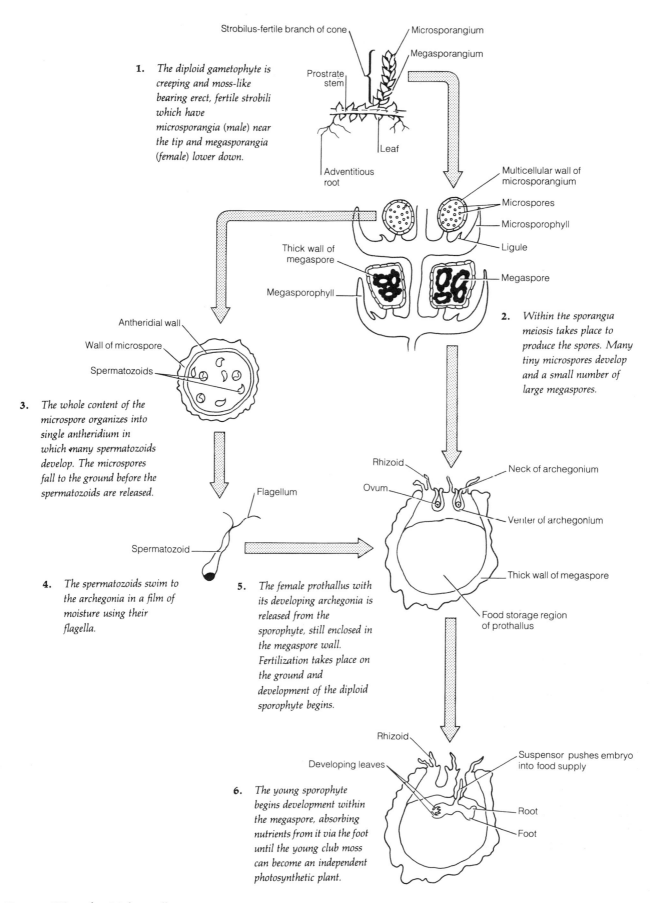

1. The diploid gametophyte is creeping and moss-like bearing erect, fertile strobili which have microsporangia (male) near the tip and megasporangia (female) lower down.

Strobilus-fertile branch of cone
Microsporangium
Megasporangium
Prostrate stem
Leaf
Adventitious root

Multicellular wall of microsporangium
Microspores
Microsporophyll
Ligule
Thick wall of megaspore
Megaspore
Megasporophyll

2. Within the sporangia meiosis takes place to produce the spores. Many tiny microspores develop and a small number of large megaspores.

Antheridial wall
Wall of microspore
Spermatozoids

3. The whole content of the microspore organizes into single antheridium in which many spermatozoids develop. The microspores fall to the ground before the spermatozoids are released.

Flagellum
Spermatozoid

Rhizoid
Ovum
Neck of archegonium
Venter of archegonium
Thick wall of megaspore
Food storage region of prothallus

4. The spermatozoids swim to the archegonia in a film of moisture using their flagella.

5. The female prothallus with its developing archegonia is released from the sporophyte, still enclosed in the megaspore wall. Fertilization takes place on the ground and development of the diploid sporophyte begins.

Rhizoid
Developing leaves
Suspensor pushes embryo into food supply
Root
Foot

6. The young sporophyte begins development within the megaspore, absorbing nutrients from it via the foot until the young club moss can become an independent photosynthetic plant.

Fig. 8.2 Life cycle of Selaginella

Equisetum with cone

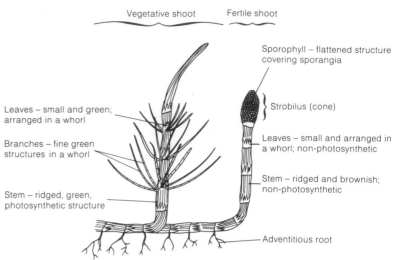

Fig. 8.3 Equisetum — *external features of sporophyte*

8.4 Filicinophyta (ferns)

Ferns have large leaves called fronds which are coiled in bud. Their sporangia are grouped in clusters called **sori**. Fig. 8.4 shows the life cycle of a typical fern. Most living ferns are quite small and have no direct economic importance to man, although they are significant groundcover plants in moist areas. The larger ferns which formed the dominant terrestrial vegetation for about 70 million years from the Devonian to the Permian periods contributed greatly to the coal measures now so useful to man.

TABLE 8.2 **Comparison of the phyla Lycopodophyta, Sphenophyta and Filicinophyta.**

	In all 3 groups:— Sporophyte generation is dominant Gametophyte reduced to tiny prothallus Stem, root and leaves present Vascular tissue present		
Lycopodophyta – clubmosses	**Sphenophyta – horsetails**	**Filicinophyta – ferns**
Small, spirally arranged leaves	Fine leaves in whorls around stem	Large leaves called fronds
Homosporous and heterosporous forms	Homosporous	Mostly homosporous
Sporangia in cones called strobili	Sporangia in cones on leaf-like sporangiophores	Sporangia in groups called sori
e.g. Selaginella (heterosporous form) *Lycopodium* (homosporous form)	*e.g. Equisetum* – horse-tail (only living genus)	*e.g. Pteridium* – bracken *Dryopteris* – fern

1. *Dominant sporophyte is diploid. It consists of a number of fronds growing from an underground rhizome. It can photosynthesize, and the roots absorb water from the soil.*

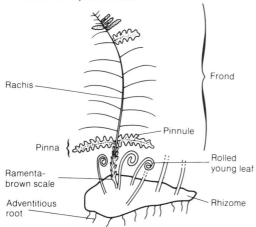

2. *In late summer, brown sori are visible on the underside of each pinnule. Each sorus is made up of a number of sporangia covered by the indusium.*

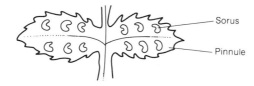

3. *The sporangia are found in groups. Within each sporangium meiosis occurs to produce haploid spores.*

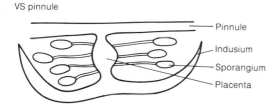

4. *As the spores mature the indusium falls off. The exposed sporangia dry out. The uneven thickening in the annulus sets up tension in the stomium whose cells rupture suddenly to release the spores.*

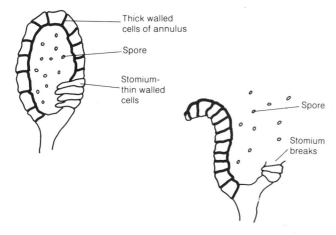

5. *The spores germinate in moist conditions and develop into a tiny plate of cells called the prothallus. This is the haploid gametophyte stage bearing archegonia and antheridia. It has rhizoids for anchorage and it is photosynthetic.*

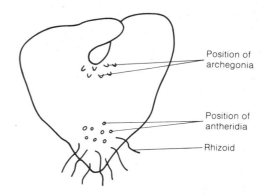

6. *Antheridia develop on the under surface of the prothallus near the rhizoids. Within them spiral, multiflagellate spermatozoids develop.*

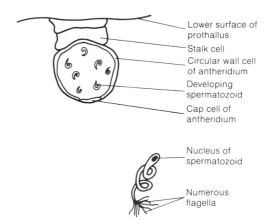

7. *The archegonia develop after the antheridia so self-fertilization is rare. The venter of the archegonium is embedded in the prothallus and the neck is short.*

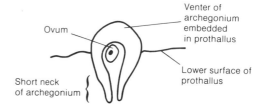

8. *Following fertilization the young, diploid sporophyte plant grows to become an independent plant. At first it has a foot which absorbs nutrients from the prothallus.*

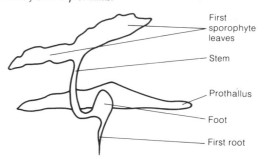

Fig. 8.4 Life cycle of a typical fern

Pinus male cones (*top*) and female cones (*bottom*)

8.5 Coniferophyta (conifers)

In conifers the retention of the gametophyte within the sporophyte plant means that it is not as susceptible to desiccation and the group therefore has colonised drier regions. The seeds which result from fertilization are supplied with food, have a resistant coat and are able to withstand adverse conditions. All coniferophyta lack flowers and fruits, but instead bear their seeds on cones. The majority have narrow leaves which may be needle-like as in *Pinus*. They show many xeromorphic features since they often live in habitats where water is relatively inaccessible. An account of the morphology and reproduction of *Pinus sylvestris* is given in Fig. 8.5.

8.5.1 Economic importance of the Coniferophyta

Many conifers are relatively fast growing trees and so provide an important source of timber. They are also grown extensively for woodpulp, needed to make paper, and are important for resin and turpentine. They are often planted to prevent land erosion and also as ornamentals. Fossilized conifers are an important component of coal measures.

TABLE 8.3 **Comparison of the phyla Coniferophyta and Angiospermophyta**

In both groups	Sporophyte is dominant generation; gametophyte is very reduced Heterosporous: microspore = pollen grain megaspore = embryo sac Seed results from fertilization Non-motile male gametes
Coniferophyta	**Angiospermophyta**
Seeds not enclosed in ovary	Seeds enclosed in ovary
Normally forms cones, not flowers	Flowers formed
No fruit	Fruit develops from ovary
No xylem vessels or phloem companion cells	Xylem vessels and phloem companion cells present
e.g. Pinus – pine　　*Taxus* – yew　　*Larix* – larch	*e.g. Lilium* – lily　　*Bellis* – daisy　　*Quercus* – oak

The trunk and branches of a pine tree are shoots of unlimited growth. Leaves develop in pairs on dwarf shoots. Male and female cones are borne on separate trees.

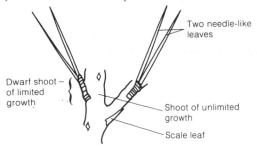

The male cones develop in clusters near the base of new shoots. Each cone is a spiral arrangement of microsporophylls, each bearing two microsporangia or pollen sacs. Within these the microspores (pollen grains) develop by meiosis.

Fig. 8.5 Life cycle of Pinus sylvestris

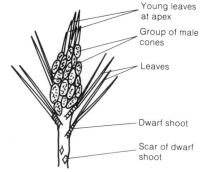

Mature pollen grains complete development in one year and when liberated have two wing-like air sacs.

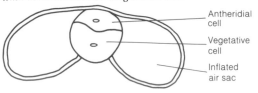

(cont.)

Female cones take three years to complete development. Various stages may be present on the same tree. Female cones have a central axis and numerous ovuliferous scales, each bearing two ovules.

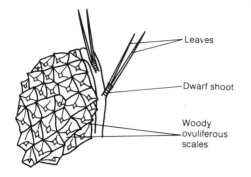

Within each ovule meiosis gives rise to four megaspores one of which enlarges and divides to form archegonia each containing an ovum. At the end of the first year pollination takes place and then the scales of the cone close for two years during which time fertilization and development are completed.

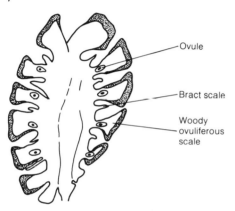

During the third year the woody scales of the female cone open and expose winged seeds which are dispersed by the wind.

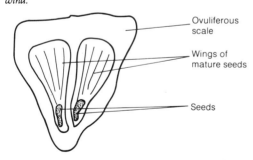

Fig. 8.5 (cont.) *Life cycle of* Pinus sylvestris

8.6 Angiospermophyta

Angiosperms form the dominant terrestrial vegetation today, having taken over from the Coniferophyta more than 100 million years ago. They are found in a wide range of habitats and have even re-established themselves in fresh water and the sea. Their evolution has closely paralleled that of the insects on which many species depend for pollination. They are extremely well suited to life on land both in their morphology, e.g. efficient water-carrying xylem vessels, and in their reproduction, e.g. seeds enclosed in an ovary. It should be noted that they still show alternation of sporophyte and gametophyte generations although the latter is severely reduced to no more than a few cells retained within the sporophyte.

A detailed account of the reproduction of a flowering plant is given in Chapter 18.

The two angiosperm classes, **monocotyledoneae** and **dicotyledoneae**, differ in a number of respects, the most significant of which are shown in Table 8.4.

TABLE 8.4 **Comparison of monocotyledonae and dicotyledonae**

Monocotyledoneae	Dicotyledoneae
Embryo has one cotyledon	Embryo has two cotyledons
Narrow leaf with parallel venation	Broad leaf with net-like venation
Scattered vascular bundles in stem	Ring of vascular bundles in stem
Rarely vascular cambium present and normally no secondary growth	Vascular cambium present which can lead to secondary growth
Many xylem groups in root	Few xylem groups in root
Flower parts usually in threes	Flower parts usually in fours or fives
Calyx and corolla not easily distinguishable	Usually distinct calyx and corolla
Often wind pollinated	Often insect pollinated
e.g. Avena – oats *Iris* *Triticum* – wheat *Lilium* – lily	*e.g. Ranunculus* – buttercup *Lamium* – nettle *Cheiranthus* – wallflower *Bellis* – daisy

8.6.1 Economic importance of the Angiospermophyta

As the dominant vegetation on land many angiosperms are economically important to man. Grasses and all the cereal crops are monocotyledons. Fruit trees, most vegetable crops and broad-leaved forest trees are dicotyledons. In addition, since they cover such a large proportion of the land they are significant primary producers and replenish the oxygen in the atmosphere.

1. (a) What is meant by the term *alternation of generations* in plants? (*2 marks*)
   (b) Describe, with the aid of suitable diagrams, the life cycle of EITHER a moss OR a fern (state which), illustrating the alternation of generations. Your answer should include details of the chromosome number at the different stages and the processes (including the types of cell division) involved in progression from each stage to the next. (*7 marks*)
   (c) Contrast this with the alternation of generations in an angiosperm (flowering plant). (*3 marks*)
   (d) (i) What problems for plant reproduction are posed by a terrestrial way of life?
       (ii) How are these problems overcome in your two examples (given in parts (b) and (c) above)?
       (iii) How do these considerations restrict the range of habitats in which your two examples can live? (*8 marks*)
       (*Total 20 marks*)

Northern Ireland Board June 1985, Paper II, No. 1

2. (a) What is a gametophyte? (*4 marks*)
   (b) By means of well annotated diagrams only, show the main structural features of the gametophytes of: (i) a moss; (ii) a homosporous fern; (iii) *Selaginella*; (iv) a flowering plant. Indicate the approximate size in each case. (*20 marks*)
   (c) What important evolutionary trends may be illustrated by these four gametophytes? (*6 marks*)
       (*Total 30 marks*)

Oxford Local 1987 Specimen Paper II, No. 1

3. Describe the structure and life cycle of *Funaria* sp., and explain why it is regarded as a more complex organism than *Chlamydomonas* sp. (*20 marks*)

Welsh Joint Education Committee June 1984, Paper AII, No. 6

4. (a) What are the distinguishing features of each of the following:
       (i) a green alga,
       (ii) a bryophyte and
       (iii) a pteridophyte? (*10 marks*)
   (b) What does the study of these organisms suggest about the evolution of terrestrial plants from primitive aquatic plants? (*8 marks*)
       (*Total 18 marks*)

Cambridge Board November 1984, Paper I, No. 6

5. The diagram shows the life cycle of a liverwort, (a bryophyte).

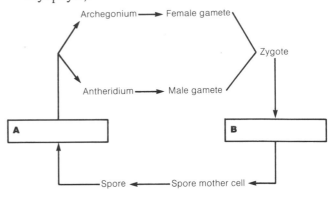

   (a) What stage in the life cycle is represented by
       (i) box A,
       (ii) box B?
       Write your answers in the boxes on the diagram. (*2 marks*)
   (b) Mark the diagram with a cross (×) to show where meiosis takes place. (*1 mark*)
   (c) On the diagram, circle the stage in the life cycle which has both the haploid number of chromosomes and possesses chlorophyll. (*1 mark*)
       (*Total 4 marks*)

Associated Examining Board June 1989, Paper I, No. 3

6. By means of diagrams compare the life-history of a bryophyte (moss or liverwort) with that of a flowering plant. What is the basis for the view that modern bryophytes and flowering plants are descended from a common ancestor? (*8 marks*)

Southern Universities Joint Board June 1985, Paper I, No. 4

7. The diagram below illustrates the life cycle of a bryophyte.

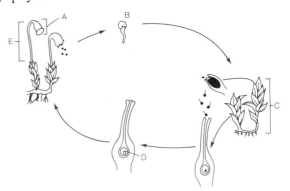

   (a) Name the structures labelled A-E. (*5 marks*)
   (b) State clearly at which stage in the life cycle meiosis occurs. (*2 marks*)

(c) (i) What is meant by the term 'alternation of generations' in relation to this life cycle?

(2 marks)

(ii) State **two** ways in which the alternation of generations shown by this cycle differs from that found in flowering plants. (2 marks)

(Total 11 marks)

London Board June 1985, Paper I, No. 5

9 Classification of animals

9.1 Phylum Porifera

TABLE 9.1 **The Porifera (sponges)**

PHYLUM PORIFERA	Multicellular without tissues or organs Lack of symmetry Body perforated by many openings Choanocytes present
Examples	*Leucosolenia* *Grantia* *Sycon*

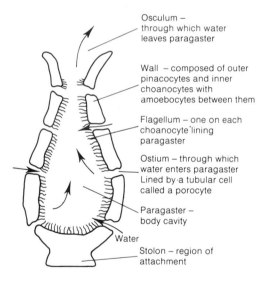

Osculum –
through which water
leaves paragaster

Wall – composed of outer
pinacocytes and inner
choanocytes with
amoebocytes between them

Flagellum – one on each
choanocyte lining
paragaster

Ostium – through which
water enters paragaster
Lined by a tubular cell
called a porocyte

Paragaster –
body cavity

Water

Stolon – region of
attachment

Fig. 9.1 Simplified diagram of the structure of a sponge such as Leucosolenia

Sponges are the most primitive multicellular animals and are almost exclusively marine. They have neither organs nor tissues and their cells demonstrate a high degree of independence. However, they do show division of labour with a wall containing outer contractile cells and inner **choanocytes** (collar cells) producing a respiratory/feeding current. Between these are **amoebocytes**, which store food and give rise to reproductive cells. The wall also contains skeletal structures called **spicules**. Four classes are recognized based on the structure of the skeleton.

9.1.1 Locomotion

Sponges are sedentary.

9.1.2 Nutrition

Flagellated choanocytes beat to cause a current of water to flow through the sponge, entering the **ostia** and leaving the **osculum**. At the base of each flagellum is a fine mesh-like structure which filters small particles from the water. These are digested intracellularly and undigested waste leaves in the exhalent current of water. Food may be stored in amoebocytes as glycogen or lipoprotein.

9.1.3 Gaseous exchange

Most sponges require a high level of oxygen in the water. The feeding current maintained by the choanocytes also acts as a respiratory current and gases are exchanged by diffusion between the water and the sponge cells.

9.1.4 Excretion and osmoregulation

The excretory product ammonia leaves the sponge via the exhalent water current. Some cells of freshwater sponges have contractile vacuoles but these are not present in marine species.

9.1.5 Nervous system

There is no nervous system present in sponges so any sensitivity they show is slow and uncoordinated – amoebocytes behave rather like individual *Amoeba*.

9.1.6 Reproduction

Sponges have remarkable powers of asexual reproduction, buds developing from any region of the body. They can also regenerate from fragments of the parent sponge. All sponges are capable of sexual reproduction and, although they are hermaphrodite, cross fertilization takes place. Following fertilization a free-swimming larva forms which soon settles and develops into an adult sponge.

9.1.7 Economic importance

Sponges form a food source for a number of other animals and provide the traditional bath sponge.

9.2 Phylum Cnidaria

Animals in this phylum are all aquatic and predominantly marine. They are **diploblastic** with a body wall composed of two cell layers, an inner **gastrodermis** and an outer **epidermis**. Between these lies a jelly-like **mesogloea** which may contain cells derived from the other two layers. The body is organized around a central cavity, the **gastro-vascular cavity**, which has one opening serving as both mouth and anus. This opening is surrounded by tentacles bearing specialized stinging cells or **nematocysts**.

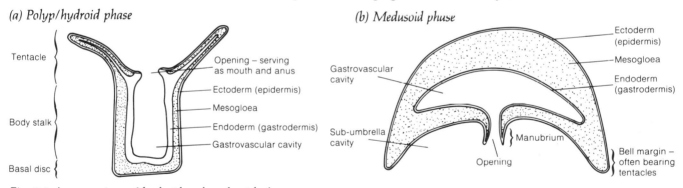

Fig. 9.2 A comparison of hydroid and medusoid phases

TABLE 9.2 **Classification of the Cnidaria**

PHYLUM CNIDARIA		Diploblastic Single body cavity with one opening surrounded by tentacles Radial symmetry Polymorphism with free swimming medusoid and/or sedentary polyps Nematocysts Planula larva	
Class	**Hydrozoa**	**Scyphozoa**	**Anthozoa**
	Dominant polyp	Reduced polyp	Only polyp
	Reduced medusa	Dominant medusa	No medusa
	No mesenteries (divisions in gastrovascular cavity)	Mesenteries in young polyp	Large mesenteries
Examples	*Obelia* – colonial; marine *Hydra* – solitary; freshwater *Physalia* – Portuguese Man-of-war	*Aurelia* – solitary, free-swimming marine jellyfish	*Actinia* – solitary; marine sea anemone *Corallium* – colonial, marine coral

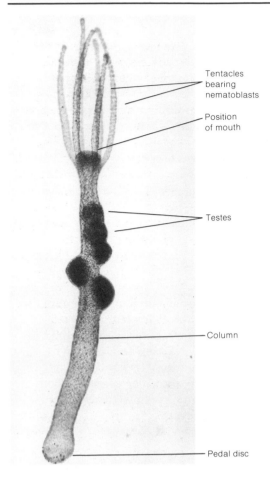

Hydra

The whole body is radially symmetrical and occurs in two main forms: a jellyfish-like medusoid phase and a hydroid or polyp phase. When the same organism can exist in a number of morphologically distinct forms it is said to show **polymorphism**. Some species have both medusoid and polyp phases in their life history, others only one.

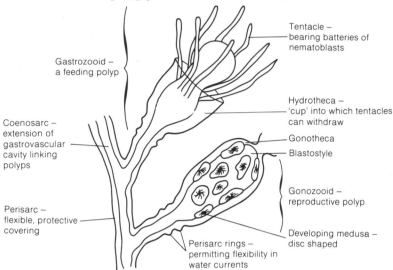

Fig. 9.3 Part of a colony of Obelia
All canals are part of the gastrovascular cavity.

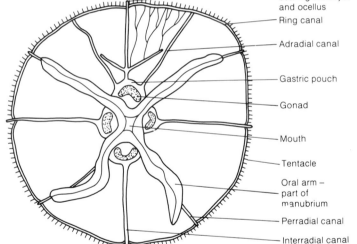

Fig. 9.4 Aurelia — oral view

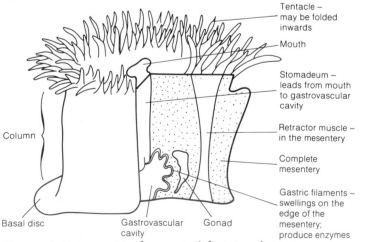

Fig. 9.5 Actinia — cut to show part of the internal structure

9.2.1 Locomotion

Most polyps are sedentary but some, like *Hydra*, move around in a series of somersaults. *Hydra* is normally attached by a basal disc but moves by extending the body, bending over and attaching its tentacles to the substrate. The basal disc is then freed and swung over to a new point of attachment. This movement is brought about by the contraction of musculo-epithelial cells in the epidermis. Each of these cells contains contractile fibrils of actin and myosin. The gastrovascular cavity acts as a hydrostatic skeleton (see Section 39.4.1) and the elastic nature of the mesogloea helps to maintain the shape of the body. Other polyps, e.g. *Actinia*, show limited movement, 'walking' on their tentacles or creeping on the basal disc. Medusae such as *Aurelia* are free-swimming and have a more highly developed muscular system than the polyps with definite muscle tracts.

Around the bell margin (see Fig. 9.2(b)) there is a circular band of muscle fibres whose contraction decreases the volume of the sub-umbrella cavity and forces water out. The shape is regained due to the elastic properties of the mesogloea. These actions result in a predominantly vertical movement, horizontal movement being largely due to water currents.

9.2.2 Nutrition

Most cnidarians are carnivorous, feeding on any suitably sized animals coming into contact with their tentacles. These tentacles bear batteries of nematocysts whose threads may either entrap the prey or penetrate it and inject paralysing toxin. (See Fig. 9.6.)

Nematocysts can only be used once and after discharge must be replaced, a process taking about 24 hours. Most medusoid forms feed on animals which become entangled with their oral arms and the flexible manubrium collects the food. *Aurelia* is a suspension feeder. Planktonic organisms become trapped in mucus on the surface of *Aurelia* as it sinks or swims. This is moved by cilia to the edge of the bell and into a ciliated groove leading to the mouth.

Extracellular digestion of the ingested food occurs in the gastro-vascular cavity where trypsin-like enzymes are produced. Further digestion of proteins and fats occurs intracellularly in food vacuoles and undigested food is removed via the mouth when the body contracts. Food may be stored as fat or glycogen.

9.2.3 Gaseous exchange and excretion

No special organs are required for gaseous exchange or excretion since, with the presence of the gastro-vascular cavity, no cells are far from the circulating water. Oxygen and carbon dioxide diffuse readily through the plasma membranes of individual cells. The excretory product is soluble ammonia and this can also diffuse readily through the general body surface.

9.2.4 Nervous system

Polyps have a net-like arrangement of multipolar nerve cells, especially concentrated around the mouth. Impulses are transmitted to the net from sense organs but their passage is slow. Free swimming medusae have a more highly developed nervous system with a multipolar nerve net as found in the polyps together with a giant

(a) Before discharge

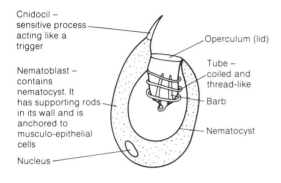

Cnidocil – sensitive process acting like a trigger

Nematoblast – contains nematocyst. It has supporting rods in its wall and is anchored to musculo-epithelial cells

Nucleus

Operculum (lid)

Tube – coiled and thread-like

Barb

Nematocyst

(b) After discharge

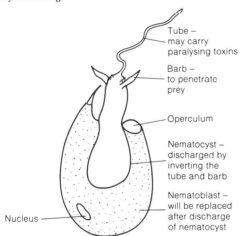

Tube – may carry paralysing toxins

Barb – to penetrate prey

Operculum

Nematocyst – discharged by inverting the tube and barb

Nematoblast – will be replaced after discharge of nematocyst

Nucleus

Fig. 9.6 Nematoblasts before and after discharge

fibre system of bipolar cells responsible for coordinating swimming. Medusae also have **statocysts** which detect changes in orientation and may have **photoreceptor cells** in the form of a simple **ocellus**.

9.2.5 Reproduction

Polyps such as *Hydra* reproduce asexually by budding in favourable conditions but in the autumn sperm and ova are produced. The ova are fertilized within the ovary and an embryo, enclosed in a chitinous shell, is released to overwinter. More typical hydrozoan polyps like *Obelia* produce **medusae** by budding and these reproduce sexually to give rise to a ciliated **planula larva** which settles and buds to give rise to new polyps. Scyphozoan medusae also reproduce sexually and the planula larvae they form develop into **scyphistoma larvae** which bear a superficial resemblance to *Hydra*. New medusae are produced by transverse fission. Sea anemones, such as *Actinia*, have gonads on the mesenteries of the body cavity; these produce ova and sperm which on fertilization develop into the planula larva. After a brief free-swimming existence this larva attaches to the substrate and develops into a mature polyp. *Actinia* may also reproduce asexually by longitudinal binary fission.

9.2.6 Economic importance

The nematocysts of many jellyfish cause painful stings and that of the Portuguese Man-of-war (*Physalia*) may be fatal. Most anthozoan cnidarians are corals and those living in warm shallow water deposit skeletons of calcium carbonate which are of great economic importance. They provide habitats for many other animals and in the Indian and Pacific Oceans may form coral islands, atolls and fringing reefs. The red coral *Corallium*, from the Mediterranean and off Japan, is used for making jewellery.

TABLE 9.3 **Classification of the Platyhelminthes**

PHYLUM PLATYHELMINTHES	Triploblastic Acoelomate Dorso-ventrally flattened One opening serving as mouth and anus Gut, if present, highly branched Excretion by flame cells		
Class	**Turbellaria**	**Trematoda**	**Cestoda**
	All free-living	Parasitic, mostly within the host, i.e. endoparasitic	All endoparasitic
	Ciliated epidermis but no cuticle	Thick cuticle; no cilia	Thick cuticle; no cilia
	Mouth and gut present	Mouth and gut present	Mouth and gut absent
	Suckers rare	Suckers present	Suckers and hooks present
Examples	*Planaria* *Dugesia* (See Fig. 9.7)	*Schistosoma* – blood fluke *Polystoma* *Fasciola* – liver-fluke (See Fig 9.8)	*Taenia* – tapeworm (See Fig. 9.9)

The testes and parts of the female reproductive system are very diffuse and if included on the diagram would obliterate the digestive system.

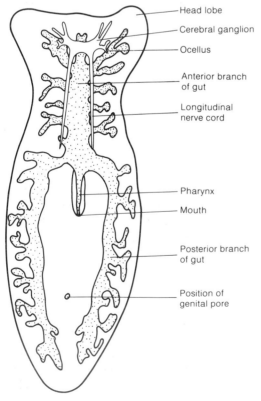

- Head lobe
- Cerebral ganglion
- Ocellus
- Anterior branch of gut
- Longitudinal nerve cord
- Pharynx
- Mouth
- Posterior branch of gut
- Position of genital pore

Fig. 9.7 A planarian

The excretory and reproductive systems are very diffuse, and cover the digestive system.

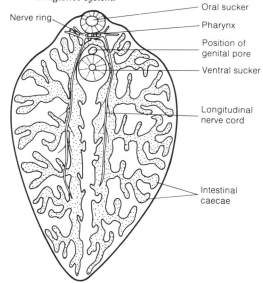

Fig. 9.8 Fasciola

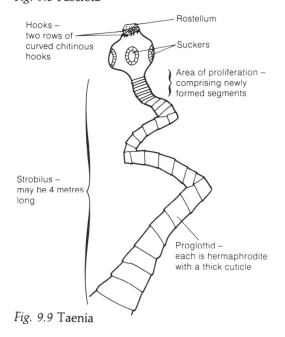

Fig. 9.9 Taenia

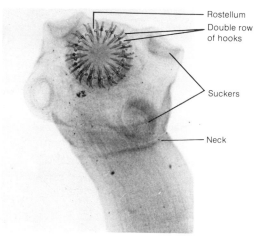

Scolex of *Taenia* (tapeworm)

9.3 Phylum Platyhelminthes

The Platyhelminthes are a group of flatworms with a definite head region and bilateral symmetry. They are **triploblastic** with a body wall composed of an outer epidermis and an inner gastrodermis separated by a relatively undifferentiated region of mesoderm called the **mesenchyme**. The phylum contains around 14 000 different species many of which are parasitic and of great economic importance.

9.3.1 Locomotion

Free-living Platyhelminthes such as *Dugesia* produce mucus from the epidermis and glide slowly over this by means of cilia. They may also move by producing a series of muscular contractions down the length of the body. Parasitic flatworms rarely move from place to place; on the contrary, they are often attached by suckers and/or hooks to their host. However, they do produce free-swimming ciliated larvae for dispersal.

9.3.2 Nutrition

Planarians are predatory or scavenging carnivores, trapping prey in the mucus they produce. The pharynx may be used to engulf small prey or protruded into larger prey aided by the action of proteolytic enzymes. Extracellular digestion in the gut is followed by intracellular digestion in food vacuoles. Trematodes have a well developed gut and a sucking pharynx used to take in cells, blood or tissue fluids from the host. Adult tapeworms are surrounded by the predigested food of their host which is absorbed directly over the body surface by diffusion; they do not need their own digestive system.

9.3.3 Gaseous exchange

The flattened shape of animals in this phylum provides them with a large surface area to volume ratio and simple diffusion is therefore adequate for gaseous exchange. No special organs are required.

9.3.4 Excretion and osmoregulation

Nitrogenous waste is collected by a number of blind-ending structures called **flame cells** which drain via a series of ducts to one or more pores. This system is better developed in the freshwater planarians than marine ones and is therefore presumed to have an osmoregulatory function as well.

9.3.5 Nervous system

All Platyhelminthes have a concentration of nervous tissue at the anterior end of the body and two longitudinal nerve cords. Turbellarians have the most highly developed sense receptors of the group, associated with their free-living existence. There are receptors for touch, chemicals and light. The sensory structures are quite well developed in trematodes, especially around the suckers and pharynx. The system is much reduced in cestodes.

9.3.6 Reproduction

Planarians show remarkable powers of regeneration and are often used experimentally. The reproductive systems of all Platyhelminthes are complex and show considerable variation between species. Most are hermaphrodite and, although self-fertilization is possible, cross-fertilization is more usual. Fertilization is internal and normally results in the production of a free-swimming ciliated larva, although some turbellarians show direct development (i.e. no larval form). Parasitic platyhelminthes have complex life cycles normally involving one or more intermediate hosts before the new adult develops in the primary host (Section 25.2.4).

TABLE 9.4 **The parasitic flatworms**

Class	Parasite	Primary host	Intermediate host	Harm caused to primary host
Trematoda	*Fasciola hepatica* − liver-fluke	Sheep and cattle	Snails	Liver rot
	Clonorchis sinensis − Chinese liver-fluke	Man	Aquatic snail and freshwater fish	Damage to liver where they feed on blood; large numbers block bile ducts
	Schistosoma − blood-fluke	Man	Freshwater snails	Bilharzia (schistosomiasis); damage to lungs and liver and localized swellings; hepatitis
Cestoda	*Taenia solium* − pork tapeworm	Man	Pig	Anaemia; diarrhoea; loss of weight; intestinal pains, heavy infestations may block gut and its associated ducts
	Diphyllobothrium latum − fish tapeworm	Man and carnivores	Copepod (crustacean) and freshwater fish	
	Echinococcus granulosus	Man, sheep and cattle	Dog	Hydatid cysts − 70% in liver, 20% in lungs and rest elsewhere

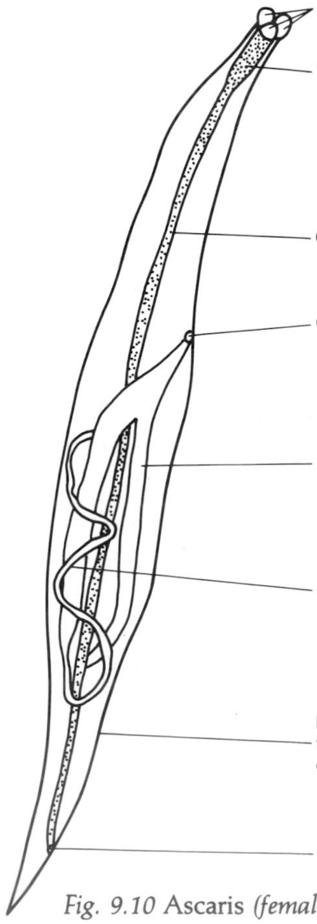

Fig. 9.10 Ascaris *(female)*

Labels: Three lips guarding mouth; Pharynx; Gut; Genital pore; Uterus; Ovary; Body wall − tough semi-transparent cuticle; Anus

9.3.7 Economic importance

The parasitic flatworms are of great significance, causing infestations of man and his domesticated animals. Some examples are given in Table 9.4.

9.4 Phylum Nematoda

Nematoda are found in very large numbers in a wide range of habitats. It is estimated that the 10 000 known species of nematodes represent only about 2 per cent of their full number. Most are free-living although the parasitic ones are best known. The animals are circular in cross-section, have very few cells in their structure and lack both cilia and flagella.

TABLE 9.5 **The Nematoda**

PHYLUM NEMATODA	Unsegmented, cylindrical body No cilia Cuticle of protein Pseudocoelom (body spaces but no true coelom − Section 6.3.2) Unbranched gut from mouth to anus

9.4.1 Locomotion

Alternate contraction of dorsal and ventral longitudinal muscles cause waves down the body allowing the nematode to move. There are no circular muscles. The **pseudocoelom** acts as a hydrostatic skeleton and the cuticle enables the animal to maintain its shape.

9.4.2 Nutrition

A pump is needed to fill the intestine with food because the high pressure of the body fluids tends to collapse the tube. The pharynx is the pump. Nematodes have a wide variety of food sources, including organic debris, algae and animals. Many are either saprophytes or parasites.

9.4.3 Gaseous exchange

There are no special organs for gaseous exchange. Some nematodes live where there is very high oxygen tension, e.g. plant parasites and parasites in blood. Others may normally live in anaerobic conditions, e.g. small intestine or mud deposits.

9.4.4 Excretion and osmoregulation

These processes are not fully understood in the nematodes but the excretory product is ammonia. The system is unusual in that its two longitudinal canals are derived from just two cells.

9.4.5 Nervous system

There is a series of longitudinal nerves attached to an anterior ring. The main sensory structures are **chemoreceptors** (especially well developed in parasites) and **tactile bristles**. Some free-living forms have simple eyes.

9.4.6 Reproduction

In most nematodes the sexes are separate, the males generally being smaller than the females. The sperm are unusual in being amoeboid and lacking flagella. Fertilization is internal and the juveniles which hatch are very much like the adults. Parasitic species produce large numbers of eggs and the life history may involve one or more intermediate hosts.

TABLE 9.6 **The parasitic nematodes**

Parasite	Ecto- or endo- parasite	Primary host	Main site of infection	Disease/damage caused	Intermediate host
Longidorus	Ecto	Higher plants	Root	Damage to roots and carries tomato black ring virus	None
Heterodora rostochiensis (potato root eelworm)	Endo	Potato	Root	Reduced growth; leaves die; tubers few and small	None
Ascaris lumbricoides	Endo	Man	Intestine	Cause damage to respiratory and other systems as they migrate through the body	None
Wucheria bancrofti	Endo	Man	Lymphatic system	Elephantiasis	Mosquito
Dracunculus medinensis (Guinea worm)	Endo	Man	Connective tissue and body spaces	Damage to internal organs and severe skin ulcerations on release of larvae	*Cyclops* (crustacean)

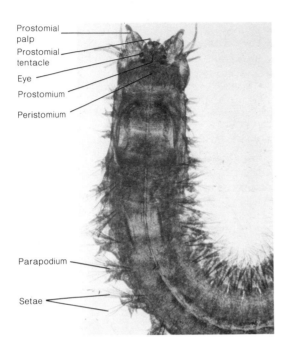

Prostomial palp
Prostomial tentacle
Eye
Prostomium
Peristomium
Parapodium
Setae

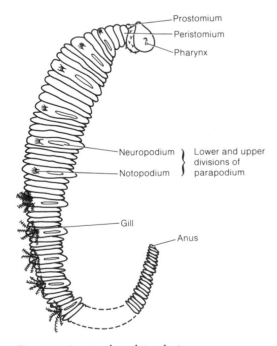

Prostomium
Peristomium
Pharynx
Neuropodium } Lower and upper divisions of parapodium.
Notopodium }
Gill
Anus

Fig. 9.11 Arenicola – lateral view

9.4.7 Economic importance

Nematode parasites are of considerable importance and just a few examples are given in Table 9.6.

9.5 Phylum Annelida

The annelids or 'true worms' are coelomate animals showing metameric segmentation (Section 6.3.3). There are about 9000 species living in the sea, fresh water or moist soil.

TABLE 9.7 **Classification of the Annelida**

PHYLUM ANNELIDA	Metameric segmentation Non-chitinous cuticle Chaetae (bristles)		
Class	**Polychaeta**	**Oligochaeta**	**Hirudinea**
	Parapodia	No parapodia	No parapodia
	Many chaetae	Few Chaetae	No Chaetae
	Outer rings correspond to inner septa	Outer rings correspond to inner septa	Outer rings more numerous than inner septa
	Distinct head	No distinct head	No distinct head
	No suckers	No suckers	Suckers (ectoparasite)
	Separate sexes	Hermaphrodite	Hermaphrodite
	Larvae	No larvae	No larvae
Examples	*Nereis* – ragworm *Arenicola* – lugworm (See Fig. 9.11)	*Lumbricus* – earthworm (See Fig. 9.12)	*Hirudo* – leech (See Fig. 9.13)

9.5.1 Locomotion

The method of locomotion varies between species. Some polychaetes such as *Nereis* may move slowly on their **parapodia** which are moved like levers by muscles. They may also swim or creep more rapidly by alternately contracting and relaxing the longitudinal muscles to produce a series of body undulations. Earthworms progress by alternately extending the body and anchoring it by chaetae. This movement involves the contraction and relaxation of longitudinal and circular muscles. Leeches may swim but normally move by a looping action similar to that of *Hydra*, attaching the body by the anterior sucker followed by the posterior one.

9.5.2 Nutrition

Polychaetes may feed on a wide variety of organisms and organic debris, ingested by means of an eversible (protruding) pharynx. Others use cilia to create a water current and organic particles which are filtered out stick to mucus before being transferred to the mouth. The many varieties of earthworm feed on decaying vegetation. Leeches attach themselves by means of a sucker and use chitinous jaws to wound the vertebrate host. Anticoagulants and anaesthetics are produced and blood is pumped into the pharynx.

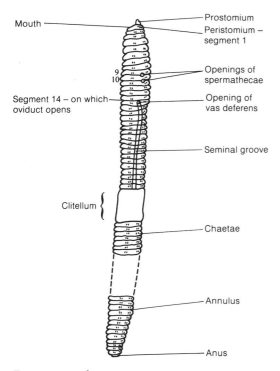

Mouth — Prostomium
— Peristomium – segment 1
9 — Openings of spermathecae
10
Segment 14 – on which oviduct opens — Opening of vas deferens
— Seminal groove
Clitellum {
— Chaetae
— Annulus
— Anus

Fig. 9.12 Lumbricus

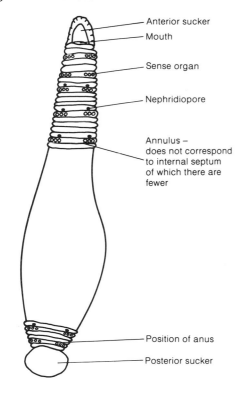

— Anterior sucker
— Mouth
— Sense organ
— Nephridiopore
— Annulus – does not correspond to internal septum of which there are fewer
— Position of anus
— Posterior sucker

Fig. 9.13 Hirudo — *ventral view*

9.5.3 Gaseous exchange

Most annelids lack specialized structures for gaseous exchange, diffusion taking place over the whole body surface e.g. *Lumbricus* (earthworm). Some annelids have parapodia e.g. *Nereis*. These, with their large surface area to volume ratio, are especially useful respiratory surfaces. Some polychaetes, such as *Arenicola*, also have gill tufts. Oxygen is carried around the body by haemoglobin enclosed in blood vessels.

9.5.4 Excretion and osmoregulation

All annelids have segmentally arranged **nephridia** to extract waste products, especially ammonia, from the body fluids and pass them to the exterior. The nephridia also function in osmoregulation.

9.5.5 Nervous system

The basic annelid nervous system comprises **cerebral ganglia**, a solid, **ventral nerve cord** and segmentally arranged **ganglia**. In addition, many polychaetes have a series of **giant fibres** which innervate the muscles permitting escape responses. Polychaetes have well-developed sense organs, the sense of touch being particularly important. They also have eyes, which, although more advanced than those of the Platyhelminthes, are unable to produce an image. The oligochaetes lack eyes and have only very simple epidermal sensory cells, chemo- and photo-receptors. The simple sensory structures of leeches are extremely effective in locating a host.

9.5.6 Reproduction

Some polychaetes, such as *Nereis*, only produce gonads in the breeding season. Fertilization is external. In most cases a free-swimming, ciliated **trochophore** larva develops which metamorphoses into the adult. Cross fertilization takes place in the hermaphrodite oligochaetes and the Hirudinea and the eggs are laid in a cocoon. There is no larval stage. Polychaetes are capable of regeneration and also asexual reproduction by budding or fragmentation.

9.5.7 Economic importance

Economically the most significant annelids are the earthworms which contribute to soil formation and improvement in the following ways:

1. Tunnels improve aeration and drainage.

2. Dead vegetation is pulled into the soil where decay by saprophytes takes place.

3. Mixing of soil layers.

4. Addition of organic matter by excretion and death.

5. Secretions of gut neutralize acid soils.

6. Improving tilth by passing soil through gut.

Annelids in general contribute to food chains and leeches used to be of medical importance. None of the parasites cause major infestations of man or of his domesticated animals.

TABLE 9.8 **Phylum Arthropoda**

PHYLUM ARTHROPODA	Exoskeleton, mainly comprising a chitinous cuticle Jointed appendages Dorsal heart and open blood system Growth in stages after moulting **(ecdysis)**

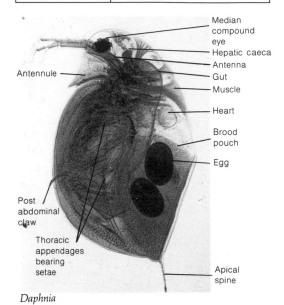

Daphnia

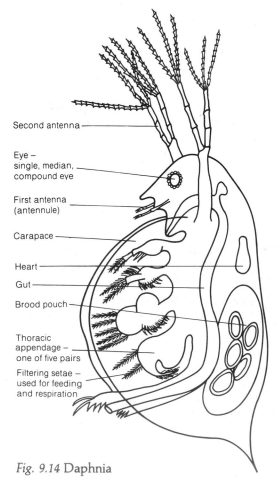

Fig. 9.14 Daphnia

9.6   Phylum Arthropoda

Arthropods make up about three quarters of living animal species. They have adapted to live successfully in both aquatic and terrestrial habitats and may be free-living or parasitic. **Trilobites** are an extinct group of arthropods, nearly 4000 species of which have been described from fossils. They became extinct at the end of the Palaeozoic era.

Because of the size and importance of this phylum it will be studied a class at a time.

9.6.1 Crustacea

There are about 26 000 species known. They are extremely abundant and many are relatively large.

TABLE 9.9 **Classification of the Crustacea**

Superclass Crustacea	Cephalothorax – formed by fusion of head and thorax Strong exoskeleton – impregnated with calcium carbonate Two pairs of antennae Three pairs of mouthparts	
Class	**Branchiopoda**	**Malacostraca**
	One pair of compound eyes which may be fused to form a single eye	One pair of stalked compound eyes
	Thoracic appendages with bristles for filter feeding	Eight pairs of thoracic appendages for walking and feeding
	No abdomen	Abdomen with appendages for swimming
	Body enclosed in carapace of two pieces	Carapace covers thorax
Examples	*Daphnia* – water flea (See Fig. 9.14)	*Carcinus* – crab *Astacus* – crayfish *Oniscus* – woodlouse *Leander* – prawn

9.6.2 Locomotion

Most crustaceans are aquatic and capable of swimming using specialized thoracic or abdominal appendages. However, many crustaceans have evolved heavy limbs adapted for burrowing or walking on the sea bottom.

9.6.3 Nutrition

Some crustaceans, such as *Daphnia*, use hair-like bristles called **setae** to filter organic particles from the water, the current being produced by the beating thoracic appendages. Other species may use modified antennae or mandibles for filter feeding. Woodlice feed on decaying plant and animal matter. Many larger crustaceans such as *Astacus* feed on almost any organic matter, dead or alive, using a series of anterior appendages modified as mouthparts. There are a few groups of parasitic crustaceans.

9.6.4 Gaseous exchange

The presence of the exoskeleton prevents gaseous exchange occurring over the whole body surface. Most crustaceans have gill-like extensions covered only by thin cuticle which permits exchange of gases with the environment.

9.6.5 Excretion and osmoregulation

Most crustaceans excrete nitrogenous waste, in the form of ammonia, via the gills. The salt and water balance of the animal is maintained by **green (antennary) glands** in the head.

9.6.6 Nervous system

Within this class the development of the nervous system varies. In many of the malacostraca, like crabs and crayfish, the nervous system is very similar to that of the annelids, comprising a double ventral nerve cord with cerebral and segmental ganglia. Active crustaceans have very well-developed compound eyes as well as efficient **statocysts** for balance, setae for touch and various chemoreceptors.

9.6.7 Reproduction

The majority of the Crustacea have separate sexes and lay eggs which develop through a series of larval stages to the adult form. The freshwater crayfish is unusual in having no larval stages and the water flea, *Daphnia*, often reproduces **parthenogenetically**. Unfertilized eggs pass into a brood pouch from which they are released as fully developed young *Daphnia*.

9.6.8 Economic importance

The numerous larval stages of crustaceans make a considerable contribution to the zooplankton fed on by many other aquatic animals. Many of the adults are also links in aquatic food webs and, as krill, even provide food for whales. Decapods are an important human food source, together with molluscs, comprising the 'shellfish'. Barnacles cause serious fouling of ships and quays.

9.6.9 Insecta

The Insecta comprise more than 750 000 species which show great diversity. They are extremely successful terrestrial animals in terms of numbers of species, individuals and habitats.

There are a few insects such as springtails and silverfish which do not develop wings. The remainder may be subdivided as follows:

TABLE 9.10 **The Insecta**

Class Insecta	Body comprises head, thorax and abdomen
	Three pairs of thoracic legs
	Two pairs of thoracic wings (except Apterygota) – sometimes modified
	One pair of antennae
	One pair of compound eyes
	Respiration by tracheae

TABLE 9.11 **The winged insects**

	Exopterygota	Endopterygota
	Wings develop externally Metamorphosis incomplete (**hemimetabolous**) Egg → nymphal stages → adult	Wings develop internally Metamorphosis complete (**holometabolous**) Egg → larval stages → pupa → adult
Order	Orthoptera, e.g. *Locusta* – locust (See Fig. 9.15) Odonata, e.g. *Libellula* – dragonfly Dictyoptera, e.g. *Periplaneta* – cockroach	Lepidoptera, e.g. *Pieris* – cabbage white butterfly Diptera, e.g. *Musca* – house-fly Hymenoptera, e.g. *Apis* – honey-bee

Cockroach

Larva

Pupa

Apis (honey-bee), brood stages

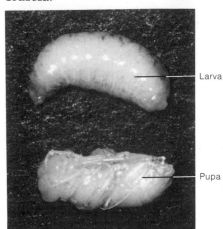

Apis adults – Queen (*top*), drone (*middle*) and worker (*bottom*)

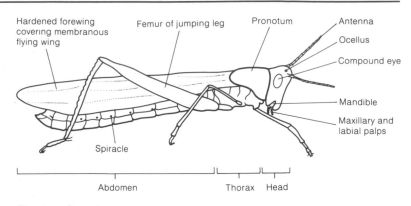

Hardened forewing covering membranous flying wing

Femur of jumping leg

Pronotum

Antenna

Ocellus

Compound eye

Mandible

Maxillary and labial palps

Spiracle

Abdomen

Thorax

Head

Fig. 9.15 Locust

9.6.10 Locomotion

Insects, being primarily terrestrial, must cope with a medium which does not provide the support offered by water. When walking, insects use a 'tripod' system so that at any one time three legs are on the ground, the first and third of one side and the middle one of the other. The reduced number of legs present allows for their elongation and the animal can move rapidly without dragging the body on the ground. Some insects, such as locusts, have highly developed hind legs for jumping. Insects are the only invertebrates capable of flight and the wing beat varies between different species, e.g. locust 4–20 beats per second, honey-bees 190 beats per second and gnats 1000 beats per second. The fore and hind-wings may be hooked together and work as one unit. In flies (Diptera) the hind-wings form halteres or balancing organs. The fore-wings may form hard, protective wing cases as in the beetles (Coleoptera).

9.6.11 Nutrition

Insects exhibit a variety of feeding mechanisms with mouthparts adapted for biting and chewing (e.g. locusts; cockroaches; bees) or for piercing and sucking (e.g. bugs; butterflies; true flies). There are also many parasitic forms, notably the fleas and lice. Some of these mechanisms are considered in Section 24.3.

9.6.12 Gaseous exchange

The exoskeleton prevents gaseous exchange occurring over the whole body surface but air enters and leaves via pores called **spiracles**. These open into a series of **tracheal tubes** which branch throughout the body carrying respiratory gases directly to and from the muscles (Section 31.2.3).

9.6.13 Excretion and osmoregulation

As terrestrial animals, most insects are constantly faced with the problem of desiccation. The development of **Malpighian tubules** (Section 34.3.2) allows water to be extracted from the nitrogenous waste which is excreted as the semi-solid uric acid.

9.6.14 Nervous system

This has a double ventral nerve cord and an arrangement of ganglia resembling that of the Crustacea such as crayfish. The sense organs of all insects are extremely well developed with large compound eyes and a variety of mechano- and chemo-receptors. Many insects are also capable of detecting sound.

9.6.15 Reproduction

In insects the sexes are separate and, in common with most terrestrial animals, fertilization is internal. In hemimetabolous insects such as a locust the eggs develop through a series of nymphal stages, which resemble the adult, to the mature imago (adult). Eggs of holometabolous insects hatch into a larval stage which bears little resemblance to the adult (e.g. caterpillar of a butterfly; maggot of a fly). The larval stage feeds, moults several times and eventually pupates to emerge as a sexually mature **imago**.

9.6.16 Economic importance

Insects are important components of terrestrial and fresh-water food webs and many of them are important scavengers removing organic waste. Many insects, e.g. bees, are pollinators of fruits and crops useful to man, e.g. clover and apples. However, insects may also be pests of man, his cultivated plants and domesticated animals. They may cause damage or disease directly, e.g. locusts, cotton bollweevils, wireworms, lice and warble flies. They may be vectors of other disease-causing organisms, e.g.:

Mosquito carries *Plasmodium*, the protozoan causing malaria.

Tsetse fly carries *Trypanosoma*, the protozoan causing sleeping sickness.

Rat flea carries the bacterium causing bubonic plague.

Other insects are vectors of yellow fever, typhoid, dysentery and typhus.

Insects may also damage stored food (e.g. cockroaches) and clothing, furnishings and wooden houses (e.g. termites).

Insects are also used in the biological control of pests, albeit often against other insects, e.g. ladybirds eat aphids. They are also used in scientific experiments, e.g. *Drosophila* (fruit fly) in the study of genetics. They provide sources of products useful to man such as cochineal, beeswax, honey and silk.

TABLE 9.12 **The Diplopoda, Chilopoda and Arachnida**

Class	Diplopoda	Chilopoda	Arachnida
	Simple eyes (if present)	Simple eyes (if present)	Simple eyes only
	One pair of antennae	One pair of antennae	No antennae
	Numerous body segments	Numerous body segments	Body divided into two main regions: prosoma and opisthosoma
	Two pairs of legs per body segment	One pair of legs per body segment	Four pairs of walking legs
Examples	*Iulus* – millipede	*Lithobius* – centipede	*Epeira* – garden spider *Scorpio* – scorpion *Ixodes* – tick

9.6.17 Other arthropods

The remaining classes of arthropods will not be dealt with in detail but their main characteristics are listed in Table 9.12.

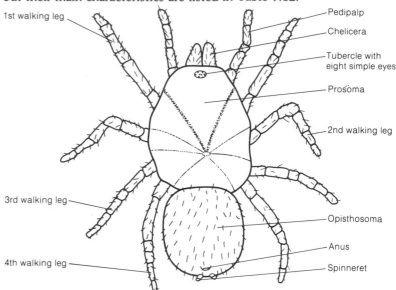

Fig. 9.16 Spider – dorsal view

9.7 Phylum Mollusca

This phylum is the second largest in the animal kingdom, comprising about 100 000 living species. There is also a very long fossil record of molluscs stretching back to the Pre-Cambrian period. Most living species are marine and examples include octopus and squid as well as members of the classes shown in Table 9.13.

TABLE 9.13 **Classification of the Mollusca**

PHYLUM MOLLUSCA	Body divided into head, muscular foot and visceral mass Mantle may secrete a calcareous shell Bilateral symmetry, but torsion may lead to asymmetry	
Class	**Gastropoda**	**Pelycopoda (Bivalves)**
	Asymmetrical (due to torsion of visceral mass)	Bilateral symmetry
	Single shell, usually coiled	Shell in two valves
	Well developed head with tentacles and eyes	Reduced head and no tentacles
	Radula (rasping tongue) for feeding	Gills with cilia used for filter feeding
Examples	*Littorina* – winkle *Limax* – slug *Helix* – snail (See Fig. 9.17)	*Mytilus* – mussel *Anodonta* – freshwater mussel (See Fig. 9.18)

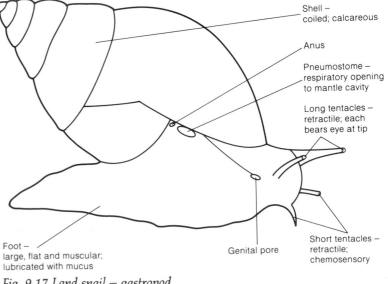

Fig. 9.17 Land snail – gastropod

9.7.1 Locomotion

Pelycopods, such as *Mytilus*, usually remain in one place attached by byssal threads. Snails, like *Helix*, move slowly using a complicated series of muscular contractions within the foot and gliding on a film of mucus.

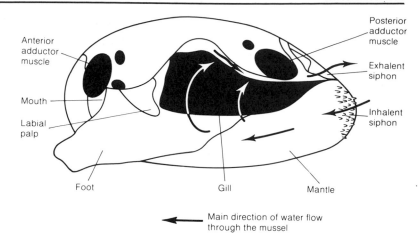

Fig. 9.18 Anodonta — *viewed from left side with left valve and one gill removed*

9.7.2 Nutrition

Gastropods have a ribbon-like structure ridged with chitin, called a **radula**, and this is used to scrape the surface of vegetation prior to ingestion. Pelycopods on the other hand are filter feeders using enlarged gill filaments to strain organic matter from the water around them. The particles are trapped in mucus and moved by cilia to the mouth.

9.7.3 Gaseous exchange

In gastropods, such as *Helix*, the mantle cavity serves as a lung. Air enters the pneumostome, which can be closed by a valve, and gases diffuse from the mantle cavity into the blood vessels of the mantle. In bivalves the large gills absorb oxygen from the surrounding water. When the molluscs are exposed at low tide the valves of the shell close, retaining moisture and thus protecting the gills from desiccation.

9.7.4 Excretion and osmoregulation

Molluscs have one or more kidneys. Marine forms excrete ammonia, to which they have a high tolerance, while terrestrial forms excrete uric acid, which requires less water for its elimination.

9.7.5 Nervous system

The nervous system of molluscs contains very few ganglia when compared to the annelids or arthropods. Pelycopods, being sedentary animals, also have few sense organs. Gastropods on the other hand have a wide range of sensory structures, including eyes and chemoreceptors to 'test' the air entering the mantle cavity.

9.7.6 Reproduction

Snails, such as *Helix*, are hermaphrodite. A mutual exchange of sperm occurs between two individuals and eggs are laid. Larval stages develop within the egg, which enlarges during this period. Young snails emerge complete with an immature reproductive system. Pelycopods like *Mytilus* have separate sexes and fertilization is external. The young develop through a series of free-swimming larvae before finally settling in the adult form.

9.7.7 Economic importance

As 'shellfish', mussels, oysters, whelks etc. are an important food source for man and have been for thousands of years. Molluscs have been used for jewellery, currency and tools. The adults and the larvae are important links in food webs. Snails are intermediate hosts for a number of economically important parasites such as *Schistosoma*, causing bilharzia, and *Fasciola*, causing liver rot in sheep (Section 25.2.4). The shipworm *Teredo* is a pelycopod which bores holes in submerged wooden structures such as pilings and boats.

9.8 Phylum Echinodermata

This phylum comprises just over 5000 species which are exclusively marine and mainly bottom dwellers. There are over 20 000 fossil species described, dating back to the Cambrian period. Embryologically, echinoderms resemble chordates more closely than any other invertebrate phylum. Apart from the starfish, other examples include sea urchins and sea cucumbers.

TABLE 9.14 **The Echinodermata**

PHYLUM ECHINODERMATA	Pentamerous radial symmetry Spiny surface Tube feet Water-vascular system with opening called a madreporite
Class Stelleroidea	Star-shaped and flattened
	Few calcareous plates, therefore flexible
	Madreporite on aboral surface (opposite side from mouth)
	Pedicellariae (small, pincer-like structures)
Example	*Asterias* – starfish

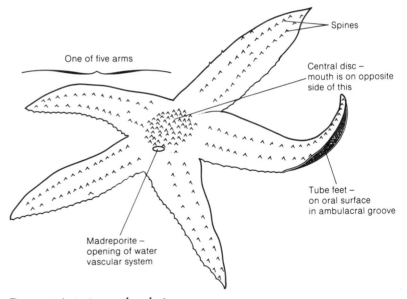

Fig. 9.19 Asterias – *aboral view*

9.8.1 Locomotion

Tube feet have a muscular wall and a cavity into which water may be pushed from the water-vascular system. There may be a sucker on the end of each 'foot'. The combined movements of many tube feet bring about movement of the whole animal.

9.8.2 Nutrition

Some starfish are suspension feeders but the majority are carnivorous, feeding on molluscs, crustaceans and annelids. *Asterias* can open bivalve shells and it then everts part of its stomach into the prey.

9.8.3 Gaseous exchange

This takes place over the thin-walled tube feet.

9.8.4 Excretion and osmoregulation

Ammonia diffuses through the wall of the tube feet. Since echinoderms are isotonic with sea water there is no need for osmoregulation.

9.8.5 Nervous system

Most of the nervous system is closely associated with the epidermis, comprising mainly a circumoral nerve ring with separate nerves into each arm. There are groups of sensory cells dotted over the surface of the animal with a concentration of photoreceptors at the tip of each arm.

9.8.6 Reproduction

Starfish show remarkable powers of regeneration and some have regular asexual reproduction. Most starfish have separate sexes and fertilization is external. The eggs develop into free-swimming, bilaterally symmetrical larvae.

9.8.7 Economic importance

Echinoderms have little economic importance although sea urchins and sea cucumbers are eaten in some parts of the world. However, starfish do damage commercial mussel and oyster beds and the Crown of Thorn Starfish (*Acanthaster*) has destroyed large areas of the Great Barrier Reef.

9.9 Phylum Chordata

This phylum includes the vertebrates which have evolved during the past 500 million years to become the dominant animals of land, sea and air. In number of species and individuals they do not rival the arthropods but their biomass and ecological dominance are much greater.

TABLE 9.15 **The chordata**

PHYLUM CHORDATA	Gill-slits present in pharynx Post-anal tail, at some stage in development Notochord Dorsal, tubular nerve cord
SUB-PHYLUM VERTEBRATA (Craniata)	Well-developed head with brain encased in cranium Vertebral column replaces notochord

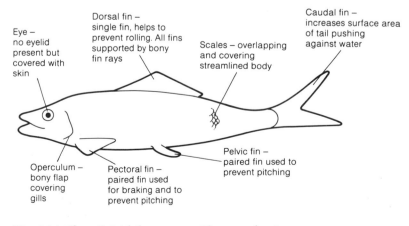

Eye – no eyelid present but covered with skin

Dorsal fin – single fin, helps to prevent rolling. All fins supported by bony fin rays

Scales – overlapping and covering streamlined body

Caudal fin – increases surface area of tail pushing against water

Operculum – bony flap covering gills

Pectoral fin – paired fin used for braking and to prevent pitching

Pelvic fin – paired fin used to prevent pitching

Fig. 9.20 Class Osteichthyes, genus Clupea *— herring*

TABLE 9.16 **Classification of the vertebrata**

Class	Characteristics	Examples		
Chondrichthyes	Cartilaginous endoskeleton No operculum Heterocercal tail fin No swim bladder	*Scyliorhinus* – dogfish *Raja* – ray		
Osteichthyes	Bony endoskeleton Bony scales Operculum over gills Homocercal tail fin Swim bladder present	*Clupea* – herring *Gasterosteus* – stickleback *Salmo* – salmon		
Amphibia	No scales Tympanum (eardrum) visible Lungs in adult Aquatic larvae Metamorphosis	*Rana* – frog *Bufo* – toad *Triturus* – newt		
Reptilia	Dry skin with horny scales Teeth – all the same type (homodont) Eggs with yolk and leathery shell No gills No larval stages	*Lacerta* – lizard *Natrix* – grass snake *Chelonia* – turtle *Crocodilus* – crocodile		
Aves	Endothermic (warm- blooded) Feathers Beak (no teeth) Forelimbs modified into wings Air sacs in light bones	*Columba* – pigeon		
Mammalia	Endothermic Hair Sweat and sebaceous glands Mammary glands Pinna (external ear) Heterodont (different types of teeth) Diaphragm	Order	Insectivora Carnivora Cetacea Chiroptera Rodentia Primates	*Talpa* – mole *Canis* – dog *Delphinus* – dolphin *Desmodus* – vampire bat *Rattus* – rat *Pan* – chimpanzee

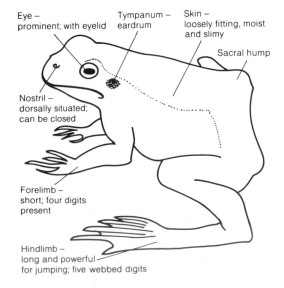

Eye –
prominent; with eyelid

Tympanum –
eardrum

Skin –
loosely fitting, moist
and slimy

Sacral hump

Nostril –
dorsally situated;
can be closed

Forelimb –
short; four digits
present

Hindlimb –
long and powerful
for jumping; five webbed digits

Fig. 9.21 Class Amphibia, genus Rana *– frog*

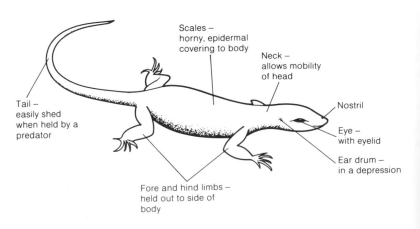

Scales –
horny, epidermal
covering to body

Neck –
allows mobility
of head

Nostril

Eye –
with eyelid

Ear drum –
in a depression

Tail –
easily shed
when held by a
predator

Fore and hind limbs –
held out to side of
body

Fig. 9.22 Class Reptilia, genus Lacerta *– lizard*

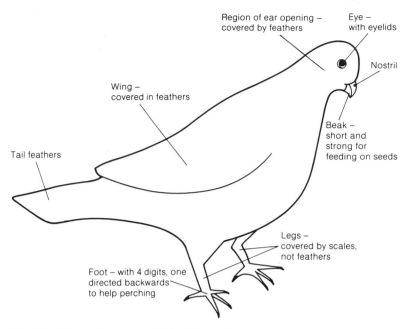

Fig. 9.23 Class Aves, genus Columba *— pigeon*

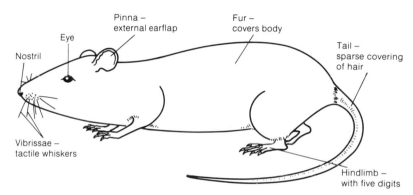

Fig. 9.24 Class Mammalia, genus Rattus *— rat*

It is not proposed to deal with the 'characteristics of life' section by section for the chordates since the most important aspects will be covered in later chapters of this book.

9.9.1 Economic importance

The importance to man of other vertebrates is enormous. As well as being parts of food webs in almost every habitat, many are a direct food source for man and may be 'farmed'. The most important are fish, some birds and mammals such as cattle, sheep and pigs. Birds and small mammals consume large quantities of insect pests and weed seeds. However, rodents, especially rats, may damage crops and stored food. Mammals also carry disease, for example bubonic plague transmitted by fleas of rats. They may also be secondary hosts for parasites such as tapeworms which infect man and his domesticated animals. Some snakes are poisonous to man. Useful products of vertebrates include leather, fur and wool for clothing, as well as such things as glue and fertilizer.

9.10 Questions

1. (a) Give three reasons why a frog and a mammal are classified in the same phylum. (3 marks)
   (b) Give three ways in which a frog and a lizard differ structurally from each other. (3 marks)
   (c) State three features of birds that have contributed to the success of this group. (3 marks)
   (Total 9 marks)

   London Board June 1982, Paper I, No. 1

2. Using examples that you have studied, discuss the major evolutionary changes that are shown in the organization of invertebrate body structure. Discuss the advantages of each change.

   Welsh Joint Education Committee June 1983, Paper AII, No. 9

3. (a) Name a terrestrial annelid and describe **one** external feature which is an adaptation to its environment.
   (b) Name a xerophyte and describe **one** external feature which is an adaptation to its environment.
   (c) Name a terrestrial insect and describe **one** external feature of the imago (adult) which is an adaptation to its environment.
   (d) Name an insect with an aquatic larva and describe **one** external feature of the larva which is an adaptation to its environment. (4 marks)

   Associated Examining Board June 1984, Paper I, No. 3

4. (a) (i) Arrange the groups shown below in the correct sequence with the largest group first:
   order   species   phylum   class   genus
   (ii) Define the term 'species'.
   (iii) Describe **two** factors which favour the evolution of new species.

(b) The drawings show four animals which belong to the same taxonomic group.
   (i) Name the taxonomic group to which they all belong.
   (ii) Give **one** feature which they all have in common. (6 marks)

   Associated Examining Board June 1984, Paper 1, No. 1

5. Complete the table with a tick ($\sqrt{}$) if the feature is present or a cross ($\times$) if it is absent.

Phylum	Hollow dorsal nerve cord	Tubular gut with mouth and anus	Exo-skeleton	Coelom	Radial sym-metry
Cnidaria (Coelenterata)					
Platyhel-minthes					
Annelida					
Arthropoda					
Chordata					

(5 marks)

Associated Examining Board June 1989, Paper I, No. 2

6. The following is a list of animal groups.
   A   Platyhelminthes
   B   Crustacea
   C   Annelida
   D   Cnidaria (Coelenterata)
   E   Mollusca

From the list **A** to **E**, above, choose the taxonomic group which best fits the descriptions given below. Each letter may be used once, more than once, or not at all.

(a) A group comprising externally-segmented animals with a circulatory system where blood is pumped through a closed system of vessels.
(b) A group in which many species are protected by shells made almost entirely of limestone.
(c) A group in which the body is protected by an external chitinous skeleton.
(d) A group in which the organisms show radial symmetry.

Joint Matriculation Board June 1988, Paper IIA, Nos. 23–26

10 Populations and communities

In previous chapters we reviewed the variety of organisms. These organisms live, not in isolation, but as part of populations and communities.

A **population** is a group of individuals of the same species, all occupying a particular area at the same time.

A **community** comprises all the plants and animals which occupy a particular area. Communities therefore consist of a number of populations.

10.1 Population growth

Provided the birth rate exceeds the death rate, a population will grow in size. If only a few individuals are present initially, the rate of growth will be very slow. This is called the **lag phase**. As numbers increase, more individuals become available for reproduction and the population grows at an ever increasing rate, provided no factor limits growth. This is called the **exponential phase**. Growth cannot continue indefinitely because there is a limit to the number of individuals any area can support. This limit is called the **carrying capacity** of the area. Beyond this point certain factors limit further population growth. The size of the population may then stabilize at a particular level. This is called the **stationary phase**. The high population level may, however, cause the carrying capacity of the environment to decline. In these circumstances the population level falls. This is called the **death phase**.

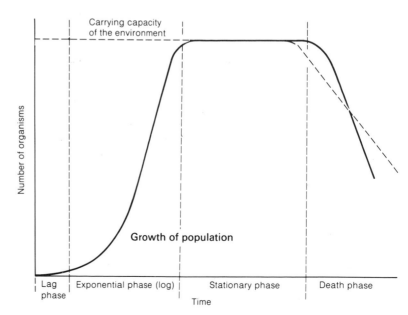

Fig. 10.1 Growth of a population

10.1.1 Environmental resistance

The factors which limit the growth of a particular population are collectively called the **environmental resistance**. Such factors include predation, disease, the availability of light, food, water, oxygen and shelter, the accumulation of toxic waste and even the size of the population itself.

10.1.2 Density-dependent growth

In this type of growth a population reaches a certain size and then remains stable. It is referred to as density-dependent because the size (or density) of the population affects its growth rate. Typical density-dependent factors are food availability and toxic waste accumulation. In a small population, little food is used up and only small amounts of waste are produced. The population can continue to grow. At high population densities the availability of food is reduced and toxic wastes build up. These cause the growth of the population to slow, and eventually stabilize at a particular level.

10.1.3 Density-independent growth

In this type of growth a population increases until some factor causes a sudden reduction in its size. Its effect is the same regardless of the size of the population, i.e. it is independent of the population density. A typical density-independent factor is temperature. A sudden fall in temperature may kill large numbers of organisms regardless of whether the population is large or small at the time. Environmental catastrophes such as fires, floods or storms are other density-independent factors.

10.1.4 Regulation of population size

The maximum possible number of offspring varies considerably from species to species. It may be as little as one offspring in two years in some mammals, or as great as one million eggs in a single laying in certain molluscs like the oyster. The term **fecundity** is used to describe the reproductive capacity of individual females of a species. In mammals the **birth rate** or **natality** is used to measure the fecundity. On the other hand the number of individuals of a species which die from whatever cause, is called the **death rate** or **mortality.** Clearly the size of a population is regulated by the balance between its fecundity and its mortality. However, there are other influences on the size of a population. Two such influences are immigration and emigration.

Immigration occurs when individuals join a population from neighbouring ones. **Emigration** occurs when individuals depart from a population. The emigrants may either enter an existing neighbouring population or, as in the swarming of locusts and bees, they may form a new population. Factors such as overcrowding often act as a stimulus for emigration. Unlike the periodic seasonal movements which occur in migration, emigration is a non-reversible, one-way process. The size of a population may fluctuate on a regular basis, called a **cycle.** These fluctuations are normally the consequence of regular seasonal changes, such as temperature or rainfall. At other times the population may be subject to sudden and unexpected fluctuations. While both types of fluctuation are usually due to a number of factors, there is often one, called the **key factor**, which is paramount in bringing about the change.

10.1.5 Control of human populations

Most animal populations are kept in check by food availability, climate, disease or predators. Populations frequently increase in size rapidly and then undergo a sudden 'crash' during which there is a dramatic reduction in numbers.

In increasingly more regions of the world man's knowledge, expertise and technology are succeeding in reducing the impact of the climate and disease. As the top organism in many food chains, man has little to fear from predators. Even in food production man has made considerable advances, although in parts of South America, Asia and Africa famine remains a major check to population growth. As a result, the human population as a whole has grown virtually unchecked in recent times. Figures for the past rate of increase in human populations can only be estimated, but it seems probable that prior to 1600 it had taken around 2000 years to double the world's population. By 1850, it had doubled again. It took just 80 years to complete the next doubling in 1939 and a mere 50 years to complete the latest one. At present the world population increases by about 1 500 000 every week, equivalent to about 150 people every minute.

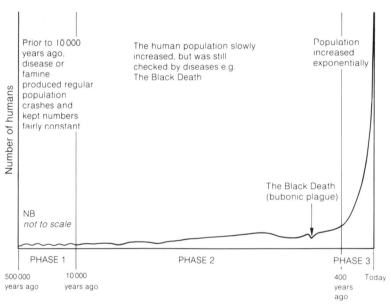

Fig. 10.2 The growth of the human population

In 1798, Thomas Robert Malthus, an English economist, published an essay on population in which he suggested that while the world's food supply would increase arithmetically, the human population would do so geometrically. The so-called **Malthusian principle** suggested famine as the inevitable consequence of this state of affairs. Despite many important agricultural advances, much of the world's population is still undernourished. It is inconceivable that the present rate of population can be sustained for much longer. War, famine or disease will inevitably curb further increases unless man reduces his birth rate by appropriate forms of birth control. The variety of birth control methods available are given in Chapter 19 (Table 19.1). The solution may seem simple enough, but opposition to birth control is often deeply rooted in personal, social, religious or traditional belief. Perhaps only the awful consequences of unlimited growth in the human population will persuade people of the necessity of population control.

10.2 Interactions between populations

10.2.1 Competition

Individuals of species in a population are continually competing with each other, not only for nutrients but also for mates and breeding sites. This competition between individuals of the same species is called **intra-specific competition.** Individuals are also in continual competition with members of different species for such factors as nutrients, space and shelter. Competition between individuals of different species is called **inter-specific competition.**

10.2.2 Predation

Predator-prey relationships are important in producing cyclic changes in the size of a population. By eating their prey, predators remove certain members of a population and so reduce their numbers. As the size of the prey population diminishes, the predators experience greater competition with each other for the remaining prey. The predator population therefore diminishes as some individuals are unable to obtain enough food to sustain them. The reduction in the predator population results in fewer prey being taken and so allows their numbers to increase again. This increase in its turn leads to an increase in the predator population.

The number of predators is usually less than the number of prey.
The shape of the two curves is similar, but there is a time lag between the two; the curve for the predators lags behind that of the prey.

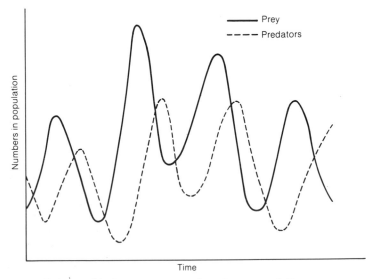

Fig. 10.3 Relationship between prey and predator populations

Typical examples of the relationship, shown in Fig. 10.3, include the lynx preying upon the Canadian snowshoe hare and *Hydra* preying upon the water flea *Daphnia*. It must be said that predator-prey relationships alone are not responsible for fluctuations in the numbers in a population; disease and climatic factors also play a rôle. Nevertheless, predation is significant in the regulation of natural populations. The type of cyclic fluctuation shown in Fig. 10.3 plays an important rôle in evolution. The periodic population crashes create selection pressure whereby only those individuals who are able to escape predation, or withstand disease or adverse climatic conditions, will survive to reproduce. The population thereby evolves to be better adapted to the prevailing conditions.

10.2.3 Competitive exclusion principle

In 1934, a Russian biologist, C. F. Gause, experimented on two species of *Paramecium*. He grew *P. caudatum* and *P. aurelia* both separately and together. When grown together, the two species competed for the available food. After a few days the population of *P. caudatum* began to decline, and after three weeks all its members had died. It seemed that the two species were in such close competition that only one could survive. This became known as the **competitive exclusion principle** or **Gause's principle.** It states that only one species (population) in a given community can occupy a given ecological niche at any one time. Fig. 10.4 summarizes the results of Gause's experiments with *Paramecium*. Although *P. aurelia* survives, its population is reduced in the presence of *P. caudatum*, compared to its population when grown alone.

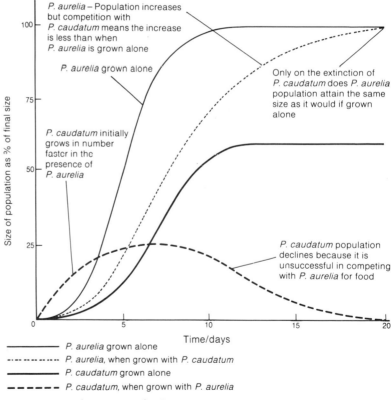

Fig. 10.4 *Population growth of two species of* Paramecium *grown separately and together*

Why *P. aurelia* is more successful in the long term may be because of its smaller size. Being smaller, it requires less food and is better able to survive when food is scarce. Its success may, however, be due to a faster reproductive rate or greater efficiency in obtaining its food. The reasons for success are hard enough to isolate in the laboratory; it is considerably more difficult to do so for a wild population.

TABLE 10.1 **Some examples of biological control**

Target (pest)	Harmful effects of pest	Control agent	Method of action
Scale insect (*Icerya*)	Kills citrus fruit trees	Ladybird (*Rodolia*)	Ladybird uses scale insect as a food source
Codling moth (*Crypto-phlebia*)	Ruins orange crop	African wasp (*Tricho-gamatoidea*)	Wasp parasitises moth eggs
Mosquito (*Anopheles*)	Vector of malarial parasite (*Plasmo-dium*)	Hydra (*Chloro-hydra*)	Hydra is a predator of mosquito larvae
Snail (*Biomphal-aria*)	Vector of *Schistosoma* which causes bilharzia	Snail (*Marisa*)	Control agent snail is predatory on the snail vector
Prickly pear (*Opuntia*)	Makes land difficult to farm by restricting access	Cochineal insect (*Dactylo-pius*)	*Opuntia* is a food source for the insect
Larvae of many butterflies and moths	Consume the foliage of many economic-ally important plants	Bacterium (HD-1 strain of *Bacillus thuringiensis*)	Bacterium parasitises the larvae of moths and butterflies

Fig. 10.5 General relationship between pest and control agent populations in biological control

10.2.4 Biological control

The effect of the predator-prey relationship in regulating populations has been exploited by man to enable him to control various pests. Biological control is a means of managing populations of organisms which compete for his food or damage his health or that of his livestock. The aim is to bring the population of a pest down to a tolerable level by use of its natural enemies. A beneficial organism (the **agent**) is deployed against an undesirable one (the **target**). A typical situation is where a natural predator of a harmful organism is introduced in order to reduce its numbers to a level where they are no longer harmful. The aim is not to eradicate the pest; indeed, this could be counter-productive. If the pest was reduced to such an extent that it no longer provided an adequate food source for the predator, then the predator in its turn would be eradicated. The few remaining pests could then increase their population rapidly, in the absence of the controlling agent. The ideal situation is where the controlling agent and the pest exist in balance with one another, but at a level where the pest has no major detrimental effect.

Biological control was originally used against insect and weed pests of economically important crops. In more recent times its use has broadened to include medically important pests such as snails and even vertebrate pests. In the same way the type of controlling agent has become more diverse and the following are now employed: bacteria, viruses, fungi, protozoans, nematodes, insects and even amphibians and birds. Table 10.1 lists some examples of biological control.

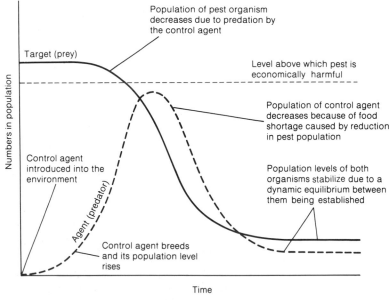

One interesting and unusual form of biological control takes place in Australia. Cattle dung there presents a problem because two major pests, the bush fly and buffalo fly, lay their eggs in it. In addition, the dung carries the eggs of worms which parasitize the cattle. The indigenous dung beetles which are adapted to coping with the fibrous wastes of marsupials are ineffective in burying the soft dung of cattle. The introduction of African species of dung beetle, which bury the dung within forty-eight hours, has been effective in controlling the flies. By burying the dung before the flies can mature, or before the parasitic worm can develop and reinfect cattle, they have controlled the populations of these pests.

10.3 Ecological concepts

10.3.1 Communities

A community is all the plants and animals which occupy a particular area. The individual populations within the community interact with one another. The community is a constantly changing dynamic unit, which passes through a number of stages from its origin to its climax. The transition from one stage to the next is called **succession**.

10.3.2 Succession

Imagine an area of bare rock. One of the few kinds of organisms capable of surviving on such an inhospitable area are the lichens. The symbiotic relationship between an alga and a fungus which makes up a lichen, allows it to survive considerable drying out. As the first organisms to bring about **colonization** of a new area, the lichens are called **pioneers** or the **pioneer community**.

The weathering of any rock produces a sand or soil, but in itself this is inadequate to support other plants. With the decomposing remains of any dead lichen, however, sufficient nutrients are made available to support a community of small plants. Mosses are typically the next stage in the succession, followed by ferns. With the continuing erosion of the rock and the increasing amounts of organic material available from these plants, a thicker layer of soil is built up. This will then support smaller flowering plants such as grasses and, by turn, shrubs and trees. In Britain the ultimate community is most likely to be deciduous oak woodland. The stable state thus formed comprises a balanced equilibrium of species with few, if any, new varieties replacing those established. This is called the **climax community**. This community consists of animals as well as plants. The animals have undergone a similar series of successional stages, largely dictated by the plant types available. Within the climax community there is normally a **dominant** plant and animal species, or sometimes two or three **co-dominant** species. The dominant species is normally very prominent and has the greatest biomass.

The succession described above, where bare rock or some other barren terrain is first colonized, is called **primary succession.** If, however, an area previously supporting life is made barren, the subsequent recolonization is called **secondary succession**. Secondary succession occurs after a forest fire or the clearing of agricultural land. Spores, seeds and organs of vegetative propagation may remain viable in the soil, and there will be an influx of animals and plants through dispersal and migration from the surrounding area. In these circumstances the succession will not begin with pioneer species but with organisms from subsequent successional stages.

A series of successional stages is called a **sere.** There are a number of different seres according to the environment being colonized. A **hydrosere** refers to a series of successions in an aquatic environment and a **halosere** to one in a saltmarsh.

An oak wood to show a climax community

10.4 Questions

1. Fig. 1 is a generalized graph of population growth.
   (a) Why does the initial accelerating phase start slowly?
   (b) List **three** environmental factors other than shortage of food, oxygen or water which may cause the growth of the population to decelerate.
   (c) How do you explain the final phase in which the number of individuals in the population remains constant?

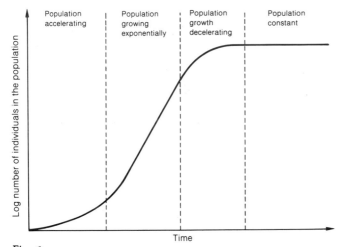

Fig. 1

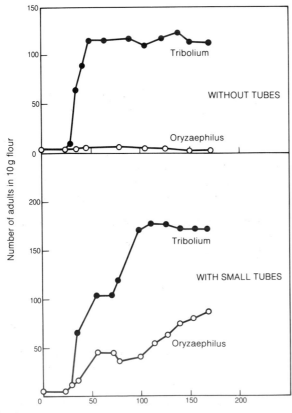

Fig. 2

Fig. 2 gives some results of a population study on two genera of beetles, *Tribolium* and *Oryzaephilus*, living in flour that is continually replenished. *Tribolium* rapidly eliminates *Oryzaephilus* unless small glass tubes are put into the flour. The larvae of *Oryzaephilus* hide in these away from predation by *Tribolium*. The adults can coexist.
   (d) What principle do these results illustrate?

(*Total 7 marks*)

Cambridge Board November 1983, Paper II, No. 7

2. Discuss the ways in which biotic factors, excluding human activities, may influence the size of an animal population. (*Total 20 marks*)

London Board June 1985, Paper II, No. 9

3. (a) Fig. 1 shows an exponential growth curve. This does not apply to real populations, such as yeast in culture or herbivores living on grassland. Explain why this is so. (*1 mark*)

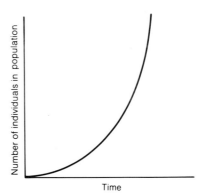

Fig. 1

(b) Fig. 2 shows a more common population curve. For each stage, A–D, describe and explain what is happening to the population. (*6 marks*)

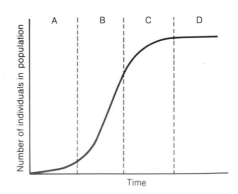

Fig. 2

(c) Fig. 3 shows the change in biomass during the years following clearance of an oak/pine forest in Brookhaven, New York State, USA.

    (i) What type of vegetation would you expect to predominate at 2 years, 20 years and 150 years?

    (ii) Describe how the succession of plants leads to a climax community as illustrated by Fig. 3.   *(4 marks)*

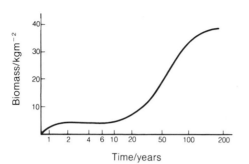

Fig. 3

(d) On a journey along the length of the Mississippi River Basin in the USA a biologist noticed that there was a considerable variation in the tree species from one place to another, although there were no abrupt changes. What does this imply about the climax community in the Mississippi River Basin?   *(4 marks)*

Total 15 marks)

Northern Ireland Board June 1985, Paper I, No. 6

4. (a) Define the term *population*.   *(2 marks)*

(b) With reference to examples of **named** organisms, discuss the factors which limit the sizes of populations in nature.   *(11 marks)*

(c) Describe, with examples, how parasites and predators are used in attempts to control pests.   *(5 marks)*

(Total 18 marks)

Cambridge Board June 1988, Paper I, No. 9

5. In some parts of the world, maize is the staple food. After harvest, it is stored in simple mud-walled, thatched granaries. Insects may cause considerable loss of the stored grain.

Fig. 1 is a graph showing variations in the population density of a grain pest, *Tribolium castaneum*, in a maize store.

(a) Calculate the rate of increase of the population density between:

    (i) 1 and 4 months;   *(2 marks)*

    (ii) 6 and 7 months.   *(2 marks)*

(b) Suggest **two** reasons for the relatively slow increase in the population density in the first five months.   *(2 marks)*

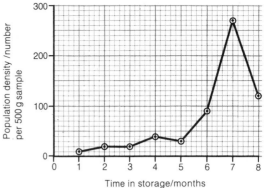

Fig. 1

(c) Suggest **two** reasons for the change in the population density which took place after 7 months.   *(2 marks)*

In the sample of grain from which the results shown in Fig. 1 were obtained, only *Tribolium* was present. In another maize store, two species of pest were found, *Tribolium castaneum* and *Sitophilus zeamais*. Fig. 2 shows the variations in the densities of the two populations in this second maize sample.

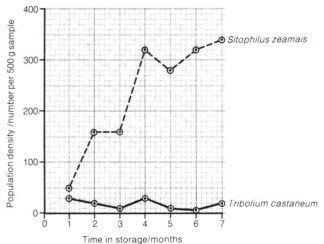

Fig. 2

(d) What is the evidence from Figs. 1 and 2 that there is competition between *Tribolium castaneum* and *Sitophilus zeamais*?   *(2 marks)*

Samples of grain were also collected each month to determine the amount of damage done by insect pests. A standard formula was used:

$$\frac{UNd - DNd}{U(Nd + Nu)} \times 100\%$$

where U = the average mass of an undamaged grain;

    D = the average mass of a damaged grain;

    Nd = the number of damaged grains in the sample;

    Nu = the number of undamaged grains in the sample.

(e) In order to draw valid conclusions, give **two** precautions that should be taken in collecting the grain samples. (*2 marks*)

(f) Explain precisely what is being measured by the expressions
  (i) UNd − DNd;
  (ii) U(Nd + Nu). (*2 marks*)

(g) Construct a table suitable for collecting the raw data for this determination from a series of grain samples. (*2 marks*)

(h) Outline a suitable method of determining moisture content of two different grain samples. (*4 marks*)

An investigation was carried out into the effectiveness of different insecticides in controlling *Tribolium*. A number of granaries were selected for the trial. Before the start of the experiment, farmers were asked to clean them thoroughly and plaster the walls with fresh mud. The granary was filled with maize treated with the appropriate insecticide.

(j) What was the purpose of plastering the walls with fresh mud before introducing the grain? (*1 mark*)

The table shows the effects of three different insecticides on the mean population density of *Tribolium* in 500 g samples of maize.

Treatment	Population density after	
	3 months	7 months
Malathion	0.9	2.5
Tetrachlorvinfos	1.1	0.1
Etrimfos	0.9	0.1
Untreated	1.0	10.5

(k) Draw a barchart to show the effect of these insecticides on the population densities of *Tribolium* after 3 months and after 7 months. (*5 marks*)
(*Total 24 marks*)

Associated Examining Board June 1989, Paper II, No. 4

6. (a) State what is meant by the following terms:
  (i) A population (*2 marks*)
  (ii) Interspecific competition. (*2 marks*)

(b) A small number of yeast cells were introduced into a large volume of well-aerated culture solution and kept at 25 °C.
The graph below shows the pattern of population change during a period of 20 hours.

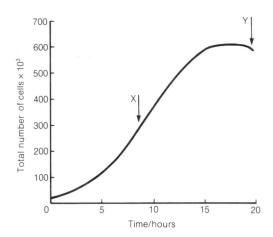

(i) Assuming environmental conditions to be optimal, state *two* factors that determine the rate of population change at point X. (*2 marks*)

(ii) Suggest *two* reasons for the decline in population at point Y. (*2 marks*)
(*Total 8 marks*)

London Board January 1989, Paper I, No. 5

Part of a model of the DNA double helix (opposite)

Part II
The Continuity of Life

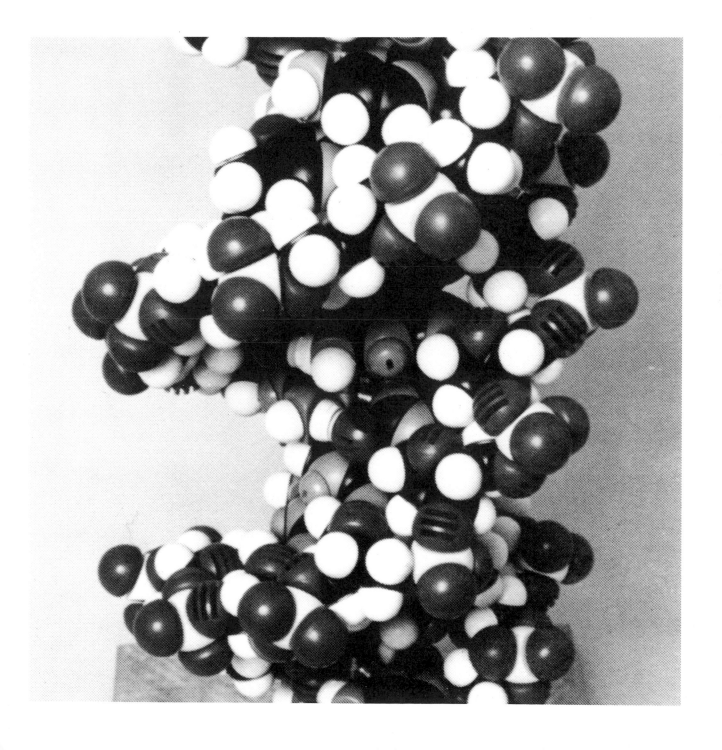

11 *Inheritance in context*

The obvious similarities between children and their parents, or sometimes their grandparents, has long been recognized. Despite many attempts to explain this phenomenon, it is only in recent years that our knowledge of the process of heredity has enabled us to fully understand the mechanism.

The fact that, in some species, both male and female are needed to produce offspring was realized from early times. The rôle each sex played was, however, a matter of argument. Aristotle believed that the male's semen was composed of an incomplete blend of ingredients which upon mixing with the menstrual fluid of the female, gained form and power and became the new organism. Apart from minor refinements this belief was generally accepted until the seventeenth century. When Anton van Leeuwenhoek observed sperm in human semen, the idea arose that these contained a miniature human. When these sperm were introduced into the female, one would implant in the womb and develop there, the female's rôle being nothing more than a convenient incubator. Around this period Regnier de Graaf discovered in ovaries what was later to be called the Graafian follicle. This was thought by another group of scientists to contain the miniature human, the sperm simply acting as a stimulus for its development.

The problem with both these beliefs was that it could easily be observed that any offspring tended to show characteristics of both parents rather than just one. This led, in the last century, to the idea that both parents contributed hereditary characteristics and the offspring was merely an intermediate blend of both. While closer to present thinking, it too had one flaw. Logically the offspring of a cross between a red flower and a white flower should have pink flowers and the children of a tall father and a short mother should be of medium height. It took the rediscovery of the work of Mendel at the beginning of this century to provide what is now an accepted explanation. Both parents do provide hereditary material within the sperm and ovum. The offspring therefore has two sets of genetic information – one from the mother and one from the father. For any individual characteristic, e.g. eye colour, only one of the two factors expresses itself. An individual with one factor for blue eyes and one for brown eyes will always have brown eyes. A few characters do show an intermediate state between two contrasting factors, but this is relatively rare. Only one of each pair of factors will be present in any one gamete.

Alongside these changes in the last century, another development took place. It was originally believed that new species arose spontaneously in some manner. By the end of the century it was more or less accepted that they were formed by adaptation of existing forms. Natural selection is considered to be the mechanism by which these changes arise and depends upon there being much variety among individuals of a species. Without this variety and consequent selection of the types best suited to the present conditions, species could not adapt and evolve to meet the changing demands of the environment.

If this theory of evolution is accepted, then the process of inheritance must permit variety to occur. At the same time, if the offspring are to be supplied with the same genetic information as the parents, the genetic material must be extremely stable. This stability is especially important to ensure that favourable characteristics are passed on from one generation to the next. This then is the paradox of inheritance – how to reconcile the genetic stability needed to preserve useful characteristics with the genetic variability necessary for evolution. To satisfy both requirements it is necessary to have hereditary units which are in themselves exceedingly stable, which can be reassorted in an almost infinite variety of ways. The idea can be likened to a pack of playing cards. The cards themselves are stable, fixed units, but the number of different possible combinations in a typical hand of thirteen cards is immense. Imagine how much greater are the possible combinations of the thousands of hereditary units in a typical organism.

A summary of the historical events which contributed to our current understanding of heredity is given in Table 11.1.

TABLE 11.1 **Historical review of events leading to present-day knowledge of reproduction and heredity**

Name	Date	Observation/discovery	Name	Date	Observation/discovery
Aristotle	384–322 BC	Mixing of male semen and female semen (menstrual fluid) was like blending two sets of ingredients which gave 'life'.	Morgan	early 1900s	Pioneered use of *Drosophila* in genetics experiments and described linkage
General scientific belief	Up to 17th century	Simple organisms arose spontaneously out of non-living material	Garrod	1908	Postulated mutations as sources of certain hereditary diseases
van Leeuwenhoek	1677	Discovered sperm – it was generally believed that these contained miniature organisms which only developed when introduced into a female.	Johannsen	1909	Coined term 'gene' as hereditary unit
			Janssens	1909	Observed chiasmata and crossing over
			Sturtevant	1913	Mapped genes on chromosomes of *Drosophila*
de Graaf	1670s	Described the ovarian follicle (later called Graafian follicle)	Muller	1920s	Observed mutagenic effect of X-rays
			Oparin	1923	Suggested theory of origin of life
Lamarck	1809	Proposed theory of evolution based on inheritance of acquired characteristics	Griffith	1928	Produced evidence suggesting that a chemical 'transforming principle' was responsible for carrying genetic information
Darwin	1859	*On the Origin of Species by Means of Natural Selection* published	Beadle and Tatum	1941	Produced evidence supporting the one gene, one enzyme hypothesis
Pasteur	1864	Experimentally disproved the theory of spontaneous generation	Avery, McCarty and McCleod	1944	Showed nucleic acid to be the chemical which carried genetic information
Mendel	1865	Experiments on the genetics of peas and formulation of his two laws	Hershey and Chase	1952	Showed DNA to be the hereditary material
Hertwig	1875	Witnessed fusion of nuclei during fertilization	Watson and Crick	1953	Formulated the detailed structure of DNA
Flemming	1882	Described all stages of mitosis	Kornberg	1956	Produced DNA copies from single DNA template using DNA polymerase
de Vries	1900	Rediscovery of the significance of Mendel's 1865 experiment			
Sutton	1902	Observed pairing of homologous chromosomes during meiosis and suggested these carried genetic information	Meselsohn and Stahl	1959	Described mechanism of semi-conservative replication in DNA
			Jacob and Monod	1961	Postulated existence of mRNA in theory on control of protein synthesis

12 DNA and the genetic code

12.1 Evidence that the nucleus contains the hereditary material

The universal occurrence of a nucleus at some stage of the life cycle of cells suggests that it performs an essential rôle. The functions of the nucleus are listed in Chapter 4 Section 3.4. The fundamental rôle of the nucleus in determining the features of a cell was established by Hämmerling. Working with individual cells is normally a difficult task, not least because of their small size. Hämmerling, however, used unusually large single-celled algae belonging to the genus *Acetabularia*. Each cell is up to 5 cm in length, making the sectioning of it relatively easy.

Fig. 12.1, over the page, gives a summary of the experiments using two species of *Acetabularia*, which show the nucleus to contain the hereditary material. The experiments are based on those of Hämmerling although they incorporate some refinements made possible by modern techniques.

The evidence that **DNA** (deoxyribonucleic acid) is the hereditary material comes from seven sources:

1. Chromosome analysis.

2. Metabolic stability of DNA.

3. Constancy of DNA in a cell.

4. Correlation between mutagens and their effects on DNA.

5. Experiments on bacterial transformation (Griffith 1928).

6. Experiments to identify the transforming principle (Avery, McCarty and McCleod 1944).

7. Transduction experiments (Hershey and Chase 1952).

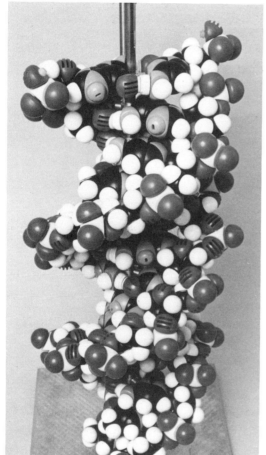

Molecular model of DNA

12.2 Evidence that DNA is the hereditary material

12.2.1 Chromosome analysis

With the nucleus having been shown to contain the hereditary material, attention focused on determining the precise nature of this material. As **chromosomes** only become visible during cell division, it was hardly surprising that they quickly attracted attention. Chromosomes were shown to be made up of protein and DNA. Of the two, protein was thought a more likely candidate as it was known to be a complex molecule existing in an almost infinite number of forms – a necessary characteristic of a material which must carry an immense diversity of information. Later work showed this not to be the case and research centred on the DNA.

METHOD		RESULTS	CONCLUSION	EXPLANATION IN LIGHT OF PRESENT KNOWLEDGE
Experiment 1 *A. mediterranea* is cut into two approximately equal halves 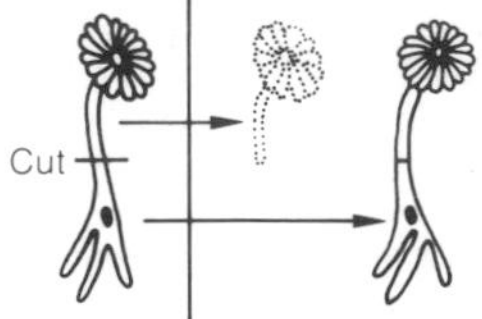	Cut	The portion without the nucleus degenerates. The portion with the nucleus regenerates a new cap of the same type	The information for the regeneration of the cap is contained in, or produced by, the lower portion which contains the nucleus	The DNA in the nucleus produces mRNA which enters the cytoplasm where it provides the instructions for the formation of the enzymes needed in the production of a new cap. In the absence of a nucleus, the upper portion cannot do this
Experiment 2 *A. mediterranea* is cut to isolate the stalk section which does not contain the nucleus 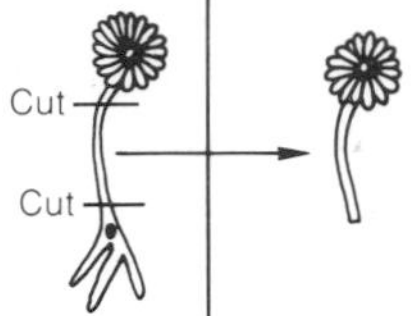	Cut Cut	A new cap is regenerated from the stalk section	The information on how to regenerate the cap is present in the stalk	As the nucleus produces a constant supply of mRNA there is sufficient in the cytoplasm of the stalk to provide instructions on how to form the enzymes necessary for the regeneration of the cap
Experiment 3 The regenerated cap from the previous experiment is again removed 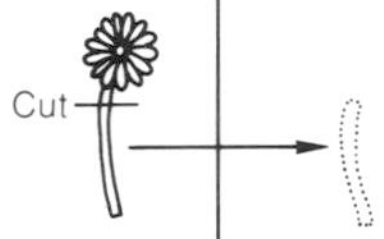	Cut	The stalk does not regenerate a cap for a second time	The information on how to regenerate the cap, which is contained in the stalk, must be used up and so cannot effect a second regeneration	The mRNA is broken down once its role in regenerating the cap is complete. It is therefore not available for the cap to be regenerated a second time. In the absence of a nucleus, there is no new source of mRNA
Experiment 4 The stalk of *A. mediterranea* is grafted onto the base portion (which contains the nucleus) of *A. crenulata* – a species possessing a different shaped cap 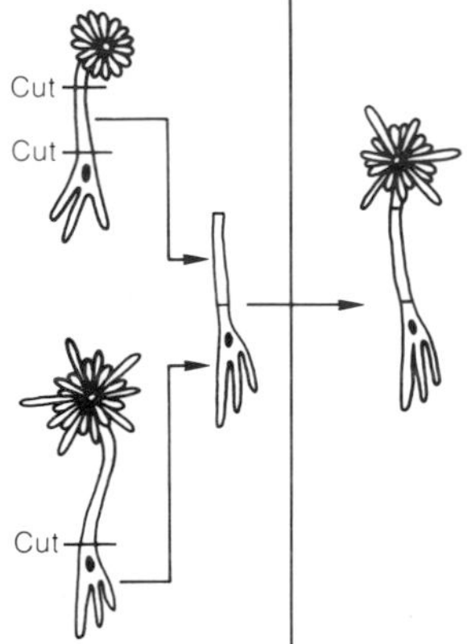	Cut Cut Cut	The cap of *A. crenulata* is regenerated	The influence on cap regeneration of the base portion (with nucleus) is greater than the influence of the stalk portion (without the nucleus)	With the nucleus of *A. crenulata* present, a constant supply of mRNA is available to regenerate this type of cap. The mRNA from *A. mediterranea* is limited to that present in the stalk when it was separated from its nucleus. The influence of the mRNA from *A. crenulata* is therefore greater
Experiment 5 The nucleus from a decapitated *A. crenulata* is removed and replaced with a transplanted nucleus from *A. mediterranea* 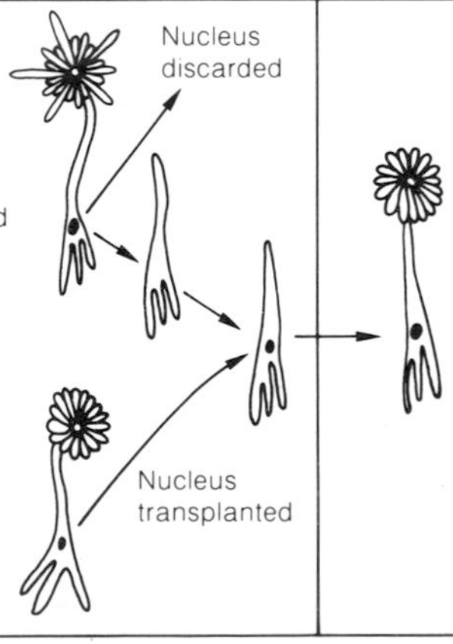	Nucleus discarded Nucleus transplanted	The cap regenerated is of the *A. mediterranea* type	As the only part of *A. mediterranea* which is present is the nucleus, it alone must contain the instructions on how to regenerate the cap	The situation is similar to that in experiment 4 except that it is the mRNA of *A. mediterranea* which is present in greater quantities, because its nucleus is present and forms a constant supply of mRNA

Fig. 12.1 *Summary of experiments to show that the nucleus contains hereditary material*

12.2.2 Metabolic stability of DNA

Any material which is responsible for transferring information from one generation to another must be extremely stable. If it were altered to any extent imperfect copies would be made. Unlike protein, DNA shows remarkable metabolic stability. If DNA is labelled with a radioactive isotope it can be shown that its rate of disappearance from the DNA is very slow. This suggests that, once formed, a DNA molecule undergoes little if any alteration.

12.2.3 Constancy of DNA within a cell

Almost all the DNA of a cell is associated with the chromosomes in the nucleus. Small amounts do occur in cytoplasmic organelles such as mitochondria, but this represents a small proportion of the total. Changes in the DNA content of a cell may therefore be attributed to changes in chromosomal DNA. Analysis shows that the amount of DNA remains constant for all cells within a species except for the gametes (or spores in mosses and ferns), which have almost exactly half the usual quantity. Prior to cell division the amount of DNA per cell doubles. This is shared equally between the two daughter cells which therefore have the usual quantity. These changes are consistent with those expected of hereditary material which is being transmitted from cell to cell during division.

12.2.4 Correlation between mutagens and their effects on DNA

Mutagens are agents which cause **mutations** in living organisms. A mutation is an alteration to an organism's characteristics which is inherited. Many agents are known mutagens; they include X-rays, nitrous acid and various dyes. It can be shown that these mutagens all alter the structure of DNA in some way. A typical example is ultra-violet light of wavelength 260 nm. It both causes mutations and alters the structure of the pyrimidine bases of which DNA is made. This suggests that it is this alteration of DNA which is the source of the mutation and DNA must therefore be the hereditary material.

12.2.5 Experiments on bacterial transformation

The most convincing evidence for the genetic rôle of DNA was provided by Griffith in 1928. He experimented on the bacterium *Pneumococcus* which causes pneumonia. It exists in two forms:

1. The harmful form - a virulent (disease-causing) type which has a gelatin coat. When grown on agar it produces **shiny, smooth** colonies and is therefore known as the S-strain.

2. The safe form – a non-virulent (does not cause disease) type which does not have a gelatin coat. When grown on agar it produces **dull, rough** colonies and is therefore known as the R-strain. Griffith's experiments may be summarized thus:

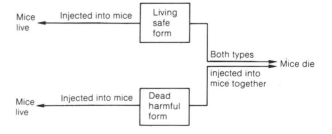

The living safe form and dead harmful form, while not causing pneumonia when injected separately, did so when injected together. The resulting dead mice were found to contain living harmful forms of *Pneumococcus*. If one discounts the improbable explanation that the dead harmful forms have been resurrected, how then could the living safe forms suddenly have acquired the ability to form a gelatin coat, produce rough colonies and cause pneumonia? It is possible that the safe form had mutated into the harmful form, but this is unlikely. Furthermore, the experiment can be repeated with similar results and the likelihood of the same mutation arising each and every time is so improbable that it can be discounted.

If pneumonia is caused by some toxin produced by *Pneumococcus*, then the harmful type must have the ability to produce it, whereas the safe type does not. The explanation could therefore be that the dead harmful type has the information on how to make the toxin but, being dead, is unable to manufacture it. The safe type, being alive, is potentially able to make the toxin but lacks the information on how to go about it. If then the recipe for the toxin can in some way be transferred from the dead harmful to the living safe variety, the toxin can be manufactured and pneumonia will result. As the substance was able to transform one strain of *Pneumococcus* into another, it became known as the **transforming principle.**

12.2.6 Experiments to identify the transforming principle

The identity of the transforming principle was determined by Avery, McCarty and McCleod in 1944. In a series of experiments they isolated and purified different substances from the dead harmful types of *Pneumococcus*. In turn they tested the ability of each to transform living safe types into harmful ones. Purified DNA was shown to be capable of bringing about transformation, and this ability ceased when the enzyme which breaks down DNA (deoxyribonuclease) was added.

12.2.7 Transduction experiments

In 1952 Hershey and Chase performed a series of experiments involving the bacterium *Escherichia coli* and a bacteriophage (T_2 phage) which attacks it. (Details of a phage life cycle are given in Section 7.1.1.) T_2 phage transfers to *E. coli* the necessary hereditary material needed to make it manufacture new T_2 phage viruses. As the T_2 phage virus is composed of just DNA and protein, one or the other must constitute the hereditary material. Hershey and Chase carefully labelled the protein of one phage sample with radioactive sulphur (^{35}S) and the DNA of another phage sample with radioactive phosphorus (^{32}P). They then separately introduced each sample into a culture of *E. coli* bacteria. At a critical stage, when the viruses had transferred their hereditary material into the bacterial cells, the two organisms were separated mechanically and each culture of bacteria was examined for radioactivity. The culture injected with radioactive DNA contained radioactive bacteria, while that injected with radioactive protein did not. The evidence was conclusive; DNA was the hereditary material but if further proof were needed this was provided by electron microscope studies which actually traced the movement of DNA from viruses into bacterial cells.

12.3 Nucleic acids

12.3.1 Structure of nucleotides

Individual nucleotides comprise three parts:

1. **Phosphoric acid** (phosphate H_3PO_4). This has the same structure in all nucleotides.

2. Pentose sugar. Two types occur, ribose ($C_5H_{10}O_5$) and deoxyribose ($C_5H_{10}O_4$).

3. Organic base. There are five different bases which are divided into two groups, described on the next page.

NAME OF MOLECULE	CHEMICAL STRUCTURE	REPRESENTATIVE SHAPE
Phosphate	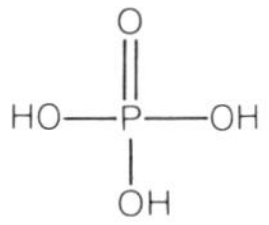	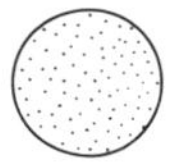
Ribose	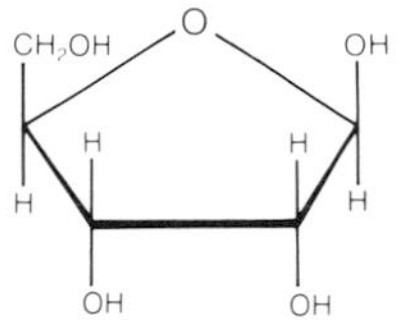	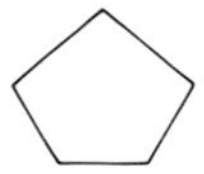
Deoxyribose	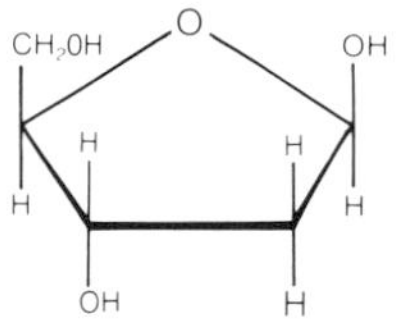	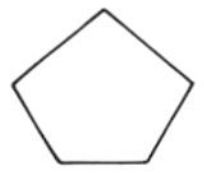
Adenine (a purine)	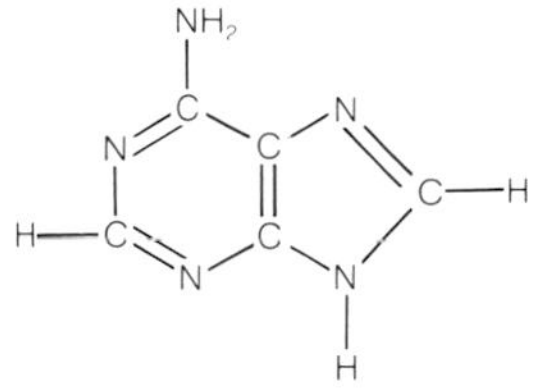	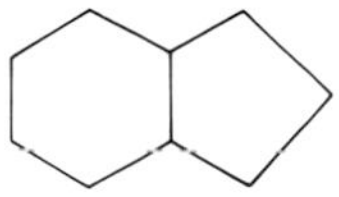 Adenine
Guanine (a purine)	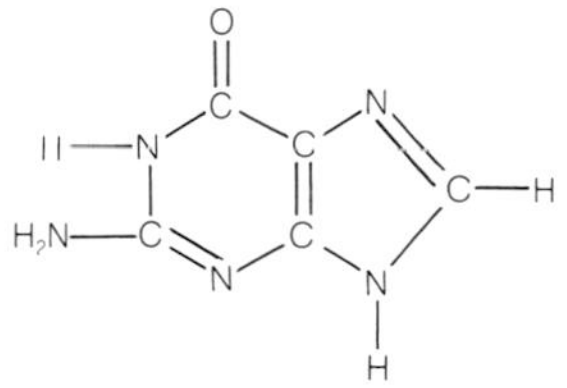	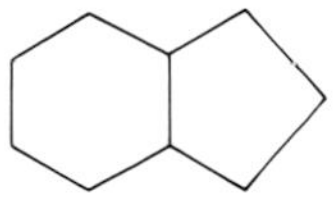 Guanine
Cytosine (a pyrimidine)	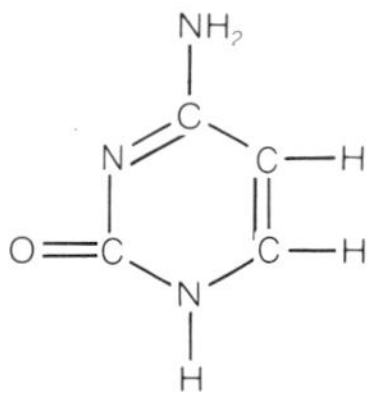	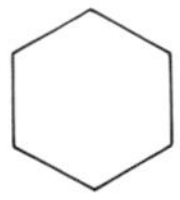 Cytosine
Thymine (a pyrimidine)	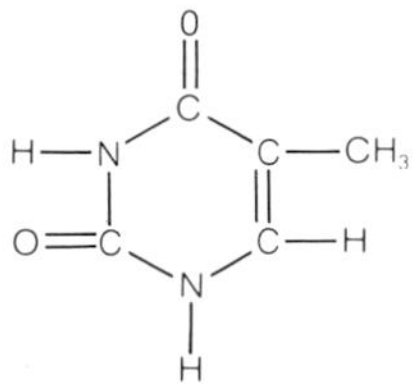	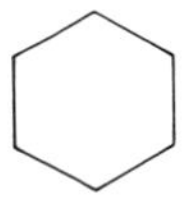 Thymine
Uracil (a pyrimidine)	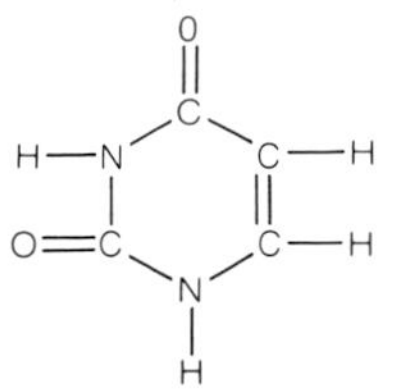	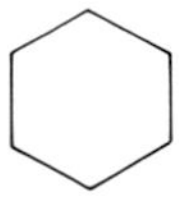 Uracil

Fig. 12.2 Structure of molecules in a nucleotide

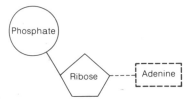

Adenosine monophosphate (adenylic acid)

Fig. 12.3(a) Structure of a typical nucleotide

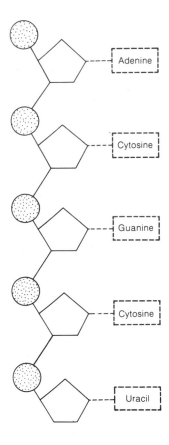

Fig. 12.3(b) Structure of a section of polynucleotide, e.g. RNA

(a) **Pyrimidines** – these are single rings each with six sides. Examples found in nucleic acids are: **cytosine, thymine** and **uracil.**

(b) **Purines** – these are double rings comprising a six-sided and a five-sided ring. Two examples are found in nucleic acids: **adenine** and **guanine.**

The three components are combined by condensation reactions to give a nucleotide, the structure of which is shown in Fig. 12.3(a). By a similar condensation reaction between the sugar and phosphate groups of two nucleotides, a **dinucleotide** is formed. Continued condensation reactions lead to the formation of a **polynucleotide** (Fig. 12.3(b)).

The main function of nucleotides is the formation of the nucleic acids **RNA** and **DNA** which play vital rôles in protein synthesis and heredity. In addition they form part of other metabolically important molecules. Table 12.1 gives some examples.

TABLE 12.1 **Biologically important molecules containing nucleotides, and their functions**

Molecule	Abbreviation	Function
Deoxyribonucleic acid	DNA	Contains the genetic information of cells
Ribonucleic acid	RNA	All three types play a vital rôle in protein synthesis
Adenosine monophosphate Adenosine diphosphate Adenosine triphosphate	AMP ADP ATP	Coenzymes important in making energy available to cells for metabolic activities, osmotic work, muscular contractions etc.
Nicotinamide adenine dinucleotide Flavine adenine dinucleotide	NAD FAD	Electron (hydrogen) carrier important in respiration in transferring hydrogen atoms from the Krebs cycle along the respiratory chain
Nicotinamide adenine dinucleotide phosphate	NADP	Electron (hydrogen) carrier important in photosynthesis for accepting electrons from the chlorophyll molecule and making them available for the photolysis of water
Coenzyme A	CoA	Coenzyme important in respiration in combining with pyruvic acid to form acetyl coenzyme A and transferring the acetyl group into the Krebs cycle

12.3.2 Ribonucleic acid (RNA)

RNA is a single-stranded polymer of nucleotides where the pentose sugar is always ribose and the organic bases are adenine, guanine, cytosine and uracil, but never thymine. Its basic structure is given in Fig. 12.3(b). There are three types of RNA found in cells all of which are involved in protein synthesis.

Ribosomal RNA (rRNA) is a large, complex molecule made up of both double and single helices. Although it is manufactured by the DNA of the nucleus, it is found in the cytoplasm where it makes up more than half the mass of the ribosomes. It comprises more than

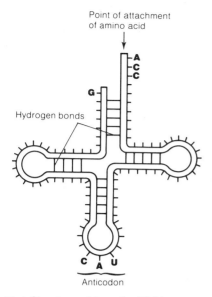

Point of attachment
of amino acid

Hydrogen bonds

Anticodon

Fig. 12.4 Structure of transfer RNA

TABLE 12.2 **Differences between RNA and DNA**

RNA	DNA
Single polynucleotide chain	Double polynucleotide chain
Smaller molecular mass (20 000–2 000 000)	Larger molecular mass (100 000–150 000 000)
May have a single or double helix	Always a double helix
Pentose sugar is ribose	Pentose sugar is deoxyribose
Organic bases present are adenine, guanine, cytosine and uracil	Organic bases present are adenine, guanine, cytosine and thymine
Ratio of adenine and uracil to cytosine and guanine varies	Ratio of adenine and thymine to cytosine and guanine is one
Manufactured in the nucleus but found throughout the cell	Found almost entirely in the nucleus
Amount varies from cell to cell (and within a cell according to metabolic activity)	Amount is constant for all cells of a species (except gametes and spores)
Chemically less stable	Chemically very stable
Maybe temporary – existing for short periods only	Permanent
Three basic forms: messenger, transfer and ribosomal RNA	Only one basic form, but with an almost infinite variety within that form

half the mass of the total RNA of a cell and its base sequence is similar in all organisms.

Transfer RNA (tRNA) is a small molecule (about eighty nucleotides) comprising a single strand. Again it is manufactured by nuclear DNA. It makes up 10–15% of the cell's RNA and all types are fundamentally similar. It forms a clover-leaf shape (Fig. 12.4), with one end of the chain ending in a cytosine–cytosine–adenine sequence. It is at this point that an amino acid attaches itself. There are at least twenty types of tRNA, each one carrying a different amino acid. At an intermediate point along the chain is an important sequence of three bases, called the **anti-codon.** These line up alongside the appropriate codon on the mRNA during protein synthesis (Section 12.6.4).

Messenger RNA (mRNA) is a long single-stranded molecule, of up to thousands of nucleotides, which is formed into a helix. Manufactured in the nucleus, it is a mirror copy of part of one strand of the DNA helix. There is hence an immense variety of types. It enters the cytoplasm where it associates with the ribosomes and acts as a template for protein synthesis (Section 12.6.4). It makes up less than 5% of the total cellular RNA. It is easily and quickly broken down, sometimes existing for only a matter of minutes.

12.3.3 Deoxyribonucleic acid (DNA)

DNA is a double-stranded polymer of nucleotides where the pentose sugar is always deoxyribose and the organic bases are adenine, guanine, cytosine and thymine, but never uracil. Each of these polynucleotide chains is extremely long and may contain many million nucleotide units.

By the early 1950s, information on DNA from a variety of sources had been collected, but no molecular structure had been agreed. The available facts about DNA included:

1. It is a very long, thin molecule made up of nucleotides.
2. It contains four organic bases: adenine, guanine, cytosine and thymine.
3. The amount of guanine is usually equal to that of cytosine.
4. The amount of adenine is usually equal to that of thymine.
5. It is probably in the form of a helix whose shape is maintained by hydrogen bonding.

Using the accumulated evidence, James Watson and Francis Crick in 1953 suggested a molecular structure which proved to be one of the greatest milestones in biology. They postulated a double helix of two nucleotide strands, each strand being linked to the other by pairs of organic bases which are themselves joined by hydrogen bonds. The pairings are always cytosine with guanine and adenine with thymine. This was not only consistent with the known ratio of the bases in the molecule, but also allowed for an identical separation of the strands throughout the molecule, a fact shown to be the case from X-ray diffraction patterns. As the purines, adenine and guanine, are double ringed structures (Fig. 12.2) they form much longer links if paired together than the two single ringed pyrimidines, cytosine and thymine. Only by pairing one purine with one pyrimidine can a consistent separation of three rings' width be achieved. In effect, the structure is like a ladder where the deoxyribose and phosphate units form the uprights and the organic base pairings form the rungs.

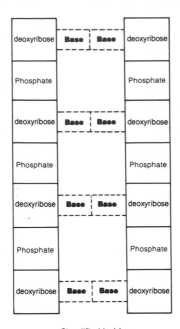

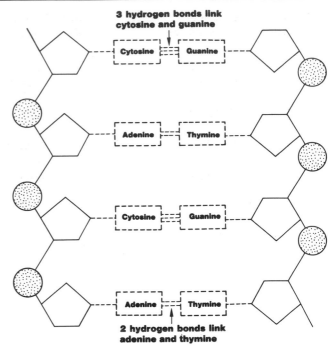

Fig. 12.5(a) Basic structure of DNA

Simplified ladder
D.N.A. structure may be likened to a ladder where alternating phosphate and deoxyribose molecules make up the 'uprights' and pairs of organic bases comprise the 'rungs'.

Molecular arrangement
Note the base pairings are always cytosine-guanine and adenine-thymine this ensures a standard 'rung' length. Note also that the 'uprights' run in the opposite directions to each other.

The uprights are composed of deoxyribose-phosphate molecules, the rungs of pairs of bases

3.4 nm
10 base pairs

2.0 nm

Fig. 12.5(b) The DNA double helix structure

However, this is no ordinary ladder; instead it is twisted into a helix so that each upright winds around the other. The two chains that form the uprights run in opposite directions, i.e. are **antiparallel.** The structure of DNA is shown in Fig. 12.5.

The structure postulated both fitted the known facts about DNA and was consistent with its biological rôle. Its extreme length (around 2.5 billion base pairs in a typical mammalian cell) permitted a very long sequence of bases which could be almost infinitely various, thus providing an immense store of genetic information. In addition its structure allowed for its replication. The separation of the two strands would result in each half attracting its complementary nucleotide to itself. The subsequent joining of these nucleotides would form two identical DNA double helices. This fitted the observation that DNA content doubles prior to cell division. Each double helix could then enter one of the daughter cells and so restore the normal quantity of DNA.

12.3.4 Differences between RNA and DNA

Despite the obvious similarities between these two nucleic acids, a number of differences exist and these are listed in Table 12.2 on the previous page.

12.4 DNA replication

The Watson-Crick model for DNA allows for a relatively simple method by which the molecule can make exact copies of itself, something which must occur if genetic information is to be transmitted from cell to cell and from generation to generation. Replication is controlled by the enzyme DNA polymerase and an illustrated description is given in Fig. 12.6.

1. *A representative portion of DNA, which is about to undergo replication, is shown.*

2. *DNA polymerase causes the two strands of the DNA to separate.*

3. *The DNA polymerase completes the splitting of the strand. Meanwhile free nucleotides are attracted to their complementary bases.*

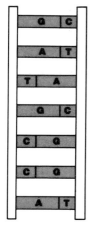
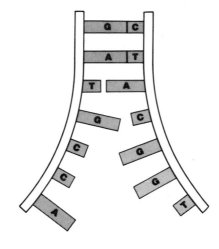
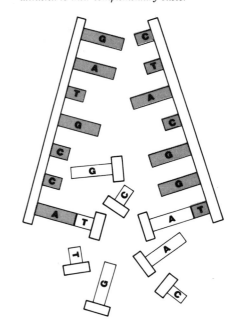

4. *Once the nucleotides are lined up they join together (bottom 3 nucleotides). The remaining unpaired bases continue to attract their complementary nucleotides.*

5. *Finally all the nucleotides are joined to form a complete polynucleotide chain. In this way two identical strands of DNA are formed. As each strand retains half of the original DNA material (shaded), this method of replication is called the semi-conservative method.*

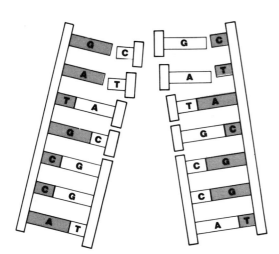
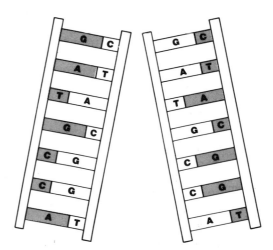

Fig. 12.6 The replication of DNA

Evidence for **semi-conservative replication** came from experiments by Meselsohn and Stahl. They grew successive generations of *Escherichia coli* in a medium where all the available nitrogen was in the form of the isotope ^{15}N (heavy nitrogen). In time, all the nitrogen in the DNA of *E. coli* was of the heavy nitrogen type. As DNA contains much nitrogen, the molecular weight of this DNA was measurably greater than that of DNA with normal nitrogen (^{14}N).

The *E. coli* containing the heavy DNA were then transferred into a medium containing normal nitrogen (^{14}N). Any new DNA produced would need to use this normal nitrogen in its manufacture. The question was, would the new DNA all be of the light type

151

DNA extracted from *E. coli* grown in a medium containing normal nitrogen (¹⁴N)

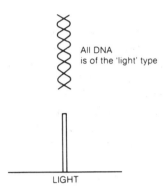

All DNA is of the 'light' type

LIGHT

DNA extracted from *E. coli* grown in a medium containing heavy nitrogen (¹⁵N) and then transferred to a medium containing normal nitrogen (¹⁴N)

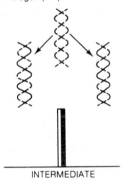

INTERMEDIATE

DNA extracted from *E. coli* grown in a medium containing heavy nitrogen (¹⁵N)

HEAVY

Relative weight of DNA as determined by centrifugation

Fig. 12.7 Interpretation of Meselsohn–Stahl experiments on semi-conservative replication of DNA

(contain only ¹⁴N) or would it, as the semi-conservative replication theory suggests, be made up of one original strand of heavy DNA and one new strand of light DNA? In the latter case its weight would be intermediate between the heavy and light types. To answer this they allowed *E. coli* to divide once and collected all the first generation cells. The DNA from them was then extracted and its relative weight determined by special techniques involving centrifugation with caesium chloride. As the results depicted in Fig. 12.7 show, the weight was indeed intermediate between the heavy and light DNA types, thus confirming the semi-conservative replication theory.

If a second generation of *E. coli* is grown from the first generation it is found to comprise half light and half intermediate weight DNA. Can you explain this?

Analysis shows that the replication of DNA takes place during interphase, shortly before cell division. Thus when the chromatids appear during prophase each has a double helix of DNA.

12.5 The genetic code

Once the structure of DNA had been elucidated and its mechanism of replication discovered, one important question remained: how exactly are the genetic instructions stored on the DNA in such a way that they can be used to mastermind the construction of new cells and organisms? Most chemicals within cells are similar regardless of the type of cell or species of organism. It is in their proteins and DNA that cells and organisms differ. It seems a reasonable starting point, therefore, to assume that the DNA in some way provides a 'code' for an organism's proteins. Moreover, most chemicals in cells are manufactured with the aid of enzymes, and all enzymes are proteins. Therefore by determining which enzymes are produced, the DNA can determine an organism's characteristics. Every species possesses different DNA and hence produces different enzymes. The DNA of different species differs not in the chemicals which it comprises, but in the sequence of base pairs along its length. This sequence must be a code that determines which proteins are manufactured.

Proteins show almost infinite variety. This variety likewise depends upon a sequence, in this instance the sequence of amino acids in the protein (Section 3.9). There are just twenty amino acids which regularly occur in proteins, and each must presumably have its own code of bases on the DNA. With only four different bases present in DNA, if each coded for a different amino acid, only four different amino acids could be coded for. Using a pair of bases, sixteen different codes are possible – still inadequate. A triplet code of bases produces sixty-four codes, more than enough to satisfy the requirements of twenty amino acids. This is called the **triplet code**.

The next problem was to determine the precise codon for each amino acid. Nirenberg devised a series of experiments towards the end of the 1950s which allowed him to break the code. He synthesized mRNA which had a triplet of bases repeated many times, e.g. GUA, GUA, GUA etc. He prepared test tubes which contained cell-free extracts of *E. coli*, i.e. they possessed all the necessary biochemical requirements for protein synthesis. Twenty tubes were set up, each with a different radioactively labelled amino acid. His synthesized mRNA was added to each tube and the presence of a polypeptide was looked for. Only in the test tube containing valine was a polypeptide found, indicating that GUA codes for valine. By repeating the process for all sixty-four possible combinations of bases, Nirenberg was able to determine which amino acid each coded for.

In some cases only the first two bases of the codon are relevant. Valine for instance is coded for by GU*, where * can be any of the four bases. Some amino acids have up to six codons. Arginine, for example, has CGU, CGC, CGA, CGG, AGA and AGG. At the other extreme, methionine, with AUG, and tryptophan, with UGG, have only one codon each. As there is more than one triplet for most amino acids it is called a **degenerate code** (a term derived from cybernetics). There are three codons UAA, UAG and UGA which are not amino acid codes. There are **stop** or **nonsense codons** and their importance is discussed in Section 12.6.4.

All the codons are **universal**, i.e. they are precisely the same for all organisms.

The code is also **non-overlapping** in that each triplet is read separately. For example, CUGAGCUAG is read as CUG–AGC–UAG and not CUG–UGA–GAG–AGC etc., where each triplet overlaps the previous one, in this case by two bases. Overlapping would allow more information to be provided by a given base sequence, but it limits flexibility. Some viruses, with limited amounts of DNA, may use overlapping codes, but this is very rare.

12.6   Protein synthesis

If the triplet code on the DNA molecule determines the sequence of amino acids in a given protein, how exactly is the information transferred from the DNA and how is the protein assembled? There are four main stages in the formation of a protein:

1. Synthesis of amino acids.

2. Transcription (formation of mRNA).

3. Amino acid activation.

4. Translation.

12.6.1 Synthesis of amino acids

In plants, the formation of amino acids occurs in mitochondria and chloroplasts in a series of stages:

(a) absorption of nitrates from the soil (Section 33.6.1);

(b) reduction of these nitrates to the amino group (NH_2);

(c) combination of these amino groups with a carbohydrate skeleton (e.g. a-ketoglutaric acid from Krebs cycle);

(d) transfer of the amino groups from one carbohydrate skeleton to another by a process called **transamination**. In this way all twenty amino acids can be formed.

Animals usually obtain their supply from the food they ingest, although they have some capacity to synthesize their own amino acids (called **non-essential amino acids**). The remaining nine – **essential amino acids** – must be provided in the diet.

12.6.2 Transcription (formation of messenger RNA)

Transcription is the process by which a complementary mRNA copy is made of the specific region (= **cistron**) of the DNA molecule which codes for a polypeptide (about 17 base pairs). A specific region of the DNA molecule, called a cistron, unwinds. This unwinding is the result of hydrogen bonds between base pairs in the DNA double helix being broken. This exposes the bases along each strand. Each base along one strand attracts its complementary RNA nucleotide, i.e. a free guanine base on the DNA will attract an RNA nucleotide with a cytosine base. It should be remembered, however, that uracil, and not thymine, is attracted to adenine (Fig. 12.8).

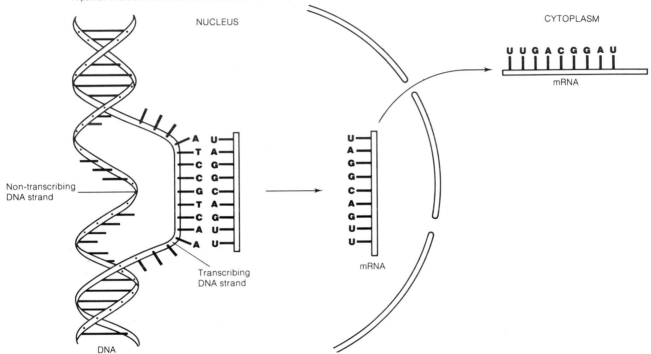

A portion of a cistron unwinds. One strand acts as a template for the formation of mRNA

Fig. 12.8 Transcription

The enzyme **RNA polymerase** moves along the DNA adding one complementary RNA nucleotide at a time to the newly unwound portion of DNA. The region of base pairing between the DNA and the RNA is only around 12 base pairs at any one time as the DNA helix reforms behind the RNA polymerase. The DNA thus acts as a **template** against which mRNA is constructed. A number of mRNA molecules may be formed before the RNA polymerase leaves the DNA, which closes up reforming its double helix. Being too large to diffuse across the nuclear membrane, the mRNA leaves instead through the nuclear pores. In the cytoplasm it is attracted to the ribosomes. Along the mRNA is a sequence of triplet codes which have been determined by the DNA. Each triplet is called a **codon**.

12.6.3 Amino acid activation

Activation is the process by which amino acids combine with tRNA using energy from ATP. Fig. 12.4 shows the structure of a tRNA molecule. Each type of tRNA binds with a specific amino acid which means there must be at least twenty types of tRNA. Each type differs, among other things, in the composition of a triplet of bases called the **anticodon.** What all tRNA molecules have in common is a free end which terminates in the triplet CCA. It is to this free end that the individual amino acids become attached, although how each specific amino acid is specified is not known (Fig. 12.9). The tRNA molecules with attached amino acids now move towards the ribosomes.

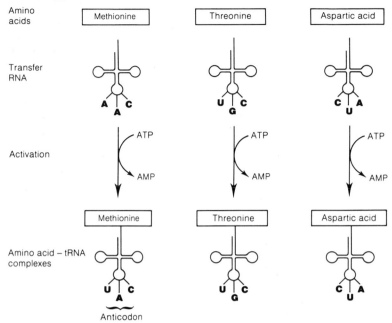

Fig. 12.9 Activation

12.6.4 Translation

Translation is the means by which a specific sequence of amino acids is formed in accordance with the codons on the mRNA. A group of ribosomes becomes attached to the mRNA to form a structure called a **polysome**. The complementary anticodon of a tRNA-amino acid complex is attracted to the first codon on the

Many ribosomes may move along the mRNA at the same time, thus forming many identical polypeptides.

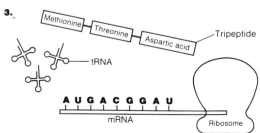

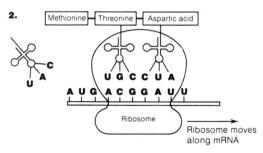

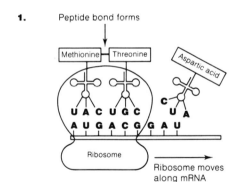

Fig. 12.10(a) Translation

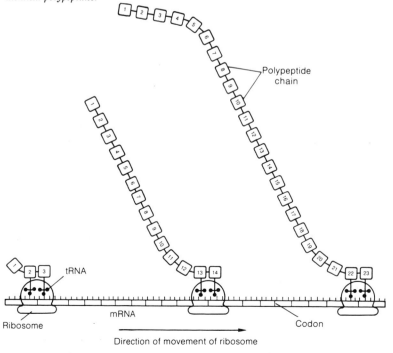

Fig. 12.10(b) Polypeptide formation

mRNA. The second codon likewise attracts its complementary anticodon. The ribosome acts as a framework which holds the mRNA and tRNA amino acid complex together until the two amino acids form a peptide bond between each other. Once they have combined, the ribosome will move along the mRNA to hold the next codon-anticodon complex together until the third amino acid is linked with the second. In this way a polypeptide chain is assembled, by the addition of one amino acid at a time. Second and subsequent ribosomes may pass along the mRNA immediately behind the first. In this way many identical polypeptides are produced simultaneously.

Once each amino acid is linked, the tRNA which carried it to the mRNA is released back into the cytoplasm. It is again free to combine with its specific amino acid. The ribosome continues along the mRNA until it reaches one of the nonsense codes (Section 12.5) at which point the polypeptide is cast off. The process of translation is summarized in Fig. 12.10, on the previous page.

The polypeptides so formed must now be assembled into proteins. This may involve the spiralling of the polypeptide to give a secondary structure, its folding to give a tertiary structure and its combination with other polypeptides and/or prosthetic groups to give a quaternary structure (see Fig. 3.13).

12.7   Gene expression and control

The part of the DNA molecule which specifies a polypeptide is termed a **cistron** by the molecular biologist. The geneticist, however, terms a similar functional unit of DNA a **gene.** Experiments have confirmed the theory that **one gene specifies one polypeptide.** Some polypeptides are required continually by a cell and these must therefore be produced continuously. Others are only required in certain circumstances and need not be produced all the time, indeed it would be needlessly wasteful to do so. How then are these genes switched on or off? Jacob and Monod studied the problem using *E. coli.* They considered two different methods of control:

1. Enzyme induction – Some genes are only switched on when the enzymes they code for are needed. *E. coli* normally respires glucose, but if grown on a medium containing lactose it will start to produce two enzymes, one enabling the lactose to be absorbed and the other to allow the lactose to be respired. The presence of the lactose in some way switches on the appropriate genes needed to produce these enzymes.

2. Enzyme repression – Some genes are normally switched on but may be switched off in special circumstances. *E. coli* manufactures its own amino acid, tryptophan, using a group of enzymes called tryptophan synthetase. If, however, tryptophan is added to the growing medium, it is absorbed by the bacterium, which consequently stops the production of tryptophan synthetase. The presence of tryptophan must switch off the appropriate gene.

Jacob and Monod put forward the concept of an **operon** – a group of adjacent genes which act together. The operon has two main sections:

1. The structural genes – These are the genes responsible for the production of the polypeptides which make up an enzyme or group of enzymes.

2. The operator gene – The gene which regulates the structural genes, in effect switching them on or off.

A third gene, the **regulator gene,** is involved. This is not part of the operon and may be some distance from it on the DNA. The regulator gene codes for a protein, called the **repressor**. This repressor can bind with the operator gene and prevent it switching on the structural genes, i.e. it represses the production of those particular enzymes.

Enzyme induction and repression can both be explained in terms of this system, differing only in their effects on the repressor. In the case of enzyme induction, the inducer combines with the repressor in such a way that it prevents it binding with the operator gene. In the absence of the repressor, the operator gene switches on the structural genes which begin mRNA production which in turn synthesizes the relevant enzymes. To take the earlier example, lactose combines with the repressor and prevents it binding with the operator gene, which thus switches on the structural genes. The result is the manufacture of the enzymes which allow the lactose to be absorbed and respired.

(a) Enzyme induction by lactose

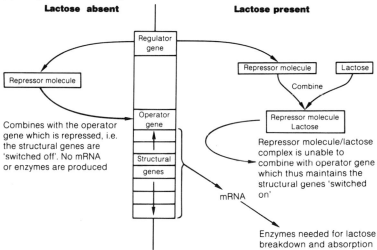

In the case of enzyme repression, the **co-repressor molecule,** as it is called, combines with the repressor. Together they bind with the operator gene and cause it to switch off the structural genes. These cease to produce mRNA and enzyme production ceases. To

(b) Enzyme repression by tryptophan

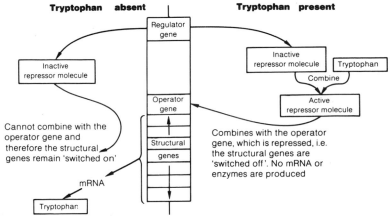

Fig. 12.11 Jacob–Monod hypothesis of control of enzyme synthesis

use the earlier example, tryptophan combines with the repressor allowing it in turn to bind with the operator gene. The operator gene, being repressed, switches off the structural genes and the production of tryptophan synthetase ceases.

This system of control helps to maintain a steady state within cells despite fluctuations in the supply of materials. As such it is an example of **cellular homeostasis** (Section 36.6).

12.8   Genetic Engineering

Perhaps the most significant scientific advance in recent years has been the development of technology which allows genes to be manipulated, altered and transferred from organism to organism — even to transform DNA itself. This has enabled us to use rapidly reproducing organisms such as bacteria as chemical factories producing useful, often life-saving, substances. The list of these substances expands almost daily and includes hormones, antibiotics, interferon and vitamins. Details of the production of some of these chemicals are given in Chapter 41.

12.8.1 Recombinant DNA technology

It has been known that a number of human diseases are the result of individuals being unable to produce for themselves chemicals which have a metabolic rôle. Many such chemicals, e.g. insulin and thyroxine, are proteins and therefore the product of a specific portion of DNA. The treatment of such deficiencies had previously been to extract the missing chemical from either an animal or human donor. This has presented problems. While the animal extracts may function effectively, subtle chemical differences in their composition have been detected by the human immune system, which has responded by producing antibodies which destroy the extract. Even chemically compatible extracts from human donors present a risk of infection from other diseases, as the transmission of the HIV virus to haemophiliacs illustrates only too well. Whether from animals or humans, the cost of such extracts are considerable.

It follows that there are advantages in producing large quantities of 'pure' chemicals from non-human sources. As a result, methods have been devised for isolating the portion of human DNA responsible for the production of insulin and combining it with bacterial DNA in such a way that the microorganism will continually produce the substance. This DNA, which results from the combination of fragments from two different organisms, is called **recombinant DNA**.

12.8.2 Techniques used to manipulate DNA

The manipulation of DNA involves three main techniques each using a specific enzyme or group of enzymes:

1. Cutting of DNA into small sections using restriction endonucleases

These enzymes are used to cut DNA between specific base sequences which the enzyme recognizes. For example, Hae III nuclease recognizes a four base-pair sequence and cuts it as shown by the arrow:

G G │ C C ⎞ four complementary base-pairs
C C │ G G ⎠ on DNA double helix

The Hind III nuclease however recognises a six base sequence, cutting it as shown by the arrow below:

A │ A G C T T ⎞ six complementary base-
T T C G A │ A ⎠ pairs on DNA double helix

As any sequence of four base-pairs is likely to occur more frequently than a six base-pair sequence, the nucleases recognizing four base-pairs cut DNA into smaller sections than those recognizing six base-pairs. The latter group are, however, more useful as the longer sections they produce are more likely to contain an intact gene.

2. Production of copies of DNA using either plasmids or reverse transcriptase

In bacterial cells there are small circular loops of DNA called **plasmids**. Plasmids are distinct from the larger circular portions of DNA which make up the bacterial chromosome. Bacteria replicate their plasmid DNA so that a single cell contains many copies. If a portion of DNA from, say, a human cell, is inserted into a plasmid and it is reintroduced into the bacterial cell, replication of the plasmid will result in up to 200 identical copies of the human DNA being made. A population of bacteria containing this human DNA can now be grown to provide a permanent source of it. By repeating the process for other DNA portions, a complete library of human DNA can be maintained. Geneticists can then select as required, any gene (a portion of DNA) they require for further investigation or use, in much the same way as a book is selected from a conventional library. Selection is, however, more complex requiring the use of DNA probes or specific antibodies. This collection of genetic information is called a **genome library**.

A second method of duplicating particular portions of DNA is appropriate where the protein for which it codes is synthesized in a specific organ. Thyroxine, for example, is produced in the thyroid gland and therefore cells from this gland would be expected to contain a relatively large amount of messenger RNA which codes for thyroxine. Reverse transcriptase (Section 7.1.2) can be used to synthesize DNA, called **copy DNA (cDNA)**, from the mRNA in thyroid cells. A large proportion of the cDNA produced is likely to code for thyroxine and it can be isolated using the techniques described in Section 12.8.3.

3. Joining together portions of DNA using DNA ligase

The recombination of pieces of DNA, e.g. the addition of cDNA into bacterial plasmid DNA, is carried out with the aid of the enzyme **DNA ligase**.

12.8.3 Gene cloning

The techniques described in the previous section are utilized in the process of gene cloning in which multiple copies of a specific gene are produced which may then be used to manufacture large quantities of valuable products.

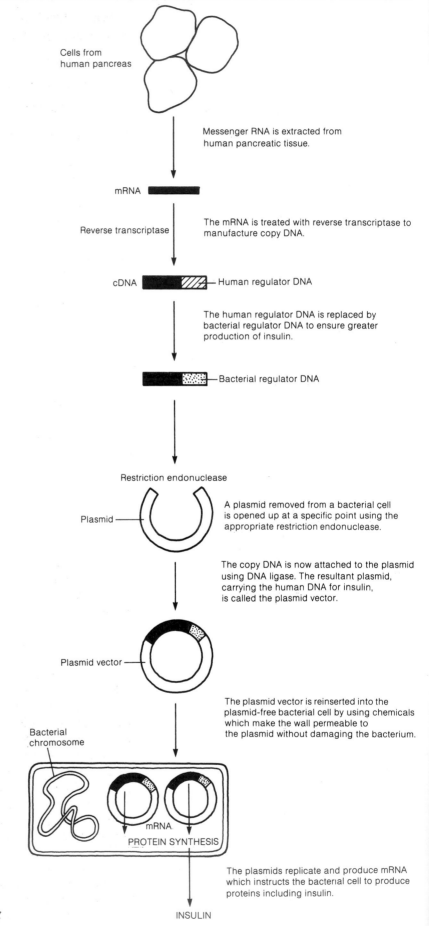

Cells from
human pancreas

Messenger RNA is extracted from
human pancreatic tissue.

mRNA

Reverse transcriptase

The mRNA is treated with reverse transcriptase to
manufacture copy DNA.

cDNA — Human regulator DNA

The human regulator DNA is replaced by
bacterial regulator DNA to ensure greater
production of insulin.

— Bacterial regulator DNA

Restriction endonuclease

Plasmid

A plasmid removed from a bacterial cell
is opened up at a specific point using the
appropriate restriction endonuclease.

The copy DNA is now attached to the plasmid
using DNA ligase. The resultant plasmid,
carrying the human DNA for insulin,
is called the plasmid vector.

Plasmid vector

The plasmid vector is reinserted into the
plasmid-free bacterial cell by using chemicals
which make the wall permeable to
the plasmid without damaging the bacterium.

Bacterial
chromosome

mRNA

PROTEIN SYNTHESIS

The plasmids replicate and produce mRNA
which instructs the bacterial cell to produce
proteins including insulin.

Fig. 12.12 Use of plasmid vector in gene cloning

INSULIN

Manufacture involves the following stages:

1. Identification of the required gene.

2. Isolation of that gene.

3. Insertion of the gene into a vector.

4. Insertion of the vector into a host cell.

5. Multiplication of the host cell.

6. Synthesis of the required product by the host cell.

7. Separation of the product from the host cell.

8. Purification of the product.

Figure 12.12 illustrates how gene cloning is used in the production of insulin. The bacteria produced in this way can be grown in industrial fermenters using a specific nutrient medium under strictly controlled conditions. The bacteria may then be collected and the insulin extracted from them by suitable means. Alternatively, it is possible to engineer bacteria which secrete the insulin and this can be extracted from the medium which is periodically drawn off. Details of these, and other, fermentation techniques are given in Chapter 41.

12.8.4 Applications of genetic engineering

The techniques illustrated above may be utilized to manufacture a range of materials which can be used to treat diseases and disorders. In addition to insulin, human growth hormone is now produced by bacteria in sufficient quantity to allow all children in this country requiring it to be treated. Among other hormones being produced in this manner are erythropoietin, which controls red blood cell production and calcitonin which regulates the levels of calcium in the blood (Section 37.4). Much research is taking place into the production of antibiotics and vaccines through recombinant DNA technology and already interferon, a chemical produced in response to viral infection, is in production.

A form of abnormal haemoglobin is produced by a defective gene causing a disease called **thalassaemia**. It is likely that genetic engineering will provide a cure for the disease, by the transference of a normal gene for haemoglobin into patients afflicted by the disease.

The scope of recombinant DNA technology is not restricted to the field of medicine. In agriculture, it is now possible to transfer genes which produce toxins with insecticidal properties from bacteria to higher plants such as potatoes and cotton. In this way these plants have 'built-in' resistance to certain insect pests. The saving in time and money by not having to regularly spray such crops with insecticides is obvious, to say nothing of avoiding killing harmless or beneficial insect species which inevitably happens however carefully spraying is carried out. It may prove possible to transfer genes from nitrogen-fixing bacteria to cereal crops, to enable them to fix their own nitrogen. There would then be less need to apply expensive nitrogen fertilizers thus reducing the pollution problems of 'run-off' (Section 29.3.2).

There would seem to be no end of possibilities – transfer of genes conferring resistance to all manner of diseases, development of plants with more efficient rates of photosynthesis, the control of weeds and the development of oil-digesting bacteria to clear up oil spillages are just some of the potential uses of recombinant DNA technology.

There are however ethical as well as practical problems to be overcome before many of these ideas can be brought to fruition. Some of these problems are discussed in the following section.

12.8.5 Implications of genetic engineering

The benefits of genetic engineering are obvious but it is not without its hazards. It is impossible to predict with complete accuracy what the ecological consequences might be of releasing genetically engineered organisms into the environment. It is always possible that the delicate balance that exists in any habitat may be irretrievably damaged by the introduction of organisms with new gene combinations. It is also possible that organisms designed for use in one environment may escape to others with harmful consequences. We know that viruses can transfer genes from one organism to another. Advantageous genes added to our domestic animals or crop plants may be transferred in this way to their competitors making them even greater potential dangers. The escape of a single pathogenic bacterium into a susceptible population could result in considerable damage to a species. Perhaps more sinister is the fear that the ability to manipulate genes could allow human characteristics and behaviour to be modified. In the wrong hands this could be used by individuals, groups or governments in order to achieve certain goals, control opposition or gain ultimate power.

Even without these dangers there are still ethical issues which arise from the development of recombinant DNA technology. Is it right to replace a 'defective' gene with a 'normal' one? Is the answer the same for a gene which causes the bearer pain, as it is where the gene has a merely cosmetic effect? Who decides what is 'defective' and what is 'normal'? A 'defective' gene may actually confer some other advantage, e.g. sickle cell gene (Section 15.4.3). Is there a danger that we shall in time reduce the variety so essential to evolution, by the progressive removal of unwanted genes or, by combining genes from different species, are we actually increasing variety and favouring evolution? Where a gene probe detects a fatal abnormality, what criteria, if any, should be applied before deciding whether to carry out an abortion?

It is inevitable that we shall remain inquisitive about the world in which we live and, in particular, about ourselves. Scientific research will therefore continue. The challenge is to develop regulations and safeguards within moral boundaries which permit genetic engineering to be used in a safe and effective way to the benefit of both individuals in particular and mankind in general.

12.9   Genetic fingerprinting

The pattern of dermal ridges and furrows which constitute the fingerprint not only persist unchanged throughout our lives, but are also unique to each one of us (identical twins excepted). For this reason they have long been used to help solve crimes by comparing the fingerprint pattern of the suspect with the impressions left, as a result of the furrow's oily secretions, at the scene of the crime. To this well-tried and successful forensic technique has now been added another – **genetic fingerprinting**.

While having nothing to do with either fingers or printing, the

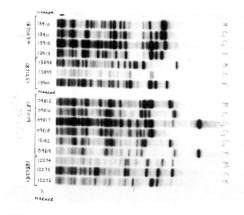

DNA fingerprint
Credit: David Parker/Science Photo Library

technique is equally, if not more, successful in identifying individuals from 'information' they provide. This 'information' is contained in a spot of blood, a sample of skin, a few sperm – in fact almost any cell of the body.

The technique, developed by Alec Jeffreys of Leicester University, takes around six days and involves the following stages:

1. The DNA is separated from the sample.

2. Restriction endonucleases are used to cut the DNA into sections.

3. The DNA fragments are separated in an agarose gel using electrophoresis.

4. The fragments are transferred to a nitrocellulose (or nylon) membrane – a process called **Southern Blotting** after its inventor, Professor Southern.

5. Radioactive DNA probes are used to bind to specific portions of the fragments known as the core sequences.

6. The portions of the DNA not bound to the radioactive probes are washed off.

7. The remaining DNA still attached to the nylon membrane is placed next to a sheet of X-ray film.

8. The radioactive probes on this DNA expose the film, revealing a pattern of light and dark bands when it is developed. The pattern makes up the genetic fingerprint.

The patterns, like fingerprints, are unique to each individual (except identical twins) and remain unchanged throughout life. Unlike fingerprints, however, the pattern is inherited from both parents. The scope for genetic fingerprinting, therefore, extends beyond catching criminals, it can also be used in paternity suits for example (Section 14.5.2). To do this, white blood cells are taken from the mother and the possible father. From the pattern of bands of the child are subtracted those bands which correspond to the mother's bands. If the man is truly the parent, he must possess all the remaining bands in the child's genetic fingerprint.

As sperm contain DNA, they too can be used to provide a genetic fingerprint, leading to a remarkably accurate method of determining guilt, or otherwise, of an accused rapist. The method has also been successfully applied in immigration cases where the relationship of an immigrant to someone already resident in a country, is in dispute. Confirming the pedigree of animals, detecting some inherited diseases and monitoring bone-marrow transplants are other applications of the technique.

Despite the fact that we are, as yet, unclear as to what exactly the dark bands of the genetic fingerprint represent, the chances of two individuals (other than identical twins) having identical patterns is so small, that the technique is widely used and its results accepted as accurate.

1. Give an illustrated account of the mechanisms by which the living cell is believed to synthesise proteins.
(20 marks)

Welsh Joint Education Committee June 1988, Paper A, No. 1

2. Describe the properties of DNA that allow self-replication to take place, together with one piece of experimental evidence that indicates how this process occurs. *(14 marks)*

In sickle-cell anaemia one of the DNA codons for glutamic acid is changed to that for valine, so the haemoglobin formed does not function in the normal way. From your knowledge of the sequence of events involved in protein synthesis explain why the mutant haemoglobin would be synthesized. *(6 marks)*

Oxford and Cambridge Board July 1984, Paper II, No. 1

3. (a) Outline the chemical structure of DNA.
(5 marks)
(b) Describe the experimental evidence which confirmed that DNA replication is semi-conservative. *(8 marks)*
(c) Discuss some of the evidence which suggests that DNA is the genetic material. *(5 marks)*
(Total 18 marks)

Cambridge Board November 1988, Paper I, No. 1

4. Indicate whether the following statements are true or false by placing a tick in the appropriate box.

	TRUE	FALSE
(a) DNA is a protein.	☐	☐
(b) DNA is a double helix.	☐	☐
(c) The molecular shape of DNA is maintained by hydrogen bonding between complementary base pairs.	☐	☐
(d) DNA contains four bases, adenine, guanine, cytosine, and uracil.	☐	☐
(e) Transcription, the transfer of information from DNA to RNA, occurs only when DNA is replicating.	☐	☐
(f) In messenger RNA (mRNA) there are codons, each consisting of three bases.	☐	☐
(g) Transfer RNA (tRNA) carries specific amino acids to ribosomes where complementary anticodons on tRNA match codons on mRNA.	☐	☐
(h) Ribosomes attach adjacent amino acids by hydrogen bonding to form polypeptide chains.	☐	☐
(i) mRNA and tRNA cannot be re-used after being released by the ribosome.	☐	☐
(j) Growth of the polypeptide chain ultimately automatically forms a helix, and this is known as a primary protein.	☐	☐

(10 marks)

Welsh Joint Education Committee June 1983, Paper AI, No. 2

5. (a) DNA (deoxyribonucleic acid) is a polynucleotide, i.e. a linear polymer of nucleotide monomers.
  (i) What are the three chemical groupings within each nucleotide monomer?
  (ii) What is the difference between the four types of nucleotide?
  (iii) Briefly describe the role of each of the three chemical groupings within the nucleotide monomers in the function of the DNA molecule. *(3 marks)*
(b) Explain the term semi-conservative replication, as applied to DNA. *(1 mark)*
(c) What is the evidence for the semi-conservative nature of DNA replication? *(4 marks)*
(d) (i) Why is it important that DNA replication should produce two exact copies of the original DNA molecule?
  (ii) What is the result of occasional mistakes in replication?
  (iii) What are the implications of such mistakes for the organism, the species and evolution? *(7 marks)*
(Total 15 marks)

Northern Ireland Board June 1983, Paper I, No. 5

6. Meselson and Stahl performed the following experiment with the bacterium *Escherichia coli.* They grew cells for several generations in the presence of the heavy isotope of nitrogen, ^{15}N. They labelled all the DNA of the progeny that were produced from the initial small inoculum with ^{15}N. The ^{15}N-labelled cells were then placed in ^{14}N-containing medium. The DNA produced in subsequent generations was extracted. The DNA from the different generations was compared using the technique of differential centrifugation using a caesium chloride density gradient to separate the DNA containing ^{15}N from DNA containing ^{14}N, as ^{15}N-labelled DNA is more dense than ordinary ^{14}N DNA. The position of the ^{15}N and ^{14}N DNA absorbance was measured in ultraviolet light at 260 nm. The results obtained are shown in Fig. 1(a).

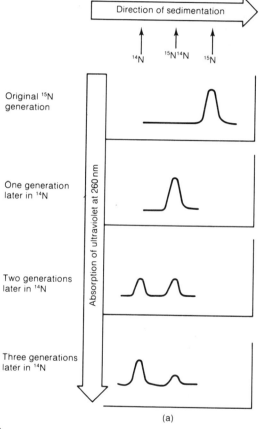

Fig. 1

(a) Do you consider that the data presented in Fig. 1(a) suggests that replication of DNA in *Escherichia coli* is conservative or semi-conservative? Give your reasons.

(b) Assume that the DNA molecule is a double coil and use a coloured pen or thicker line for ^{15}N to indicate on Fig. 1(b) the isotope labelling pattern for ^{15}N in the four generations. (Your answer must be given on Fig. 1(b), at the top of the next column.)

(Total 6 marks)

Cambridge Board November 1984, Paper II, No. 3

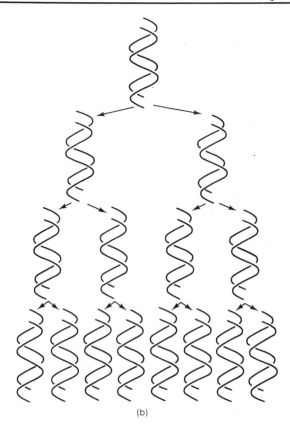

(b)

7. Read through the following account of DNA and protein synthesis and then write on the dotted lines the most appropriate word or words to complete the account.

The DNA molecule is composed of .... sugar, phosphoric acid and four types of .... base. Within this molecule the bases are arranged in pairs held together by .... bonds. For example, adenine is paired with .... Adenine and guanine are examples of a group of bases called .... The two strands of nucleotides are twisted around one another to form a double ...., and in each turn of the spiral there are .... base pairs. The DNA controls protein synthesis by the formation of a template known as .... Compared with DNA, the sugar component of this template is ...., the base .... occurs instead of .... and the molecule consists of a .... chain of nucleotides. The template is stored temporarily in the ...., before passing out into the cytoplasm through the .... It becomes associated with organelles called ...., which supply the .... required for protein synthesis. Transfer RNA molecules, each with an attached ...., are lined up on the surface of the template according to their .... of .... bases. The amino acids are joined

in a chain by .... links to form a
polypeptide molecule.

(*Total 20 marks*)

London Board June 1985, Paper I, No. 1

8. A space probe brings back some living material from a distant planet. Proteins very similar to those found on earth are present but the genetic material (DNA), although composed of a double helix of nucleotides, contains **six** different nucleic acid bases, identified as H, I, J, K, L and M. Of these, J, L and M were found to be pyrimidine bases.

Studies of the DNA from the alien material gave the following results.

Ratio of bases in double-stranded DNA	Numerical value
H/J	1.00
H+I/J+M	1.02
H+K/J+M	1.14
H+K/J+L	1.01

(*a*) Assuming that base-pairing rules similar to those discovered by Watson and Crick apply, what can be concluded from these data? (*3 marks*)

(*b*) Further work on the alien DNA revealed the sequence of bases in one particular gene which coded for a peptide composed of a total of eleven amino acids. The sequence is shown below, with the bases coding for the N-terminal amino acid on the left.

H–H–J–I–J–I–K–J–L–L–M–J–L–L–J–I–L–
L–L–L–M–J

When the peptide for which this gene codes was hydrolysed, it yielded the amino acids in the proportions shown in the table below.

Identity of amino acid	Number of amino acid residues per peptide
v	1
w	3
x	2
y	1 (at N-terminal end)
z	4

Using this information answer the following questions concerning the alien genetic coding system.

(i) How many bases are present in a codon specifying a single amino acid? Explain your answer.

(ii) Work out the actual sequence of amino acids in the peptide and write them in correct sequence in the boxes provided.

N-terminal
amino acid
↓

y										

(*6 marks*)
(*Total 9 marks*)

Joint Matriculation Board June 1985, Paper IA, No. 7

9. Read through the following passage and then answer the questions.

Genetic engineering is not new: the importance of DNA for all living processes has long been appreciated. It provides the blueprints for life, not only determining by a precise and reproducible arrangement of purine and
5 pyrimidine bases in a double helix, the heredity of the cell, be it of bacterial or human origin, but at the same time coding for every function of the cell including the deployment of enzyme systems which will allow the DNA double-helix to repair itself if it becomes damaged.
10 As early as 1967 an enzyme, DNA-ligase, had been discovered which joined the breaks in DNA chains. In 1972 some workers were making artificial genes – chains of DNA constructed entirely in the test-tube and suitably, and specifically, joined together. The phenomenon of
15 'restriction' had been known earlier. Although certain small viruses could invade and multiply in certain bacteria, others could not. This 'foreign' DNA was, in fact, cut into pieces by bacterial enzymes called restriction endonucleases, and so rendered non-infective. A score or
20 so of different bacterial restriction enzymes have now been identified. DNA strands cannot only be chopped up into smaller pieces but methods are available for joining the ends together again, either to each other or to strange DNA. If the 'new' DNA was to be produced in quantity,
25 then suitable DNA vehicles, or carriers, were required.

Adapted from *Commonsense in Genetic Engineering* by
R. Harris in *Biologist* (1977)

(*a*) What is meant by 'a precise and reproducible arrangement of purine and pyrimidine bases' (lines 4–5)? (*4 marks*)

(*b*) Indicate how DNA is involved in 'coding for every function of the cell' (line 7). (*5 marks*)

(*c*) Suggest how enzymes such as 'DNA-ligase' (line 10) and 'restriction endonucleases' (line 19) could be used in genetic engineering.
(*3 marks*)

(*d*) Describe *in outline* how 'viruses could invade and multiply in' cells (line 16). (*4 marks*)

(e) Suggest *one* potential danger and *one* potential benefit of genetic engineering to humans.

(*4 marks*)

London Board June 1988, Paper I, No. 13

10. Read through the following account of nucleotide structure and then replace the dotted lines with the most appropriate word or words to complete the account.

Nucleotides are organic compounds containing the elements carbon, hydrogen, oxygen, ... and .... A molecule of the mononucleotide ATP includes the organic base .... and a sugar called .... which has .... carbon atoms. The organic base in ATP is a double-ringed molecule of the kind called a .... RNA and DNA both contain a nucleotide which includes the organic base guanine, but the nucleotide in RNA contains one more atom of .... In both RNA and DNA the nucleotide containing guanine pairs with the nucleotide containing .... by means of .... bonding.

(*Total 9 marks*)

London Board January 1989, Paper I, No. 2

13 *Cell division*

Modern cell theory, as described in Chapter 4, states that 'all new cells are derived from other cells'. The process involved is **cell division**. All 10^{14} cells which comprise a human are derived, through cell division, from the single zygote formed by the fusion of two gametes. These gametes in turn were derived from the division of certain parental cells. It follows that all cells in all organisms have been formed from successive divisions of some original ancestral cell. The remarkable thing is that, while cells and organisms have diversified considerably over millions of years, the process of cell division has remained much the same.

There are two basic types:

Mitosis which results in all daughter cells having the same number of chromosomes as the parent.

Meiosis which results in the daughter cells having only half the number of chromosomes found in the parent cell.

13.1 Chromosomes

13.1.1 Chromosome structure

Chromosomes carry the hereditary material DNA (15%). In addition they are made up of protein (70%) and RNA (10%). Individual chromosomes are not visible in a non-dividing (resting) cell, but the chromosomal material can be seen, especially if stained. This material is called **chromatin**. It is only at the onset of cell division that individual chromosomes become visible. They appear as long, thin threads between 0.25 μm and 50 μm in length. Each chromosome is seen to consist of two threads called **chromatids** joined at a point called the **centromere** (Fig. 13.1) Chromosomes vary in shape and size, both within and between species.

13.1.2 Chromosome number

The number of chromosomes varies from one species to another but is always the same for normal individuals of one species. Table 13.1 gives some idea of the range of chromosome number in different species. It can be seen that the numbers are not related to either the size of the organism or to its evolutionary status; indeed it is quite without significance.

Although the chromosome number of a cell varies from two to 300 or more, the majority of organisms have between ten and forty chromosomes in each of their cells. With well over one million different species, it follows that many share the same chromosome number, twenty-four being the most common.

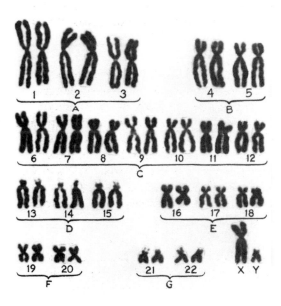

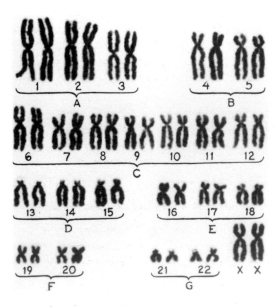

Human karyotypes (male and female)

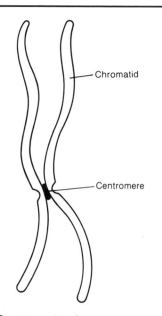

Fig. 13.1 Structure of a chromosome

TABLE 13.1 **The chromosome number of a range of species**

Species	Chromosome number
Certain roundworms	2
Crocus (*Crocus balansae*)	6
Fruit fly (*Drosophila melanogaster*)	8
Onion (*Allium cepa*)	16
Maize (*Zea mays*)	20
Locust (*Locusta migratoria*)	24
Lily (*Lilium longiflorum*)	24
Tomato (*Solanum lycopersicum*)	24
Mouse (*Mus musculus*)	40
Human (*Homo sapiens*)	46
Potato (*Solanum tuberosum*)	48
Certain Protozoa	300 +

Nuclear division, or mitosis, typically occupies 5–10% of the total cycle. The cycle may take as little as 20 minutes in a bacterial cell, although it typically takes 8–24 hours.

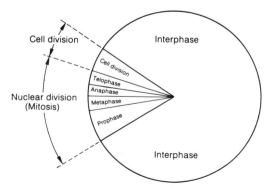

Fig. 13.2 The cell cycle

13.2 Mitosis

Dividing cells undergo a regular pattern of events, known as the **cell cycle**. This cycle may be divided into two basic parts –

Interphase – when the nucleus is mechanically inactive although chemically very active.

Mitosis – when the nucleus is mechanically active as it divides.

In single-celled organisms and actively dividing cells, the cycle is continuous, i.e. the cells continue to divide regularly. In some cells, like those of the liver, division ceases after a certain time and only resumes if damaged or lost tissue needs replacing. In specialized tissues, such as nerves, division ceases completely once the cells are mature.

Interphase
Although often termed the **resting phase** because the chromosomes are not visible, interphase is in fact a period of considerable metabolic activity. It is during this phase that the DNA content of the cell is doubled. Duplication of the cell organelles also takes place at this time.

Prophase
The chromosomes become visible as long, thin tangled threads. Gradually they shorten and thicken, and each is seen to comprise two chromatids joined at the centromere. With the exception of higher plant cells which lack them, the centrioles migrate to opposite ends or **poles** of the cell. From each centriole, microtubules develop and form a star-shaped structure called an **aster**. Some of these microtubules, called **spindle fibres**, span the cell from pole to pole. Collectively they form the **spindle**. The nucleolus disappears and finally the nuclear envelope disintegrates, leaving the chromosomes within the cytoplasm of the cell.

Metaphase
The chromosomes arrange themselves at the centre or **equator** of the spindle, and become attached to certain spindle fibres at the centromere. Contraction of these fibres draws the individual chromatids slightly apart.

Anaphase
Further shortening of the spindle fibres causes the two chromatids of each chromosome to separate and migrate to opposite poles. The energy for this contraction is provided by mitochondria which are observed to collect around the spindle fibres.

Telophase
The chromatids reach their respective poles and a new nuclear envelope forms around each group. The chromatids uncoil and lengthen, thus becoming invisible again. The spindle fibres disintegrate and a nucleolus reforms in each new nucleus.

13.2.1 Differences between mitosis in plant and animal cells

The cells of higher plants lack centrioles and do not form asters as in animal cells. Spindle formation still takes place, however, and therefore the centrioles would not seem to be the centre of spindle synthesis.

In animal cells, cell division or **cytokinesis** occurs by the constriction of the centre of the parent cell from the outside inwards

(a) Interphase

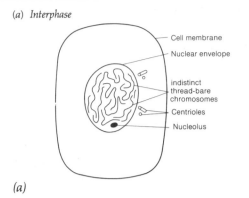

- Cell membrane
- Nuclear envelope
- indistinct thread-bare chromosomes
- Centrioles
- Nucleolus

(a)

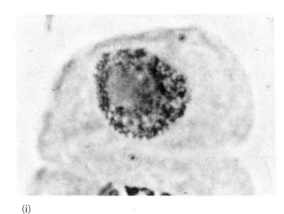

(i)

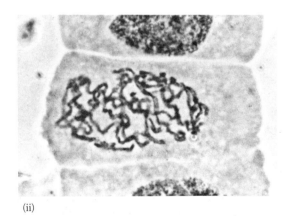

(ii)

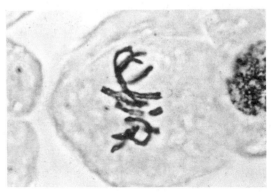

(iii)

(Fig. 13.3h). In plant cells, however, the process occurs by the growth of a **cell plate** across the equator of the parent cell from the centre outwards. The plate is formed from the fusion of vesicles produced by the dictyosome. Cellulose is laid down on this plate to form the cell wall.

Whereas most animal cells are, if the need arises, capable of mitosis, only a specialized group of plant cells, called **meristematic cells**, are able to do so.

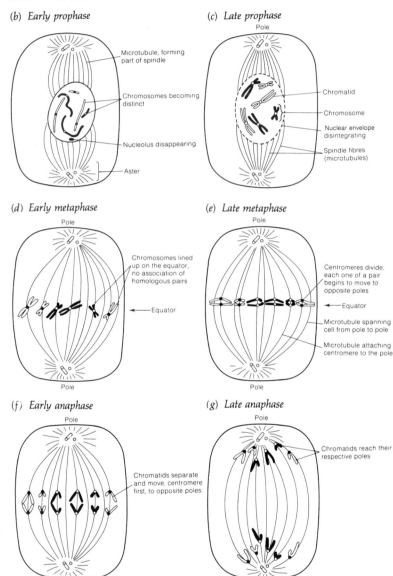

(b) Early prophase

- Microtubule, forming part of spindle
- Chromosomes becoming distinct
- Nucleolus disappearing
- Aster

(c) Late prophase

Pole

- Chromatid
- Chromosome
- Nuclear envelope disintegrating
- Spindle fibres (microtubules)

(d) Early metaphase

Pole

- Chromosomes lined up on the equator, no association of homologous pairs
- Equator

(e) Late metaphase

Pole

- Centromeres divide, each one of a pair begins to move to opposite poles
- Equator
- Microtubule spanning cell from pole to pole
- Microtubule attaching centromere to the pole

Pole

(f) Early anaphase

Pole

- Chromatids separate and move, centromere first, to opposite poles

Pole

(g) Late anaphase

Pole

- Chromatids reach their respective poles

Pole

(h) Early telophase

- Constriction begins to separate cells in animals only

(i) Late telophase

- Nuclear envelope reforms
- Cell divides by constriction
- Chromosomes become indistinct
- Centrioles divide

Fig. 13.3 Stages of mitosis

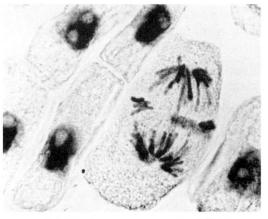

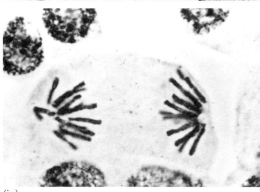

(iv)

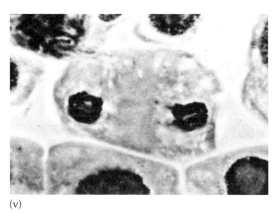

(v)

The main stages of mitosis (× 500 approx.): (i) interphase (ii) prophase-chromosomes become visible (iii) metaphase – chromosomes line up on the equator (iv) anaphase – chromatids migrate to opposite poles (v) telophase – daughter nuclei form at opposite poles

13.2.2 Experiment to demonstrate mitosis in root tips

The growing apical meristems of roots provide an abundant source of cells in various stages of mitotic division.

Requirements

Apparatus	*Chemicals*
Microscope	2 cm³ acetic orcein stain
Spirit lamp or Bunsen burner	1 cm³ molar hydrochloric acid
Watch glass	
Microscope slides	*Living material*
Coverslips	Onion or broad bean root tips
Dropping pipette	
Filter (or blotting) paper	
Small paintbrush	
Scalpel	

Precautions

Molar hydrochloric acid is caustic and causes burns. If any reaches the skin the affected part should be washed thoroughly with water.

Method

1. Pipette 10 drops of acetic orcein stain into a watch glass and add one drop of molar hydrochloric acid.

2. Cut the terminal 1 cm off two onion roots and place the tips in the watch glass.

3. Warm the watch glass *gently* over a spirit burner or very small Bunsen flame. *Do not allow the liquid to boil.* Maintain the temperature for five minutes.

4. Using a small paintbrush, place the root tips on a clean microscope slide. Cut off the 2 mm nearest the tip with a scalpel and discard the rest.

5. Add 2 drops of acetic orcein stain and gently tease apart the cells, keeping them in the same relative positions as far as possible.

6. Place a coverslip over the preparation and cover with several layers of filter (blotting) paper.

7. Apply a little downward pressure to the coverslip with a finger, taking care not to allow any lateral movement.

8. Warm the preparation over a flame for a few seconds and examine it under a microscope, using an oil immersion lens if possible.

9. Make fully labelled drawings of each stage of mitosis.

10. Scan the slide and record the number of cells which exhibit each stage of mitosis. How can these results be used to determine the relative length of each stage in the process of mitosis?

13.3 Meiosis (reduction division)

Meiosis involves one division of the chromosomes followed by two divisions of the nucleus and cell. The result is that the number of chromosomes in each cell is reduced by half. The **diploid (2n)** parent cell gives rise to **four haploid (n)** daughter cells. Meiosis occurs in the formation of gametes, sperm and ova, in animals, and in the production of spores in most plants.

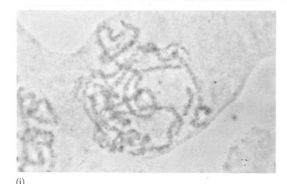

(i)

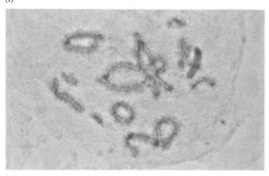

(ii)

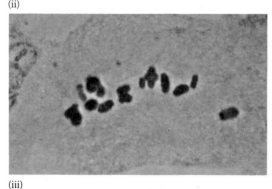

(iii)

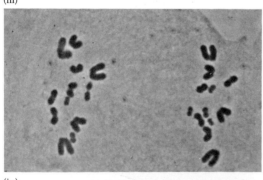

(iv)

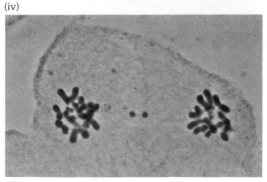

(v)

Meiosis comprises two divisions –

1. **First meiotic division** – similar to mitosis except for a highly modified prophase stage.

2. **Second meiotic division** – a typically mitotic division.

The process is continuous but for convenience is divided into the same stages as mitosis. The symbols I and II indicate the first and second meiotic divisions respectively.

Prophase I

Organisms have two sets of chromosomes, one derived from each parent. Any two chromosomes which determine the same characteristics, e.g. eye colour, blood groups etc., are called an **homologous pair**. Although each chromosome of a pair determines the same characteristics, they need not be identical. For instance, while one of the pair may code for blue eyes, the other may code for brown eyes.

Prophase I of meiosis is similar to prophase in mitosis, in that the chromosomes become visible, shorten and fatten, but differs in that they associate in their homologous pairs. They come together by a process termed **synapsis** and each pair is called a **bivalent**.

Each chromosome of the pair is seen to comprise two chromatids. These chromatids wrap around each other. The chromatids of the pair partially repel one another although they remain joined at certain points called **chiasmata** (singular – **chiasma**). It is at these points that chromatids may break and recombine with a different chromatid. This swapping of portions of chromatids is termed **crossing over**. The chromatids continue to repel one another although at this stage they still remain attached at the chiasmata. The nucleolus disappears and the nuclear envelope breaks down. Where present, the centrioles migrate to the poles and the spindle forms.

Metaphase I

The bivalents arrange themselves on the equator of the cell with each of a pair of homologous chromosomes orientated to opposite poles. This arrangement is completely random relative to the orientation of other bivalents. The genetic significance of this will be discussed later. The spindle fibres attached to the centromeres contract slightly, pulling the chromosomes apart as much as the chiasmata allow.

Anaphase I

The spindle fibres, which are attached to the centromeres, contract and pull the homologous chromosomes apart. One of each pair is pulled to one pole, its sister chromosome to the opposite one.

Telophase I

The chromosomes reach their opposite poles and a nuclear envelope forms around each group. In most cells the spindle fibres disappear and the chromatids uncoil. Cell division, or **cleavage**, may follow. The nucleus may enter interphase although no replication of the DNA takes place. In some cells this stage does not occur and the cell passes from anaphase I directly into prophase II.

Prophase II

In those cells where telophase and interphase take place, the nucleolus disappears and the nuclear envelope breaks down. Where centrioles are present these divide and move to opposite poles. The poles on this occasion are at right angles to the plane of the previous cell division and therefore the spindle fibres develop at right angles to the spindle axis of the first meiotic division.

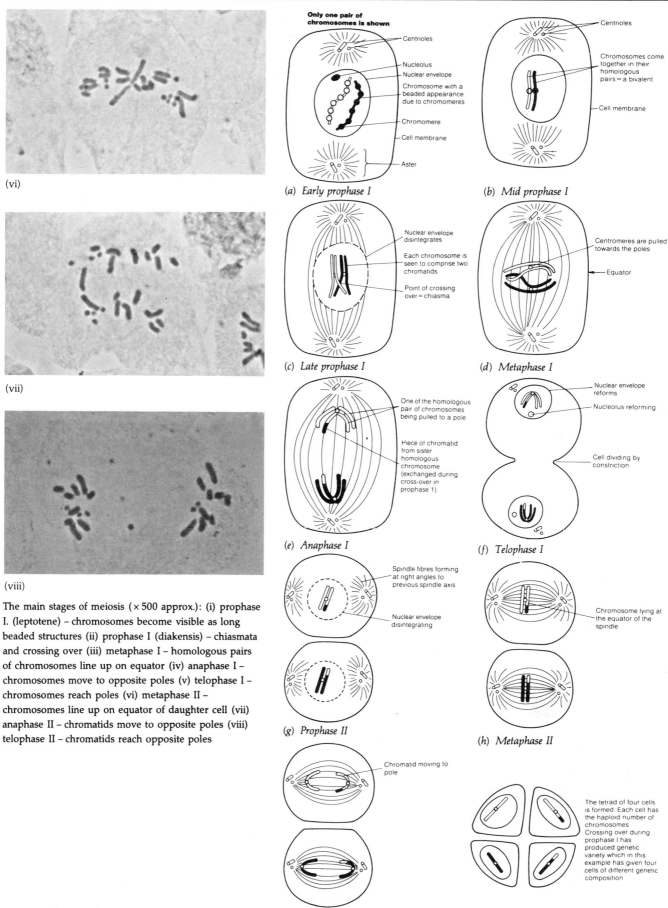

(vi)

(vii)

(viii)

The main stages of meiosis (× 500 approx.): (i) prophase I. (leptotene) – chromosomes become visible as long beaded structures (ii) prophase I (diakensis) – chiasmata and crossing over (iii) metaphase I – homologous pairs of chromosomes line up on equator (iv) anaphase I – chromosomes move to opposite poles (v) telophase I – chromosomes reach poles (vi) metaphase II – chromosomes line up on equator of daughter cell (vii) anaphase II – chromatids move to opposite poles (viii) telophase II – chromatids reach opposite poles

Only one pair of chromosomes is shown

Centrioles
Nucleolus
Nuclear envelope
Chromosome with a beaded appearance due to chromomeres
Chromomere
Cell membrane
Aster

(a) *Early prophase I*

Centrioles
Chromosomes come together in their homologous pairs = a bivalent
Cell membrane

(b) *Mid prophase I*

Nuclear envelope disintegrates
Each chromosome is seen to comprise two chromatids
Point of crossing over = chiasma

(c) *Late prophase I*

Centromeres are pulled towards the poles
Equator

(d) *Metaphase I*

One of the homologous pair of chromosomes being pulled to a pole
Piece of chromatid from sister homologous chromosome (exchanged during cross-over in prophase 1)

(e) *Anaphase I*

Nuclear envelope reforms
Nucleolus reforming
Cell dividing by constriction

(f) *Telophase I*

Spindle fibres forming at right angles to previous spindle axis
Nuclear envelope disintegrating

(g) *Prophase II*

Chromosome lying at the equator of the spindle

(h) *Metaphase II*

Chromatid moving to pole

(i) *Anaphase II*

The tetrad of four cells is formed. Each cell has the haploid number of chromosomes. Crossing over during prophase I has produced genetic variety which in this example has given four cells of different genetic composition

(j) *Telophase II*

Fig. 13.4 Stages of meiosis

Metaphase II
The chromosomes arrange themselves on the equator of the new spindle. The spindle fibres attach to the centromere of each chromosome.

Anaphase II
The centromeres divide and are pulled by the spindle fibres to opposite poles, carrying the chromatids with them.

Telophase II
Upon reaching their opposite poles, the chromatids unwind and become indistinct. The nuclear envelope and the nucleolus are reformed. The spindle disappears and the cells divide to give four cells, collectively called a **tetrad**.

13.4 The significance of cell division

13.4.1 Significance of mitosis

The significance of mitosis is its ability to produce daughter cells which are exact copies of the parental cell. It is important in three ways.

(a) Growth
If a tissue is to extend by growth it is important that the new cells are identical to the existing cells. Cell division must therefore be by mitosis.

(b) Repair
Damaged cells must be replaced by exact copies of the originals if the repair is to return a tissue to its former condition. Mitosis is the means by which this is achieved.

(c) Asexual reproduction
If a species is successful in colonizing a particular habitat, there is little advantage, in the short term, in producing offspring which differ from the parents, because these may be less successful. It is better to quickly establish a colony of individuals which are similar to the parents. In simple animals and most plants this is achieved by mitotic divisions.

13.4.2 Significance of meiosis

The long-term survival of a species depends on its ability to adapt to a constantly changing environment. It should also be able to colonize a range of new environments. To achieve both these aims it is necessary for offspring to be different from their parents as well as different from each other.

There are three ways in which this variety is brought about with the aid of meiosis.

(a) Production and fusion of haploid gametes
Variety of offspring is increased by mixing the genotype of one parent with that of the other. This is the basis of the sexual process in organisms. It involves the production of special sex cells, called gametes, which fuse together to produce a new organism. Each gamete must contain half the number of chromosomes of the adult if the chromosome number is not to double at each generation. It is

therefore essential that meiosis, which halves the number of chromosomes in daughter cells, occurs at some stage in the life cycle of a sexually reproducing organism. Meiosis is thus instrumental in permitting variety in organisms, and giving them the potential to evolve.

(b) The creation of genetic variety by the random distribution of chromosomes during metaphase I

When the pairs of homologous chromosomes arrange themselves on the equator of the spindle during metaphase I of meiosis, they do so randomly. Although each one of the pair determines the same general features, they differ in the detail of these features. The random distribution and consequent independent assortment of these chromosomes produces new genetic combinations. A simple example is shown in Fig. 13.5.

(c) The creation of genetic variety by crossing over between homologous chromosomes

During prophase I of meiosis, equivalent portions of homologous chromosomes may be exchanged. In this way new genetic combinations are produced and linked genes separated.

The variety which meiosis brings about is essential to the process of evolution. By providing a varied stock of individuals it permits the natural selection of those best suited to the existing conditions and so ensures that species constantly change and adapt when these conditions alter. This is the main significance of meiosis.

In arrangement 1, the two pairs of homologous chromosomes orientate themselves on the equator in such a way that the chromosome carrying the allele for brown eyes and the one carrying the allele for blood group A migrate to the same pole. The alleles for blue eyes and blood group B migrate to the opposite pole. Cell 1 therefore carries the alleles for brown eyes and blood group A while cell 2 carries the ones for blue eyes and blood group B.

In arrangement 2, the left hand homologous pair of chromosomes is shown orientated the opposite way around. As this orientation is random this arrangement is equally as likely as the first one. The result of this different arrangement is that cell 3 carries the alleles for blue eyes and blood group A whereas cell 4 carries ones for brown eyes and blood group B.

All four resultant cells are different from one another. With more homologous pairs the number of possible combinations becomes enormous. A human, with 23 such pairs, has the potential for $2^{23} = 8\,388\,608$ combinations.

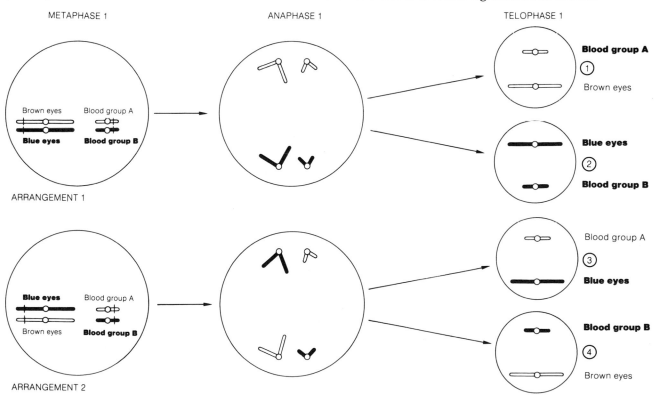

Fig. 13.5 Variety brought about by meiosis

13.5 Comparison of mitosis and meiosis

The process of nuclear division is basically the same in mitosis and meiosis. The appearance and behaviour of the chromosomes is similar, but there are nevertheless some differences. These are listed in Table 13.2.

TABLE 13.2 **Differences between mitosis and meiosis**

Mitosis	Meiosis
A single division of the chromosomes and the nucleus	A single division of the chromosomes but a double division of the nucleus
The number of chromosomes remains the same	The number of chromosomes is halved
Homologous chromosomes do not associate	Homologous chromosomes associate to form bivalents in prophase I
Chiasmata are never formed	Chiasmata may be formed
Crossing over never occurs	Crossing over may occur
Daughter cells are identical to parent cells (in the absence of mutations)	Daughter cells are genetically different from parental ones
Two daughter cells are formed	Four daughter cells are formed, although in females only one is usually functional
Chromosomes shorten and thicken	Chromosomes coil but remain longer than in mitosis
Chromosomes form a single row at the equator of the spindle	Chromosomes form a double row at the equator of the spindle during metaphase I
Chromatids move to opposite poles	Chromosomes move to opposite poles during the first meiotic division

13.6 Differences between nuclear division and cell division

The terms 'nuclear division' and 'cell division' are frequently confused. Nuclear division is the division of the nucleus alone. Mitosis and meiosis are two types of nuclear division. Cell division, however, includes nuclear division *and* many processes which follow it, e.g. replication of cell organelles and the division of the whole cell into two. Whereas cell division is preceded by nuclear division, the reverse is not always the case. A nucleus may divide without subsequent division of the cell; it is in this way that multinucleate cells are produced. Whereas nuclear division is similar in plants and animals, cell division is very different. Plant cells divide by the formation of a cell plate, whereas animal cells do so by constriction. These and other differences are summarized in Table 13.3.

TABLE 13.3 **Differences between nuclear and cell division**

Nuclear division	Cell division
Similar in plants and animals	Different in plants and animals
Involves duplication of chromosomes	Involves duplication of cell organelles
Daughter nuclei may be similar (mitosis) or dissimilar (meiosis)	Daughter cells are always similar
Spindle formation occurs	No spindle formation
Often, but not always, followed by cell division	Always preceded by nuclear division

13.7 Cancer – a breakdown of control of cell division

Cancer is the name given to a group of disorders in which certain cells no longer respond to the normal controls which determine the extent of growth of a tissue. In the absence of these controls the tissue grows, invading and crowding out other tissues, sometimes destroying them. There would appear to be some breakdown in the normal mechanism which controls when a cell divides by mitosis. The cause of this breakdown is not definitely known, but it seems likely that there may be more than one explanation. It could be a mutation. The fact that certain chemicals and forms of radiation are known to cause cancer would tend to support this view. Another possible cause, for certain cancers at least, is a viral infection, although, as cancer is not infectious, its effects are obviously complex. (See Section 7.1.2.)

1. Where does the process of mitosis occur and what is its significance? *(5 marks)*
With clear annotated diagrams explain the stages of mitosis in animals. *(20 marks)*
(Total 25 marks)

Southern Universities Joint Board June 1986, Paper II, No. 2

2. (a) (i) With the aid of diagrams, describe mitosis in an animal cell where the diploid number of chromosomes is 4.
(ii) What is the significance of mitotic divisions?
(iii) List the ways in which mitosis differs from meiosis. *(10 marks)*
(b) Compare the sites of cell division in mammals and flowering plants, explaining clearly the effect of any differences upon growth and reproduction. *(5 marks)*
(c) (i) What advantage is it for a species to have a constant number of chromosomes?
(ii) How do mammalian sex chromosomes differ in form and function from each other and from other chromosomes? *(5 marks)*
(Total 20 marks)

Joint Matriculation Board June 1983, Paper IIB, No. 2

3. Show, by placing a tick in the appropriate column, which of the following are produced as a direct result of meiosis and those which are not produced as a direct result of meiosis.

	Produced as a direct result of meiosis	Not produced as a direct result of meiosis
Pollen grains		
Antherozoids of a fern		
Antheridia of a moss		
Obelia medusae		
Pollen tube		
Secondary oocyte of a mammal		
'Seed' potatoes (as bought in order to raise next year's crop)		
Egg cell and synergidae of a flowering plant		
Moss spores		
Blood cells		

(Total 10 marks)
London Board June 1983, Paper I, No. 10

4. A successful return flight from another planet brought back some 'organisms' for investigation. Chromosome-like structures were identified in the cells and counted. The table below gives the results of these counts.

'Organisms'	Cells investigated	'Chromosome' number
Stunted tood	All cells	5
Bristly granet	Most cells	24
	A few cells	12
Slimy munger	1 specimen all cells	8
	1 specimen all cells	10
	1 specimen all cells	7
Flighted bobbler	6 specimens all cells	24
	5 specimens all cells	23

Decide for each whether these could be identified as organisms similar to ones found on earth, giving your reasons for each decision. *(Total 12 marks)*

London Board June 1982, Paper I, No. 12

5. The diagram (on next page) shows the chromosomal constitution of a human individual as seen during one of the stages of mitosis in a lymphocyte. In normal humans $2n = 46$. The chromosomal constitution shown here has several abnormalities.

The preparation of such material is done as follows: *Heparin* (or citrate) is added to a sample of blood which is then incubated for 48 hours. A *spindle inhibitor* is added, and several hours later the cells are centrifuged out. A squash preparation of the cells is made and stained. The chromosomes show a characteristic banding on examination under a microscope.

(a) (i) Suggest briefly the reason for using each of the two substances in italics in the description given.
(ii) Why does each chromosome appear as a double structure? *(3 marks)*
(b) This section concerns the six largest chromosome pairs, numbered 1 to 6 on the diagram.
(i) One of the number 6 homologues has had an addition to it from another of the large chromosomes. **Draw a ring** round the chromosome from which the addition came.
(ii) Another chromosome has become a dicentric (with two centromeres) by addition. **Draw**

boxes round the added portion and the fragment which remains. (*3 marks*)

(c) This section concerns the pairs of medium-sized chromosomes, numbers 10 to 12. Instead of a number, these are shown on the diagram by a circle nearby. Distinguish the homologous pairs by **writing letters** in the circles. Write *A* in each of the circles of one pair of chromosomes. Similarly, use *B* for a second pair and *C* for the third pair. (*2 marks*)

(d) (i) Four of the small chromosomes have been given boxes but no identification. Identify each chromosome by writing in the box shown near to it.

(ii) On the basis of these four chromosomes, what would be the sex and condition of this individual? (*4 marks*)

(*Total 12 marks*)

Joint Matriculation Board June 1982, Paper IA, No. 7

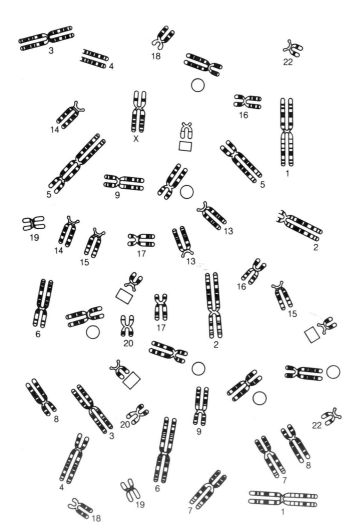

6. The simplified diagram below shows a nucleus from a cell of an insect.

(a) Draw diagrams to show the two nuclei which would be formed if the cell divided by mitosis. (*1 mark*)

(b) Draw diagrams to show **three** different nuclei which could be formed if the cell divided by meiosis. (Assume that no crossing over has taken place.)

(c) In an actively dividing cell the process of mitosis lasts, on average, for an hour and is repeated every 12 hours.
   (i) Name the period in the cell cycle between two successive mitotic divisions.
   (ii) What percentage of the cell cycle is occupied by this period?
   (iii) State **three** characteristic events which occur during this period. (*6 marks*)

(d) Certain simple organisms are known to be haploid in the adult stage of their life cycles.
   (i) Name the type of cell division which will occur during gamete formation in such organisms.
   (ii) State and explain the chromosome number in the zygotes of these organisms.
   (iii) Name the type of cell division likely to immediately follow zygote formation and comment on its significance. (*5 marks*)

(*Total 12 marks*)

Welsh Joint Education Committee June 1984, Paper AI, No. 3

7. Although mitosis is a continuous process, it is conventionally divided into the following sequence of stages:

> Interphase;
> Prophase;
> Metaphase;
> Anaphase;
> Telophase.

(a) Name the stage of mitosis during which:
   (i) the chromosomes align on the equator of the spindle;
   (ii) the chromosomes become visible and the spindle apparatus forms;
   (iii) the nuclear membrane reforms and cytoplasmic cleavage takes place.

(*3 marks*)

(b) If the amount of DNA present in the cell at metaphase is 10 units, how much DNA will be present in each nucleus:
(i) at the start of prophase;
(ii) immediately following telophase? (*2 marks*)
(*Total 5 marks*)

Associated Examining Board June 1989, Paper I, No. 15

8. (a) What is a chromosome? (*3 marks*)
(b) Why are chromosomes visible only in actively dividing cells? (*2 marks*)
(c) What is the significance of mitosis in a meristematic cell? (*2 marks*)
(d) Describe the function of a centromere in mitosis. (*2 marks*)
(*Total 9 marks*)

London Board January 1989, Paper I, No. 7

9. The figure shows parts of two cells from the same flowering plant, in the process of division.

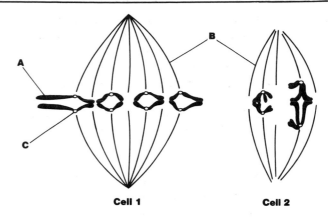

Cell 1   **Cell 2**

(a) Name the structures labelled **A, B** and **C**.
(b) Name the type of nuclear division and the stage of division in each cell as precisely as you can.
(c) Name **two** types of cell in a flowering plant which you would expect to find undergoing division as shown in **Cell 2**.
(d) Explain the functions of the two types of nuclear division occurring in **Cells 1** and **2**.
(*Total 8 marks*)

Cambridge Board June 1988, Paper III, No. 1

14 *Heredity and genetics*

Gregor Mendel

It has been estimated that all the ova from which the present human population was derived could be contained within a five-litre vessel. All the sperm which fertilized these ova could be contained within a thimble. Put another way, all the information necessary to produce in excess of 5 000 000 000 different humans can be stored within this relatively tiny volume. Indeed, we now know that the genetic information itself occupies only a small proportion of the ovum. Imagine the space occupied if all the characteristics of every living human being were printed in book form. Whatever else it may be, the genetic information is remarkably condensed. In Chapter 12 we described the chemical nature of this genetic information, namely DNA. In this chapter we shall concern ourselves with the means by which it is transmitted from generation to generation.

Genetics is a fundamental and increasingly important branch of biology. Man has unknowingly used genetic principles in the breeding of his animals and plants for many thousands of years. An understanding of the underlying genetic principles is, however, fairly recent. In fact the term genetics was first used at the beginning of this century. The understanding of the chemical foundations of heredity and genetics came even more recently with the discovery of the structure of DNA by Watson and Crick in 1953.

Observation of individuals of the same species shows them all to be recognizably similar. This is **heredity**. Closer inspection reveals minor differences by which each individual can be distinguished. This is **variation**.

The genetic composition of an organism is called the **genotype**. This often sets limits within which individual characteristics may vary. Such variation may be due to the effect of environmental influences, e.g. the genotype may determine a light-coloured skin, but the precise colour of any part of the skin will depend upon the extent to which it is exposed to sunlight. The **phenotype**, or set of characteristics, of an individual is therefore determined by the interaction between the genotype and the environment. Any change in the genotype is called a **mutation** and may be inherited. Any change in the phenotype only is called a **modification** and it is not inherited.

14.1 Mendel and the laws of inheritance

Gregor Mendel (1822–84) was an Austrian monk and teacher. He studied the process of heredity in selected features of the garden pea *Pisum sativum*. He was not the first scientist to study heredity, but he was the first to obtain sufficiently numerous, accurate and detailed data upon which sound scientific conclusions could be based. Partly by design and partly by luck, Mendel made a suitable choice of characteristics for study. He isolated pea plants which were **pure-breeding**. That is, when bred with each other, they produced consistently the same characteristics over many generations. He referred to each character as a **trait**. He chose traits which had two contrasting features, e.g. he chose stem length, which could be either long or short and flower colour which could be red or white. It must be remembered that Mendel began his ten-year-long experiments in 1856, when the nature of chromosomes and genes was yet to be discovered.

14.1.1 Monohybrid inheritance (Mendel's Law of Segregation)

Monohybrid inheritance refers to the inheritance of a single character only. One trait which Mendel studied was the shape of the seed produced by his pea plants. This showed two contrasting forms, round and wrinkled. When he crossed plants which were pure-breeding for round seed with ones pure-breeding for wrinkled seed, all the resulting plants produced round seed. The first generation of a cross is referred to as the **first filial generation (F_1)**. When individuals of the F_1 generation were intercrossed the resulting **second filial generation (F_2)** produce 7324 seeds, 5474 of which were round and 1850 wrinkled. This is a ratio of 2.96:1.

In all his crosses, Mendel found that one of the contrasting features of a pair was not represented in the F_1 generation. This feature reappeared in the F_2 generation where it was consistently outnumbered 3 to 1 by the contrasting feature.

The significance of these findings was that the F_1 seeds were not intermediate between the two parental types, i.e. partly wrinkled, partly smooth. This shows that there was no blending or mixing of the features. It also indicated that as only one of the features expressed itself in the F_1, this feature was **dominant** to the other. The feature which does not express itself in the F_1 is said to be **recessive**. In the example given, round is dominant and wrinkled is recessive.

In interpreting his results, Mendel concluded that the features were passed on from one generation to the next via the gametes. The parents he decided must possess two pieces of information about each character. However, only one of these pieces of information was found in an individual gamete. On the basis of this he formulated his first law, the **Law of Segregation**, which states:

The characteristics of an organism are determined by internal factors which occur in pairs. Only one of a pair of such factors can be represented in a single gamete.

We know that Mendel's 'factors' are specific portions of a chromosome called **genes**. We also know that the process which produces gametes with only one of each pair of factors is meiosis. On the basis of his results, Mendel had effectively predicted the existence of genes and meiosis.

14.1.2 Representing genetic crosses

Genetic crosses are usually represented in a form of shorthand. There is more than one system of this shorthand, but the following one has been adopted here because it is both quick and less liable to errors, especially under the pressures of an examination.

TABLE 14.1

Instruction	Reason/notes	Example [round and wrinkled seed]
Choose a single letter to represent each characteristic	An easy form of shorthand. In some conventional genetic crosses, e.g. in *Drosophila*, there are set symbols, some of which use two letters	—
Choose the first letter of one of the contrasting features	When more than one character is considered at one time such a logical choice means it is easy to identify which letter refers to which character	Choose either R (round) or W (wrinkled)
If possible, choose the letter in which the higher and lower case forms differ in shape as well as size	If the higher and lower case forms differ it is almost impossible to confuse them regardless of their size	Choose R, because the higher case form (R) differs in shape from the lower case form (r), whereas W and w differ only in size, and are more likely to be confused
Let the higher case letter represent the dominant feature and the lower case letter the recessive one. Never use two different letters where one character is dominant. Always state clearly what feature each symbol represents	The dominant and recessive features can easily be identified. Do *not* use two different letters as this indicates incomplete dominance or codominance	Let R = round and r = wrinkled Do *not* use R for round and W for wrinkled
Represent the parents with the appropriate pairs of letters. Label them clearly as 'parents' and state their phenotypes	This makes it clear to the reader what the symbols refer to	Round seed / Wrinkled seed; Parents RR v rr
State the gametes produced by each parent. Label them clearly, and encircle them. Indicate that meiosis has occurred	This explains why the gametes only possess one of the two parental factors. Encircling them reinforces the idea that they are separate	Meiosis / Meiosis; Gametes (R) (r)
Use a type of chequerboard or matrix, called **a Punnett square**, to show the results of the random crossing of the gametes. Label male and female gametes even though this may not affect the results	This method is less liable to error than drawing lines between the gametes and the offspring. Labelling the sexes is a good habit to acquire – it has considerable relevance in certain types of crosses, e.g. sex-linked crosses	♂ gametes: R, R / ♀ gametes: r → Rr, Rr; r → Rr, Rr
State the phenotype of each different genotype and indicate the numbers of each type. Always put the higher case (dominant) letter first when writing out the genotype.	Always putting the dominant feature first can reduce errors in cases where it is not possible to avoid using symbols with the higher and lower case letters of the same shape	All offspring are plants producing round seeds (Rr)

NB Always carry out the above procedures in their entirety. Once you have practised a number of crosses, it is all too easy to miss out stages or explanations. Not only does this lead to errors, but often makes your explanations impossible for others to follow. *You* may understand what you are doing, but if the reader cannot follow it, it isn't much use, neither will it bring full credit in an examination.

14.1.3 Genetic representation of the monohybrid cross

Using the principles outlined above, the full genetic explanation of one of Mendel's experiments is shown below:

Let R = allele for round seed
r = allele for wrinkled seed

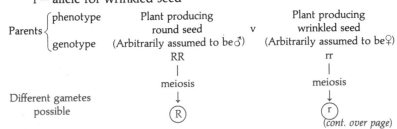

(cont. over page)

(cont. from previous page)

F₁ generation:

		♂ gametes	
		R	R
♀ gametes	r	Rr	Rr
	r	Rr	Rr

All offspring are plants producing round seed (Rr)

F₁ intercross

| phenotype | Plant producing round seed | v | Plant producing round seed |
| genotype | Rr | | Rr |

meiosis   meiosis

Gametes: Ⓡ   ⓡ   Ⓡ   ⓡ

F₂ generation:

		♂ gametes	
		R	r
♀ gametes	R	RR	Rr
	r	Rr	rr

3 plants producing round seeds (1 × RR + 2 × Rr)
1 plant producing wrinkled seeds (rr)

Mendel's actual results gave a ratio of 2.96:1, a very good approximation to the 3:1 ratio which the theory suggests should be achieved. Any discrepancy is due to statistical error. Such errors are inevitable. Imagine for instance tossing a coin ten times — it should in theory come down heads five times and tails five times. More often than not, some other ratio is achieved in practice. The actual results are rarely exactly the same as predicted by the theory. The larger the sample, the more nearly the results approximate to the theoretical value. This was an essential aspect of Mendel's experiments. Probably because he was trained partly as a mathematician, he appreciated the need to collect large numbers of offspring if he was to draw meaningful conclusions from his experiments.

The use of 'F₁ generation' should be limited to the offspring of homozygous parents. Similarly, 'F₂ generation' should refer only to the offspring of the F₁ generation. In all other cases 'offspring(1)' should replace 'F₁ generation', and 'offspring (2)' should replace 'F₂ generation'. The complete set of headings in order therefore will be:

Parents phenotypes

Parents genotypes

Gametes

Offspring (1) genotypes

Offspring (1) phenotypes

Gametes

Offspring (2) genotypes

Offspring (2) phenotypes

Whether the variation from an expected ratio is the result of statistical chance or not can be tested for mathematically using the chi-squared test. Details of this are given in Section 15.3.

14.1.4 Genes and alleles

A character such as the shape of the seed coat in peas is determined by a single gene. The gene is therefore the basic unit of inheritance. It is a region of the chromosome or, more specifically, a length of the DNA molecule, which has a particular function (see Section 12.7). Each gene may have two, occasionally more, alternative forms. Each form of the gene is called an **allele**. The gene for the shape of the seed coat in peas has two alleles, one determining round shape, the other wrinkled. The position of a gene within a DNA molecule is called the **locus**. When two identical alleles occur together at the same locus on a chromosome, they are said to be **homozygous**, e.g. when two alleles for round seeds occur together (RR) they are said to be **homozygous dominant**. Similarly, the two alleles for wrinkled seeds (rr) are referred to as **homozygous recessive**. Where the two alleles differ (Rr) they are termed **heterozygous**.

14.1.5 Dihybrid inheritance (Mendel's Law of Independent Assortment)

Dihybrid inheritance refers to the simultaneous inheritance of two characters. In one of his experiments Mendel investigated the inheritance of seed shape (round v. wrinkled) and seed colour (green v. yellow) at the same time. He knew from his monohybrid crosses that round seeds were dominant to wrinkled ones and yellow seeds were dominant to green. He chose to cross plants with both dominant features (round and yellow) with ones that were recessive for both (wrinkled and green). The F_1 generation yielded plants all of which produced round, yellow seeds — hardly surprising as these are the two dominant features.

TABLE 14.2 **Results of Mendel's dihybrid cross**

Parents: round, yellow seeds v. wrinkled, green seeds

F_1 generation: all round, yellow seeds

F_2 generation:

		Seed shape			
		Round	Wrinkled	Total	Approx. ratio
Seed colour	Yellow	315	101	416	3 yellow
	Green	108	32	140	1 green
	Total	423	133		
	Approx. ratio	3 round	1 wrinkled		

Approx. ratio: round, yellow (2 dominants) : round, green (dominant + recessive) : wrinkled, yellow (recessive + dominant) : wrinkled, green (2 recessives)

9 : 3 : 3 : 1

Mendel planted the F_1 seeds, raised the plants and allowed them to self-pollinate. He then collected the seeds. Of the 556 seeds produced, the majority, 315, possessed the two dominant features — round and yellow. The smallest group, 32, possessed the two recessive features — wrinkled and green. The remaining 209 seeds were of types not previously found. They combined one dominant and one recessive feature; 108 were round (dominant) and green (recessive) and 101 were wrinkled (recessive) and yellow (dominant). At first inspection these results may appear to contradict those obtained in the monohybrid cross, but as Table 14.2 shows, the ratio of dominant to recessive for each feature is still 3:1, as expected.

The significance of these findings was that as the features of seed shape and colour had each produced a 3:1 ratio (dominant: recessive), the two features had behaved completely independently of one another. The presence of one had not affected the behaviour of the other. On the basis of these findings, Mendel formulated his second law, the Law of Independent Assortment, which states:

Each of a pair of contrasted characters may be combined with either of another pair.

With our present knowledge of genetics the law could now be rewritten as:

Each member of an allelic pair may combine randomly with either of another pair.

14.1.6 Genetic representation of the dihybrid cross

Using the principles outlined in Table 14.1, the full genetic explanation of this dihybrid cross is shown below.

Let R = allele for round seed
    r = allele for wrinkled seed
    G = allele for yellow seed
    g = allele for green seed

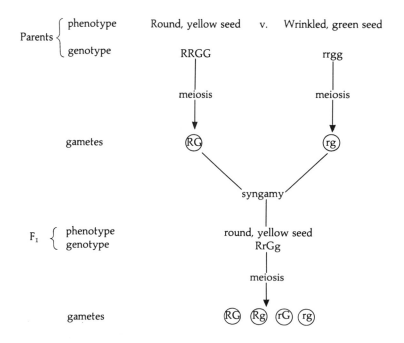

As the plants are self-pollinated the male and female gametes are of the same types. The offspring of this cross may therefore be represented in the following Punnett square.

		♂ gametes			
		RG	Rg	rG	rg
♀ gametes	RG	RRGG	RRGg	RrGG	RrGg
	Rg	RRGg	RRgg	RrGg	Rrgg
	rG	RrGG	RrGg	rrGG	rrGg
	rg	RrGg	Rrgg	rrGg	rrgg

In the following list, '–' represents either the dominant or recessive allele.

		Total
R–G– = round, yellow seed		9 (315)
R–gg = round, green seed		3 (108)
rrG– = wrinkled, yellow seed		3 (101)
rrgg = wrinkled, green seed		1 (32)

Allowing for statistical error, Mendel's results (shown in brackets) were a reasonable approximation to the expected 9:3:3:1 ratio.

14.2   The test cross

One common genetic problem is that an organism which shows a dominant character can have two possible genotypes. For example, a plant producing seeds with round coats could either be homozygous dominant (RR) or heterozygous (Rr). The appearance of the seeds (phenotype) is identical in both cases. It is often necessary, however, to determine the genotype accurately. This may be achieved by crossing the organism of unknown genotype with one whose genotype is accurately known. One genotype which can be positively identified from its phenotype alone is one which shows the recessive feature. In the case of the seed coat, any pea seed with a wrinkled coat must have the genotype rr. By crossing the dominant character, the unknown genotype can be identified. To take the above example:

Let R = allele for round seeds
    r = allele for wrinkled seeds

If the plant producing round seed has the genotype RR:

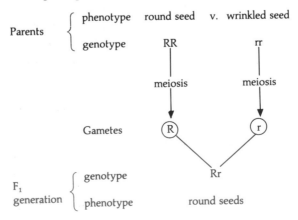

The only possible offspring are plants which produce round seeds.

If the plant producing round seeds has the genotype Rr:

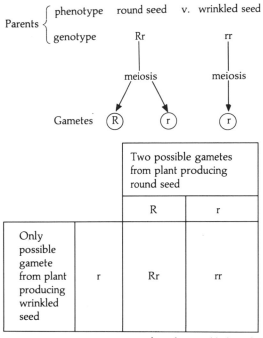

round seed   wrinkled seed

The offspring comprise equal numbers of plants producing round seeds and ones producing wrinkled seeds.

If some of the plants produce seeds with wrinkled coats (rr), then the unknown genotype must be Rr. An exact 1:1 ratio as above is not often achieved in practice, but this is unimportant as the presence of a single plant producing wrinkled seeds is proof enough (the possibility of a mutation must be discounted as it is highly unlikely and totally unpredictable). Such a plant with its rr genotype could only be produced if both parents donated an r gamete. The only way a plant which produces round seed could donate such a gamete would be if it is heterozygous (Rr).

If all the offspring of our test cross were plants producing round seed, then no definite conclusions could be drawn, since both parental genotypes (Rr and RR) are capable of producing such offspring. However, provided a large enough number of offspring are produced, the absence of ones producing wrinkled seeds would strongly indicate that the unknown genotype was RR. Had it been Rr, half the offspring should produce wrinkled seeds. While it would be theoretically possible for no wrinkled seeds to arise, this would be highly improbable where the sample was large.

It is possible to perform a dihybrid test cross. A plant which produces round, yellow seeds has four possible genotypes, namely: RRGG, RrGG, RRGg and RrGg. To determine the genotype of such a plant, it must be crossed with one producing wrinkled, green seeds. Such a plant has only one possible genotype, rrgg, and produces only one type of gamete, namely rg. The outcome of each of the crosses is shown in Table 14.3.

From the table it can be seen that the unknown genotypes can be identified from the results of the test cross as follows. If the offspring contain at least one plant producing wrinkled, yellow seeds, the unknown genotype is RrGG; if round, green seeds it is RRGg, and

TABLE 14.3 **Dihybrid backcross**

Possible genotypes of plant producing round, yellow seeds	Possible gametes	Genotypes of offspring crossed with plant producing wrinkled, green seeds (gamete = rg)	Phenotype (type of seeds produced)
RRGG	RG	RrGg	All round and yellow
RrGG	RG	RrGg	$\frac{1}{2}$ round and yellow
	rG	rrGg	$\frac{1}{2}$ wrinkled and yellow
RRGg	RG	RrGg	$\frac{1}{2}$ round and yellow
	Rg	Rrgg	$\frac{1}{2}$ round and green
RrGg	RG	RrGg	$\frac{1}{4}$ round and yellow
	Rg	Rrgg	$\frac{1}{4}$ round and green
	rG	rrGg	$\frac{1}{4}$ wrinkled and yellow
	rg	rrgg	$\frac{1}{4}$ wrinkled and green

if wrinkled, green seeds it is RrGg. If the number of offspring is large and *all* produce round, yellow seeds it is highly probable that the genotype is RRGG.

14.3 Sex determination

In humans there are twenty-three pairs of chromosomes. Of these, twenty-two pairs are identical in both sexes. The twenty-third pair, however, is different in the male from the female. The twenty-two identical pairs are called **autosomes** whereas the twenty-third pair are referred to as **sex chromosomes** or **heterosomes**. In females, the two sex chromosomes are identical and are called **X chromosomes**. In males, an X chromosome is also present, but the other of the pair is smaller in size and called the **Y chromosome**. Unlike other features of an organism, sex is determined by chromosomes rather than genes.

As sexual reproduction can only occur between a male and a female, there is only one possible genetic cross:

It can be seen that in humans the female produces gametes which all contain an X chromosome and are therefore the same. She is called the **homogametic sex** ('same gametes'). The male, however, produces gametes of two genetic types: one which contains an X chromosome, the other a Y chromosome. The male is called the **heterogametic sex** ('different gametes').

Sex determination differs in other organisms. In birds, most reptiles, some fish and all butterflies, the male is the homogametic sex (XX) and the female is the heterogametic sex (XY). In some insects, while the female is XX, the Y chromosome is absent in the male, which is therefore XO. In the fruit fly *Drosophila*, the female is XX and the male XY; however, the Y chromosome is not smaller, as in humans, but simply a different shape.

Parents	phenotype	male ♂	female ♀

Sex ratio   1 female:1 male

14.4 Linkage

For just twenty-three pairs of chromosomes to determine the many thousands of different human characteristics, it follows that each chromosome must possess many different genes. Any two genes which occur on the same chromosome are said to be **linked**. All the genes on a single chromosome form a **linkage group**.

Under normal circumstances, all the linked genes remain together during cell division and so pass into the gamete, and hence the offspring, together. They do not therefore segregate in accordance with Mendel's Law of Independent Assortment. Fig. 14.1 shows the different gametes produced if a pair of genes A and B are linked rather than on separate chromosomes.

14.4.1 Crossing over and recombination

It is known that genes for flower colour and fruit colour in tomatoes are on the same chromosome. Plants with yellow flowers bear red fruit, those with white flowers bear yellow fruit. If the two types are crossed, the following results are obtained.

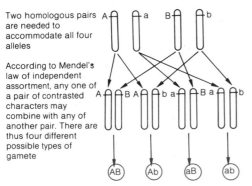

Fig. 14.1 Comparison of gametes produced by an organism heterozygous for two genes A and B, when they are linked and not linked

Let R = allele for red fruit (dominant) and
    r = allele for yellow fruit (recessive)
    W = allele for yellow flowers (dominant) and
    w = allele for white flowers (recessive)

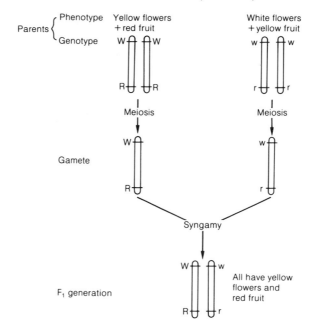

If the F_1 generation is intercrossed (i.e. self-pollinated), the following results would be expected:

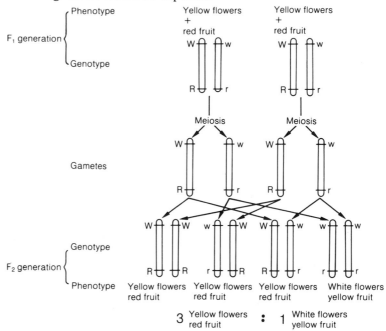

When the actual cross is performed, however, the following results are typical if 100 F_2 plants are produced:

Yellow flowers and red fruit   68
Yellow flowers and yellow fruit   7
White flowers and red fruit   7
White flowers and yellow fruit   18

What then is the explanation? Could it be that the two characters are not linked, but occur on separate chromosomes? If this were so, it would be a normal dihybrid cross, and a 9:3:3:1 ratio should be found. For 100 plants this would mean a 56:19:19:6 distribution. This is sufficiently different from the actual ratio of 68:7:7:18 which was obtained for it to be discounted. For the answer, we have to go back to Section 13.3 and the events in prophase I of meiosis. During this stage portions of the chromatids of homologous chromosomes were exchanged in the process called crossing over. Could this be the explanation as to how the two unexpected phenotypes (yellow flowers/yellow fruit and white flowers/red fruit) came about? To find out, let us consider the same F_1 intercross as before but assume that in one parent crossing over took place and this plant was subsequently self-pollinated (see opposite page).

The new combinations are thus the result of crossing over in prophase I of meiosis. These new combinations are called **recombinants**. As shown, this cross produces a 9:3:3:1 ratio. However, in practice, crossing over will not always occur between the two genes. In some cases it may not occur at all; in others it may occur in such a way that the two genes are not separated (Fig. 14.2). In these circumstances the only gametes are WR and wr. For this reason plants with yellow flowers and red fruit, and those with white flowers and yellow fruit, occur in greater numbers than expected.

14.4.2 Cross-over values

When crossing over occurs, recombinants are formed. The number of recombinants depends on the proximity of the linked genes.

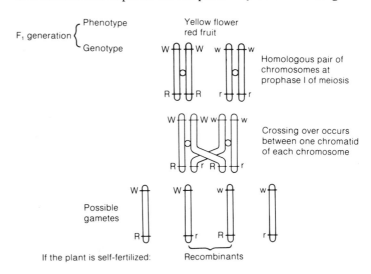

F₂ generation:

		♂ gametes			
		WR	Wr	wR	wr
♀ gametes	WR	WWRR	WWRr	WwRR	WwRr
	Wr	WWRr	WWrr	WwRr	Wwrr
	wR	WwRR	WwRr	wwRR	wwRr
	wr	WwRr	Wwrr	wwRr	wwrr

New combinations WWrr and Wwrr (yellow flowers, yellow fruit)
wwRR and wwRr (white flowers, red fruit)

Crossing over can occur at any point along a chromosome (Fig. 14.2). It follows that if two genes are close together on a chromosome, the chances of them being separated by any one cross-over is smaller than if they were far apart. The distance apart of genes can therefore be determined by the number of recombinants. The further two genes are apart, the greater the statistical chance that crossing over will separate them, and the greater the number of recombinants that will be formed.

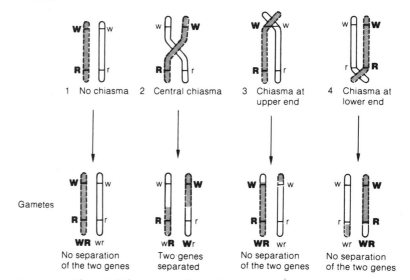

Fig. 14.2 *Effect of differences in the position of the chiasmata on the separation of two linked genes (only one chromatid is shown to simplify the diagrams)*

The proportion of recombinants in a group of offspring can be used to calculate the **cross-over value (COV)** or **recombination frequency** as follows:

$$\frac{\text{Number of offspring showing recombination}}{\text{Total number of offspring}} \times 100$$

In our earlier cross using tomatoes, the following results were obtained:

Yellow flowers and red fruit	68
Yellow flowers and yellow fruit	7
White flowers and red fruit	7
White flowers and yellow fruit	18

Applying the formula, the cross-over value is calculated as:

$$\frac{7+7}{68+7+7+18} \times 100$$

$$= \frac{14}{100} \times 100$$

$$= 14\%$$

14.4.3 Mapping of chromosomes

As the cross-over value for any two genes is always constant, it is possible to map the exact location of all genes on a chromosome, simply by calculating the cross-over values between all possible pairs. Take for example the following values for hypothetical genes A, B, C and D.

Linked gene pair	Cross-over value (%)
AB	60
AC	10
AD	20
BC	70
BD	40
CD	30

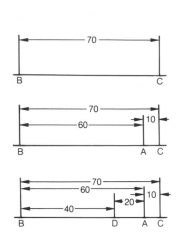

As the COV for genes B and C is the largest, then these two genes must be farthest apart.

To locate gene A
As the COV for genes A and B is 60% these must also be well separated. As the value for A and C is 10%, A must lie near C.

To locate gene D
As the COV for genes B and D is 40% and for A and D is 20%, D must lie nearer to A than B. Its position is confirmed by the COV for genes C and D, which is 30%.

By carrying out appropriate crosses, it is possible to map chromosomes accurately. If two genes have a COV of 10%, they are said to be 10 units apart. In the fruit fly *Drosophila*, all eight chromosomes have been accurately mapped. Given opposite are the details for chromosome number 1 (the X chromosome).

Double cross-overs may arise in which two chiasmata occur between two linked genes. A double cross-over results in genes which are separated by the first cross-over being joined again by a second cross-over. It will appear that cross-over has not taken place when in fact it has – twice. Double cross-overs are more likely to arise when two genes are widely separated. It leads to an underestimate of their cross-over value.

14.4.4 Sex linkage

Sex linkage refers to the carrying of genes on the sex chromosomes. These genes determine body characters and have nothing to do with sex. The X chromosome carries many such genes, the Y chromosome has very few. Features linked on the Y chromosome will only arise in the heterogametic (XY) sex, i.e. males in mammals, females in birds. Features linked on the X chromosome may arise in either sex.

White eye colour is a sex-linked character in the fruit fly *Drosophila*. It is carried on the X chromosome and the male is the heterogametic sex. To represent sex-linked crosses, the same principles which were laid down in Table 14.1 should be followed. The letter representing each allele should, however, be attached to the letter X to indicate it is linked to it. No corresponding allele is found on the Y chromosomes, which therefore have no attached letter.

Below are the expected results of a cross between a white-eyed male mutant and a **wild-type** (red-eyed) female. 'Wild-type' is a term used to describe an organism as it normally occurs in nature. The **reciprocal cross** is also shown. A reciprocal cross is one where the same genetic features are used, but the sexes are reversed. In this case the reciprocal cross is between a white-eyed mutant female and a wild-type (red-eyed) male.

Red eyes are dominant over white eyes.

Gene	Units
Yellow body	0.0
White eyes	1.5
Facet eyes	3.0
Ruby eyes	7.5
Cut wings	20.0
Singed bristles	21.0
Lozenge eyes	27.7
Vermillion eyes	33.0
Miniature wings	36.1
Sable body	43.0
Garnet eyes	44.0
Forked bristles	56.7
Fused veins	59.5
Carnation eyes	62.5
Bobbed hairs	66.0

Therefore let R represent the allele for red eyes and
let r represent the allele for white eyes

As the genes for eye colour are carried on the X chromosome the alleles are represented as X^R and X^r respectively. In *Drosophila* the male is the heterogametic sex (XY) and the female is the homogametic sex (XX).

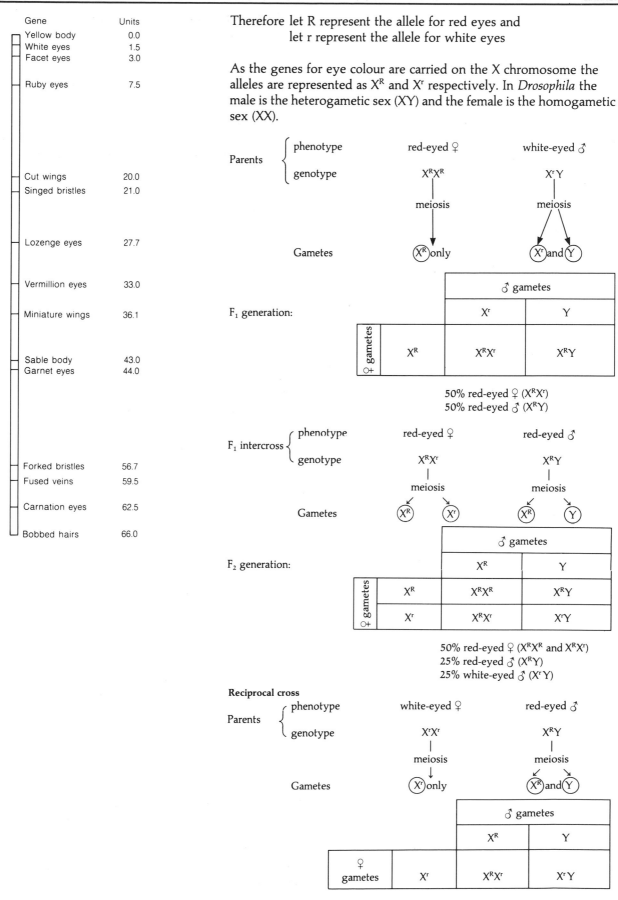

Parents { phenotype — red-eyed ♀ — white-eyed ♂
genotype — $X^R X^R$ — $X^r Y$

meiosis — meiosis

Gametes — X^R only — X^r and Y

F_1 generation:

		♂ gametes	
		X^r	Y
♀ gametes	X^R	$X^R X^r$	$X^R Y$

50% red-eyed ♀ ($X^R X^r$)
50% red-eyed ♂ ($X^R Y$)

F_1 intercross { phenotype — red-eyed ♀ — red-eyed ♂
genotype — $X^R X^r$ — $X^R Y$

meiosis — meiosis

Gametes — X^R — X^r — X^R — Y

F_2 generation:

		♂ gametes	
		X^R	Y
♀ gametes	X^R	$X^R X^R$	$X^R Y$
	X^r	$X^R X^r$	$X^r Y$

50% red-eyed ♀ ($X^R X^R$ and $X^R X^r$)
25% red-eyed ♂ ($X^R Y$)
25% white-eyed ♂ ($X^r Y$)

Reciprocal cross

Parents { phenotype — white-eyed ♀ — red-eyed ♂
genotype — $X^r X^r$ — $X^R Y$

meiosis — meiosis

Gametes — X^r only — X^R and Y

		♂ gametes	
		X^R	Y
♀ gametes	X^r	$X^R X^r$	$X^r Y$

50% red-eyed ♀
50% white-eyed ♂
(*cont. over page*)

(cont. from previous page)

F₁ intercross
{
phenotype   red-eyed ♀   white-eyed ♂

genotype $X^R X^r$ $X^r Y$
}

meiosis   meiosis

Gametes X^R and X^r X^r Y

F₂ generation:

	♂ gametes	
	X^r	Y
♀ gametes X^R	$X^R X^r$	$X^R Y$
X^r	$X^r X^r$	$X^r Y$

25% red-eyed ♀   25% red-eyed ♂
25% white-eyed ♀   25% white-eyed ♂

    Two well known sex-linked genes in humans are those causing haemophilia and red-green colour-blindness. Both are linked to the X chromosome and both occur almost exclusively in males. For the condition to arise in females requires the double recessive state and as the recessive gene is relatively rare in the population this is unlikely to occur. In females the recessive gene is normally masked by the appropriate dominant gene which occurs on the other X chromosome. These heterozygous females are not themselves affected but are capable of passing the recessive gene to their offspring. For this reason such females are termed **carriers**. When the recessive gene occurs in males it expresses itself because the Y chromosome cannot carry any corresponding dominant gene. The inheritance of red-green colour-blindness is illustrated below.

Normal sight is dominant over red-green colour-blindness.
Therefore let B represent the allele for normal sight and
       let b represent the allele for colour-blindness
    As this gene is carried on the X chromosome, its alleles are represented as X^B and X^b respectively. In humans the male is the heterogametic sex (XY) and the female is the homogametic sex (XX).

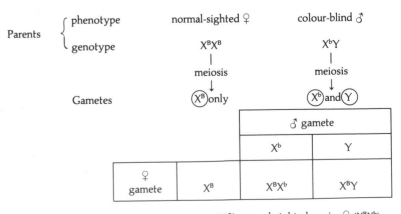

Parents
{
phenotype   normal-sighted ♀   colour-blind ♂

genotype $X^B X^B$ $X^b Y$
}

meiosis   meiosis

Gametes X^B only X^b and Y

	♂ gamete	
	X^b	Y
♀ gamete X^B	$X^B X^b$	$X^B Y$

50% normal-sighted carrier ♀ ($X^B X^b$)
50% normal-sighted ♂ ($X^B Y$)

F_1 intercross $\left\{\begin{array}{l}\text{phenotype}\\\text{genotype}\end{array}\right.$

normal-sighted carrier ♀   normal-sighted ♂

$X^B X^b$ $X^B Y$

| Meiosis | Meiosis |

gametes   Ⓧ^{B} Ⓧ^{b} Ⓧ^{B} Ⓨ

		♂ gametes	
F_2 generation		X^B	Y
♀ gametes	X^B	$X^B X^B$	$X^B Y$
	X^b	$X^B X^b$	$X^b Y$

25% normal-sighted ♀ ($X^B X^B$)   25% normal-sighted ♂ ($X^B Y$)
25% normal-sighted, carrier ♀ ($X^B X^b$)   25% colour-blind ♂ ($X^b Y$)

A study of the crosses reveals that the recessive gene causing colour-blindness is exchanged from one sex to the other at each generation. The father passes it to his daughters, who thus become carriers. The daughters in turn may pass it to their sons, who are thus colour-blind. This pattern of inheritance is perhaps more obvious when viewed another way. As the male is XY, his Y chromosome must have been inherited from his father as the mother does not possess a Y chromosome. The X chromosome and hence colour-blindness must therefore have been inherited from the mother. The colour-blind male can only donate his X chromosome to his daughters as it is bound to fuse with another X chromosome — the only type the mother produces. Colour-blind females can only arise from a cross between a carrier female and a colour-blind male. As both types are rare in the population, the chance of this happening is very small indeed. Even then, there is only a one in four chance of any single child of such a cross being a colour-blind female.

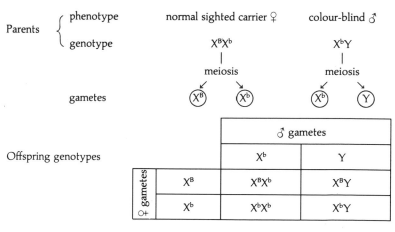

Parents $\left\{\begin{array}{l}\text{phenotype}\\\text{genotype}\end{array}\right.$

normal sighted carrier ♀   colour-blind ♂

$X^B X^b$ $X^b Y$

| meiosis | meiosis |

gametes   Ⓧ^{B} Ⓧ^{b} Ⓧ^{b} Ⓨ

		♂ gametes	
Offspring genotypes		X^b	Y
♀ gametes	X^B	$X^B X^b$	$X^B Y$
	X^b	$X^b X^b$	$X^b Y$

25% normal-sighted carrier ♀ ($X^B X^b$)   25% colour-blind ♂ ($X^b Y$)
25% normal-sighted ♂ ($X^B Y$)   25% colour-blind ♀ ($X^b X^b$)

The inheritance of haemophilia follows a similar pattern to that of colour-blindness. Haemophilia is the inability of the blood to clot leading to slow and persistent bleeding, especially in the joints. Unlike colour-blindness it is potentially lethal. For this reason, the recessive gene causing it is even rarer in the population. Haemophiliac females are thus highly improbable, and in any case are unlikely to have children as the onset of menstruation at puberty is often fatal.

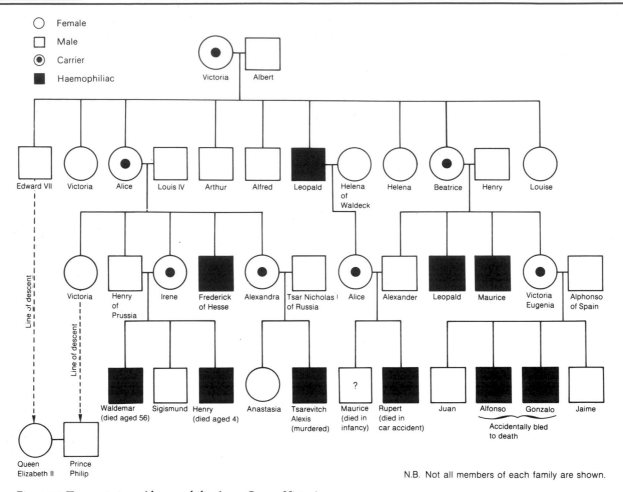

Fig. 14.3 Transmission of haemophilia from Queen Victoria

Haemophilia is the result of an individual being unable to produce one of the many clotting factors, namely **factor 8** or **anti-haemophiliac globulin (AHG).** The extraction of this factor from donated blood now permits haemophiliacs to lead near-normal lives, although they still run the same risk of conveying the disease to their children.

Any mutant recessive gene, such as that causing haemophilia, is normally rapidly diluted among the many normal genes in a population. Its expression is thus a rare event. If, however, there is close breeding between members of a family in which a mutant gene exists, the chances of it expressing itself are enhanced. This accounts for the higher than normal occurrence of the haemophiliac gene among members of various European royal families. The origin of this particular gene can be traced back to England's Queen Victoria, who had a haemophiliac son, Leopold Duke of Albany. Prior to this there was no history of the gene among the royal family, although it existed elsewhere in the population. In order to marry someone of similar status, members of the European royal families were limited in their choice of partners and tended to marry within a relatively small circle. In effect, the gene pool was very restricted. As a result there was a disproportionately large number of haemophiliacs in these families. Fig. 14.3 traces the inheritance of this gene. The present English royal family is unaffected, as it is descended from Edward VII who did not inherit the haemophilia gene. The chart also illustrates another method of representing sex-linked crosses.

It is unusual to find dominant mutant genes linked to the X chromosome in humans, but one example is the congenital absence of incisor teeth. These conditions occur in both sexes but are more common in females as they have two X chromosomes. Genes linked to the Y chromosome are very rare, but hairy ear rims are an example. In these cases the condition is limited to males and is always inherited, whether it is dominant or recessive.

14.5 Allelic interaction

We have so far dealt with situations where a gene comprises two alleles which may occur at the locus of a chromosome. One of the two alleles is dominant, the other recessive. Sometimes, however, neither allele completely dominates the other. Such a condition is termed **codominance**. Sometimes more than two alleles exist for a given gene, but only two alleles exist at a single locus at any one time. This condition is called **multiple alleles**.

These are all examples of the way in which genes may interact to produce phenotypic characteristics.

14.5.1 Codominance

One example of codominance occurs when a snapdragon (*Antirrhinum*) with red flowers is crossed with one with white flowers. All the F_1 generation produce flowers of intermediate colour, namely pink. The F_2 generation produces red, pink and white flowers in the ratio 1:2:1. The cross may be represented by the procedure shown below:

Let C^R = the allele for red flowers

C^W = the allele for white flowers

F₂ generation:

	♂ gametes	
	C^R	C^W
♀ gametes C^R	C^R C^R	C^R C^W
♀ gametes C^W	C^R C^W	C^W C^W

25% red flowers (C^R C^R)
50% pink flowers (C^R W)
25% white flowers (C^W C^W)

N.B. Where codominance is involved, it is normal to use different letters to represent each allele, e.g. R to represent red flowers and W to represent white flowers. The use of higher and lower cases of one letter, e.g. R and r or W and w, would imply dominance and be confusing. It is also usual to assign the gene an upper case letter (in the above case C = colour) and use superscript upper case letters to designate the different alleles.

This type of inheritance, where an intermediate form arises, is called **blending** or **epistasis**. An example in humans is the inheritance of skin pigmentation which is controlled by two genes A and B. An individual with the genotype AABB produces darkly pigmented skin whereas an individual with the genotype aabb has white unpigmented skin. A mating between these two types produces an intermediate skin colour (genotype AaBb). In the F₂ generation skin colour varies from dark (AABB) through dark brown (AABb or AaBB), half-coloured (AAbb or AaBb or aaBB), light brown (Aabb or aaBb) to white (aabb).

Epistasis does not always result in the blending of features to produce intermediates; it can create entirely new features. Take the case of comb shape in poultry. When a pure-breeding fowl with a rose-type comb is crossed with a pure-breeding pea-type fowl, all the offspring produced have walnut-type combs. All three types are represented in the F₂ generation along with a fourth variety – the single comb, as shown at the top of the next page.

Let R represent the gene for rose-type comb and
P represent the gene for pea-type comb

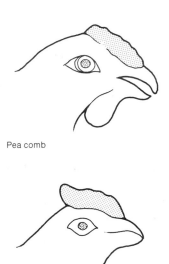
Pea comb

Rose comb

Walnut comb

Single comb

Fig. 14.4 Types of chicken comb

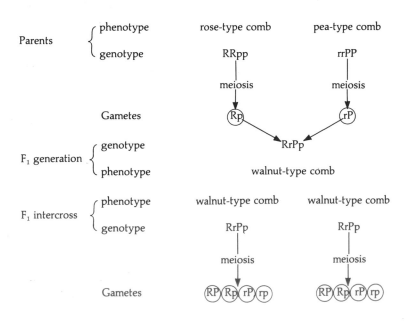

Parents	phenotype	rose-type comb	pea-type comb
	genotype	RRpp	rrPP
		meiosis	meiosis
Gametes		Rp	rP
		RrPp	
F₁ generation	genotype		
	phenotype	walnut-type comb	
F₁ intercross	phenotype	walnut-type comb	walnut-type comb
	genotype	RrPp	RrPp
		meiosis	meiosis
Gametes		RP Rp rP rp	RP Rp rP rp

F₂ generation

		♂ gametes			
		RP	Rp	rP	rp
♀ gametes	RP	RRPP	RRPp	RrPP	RrPp
	Rp	RRPp	RRpp	RrPp	Rrpp
	rP	RrPP	RrPp	rrPP	rrPp
	rp	RrPp	Rrpp	rrPp	rrpp

9 walnut-type comb (R–P–)
3 rose-type comb (R–pp)
3 pea-type comb (rrP–)
1 single-type comb (rrpp)

14.5.2 Multiple alleles

In humans the inheritance of the ABO blood groups is determined by a gene I which has three different alleles. Any two of these can occur at a single locus at any one time.

Allele A causes production of antigen A on red blood cells.

Allele B causes production of antigen B on red blood cells.

Allele O causes no production of antigens on red blood cells.

Alleles A and B are codominant and allele O is recessive to both.

TABLE 14.4 **Possible genotypes of blood groups in the ABO system**

Blood group	Possible genotypes
A	$I^A I^A$ or $I^A I^O$
B	$I^B I^B$ or $I^B I^O$
AB	$I^A I^B$
O	$I^O I^O$

The transmission of these alleles occurs in normal Mendelian fashion.

A cross between an individual of group AB and one of group O therefore gives rise to individuals none of whom possess either parental blood group.

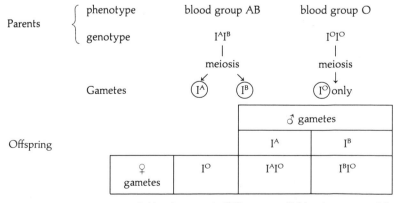

A cross between certain individuals of blood group A and certain individuals of blood group B may produce offspring with any one of the four blood groups.

Parents
{
phenotype   blood group A   blood group B

genotype I^AI^O I^BI^O
}

  meiosis   meiosis

Gametes I^A and I^O I^B and I^O

Offspring

		♂ gametes	
		I^A	I^O
♀ gametes	I^B	I^AI^B	I^BI^O
	I^O	I^AI^O	I^OI^O

25% blood group A (I^AI^O)   25% blood group AB (I^AI^B)
25% blood group B (I^BI^O)   25% blood group O (I^OI^O)

Paternity suits

Although blood groups cannot prove who is the father of a child, it is possible to use their inheritance to show that an individual could not possibly be the father. Imagine a mother who is blood group B having a child of blood group O. She claims the father is a man whose blood group is found to be AB. As the child is group O its only possible genotype is I^OI^O. It must therefore have inherited one I^O allele from each parent. The mother, if I^BI^O, could donate such an allele. The man with blood group AB can only have the genotype I^AI^B. He is unable to donate an I^O allele and cannot therefore be the father.

Dominance series

Coat colour in rabbits is determined by a gene C which has four possible alleles:

Allele C^F determines full coat colour and is dominant to
Allele C^{CH} which determines chinchilla coat and is in turn dominant to
Allele C^H which determines Himalayan coat and is in turn dominant to
Allele C^A which determines albino coat colour.

There is therefore a dominance series, and each type has a range of possible genotypes.

Inheritance is once again in normal Mendelian fashion.

TABLE 14.5 **Possible genotypes of rabbits with different coat colour**

Coat colour	Possible genotypes
Full	C^FC^F or C^FC^{CH} or C^FC^H or C^FC^A
Chinchilla	$C^{CH}C^{CH}$ or $C^{CH}C^H$ or $C^{CH}C^A$
Himalayan	C^HC^H or C^HC^A
Albino	C^AC^A

14.6   Lethal genes

A lethal gene is one that, when present in the homozygous condition, will cause the death of the offspring. Such genes are found in both plants and animals but the classic example is that found in mice. The yellow race of the house mouse (*Mus musculus*) is heterozygous. Whenever two yellow mice are bred together the offspring are always in the ratio of 2 yellow to 1 agouti (grey). The expected ratio is 3 yellow to 1 agouti. Examination of pregnant yellow mice reveals that the homozygous yellow embryo always dies. The precise cause of this lethal condition is not known.

Let Y represent the dominant allele for yellow fur
and y represent the recessive allele for agouti fur

Parents
{ phenotype   yellow fur   yellow fur
    genotype   Yy   Yy

meiosis   meiosis

Gametes   (Y) and (y)   (Y) and (y)

	♂ gametes	
	Y	y

Offspring

♀ gametes	Y	YY	Yy
	y	Yy	yy

2 yellow fur (Yy)   YY is a lethal condition, resulting
1 agouti fur (yy)   in death as an embryo

Another lethal dominant gene occurs in the creeper fowl which is normal when homozygous recessive, has short legs when heterozygous but the chicks die before hatching when homozygous dominant. Recessive lethal genes are more common than dominant genes, and cause death in the homozygous recessive condition.

14.7   Choice of species for genetic crosses

Much of our knowledge about genetics has been derived from genetic experiments using living organisms. By crossing individuals, collecting and analysing the offspring and then carrying out further crosses, much can be learned of the mechanism of heredity. The choice of species for such experiments depends on a number of factors.

1. Easy to breed – the species must readily produce offspring and not be particular with whom they breed.

2. Readily grown/cultured/reared – the organisms should be convenient and easy to keep.

3. Cheap and easy to feed – they should not have highly specific nutritional requirements.

4. Small size – it follows that the smaller the organism the more likely the previous conditions are to be met.

5. Short life cycle – this allows many generations to be investigated in a short period.

6. Production of many offspring – to give statistically accurate results large numbers of offspring need to be produced from each mating.

7. Early sexual maturity – this allows more rapid production of subsequent generations.

8. Obviously recognizable features – genetic differences should be easy to observe.

9. Sexual dimorphism – it is helpful if the male and female of the species are quickly and easily distinguished.

The organisms favoured for genetic research such as *Drosophila melanogaster* (fruit fly), mice and maize plants combine most, if not all, the above features.

14.8 Questions

1. (a) Describe how you would carry out and record the results of a dihybrid cross (to obtain F_1 and F_2 generations) in a **named** organism, emphasizing the reasons for the various procedures. (5 marks)

   (b) Consider an F_1 generation of a dihybrid cross. Explain how the results in the F_2 are dependent on the behaviour of chromosomes during meiosis in the F_1. (10 marks)

   (c) Typically, a 9:3:3:1 phenotypic ratio is obtained in the F_2 of a dihybrid cross. What effect does (i) linkage and (ii) incomplete dominance have on this ratio? Explain your answer. (5 marks)

   (Total 20 marks)

   Joint Matriculation Board June 1985, Paper IIB, No. 4

2. In the fruit fly *Drosophila*, vestigial wing (vg) is recessive to normal and white eye colour (w) recessive to the normal red. These genes are on the X-chromosome and in *Drosophila* the heterogametic sex is male.

   (a) Briefly explain the terms 'heterogametic' and 'sex-linkage' and describe how you would distinguish between male and female offspring. (5 marks)

   (b) What phenotypes would be expected in the F_1 of a cross between a vestigial winged, red-eyed male and a homozygous normal winged white eyed female? (8 marks)

   (c) What phenotypes would be expected in the F_2 generation when F_1 flies interbreed? Show clearly all your working (from (b)). (12 marks)

   (Total 25 marks)

   Southern Universities Joint Board June 1986, Paper II, No. 3

3. (a) Distinguish between the terms 'gene' and 'allele'. (4 marks)

   (b) (i) In maize plants, normal size is dominant to pygmy size, and normal leaf shape is dominant to crinkly leaf shape.
   A plant heterozygous for both these genes was self-pollinated.
   Its seeds were collected and 320 plants subsequently grew. Assuming that the genes are not linked, what phenotypes and how many of each type would you expect to appear in these plants? Give a full explanation for your answer. (12 marks)

   (ii) What differences would you expect in the results if the genes had been linked? (4 marks)

   (Total 20 marks)

   London Board January 1989, Paper II, No. 3

4. (a) Explain the differences between the members of **each** of the following pairs of genetical terms and give **one** example of **each** term to illustrate your answer.
   (i) complete and incomplete dominance
   (ii) continuous and discontinuous variation
   (iii) chromosomal mutation and crossing-over
   (iv) polyploidy and haploidy (12 marks)

   (b) Crosses between ginger female cats and black male cats produce only tortoiseshell females and ginger-coloured males. A single gene controls expression of colour in cats.
   (i) Give a reasoned explanation of these results and show the genotypes of the parents, their gametes and the offspring produced in these crosses.
   (ii) Is it possible to have tortoiseshell male cats? Explain your answer. (8 marks)

   (Total 20 marks)

   Joint Matriculation Board June 1989, Paper IIB, No. 1

5. The human ABO blood groups are determined by three alleles; A, B and O (sometimes represented as I_A, I_B, I_O). The A and B alleles show co-dominance (incomplete dominance) when present in an individual. Both of these alleles are dominant to the O allele.

   Another pair of alleles, present at a separate (unlinked) gene locus, control the Rhesus blood group. The Rhesus positive allele (Rh+) is dominant to the Rhesus negative allele (Rh−).

   Below are given the blood groups (phenotypes) of four sets of parents, along with the blood groups of the first two children of each couple.

CROSS	PARENTAL BLOOD GROUPS	
	FATHER	MOTHER
1	O Rhesus +	O Rhesus −
2	O Rhesus +	A Rhesus −
3	AB Rhesus +	A Rhesus +
4	A Rhesus +	A Rhesus −

CROSS	OFFSPRING BLOOD GROUPS	
	1st CHILD	2nd CHILD
1	O Rhesus +	O Rhesus −
2	A Rhesus −	O Rhesus +
3	A Rhesus +	B Rhesus −
4	O Rhesus +	O Rhesus −

Taking **each** of crosses 1–4 above **in turn**, deduce:

(a) the genotypes of the parents;

(b) the genotypes of the first two children;

(c) the possible phenotypes (blood groups) and genotypes of any other children of the parents.

(Note that in each case you should show all your working and explain your reasoning and that you should also state any assumptions you have made.)

(7, 7, 8, 8 marks)

(Total 30 marks)

Oxford Local June 1989, Paper II, No. 9

6. In dogs, coat colour is determined by a series of multiple alleles.

The allele A^S produces a uniformly dark coat, the allele a^y produces a tan coat colour and the allele a^t produces a spotted coat.

The dominance hierarchy is $A^S > a^y > a^t$, which means that A^S is dominant to both a^y and a^t, but a^y is dominant only to a^t.

A family tree for dogs showing these coat colours is given below.

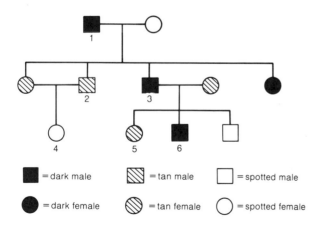

■ = dark male   ▨ = tan male   □ = spotted male

● = dark female   ◒ = tan female   ○ = spotted female

(a) State the genotypes of each of the individuals numbered 1–5. (5 marks)

(b) By means of genetic diagrams, deduce the possible genotypes and phenotypes of the puppies which could be produced by a mating between individuals 4 and 6.

(8 marks)

(Total 13 marks)

London Board June 1985, Paper I, No. 12

7. In this question, phenotypes may be described by showing the alleles which are expressed. e.g. Phenotypes with the alleles AABBCC and phenotypes with the alleles AaBbCc would **both** be shown as ABC.

Three pairs of alleles, A/a, B/b and C/c are carried on homologous chromosomes as indicated in the diagram.

When an individual with this genotype was crossed with another which was recessive for all three genes, the following offspring were produced:

Offspring Phenotype	Number
ABC	284
Abc	50
ABc	76
AbC	2
aBc	3
abC	81
aBC	44
abc	260

(a) Give the phenotypes that would result from a cross-over between:

(i) the A/a and B/b loci;

(ii) the B/b and C/c loci. (2 marks)

(b) Explain how the aBc and the AbC phenotypes could arise. (2 marks)

(c) How many of the offspring show crossing over between:

(i) the A/a and B/b loci;

(ii) the B/b and C/c loci? (4 marks)

(d) Calculate the cross-over value between:

(i) the A/a and B/b loci;

(ii) the B/b and C/c loci. (2 marks)

(Total 10 marks)

Oxford Local June 1988, Paper I, No. 9

8. A maize plant homozygous for smooth, coloured grain was cross-pollinated with a plant homozygous for wrinkled, colourless grain. The F_1 plants all produced smooth, coloured grain. On cross-pollinating the F_1 plants, it was found that most of the F_2 generation resembled the original plants, 73% producing smooth, coloured grain and 22% producing wrinkled, colourless grain.

(a) (i) Which of the above characteristics are

1. dominant

2. recessive?

(ii) What else can you deduce about the alleles for texture and colour of the grains?

(3 marks)

(b) (i) Give the probable phenotypes of the 5% F₂ plants **not** described above.

(ii) State how these phenotypes arose.

(2 marks)

(c) Using appropriate symbols, give a genotype of an

F₁ plant

F₂ plant you described in (b)(i).   (2 marks)

(Total 7 marks)

Welsh Joint Education Committee June 1988, Paper A2, No. 2

9.   (a) Define sex-linkage.   (2 marks)

(b) Sex-linkage was first demonstrated by Thomas Hunt Morgan using *Drosophila*. He was looking at the eye colour characteristic in which red (wild-type) is dominant and white-eye is recessive. He crossed a white-eyed male with a red-eyed female and all of the F₁ generation were red-eyed. Then the F₁ generation was interbred with this result:

Red-eyed females   2459
White-eyed females   0
Red-eyed males   1011
White-eyed males   782

Morgan then crossed the original white-eyed male with one female of the F₁ generation. These were his results:

Red-eyed females   129
White-eyed females   88
Red-eyed males   132
White-eyed males   86

(He later found that white-eyed flies are more likely to die before they hatch than red-eyed flies and this explains the lower number of white-eyed flies than expected in some crosses.)

(i) Using R for the red-eyed allele and r for the white-eyed allele; state the genotypes of the parental white-eyed male and the red-eyed female. (2 marks)

(ii) Give details of the crosses to explain the experimental results.   (6 marks)

(iii) What evidence is there to suggest that the factor is sex-linked?   (2 marks)

(c) Haemophilia, a human disease in which blood is slow to clot, is also sex-linked. The gene involved is carried on the X chromosome.

In the pedigree table shown below, females are represented by circles and males by rectangles. Shaded rectangles represent males suffering from haemophilia.

Give the genotypes of the individuals labelled A–F and state whether each is normal, a carrier or a haemophiliac.   (6 marks)

(Total 18 marks)

Oxford and Cambridge Board June 1989, Paper I, No. 4

10. A cross was made between two varieties of tobacco (*Nicotiana affinis*) which differed markedly in flower size. One parental form had a long-tubed corolla and the other a short-tubed corolla. The corolla length of both parental forms, the F₁ and the F₂ generations were all measured and the measurements grouped into size classes of 3 mm. The frequency distribution of the individuals with different corolla lengths is shown in the figure.

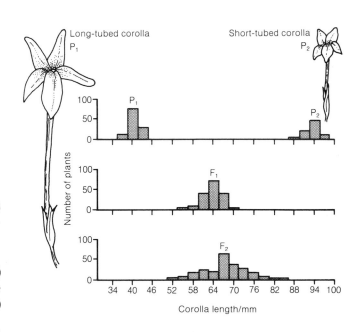

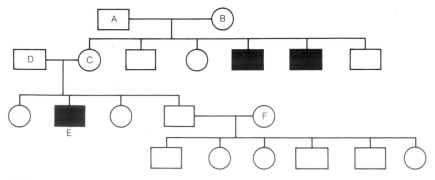

(a) The data presented in the figure indicate that both parents showed variation in corolla length. The variation in the F_1 generation is similar to the parental variation but the variation in the F_2 generation is greater than that found within the parental forms or the F_1 generation. Suggest an explanation for this.

(b) Explain the different degrees of expression of this characteristic.

(c) State a term used to describe the range of the phenotypes found in the F_2 generation.

(*Total 10 marks*)

Cambridge Board June 1988, Paper III, No. 4

11. Humans may possess one of the following blood groups: A, B, AB or O.

(a) Describe these phenotypes in terms of
   (i) the constituents of the red blood cell membrane,
   (ii) the plasma antibodies,
   (iii) their possible genotypes. (*6 marks*)

(b) Describe the biochemical linkage between the possession of the alleles and the phenotypes, explaining the term multiple allele. (*5 marks*)

(c) Some aspects of tail length in Columbian gerbils are inherited in the same way as blood groups. Three alleles control four phenotypes; long tailed, short tailed, intermediate tailed and tailless.

A group of three of these animals was kept together. There was one intermediate tailed female, one long tailed male and one short tailed male. Three young were eventually born in a single litter, one of each tailed phenotype.

   (i) Using the symbols L (long), S (short) and O (tailless) explain these phenotypes in terms of the possible genotypes of parents and offspring. (*4 marks*)

   (ii) What test crosses could be performed on the offspring to determine the most probable paternal genotype? What would you expect the results to be for each possible parent? (*5 marks*)

(*Total 20 marks*)

Northern Ireland Board June 1988, Paper II, No. 6X

15 *Genetic change and variation*

Within any given population there are variations among individual organisms. It is this variation which forms the basis of the evolutionary theory of Darwin. There are two basic forms of variation: **continuous variation**, where the individuals in a population show a gradation from one extreme to the other, and **discontinuous (discrete) variation**, where there is a limited number of distinct forms within the population. Any study of variation inevitably involves the collection of large quantities of data.

15.1 Methods of recording variation

The investigation of variation within a population may involve recording the number of individuals which possess a particular feature, e.g. black fur. On the other hand, it may involve recording the number of individuals which fall within a set range of values, e.g. those weighing between 1 kg and 10 kg. What is being measured in both cases is the **frequency distribution**. This may be presented in a number of ways. To illustrate each method the same set of data is used throughout, although it is not really suitable for some methods of presentation.

15.1.1 Table of data

Tabulation is the simplest means of presenting data. It is a useful method of recording information initially but is less useful for demonstrating the relationship between two variables.

15.1.2 Line graph

A graph typically has two axes, each of which measures a variable. One variable has fixed values which are selected by the experimenter. This is called the **independent variable**. The other variable is the measurement taken and as such is not selected by the experimenter. This is called the **dependent variable**. In Table 15.1, 'height' is the independent variable and 'frequency' the dependent variable. The values of the independent variable are plotted along the horizontal axis (also known as the x axis or abscissa) and the values of the dependent variable are plotted along the vertical axis (also known as the y axis or ordinate). The corresponding values of the two variables can be plotted as points on the graph known as **coordinates**. These points may then be joined to give a line or smooth curve, as in Fig. 15.1(a).

15.1.3 Histogram

Axes are drawn in much the same way as for a line graph. The values for the independent variable on the x axis are, however, normally reduced, often by grouping the data into convenient classes. For example, the twenty-six values for height used on the line graph may be reduced to ten by grouping the heights into sets of 5 cm, e.g. 140–145, 145–150 etc. Instead of plotting points, vertical columns are drawn. The method is illustrated in Fig. 15.1(b).

15.1.4 Bar graph

This is similar to a histogram except that a non-numerical value is plotted on the y axis. Let us suppose the sample population is divided into non-numerical sets such as racial groups and sex. These can be plotted along the y axis, with average height being plotted along the x axis. The resultant bar graph is shown in Fig. 15.1(c).

TABLE 15.1 **Frequency of heights (measured to the nearest 2 cm) of a sample of humans**

Height/cm	Frequency
140	0
142	1
144	1
146	6
148	23
150	48
152	90
154	175
156	261
158	352
160	393
162	462
164	458
166	443
168	413
170	264
172	177
174	97
176	63
178	46
180	17
182	7
184	4
186	0
188	1
190	0

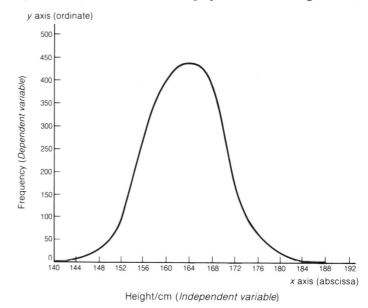

Fig. 15.1(a) Graph of frequency against height/cm for a sample of humans

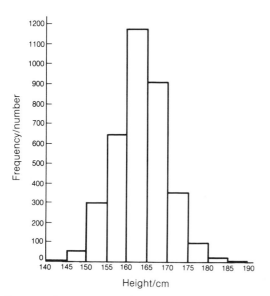

Fig. 15.1(b) Histogram showing height frequencies in a sample human population

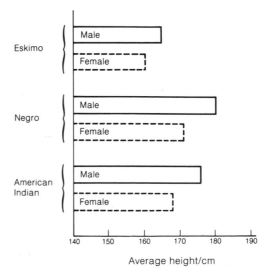

Fig. 15.1(c) Bar graph showing average height variation according to racial group and sex

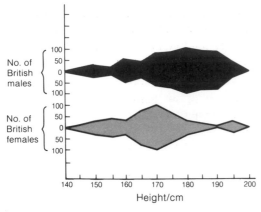

Fig. 15.1(d) Kite graphs to show the height frequency for British males and females in the population sample

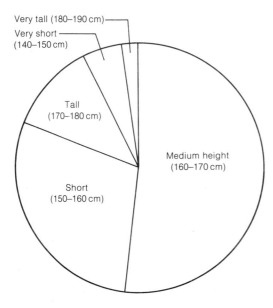

Fig. 15.1(e) Pie chart to show the relative proportions of five height categories in a population sample

15.1.5 Kite graph

This is a form of bar graph, which gives more detailed information on the frequency of a non-numerical variable. To take the information given in Fig. 15.1(c), it simply reveals the average height of each group and not the frequency at different heights. In a kite graph, the frequency of particular heights is plotted vertically for certain non-numerical variables, e.g. males and females. A kite graph is shown in Fig. 15.1(d).

15.1.6 Pie chart

Pie charts are a simple and clearly visible means of showing how a whole sample is divided up into specified parts. A circle (the pie) represents the whole and it is subdivided into different sized sections according to the relative proportions of each constituent part. To be effective the pie chart should not be divided into a large number of portions, nor should it be used when it is necessary to read off precise information from the chart. It is simply a means of giving an idea of relative proportions. Fig. 15.1(e) shows a pie chart depicting the proportions of our sample which fall within broad height ranges.

15.2 Types of variation

15.2.1 Continuous variation

Certain characteristics within a population vary only very marginally between one individual and the next. This results in a gradation from one extreme to the other, called continuous variation. The height and weight of organisms are two characteristics which show such a gradation. If a frequency distribution for such a characteristic is plotted, a bell-shaped graph similar to that in Fig. 15.1(a) is obtained. This is called a **normal distribution curve** or **Gaussian curve** (after the mathematician Fredrick Gauss). It is discussed further in Section 15.2.2.

Characteristics which show continuous variation are controlled not by one, but by the combined effect of a number of genes, called **polygenes**. Thus any character which results from the interaction of many genes is called a **polygenic character**. The effect of an individual gene is small, but their combined effect is marked. The random assortment of the genes during prophase I of meiosis ensures that individuals possess a range of genes from any polygenic complex. Where a group of genes all favouring the development of a tall individual combine, a very tall individual results. A combination of genes favouring small size results in a very short individual. These extremes are rare because it is probable that an individual will possess genes from both extremes. The combined effect of these genes produces individuals of intermediate height.

15.2.2 The normal distribution (Gaussian) curve

Fig. 15.2(a) shows a normal distribution curve; its bell-shape is typical for a feature which shows continuous variation, e.g. height in humans. The graph is symmetrical about a central value. Occasionally the curve is shifted slightly to one side. This is called a skewed distribution and is illustrated in Fig. 15.2(b). There are three main

Family	Number of children
A	0
B	1
C	1
D	1
E	2
F	2
G	3
H	4
I	6
J	6
K	7

terms used in association with normal distribution curves, whether skewed or not. To illustrate these terms let us consider the following set of values for the number of children in eleven different families.

The mean (arithmetic mean)
This is the average of a group of values. In our example opposite this is found by totalling the number of children in all families and dividing it by the number of families.
Total children in all families
$= 0 + 1 + 1 + 1 + 2 + 2 + 3 + 4 + 6 + 6 + 7 = 33$
Total number of families A–K $= 11$

Mean $= 33 \div 11 = 3$

The mode
This is the single value of a group which occurs most often. In our example more families have one child than any other number. The mode is therefore equal to 1.

The median
This is the central or middle value of a set of values. In our example the values are already arranged in ascending order of the number of children in each family. There are eleven families. The sixth family in the series (family F) is therefore the middle family of the group. There are five families (A–E) with the same number or fewer children, and five families (G–K) with more children. As family F has two children the median is 2.

Fig. 15.2(a) shows a typical symmetrical normal distribution curve in which the mean and mode (and often the median) have the same value.

Fig. 15.2(b) shows a skewed distribution in which the mean, mode and median all have different values.

The mean height of a sample population gives a good indication of its relative height compared to other sample populations. It does not, however, give any indication of the distribution of height within the sample. Indeed, the mean can be misleading. A population made up of individuals who were either 140 or 180 cm tall would have a mean of 160 cm, and yet no single individual would be anywhere near this height. It is therefore useful to have a value which gives an indication of the range of height either side of the mean. This value is called the **standard deviation (SD)**. It is calculated as follows:

$$SD = \sqrt{\frac{\Sigma d^2}{n}}$$

$\Sigma =$ the sum of
$d =$ difference between each value in the sample and the mean
$n =$ the total number of values in the sample

How then does the standard deviation provide information on the range within a sample? Let us suppose the mean height of a sample human population is 170 cm and its standard deviation is ± 10 cm. This means that over two thirds (68%) of the sample have heights which are within 10 cm of 170 cm, i.e. 68% of the sample have heights between 160 cm and 180 cm. Furthermore we can say that 95% of the sample lie within two standard deviations of the mean. In our example two standard deviations $= 2 \times 10 = 20$ cm. In other words, 95% of the sample have heights between 150 cm and 190 cm. Fig. 15.2(c), at the top of the next page, illustrates these values.

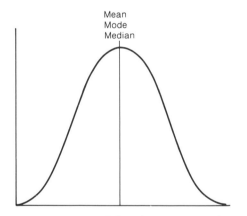

Mean
Mode
Median

Fig. 15.2(a) A normal distribution curve where the mean, mode and median have the same value

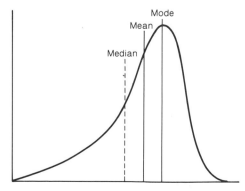

Mode
Mean
Median

Fig. 15.2(b) A skewed distribution where the mean, mode and median have different values

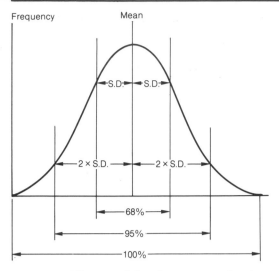

Fig. 15.2(c) *The normal distribution curve showing the values for standard deviation*

15.2.3 Discontinuous (discrete) variation

Certain features of individuals in a population do not show a gradation between extremes but instead fall into a limited number of distinct forms. There are no intermediate types. For example, humans may be separated into distinct sets according to their blood groups. In the ABO system there are just four groups: A, B, AB and O. Unlike continuous variation, which is controlled by many genes (polygenes), a feature which exhibits discontinuous variation is normally controlled by a single gene. This gene may have two or more alleles. Features exhibiting discontinuous variation are normally represented on histograms, bar graphs or pie charts.

15.3 The chi-squared test

Imagine tossing a coin 100 times. It is reasonable to expect it to land heads on 50 occasions and tails on 50 occasions. In practice it would be unusual if these results were obtained (try it if you like!). If it lands heads 55 times and tails only 45 times, does this mean the coin is weighted or biased in some way, or is it purely a chance deviation from the expected result?

The **chi-squared test** is the means by which the statistical validity of results such as these can be tested. It measures the extent of any deviation between the expected and observed results. This measure of deviation is called the **chi-squared value** and is represented by the Greek letter chi, shown squared, i.e. χ^2. To calculate this value the following equation is used:

$$\chi^2 = \sum \frac{d^2}{x}$$

where $\sum$ = the sum of

d = difference between observed and expected results (the deviation)

x = the expected result

Using our example of the coin tossed 100 times, we can calculate the chi-squared value. We must first calculate the deviation from the expected number of times the coin should land heads:

Expected number of heads in 100 tosses of the coin (x) = 50

Actual number of heads in 100 tosses of the coin $\quad$ = 55

Deviation (d) $\quad$ 5

Therefore $\dfrac{d^2}{x} = \dfrac{5^2}{50} = \dfrac{25}{50} = 0.5$

We then make the same calculation for the coin landing tails.

Expected number of tails in 100 tosses of the coin (x) = 50

Actual number of tails in 100 tosses of the coin $\quad$ = 45

Deviation (d) $\quad$ 5

Therefore $\dfrac{d^2}{x} = \dfrac{5^2}{50} = \dfrac{25}{50} = 0.5$

The chi-squared value can now be calculated by adding these values:

Therefore $\chi^2 = 0.5 + 0.5 = 1.0$.

The whole calculation can be summarized thus:

$$\chi^2 = \sum \frac{d^2}{x}$$

$$\chi^2 = \overset{\text{Heads}}{\left[\frac{(55-50)^2}{50}\right]} + \overset{\text{Tails}}{\left[\frac{(50-45)^2}{50}\right]}$$

$$= \left[\frac{(5)^2}{50}\right] + \left[\frac{(5)^2}{50}\right]$$

$$= \frac{1}{2} + \frac{1}{2}$$

$$= 1.0$$

To find out whether this value is significant or not we need to use a chi-squared table, part of which is given in Table 15.2. Before trying to read these tables it is necessary to decide how many **classes of results** there are in the investigation being carried out. In our case there are two classes of results, 'heads' and 'tails'. This corresponds to one degree of freedom. We now look along the row showing 2 classes (i.e. one degree of freedom) for our calculated value of 1.0. This lies between the values of 0.45 and 1.32 on the table. Looking down this column we see that this corresponds to a probability between 0.50 (50%) and 0.25 (25%). This means that the probability that chance alone could have produced the deviation is between 0.50 (50%) and 0.25 (25%). If this probability is greater than 0.05 (5%), the deviation is said to be **not significant**. In other words the deviation is due to chance. If the deviation is less than 0.05 (5%), the deviation is said to be **significant**. In other words, some factor other than chance is affecting the results. In our example the value is greater than 0.05 (5%) and so we assume the deviation is due to chance. Had we obtained 60 heads and 40 tails, a chi-squared value of slightly less than 0.05 (5%) would be obtained, in which case we would question the validity of the results and assume the coin might be weighted or biased in some way. This test is especially useful in genetic experiments.

In *Drosophila*, normal (wild-type) wings are dominant to vestigial wings. Suppose we cross two normal-winged individuals both believed to be heterozygous for this character. We should expect a 3:1 ratio of normal wings to vestigial wings. In practice, of 48 offspring produced, 30 have normal wings and 18 have vestigial wings. Is this close enough to a 3:1 ratio to justify the view that both parents were heterozygous?

Applying the chi-squared test:

	Normal wings	Vestigial wings
Expected number of *Drosophila* (x)	36	12
Actual number of *Drosophila*	30	18
Deviation (d)	6	6
$d^2 =$	36	36

$$\chi^2 = \sum \frac{d^2}{x}$$

$$= \frac{36}{36} + \frac{36}{12}$$

$$= 1.0 + 3.0$$

$$= 4.0$$

With two classes of results (vestigial and normal wings) there is just one degree of freedom. Using the relevant row on the chi-squared table we find that the value of 4.0 lies between 3.84 and 5.41, i.e. 0.05 (5%) and 0.02 (2%), which means that the possibility that the deviation is due to chance is less than 5%. The deviation is therefore significant and we cannot assume the parents are heterozygous.

In another experiment domestic fowl with walnut combs were crossed with each other. The expected offspring ratio of comb types was 9 walnut, 3 rose, 3 pea and 1 single. In the event, the 160 offspring produced 93 walnut combs, 24 rose combs, 36 pea combs and 7 single combs. Applying the chi-squared test:

	Walnut	Rose	Pea	Single
Expected number of comb types (χ)	90	30	30	10
Actual number of comb types	93	24	36	7
Deviation (d)	3	6	6	3
$d^2 =$	9	36	36	9

$$\chi^2 = \sum \frac{d^2}{x}$$

$$\therefore \chi^2 = \frac{9}{90} + \frac{36}{30} + \frac{36}{30} + \frac{9}{10}$$

$$= \frac{1}{10} + \frac{12}{10} + \frac{12}{10} + \frac{9}{10}$$

$$= \frac{34}{10}$$

$$= 3.4$$

In this instance there are four classes of results (walnut, rose, pea and single) and this is equivalent to three degrees of freedom. We must

TABLE 15.2 **Part of a χ^2 table (based on Fisher)**

Degrees of freedom	Number of classes	χ^2							
1	2	0.00	0.10	0.45	1.32	2.71	3.84	5.41	6.64
2	3	0.02	0.58	1.39	2.77	4.61	5.99	7.82	9.21
3	4	0.12	1.21	2.37	4.11	6.25	7.82	9.84	11.34
4	5	0.30	1.92	3.36	5.39	7.78	9.49	11.67	13.28
5	6	0.55	2.67	4.35	6.63	9.24	11.07	13.39	15.09
Probability that deviation is due to chance alone		0.99 (99%)	0.75 (75%)	0.50 (50%)	0.25 (25%)	0.10 (10%)	0.05 (5%)	0.02 (2%)	0.01 (1%)

therefore use this row to determine whether the deviations are significant. The value lies between 2.37 and 4.11 which is equivalent to a probability of 0.5 (50%) to 0.25 (25%). This deviation is not significant and is simply the result of statistical chance.

15.4 Origins of variation

Variation may be due to the effect of the environment on an organism. For example, the action of sunlight on a light-coloured skin may result in its becoming darker. Such changes have little evolutionary significance as they are not passed from one generation to the next. Much more important to evolution are the inherited forms of variation which result from genetic changes. These genetic changes may be the result of the normal and frequent reshuffling of genes which occurs during sexual reproduction, or as a consequence of mutations.

15.4.1 Environmental effects

We saw in the previous chapter that the final appearance of an organism (phenotype) is the result of its genotype and the effect of the environment upon it. If organisms of identical genotype are subject to different environmental influences, they show considerable variety. If one of a pair of genetically identical plants is grown in a soil deficient in nitrogen, it will not attain the height of the other grown in a soil with sufficient nitrogen. Because environmental influences are themselves very various, and because they often form gradations, e.g. temperature, light intensity, they are largely responsible for continuous variation within a population.

15.4.2 Reshuffling of genes

The sexual process in organisms has three inbuilt methods of creating variety:

1. The mixing of two different parental genotypes where cross-fertilization occurs.

2. The random distribution of chromosomes during metaphase I of meiosis.

3. The crossing over between homologous chromosomes during prophase I of meiosis.

These changes, which were dealt with in more detail in Section 13.4.2, do not bring about major changes in features but rather create new combinations of existing features.

Mutations
Any change in the structure or the amount of DNA of an organism is called a **mutation**. Most mutations occur in somatic (body) cells and are not passed from one generation to the next. Only those mutations which occur in the formation of gametes can be inherited. These mutations produce sudden and distinct differences between individuals. They are therefore the basis of discontinuous variation.

TABLE 15.3

Message on telegram	Equivalent form of gene mutation	Likely result of receiving the message
Meet station 2100 hours today	Normal	Individuals meet as arranged.
Meet Met station 2100 hours today	Duplication	Individuals arrive at correct time, one at the prearranged station, the other at the nearest station on the Metropolitan line (or London police station)
Met station 2100 hours today	Deletion	
Meet bus station 2100 hours today	Addition	Individuals arrive on time but one at the bus station, the other at the prearranged station
Meet station 1200 hours today	Inversion	Individuals arrive at correct place – but 9 hours apart
Meet station 1100 hours today	Substitution	Individuals arrive at correct place but 12 hours apart
Meet sat on it 2100 hours today	Inversion (including translocation)	Message incomprehensible

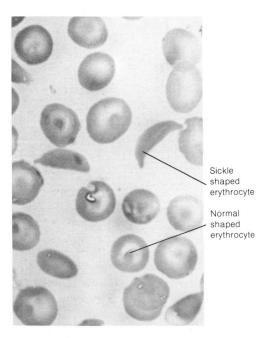

Blood with sickle cells (× 1000)

15.4.3 Changes in gene structure (point mutations)

A change in the structure of DNA which occurs at a single locus on a chromosome is called a **gene mutation** or **point mutation**. In Section 12.5 we saw that the genetic code, which ultimately determines an organism's characteristics, is made up of a specific sequence of nucleotides on the DNA molecule. Any change to one or more of these nucleotides, or any rearrangement of the sequence, will produce the wrong sequence of amino acids in the protein it makes. As this protein is often an enzyme, it may result in it having a different molecular shape and hence prevent it catalysing its reaction. The result will be that the end product of that reaction cannot be formed. This may have a profound effect on the organism. For example, a gene mutation may result in the absence of pigments such as melanin. The organism will be unpigmented, i.e. an albino. There are many forms of gene mutation.

1. Duplication – a portion of a nucleotide chain becomes repeated.

2. Addition (insertion) – an extra nucleotide sequence becomes inserted in the chain.

3. Deletion – a portion of the nucleotide chain is removed from the sequence.

4. Inversion – a nucleotide sequence becomes separated from the chain. It rejoins in its original position, only inverted. The nucleotide sequence of this portion is therefore reversed.

5. Substitution – one of the nucleotides is replaced by another which has a different organic base.

To illustrate these different types, let us imagine each nucleotide is equivalent to a letter of the alphabet. The sequence of nucleotides therefore makes up a sentence or groups of sentences which can be understood by the cell's chemical machinery as the instructions for making specific proteins. If a mutation results in the instructions being incomprehensible, the cell will be unable to make the appropriate protein. In most cases this will result in the death of the cell or organism at an early stage. Sometimes, however, the mutation will result in inaccurate, and yet comprehensible, instructions being given. A protein may well be produced, but it is the wrong one. The defect may create some phenotypic change, but not of sufficient importance to cause the death of the organism.

Imagine a telegram to confirm the details of an earlier arrangement to meet at a pre-arranged station. If we alter just one or two letters each time, the message may either be totally incomprehensible or it may be understood by the receiver but not in the way intended by the sender. Table 15.3, above, gives some examples.

A gene mutation in the gene producing haemoglobin results in a defect called **sickle-cell anaemia**. The replacement of just one base in the DNA molecule results in the wrong amino acid being incorporated into two of the polypeptide chains which make up the haemoglobin molecule. The abnormal haemoglobin causes red blood cells to become sickle-shaped, resulting in anaemia and possible death. The detailed events are illustrated in Fig. 15.3.

The mutant gene causing sickle-cell anaemia is recessive. In the homozygous recessive state, the individual suffers the disease and frequently dies. In the heterozygous state, the individual has 30–40% sickle cells, the rest being normal. This is called the sickle-cell trait. These individuals suffer less severe anaemia and rarely die from the

1. *The DNA molecule which codes for the beta amino acid chain in haemoglobin has a mutation whereby the base adenine replaces thymine.*

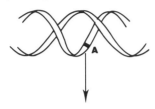

2. *The mRNA produced has the triplet codon GUA for amino acid valine) rather than GAA (for amino acid glutamic acid).*

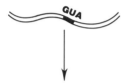

3. *The beta amino acid chain produced has one glutamic acid molecule replaced by a valine molecule.*

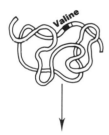

4. *The haemoglobin molcule containing the abnormal beta chains forms abnormal long fibres when the oxygen level of the blood is low. This haemoglobin is called haemoglobin-S.*

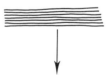

5. *Haemoglobin-S causes the shape of the red blood cell to become crescent (sickled) shaped.*

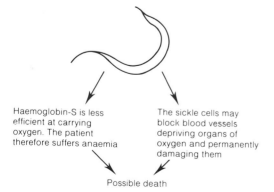

Haemoglobin-S is less efficient at carrying oxygen. The patient therefore suffers anaemia

The sickle cells may block blood vessels depriving organs of oxygen and permanently damaging them

Possible death

Fig. 15.3 Sequence of events whereby a gene mutation causes sickle cell anaemia

condition. As they still suffer some disability, it might be expected that the disease would be very rare, if not completely eliminated, by natural selection. In parts of Africa, however, it is very common. The reason is that the malarial parasite, *Plasmodium*, cannot easily invade sickle cells. Individuals with either sickle-cell condition are therefore more resistant to severe attacks of malaria. In the homozygous recessive condition, this resistance is still sufficient to offset the considerable disadvantage of having sickle-cell anaemia. In the heterozygous condition (sickle-cell trait), however, the advantage of being resistant to severe malarial attacks outweighs the disadvantage of the mild anaemia the individual suffers. In malarial regions of the world the mutant gene is selected in favour of the one producing normal haemoglobin. Outside malarial regions, there is no advantage in being resistant to malaria, and the disadvantage of suffering anaemia results in selection *against* the mutant gene.

One relatively common gene mutation in European countries causes **cystic fibrosis**, which is the result of a recessive gene. Around 1 in 20 people in Europe are heterozygous for the condition, i.e. they are carriers. They are usually perfectly healthy. For this reason, and because the mutant gene is very stable, it is unlikely to disappear from the population. At present around 1 in 2000 babies is born with cystic fibrosis. The disease causes the mucus secretions of the pancreas, the intestines and lungs to be more viscous than normal. These secretions easily dry up and block glands and ducts causing the glands to be destroyed.

Dominant gene mutations are rarer but include **Huntington's chorea**. The latter is characterized by involuntary muscular movement and progressive mental deterioration. The mutant gene is so rare (around 1 in 100 000 people carry it) that it occurs almost exclusively in the heterozygous state. It would most probably be lethal in the homozygous condition.

15.4.4 Changes in whole sets of chromosomes

Sometimes organisms occur that have additional whole sets of chromosomes. Instead of having a haploid set in the sex cells and a diploid set in the body cells, they have several complete sets. This is known as **polyploidy**. Where three sets of chromosomes are present, the organism is said to be **triploid**. With four sets, it is said to be **tetraploid**.

Polyploidy can arise in several different ways. If gametes are produced which are diploid and these self-fertilize, a tetraploid is produced. If instead the diploid gamete fuses with a normal haploid gamete, a triploid results. Polyploidy can also occur when whole sets of chromosomes double after fertilization.

Tetraploid organisms have two complete sets of homologous chromosomes and can therefore form homologous pairings during gamete production by meiosis. Triploids, however, cannot form complete homologous pairings and are usually sterile. They can only be propagated by asexual means. The type of polyploidy whereby the increase in sets of chromosomes occurs within the same species is called **autopolyploidy**. The actual number of chromosomes in an autopolyploid is always an exact multiple of its haploid number. Autopolyploidy can be induced by a chemical called **colchicine** which is extracted from certain crocus corms. Colchicine inhibits spindle formation and so prevents chromosomes separating during anaphase.

Sometimes hybrids can be formed by combining sets of chromosomes from species with different chromosome numbers. These hybrids are ordinarily sterile because the total number of chromosomes does not allow full homologous pairing to take place. If, however, the hybrid has a chromosome number which is a multiple of the original chromosome number, a new fertile species is formed. The species of wheat used today to make bread was formed in this way. The basic haploid number of wild grasses is seven. A tetraploid with 28 chromosomes called emmer wheat was accidentally cross-fertilized with a wild grass with 14 chromosomes. The resultant wheat with 42 chromosomes is today the main cultivated variety. Having a chromosome number which is a multiple of the original haploid number of 7, it is fertile. This form of polyploidy is called **allopolyploidy**.

Polyploidy is rare in animals, but relatively common in plants. Almost half of all flowering plants (angiosperms) are polyploids, including many important food plants. Wheat, coffee, bananas, sugar cane, apples and tomatoes all have polyploid forms. The polyploid varieties often have some advantage. Tetraploid apples, for example, form larger fruits and tetraploid tomatoes produce more vitamin C.

15.4.5 Changes in chromosome number

Sometimes it is an individual chromosome, rather than a whole set, which fails to separate during anaphase. If, for example, in humans one of the 23 pairs of homologous chromosomes fails to segregate during meiosis, one of the gametes produced will contain 22 chromosomes and the other 24, rather than 23 each. This is known as **non-disjunction** and is often lethal. The condition where an organism possesses an additional chromosome is represented as $2n + 1$; where one is missing, $2n - 1$. Where two additional chromosomes are present it is represented as $2n + 2$ etc.

One frequent consequence of non-disjunction in humans is **Down's syndrome** (mongolism). In this case the 21st chromosome fails to segregate and the gamete produced possesses 24 chromosomes. The fusion of this gamete with a normal one with 23 chromosomes results in the offspring having 47 ($2n + 1$) chromosomes. Non disjunction does occur with other chromosomes but these normally result in the foetus aborting or the child dying soon after birth. The 21st chromosome is relatively small, and the offspring is therefore able to survive. Down's syndrome children have disabilities of varying magnitude. Typically they have a flat, broad face, squint eyes with a skin fold in the inner corner and a furrowed and protruding tongue. They have a low IQ and a short life expectancy.

Non-disjunction in the case of Down's syndrome appears to occur in the production of ova rather than sperm. Its incidence is related to the age of the mother. The chance of a teenage mother having a Down's syndrome child is only one in many thousands. A forty-year-old mother has a one in a hundred chance and by forty-five the risk is three times greater. The risk is unaffected by the age of the father. Non-disjunction of the sex chromosomes can occur. One example is **Klinefelter's syndrome**. This may result in individuals who have the genetic constitution XXY, XXXY or XXXXY. These individuals are phenotypically male but have small testes and no sperm in the ejaculate. There may be abnormal breast development and the body proportions are generally female. The greater the number of Xs the more marked is the condition. As individuals are

47 XX + 21

Karyotype of Down's Syndrome

phenotypically male, this indicates that the presence of a Y chromosome is the cause of maleness. This is borne out by a second abnormality of the sex chromosomes. Individuals with **Turner's syndrome** have one missing X chromosome. Their genetic constitution is therefore XO and they have only 45 (2n − 1) chromosomes. Individuals with this condition often do not survive pregnancy and are aborted. Those that do are phenotypically female, but small in stature and sexually immature. Despite having a single X chromosome, like males, they are female, indicating again that the Y chromosome is the cause of maleness.

15.4.6 Changes in chromosome structure

During meiosis it is normal for homologous pairs of chromosomes to form chiasmata. The chromatids break at these points and rejoin with the corresponding portion of chromatid on its homologous partner. It is not surprising that from time to time mistakes arise during this process. Indeed, it is remarkable that these chromosome mutations do not occur more frequently. There are four types:

1. Deletion – a portion of a chromosome is lost (Fig. 15.4a). As this involves the loss of genes, it can have a significant effect on an organism's development, often proving lethal.

2. Inversion – a portion of chromosome becomes deleted, but becomes reattached in an inverted position. The sequence of genes on this portion are therefore reversed (Fig. 15.4b). The overall genotype is unchanged, but the phenotype may be altered. This indicates that the sequence of genes on the chromosome is important.

3. Translocation – a portion of chromosome becomes deleted and rejoins at a different point on the same chromosome or with a different chromosome (Fig. 15.4c). The latter is equivalent to crossing over except that it occurs between non-homologous chromosomes.

4. Duplication – a portion of chromosome is doubled, resulting in repetition of a gene sequence (Fig. 15.4d).

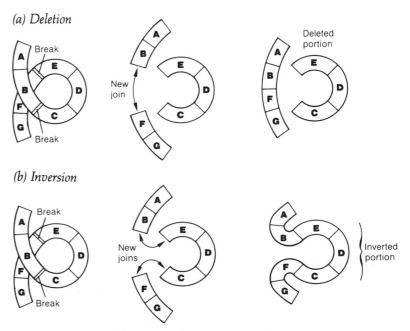

(a) Deletion

(b) Inversion

Fig. 15.4 *Diagrams illustrating the four types of chromosome mutation*

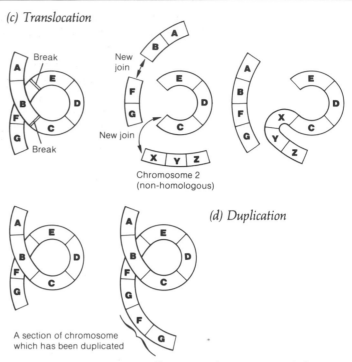

(c) Translocation

(d) Duplication

A section of chromosome which has been duplicated

Fig. 15.4 (cont.) Diagrams illustrating the four types of chromosome mutation

15.5   Causes of mutations

Mutations occur continually. There is a natural mutation rate which varies from one species to another. In general, animals with shorter life cycles, and therefore more frequent meiosis, show a greater rate of mutation. A typical rate of mutation is 1 or 2 new mutations per 100 000 genes per generation.

This natural mutation rate can be increased artificially by certain chemicals or energy sources. Any agent which induces mutations is called a **mutagen**. Most forms of high energy radiation are capable of altering the structure of DNA and thereby causing mutations. These include ultra-violet light, X-rays and gamma rays. High energy particles such as α and β particles and neutrons are even more dangerous mutagens.

A number of chemicals also cause mutations. We saw in Section 15.4.4 that colchicine inhibits spindle formation and so causes polyploidy. Other chemical mutagens include formaldehyde, nitrous acid and mustard gas.

15.6   Genetic screening and counselling

As our knowledge of inheritance has increased, more and more disabilities have been found to have genetic origins. Some of these disabilities cannot be predicted with complete accuracy. In Down's syndrome, for example, it is impossible to give a precise prediction of its occurrence for any individual. The risk for a mother of a particular age can, however, be calculated. Other disabilities like haemophilia, cystic fibrosis, some forms of muscular dystrophy and Huntington's chorea can be predicted fairly accurately, provided

Let F = gene for normal mucus development
f = gene for abnormal mucus development associated with cystic fibrosis

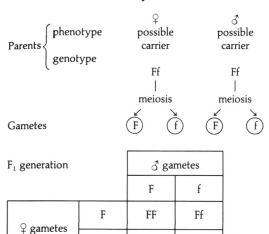

25% normal children
50% normal children, but carriers of the gene
25% children with cystic fibrosis

enough information on the history of the disease in the family is known. Genetic counselling has developed in order to research the family history of inherited disease and to advise parents on the likelihood of it arising in their children. Imagine a mother, whose family has a history of cystic fibrosis. If she herself is unaffected but possesses the gene, she can only be heterozygous for the condition. Suppose she wishes to produce children by a man with no history of the disease in his family. It must be assumed that he does not carry the gene for the disease and therefore none of the children will suffer from it, although they may be carriers. If, on the other hand, the potential father's family has a history of the disease, it is possible he too carries the gene. As we see from the genetic diagram opposite, it is possible to advise the parents that there is a one in four chance of their children being affected. The gene for cystic fibrosis is recessive and autosomal (i.e. not sex-linked).

On the basis of this advice the parents can choose whether or not to have children. With a very detailed knowledge of the disease in each of the parent's families it may even be possible to establish for certain whether they are carriers or not. It is now possible to carry out tests to establish with some accuracy whether an individual is heterozygous for the gene, and so make precise predictions about the likelihood of having a child with cystic fibrosis.

What then of the parents who have children knowing that there is a greater than usual possibility of them inheriting a genetic defect? Is there any means of establishing whether a child is affected, before it is born? The answer is yes, for some defects at least. Doctors can now diagnose certain genetic defects in a foetus, by studying samples of cells taken from the amniotic fluid which surrounds the foetus. In a process called **amniocentesis** a little of the fluid is removed using a hypodermic syringe which is inserted into the uterus through the abdominal wall. In the amniotic fluid are skin cells which are continually lost from the foetus as the skin grows. The fluid removed is centrifuged to separate the skin cells. These are then stained to show up the chromosomes and examined under the microscope. Certain genetic defects such as Down's syndrome can be detected directly as the additional chromosomes are easily seen. Biochemical tests on the cells, or even the amniotic fluid (which contains much foetal urine), may reveal other genetic defects. On the basis of these tests the parents can decide whether or not to have the pregnancy terminated.

15.6.1 Gene tracking

Gene tracking has proved a useful aid in genetic counselling. It is first necessary to find out on which chromosome a defective gene is located, something achieved through mapping chromosomes (Section 14.4.3). The technique of making selected crosses of pure-breeding parents and collecting thousands of F_1 offspring in order to determine cross-over values and hence map chromosomes, may be appropriate with *Drosophila*, but can hardly be used with humans. Instead the inheritance of other, easily distinguished features such as blood groups, are traced in families to act as **genetic markers**. Study of the correlation between certain blood group alleles and the occurrence of a genetic disease can determine whether or not the gene for the disease is on the same chromosome as that for blood groups. If one genetic marker is not linked to the disease in question another must be tried and so on until the one which shows linkage with the disease is found. Linked markers are then used to work out whether or not someone carries a disease – this is the process of gene tracking. The technique is used with Huntington's disease (also called Huntington's chorea), a neuropsychiatric disorder leading to the loss of control of movements. It is caused by an autosomal dominant allele and sufferers are usually heterozygous. As the disease often fails to manifest itself until those affected are in their 40's or 50's, it is usually passed on to their children even before they realize they carry the gene. Any technique which allows the presence of the gene to be detected early helps individuals to decide whether or not to have children, and also shows the size of the risk of those children carrying the gene. Gene tracking has not only achieved this, but also established that some sufferers of Huntington's disease are homozygous for the defect.

15.7 Questions

1. The diagram below represents the molecular structure of a nucleic acid.

(a) (i) Where in a eukaryotic cell would you expect to find the molecule represented by the diagram? *(1 mark)*

(ii) How would the location of the molecule differ in a prokaryotic cell? *(1 mark)*

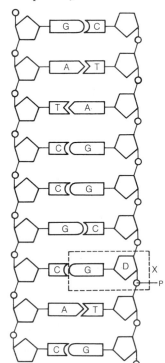

(b) Name the part of the molecule inside the dotted line labelled X in the diagram. *(1 mark)*

(c) What do the following letters in the diagram represent?

D P C G A T *(3 marks)*

(d) During replication of the molecule, the letters GAT CCG CAC might be copied incorrectly, the 'T' being omitted. The 'code' would then be read as GAC CGC AC.

(i) What is such a change called? *(1 mark)*

(ii) Suggest *two* possible effects of such a change in a cell. *(2 marks)*

(iii) How may such changes contribute to evolution? *(2 marks)*

(Total 11 marks)

London Board June 1989, Paper I, No. 4

2. A cross was made between two maize (*Zea mays*) varieties called Tom Thumb and Black Mexican which differed markedly in ear length. The ear length of both parents, the F_1 and the F_2 generations, was measured to the nearest centimetre. This is given in the table at the top of the next page, with the number of ears in each length category (for example, 14 of the F_1 plants produced ears 12 cm in length).

(a) Both parents showed variation in ear length. The variation in the F_1 is comparable to the average of the parental variations while the variation in the F_2 is greater than that found within the parental lines or the F_1. How would you explain this?

(b) How would you explain the whole spectrum of different degrees of expression of this particular characteristic?

(c) What is the term used to describe the range of the phenotypes in the above example?

(Total 7 marks)

Cambridge Board November 1983, Paper II, No. 4

Ear length (cm)																
	5	6	7	8	9	10	11	12	13	14	15	16	17	18	19	20	21
Black Mexican parents									3	11	12	15	26	15	10	7	2
Tom Thumb parents	4	21	24	8													
F₁					1	12	12	14	17	9	4						
F₂			1	10	19	26	47	73	68	68	39	25	15	9	1		

3. In people who suffer from the genetic condition known as sickle-cell anaemia, up to 40% of the red blood cells are abnormal in shape and also function inefficiently. The condition is widespread in African Negroes and is caused by a single gene (H^s) which is equal in dominance to its normal allele (H). The heterozygous genotype is usually encountered because in a high proportion of cases the homozygous condition is lethal. The homozygous (H^sH^s) genotype should be considered lethal in answering the following questions.

(a) Show diagrammatically the proportion of offspring carrying the sickle-cell gene in the F₁ of a cross between
  (i) a normal parent and one carrying the sickle-cell gene;
  (ii) two parents, both carrying the sickle-cell gene.

(b) The percentages of heterozygotes occurring in different areas are shown in circles on the map of parts of Uganda and Kenya given below. The distribution of the malaria parasite, *Plasmodium falciparum*, is also shown.
Explain fully the relationship between the distribution of the sickle-cell anaemia gene and that of *P. falciparum*.

(c) *P. falciparum* is transmitted to man by the bite of the mosquito. Suggest an explanation for the distribution of *P. falciparum* shown on the map.

Welsh Joint Education Committee June 1983, Paper I, No. 6

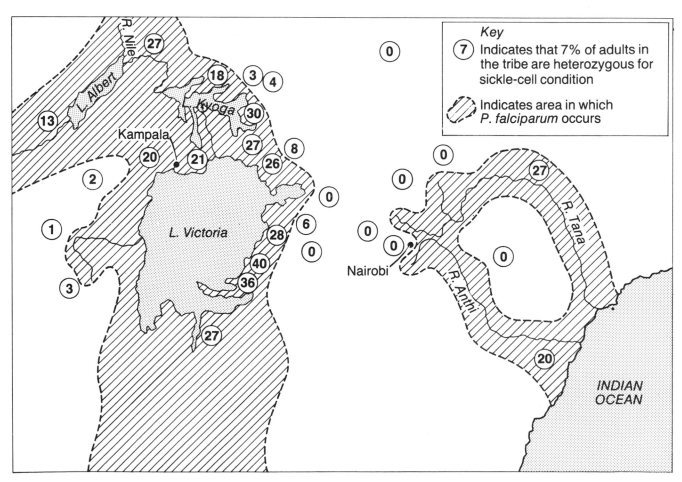

4. (a) In what different ways can DNA change when a gene mutates? Name **two** agents which could promote such a mutation. *(4 marks)*

(b) Explain how the changed information in such mutant DNA results in the synthesis of a new protein. *(8 marks)*

(c) Suppose the original protein is an enzyme and that after mutation it has an altered sequence of amino acids. Explain how and why this altered sequence of amino acids could affect the activity of the enzyme. *(4 marks)*

(d) Assume that the new mutant allele is recessive and advantageous to an organism in its normal environment. Describe how natural selection could lead to the spread of this allele through a population. *(4 marks)*

(Total 20 marks)

Joint Matriculation Board June 1984, Paper IIB, No. 1

5. The typical form of the European Swallow-tail butterfly has yellow patches on its wings, but in a rare variety called *nigra* these areas are shaded black. A cross between a typical male and a *nigra* female produced 14 typical offspring and 6 *nigra*.

(a) As a first hypothesis to explain this result it was suggested that the *nigra* variety is caused by a recessive allele of a single gene, which is not sex-linked.

(i) On the basis of this hypothesis, a 1:1 ratio would be expected in the offspring. Construct a genetic diagram to show the genotypes of the parents in this cross, and to show how this ratio could be obtained. Use the letters **N** and **n** to represent the two alleles of the gene. *(3 marks)*

(ii) Complete the following table, filling in the expected numbers (E) of each phenotype on the basis of this hypothesis, and the differences between the observed and expected numbers (O–E).

Phenotype	Observed number O	Expected number E	Difference $O–E$
typical	14		
nigra	6		

(2 marks)

(iii) Using the following formula, calculate the value of chi-squared (χ^2). Show your working.

$$\chi^2 = \sum \frac{(O-E)^2}{E}$$

(3 marks)

(iv) Probability levels (P) corresponding to some values of χ^2 in this case are shown below.

χ^2	0.004	2.71	3.85	6.63
P(%)	95	10	5	1

What does your calculated value for χ^2 indicate in relation to this particular cross? *(1 mark)*

(b) A second hypothesis was suggested, in which the *nigra* variety is produced by interaction between two unlinked genes **A** and **B**.

One of the genes has alleles **A** and **a** and the other gene has alleles **B** and **b**.

The *nigra* phenotype is seen only in individuals which are homozygous for the **a** allele and *either* homozygous *or* heterozygous for the **B** allele (*either* **aa BB** *or* **aa Bb**).

This second hypothesis suggested that the genotypes of the original parents in the above cross were as follows.

The typical parent   **Aa Bb**
The *nigra* parent   **aa Bb**

Construct a genetic diagram to show the genotype and phenotype ratios expected from this cross on the basis of the second hypothesis. *(4 marks)*

(c) (i) Comparison of the observed results with those expected from the second hypothesis gives a value for χ^2 of 0.48. Compare this with the value calculated for the first hypothesis and suggest which hypothesis has the greater probability of being correct. Justify your answer. *(2 marks)*

(ii) State *two* aspects of the data given that might reduce the reliability of the conclusion drawn in (c) (i). *(2 marks)*

(d) Using only the *nigra* offspring from the original cross, suggest a further investigation to test the two hypotheses. Explain how the results of the investigation might show one or the other of the hypotheses to be incorrect. *(3 marks)*

(Total 20 marks)

London Board June 1989, Paper I, No. 16

6. (a) Photograph A shows a set of human chromosomes in which a mutation has occurred.
(i) What is the nature of this mutation?
(ii) What appears to be the criterion by which the complete set of chromosomes has been arranged?

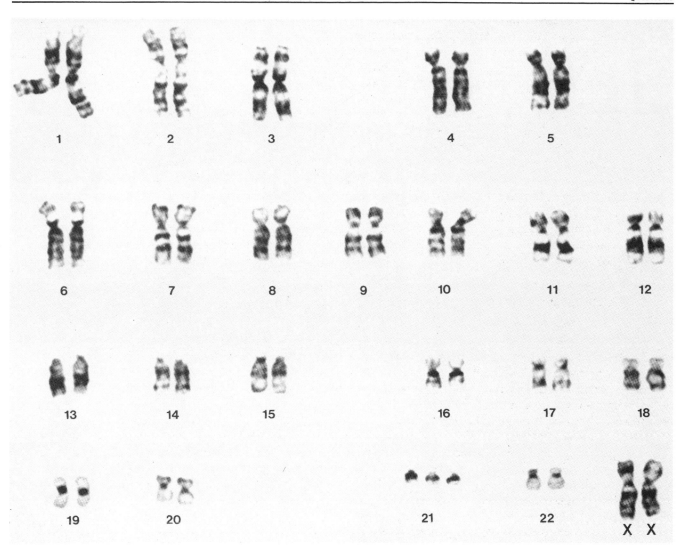

Photograph A

(iii) What is the sex of the person from whom this chromosome set has been taken? Give the reason for your answer.

(b) Photograph B shows a cell at mitosis in which a mutation has occurred.

(i) Name the stage of mitosis shown in the photograph.

(ii) What is the evidence in the photograph that a mutation has occurred?   (*Total 6 marks*)

Associated Examining Board June 1984, Paper I, No. 11

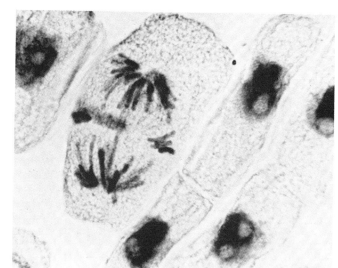

Photograph B

16 *Evolution*

Evolution is the process by which new species are formed from pre-existing ones over a period of time. It is not the only explanation of the origins of the many species which exist on earth, but it is the one generally accepted by the scientific world. Evolution is, in effect, the continuous change from simple to complex organisms.

16.1 The origins of life

There are a number of theories on the origin of life:

1. Steady state theory – This suggests that the earth and the species upon it have always existed. Life therefore had no origin.

2. Creation theory – This is the belief that the earth and the species upon it were created by a single event initiated by a 'super-being' or 'God'.

3. Cosmozoan (Panspermian) theory – This theory states that life arose elsewhere in the universe and arrived on earth by some means, e.g. UFOs.

4. Spontaneous generation theory – This theory contends that life arose from non-living material on a number of separate occasions.

5. Biochemical evolution theory – This theory suggests that life arose from the combination of simple molecules into complex ones and their evolution, via coacervates, into cells.

Of these theories, that of biochemical evolution is the most widely accepted by present-day scientists.

16.1.1 The biochemical evolution theory of the origin of life

In the 1920s, A. I. Oparin in the USSR and J. B. S. Haldane in Britain put forward the theory that the origin of life occurred in three stages:

1. The formation of small molecules like amino acids, organic bases and monosaccharides.

2. The formation of large polymers of these small molecules.

3. The integration of the polymers to form cells.

The formation of small molecules
We now believe that the earth originated about 5000 million years ago. Conditions at that time were vastly different from today; the gases present in the atmosphere included carbon dioxide, methane, hydrogen, ammonia and water. At this time the ozone belts, which now protect the earth from the sun's ultra-violet radiation, had not

Major events in the biochemical theory of the origin of life

Atmosphere of carbon dioxide, methane, hydrogen, ammonia and water on the earth soon after its formation

Ultra-violet radiation from the sun

Electric discharge from storms

Simple organic molecules such as amino acids, adenine and ribose are formed

The soup of organic molecules floating on oceans is concentrated in estuarine mud

Simple organic molecules become polymerized and coacervates are formed

Enzymes catalyse further polymerization and coacervates increase in size, then break up into smaller portions

Lipid layer forms around coacervates which contain self-replicating molecules

Primitive anaerobic prokaryotic cells develop which do not produce oxygen

Oxygen-producing anaerobic autotrophs develop

Aerobic cells develop

Eukaryotic cells develop

Colonial/syncitial forms develop

Multicellular organisms evolve from colonial/syncitial ones

Adaptive radiation and natural selection gives rise to numerous different species including ones capable of colonizing land

formed. It is thought that this ultra-violet radiation could have provided the source of energy by which these molecules combined into more complex ones. In 1953, Urey and Miller carried out a series of experiments to test this hypothesis. They prepared a mixture of gases similar to that originally present on the earth. By passing electric sparks through the mixture, they produced important biological molecules such as amino acids. Similar experiments performed recently using a variety of gas mixtures and ultra-violet radiation have produced similar results. In one experiment very simple nucleic acids, up to six nucleotides long, were formed.

The formation of polymers

The organic molecules so formed probably floated as a thin layer on the surface of the oceans. Wind may have caused this layer to drift and concentrate especially on the mud of estuaries. The concentration may have been great enough for polymers to form by condensation reactions. Special droplets which were rich in these polymers could have formed. These are called **coacervates**. Among these polymers there may have been enzymes which catalysed the formation of more polymers, e.g. starch. The droplets hence increased in size and when they became large enough broke up into smaller coacervates. These in turn increased in size and again broke up. The coacervates in effect replicated themselves.

The formation of cells

The final stage involved the formation of a lipid layer around the coacervate. This was a primitive membrane. Within the coacervate were molecules such as nucleic acids which were capable of replicating themselves. In the absence of oxygen these primitive cells were anaerobic and probably fed by the absorption of molecules from the primeval soup around them. They were prokaryotic in that they possessed no membrane-bounded organelles. The origins of eukaryotic cells are discussed in Section 4.2.2.

These heterotrophic cells which fed on the primeval soup would in time incorporate into themselves all available molecules from their surroundings, and no further evolution would have been possible. However, photosynthetic (autotrophic) organisms must have arisen. These utilized the earth's radiation to build complex organic polymers from simple inorganic molecules. In this way the supply of food for the heterotrophs was constantly replenished. It is unlikely that these early autotrophs formed oxygen, but in time oxygen-producing cells, similar to present-day blue-green bacteria, developed. The oxygen they produced led to the formation of ozone layers above the earth, which prevented the potentially harmful ultra-violet radiation reaching its surface. At the same time aerobic forms of life evolved in the presence of this newly formed oxygen.

16.2 Population genetics

To a geneticist, a population is an interbreeding group of organisms. In theory, any individual in the population is capable of breeding with any other. In other words, the genes of any individual organism are capable of being combined with the genes of any other. The genes of a population are therefore freely interchangeable. The total of all the alleles of all the genes in a population is called the **gene pool**. Within the gene pool the number of times any one allele occurs is referred to as its frequency.

16.2.1 Heterozygotes as reservoirs of genetic variation (the Hardy–Weinberg principle)

If one looks at a particular characteristic in a population, it is apparent that the dominant form expresses itself more often than the recessive one. In almost all human populations, for example, brown eyes occur more frequently than blue. It might be thought, therefore, that in time the dominant form would predominate to the point where the recessive type disappeared from the population completely. The proportion of dominant and recessive alleles of a particular gene remains the same, however. It is not altered by interbreeding. This phenomenon is known as the **Hardy–Weinberg principle**. It is a mathematical law which depends on four conditions being met:

1. No mutations arise.

2. The population is isolated, i.e. there is no flow of genes into, or out of, the population.

3. There is no natural selection.

4. The population is large and mating is random.

While these conditions are probably never met in a natural population, the Hardy–Weinberg principle nonetheless forms a basis for the study of gene frequencies.

To help understand the principle, consider a gene which has a dominant allele A and a recessive one a.

Let p = the frequency of allele A and
q = the frequency of allele a

1st allele	2nd allele	Frequency
A	A	$p \times p = p^2$
A	a	$p \times q$
a	A	$q \times p$ $\Big\}$ 2pq
a	a	$q \times q = q^2$

In diploid individuals these alleles occur in the combinations given opposite.

As the homozygous dominant (AA) combination is 1/4 of the total possible genotypes, there is a 1/4 (25%) chance of a single individual being of this type. Similarly, the chance of it being homozygous recessive (aa) is 1/4 (25%) whereas there is a 1/2 (50%) chance of it being heterozygous. There is a 1/1 (100%) chance of it being any one of these three types. In other words:

homozygous dominant (1/4) + heterozygous (1/2) + homozygous recessive (1/4) = 1.0 (100%)

thus   AA + 2Aa + aa = 1.0 (100%)

and $p^2 + 2pq + q^2$ = 1.0 (100%)

The Hardy-Weinberg principle is expressed as:
$$p^2 + 2pq + q^2 = 1.0$$

(where p and q represent the respective frequencies of the dominant and recessive alleles of any particular gene).

The formula can be used to calculate the frequency of any allele in the population. For example, imagine that a particular mental defect is the result of a recessive allele. If the number of babies born with the defect is one in 25 000, the frequency of the allele can be calculated as follows:

The defect will only express itself in individuals who are homozygous recessive. Therefore the frequency of these individuals (q^2) = 1/25 000 or 0.00004.
The frequency of the allele (q) is therefore $\sqrt{0.00004}$
$$= 0.0063 \text{ approx.}$$

As the frequency of both alleles must be 1.0, i.e. p + q = 1.0, then the frequency of the dominant allele (p) can be calculated.

$$p + q = 1.0$$
$$\therefore p = 1.0 - q$$
$$\therefore p = 1.0 - 0.0063$$
$$\therefore p = 0.9937$$

The frequency of heterozygotes can now be calculated.

From the Hardy-Weinberg formula, the frequency of heterozygotes is 2pq, i.e. $2 \times 0.9937 \times 0.0063 = 0.0125$.

In other words, 125 in 10 000 (or 313 in 25 000) are carriers (heterozygotes) of the allele.

This means that in a population of 25 000 individuals, just one individual will suffer the defect but around 313 will carry the allele. The heterozygotes are acting as a reservoir of the allele, maintaining it in the gene pool. As these heterozygotes are normal, they are not specifically selected against, and so the allele remains. Even if the defective individuals are selectively removed, the frequency of the allele will hardly be affected. In our population of 25 000, there is one individual who has two recessive alleles and 313 with one recessive allele – a total of 315. The removal of the defective individual will reduce the number of alleles in the population by just 2, to 313. Even with the removal of all defective individuals it would take thousands of years just to halve the allele's frequency.

Occasionally, as in sickle cell anaemia (Section 15.4.3), the heterozygote individuals have a selective advantage. This is known as **heterozygote superiority**.

16.2.2 Genetic drift

The Hardy-Weinberg principle is only applicable to large populations. In small populations a situation called **genetic drift** arises. Consider an allele which occurs in 1% of the members of a species. In a population of one million, 10 000 individuals may be expected to possess this allele. Even if some of these fail to pass it on to their offspring, the vast majority are likely to do so. The proportion of individuals with the allele will not be significantly altered in the next generation. If, however, the population is much smaller, say 1000 individuals, only 10 will carry the allele. The effect of some of these failing to pass it on will have a marked effect on its frequency in the next generation. This drift in the frequency of an allele is greater the smaller the population. Indeed, in an extreme example, a population of just 100 will have only a single individual with the allele. If this individual fails to breed, the allele will be lost from the population altogether.

16.3 Theories of evolution

Prior to the last century, popular belief held that new species arose by some form of special creation. During the last century two main theories were put forward, both based on the evolution of new species by the gradual adaptation of existing ones. The theories differed, however, in the mechanism by which this gradual adaptation occurred.

16.3.1 The inheritance of acquired characters (Lamarck)

Jean-Baptiste Lamarck (1744–1829) was a French biologist who put forward the theory that evolutionary change was the result of minor alterations which individuals acquired during their lifetime. These changes were passed on to their offspring. Thus the ancestor of the present giraffe, stretching to reach foliage high up in trees, would develop a slightly longer neck which would be inherited by its offspring. They in their turn, in reaching for food, would acquire still longer necks which they would transmit to their offspring. The present long neck of a giraffe is thus the cumulative result of many generations of reaching up for food. Lamarck believed that the disuse of any character would in time result in its disappearance. On the other hand, the constant use of any character would lead to its further development. While his views have never been widely accepted, he did influence the thinking of Darwin to some degree.

16.3.2 Evolution through natural selection (Darwin/Wallace)

Charles Darwin (1809–1882) became the naturalist on HMS Beagle which sailed in 1832 to South America and Australasia. On the voyage he had an excellent opportunity to examine a wide variety of living plants and animals and his knowledge of geology was invaluable for studying fossils he came across. He was struck by the remarkable likeness between the fossils he found and present-day organisms. At the same time he observed the differences in certain characters that occurred when otherwise similar animals lived in different environments. He noted such differences between the organisms of different continents and between the east and west coast of South America. What impressed him most were the distinct variations between the species which inhabited different islands in a small group 580 miles off the coast of Ecuador. These were the Galapagos Islands. In particular he studied the finches which inhabited each of the islands. While they all had a general resemblance to those on the mainland of Ecuador, they nevertheless differed in certain respects, such as the shape of their beaks. He considered that originally a few finches had strayed from the mainland to these volcanic islands, shortly after their formation. Encountering, as they did, a range of different foods, each type of finch developed a beak which was adapted to suit their diet. Following the five-year voyage, Darwin set about developing his views on the mechanism by which these changes occurred.

Charles Darwin

Quite independently of Darwin, Alfred Wallace had drawn his own conclusions on the mechanism of evolution. Wallace sent Darwin a copy of this theory and Darwin realized that they were in essence the same as his own. As a result, they jointly presented their findings to the Linnaean Society in 1858. A year later Darwin published his book *On the Origin of Species by Means of Natural Selection and the Preservation of Favoured Races in the Struggle for Life*. The essential features of the theory Darwin put forward are:

1. Overproduction of offspring

All organisms produce large numbers of offspring which, if they survived, would lead to a geometric increase in the size of any population.

2. Constancy of numbers

Despite the tendency to increase numbers due to overproduction of offspring, most populations actually maintain relatively constant numbers. The majority of offspring must therefore die, before they are able to reproduce.

3. Struggle for existence

Darwin deduced on the basis of **1** and **2** that members of a species were constantly competing with each other in an effort to survive. In this struggle for existence only a few would live long enough to breed.

4. Variation among offspring

The sexually produced offspring of any species show individual variations (Section 17.1) so that generally no two offspring are identical.

5. Survival of the fittest by natural selection

Among the variety of offspring will be some better able to withstand the prevailing conditions than others. That is, some will be better adapted ('fitter') to survive in the struggle for existence. These types are more likely to survive long enough to breed.

6. Like produces like

Those which survive to breed are likely to produce offspring similar to themselves. The advantageous characteristics which gave them the edge in the struggle for existence are likely to be passed on to the next generation.

7. Formation of new species

Individuals lacking favourable characteristics are less likely to survive long enough to breed. Over many generations their numbers will decline. The individuals with favourable characteristics will breed, with consequent increase in their numbers. The inheritance of one small variation will not, by itself, produce a new species. However, the development of a number of variations in a particular direction over many generations will gradually lead to the evolution of a new species.

In a population of a particular mammal fur length shows continuous variation.

1. *When the average environmental temperature is 10°C, the optimum fur length is 1.5 cm. This then represents the mean fur length of the population.*

2. *A few individuals in the population already have a fur length of 2.0 cm or greater. If the average environmental temperature falls to 5°C these individuals are better insulated and so are more likely to survive to breed. There is a selection pressure favouring individuals with longer fur.*

3. *The selection pressure causes a shift in the mean fur length towards longer fur over a number of generations. The selection pressure continues.*

4. *Over further generations the shift in the mean fur length continues until it reaches 2.0 cm — the optimum length for the prevailing average environmental temperature of 5°C. The selection pressure now ceases.*

1.
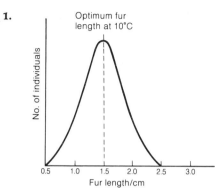

2.

3.

4.
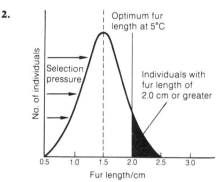
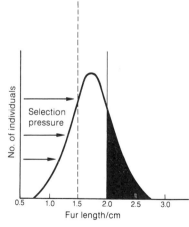
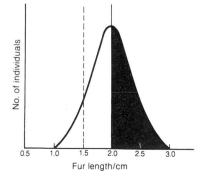

Fig. 16.1(a) Directional selection

16.4   Natural selection

The evolutionary theory of Darwin and Wallace is based on the mechanism of natural selection. Let us look more closely at exactly how this process operates.

Selection is the process by which organisms that are better adapted to their environment survive and breed, while those less well adapted fail to do so. The better adapted organisms are more likely to pass their characteristics to succeeding generations. Every organism is therefore subjected to a process of selection, based upon its suitability for survival given the conditions which exist at the time. The organism's environment exerts a **selection pressure**. The intensity and direction of this pressure varies in both time and space. Selection pressure determines the spread of any allele within the gene pool.

16.4.1   Types of selection

There are three types of selection which operate in a population of a given species.

Directional selection

When environmental conditions change, there is a selection pressure on a species causing it to adapt to the new conditions. Within a population there will be a range of individuals in respect of any one character. The continuous variation among individuals forms a normal distribution curve, with a mean which represents the optimum for the existing conditions. When these conditions change, so does the optimum necessary for survival. A few individuals will possess the new optimum and by selection these in time will predominate. The mean for this particular character will have shifted. An example is illustrated in Fig. 16.1(a).

Stabilizing selection

This occurs in all populations and tends to eliminate the extremes within a group. In this way it reduces the variability of a population and so reduces the opportunity for evolutionary change.

In our earlier example, we see that at 10 °C there was an optimum coat length of 1.5 cm. Individuals within the population, however, had a range of coat lengths from 0.5 cm to 3.0 cm. Under normal climatic circumstances the average temperature will vary from one year to the next. In a warm year with an average temperature of 15 °C the individuals with shorter fur may be at an advantage as they can lose heat more quickly. In these years the numbers of individuals with short fur increase at the expense of those with long fur. In cold years, the reverse is true and individuals with long fur increase at the expense of their companions with shorter coats. The periodic fluctuations in environmental temperature thus help to maintain individuals with very long and very short fur.

Imagine that the average environmental temperature was 10 °C every year and there were no fluctuations. Without the warmer years to give them an advantage in the competition with others in the population, the individuals with short hair would decline in numbers. Likewise the absence of colder years would reduce the number of long-haired individuals. The mean fur length would

1.   *Initially there is a wide range of fur length about the mean of 1.5 cm. The fur lengths of individuals in the shaded areas are maintained by rapid breeding in years when the average temperature is much warmer or colder than normal.*

2.   *When the average environmental temperature is consistently around 10°C with little annual variation, individuals with very long or very short hair are eliminated from the population over a number of generations.*

1.

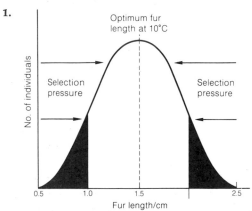

2.

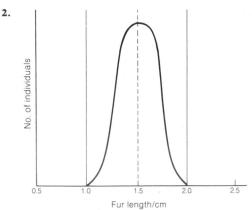

Fig. 16.1(b) Stabilizing selection

1. *When there is a wide range of temperatures throughout the year, there is continuous variation in fur length around a mean of 1.5 cm.*

2. *Where the summer temperature is static around 15 °C and the winter temperature is static around 5 °C, individuals with two distinct fur lengths predominate: 1.0 cm types which are active in summer and 2.0 cm types which are active in winter.*

3. *After many generations two distinct sub-populations are formed.*

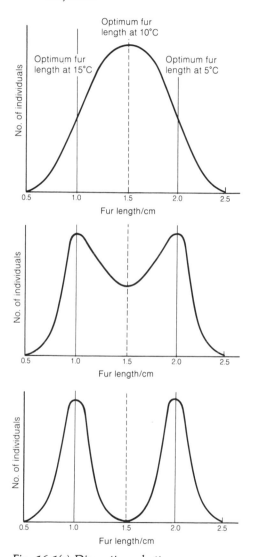

Fig. 16.1(c) Disruptive selection

remain at 1.5 cm but the distribution curve would show a much narrower range of lengths (Fig. 16.1b).

Disruptive selection

Although much less common, this form of selection is important in achieving evolutionary change. Disruptive selection may occur when an environmental factor takes a number of distinct forms. To take our hypothetical example, suppose the environmental temperature alternated between 5 °C in the winter and 15 °C in the summer, with no intermediate temperatures occurring. These conditions would favour the development of two distinct phenotypes within the population. One with a fur length of 2.0 cm (the optimum of an environmental temperature of 5 °C), the other with a fur length of 1.0 cm (optimum length at 15 °C). (Fig. 16.1(c)

It is possible that the group with 2.0 cm fur would aestivate or migrate in summer to avoid the problem of overheating. The other group might hibernate or migrate in winter to avoid the problem of retaining heat. In this way reproduction between the two groups might be interrupted and the flow of genes between them prevented. Each population might then become a separate species. Where interbreeding continues but a number of distinct phenotypic forms remains, a condition called **polymorphism** exists.

16.4.2 Polymorphism

Polymorphism (*poly* – 'many', *morph* – 'form') is the word used to describe the presence of clear-cut, genetically determined differences between large groups in the same population. One well known example is the A B O blood grouping system found in human populations; another is the colour and banding patterns which arise in certain species of land snail.

Polymorphism in the land snail (Cepaea nemoralis)

The shells of the land snail *Cepaea nemoralis* have a variety of distinct colours including yellow, pink and brown. The snail shells may also be marked with dark bands, ranging in number from 1 to 5. Colour and banding are both genetically determined.

The snail is an important source of food for the song thrush. The thrush uses rocks as anvils upon which to smash the shells to enable them to eat the soft parts within. Examination of the smashed shells reveals that the proportions of each type differ according to the habitat around. Where the surrounding area is relatively uniform, e.g. grassland, the proportion of banded shells is greater among those smashed by thrushes. Where the surrounding area is less uniform, e.g. hedgerows, the proportion of unbanded shells is greater. How can we interpret these results?

It must be assumed that on uniform backgrounds, banded shells are more conspicuous and so more easily seen by predators such as thrushes. These shells form a disproportionately large number of smashed shells found around 'anvils' in this type of habitat. In less uniform backgrounds, bands on the shells tend to break up the shape of the animal, thus providing a form of disruptive camouflage. By contrast, the unbanded ones are not camouflaged. Being more conspicuous these forms are more frequently captured by thrushes and are consequently more common around nearby 'anvils'. As a result of this selection, the variety of snail shell commonly found around any anvil is correspondingly rare in the surrounding habitat.

Polymorphism in *Cepaea nemoralis*

Melanic and normal forms of *Biston betularia* against a natural background

In a similar way, there is a correlation between shell colour and the predominate colour of the habitat background. Where the habitat is light in colour yellow-shelled snails are most common in the area, but brown ones occur more frequently around 'anvils'. The reverse is true of habitats with dark backgrounds where brown snails predominate overall, but yellows are more common around 'anvils'. Clearly the thrushes find brown snails easier to spot against a light background and yellow snails easier to spot against a dark one. In time one might expect this selection to eliminate the more conspicuous variety from any habitat. It appears, however, that the genes for colour and banding are closely linked with other genes which may confer advantages. In this way any evolutionary disadvantage due to colour or banding, is offset by these advantages, in a way comparable with the sickle cell gene (Section 15.4.3). When particular gene loci are so close together that crossing over almost never separates them, then these genes effectively act as a single unit. This unit is called a **super-gene** and is the possible explanation for the continued existence of certain polymorphic forms of *Cepaea* in particular habitats.

Polymorphism in the peppered moth (Biston betularia) (*Industrial melanism*)

Another example of polymorphism occurs in the peppered moth *Biston betularia*. It existed only in its natural light form until the middle of the last century. Around this time a melanic (black) variety arose as a result of a mutation. These mutants had doubtlessly occurred before (one existed in a collection made before 1819) but they were highly conspicuous against the light background of lichen-covered trees and rocks on which they normally rest. As a result, the black mutants were subject to greater predation from insect-eating birds, e.g. robins and hedge sparrows, than were the better camouflaged, normal light forms.

When in 1848 a melanic form of the peppered moth was captured in Manchester, most buildings, walls and trees were blackened by the soot of 50 years of industrial development. The sulphur dioxide in smoke emissions killed the lichens that formerly covered trees and walls. Against this black background the melanic form was less, not more conspicuous than the light natural form. As a result, the natural form was taken by birds more frequently than the melanic form and, by 1895, 98% of Manchester's population of the moth was of the melanic type.

Dr H. B. D. Kettlewell attempted to show that this change in gene frequency was the result of natural selection. He bred large stocks of both varieties of the moth. He then marked them and released them in equal numbers in two areas:

1. Birmingham (an area where soot pollution and high sulphur dioxide levels had resulted in 90% of the existing moths being of the melanic form).

2. Rural Dorset (where the absence of high levels of soot and sulphur dioxide meant lichen-covered trees and no record of the melanic form of the moth).

He recaptured samples of the moths using light traps and found that in polluted Birmingham over twice as many marked melanic moths were recaptured than normal light ones. The reverse was true for unpolluted Dorset where the normal light form was recaptured with twice the frequency of the melanic type.

Further studies have confirmed the findings that the melanic form has a selection advantage over the lighter form in industrial areas. In non-polluted areas the selection advantage lies with the lighter form.

16.4.3 Drug and pesticide resistance

Following the production of antibiotics in the 1940s, it was noticed that certain bacterial cells developed resistance to these drugs, i.e. the antibiotics failed to kill them in the normal way. Experiments showed that this was not a cumulative tolerance to the drug, but the result of chance mutation. This mutation in some way allowed the bacteria to survive in the presence of drugs like penicillin, e.g. by producing an enzyme to break it down. In the presence of penicillin non-resistant forms are destroyed. There is a selection pressure favouring the resistant types. The greater the quantity and frequency of penicillin use, the greater the selection pressure. The medical implications are obvious. Already the usefulness of many antibiotics has been destroyed by bacterial resistance to them. By 1950, the majority of staphylococcal infections were already penicillin-resistant.

The problem has been made more acute by the recent discovery that resistance can be transmitted between species. This means that disease-causing bacteria can become resistant to a given antibiotic even before the antibiotic is used against it. As a result, certain staphylococci are resistant to all major antibiotics.

Resistance to insecticides has come about in a similar way. Within two years of using DDT, many insects had developed resistance to it, often independently in different parts of the world. Most common insect pests are now resistant to most insecticides. In many cases the presence of the insecticide switches on the gene present in the mutant varieties. This gene initiates the synthesis of enzymes which break down the insecticide. Apart from directly harmful insects, insect vectors have also acquired resistance. Examples include mosquitoes of the genus *Aedes*, which carry yellow fever, and of the genus *Anopheles* which carry malaria.

Resistance to myxomatosis in rabbits takes two forms. In one type a mutant gene renders the myxomatosis virus ineffective in some way. In the second form a mutant gene alters the rabbits' behaviour, in that they spend more time above ground and less in their burrows. The disease is spread by a vector, the rabbit flea. Normal rabbits live in crowded warrens underground where the flea can easily be transferred between individuals. The mutant variety, spending less time underground, has a reduced chance of being affected by fleas, and hence catching the myxomatosis virus. This variety is favourably selected whereas previously it was selected against because of the increased chance of predation due to its vulnerability when above ground.

16.4.4 Heavy metal tolerance in plants

Another example of natural selection occurs on spoil heaps which contain the waste material from mining activities. These heaps contain high concentrations of certain heavy metals, e.g. tin, lead, copper and nickel. In the concentrations found, these metals are toxic to most plants. Some varieties of grasses, e.g. *Festuca ovina* and *Agrostis tenuis*, have become genetically adapted to survive high levels of these metals. These plants are less competitive where the concentration of these metals is low and so do not always survive.

233

16.5 Artificial selection

Man has cultivated plants and kept animals for about 10 000 years. Over much of this time he has bred them selectively. There have been two basic methods, each with a particular aim:

1. Inbreeding – When, by chance, a variety of plant or animal arose which possessed some useful character, it was bred with its close relatives in the hope of retaining the character for future generations. Inbreeding is still widely practised today, especially with dogs and cats.

One problem with inbreeding is that it increases the danger of a harmful recessive gene expressing itself, because there is greater risk of a double recessive individual arising (Section 14.4.4). As a result, inbreeding is not usually carried out indefinitely but new genes are introduced by outbreeding with other stock. While this makes consistent qualities harder to achieve, it can lead to stronger, healthier offspring.

2. Outbreeding – This is carried out in order to improve existing varieties. Where two individuals of a species each have their own beneficial feature they are often bred together in order to combine the two. A racehorse breeder, for example, might cross a fast mare with a strong stallion in the hope of attaining a strong, fast foal. Outbreeding frequently produces tougher individuals with a better chance of survival, especially where many generations of inbreeding have taken place. This is known as **hybrid vigour**.

Extreme examples of outbreeding occur when individuals of different species are mated. Only rarely is this successful. When it is, the resulting offspring are normally sterile. These sterile hybrids may still be useful. Mules, produced from a cross between a horse and a donkey, have strength and endurance which make them useful beasts of burden.

The improvement of the human race by the selection or elimination of specific characters is called **eugenics**. To some, the idea of such selection is offensive but, as we saw in Section 15.6, genetic counselling is now fairly commonplace. Provided the individuals involved remain free to make their own choice about whether to have children, many see no harm in providing them with statistical information which might help them reach a decision.

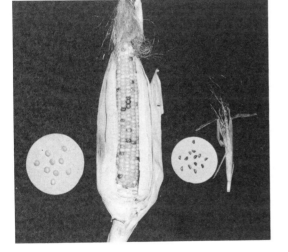

Selective breeding of corn

16.6 Isolation mechanisms

Within a population of one species there are groups of individuals which breed with one another. Each of these breeding sub-units is called a **deme**. Although individuals within the deme breed with each other most of the time, it is still possible for them to breed with individuals of separate demes. There therefore remains a single gene pool. If demes become separated in some way, the flow of genes between them may cease. Each deme may then evolve along separate lines. The two demes may become so different that, even if reunited, they would be incapable of successfully breeding with each other. They would thus become separate species each with its own gene pool. The process by which species are formed is called **speciation** and depends on groups within a population becoming isolated in some way. There are three main forms of isolation:

16.6.1 Geographical isolation

Any physical barrier which prevents two groups of the same species from meeting must prevent them interbreeding. Such barriers include mountain ranges, deserts, oceans, rivers etc. The effectiveness of any barrier varies from species to species. A small stream may separate two groups of woodlice, whereas the whole of the Pacific Ocean may fail to isolate some species of birds. A region of water may separate groups of terrestrial organisms, whereas land may isolate aquatic ones. The environmental conditions on either side of a barrier frequently differ. This leads to the group on each side adapting to suit its own environment – a process called **adaptive radiation**.

Fig. 16.2 Speciation due to geographical isolation

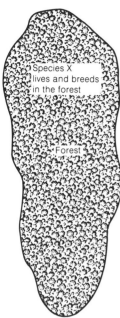

1. Species X occupies a forest area. Individuals within the forest form a single gene pool and freely interbreed.

2. Climatic changes to drier conditions reduce the size of the forest to two isolated regions. The distance between the two regions is too great for the two groups of species X to cross to each other.

3. Further climatic changes result in the northerly region (forest A) becoming colder and wetter. Group X, adapts to these new conditions. Physiological and anatomical changes occur in this group.

4. Continued adaptation leads to the eolution of a new form – group Y in forest A.

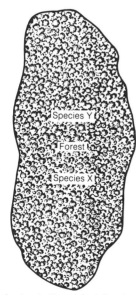

5. A return to the original climatic conditions results in regrowth of forest. Forest A and B are merged and groups X and Y are reunited. The two groups are no longer capable of interbreeding. They are now two species X and Y each with its own gene pool

Imagine, for example, that climatic changes resulted in two areas becoming separated from one another by an area of arid grassland. A possible sequence of events which could lead to a new species being formed under these conditions is illustrated in Fig. 16.2.

Speciation may occur when two groups inhabit the same region but their habitats differ. For example, two species of violet occur in the same region, but one prefers an acid soil, the other an alkaline one. They were probably originally one species, but their preferences for soils of different pH effectively isolated them and led to the formation of different species. This is sometimes called **ecological isolation**.

16.6.2 Physiological isolation

Even when two groups of individuals live side by side, they may still be unable to breed fertile offspring successfully for a number of reasons:

1. The genitalia of the two groups may be incompatible (mechanical isolation) – It may be physically impossible for the penis of a male mammal to enter the female's vagina.

235

2. The gametes may be prevented from meeting – In animals, the sperm may not survive in the females reproductive tract or, in plants, the pollen tube may fail to grow.

3. Fusion of the gametes may not take place – Despite the sperm reaching the ovum, or the pollen tube entering the micropyle, the gametes may be incompatible and so do not fuse.

4. Development of the embryo may not occur (hybrid inviability) – Despite fertilization taking place, further development may not occur, or fatal anomalies may arise during early growth.

5. The offspring may be sterile (hybrid sterility) – When individuals of different species breed, the sets of chromosomes from each parent are obviously different. These sets are unable to pair up during meiosis and so the offspring cannot produce gametes. For example, the cross between a horse (2n = 60) and an ass (2n = 66) results in a mule (2n = 63). It is impossible for 63 chromosomes to pair up during meiosis.

16.6.3 Behavioural isolation

Before copulation can take place, many animals undergo elaborate courtship behaviour. This behaviour is often stimulated by the colour and markings on members of the opposite sex, the call of a mate or particular actions of a partner. Small differences in any of these may prevent mating. If a female stickleback does not make an appropriate response to actions of the male, he ceases to court her. The beak shape in many of Darwin's finches in the Galapagos Islands is the only feature which distinguishes the species. Individuals will only mate with partners having a similar beak to themselves. The song of a bird or the call of a frog must be exact if it is to elicit the appropriate breeding response from the opposite sex. The timing of courtship behaviour and gamete production is also important. If the breeding season of two groups (demes) does not coincide, they cannot breed. Different flowering times in plants may mean that cross-pollination is impossible. These are both examples of **seasonal isolation**.

16.7 Evidence for evolution

Since Darwin and Wallace formulated their theory of evolution by natural selection, further evidence has been accumulated which would seem to substantiate their view. Much of this evidence is concerned with similarities between organisms. If the Darwin/Wallace theory is correct and organisms evolved from existing ones through gradual change, it follows that many similarities should exist between organisms, especially where they are related by evolution.

16.7.1 Palaeontology

When an organism dies it often decomposes rapidly or is consumed by scavengers, in which case no permanent evidence of its existence remains. Sometimes the animal becomes preserved in some way. The preserved remains are called **fossils**, the study of which is called **palaeontology**. Most fossils occur in rock which is formed from the slow deposition of mud and silt. This is called **sedimentary rock** and is laid down in layers called **strata**. The oldest layers are at the bottom, the newest at the top. It is possible to find out the age of each layer by a method called **radio-isotope dating**. This involves using unstable radio-active isotopes found in the rock. For example, carbon exists as an isotope called ^{14}C. This breaks down to the more common ^{12}C. The time taken for half this breakdown to occur is called the **half life** and in the case of ^{14}C is around 5700 years. By using information which includes the ratio of ^{14}C to ^{12}C in the rock, the age of it can be quite accurately determined. This process is called **carbon dating**.

It is logical to suppose that fossils in the older, deeper layers were formed earlier than those in the newer, more superficial layers. Sometimes the fossils reveal a gradual change from one form to the next as one moves upwards through the strata. Often, however, the sequence of fossils is not continuous; there are 'missing links'. These may be the result of intermediate forms not having fossilized for some reason, or simply that they have not yet been discovered. An increasing number of 'missing links' have been found, e.g. *Seymouria* which is intermediate between an amphibian and a reptile and *Archaeopteryx* which is intermediate between a reptile and a bird.

Palaeontology is not itself evidence of how evolutionary change came about, but it does support the idea of gradual and progressive change from simple to complex forms. This is consistent with the Darwin/Wallace theory.

16.7.2 Classification

Before Darwin and Wallace put forward their theory of evolution, the similarities between certain organisms had led Linnaeus to propose a system of classification based upon grouping organisms with similar features. The present-day taxonomists have devised a system of classification based upon the evolutionary relationships between organisms. It forms a natural series of groups of phyla, classes, orders, families, genera and species. It is difficult to see how this would be possible unless the organisms were related by descent.

16.7.3 Comparative embryology

Animals which are different as adults often have similar embryos. For example, the embryos of all vertebrates possess gill slits although only fish retain the gill slits as adults. Annelids and molluscs have a similar larva – the trochophore larva. Where embryos have similar features, this is taken to indicate that they evolved from a common ancestor. There seems little point otherwise in organisms having features as embryos which are not retained into adulthood. It is therefore assumed that these features are the remains of structures found in the organisms' ancestors. Haeckel (1834–1919) went as far as suggesting that during its development an organism repeated its evolutionary history – 'ontology recapitulates phylogeny'. While this rather oversimplifies the situation, it is still reasonable to assume that similarities in embryology indicate an evolutionary relationship. The greater the similarities, the closer the relationship.

16.7.4 Comparative anatomy

Detailed study of apparently unrelated organisms may reveal many structures which are fundamentally similar. These similarities may

(a) General structure of a pentadactyl limb

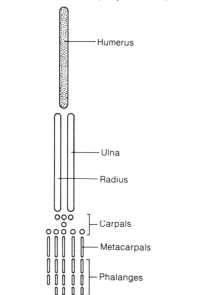

(b) Bird wing

(c) Bat wing

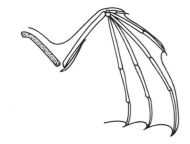

Fig. 16.3 *Pentadactyl limbs of two different mammals*

indicate common ancestry. The **pentadactyl limb** is common to all vertebrates (except fish) but during evolution has become modified to a number of functions. In birds it forms the wing for flight, in primates it forms the hand and is used for grasping whereas in whales it is modified as a paddle to aid swimming. Despite these diverse uses, these structures are all **homologous**, i.e. they have the same evolutionary origin. The presence of the pentadactyl limb in most vertebrates suggests they have a common ancestor. (See Fig. 16.3.)

16.7.5 Comparative biochemistry

In the same way that similar structures like the pentadactyl limb indicate a common ancestry, so too can similar chemicals. Obviously simple chemicals such as water, glucose, lipids etc. are common to all organisms. More complex compounds like proteins and nucleic acids are much more specific to individual species. Cytochromes, haemoglobin and ribosomal RNA have been extensively used in the search for evolutionary affinities.

The sequence of amino acids in the polypeptide portion of cytochrome c shows considerable similarity between all organisms so far studied. This suggests a common ancestry for all organisms. There are, however, slight differences. These differences are less, the more closely related the species. The chimpanzee has an identical sequence of amino acids to that in human cytochrome c. This indicates a very close affinity. The rhesus monkey's cytochrome c differs in respect of a single amino acid, indicating a slightly less close evolutionary affinity to man. Similar results from haemoglobin can be obtained. When these relationships are used to construct an evolutionary tree, the results are remarkably similar to those obtained from comparative anatomy.

Immunological research has provided further evidence of evolutionary relationships. When the serum of animal A is injected into animal B the latter forms antibodies against it. These antibodies are specific to the proteins found in animal A. If the serum from animal B (which contains the antibodies) is mixed with serum from animal C, the antibodies will combine with their specific proteins. This causes a precipitate to form. If animal C has the same proteins as animal A there will be the maximum precipitation (100%). If animal C has few similar proteins, the amount of precipitation will be less. In other words, the more closely related any two animals, the more similar are their proteins and the more precipitation occurs. Fig. 16.4 illustrates some typical results. Once again the relationships indicated by these experiments are similar to those provided by evidence from other sources.

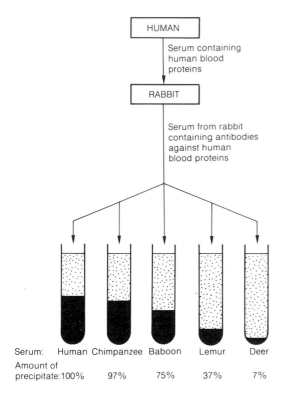

Fig. 16.4 Evolutionary affinities as shown by immunological studies

16.7.6 Geographical distribution

Plant and animal species are not evenly distributed throughout the world. Certain zones have their own characteristic fauna and flora. It might be expected that where identical conditions occur in different parts of the world, the same organisms would be found, but this is not the case. Elephants, for example, are found in Africa and India, but not in similar habitats in South America. Britain and New Zealand, while having similar climatic conditions, have very different fauna and flora. This discontinuous distribution of species can be explained as follows:

1. A species originates in a particular area.

2. Individuals continuously disperse to avoid overcrowding.

3. As they encounter new environments as a result of dispersal, they adapt to meet the new conditions (adaptive radiation).

4. Climatic, topographical and other changes create barriers between the new varieties and their ancestors.

5. This genetic isolation leads to separate gene pools and new species.

It is thought that, in this way, individual species become restricted to specific areas. But how did these barriers arise? One explanation is **continental drift**. It is thought that the continents of the earth once formed a single land mass at the south pole. This land mass broke up into sections which, floating on the earth's molten mantle, drifted apart. While land bridges remained between individual sections, members of a species could freely interbreed. Where these bridges were submerged by changes in sea level, groups became genetically isolated and new species arose. By the time the land bridges were reformed due to a fall in sea level, interbreeding between the original groups was impossible. Continental drift therefore helps to explain the discontinuity in the distribution of species.

16.8 Questions

1. (a) Define the terms *natural selection, artificial selection* and *adaptive radiation*. (6 marks)
(b) Show how two out of three of the following support the theory that evolution has taken place:
(i) industrial melanism in moths;
(ii) breeding in domestic animals;
(iii) vestigial organs. (6,6 marks)
(Total 18 marks)

Cambridge Board June 1985, Paper I, No. 1

2. Explain how each of the following provides evidence for evolution. Illustrate your answer in each case with **one** suitable example.
(a) Geographical isolation (6 marks)
(b) Chromosome changes (6 marks)
(c) Behavioural mechanisms (4 marks)
(d) Comparative physiology (4 marks)
(Total 20 marks)

London Board June 1985, Paper II, No. 6

3. Review the means by which new species are believed to evolve from existing species of organisms.

Southern Universities Joint Board June 1984, Paper II, No. 6

4. 'Evolution occurs in populations as a result of *mutation* not *reassortment* or *recombination*. The evolution of many species from an ancestral form can follow *isolation* and *adaptive radiation*.'
Explain clearly the meaning of each of the terms in italics. (25 marks)

Southern Universities Joint Board June 1986, Paper II, No. 6

5. Read the passage below, then answer the following questions:

From *The Origin of Species*
Variation under Domestication

When we compare the individuals of the same variety of our older cultivated plants and animals, it is found that they generally differ more from each other than do the individuals of any one species or variety in a state of
5 nature. We are driven to conclude that this great variability is due to our domestic productions having been raised under conditions of life not so uniform as, and somewhat different from, those to which the parent species had been exposed in nature. This variability may
10 be partly connected with excess of food. Our oldest cultivated plants, such as wheat, still yield new varieties: our oldest domesticated animals are still capable of rapid improvement or modification.

Some naturalists have maintained that all variations are
15 connected with sexual reproduction but this is certainly
an error; for I have given in another work a long list of
'sporting plants', as they are called by gardeners, that is
of plants that have suddenly produced a single bud with
a new and sometimes widely different character from
20 that of other buds on the same plant. These variations
can be propagated by graft, offsets etc. and sometimes
by seed. We clearly see that the nature of the conditions
is of subordinate importance in comparison with the
nature of the organism in determining each particular
25 form of variation.

Let us now consider the steps by which domestic races
have been produced. Some variations useful to man have
probably arisen suddenly or by one step: this probably
occurred with the teasel with its hooks and this is known
30 to have been the case with the ancon sheep. But when
we compare the cart horse with the race horse, the various
breeds of sheep fitted either for cultivated land or
mountain pasture, we cannot suppose that all the breeds
were suddenly produced as perfect and as useful as we
35 now see them. The key is accumulative selection: nature
gives successive variations, man adds them up in certain
directions useful to him.

(a) (i) Who was the author of this passage?
(1 mark)

(ii) When was it published? (1 mark)

(b) (i) How justified was the author in thinking that
more variation may appear in domestic
animals and plants than in wild ones?
(2 marks)

(ii) How may domestic conditions differ from
those in nature? (2 marks)

(c) Give two ways in which useful (or domestic)
varieties may be produced. (2 marks)

(d) Give the principles which the author thought
were involved in producing a new variety of
plant or animal with particular characters.
(2 marks)

(e) Comment briefly on the phrase 'as perfect and
as useful' (line 34). (2 marks)

(f) (i) What evidence is there in the passage that
the author had an understanding of the
mechanism of organic evolution? (4 marks)

(ii) What do you think the author meant by 'the
steps' in line 26? (2 marks)

(iii) How could variability be connected with
'excess of food' (line 10)? (2 marks)

(g) Why can 'sporting plants' be propagated by
grafts but not always by seed? (2 marks)

(h) (i) What is meant, in line 23, by 'the nature of
the conditions'? (1 mark)

(ii) What is meant in line 24, by 'the nature of
the organism'? (1 mark)

(iii) Explain why 'the nature of the conditions is
of subordinate importance in comparison
with the nature of the organism'. (2 marks)

(i) Describe in present day terms what the author
understood by 'sporting plants' (line 17).
(2 marks)

(j) Comment briefly on how the observations made
in the passage are of relevance to the evolution
of wild populations. (2 marks)
(Total 30 marks)

Oxford Local June 1983, Paper I, No. 10

6. (a) Write a short summary of Darwin's theory of
evolution. (8 marks)

(b) Explain how a knowledge of genetics adds to
our understanding of Darwin's theory of
evolution. (8 marks)

(c) A feature of Darwin's theory is that it
emphasises a process of gradual change. Explain
briefly why modern developments such as the
theory known as punctuated equilibrium pose
a challenge to the traditional view of Darwinism.
(4 marks)
(Total 20 marks)

*Joint Matriculation Board (Nuffield) June 1988,
Paper IIC, No. 4*

7. The MN blood group system in man depends upon
the inheritance of one pair of alleles. In a sample of 1100
Chinese from Peking, it was found that the number of
people with blood group M was 356, with MN, 519 and
with N, 225 respectively.

(a) Calculate the frequencies of the two alleles and
the expected Hardy-Weinberg genotypic ratios.

(b) Is the population in Hardy-Weinberg equilbrium?
(9 marks)

Cambridge Board June 1984, Paper II, No. 5

8. The shell of the land snail *Cepaea nemoralis* varies both
in its ground colour and in its banding pattern. In a
particular area of the British Isles the ground colour of
the shell is either yellow (appearing green with the live
snail inside) or pink. These shells may either possess 5
black bands (banded) or none (unbanded). These forms
are determined genetically. Pink shell is a dominant trait
as is the unbanded condition.

(a) 2000 snails were collected from a beechwood
and 320 of these were yellow, the remainder
being pink. With respect to these colour forms,
determine the allele and genotype frequencies
assuming the population to be in Hardy-
Weinburg equilibrium.

The Hardy-Weinburg equation is given by:
$$p^2 + 2pq + q^2 = 1$$
(5 marks)

(b) In a woodland there was a stable population of banded and unbanded snails. A sudden environmental change resulted in the heterozygote having an advantage over both homozygotes. Describe the outcome of this change lasting a number of years in terms of the relative frequency of banded and unbanded snails. *(3 marks)*

(c) Studies of the frequencies of colour and banding forms in different environments gave the results shown in the graph.

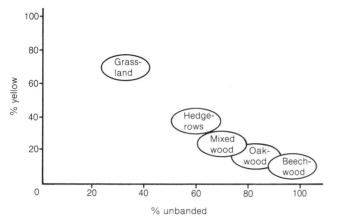

Suggest an explanation for the proportions of forms found in beechwoods which tend to have a uniform carpet of leaf litter. *(3 marks)*

(d) A study of a grassland area involved not only counting the number of coloured and banded forms of living snails but also the frequency with which the different forms were predated by the song thrush. The thrush, having captured a snail, breaks open the shell by hammering it against a stone, so that examination of the shells found at these 'anvil' sites reveal the frequencies of the forms predated. The table below gives the results of this study.

	Number yellow	Number unbanded	Total collected
Living	83	13	111
Predated	21	11	35

(i) Calculate the proportion of (a) yellow and (b) unbanded forms for both the living and the predated snails. Draw up a table in which the proportion of each form found living may be compared more directly with that found predated.

(ii) Discuss the relationship between the yellow/banded shell characters and predation in the light of both the findings from (d) (i) above and the results given in part (c). *(5 marks)*

(e) Other studies have suggested that predators tend to take proportionately more of the most common form when faced with a choice of prey items.
Explain how such predation would prevent a population of snails becoming all of one form. *(4 marks)*
(Total 20 marks)

Northern Ireland Board June 1988, Paper 2, No. 6Y

9. (a) Lamarck (1809) held the view that
   (i) organisms may acquire new characteristics during their lifetime and
   (ii) these acquired characteristics are inherited by their offspring, leading to evolution.
   For **each** part of Lamarck's view, state whether or not it is now held to be correct and give an example as evidence to support your choice. *(4 marks)*

(b) Before 1850, all reported forms of the peppered moth, *Biston betularia*, were creamy-white, speckled with dark dots. In 1850, a black form of the moth was recorded in Manchester. By 1895, as a result of selective predation, 98% of this moth population in Manchester was black.
   (i) Account for the appearance of the first black moths.
   (ii) State the meaning of the term 'selective predation'.
   (iii) Complete the sentence:
      'The presence of different phenotypes of the moth *Biston betularia* in the same population provides an example of .... variation and of ....'
   (iv) Suggest why the creamy-white form of the moth did not become totally eliminated from the Manchester population.
   (v) Offer an explanation for the fact that black forms of the moth are found in the non-industrial areas of the East Coast, but not in those of the West Country. *(7 marks)*

(c) Consider the evolutionary series:
   Invertebrates → fish → amphibia → reptiles → mammals.
   State why it would be wrong to accept this series as illustrating the evolution of advanced organisms from primitive organisms. *(2 marks)*

(d) Parasites are sometimes thought of as degenerate organisms because they lack features of their free-living relatives. Suggest why this view is misguided. *(2 marks)*
(Total 15 marks)

Welsh Joint Education Committee June 1988, Paper A2, No. 12

10. Read through the following passage and then answer the questions.

The chances that a population will be represented by its descendants many generations hence will depend upon it having sufficient genetic variation to respond to any change in conditions which produce novel selective
5 forces in the long term. Current evidence suggests that most populations contain an enormous reservoir of genetic differences, certainly in so far as they produce differences in the properties of enzymes and other proteins. In fact the number of possible genetic
10 combinations is likely to be so huge that only a very small proportion of them will actually occur in even a large population. If the various combinations were also formed through random associations of the component genes then their probabilities were also formed through
15 random associations of the component genes then their probabilities of occurrence would depend solely on the frequencies of the alleles concerned. Thus, if A and B were two unlinked genes and the frequencies of homozygous individuals of genotype **AA** and **BB** were both 50%, the
20 expected frequency of individuals of genotype **AABB** would be 25% and this genotype would almost certainly be found in any population. On the other hand, if the frequencies of **AA** individuals and of **BB** individuals were both 1%, that of the **AABB** genotype would be 0.01%
25 and it would be likely to be absent completely from most populations except very large ones. But if the environmental conditions changed so that genotypes **AA** and **BB** became favoured by natural selection, their frequencies would rise, as would the probability that
30 **AABB** would occur, and so a new genotype would arise.

Adapted from Evolution in Modern Biology *by K. J. R. Edwards (Arnold 1977)*

(a) State *two* ways by which genetic variation (line 3) may arise within a population. (*2 marks*)

(b) Explain how a population showing genetic variation can 'respond to any change in conditions which produce novel selective forces in the long term' (lines 3–5). (*4 marks*)

(c) Explain why differences between individuals in the properties of their enzymes and other proteins (lines 8–9) are indications of genetic variation in populations. (*4 marks*)

(d) Explain why, if genotypes **AA** and **BB** both had frequencies of 1%, the frequency of the **AABB** genotype would be 0.01% (lines 23–24). Assume that A and B are unlinked genes and that the Hardy–Weinburg equilibrium applies. (*2 marks*)

(e) If the frequency of genotype **AA** were 1%, and the gene concerned had only one other allele, **a**, what would be (i) the frequencies of alleles **A** and **a** (ii) the frequencies of genotypes **Aa** and **aa**? Show your working. (*5 marks*)

(f) How is it that natural selection could lead to an increase in the frequencies of genotypes **AA** and **BB** in the population (lines 27–29)? (*3 marks*)

(*Total 20 marks*)

London Board January 1989, Paper I, No. 15

17 *Reproductive strategies*

No individual can live indefinitely. Some of its cells may be worn or damaged beyond repair or it may be killed by predators, disease or other environmental factors. If a species is to survive it must therefore produce new individuals. This is achieved in two ways:

1. Asexual reproduction – Rapidly produces large numbers of individuals, usually having an identical genetic composition to each other and to the single parent from which they are derived; gametes are never involved.

2. Sexual reproduction – Less rapid, often involves two parents and produces offspring which are genetically different. The fusion of haploid gametes is always involved.

Apart from purely increasing numbers, reproduction may involve one or more of the following:

(a) a means of increasing genetic variety and therefore helping a species adapt to changing environmental circumstances;

(b) the development of resistant stages in a life cycle which are capable of withstanding periods of drought, cold or other adverse conditions;

(c) the formation of spores, seeds or larvae which may be used to disperse offspring and so reduce intraspecific competition as well as capitalizing on any genetic variety among the offspring.

17.1 Comparison of asexual and sexual reproduction

Sexual reproduction always involves the fusion of special sex cells called gametes; asexual reproduction never does. If these gametes are produced by meiosis they will show considerable genetic variety (Section 13.4.2). The offspring resulting from the fusion of gametes will likewise show genetic variability and therefore be better able to adapt to environmental change. In other words, they have the capacity to evolve to suit new conditions. As asexual reproduction rarely involves meiosis the offspring are usually identical to each other and to their parents. While this lack of variety is a disadvantage in adapting to environmental change, it has one main advantage. If an individual has a genetic make-up which is suited to a particular set of conditions, asexual reproduction is a means by which large numbers of this successful type may be built up. A localized area can be rapidly colonized, something which is of particular advantage to plants.

It must be said that the differences in the variety of offspring resulting from sexual and asexual reproduction are not the same in all organisms. In mosses and ferns, for instance, the gametes are produced by a haploid gametophyte generation. Being haploid, its gametes can only be produced mitotically. Mitosis does not introduce genetic variety (Section 13.4.1) and these gametes are therefore identical. During asexual reproduction, mosses and ferns produce spores by meiosis. These spores show genetic variability. The usual differences between asexual and sexual reproduction are therefore reversed in mosses and ferns.

In all cases of asexual reproduction only one parent is involved but in sexual reproduction it may be one or two parents. If a species has male and female sex organs on separate individuals, it is said to be **dioecious**. In this case the male produces one gamete, the **sperm**, and the female another, the **ovum**. Only fusion between a sperm and an ovum can give rise to a new individual. Two parents are therefore always necessary for reproduction in these species. In some other species, one individual is capable of producing both male and female gametes. Such species are said to be **monoecious** or **hermaphrodite**. In most of these species the sperm from one individual fuses with the ovum from a separate individual. The fusion of gametes from two separate parents is called **cross-fertilization**. Provided the two parents are genetically different, greater variety of offspring results. In some species, however, sperm and ova from the same individual fuse. This is called **self-fertilization** and involves a single parent; the degree of variety is therefore less. The process is nonetheless sexual as fusion of gametes is involved and as these are produced by meiosis some variety is still achieved.

It would be inaccurate to say that asexual reproduction always produces identical offspring. Mutations, although rare, nevertheless occur and so help create a little variety. Mutations also arise during sexual reproduction, indeed the greater complexity of the process means they arise more frequently.

To summarize, the main methods by which variety among offspring is achieved are:

1. The recombination of two different parental genotypes (genetic recombination).

2. During meiosis by:

    (a) random segregation of chromosomes on the metaphase plate (independent assortment);
    (b) crossing over during prophase I.

3. Mutations.

4. The effect of the environment on the genotype.

Numbers **1** and **2** apply only to sexual reproduction; **3** applies to both, although the frequency is usually greater in sexual reproduction and **4** applies equally to asexual and sexual reproduction. Overall the processes of sexual reproduction are more complex than those of asexual reproduction. The events of meiosis are more complex than those of mitosis, and the processes of producing and transferring gametes are often complicated. Elaborate **courtship** and **mating rituals** are frequently part of sexual reproduction and serve to increase the likelihood of gametes successfully fusing (Section 19.4). These processes may take many months in some species and it is not therefore surprising that the offspring are often cared for by the parents. Where this is the case, the number of offspring is small to permit this parental care to be effective. By contrast, the process of asexual reproduction is normally more simple and straightforward. It is rapid and involves no parental care; the number of offspring is normally large.

On account of the variety of offspring produced and the consequent evolutionary potential, almost all organisms have a sexual phase at some stage in their life cycle. While simpler animals have retained the asexual process, most complex ones have abandoned it. A major disadvantage of being totally reliant on the sexual process is that it is difficult to maintain a favourable genotype. Once an organism has adapted to a particular set of conditions, sexual reproduction will tend to produce different offspring. These may not be as well adapted as identical copies of the parents would be. At least animals, with their ability to move from place to place, can search out conditions that suit any new variety. Plants do not exhibit locomotion and individuals must remain where they are. For this reason most have retained the asexual process as part of their life cycle. Hence, once a plant has successfully established itself in a suitable environment, it uses asexual means to rapidly establish a colony of identical, and therefore equally well-suited, individuals. Such a group has the advantage of reducing competition from other plant species, although with its identical genotypes it may be vulnerable to disease.

17.2 Asexual reproduction

As we have seen, asexual reproduction requires only a single parent and haploid gametes are not involved. There are five major forms of asexual reproduction in organisms:

17.2.1 Binary or multiple fission

This occurs in single-celled organisms like bacteria and protozoans. The organism divides into two or more parts each of which leads a separate existence. Where the cell divides into two parts it is called binary fission and this typically occurs in bacteria.

The bacterial DNA replicates first and the nucleoplasm then divides into two, followed by the cell as a whole. Under favourable conditions (temperature around 20 °C and an abundant supply of food) the daughter cells grow rapidly and may themselves divide within twenty minutes. Under unfavourable conditions some species develop a thick resistant wall around each daughter cell. The **endospore** thus formed is resistant to desiccation, extremes of temperature and toxic chemicals. Only when favourable conditions return does a vegetative cell emerge to continue binary fission as before. It is these endospores which can survive certain forms of sterilization and so cause infection or disease. Multiple fission, or **schizogamy**, occurs when a cell divides into many parts rather than just two. In *Plasmodium*, the malarial parasite, for instance, the process occurs at a number of stages in the life cycle. These are described in Section 25.2.3.

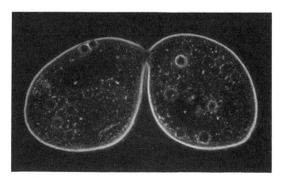

Binary fission in *Paramecium* (× 500)

17.2.2 Budding

An outgrowth develops on the parent and this later becomes detached and is then an independent organism. The process occurs in the

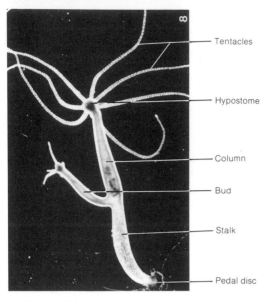

Budding in *Hydra*

Platyhelminthes (flatworms), some annelids (segmented worms) and cnidarians like *Hydra* and *Obelia*. In its simplest form it is little more than a type of binary fission, except that the two resultant cells are not of equal size but comprise a smaller bud cell, becoming detached from the larger parent cell. This occurs in *Saccharomyces* (yeast). In *Hydra* the buds arise near the centre of the parental column. They become highly differentiated, multicellular structures, and may develop their own buds before becoming detached.

In the colonial cnidarian *Obelia*, during the warmer months of the year when food is plentiful, apical growth gives rise to long stalks called **blastostyles**. Along the length of these blastostyles bell-shaped buds are formed. These buds develop into **medusae**, the sexual stage of the life cycle, before they become detached by basal constriction.

17.2.3 Fragmentation

In one sense fragmentation is no more than a form of regeneration. If certain organisms are divided into sections, each portion will regenerate the missing parts thus giving rise to new individuals. If the division occurs as a result of injury then the process is regeneration. If however an organism regularly and spontaneously divides itself up in this way the process is fragmentation. Organisms exhibiting fragmentation must have relatively undifferentiated tissues and it is therefore limited to certain algae, sponges, cnidarians and flatworms. In *Spirogyra*, for example, portions of the filamentous alga break away when the filament reaches a certain length. These drift away, attach themselves elsewhere and begin vegetative growth again.

17.2.4 Sporulation

Sporulation is the formation of small unicellular bodies called **spores**, which detach from the parent and, given favourable conditions, grow into new organisms. Spores are usually small, light and easily dispersed. They are produced in vast numbers – a single mushroom for example may produce 500 000 spores a minute at the peak of its production. Sporulation occurs in bacteria, protozoans, algae, fungi, mosses and ferns (in one sense at least, all plants produce spores). The range of spore-producing structures, called **sporangia**, is very varied and described elsewhere in the book, notably in Chapters 7 and 8.

17.2.5 Vegetative propagation

In general, vegetative propagation involves the separation of a part of the parent plant which then develops into a new individual. Almost any part – root, stem, leaf or bud – may serve the purpose. They are often highly specialized for the task and bear little resemblance to the original plant organ from which they evolved; the potato, for instance, is actually a modified stem. No plant is likely to survive long in a changing world if it relies exclusively on asexual reproduction. Plants exhibiting vegetative propagation have not therefore abandoned the sexual process but continue to produce flowers in the usual way. In some plants, to ensure the continued survival of favourable genotypes, organs of vegetative propagation also often act as **perennating organs** which lie in the soil over the winter. They are frequently swollen with excess food from the previous summer which is used to produce the new offspring the following year. One advantage is that growth can begin early in the spring using the stored food and the plant is therefore able to start photosynthesizing when there is little competition for light from other species. This is particularly important for small plants which live in deciduous woodland where they are subject to shading by their larger neighbours. Some organs of vegetative propagation are described in Table 17.1.

TABLE 17.1

Organs of vegetative propagation and perennation

Name	Example	Description of organ	Mechanism of action	Perennating organ
Bulb	Onion Garlic Daffodil Tulip	Underground, swollen, fleshy leaf – bases closely packed on a short stem, i.e. a bud	Apical and axillary buds among the leaves each give rise to more new plants	Yes
Corm	Crocus Montbretia Gladiolus Cyclamen	Underground, vertical swollen base of main stem	Buds develop in the axils of the scale leaves surrounding the corm. Each may develop into a new plant	Yes
Rhizome	Solomon's seal Iris Couch grass Canna	Underground horizontal branching stem	Stem grows and branches. At the tip of each branch a bud produces new vertical growth which gives rise to a new plant	Yes
Stem tuber	Potato Artichoke	Swollen tip of slender rhizome	Many slender rhizomes arise from axil of scale leaf. The tips swell to form a stem tuber, each giving rise to a new plant	Yes
Suckers	Mint Pear	Underground horizontal branches	A number of underground branches radiate laterally from the parent plant. The tips ultimately turn upwards out of the soil and develop into new plants	Yes
Runner	Creeping buttercup Strawberry	Thin lateral stems on the soil surface	A number of stems radiate from the parent plant. Adventitious roots arise at points along the stem and new plants arise from these	No
Offset	Leek	A short, stout lateral stem on the soil surface	Stem grows laterally along the soil and a single plant arises from a bud at the top of each stem	No
Stolon	Blackberry	A long vertical stem with little structural support	The stem grows vertically at first but then bends over until the tip touches the soil. Adventitious roots develop and at this point a new plant arises from a nearby lateral bud	No
Root Tuber	Dahlia	Swollen fibrous root	The tuber stores food but the new plant arises from an axillary bud at the base of the old stem	Yes
Leaf buds	Lily	Small, detachable bud	Buds easily break away from the parent plant and each develops into a new individual	No
Bulbil	Saxifrage *Bryophyllum*	Shoot, bud (very occasionally foliage leaves, e.g. *Bryophyllum*)	Axillary buds (or other parts) become detached and develop into new plants	No
Tap root	Carrot Parsnip	A swollen, vertical root	*Not* an organ of vegetative propagation – only a perennating organ	Yes

17.2.6 Cloning

A group of genetically identical offspring produced by asexual reproduction is called a **clone**. The nucleus of every cell of an individual contains all the genetic information needed to develop the entire organism. It is therefore possible under suitable conditions to produce a whole organism from a single cell. If a cell divides mitotically it will produce a clone. If each cell of the clone is separated and allowed to develop into the complete organism, a group of genetically identical offspring is formed. This is known as cloning. It was first performed on carrot root cells but the techniques have become so developed that it has been carried out on vertebrates such as frogs by the transplantation of nuclei from one cell to another. Cloning is already used to produce identical copies of useful plant strains (Section 41.8.2.) and may shortly be extended to economically important animals. The production of identical copies of humans is theoretically possible and may have some practical uses. However, it raises considerable moral and ethical problems.

17.2.7 Parthenogenesis

Parthenogenesis is the further development of a female gamete in the absence of fertilization. As a gamete is produced the process is a modified form of sexual reproduction, even though only a single parent is involved. The parent is always diploid and the gametes are produced by mitosis or meiosis. In **diploid parthenogenesis** the gamete is produced by mitosis and therefore the offspring are likewise diploid. Strictly speaking, the gametes are formed by a type of meiosis in which all the chromosomes show non-disjunction, although genetically its effects are the same as mitosis. In aphids, diploid parthenogenesis is used as a means of rapidly building up numbers when conditions are favourable. During the summer, colonies of wingless females are formed by this means. So rapid is the process that it has been calculated that a pair of aphids could produce 800 million tonnes of offspring in a single year – assuming of course they all survived. Normal sexual reproduction involving the fusion of haploid gametes continues to occur in order that genetic variety can be achieved. In **haploid parthenogenesis** the gametes are produced by meiosis giving rise to haploid eggs which may develop directly into haploid offspring. In the honey bee, unfertilized haploid eggs develop into males or drones, while fertilized eggs develop into female bees. If these females are fed honey and pollen they develop into workers. If they are fed a special food called royal jelly, produced by the workers, a queen results.

17.2.8 Apomixis

In some plants a diploid cell of the ovule may develop directly into an embryo without the involvement of a male gamete. As the diploid cell is not itself a gamete, the process is not strictly sexual at all, although it much more closely resembles sexual reproduction than it does any known asexual process in plants.

17.3 Life cycles

During sexual reproduction, the fusion of gametes, called **syngamy**, results in the doubling of the number of chromosomes in each cell. At some stage in the life cycle this number must be halved, otherwise further fusion of gametes will result in a doubling of chromosome number at each generation. This halving of the chromosome number is brought about by meiosis, which is therefore a feature of the life cycle of all sexually reproducing organisms. All life cycles are in effect an **alternation of generations** between a haploid and diploid phase. There are three basic life cycles which vary in the relative positions of meiosis and syngamy and hence the duration of the haploid and diploid phases. These life cycles are illustrated in Fig. 17.1.

The adults are diploid.
Gametes are produced by meiosis, and they fuse almost immediately. Occurs in most animals and some algae e.g. diatoms.

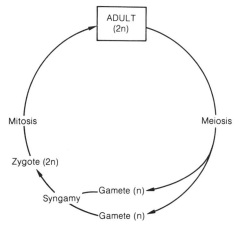

The interval between meiosis and syngamy is short.
The diploid phase is long and dominant.
The haploid phase is short.

Fig. 17.1(a) The gametic life cycle

The adults are haploid.
Gametes are produced by mitosis and fuse to give a diploid zygote. The zygote undergoes meiosis almost immediately during its development. Occurs in primitive algae e.g. Spirogyra and some fungi e.g. Rhizopus.

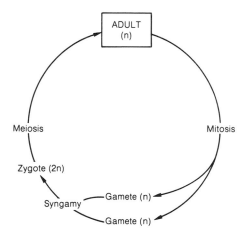

The interval between meiosis and syngamy is long.
The diploid phase is short.
The haploid phase is long and dominant.

Fig. 17.1(b) The zygotic life cycle

There are distinct adults in both haploid and diploid phases.
Gametes are produced by mitosis.
Spores are produced by meiosis.
Occurs in most plants but alternation of generation is most noticeable in mosses and ferns.

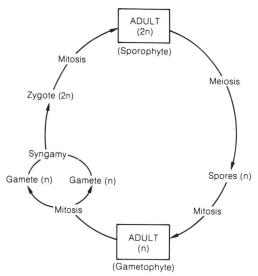

The interval between meiosis and syngamy is intermediate.
The diploid and haploid phases are both relatively long.
Either phase may be dominant, e.g. in mosses it is the haploid gametophyte, in ferns, the diploid sporophyte.

Fig. 17.1(c) The sporic life cycle

1. Give a brief account of asexual reproduction in the plant kingdom including its association with perennating organs.

Under what circumstances does asexual reproduction promote the survival of the species? Discuss how the farmer has exploited these phenomena.

Southern Universities Joint Board June 1982, Paper II, No. 2

2. By reference to the life-cycles and reproductive processes of appropriate organisms, illustrate the ways in which sexual reproduction promotes the survival of the species in adverse or changing environmental conditions.

Southern Universities Joint Board June 1983, Paper II, No. 10

3. (a) What is the significance of the differences between sexual and non-sexual reproduction?
        (6 marks)
   (b) Describe the processes of non-sexual reproduction in
      (i) a **named** protozoan;
      (ii) a **named** fungus;
      (iii) a **named** flowering plant.   *(3, 4, 7 marks)*
              (Total 20 marks)

London Board June 1986, Paper II, No. 3

4. (a) State briefly the difference between sexual and non-sexual reproduction.   *(3 marks)*
   (b) How are large increases in numbers brought about in
      (i) a **named** protozoan;
      (ii) a **named** coelenterate (cnidarian);
      (iii) a mammal, other than man;
      (iv) a fern?   *(14 marks)*
   (c) Suggest why populations of most plants and animals remain fairly constant despite the production of large numbers of offspring.
        (3 marks)
        (Total 20 marks)

London Board June 1982, Paper II, No. 3

5. Fertilization is not a universal phenomenon among higher animals since there are many animals which can reproduce parthenogenetically. However, relatively few reproduce exclusively in this way. In hymenopterous insects, such as ants, bees and wasps, fertilized eggs develop into diploid females but unfertilized eggs develop into haploid fertile males. The haploid males produce sperm by mitosis rather than meiosis. In other insects, such as aphids, the females produce only daughters during much of the breeding period. These daughters develop from unfertilized, diploid eggs.

   (a) Explain the meaning of the term *parthenogenesis*.
   (b) Why do relatively few animals reproduce exclusively parthenogenetically?
   (c) What is the biological significance of:
      (i) haploid parthenogenesis to insects such as honey-bees?
      (ii) diploid parthenogenesis to insects such as aphids?
   (d) What is the equivalent of diploid parthenogenesis in plants?   *(8 marks)*

Cambridge Board November 1983, Paper II, No. 6

6. This question is concerned with reproduction in flowering plants.
   (a) Outline any **three** methods of asexual reproduction.   *(6 marks)*
   (b) Describe the role of nuclear division in sexual and asexual reproduction.   *(6 marks)*
   (c) One population of a plant reproduces by sexual means but a second population of the same species reproduces exclusively by asexual means. Describe the advantages and disadvantages of these methods of reproduction and comment on their possible evolutionary consequences.   *(8 marks)*
        (Total 20 marks)

Joint Matriculation Board June 1989, Paper IIB, No. 8

7. (a) A, B and C represent three basic types of sexual life cycle.

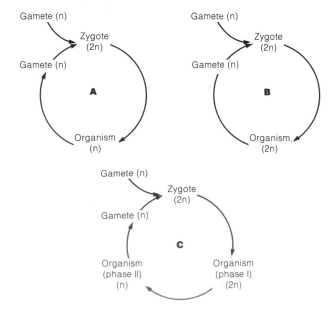

(i) Use the letters A, B or C to indicate the type of life cycle found in

Fucus   ☐   fern   ☐

flowering plant ☐   mammal. ☐

(ii) Name the type of cell division involved in the production of gametes in cycle A.

(iii) An alga, *Cladophora*, has a type C life cycle in which the two phases of the organism are identical in appearance. Suggest **two** ways in which a mature specimen in phase I could be distinguished from a mature specimen in phase II.   (*7 marks*)

(b) Briefly explain why the gametes produced by a mammal are genetically different, whereas those produced by a fern are usually genetically identical.   (*4 marks*)

(c) The diagram illustrates gametogenesis in a flowering plant.

  (i) Draw an arrow from box M to the exact point at which meiosis occurs.

  (ii) Draw an arrow from box G to the exact structure that corresponds to a mammalian spermatozoon.

  (iii) Name the structure in a mammalian testis that corresponds to the pollen grain mother cell.   (*3 marks*)

(*Total 14 marks*)

Welsh Joint Education Committee June 1988, Paper A2, No. 4

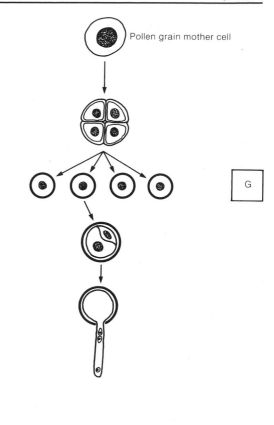

Pollen grain mother cell

M

G

18 *Reproduction in flowering plants*

Although the flower is most readily thought of as the reproductive structure in angiosperms, leaves, stems and roots may also carry out this function. For example, potatoes are often propagated from stems, African violets from individual leaves and dahlias from roots. However, a plant produced from a cutting has the same genetic constitution as its parent. Variation is introduced through sexual reproduction.

18.1 Floral structure

Sexual reproduction confers on organisms the advantages of increased variety. This variety is achieved during meiosis by:

(i) the independent assortment of chromosomes on the metaphase plate;

(ii) recombination of genes due to crossing over between homologous chromosomes at prophase I.

Both processes achieve variety even when the offspring result from self-fertilization. Cross-fertilization confers an additional source of variety, that of mixing two parental genotypes. As terrestrial plants are incapable of moving from place to place in order to transfer genetic material between individuals, this third source of variety would be denied them but for the assistance of some external agent like insects or the wind. This, however, exposes the vulnerable gamete to dangers such as drying out during its transfer. To overcome this, the Spermatophyta have enclosed their male gamete within a spore, the **microspore** or **pollen grain**, a structure resistant to desiccation.

A typical flower is made up of four sets of modified leaves: **carpels, stamens, petals** and **sepals**, all attached to a modified stem, the **receptacle**.

In the centre are the carpels which comprise a sticky **stigma** at the end of a slender stalk called the **style**. At the base of the carpel is the **ovary**, a hollow structure containing one or more **ovules**, each of which encloses the female gamete, the **egg nucleus**. The carpels may be separate as in *Ranunculus* (buttercup) or fused in groups as in *Tulipa* (tulip) where three carpels combine to give a single structure. Around the carpels are the **stamens**, each comprising a long stalk, the **filament**, at the end of which are the **anthers**. The anthers produce pollen grains which contain the male gametes.

The stamens are surrounded by the petals, brightly coloured leaf-like structures which attract insects. They may produce **nectar** and be scented. The outer ring of structures is the sepals. These are usually green and may photosynthesize, but their main function is to protect the other floral parts when the flower is a bud.

Occasionally the sepals are brightly coloured, e.g. in lilies, and help in insect attraction or may later assist in dispersal, e.g. in

TABLE 18.1 **Some basic plant terminology**

Floral part	Collective names of parts	Name given to flower if floral part is absent
Carpel	Gynoecium	Staminate, e.g. *Zea*
Stamen	Androecium	Carpellate, e.g. *Zea*
Petal	Corolla	Apetalous, e.g. *Clematis*
Sepal	Calyx	Asepalous, e.g. most of the Umbelliferae
Corolla + calyx	Perianth	e.g. wild arum

mulberry, where they become juicy and attract animals.

The four sets of floral parts are given collective names as shown in Table 18.1 (on the previous page).

Flowers with petals and sepals of similar size and shape exhibiting radial symmetry, e.g. buttercup, are called **regular** or **actinomorphic**. Flowers with unequal sepals and petals of different shapes and arranged in bilateral symmetry, e.g. white dead-nettle, are termed **irregular** or **zygomorphic**. (See Figs. 18.1–18.3.)

Fig. 18.1 Floral structure

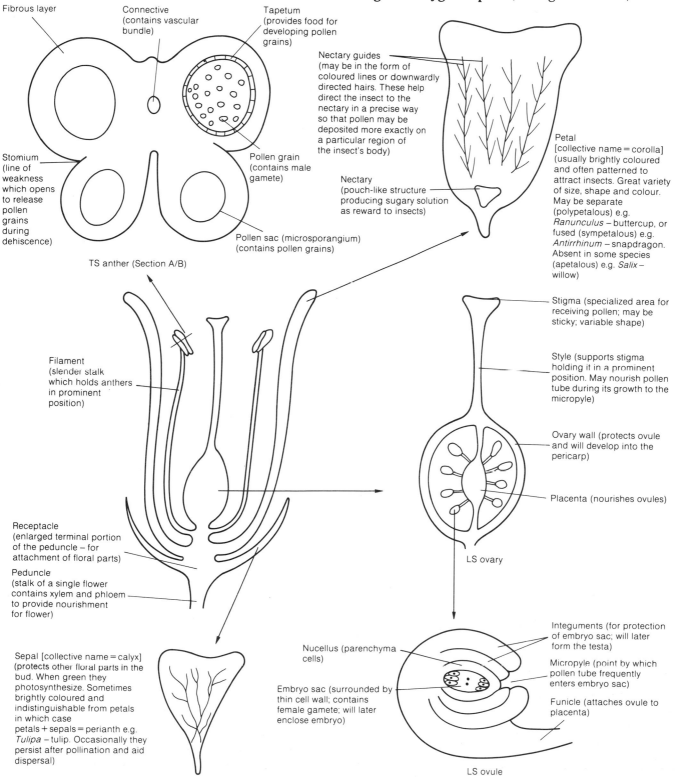

Fibrous layer

Connective (contains vascular bundle)

Tapetum (provides food for developing pollen grains)

Nectary guides (may be in the form of coloured lines or downwardly directed hairs. These help direct the insect to the nectary in a precise way so that pollen may be deposited more exactly on a particular region of the insect's body)

Stomium (line of weakness which opens to release pollen grains during dehiscence)

Pollen grain (contains male gamete)

Nectary (pouch-like structure producing sugary solution as reward to insects)

Petal [collective name = corolla] (usually brightly coloured and often patterned to attract insects. Great variety of size, shape and colour. May be separate (polypetalous) e.g. *Ranunculus* – buttercup, or fused (sympetalous) e.g. *Antirrhinum* – snapdragon. Absent in some species (apetalous) e.g. *Salix* – willow)

Pollen sac (microsporangium) (contains pollen grains)

TS anther (Section A/B)

Filament (slender stalk which holds anthers in prominent position)

Stigma (specialized area for receiving pollen; may be sticky; variable shape)

Style (supports stigma holding it in a prominent position. May nourish pollen tube during its growth to the micropyle)

Ovary wall (protects ovule and will develop into the pericarp)

Placenta (nourishes ovules)

Receptacle (enlarged terminal portion of the peduncle – for attachment of floral parts)

Peduncle (stalk of a single flower contains xylem and phloem to provide nourishment for flower)

LS ovary

Sepal [collective name = calyx] (protects other floral parts in the bud. When green they photosynthesize. Sometimes brightly coloured and indistinguishable from petals in which case petals + sepals = perianth e.g. *Tulipa* – tulip. Occasionally they persist after pollination and aid dispersal)

Nucellus (parenchyma cells)

Embryo sac (surrounded by thin cell wall; contains female gamete; will later enclose embryo)

Integuments (for protection of embryo sac; will later form the testa)

Micropyle (point by which pollen tube frequently enters embryo sac)

Funicle (attaches ovule to placenta)

LS ovule

253

The anthers comprise pollen sacs (usually 4) which contain a mass of diploid pollen mother cells.

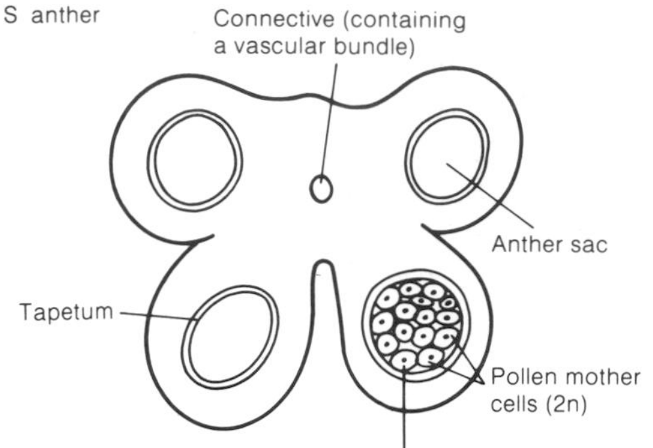

T S anther

Connective (containing a vascular bundle)

Anther sac

Tapetum

Pollen mother cells (2n)

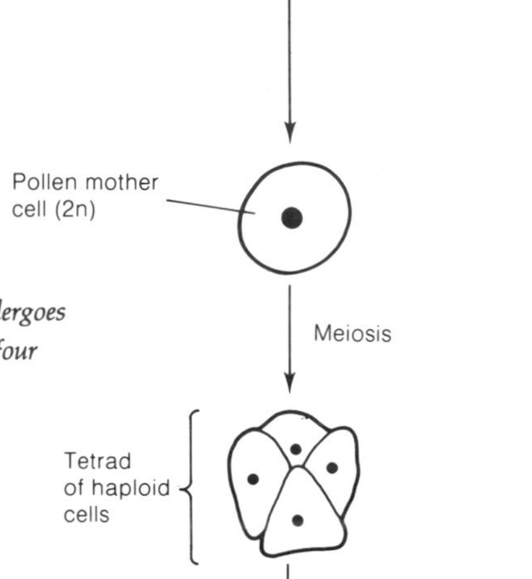

Pollen mother cell (2n)

Each pollen mother cell undergoes meiosis to form a tetrad of four haploid cells.

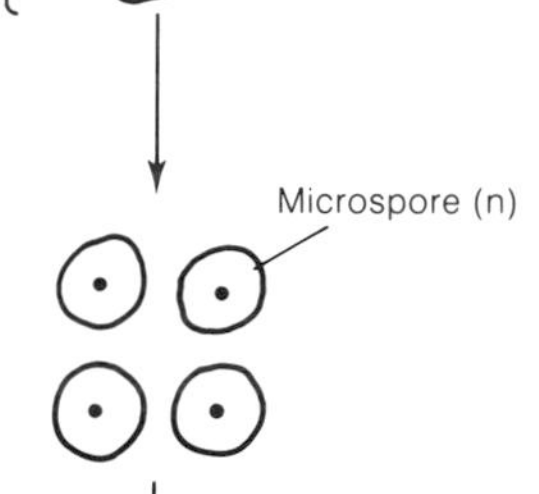

Meiosis

Tetrad of haploid cells

The cells round off and are called microspores.

Microspore (n)

Pollen grain

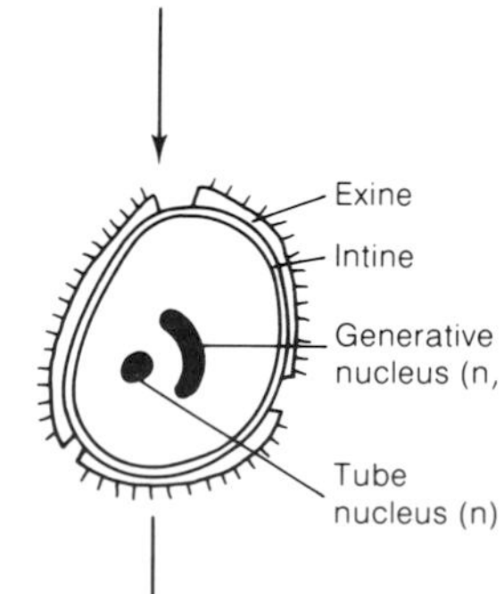

The single nucleus divides by mitosis to give the tube nucleus and the generative nucleus. The wall thickens and forms an inner layer, the intine and an often highly sculptured outer layer, the exine.

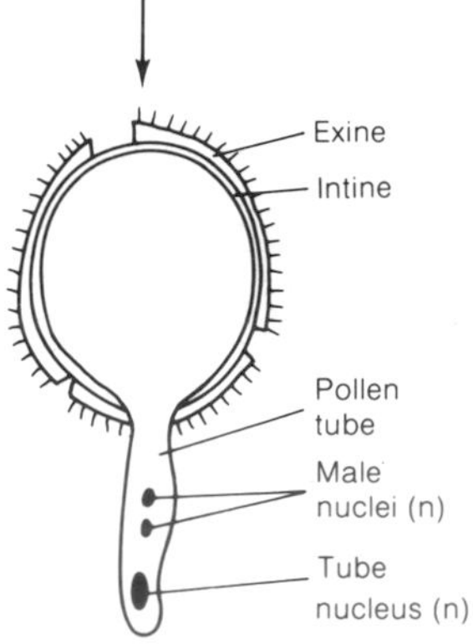

Exine

Intine

Generative nucleus (n)

Tube nucleus (n)

When transferred to the stigma of a plant of the same species the pollen grain germinates to produce a pollen tube. The tube nucleus moves down the tube first, followed by the generative nucleus which soon divides mitotically to give two male nuclei.

Exine

Intine

Pollen tube

Male nuclei (n)

Tube nucleus (n)

Fig. 18.2 Structure and development of the pollen grain

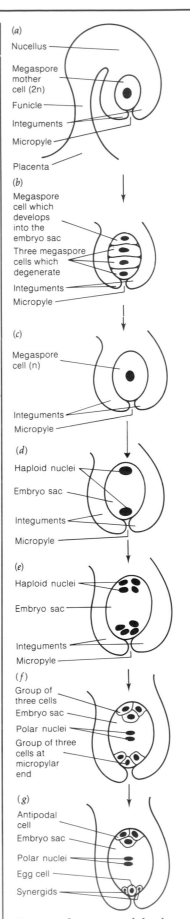

(a)
Nucellus
Megaspore mother cell (2n)
Funicle
Integuments
Micropyle
Placenta

(b)
Megaspore cell which develops into the embryo sac
Three megaspore cells which degenerate
Integuments
Micropyle

(c)
Megaspore cell (n)
Integuments
Micropyle

(d)
Haploid nuclei
Embryo sac
Integuments
Micropyle

(e)
Haploid nuclei
Embryo sac
Integuments
Micropyle

(f)
Group of three cells
Embryo sac
Polar nuclei
Group of three cells at micropylar end

(g)
Antipodal cell
Embryo sac
Polar nuclei
Egg cell
Synergids

Fig. 18.3 Structure and development of the ovule

(a) The ovule consists of a mass of cells called the **nucellus** which is carried on a short stalk called the **funicle**. The nucellus is completely surrounded by two protective **integuments** except for a narrow channel at the tip called the **micropyle**. One cell of the nucellus becomes larger and more conspicuous than the rest. This is the **embryo sac mother cell**.

(b) The embryo sac mother cell divides meiotically to give four haploid **megaspore cells**.

(c) The three cells nearest the micropyle degenerate while the remaining one enlarges to form the **embryo sac**.

(d) The embryo sac nucleus divides by mitosis and the resultant nuclei migrate to opposite poles.

(e) Each nucleus undergoes two mitotic divisions to give a group of four haploid nuclei at each pole.

(f) One nucleus from each polar group moves to the centre of the embryo sac. These are the polar nuclei. The remaining nuclei develop cytoplasm around them and become separated by cell walls leaving two groups of three cells at each pole.

(g) The three cells at the opposite end to the micropyle are called **antipodal cells** and play no further role in the process. Of the three cells at the micropyle end, one, the **egg cell**, remains, the other two, the **synergids**, degenerate.

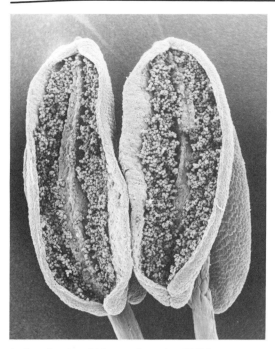

Anther showing dehiscence (scanning EM) (× 60 approx.)

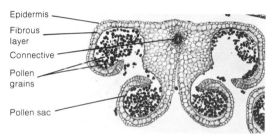

Epidermis
Fibrous layer
Connective
Pollen grains
Pollen sac

Anther (TS) showing dehiscence (× 40 approx.)

18.2 Pollination

Pollination is the transfer of pollen from anthers to stigmas. If the transfer occurs between two plants of different genetic make-up the process is **cross-pollination**. If the transfer takes place between flowers of identical genetic constitution, the process is **self-pollination**. It is common to think of self-pollination occurring within a single flower on a plant, such as garden peas, where the petals so enclose the stamens that the pollen has little chance of escaping. However, it also occurs if pollen is transferred between different flowers on the same plant. This may occur as an insect moves from flower to flower collecting nectar.

A third type occurs when pollen is transferred between flowers on two separate plants which are genetically identical. This is most common where groups of plants have arisen as a result of asexual reproduction, e.g. groups of daffodils or irises. As members of a single clone, these groups have individuals with identical genotypes. The design of any individual flower is related to the precise agent used to transfer pollen. If the plant is insect-pollinated its bright colour, patterns and scent attract potential pollinating insects. They receive nectar or excess pollen which encourages them to seek out a similar flower and thereby transfer more pollen.

As colour and scent have no bearing on wind direction, wind-pollinated flowers are dull, unattractive and without scent. Indeed as petals may shelter the reproductive structures from the wind they are frequently dispensed with altogether, leaving the anthers and stigmas exposed. (See Fig. 18.4 and Table 18.2.)

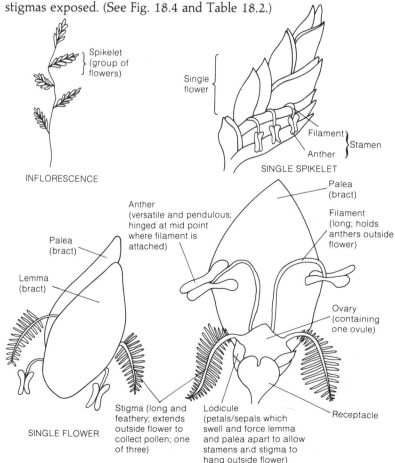

Spikelet (group of flowers)

INFLORESCENCE

Single flower

Filament
Anther
} Stamen

SINGLE SPIKELET

Palea (bract)

Anther (versatile and pendulous; hinged at mid point where filament is attached)

Filament (long; holds anthers outside flower)

Palea (bract)

Lemma (bract)

Ovary (containing one ovule)

Stigma (long and feathery; extends outside flower to collect pollen; one of three)

Lodicule (petals/sepals which swell and force lemma and palea apart to allow stamens and stigma to hang outside flower)

Receptacle

SINGLE FLOWER

FLOWER WITH LEMMA REMOVED

Fig. 18.4(a) Rye grass (Lolium perenne)

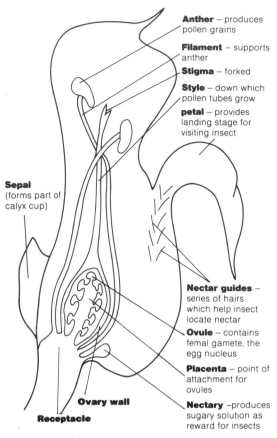

Fig. 18.4(b) Half-flower of antirrhinum, an irregular (zygomorphic) flower

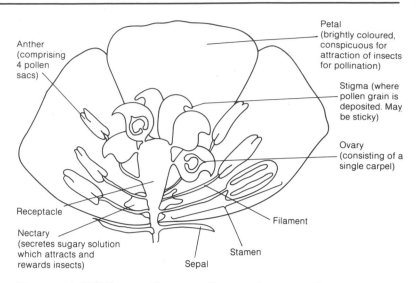

Fig. 18.4(c) Half-flower of buttercup (Ranunculus), *a regular (actinomorphic) flower*

TABLE 18.2 **Comparison of wind and insect-pollinated flowers**

Wind-pollinated flowers (anemophilous)	Insect-pollinated flowers (entomophilous)
e.g. Rye grass (*Lolium perenne*)	e.g. Buttercup (*Ranunculus repens*)
Plants often occur in dense groups covering large areas	Plants often solitary or in small groups
Flowers occur in groups (inflorescences) on the plant (e.g. Graminae)	Flowers may occur on the plant as inflorescences (e.g. apple) but may also be solitary (e.g. tulip)
Flowers are often unisexual with an excess of male flowers	Mostly bisexual (hermaphrodite) flowers
Petals are dull and much reduced in size	Petals are large and brightly coloured to make them conspicuous to insects
No scent or nectar is produced	Flowers produce scent and/or nectar to attract insects
Stigmas often protrude outside the flower on long styles	Stigmas lie deep within the corolla
Stigmas are often feathery, giving them a large surface area to filter pollen from the air	Stigmas are relatively small as the pollen is deposited accurately by the pollinating insects
Anthers dangle outside the flower on long filaments so the pollen is easily released into the air	Anthers lie inside the corolla so the pollinating insect brushes against them when collecting the nectar
Enormous amounts of pollen are produced to offset the high degree of wastage during dispersal	Less pollen is produced as pollen transfer is more precise and so entails less wastage
Pollen is smooth, light and small and sometimes has 'wing-like' extensions to aid wind transport	Pollen is larger and often bears projections which help it adhere to the insect

18.3 Fertilization

In angiosperms the female gamete is protected within the carpel and the male gamete can only reach it via the **pollen tube**. On landing upon the stigma the pollen grains absorb water and germinate to give the pollen tube. The tube pushes between the loosely packed cells of the style, the **tube nucleus** preceding the **male nuclei**. The role of the tube nucleus is to control the growth of the pollen tube and it plays no part in fertilization. While the initial growth into the style may be the result of a negative aerotropic response, it is thought that the tube then shows a positive chemotropic response to some substance produced in the embryo sac. The secretion of pectases by the pollen tube may soften the middle lamellae of the cells in the style and so assist its growth towards the micropyle – the usual point of entry to the embryo sac. Many tubes grow down the style simultaneously and where an ovary has many ovules a separate one penetrates each. On entering the embryo sac the tube nucleus, its work done, degenerates and the two male nuclei enter. One male nucleus fuses with the egg cell to give a diploid zygote; the other fuses with the two polar nuclei to form the primary endosperm nucleus, which is triploid. This double fertilization is peculiar to flowering plants.

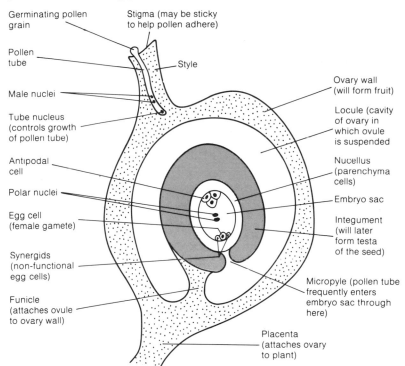

Fig. 18.5 Mature carpel at fertilization (LS)

18.4 Methods of preventing inbreeding

There can be no doubt that in some plants self-pollination occurs more or less regularly. However, there appears to be a general tendency to avoid self-pollination since this is a form of inbreeding and would very quickly reduce the variability of a population. There are four main methods of avoiding self-pollination.

1. The stamens and stigma of a flower mature at different times. If the stamens ripen before the stigma is in a condition to receive pollen, the flower is **protandrous**, e.g. white dead nettle (*Lamium*). If the stigma and ovule ripen before the stamens, the flower is **protogynous**, e.g. plantain (*Plantago*). Protandry is more common than protogyny.

2. If a plant has separate male and female flowers it is said to be **monoecious**, e.g. maize (*Zea*). This condition clearly limits the possibility of self-pollination and in *Zea* the number of seeds produced by self-pollination is reduced to less than 1%.

3. The structure of the flower itself makes self-pollination unlikely. Some flowers, e.g. *Iris*, have a stigmatic flap which is exposed to the pollen on the back of a visiting insect. The insect collects pollen from the stamens and closes the flap as it withdraws from the flower, thus protecting the stigmatic surface from its own pollen.

4. A **dioecious** species is one in which some individual plants have either all male or all female flowers. Completely dioecious plants are rare. It is more usual for a plant to be predominantly, although not completely, of one sex, e.g. plantain (*Plantago*) and ash (*Fraxinus*).

To prevent self-pollination is not the only means of preventing inbreeding. In many plants self-pollination occurs but there is a mechanism to prevent this leading to successful fertilization of the ovule and production of a seed. This is known as **incompatibility**. For example, in pears the pollen only becomes functional if the stigmatic surface on which it lands has a different genetical composition. *Primula vulgaris* (primrose) is a dimorphic plant in which there are two types of flower, pin-eyed and thrum-eyed. These differ in the length of the style (**heterostyly**) as well as in the size and chemical composition of their pollen.

These differences do not prevent self-pollination but the yield of seed produced by self-pollination is very poor.

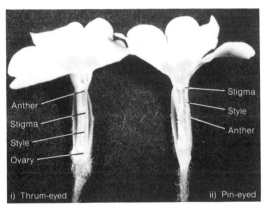

Pin-eyed and thrum-eyed *Primula* flowers (LS)

18.5 Development of fruits and seeds

Following fertilization, the zygote divides rapidly by mitosis and develops into the embryo, which then differentiates into a young shoot, called the **plumule**, a young root, the **radicle** and seed leaves known as **cotyledons**. The primary endosperm nucleus also divides mitotically to give a mass of cells, the **endosperm**. This forms the food source for the growing embryo. In some species, e.g. maize (*Zea mays*), the endosperm remains while in others, e.g. peas (*Pisum*), it is quickly absorbed by and stored in the cotyledons. Other parts develop as shown opposite. (See Fig. 18.6, on the next page.)

The most common food stores in seeds is carbohydrate. This is usually in the form of starch but some seeds, e.g. maize and peas, store quantities of sugar. Many young seeds store sugar but this changes to starch as they mature. Lipids are often stored in the cotyledons and may form a high percentage of the dry weight, e.g. 60% in walnuts and coconuts; 40% in sunflowers. Other economically important examples are peanuts, soyabeans and castor oil seeds. Proteins are found to a lesser extent in seeds but wheat has an aleurone layer and protein is stored in the cotyledons of legumes and nuts.

Before fertilization	After fertilization
Ovary and contents	Fruit
Ovary wall	Pericarp (wall of fruit)
Ovule	Seed
Integuments	Testa (seed coat)

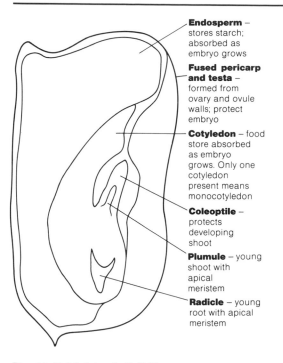

Endosperm – stores starch; absorbed as embryo grows

Fused pericarp and testa – formed from ovary and ovule walls; protect embryo

Cotyledon – food store absorbed as embryo grows. Only one cotyledon present means monocotyledon

Coleoptile – protects developing shoot

Plumule – young shoot with apical meristem

Radicle – young root with apical meristem

Fig. 18.6(a) Maize fruit (LS)

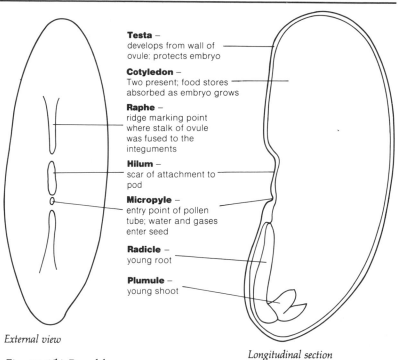

Testa – develops from wall of ovule: protects embryo

Cotyledon – Two present; food stores absorbed as embryo grows

Raphe – ridge marking point where stalk of ovule was fused to the integuments

Hilum – scar of attachment to pod

Micropyle – entry point of pollen tube; water and gases enter seed

Radicle – young root

Plumule – young shoot

External view

Longitudinal section

Fig. 18.6(b) Broad bean

Dandelion fruits showing pappus of hairs

Feathery style

Fruit

Clematis fruit showing persistent style

18.6   Fruit and seed dispersal

Sexual reproduction ensures variability. New genotypes are produced with the potential to survive in different habitats. If this potential is to be realized it is essential that fruits and seeds are dispersed to new areas with different environmental conditions. A degree of isolation also reduces the chances of backcrossing with the parents and eliminates all the problems that inbreeding incurs. Dispersal prevents overcrowding and competition and so makes species less vulnerable to epidemic attack by viruses, fungi and insects.

18.6.1 Wind dispersal

Some seeds, such as those of orchids, show no special modification but are simply small enough to be blown large distances by the wind. Some fruits, however, are modified structurally to increase their surface area; in ash and sycamore the pericarp is extended to form a wing; in lime a bract serves the same purpose; in dandelion a pappus of hair forms a 'parachute', whereas in clematis it is a long persistent style which does so. These modifications increase the fruit's surface area and so present more resistance to the wind. This delays its descent and the longer it is airborne the further it is carried.

18.6.2 Animal dispersal

The fruit or seed may develop hooks which catch in the fur of animals. These hooks may be extensions of the pericarp, e.g. goosegrass, or hooked bracts, e.g. burdock. Alternatively the fruit may be eaten by an animal and the seeds, which are resistant to digestion, later pass out in faeces some distance from the parent plant. These fruits, e.g. cherry and blackberry, are usually brightly

259

Lime fruit showing bract

Sycamore fruit showing extended pericarp

Burdock fruits showing hooked bracts

coloured with edible, sweet-tasting succulent parts. Strawberries and apples are false fruits in which the receptacle and not the pericarp forms the succulent part.

Blackberry

Sweet chestnut

18.6.3 Water dispersal

This occurs in a few species, e.g. water-lily and coconut, living in or near water. The seed is covered in a spongy or fibrous layer which traps air, making the seed or fruit buoyant.

18.6.4 Mechanical dispersal

The fruit may remain attached to the plant but open violently to expel the seeds. This is usually caused by the unequal drying of the pericarp and occurs in all legumes, e.g. pea and gorse.

18.6.5 Censer mechanisms

The fruits of the poppy and the bluebell are borne at the end of long stalks and have apertures through which the seeds are shaken as they blow in the wind. These pores may close in wet conditions.

Poppy capsule showing open pores

18.6.6 Casual mechanisms

Some seeds are not adapted to a single agent of dispersal but are opportunists, using any available means of dispersal. For example, acorns may be rolled along the ground by the wind, carried about by squirrels and even float downstream in rivers.

18.7 Dormancy

The water content of seeds at between 5–10% is very low and is the major factor in preventing them germinating. As a rule, the addition of water in the presence of oxygen and a favourable temperature is enough to break this dormancy (Section 20.3.1). Some seeds, however, still fail to germinate for one reason or another:

1. Light is necessary for the germination of certain seeds, e.g. lettuce.

2. A sustained period of cold is needed to make some seeds of temperate climates germinate (Section 40.3.3). This helps ensure that seeds do not germinate in late summer or during mild winter spells, thus making the young plant vulnerable to frosts at a later date.

3. Conversely a few seeds will not germinate unless subjected to the heat of a flash-fire.

4. A period of time is necessary to permit internal chemical changes to take place before other seeds germinate.

5. The seed coat may be impermeable to water and/or gases and time may be needed for it to decay and break. In many seeds physical abrasion or partial digestion in the intestines of an animal help break dormancy by weakening the testa.

6. Another type of dormancy is brought about by the presence of natural chemical inhibitors.

The number of dormant seeds in the soil is surprising. For example, 2.4 ha of soil at the Research Centre in Rothampstead was found to contain 300 million seeds. Periods of dormancy may last for a number of years. Lotus seeds from peat beds in Manchuria yielded high rates of germination even though they were between two hundred and one thousand years old.

1. (a) For the following list of parts of the flower briefly describe
   (i) the location, and
   (ii) the function of the
       (1) tapetum;   (4) embryo sac;
       (2) stomium;   (5) micropyle.
       (3) stigma;   (5 marks)
   (b) Flowering plants can undergo either cross-pollination or self-pollination. Give **two** disadvantages and **one** advantage of self-pollination.   (3 marks)
   (c) With reference to named examples, briefly describe **three** mechanisms by which plants encourage cross-pollination at the expense of self-pollination.   (6 marks)
   (d) In view of the importance of cross-pollination (i.e. sexual reproduction), the widespread occurrence of vegetative reproduction in flowering plants would appear anomalous.
       (i) What are the advantages of vegetative reproduction which explain this apparent anomaly?
       (ii) Describe how vegetative reproduction occurs via structures **above** the ground in a **named** flowering plant (angiosperm).
                                          (6 marks)
                                    (Total 20 marks)

   Northern Ireland Board June 1983, Paper II, No. 4

2. (a) Describe, with the aid of labelled diagrams,
       (i) the structure of a mature carpel and ovule;
       (ii) the structure of a mature anther and explain how dehiscence occurs.   (10 marks)
   (b) Give a detailed account of
       (i) the processes leading to the fertilization of an ovule after pollen has landed on the stigma, and
       (ii) the formation of a seed following fertilization of the ovule.   (10 marks)
                                    (Total 20 marks)
   Joint Matriculation Board June 1983, Paper IIB, No. 7

3. Describe the processes that occur during the production of a pollen grain in the anther of a flowering plant until the mature grain is exposed to a visiting pollinator.   (14 marks)
   By what means may such grains be prevented from reaching the receptive stigma of the same flower?
                                          (6 marks)
                                    (Total 20 marks)

   Oxford and Cambridge Board July 1984, Paper II, No. 3

4. Draw a longitudinal section through a named zygomorphic flower which you have studied. Label it and say how pollination occurs.   (8 marks)

   Southern Universities Joint Board June 1985, Paper I, No. 7

5. (a) With reference to **named** examples in each case, show how flowers are adapted for (i) wind-pollination; (ii) self-pollination.   (4,4 marks)
   (b) Describe the events which occur in the flower from pollination to the formation of a seed.
                                          (10 marks)
                                    (Total 18 marks)

   Cambridge Board June 1985, Paper I, No. 7

6. (a) Give an illustrated description of a named flower. State its method of pollination.
                                          (7 marks)
   (b) Review the typical differences which may be recognised between insect-pollinated and wind-pollinated flowers.   (8 marks)
   (c) Discuss the relative advantages and disadvantages of insect and wind pollination.
                                          (5 marks)
                                    (Total 20 marks)

   Oxford and Cambridge Board June 1989, Paper II, No. 6

7. (a) Describe the structure of a pollen grain.
                                          (4 marks)
   (b) Give a detailed account of the formation, transfer and germination of a pollen grain of a flowering plant, up to the moment of fertilisation.   (10 marks)
   (c) Relate these phases in the angiosperm life cycle to the equivalent stages in the life cycle of a **named** fern.   (4 marks)
                                    (Total 18 marks)

   Cambridge Board June 1988, Paper I, No. 8

8. (a) Why is outbreeding an advantage to flowering plants?   (2 marks)
   (b) The diagram on the next page represents a section through a flower in which there is a special mechanical device which favours outbreeding.
       (i) Suggest the sequence of events by which pollen would be transferred onto the body of a visiting insect.   (3 marks)
       (ii) How do the relative positions of the fertile anther and stigma favour outbreeding?
                                          (2 marks)

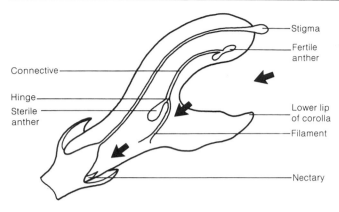

Path taken by visiting insect

(iii) In this flower the anthers ripen before the stigma becomes receptive. Explain how this favours outbreeding.　(2 marks)

(c) Name *one* condition found in certain flowering plants which makes outbreeding inevitable.
(1 mark)
(Total 10 marks)

London Board June 1988, Paper I, No. 7

9. The figure below is a diagram of a longitudinal section of part of a flower.

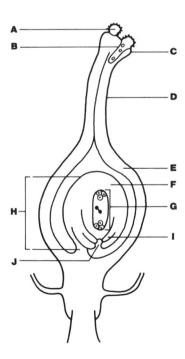

(a) Name the structures labelled **A, B, C, D, E, F, G, H, I** and **J**.

(b) Indicate how **G** developed from a spore mother cell.

(c) State what happens to the structures **E, F, H** and **I** after fertilization has occurred.
(Total 10 marks)

Cambridge Board June 1989, Paper III, No. 1

10. The diagrams below show stages in the life cycle of a flowering plant.

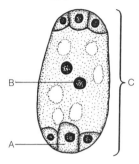
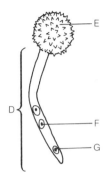

(a) Complete the table below by choosing the letter from the diagrams which refers to each of the stages given.

Stage in life cycle	Letter
Female gametophyte	
Tube nucleus	
Female gamete	
Male gamete	

(4 marks)

(b) (i) State *one* function of D.　(1 mark)
　　(ii) How has structure D enabled flowering plants to adapt to terrestrial life?
(2 marks)

(c) Comment on the surface structure of E.
(2 marks)

(d) Suggest *two* ways in which self fertilization may be avoided in flowering plants.(2 marks)
(Total 11 marks)

London Board June 1989, Paper I, No. 12

19 *Reproduction in animals*

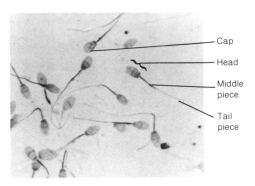

Sperm (× 600 approx.)

You saw in Chapter 17 that one major advantage of sexual reproduction is the genetic variety it creates, and that the extent of this variety is greater the more diverse the parental genotypes are. Animals, with their capacity for locomotion, are able to move far afield in their search for mates and so reproduce with individuals outside their family groups. This produces a greater degree of outbreeding, greater mixing of genes within the gene pool and hence greater variety. In animals the gametes are usually differentiated into a small motile male gamete or **sperm** which is produced in large numbers, and a larger, non-motile food-storing female gamete or **ovum** which is produced in much smaller numbers. More primitive animals may be **monoecious (hermaphrodite)** in which case a single individual is capable of producing both types of gametes. While these organisms frequently have mechanisms to avoid self-fertilization, e.g. producing male and female gametes at different times, there remains a risk of self-fertilization and hence less variety in the offspring. All higher animals are **dioecious**, producing only one type of gamete. As self-fertilization is impossible in these species the offspring show more variety.

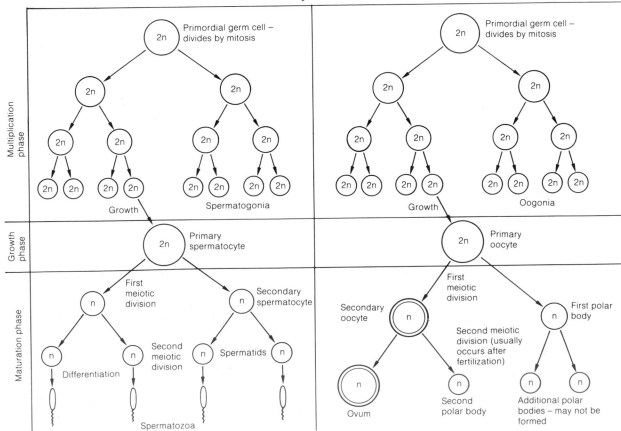

Fig. 19.1(a) Spermatogenesis — formation of sperm   Fig. 19.1(b) Oogenesis — formation of ova

The organs which produce gametes are called **gonads** and are of two types: the ovaries which produce ova and the testes which produce sperm. In some animals, e.g. mammals, the reproductive and excretory systems are closely associated with one another; in such cases they are often represented together as the urino-genital system.

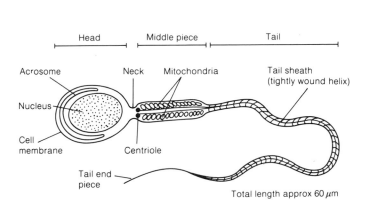

Fig. 19.2(a) Human spermatozoan based on electron micrograph

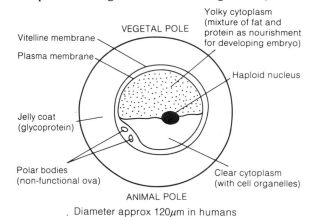

Fig. 19.2(b) A generalized egg cell

19.1 Human male reproductive system

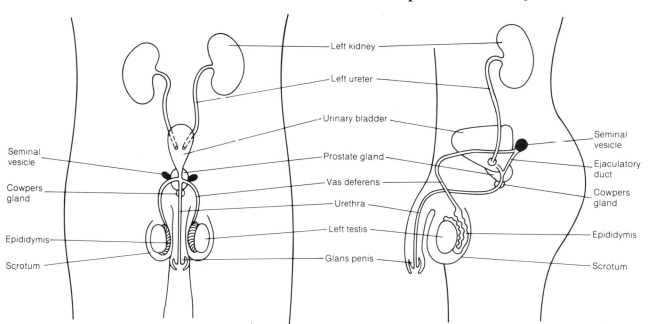

Fig. 19.3(a) Male urinogenital system (simplified) − front view

Male urinogenital system − side view

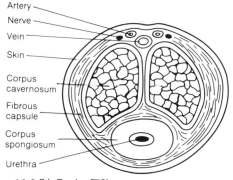

Fig. 19.3(b) Penis (TS)

The male gonads, the **testes**, develop in the abdominal cavity and descend into an external sac, the **scrotum**, prior to birth. The optimum temperature for sperm development is around 35 °C, about 2 °C below normal human body temperature. The testes can be kept at this temperature by the contraction and relaxation of muscle in the scrotal wall. When the temperature of the testes exceeds 35 °C the muscles relax, holding them away from the body to assist cooling. In colder conditions the muscles contract to bring the testes as close to the abdominal cavity as is necessary to maintain them at the optimum temperature. Each testis is suspended by a spermatic cord composed of the sperm duct or vas deferens, spermatic artery and vein, lymph vessels and nerves, bound together by connective tissue.

265

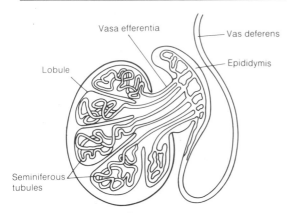

Fig. 19.3(c) Testis (LS)

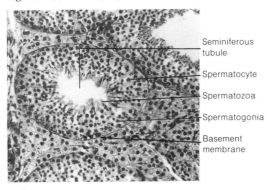

Seminiferous tubule (TS) (× 150 approx.)

Fig. 19.3(d) Photomicrograph of TS seminiferous tubule

A single testis is surrounded by a fibrous coat and is separated internally by septa into a series of lobules.

Within each lobule are convoluted **seminiferous tubules**, the total length of which is over 1 km. Between the tubules are the **interstitial cells** which secrete the hormone **testosterone**. Each seminiferous tubule is lined by **germinal epithelium** which, by a series of divisions, gives rise to sperm, a process taking 8–9 weeks.

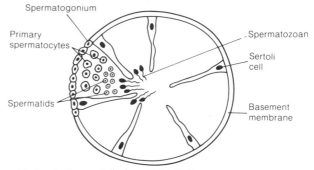

Fig. 19.3(e) Seminiferous tubule × 500 (TS)

The seminiferous tubules merge to form small ducts called the **vasa efferentia**, which in turn join up to form a six-metre-long coiled tube called the **epididymis**. The sperm are stored here, gaining motility over a period of 18 hours. From the epididymis leads another muscular tube, the **vas deferens**, which carries the sperm towards the urethra. Before it joins the urethra it combines with the duct leading from the **seminal vesicle**, forming the **ejaculatory duct**. The seminal vesicles produce a mucus secretion which aids sperm mobility. The ejaculatory duct then passes through the **prostate gland** which produces an alkaline secretion that neutralizes the acidity of any urine in the urethra as well as aiding sperm mobility. Below the prostate glands are a pair of **Cowper's glands** which secrete a sticky fluid into the urethra. The resultant combination of sperm and secretions is called **semen**. The semen passes along the **urethra** a muscular tube running through the **penis**. The penis comprises three cylindrical masses of spongy tissue covered by an elastic skin.

The end of the penis is expanded to form the **glans penis**, a sensitive region covered by loose retractable skin, the **prepuce** or **foreskin**. The foreskin is sometimes removed surgically, for medical or religious reasons, in a small operation called circumcision.

19.2 Human female reproductive system

The female gonads, the **ovaries**, lie suspended in the abdominal cavity by the ovarian ligaments. The external coat is made up of **germinal epithelium** which begins to divide to form ova while the female is still a foetus. At birth around 400 000 cells have reached prophase of the first meiotic division and are called **primary oocytes**. Each month after puberty, one of these cells completes its development into an ovum. As each cell takes some time to complete this change, the ovary consists of a number of oocytes at various stages of development. These oocytes lie in a region of fibrous tissue, the stroma, which fills the rest of the ovary. The largest and most mature of the cells are called **Graafian follicles** which are fluid-filled sacs each containing a secondary oocyte. A mature Graafian follicle can reach a diameter greater than 1 cm before it releases its ovum.

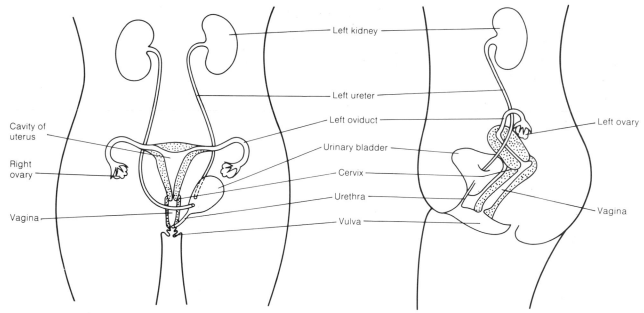

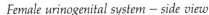

Fig. 19.4(a) Female urinogenital system — front view

Female urinogenital system — side view

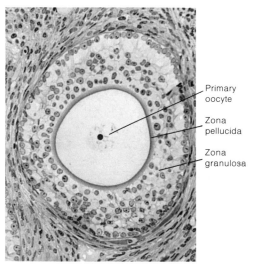

Immature ovarian follicle (× 80 approx.)

Fig. 19.4(b) Photomicrograph of section through Graafian follicle

Close to the ovary is the funnel-shaped opening of the **oviduct** or **Fallopian tube**. The opening has fringe-like edges called **fimbriae**. The oviducts are about 10 cm long and have a muscular wall lined with a mucus-secreting layer of ciliated epithelium. They open into the **uterus**, or womb, which is a pear-shaped body about 5 cm wide and 8 cm in length, held in position by ligaments joined to the pelvic girdle. It has walls of unstriated muscle and is lined internally by a mucus membrane called the **endometrium**. The uterus opens into the **vagina** through a ring of muscle, the **cervix**. The vagina has a wall of unstriated muscle with an inner mucus membrane lined by stratified epithelium. The vagina opens to the outside through the **vulva**, a collective name for the external genital organs. These consist of two outer folds of skin, the **labia majora**, covering two inner, more delicate folds, the **labia minora**. Anterior to the vaginal opening is a small body of erectile tissue, the **clitoris**, which is homologous to the penis of the male. Between the vaginal opening and the clitoris is the opening of the urethra.

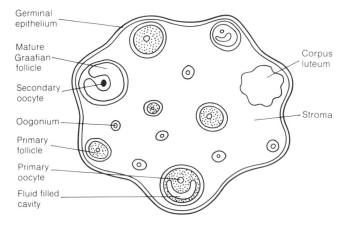

Fig. 19.4(c) Section through ovary

19.3 Sexual cycles

Many animals have cycles of sexual activity in both males and females. These cycles often occur so that fertilization takes place at

267

a time which gives the offspring the best chance of survival, e.g. the offspring are produced at a time when the climate and food availability are most favourable.

In mammals these cycles are of three main types:

1. The female undergoes a single period of sexual activity during the year, e.g. deer (**monoestrus**).

2. The female undergoes a number of periods of sexual activity during the year, each separated by a period of sexual inactivity, e.g. horses (**polyoestrus**).

3. The female has a more or less continuous cycle of activity where the end of one cycle is followed immediately by the start of the next, e.g. humans.

19.3.1 The menstrual cycle

In human females the onset of the first menstrual cycle is called **menarche** and represents the start of puberty. This takes place around the age of 12 years although the age varies widely between individuals. The menstrual cycle, which lasts about 28 days, continues until the **menopause** at the age of 45–50 years. The events of the cycle are controlled by hormones to ensure that the production of an ovum is synchronized with the readiness of the uterus to receive it, should it be fertilized. The start of the cycle is taken to be the initial discharge of blood known as **menstruation**, as this event can be easily identified. This flow of blood, which lasts about five days, is due to the lining of the uterus being shed, along with a little blood.

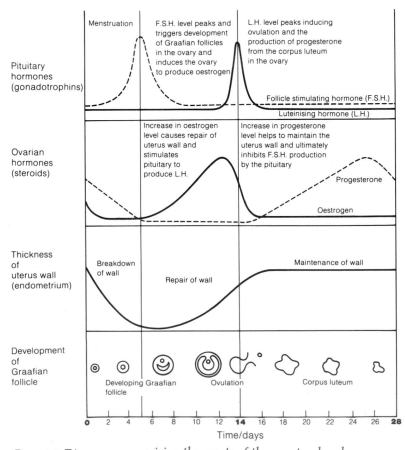

Fig. 19.5 Diagram summarizing the events of the menstrual cycle

During the following days the lining regenerates in readiness for a fertilized ovum. By day 14 it has thickened considerably and the Graafian follicle releases its ovum into the oviduct, the process being called **ovulation**. The ovum is moved down the oviduct mostly by muscular contractions of the oviduct wall, although the beating of the cilia may also assist. The journey to the uterus takes about three days during which time the ovum may be fertilized. If it is not, the ovum quickly dies and passes out via the vagina. The uterine lining is maintained for some time but finally breaks down again about 28 days after the start of the cycle.

19.3.2 Hormonal control of the menstrual cycle

The control of the menstrual cycle is an excellent example of hormone interaction. The action of one hormone is used to stimulate or inhibit the production of another. There are four hormones involved, two produced by the anterior lobe of the pituitary gland at the base of the brain, and two produced by the ovaries. The production of the hormones from the ovaries is stimulated by the pituitary hormones. These hormones are thus referred to as the gonadotrophic hormones. The two gonadotrophic hormones are **follicle stimulating hormone (FSH)** and **luteinizing hormone (LH)**. These stimulate the ovaries to produce **oestrogen** and **progesterone** respectively.

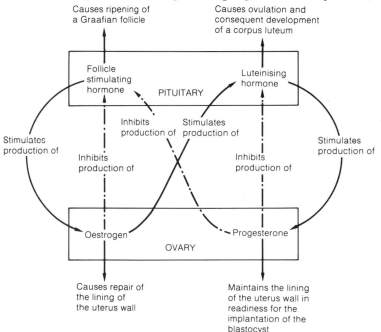

Fig. 19.6 Hormone interaction in the menstrual cycle

The functions of these hormones are as follows:

Follicle stimulating hormone
1. Causes Graafian follicles to develop in the ovary.

2. Stimulates the ovary to produce oestrogen.

Oestrogen
1. Causes repair of the uterus lining following menstruation.

2. Stimulates the pituitary to produce luteinizing hormone.

Luteinizing hormone
1. Causes ovulation to take place.

269

2. Stimulates the ovary to produce progesterone from the corpus luteum.

Progesterone
1. Causes the uterus lining to be maintained in readiness for the blastocyst (young embryo).

2. Inhibits production of FSH by the pituitary.

The hormones are produced in the following sequence: FSH, oestrogen, LH, progesterone. Progesterone at the end of the sequence inhibits the production of FSH. In turn, the production of the other hormones stops, including progesterone itself. The absence of progesterone now means that the inhibition of FSH ceases and so progesterone production commences again. In turn, all the other hormones are produced. This alternate switching on and off of the hormones produces a cycle of events – the menstrual cycle.

19.3.3 Artificial control of the menstrual cycle

The artificial control of the menstrual cycle has two main purposes: firstly as a contraceptive device by preventing ovulation and secondly as a fertility device by stimulating ovulation.

The contraceptive Pill
The Pill contains both oestrogen and progesterone and when taken daily it maintains high levels of these hormones in the blood. These high levels inhibit the production of the gonadotrophic hormones from the pituitary, and the absence of LH in particular prevents ovulation. The Pill is normally taken for 21 consecutive days followed by a period of 7 days without it, during which the uterus lining breaks down and a menstrual period occurs. The 'morning after' Pill contains the synthetic oestrogen, diethylstibestrol which is thought to prevent implantation of the fertilized ovum if it is present. Both types of Pill are very effective forms of contraception. These and other methods of contraception are reviewed in Table 19.1.

Fertility drugs
A fertility drug may induce ovulation in one of two ways:

1. It may provide gonadotrophins such as FSH which stimulate the development of Graafian follicles.

2. It may provide some chemical which inhibits the natural production of oestrogen. As oestrogen normally inhibits FSH production, the level of FSH increases and Graafian follicles develop.

Fertility drugs frequently result in multiple births.

19.3.4 Male sex hormones

Although male humans do not have a sexual cycle similar to that of females, they nonetheless produce some gonadotrophic hormones from the anterior lobe of the pituitary gland. Follicle stimulating hormone (FSH) stimulates sperm development. Luteinizing hormone (LH) stimulates the interstitial cells between the seminiferous tubules of the testis to produce **testosterone**. For this reason, in the male, LH is often called **interstitial cell stimulating hormone (ICSH)**. Testosterone is the most important of a group of male hormones or **androgens**. It is first produced in the foetus, where it controls the development of the male reproductive organs. An increase in production takes place at puberty and causes an enlargement of the reproductive organs and the development of the secondary sex characteristics. Removal of the testes (castration) prevents these changes taking place. It was used in the past as a rather drastic means of preventing choirboys' voices from breaking. Castration of animals is still practised in order to help fatten them and make the meat less tough. (See Fig. 19.7, over the page.)

TABLE 19.1 **Birth control**

Method	How it works	Effectiveness	Advantages	Disadvantages
Sterilization	**Male (vasectomy)** – The vas deferens (the duct carrying sperm from the testis to the urethra) are cut and tied off	100%	No artificial appliance is involved. Once the operation has been performed there is no further cost	Irreversible in normal circumstances
	Female (tubal ligation) – The oviducts are cut and tied off			
The contraceptive pill	**21-day Pill** – Contains progesterone and oestrogen. Prevents the production of ova	100%	Totally reliable if taken regularly	A slightly higher than normal risk of thrombosis especially in older women. Possible nausea, breast tenderness and water retention leading to an increase in weight
	28–day Pill (Mini-Pill) – Contains progesterone. Prevents the production of ova	99–100%	Very reliable. Slightly less risk of thrombosis than with the 21-day Pill	
	Morning-after Pill – Contains high level of oestrogen. Probably works by preventing implantation	Not widely used but probably 99–100%	–	
Intra-Uterine Device (IUD) (Loop, Coil)	A device usually made of plastic and/or copper which is inserted into the womb by a doctor and which prevents implantation	99–100%	Once fitted, no further action is required except for annual check-ups	Possible menstrual discomfort. The device may be displaced or rejected. Must be inserted by a trained practitioner. Only really suitable for women who have had children
Mechanical barriers	**Female (diaphragm, cap)** – A dome-shaped sheet of thin rubber with a thicker spring rim which is inserted into the vagina, over the cervix. Best used with spermicide	Very reliable	Reliable. Available for use by all women	Must be inserted prior to intercourse and should be removed 8–24 hours after intercourse. Initial fitting must be by a trained practitioner
	Male (condom, sheath) – a sheath of thin rubber unrolled onto the erect penis prior to intercourse. Semen is collected in teat at the tip. Best used with a spermicide	Very reliable	Easily available, no fitting or instruction by others needed. Available for use by all men. Gives some protection against sexually transmitted disease including AIDS	May reduce the sensitivity of the penis and so interfere with enjoyment
Spermicide	Cream, jelly or foam inserted into vagina. Only effective with a mechanical barrier. Kills sperm	Not reliable alone	Easy to obtain and simple to use	Not effective on its own. May occasionally cause irritation
Rhythm method	Refraining from intercourse during those times in the menstrual cycle when conception is most likely	Variable – not very reliable	No appliance required. Only acceptable method to some religious groups	Not reliable. Restricts times when intercourse can take place. Unsuitable for women with irregular cycles

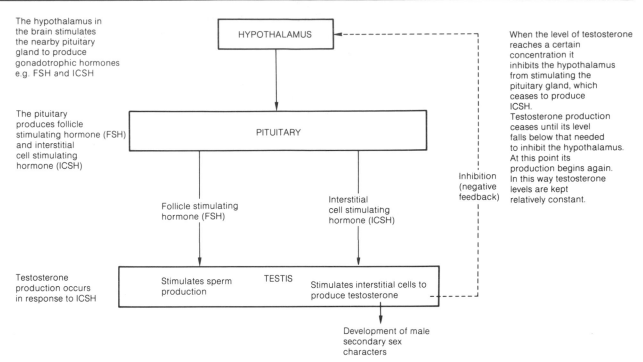

The hypothalamus in the brain stimulates the nearby pituitary gland to produce gonadotrophic hormones e.g. FSH and ICSH

HYPOTHALAMUS

When the level of testosterone reaches a certain concentration it inhibits the hypothalamus from stimulating the pituitary gland, which ceases to produce ICSH.
Testosterone production ceases until its level falls below that needed to inhibit the hypothalamus. At this point its production begins again. In this way testosterone levels are kept relatively constant.

The pituitary produces follicle stimulating hormone (FSH) and interstitial cell stimulating hormone (ICSH)

PITUITARY

Follicle stimulating hormone (FSH)

Interstitial cell stimulating hormone (ICSH)

Inhibition (negative feedback)

Testosterone production occurs in response to ICSH

TESTIS

Stimulates sperm production

Stimulates interstitial cells to produce testosterone

Development of male secondary sex characters

Fig. 19.7 Control of male hormone production

19.3.5 Factors affecting breeding cycles

In many animals environmental factors affect breeding cycles. In birds the gonads are very small outside the breeding season. This reduction in mass assists flight. The seasonal growth of the gonads prior to mating occurs in response to increasing day length, i.e. in the spring. The same stimulus promotes testosterone production in the stickleback (*Gasterosteus aculeatus*). Temperature and availability of food are other factors which affect sexual activity, e.g. sexual activity in the minnow (*Couesius plumbius*) is stimulated by a rise in temperature. In many species the production of testosterone increases in response to these factors. In male deer, for example, the increase in testosterone leads to growth of the antlers, changes in the voice and aggressive behaviour towards other males. In domestic cattle, the sight of a cow alone is sufficient to cause a massive rise in ICSH, and hence testosterone level, in the bull.

19.4   Courtship

In many species it is necessary for both partners to follow a specific pattern of behaviour before mating can occur. Courtship behaviour as it is called is developed in sexually mature individuals. In this way matings between sexually immature individuals, which cannot produce offspring, are avoided. This ensures that the often scarce sites for raising young are only occupied by pairs which have a good chance of producing offspring. On reaching sexual maturity many species develop easily recognizable features which are sexually attractive to a potential partner. These are referred to as the secondary sex characteristics. They take a variety of forms, including bright plumage in many birds, the mane of a lion, the comb and spurs of a cockerel and territory marking in dogs. In humans, secondary sex characteristics include the growth of pubic hair in both sexes, increased musculature, growth of facial hair and deepening of the voice in males and development of the breasts and broadening of the hips in females. Apart from preparing the female for child-bearing the changes in many animals help to distinguish males and females. In this way time and energy are not wasted on the fruitless courting of members of the same sex or sexually immature individuals of the opposite sex.

The females of many species undergo a cycle of sexual activity during which they are only capable of conceiving for a very brief period. Courtship behaviour is used by the male to determine whether the female is receptive or not. If she responds with the correct behavioural actions, courtship continues and is likely to result in fertilization. If she is not receptive, she exhibits a different pattern of behaviour and the male ceases to court her, turning his attentions elsewhere.

19.5 Mating

Under a variety of erotic conditions the blood supply to the genital regions increases. In females the process is slower than in males and results in the clitoris and labia becoming swollen with blood. At the same time the walls of the vagina secrete a lubricating fluid which assists the penetration of the penis. The fluid also neutralizes the acidity of the vagina which would otherwise kill the sperm. In males the increased blood supply results in the spongy tissues of the penis becoming swollen with blood, making it hard and erect. In this condition it more easily enters the vagina. By repeated thrusting of the penis within the vagina the sensory cells in the glans penis are stimulated. This leads to reflex contractions of muscles in the epididymis and vas deferens. The sperm are thus moved by peristalsis along the vas deferens and into the urethra. Here they mix with the secretions from the seminal vesicles, prostate and Cowper's glands. The resultant semen is forced out of the penis by powerful contractions of the urethra, a process called **ejaculation**. This is accompanied by a pleasant sensation known as **orgasm**. Females may experience a similar sensation due to contraction of an equivalent set of muscles, although there is no expulsion of fluid from the urethra. The process of mating is also known as **copulation** or **coitus** and in a few mammals, e.g. cats and rabbits, it actually stimulates ovulation, so increasing the likelihood of successful fertilization.

19.6 Semen

Each ejaculation consists of approximately 3 cm^3 of semen. While it contains around 500 million sperm they comprise only a tiny percentage of the total volume, the majority being made up of the fluids secreted by the seminal vesicles, prostate and Cowper's glands. The semen therefore contains:

1. **Sperm.**

2. **Sugars** which nourish the sperm and help to make them mobile.

3. **Mucus** which forms a semi-viscous fluid in which the sperm swim.

4. Alkaline chemicals which neutralize the acid conditions encountered in the urethra and vagina, which could otherwise kill the sperm.

5. Prostaglandins, hormones which help sperm reach the ovum by causing muscular contractions of the uterus and oviducts.

19.7 Fertilization

The force of ejaculation of the semen from the penis is sufficient to propel some sperm through the cervix into the uterus, with the remainder being deposited at the top of the vagina. The sperm swim up through the uterus and into the oviducts by the lashing movements of their tails. The speed with which they reach the top of the oviducts indicates that muscular contractions of the uterus and oviduct are also involved. The egg or ovum released from the Graafian follicle of the ovary is metabolically inactive and dies within 24 hours unless fertilized. The ovum is surrounded by up to 2000 **cumulus cells** which aid its movement towards the uterus by giving the cilia which line the oviduct a large mass to 'grip'. The cumulus cells may also provide nutrients to the ovum. As the journey to the uterus takes three days in humans, it follows that fertilization must take place in the top third of the oviduct if the ovum is still to be alive when the sperm reaches it. In mammals there is no evidence that the ovum attracts the sperm in any way; their meeting would appear to be largely a matter of chance. Of the 500 million sperm in the ejaculate only a few hundred reach the ovum, and only one actually fertilizes it.

The fertilized ovum is called a **zygote**. The fertilizing sperm firstly releases **acrosin**, a trypsin-like enzyme, from the acrosome. This softens the vitelline membrane which covers the ovum. Inversion of the acrosome results in a fine needle-like filament developing at the tip of the sperm and this pierces the already softened portion of the vitelline membrane. An immediate set of changes occurs which thickens the vitelline membrane. In addition it is lifted from the plasma membrane by a fluid layer which separates the two. These changes ensure that no other sperm can penetrate the egg. This is essential to prevent a 'multinucleate' fertilized egg; such cells normally degenerate after a few divisions. The thickened membrane is now called the **fertilization membrane**. The sperm discards its tail, and the head and middle piece enter the cytoplasm. The second meiotic division of the ovum nucleus normally occurs immediately following the penetration of the sperm. The sperm and ovum nuclei fuse, restoring the diploid state. A spindle forms, the two sets of chromosomes line up and the cell undergoes mitotic division at once.

If the ovum is not fertilized it quickly dies and in humans the lining of the uterus is later shed to give the menstrual flow. In the female horseshoe bat, to achieve the earliest possible fertilization in spring, mating takes place in the autumn but the female stores the sperm in a thick plug of mucus. In spring the plug dissolves releasing the sperm for fertilization. This is called **delayed fertilization** and depends on the sperm surviving considerably longer than the 2–3 days which is normal in a human.

19.7.1 In vitro fertilization and test tube babies

After first being achieved in rabbits in 1959, in vitro fertilization or IVF, was successfully performed between human sperm and ova by Drs Edwards, Bavister and Steptoe ten years later. The development of these zygotes and their successful transfer into the uterus of the mother, called **embryo transfer** or **ET**, took a number of years but finally, in 1978, the first test tube baby was born.

The success of this technique owes as much to the development of a suitable medium in which the sperm, ova and embryo can survive and grow, as to the clinical techniques of obtaining ova and implanting the embryo. Such a medium must have not only a pH, osmotic potential and ionic concentration similar to that of blood, but also contain the patient's serum as a source of protein and other macromolecules. Glucose, lactate and pyruvate are other essential components.

The process begins with a fertility drug being administered to the potential mother to increase her ova production. Around six of these are collected using a fine needle, via the vagina. Around 100 000 sperm, collected from the potential father's semen sample by centrifugation, are added to the ova in a Petri dish. When the embryo is two days old it is transferred into the mother's uterus where, if all goes well, it will develop normally.

A major cause of infertility is blocked oviducts which therefore prevent ova and sperm meeting in natural circumstances. IVF has solved this problem in some cases, allowing both parents to contribute genetically to the offspring and almost all embryo development to take place inside the natural mother. IVF clinics are now common throughout the UK and despite their low success rate, at 10%, make a major contribution to providing otherwise childless couples, with much wanted children. For those with other forms of infertility the technique is unsuitable.

19.7.2 Causes of infertility and its cures

There are a number of reasons why a couple may have difficulty conceiving a baby:

1. Blocked oviducts – These may prevent ova and sperm meeting, in which case an operation may be undertaken to unblock the tubes or in vitro fertilization can be attempted.

2. An irregular menstrual cycle – This may make the chance of fertilization remote and hormone treatment necessary to regularize the cycle.

3. Incorrect frequency and/or timing of intercourse may make conception unlikely and couples may need to be counselled on the most appropriate time (the middle of the menstrual cycle) to have sexual intercourse in order to increase the possibility of fertilization.

4. Non-production of ova – This affects a few females making it impossible for them to contribute genetically to their offspring. Adoption or the use of a donated ovum from another female for in vitro fertilization are the possible alternatives. Artificial insemination of a surrogate mother with the potential father's sperm is another option.

5. Non-production of sperm – Some men produce no sperm, or so few that there is little realistic prospect of conception. Donated semen from another male can be used to artificially inseminate the woman.

6. Impotence – Some men are unable to erect the penis and/or ejaculate semen. The cause is often psychological, or the result of prolonged drug or alcohol abuse. In these cases counselling and guidance can sometimes remedy the problem.

Some of the above causes of infertility may be the result of certain disease or infections. Sexually transmitted diseases such as gonorrhoea can cause sterility, especially in females; mumps, if contracted in adult life, sometimes makes males infertile. Even when conception occurs, a few women are not able to sustain the pregnancy because either the embryo does not implant in the uterus wall, or having implanted, is later miscarried. For some, the solution is to use in vitro fertilization (Section 19.7.1) but rather than implant the embryo into the natural mother, it is transferred to a different female. The process whereby one woman carries a fertilized egg for another through to birth, is known as **surrogacy**.

Surrogate motherhood, in vitro fertilization and artificial insemination all raise complex legal and moral issues. Should the surrogate mother or sperm donor have any legal rights over the offspring they helped produce? To what extent should the natural mother be able to influence the behaviour of the surrogate mother during pregnancy – should she be able to insist on abstinence from smoking or drinking, both of which could damage the foetus? What details, if any, should a potential mother be entitled to know about the donor of the sperm to be used in artificial insemination? Should the excess embryos which result from in vitro fertilization be used for the purposes of medical research? These are just a few of the issues which have been raised by recent scientific research into the causes of, and cures of, infertility.

19.8   Implantation

Following fertilization, the zygote divides (cleavage) mitotically until a hollow ball of cells, the **blastocyst**, is produced. It takes three days to reach the uterus and a further three or four days to become implanted in the lining of the uterus. The outer layer of cells of the blastocyst, called the **trophoblast**, develops into the embryonic membranes, the **chorion** and the **amnion**. The chorion develops villi which grow into the surrounding uterine tissue from which they absorb nutrients. These villi form part of the **placenta** which is connected to the foetus by the **umbilical cord**. The amnion develops as a membrane around the foetus and encloses the amniotic fluid, a watery liquid which protects the foetus by cushioning it from physical damage. In badgers, mating and fertilization occur in midsummer and development takes place up to the blastocyst stage. This however does not become implanted until the late winter or early spring, after which development proceeds normally. A similar process occurs in polar bears and allows offspring to be weaned when food is plentiful, regardless of the time of fertilization. This is called **delayed implantation**.

19.9   The placenta

The chorionic villi will develop about 14 days after fertilization and represent the beginning of the placenta. It rapidly develops into a disc of tissue covering 20% of the uterus. The capillaries of the mother and foetus come into close contact without actually combining.

The chorionic villi present a large surface area for the exchange of materials by diffusion across thin chorionic membrane. In some mammals the maternal and foetal bloods flow in opposite directions. This counter-current flow leads to more efficient exchange as described later, in section 31.2.5.

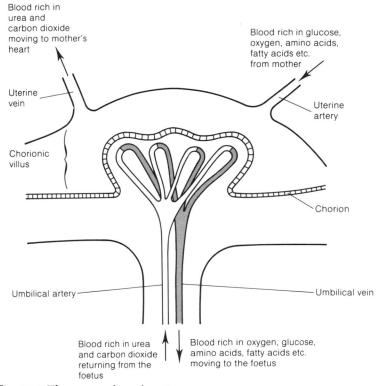

Blood rich in urea and carbon dioxide moving to mother's heart

Blood rich in glucose, oxygen, amino acids, fatty acids etc. from mother

Uterine vein

Uterine artery

Chorionic villus

Chorion

Umbilical artery

Umbilical vein

Blood rich in urea and carbon dioxide returning from the foetus

Blood rich in oxygen, glucose, amino acids, fatty acids etc. moving to the foetus

Fig. 19.8 The mammalian placenta

19.9.1 Functions of the placenta

1. It allows exchange of materials between the mother and foetus without the two bloods mixing. This is necessary as the foetal blood may be different from that of the mother due to the influence of the father's genes. If incompatible bloods mix they agglutinate (clot), causing blockage in vital organs such as the kidney, possibly resulting in death.

2. Oxygen, water, amino acids, glucose, essential minerals etc. are transferred from maternal to foetal blood to nourish the developing foetus.

3. Carbon dioxide, urea and other wastes are transferred from foetal to maternal blood to allow their excretion by the mother and prevent harmful accumulation in the foetus.

4. It allows certain maternal antibodies to pass into the foetus, providing it with some immunity against disease, particularly in the early months after birth.

5. It protects the foetus by preventing certain pathogens (disease-causing organisms) and their toxins from crossing the placenta. This protection is by no means complete. Notable exceptions include toxins of the Rubella (German measles) virus which can cross the placenta causing physical and mental damage to the foetus, and the HIV virus which can also pass into the foetus.

6. In a similar way it acts as a barrier to those maternal hormones and other chemicals in the mother's blood which could adversely affect foetal development. Again the protection is not complete and substances like nicotine, alcohol and heroin can all enter the foetus causing lasting damage.

7. As the two blood systems are not directly connected, the placenta permits them to operate at different pressures without harm to mother or foetus.

8. As the pregnancy progresses the placenta increasingly takes over the rôle of hormone production. In particular it produces progesterone which prevents ovulation and menstruation. It also secretes **human choriogonadotrophin** (HCG), a hormone whose presence in the urine of pregnant women is the basis of most pregnancy tests.

19.10 Birth (parturition)

During pregnancy the placenta continues to produce progesterone and small amounts of oestrogen. The amount of progesterone decreases during pregnancy while oestrogen increases. These changes help to trigger the onset of birth. As the end of the gestation period nears, the posterior lobe of the pituitary produces the hormone **oxytocin** which causes the uterus to contract. These contractions increase in force and frequency during labour.

The process of birth can be divided into three stages:

1. The dilation of the cervix, resulting in loss of the cervical plug ('the show') and the rupture of the embryonic membranes ('breaking of the waters').

2. The expulsion of the embryo.

3. The expulsion of the placenta ('afterbirth') which is eaten by most mammals.

19.11 Lactation

During pregnancy the hormones progesterone and oestrogen cause the development of lactiferous (milk) glands within the mammary glands. Following birth, the anterior lobe of the pituitary gland produces the hormone **prolactin** which causes the lactiferous glands to begin milk production. Suckling by the offspring causes the reflex expulsion of this milk from the nipple of the mammary glands. The first formed milk, called **colostrum**, is mildly laxative and helps the baby expel the bile which has accumulated in the intestines during foetal life. As well as essential nutrients, the milk contains antibodies which give some passive immunity to the newly born.

19.12 Parental care

Some organisms produce vast numbers of offspring, the cod (*Gadus gadus*), for example, may produce over one million eggs at a time.

In such organisms there is little or no parental care and the majority of offspring fail to reach maturity, most being consumed by predators. In birds and mammals the tendency is to reduce the number of offspring but to expend much time and energy in caring for them in order to ensure a high survival rate.

In mammals, the provision of milk is the most obvious example of parental care. As the offspring develop they are gradually introduced to other, more solid types of food, a process called weaning. In many animals the parents singly, or in pairs, collect the appropriate food for the offspring. This food may be partly digested, e.g. regurgitated from the crop in birds.

Many animals provide a nest in which to raise their young. Here the offspring may be raised in the relative safety of a warm, dry environment remote from predators.

19.13 Questions

1. (a) Describe oogenesis and fertilization in the mammal. *(10 marks)*
   (b) How is the embryo maintained until parturition?
   (8 marks)
   (Total 18 marks)

   Cambridge Board November 1984, Paper I, No. 2

2. Explain, by reference to the placenta, how materials in the blood pass between mother and foetus. What additional mechanism increases the efficiency of oxygen transfer? *(8 marks)*

   Southern Universities Joint Board June 1983, Paper I, No. 5

3. Discuss the role of hormones in relation to the reproductive activity of a female mammal. *(20 marks)*

   Welsh Joint Education Committee June 1984, Paper II, No. 5

4. (a) Relate the cyclical secretion of the human pituitary gonadotrophins to the changes occurring in
     (i) the ovary, and
     (ii) the uterus. *(12 marks)*
   (b) Explain briefly how these events are modified following the implantation of a fertilized ovum.
     (4 marks)
   (c) Some mammals (for example, rats) can reproduce at any time of the year. In others (for example, sheep) oestrous only occurs during a limited period of the year. Suggest why these

differences occur and state the possible advantages for each animal. *(4 marks)*
   (Total 20 marks)

   Joint Matriculation Board June 1982, Paper IIB, No. 5

5. The diagram shows the blood levels of the hormones involved in the control of the human menstrual cycle: luteinizing hormone (LH), follicle stimulating hormone (FSH), oestrogen and progesterone.
   (a) For each hormone state
     (i) where it is produced, and
     (ii) the organ (or organs) on which it acts.
     (4 marks)

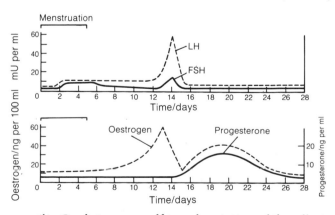

   (b) Confining yourself to a description of the effects of each of these hormones on the secretion of the others, show how negative feedback operates in this system. *(2 marks)*

(c) Following menstruation, the uterine lining undergoes repair and then proliferation of secretory tissue. Name the hormones directly responsible for initiating
  (i) repair, and
  (ii) proliferation of secretory tissue.   (2 marks)
(d) Describe the development of the Graafian follicle from the oogonium to ovulation.   (4 marks)
(e) Describe the process of implantation and formation of the placenta.   (4 marks)
(f) (i) Draw a simple sketch graph showing the changes in blood levels of oestrogen and progesterone in the event of fertilization and successful implantation occurring. You should continue your graph up to the point of parturition (birth).
  (ii) On the same sketch graph show the changes in level of one other named hormone involved in gestation and/or parturition. What is the function of the hormone you have named?   (4 marks)
  (Total 20 marks)

Northern Ireland Board June 1983, Paper II, No. 5

6. Complete the following table which summarizes some of the properties of hormones in a human female.

Name of hormone	Site of secretion	Target organ	Function
Follicle stimulating hormone (FSH)		Ovary	
		Uterus	Repair of uterine lining following menstruation.
Prolactin			

(7 marks)

Associated Examining Board June 1989, Paper I, No. 11

7. Egg formation in mammals.
(a) On the diagram, write 2N or N beside each of the cells marked with an asterisk to indicate whether it is diploid or haploid.   (2 marks)

(b) Indicate, on the diagram, where the first meiotic division and the second meiotic division occur.   (2 marks)

(c) Insert on the diagram the *names* of the cells numbered 1 to 4.   (4 marks)

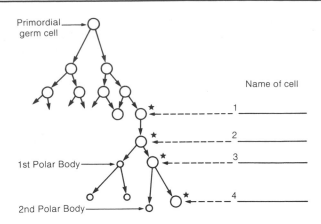

(d) Complete the table below to indicate the role of hormones in helping to control *egg release in mammals.*

	Name of hormone	Function of the hormone
(i) Two hormones released by the pituitary glands:	1.	
	2.	
(ii) Two hormones released by ovarian tissue:	1.	
	2.	
(iii) State the effect of the ovarian hormones on the release of one of the named pituitary hormones.	Name of pituitary hormone	Effect

(6 marks)
(Total 14 marks)

Oxford and Cambridge Board June 1988, Paper I, No. 3

8. (a) Describe the events in the genital tract of a human female from the development of the egg cell in an ovary to the implantation of the embryo in the wall of the uterus.   (10 marks)

(b) Give an account of the structure and functions of the placenta.   (4 marks)

(c) Explain how the loss of the lining of the uterus is prevented during pregnancy.   (4 marks)
(Total 18 marks)

Cambridge Board November 1988, Paper I, No. 3

9. Explain fully the following:
    (a) The role of the pituitary gland in the reproductive cycle of the female mammal.

    (9 marks)

    (b) The functions of the mammalian placenta.

    (6 marks)

    (c) The relationship between reproduction and population growth in animals.   *(5 marks)*

    (Total 20 marks)

    Welsh Joint Education Committee June 1988, Paper A1, No. 3

20 *Growth and development*

The growth in size of an individual cell is limited by the distance over which the nucleus can exert its control. For this reason, when single celled organisms reach a maximum size they divide to give two separate individuals. In order to attain greater size, organisms became multicellular. While being large and multicellular can present some problems, these are easily outweighed by the advantages conferred:

1. Cells may become differentiated in order to perform a particular function.

2. Specialized cells performing one particular function leads to greater efficiency.

3. It is possible to store more materials and so be better able to withstand periods when these are scarce.

4. If some cells are damaged, enough may still remain to carry out the repair.

5. Some processes require a range of conditions, e.g. digestion often has an acid and an alkaline phase. It is easier to separate regions of opposing conditions in a multicellular organism than it is in a single cell.

6. Larger organisms may have a competitive advantage, e.g. large plants compete better for light than small ones.

7. Large size may provide some protection from predators because they are simply too large to ingest.

Multicellular organisms all originate as a single cell, the zygote, and undergo three phases of development:

1. Growth: an irreversible increase in mass.

2. Differentiation: the development of cells of different types.

3. Morphogenesis: the development of organs and the organism's overall shape.

20.1 Measurement of growth

Growth is estimated by measuring some parameter (variable) over a period of time. The parameter chosen depends upon the organism whose growth is to be measured. It may be appropriate to measure the weight of a mouse, but this method would be impractical for an oak tree. Mass and length are most often used, but these may be misleading. A bush, for example, while not increasing in height, may continue to grow in size by spreading sideways. Area or volume give a more accurate indication of growth but are often impractical to measure. The measurement of mass has its problems. If an organism takes in a large amount of water its mass may increase markedly, and yet such a temporary increase could not be considered as growth. For this reason two types of mass are recognized:

1. Fresh mass — This is the mass of the organism under normal conditions. It is easy to measure and doing so involves no damage to the organism. It may, however, be inaccurate due to temporary fluctuations in water content.

2. Dry weight — This involves removing all water by drying, before weighing. It is difficult to carry out and permanently destroys the organisms involved, but does give an accurate measure of growth.

It is sometimes possible to measure one part of an organism, e.g. the girth of a tree; the length of the tail of a rat. Provided this part grows in proportion to the complete organism, increases in its size will reflect those of the individual as a whole. Groups of organisms are sometimes used rather than an individual. For instance, if the growth of peas was measured using dry mass, it would be necessary to grow a large population of the plants. Growth could be estimated by removing say ten plants every day, drying and weighing them. Provided each sample is large enough to average out individual differences in growth, a good estimate of the growth rate can be found. The growth of a population of yeast can be measured by counting the number of cells in a known, and very small, volume of the medium in which the yeast is growing.

20.2 Growth patterns

When any parameter of growth is measured against set intervals of time, a **growth curve** is produced. For many populations, organisms or organs, this curve is S-shaped and is called a **sigmoid curve**. It represents slow growth at first, because there are so few cells initially that even when dividing rapidly the actual increase in size is small.

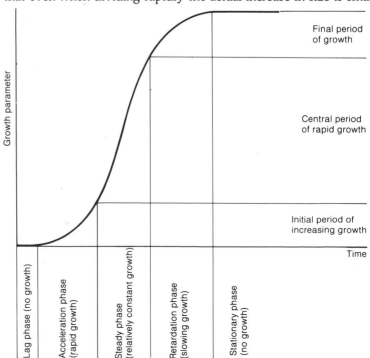

Fig. 20.1(a) The sigmoid growth curve

As the number of cells becomes larger the size increases more quickly because there are more cells carrying out division. There is a limit to this rapid phase of growth. This limit may be imposed by the genotype of the individual, which specifies a certain maximum size, or any external factors, such as shortage of food. Whatever the cause, the growth rate decreases until it ceases altogether. At this point cells are still dividing, but only at a rate which replaces those which have died. The size of the organism therefore remains constant.

While the sigmoid curve forms the basis of most growth curves, it may be modified in certain circumstances. In humans, for example, there are two phases of rapid growth; one during the early years of life, the other during adolescence. Between these two phases there is a period of relatively slow growth. The growth curve therefore resembles two sigmoid curves, one on top of the other.

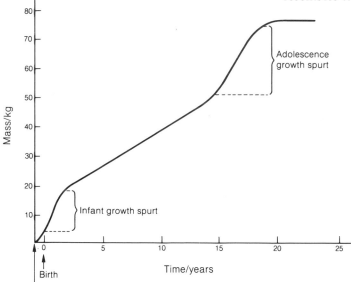

Fig. 20.1(b) Human growth curve

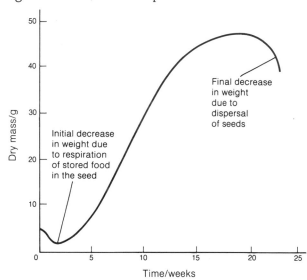

Fig. 20.1(c) Annual plant growth curve, e.g. Pea plant (Pisum sativum)

In annual plants, the growth curve is typically sigmoid, except there may be an initial decrease in mass during the early stages of germination. This occurs as the food reserves in the seed are respired in order to produce the roots and leaves. Once the leaves begin to photosynthesize growth proceeds in a sigmoid fashion. However, with the liberation of fruits and seeds at the end of the growing period, the mass of the plant may decrease prior to its death. When there is a natural limit on growth, as in annual plants, they are said to show **limited growth**. In these cases the growth curve flattens out, or even decreases prior to the organism's death.

In perennial plants, the growth pattern is an annual series of sigmoid curves. During spring when the temperature and light intensity are relatively low, there is less photosynthesis, and growth is slow. In summer, with higher temperatures and more light, the rate of photosynthesis increases and growth is rapid. The falling temperatures and lower light intensities of autumn again reduce the rate of photosynthesis and hence growth. During winter in temperate regions there is no growth in deciduous plants and so the curve flattens out. The following spring the process is repeated. The overall shape of these annual sigmoid curves is itself sigmoid, except that many perennial plants show **unlimited growth**, i.e. they grow continuously throughout their lives, and the curve therefore never flattens out.

A very different growth curve is exhibited by many arthropods. As their exoskeleton is incapable of expansion, they have to moult periodically during growth. Before a new exoskeleton has fully hardened it is capable of some expansion. During this time the insect may take up water in order to expand the exoskeleton as much as possible. This means that once it has hardened there is still some room for growth. Measuring fresh mass as the growth parameter therefore gives the unusual growth pattern shown in Fig. 20.1(e). This type of growth is called **intermittent growth**. If dry mass is measured, a normal sigmoid curve is obtained.

The annual growth follows a normal sigmoid curve. Variations occur from one year to the next according to environmental conditions. In a cold dry year for example there will be less growth than in a mild, wet one.

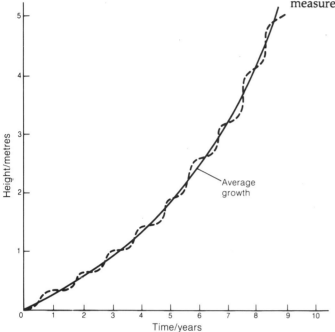

Fig. 20.1(d) Perennial plant growth curve

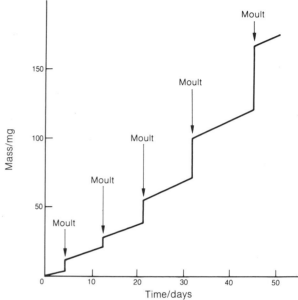

*Fig. 20.1(e) Arthropod growth curve, e.g. Water-boatman (*Notonecta glauca*)*

Certain organs of an individual grow at the same rate as the organism as a whole. This is called **isometric growth**. Other organs grow at a different rate from the entire organism. This is called **allometric growth**. The leaves of most plants exhibit isometric growth and their growth curve is typically sigmoid (Fig. 20.2(a)).

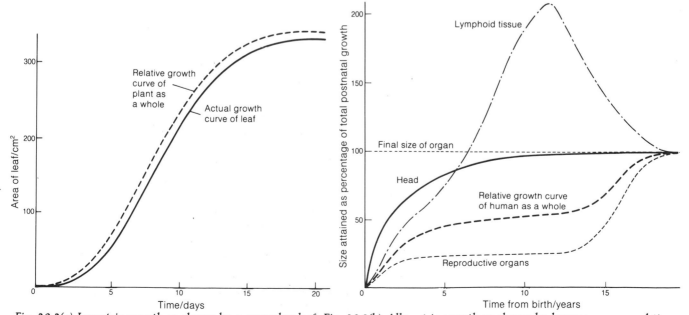

Fig. 20.2(a) Isometric growth as shown by a cucumber leaf Fig. 20.2(b) Allometric growth as shown by human organs and tissues

In animals, organs often exhibit allometric growth. Lymph tissue, which produces white blood cells to fight infection, grows rapidly in early life when the risk of disease is greater as immunity has not yet been acquired. By adult life the mass of lymph tissues is less than half of what it was in early adolescence. The reproductive organs grow very little in early life but develop rapidly with the onset of sexual maturity at puberty. Fig. 20.2(b) illustrates allometric growth in some human organs.

20.2.1 Rate of growth

The actual growth of an organism is the cumulative increase in size over a period of time. A small annual plant, for example, might grow as shown in Fig. 20.3(a), in which case a typical sigmoid growth curve results. The rate of growth is a measure of size increase over a series of equal time intervals. If instead of measuring the actual height of the plant we measure the increase in height over each three day period, a set of results like that shown in Fig. 20.3(b) is obtained. These produce a bell-shaped graph as shown.

(a)

Time/ days	Height/ mm
0	0
3	40
6	100
9	350
12	900
15	1600
18	2150
21	2400
24	2460
27	2500
30	2500

(b)

Time/days interval	Height at start of time interval	Height at end of time interval	Height increase during time interval
0–3	0	40	40
3–6	40	100	60
6–9	100	350	250
9–12	350	900	550
12–15	900	1600	700
15–18	1600	2150	550
18–21	2150	2400	250
21–24	2400	2460	60
24–27	2460	2500	40
27–30	2500	2500	0

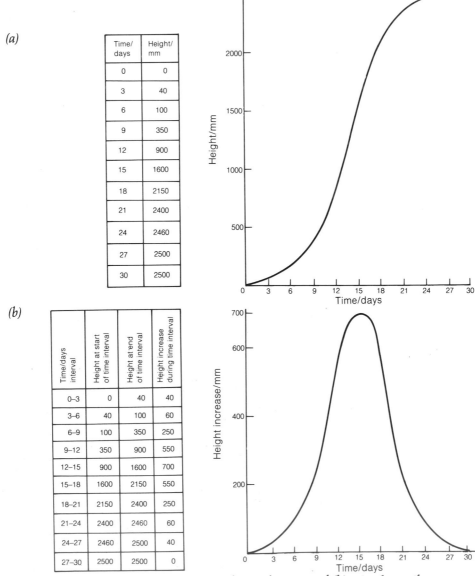

Fig. 20.3 Comparison of (a) actual growth curve and (b) rate of growth curve

20.3 Growth and development in plants

20.3.1 Germination

Prior to germination many seeds undergo a period of dormancy (Section 18.7). Germination is the onset of growth of the embryo, and requires water, oxygen and a temperature within a certain range (normally 5–40°C). In some seeds light is also required. Under these conditions the seed takes up water rapidly, initially by imbibition and later by osmosis. This water causes the seed contents to swell and so ruptures the **testa** (seed coat). At the same time the water activates enzymes in the seed which hydrolyse insoluble storage material into soluble substances which can be easily transported. In this way proteins are converted into amino acids, carbohydrates such as starch are converted into glucose, and fats are converted into fatty acids and glycerol. The soluble products of these conversions are transported to the growing point of the embryo. The glucose, fatty acids and glycerol provide respiratory substrates from which energy for growth is released. Glucose is also used in the formation of cellulose cell walls. The amino acids are used to form new enzymes and structural proteins within new cells.

Early growth results in the **plumule** (embryonic shoot) and the **radicle** (embryonic root) growing rapidly. The radicle grows downwards and the plumule upwards. The **cotyledons** (embryonic

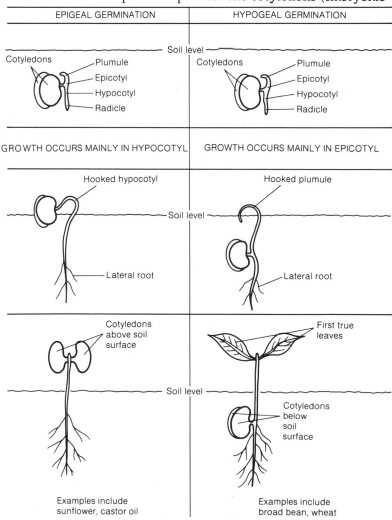

EPIGEAL GERMINATION	HYPOGEAL GERMINATION
─ Soil level ─	
Cotyledons — Plumule — Epicotyl — Hypocotyl — Radicle	Cotyledons — Plumule — Epicotyl — Hypocotyl — Radicle
GROWTH OCCURS MAINLY IN HYPOCOTYL	GROWTH OCCURS MAINLY IN EPICOTYL
Hooked hypocotyl — Soil level — Lateral root	Hooked plumule — Soil level — Lateral root
Cotyledons above soil surface — Soil level	First true leaves — Cotyledons below soil surface
Examples include sunflower, castor oil	Examples include broad bean, wheat

Fig. 20.4 To show differences between epigeal and hypogeal germination

leaves) may be carried up and out of the soil by this growth (**epigeal germination**), in which case they form the first photosynthetic structure. In some plants the cotyledons remain below the soil surface (**hypogeal germination**) (Fig. 20.4).

20.3.2 Meristems

The presence of a semi-rigid cell wall around plant cells effectively restricts their ability to divide and grow. For this reason, unlike animals, plants retain groups of immature cells which form the only actively growing tissues. These tissues are called **meristems**. Three types of meristems are generally recognized:

1. Apical meristems – These are found at the tips of roots and shoots and are responsible for primary growth of the plant. They increase its length.

2. Lateral meristems – These are found in a cylinder towards the outside of stems and roots. They are responsible for secondary growth and cause an increase in girth.

3. Intercalary meristems – These are found at the nodes in monocotyledonous plants. They allow an increase in length in positions other than the tip.

A typical meristem cell, being undifferentiated, has a simple structure. As meristems are constantly dividing it is hardly surprising that they are small in size and the cell wall is thin. Constant division means that there is much metabolic activity and for this reason the nucleus is large and the cytoplasm dense. Coupled with their small size this leaves little room for the vacuole which is therefore represented by a series of smaller ones.

The delicate growing points which comprise the apical meristem need protection. In the shoot this is afforded by the **leaf primordia** which partially envelop the tip. In the root, where physical damage is more likely because the tip is constantly being pushed through the soil, a **root cap** performs this function. The root cap is a layer of loose cells which entirely cover the root tip. As the cells are worn away, replacement cells are formed by the apical meristem.

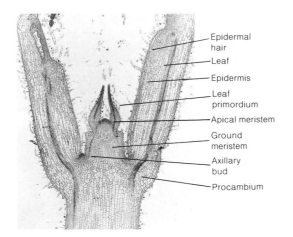

Shoot apex (LS) (× 25 approx.)

Labels: Epidermal hair, Leaf, Epidermis, Leaf primordium, Apical meristem, Ground meristem, Axillary bud, Procambium

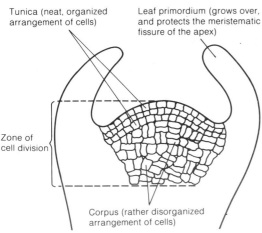

Labels: Tunica (neat, organized arrangement of cells); Leaf primordium (grows over, and protects the meristematic fissure of the apex); Zone of cell division; Corpus (rather disorganized arrangement of cells)

Fig. 20.5(a) Diagram of the meristem at the shoot apex

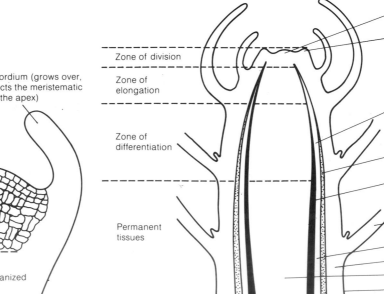

Labels: Apical meristem; Leaf primordium; Zone of division; Zone of elongation; Zone of differentiation; Permanent tissues; Procambial strand; Primary phloem; Primary xylem; Leaf; Cambium; Cortex; Pith; Epidermis

Fig. 20.5(b) Tip of a young shoot showing early development (LS)

20.3.3 Primary growth

Primary growth is the first growth to take place in a plant. In many plants it is the only form of growth which occurs. It results largely from the activity of apical meristems but intercalary meristems may also be involved.

At the apex of a stem or root there is a **zone of cell division** which extends for 1–2 mm. This is composed of an outer layer called the **tunica** which consists of neatly arranged cells. Beneath this is a less organized layer of dividing cells called the **corpus** (Fig. 20.5(a)). Behind the zone of cell division is an area where the cells rapidly increase in size. This is the **zone of cell expansion**. Further back still from the apex is a region where the cells become specialized and alter their form as they change into xylem, phloem, sclerenchyma etc. This is called the **zone of differentiation**.

During cell expansion, in meristematic cells the small vacuoles combine to form large ones and finally coalesce to give a single large central vacuole. This process is called **vacuolation**. At the same time the cell wall thickens due to more cellulose being laid down. Some of the meristem cells do not vacuolate. Instead they remain meristematic in character and form long strands extending back from the apex. This is called **procambium**. These procambial strands are the basis for the formation of vascular tissue. As they divide, the cells towards the inside of the shoot form **protoxylem**, and those to the outside form **protophloem**. Later the procambium gives rise to **metaxylem** and **metaphloem** which are larger cells and tend to crush the protoxylem and protophloem. The differentiation of procambial cells into metaxylem and metaphloem involves quite considerable changes. By contrast, differentiation of the apical meristem into parenchyma involves little more than an increase in size and vacuolation. Depending upon the extent, location and type of wall laid down, collenchyma and sclerenchyma tissues may be produced. Epidermal cells are derived from differentiation of meristematic cells of the tunica.

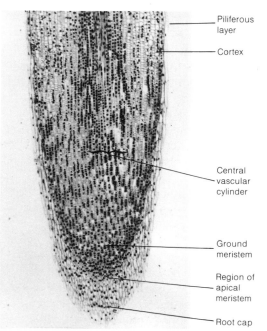

Root apex (LS) (× 30 approx.)

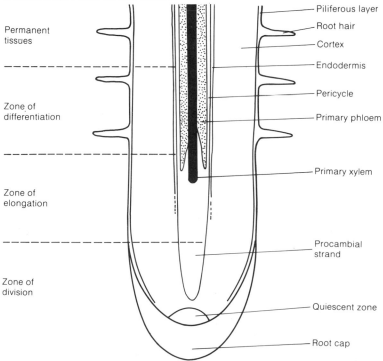

Fig. 20.5(c) Tip of a young root showing early development (LS)

289

In roots, only a single, central procambial strand is produced. Protophloem develops first from this strand, followed by protoxylem. Finally metaphloem and metaxylem respectively develop. As this differentiation occurs mostly at the centre of the procambial strand, some of it remains to the outside of the central vascular tissue it forms. This is called the **pericycle** and it retains its meristematic properties and can thus divide to give rise to lateral roots. As lateral roots develop from this pericycle, which is deep within the root tissues, their formation is described as **endogenous** (*endo* – 'inside'; *geno* – 'forming'). Lateral shoots, however, arise from the cambium which lies near the shoot surface. Their formation is described as **exogenous** (*ex* – 'out of'; *geno* – 'forming').

20.3.4 Secondary growth

Behind the zone of differentiation in shoots and roots is a region called the **permanent zone**. In this region cells are fully differentiated and each carries out its specialist rôle. Where a plant lives for a number of years, renewed growth in this region takes place annually. This helps to support the plant as it increases in size in an effort to compete successfully for light, water, mineral salts etc. This growth is termed **secondary growth** and occurs in woody perennials. It results from the activity of lateral meristems of which there are two types:

1. Vascular cambium – This produces new xylem and phloem and is responsible for most of the increase in girth.

2. Cork cambium (phellogen) – This produces a thick epidermal layer which forms part of the bark.

The first stage of secondary growth in stems occurs when the vascular cambium, located within the vascular bundle, grows to form a complete cylinder around the stem (Fig. 20.6 (*a*)). New cambium formed between the vascular bundles is called **interfascicular cambium**. The cells of this cambial cylinder divide and those on the inside differentiate into xylem while those on the outside form phloem. In this way a cylinder of **secondary xylem** and one of **secondary phloem** form either side of the cambium. As the amount of xylem formed is greater than phloem, and as xylem cells have thick, lignified walls whereas phloem cells are less rigid, the xylem cylinders push the phloem and cambium outwards. Over many years the phloem becomes crushed and normally only the previous year's growth remains functional. The xylem remains functional for many years. In an old tree there is therefore a wide cylinder of functional xylem known as the **sapwood** around the outside. In time, however, the xylem becomes blocked with various stored wastes such as tannins. This non-functional xylem lies at the centre of the trunk and is called **heartwood**.

The growth of secondary xylem is not uniform throughout the year. In spring, xylem vessels have large lumina in order to carry the heavy flow of water from the wet spring soil. By autumn, the soil is frequently drier after the relatively warm summer. The xylem vessels laid down at this time are smaller and contain a greater proportion of thick-walled sclerenchyma. This autumn wood is therefore more compact and darker in colour than the spring wood. This gives the appearance of a series of concentric rings when a trunk is cut transversely. These are called **annual rings**. Counting of these rings is an accurate means of estimating the age of a tree. In

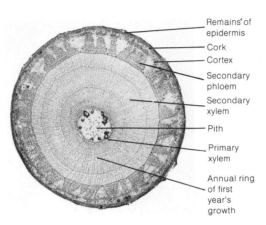

Annual growth rings (× 5 approx.)

Remains of epidermis
Cork
Cortex
Secondary phloem
Secondary xylem
Pith
Primary xylem
Annual ring of first year's growth

addition, the width of any one ring gives an estimate of how favourable the climate was for growth in that year. Under favourable conditions, e.g. warm and wet, much growth takes place and the ring is wide. In a cold and/or dry growing season the ring is much narrower.

The increased girth as a result of secondary growth inevitably ruptures the epidermis. This too must therefore undergo secondary growth in order to maintain an outer protective layer. This growth results from the division of the **cork cambium (phellogen).** This outer cylinder of meristematic cells divides to give rise to an outer layer of **cork cells**. These form an effective waterproof barrier and provide protection against the invasion of bacteria and other potentially harmful organisms. Dotted throughout this layer are regions of loose cork cells which permit gaseous exchange. These are the **lenticels**. The pattern of secondary growth in roots is very similar and is illustrated in Fig. 20.6(b). Very few monocotyledonous plants undergo secondary growth.

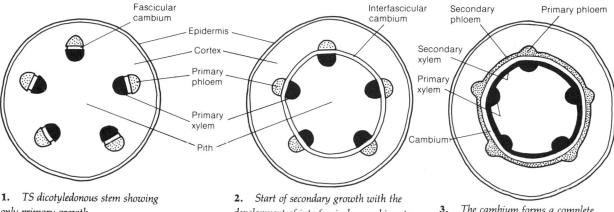

1. *TS dicotyledonous stem showing only primary growth*

2. *Start of secondary growth with the development of interfascicular cambium to form a complete ring*

3. *The cambium forms a complete phloem ring to the outside and a complete xylem ring to the inside*

Fig. 20.6(a) Diagrams to show the early stages of secondary growth in a dicotyledonous stem

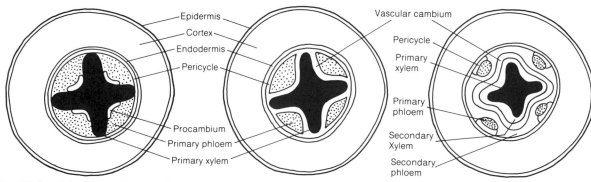

1. *TS dicotyledonous root showing only primary growth*

2. *Start of secondary growth with development of a ring of vascular cambium which merges with the pericycle*

3. *The cambium forms secondary phloem and xylem. The primary xylem remains in the centre but the primary phloem is crushed and remains, with a little pericycle, in small bundles*

Fig. 20.6(b) Diagrams to show the early stages of secondary growth in a dicotyledonous root

20.4   Growth and development in insects

20.4.1 Insect metamorphosis

During their life cycles insects undergo **metamorphosis**. This is the series of changes which take place between larval and adult forms.

They usually involve a major reorganization of larval tissues. As we saw in Section 9.6.15, there are two forms of insect metamorphosis:

1. Hemimetabolous (incomplete metamorphosis) – The eggs hatch into **nymphs** which clearly resemble the adults except that they are smaller, lack wings and are sexually immature. There are a number of nymphal stages between which moulting occurs. Examples of hemimetabolous insects include locusts and cockroaches.

2. Holometabolous (complete metamorphosis) – The eggs hatch into **larvae** which differ considerably from the adults. Each larva undergoes a series of moults until it changes its appearance and becomes a dormant stage known as a **pupa**. After much reorganization of the tissues within the pupa, the adult (**imago**) emerges. Examples of holometabolous insects include moths, butterflies and flies.

20.4.2 Control of insect metamorphosis

The control of insect metamorphosis involves two main hormones: **moulting hormone (ecdysone)** and **juvenile hormone (neotonin)**. Moulting hormone is produced by a gland in the first thoracic segment, called the **prothoracic gland**. Juvenile hormone is produced by a region behind the brain known as the **corpus allatum**. The production of both hormones is controlled by neurosecretory cells in the brain (see Section 37.8). All moults require moulting hormone. If juvenile hormone is present in high concentrations larval moults occur, which mean the insect remains as a larva. If only low concentrations of juvenile hormone are present a pupal moult occurs and the larva metamorphoses into a pupa. In the complete absence of juvenile hormone, the pupa metamorphoses into the imago (adult). These events are summarized in Fig. 20.7.

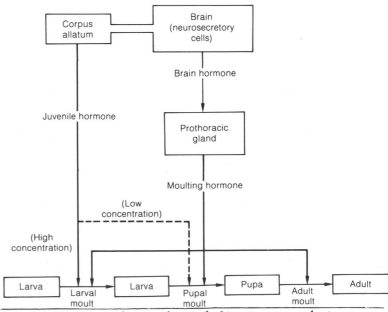

Fig. 20.7 Summary of hormonal control of insect metamorphosis

20.4.3 Larval forms

Organisms other than insects undergo metamorphosis, e.g. amphibians. In most of these organisms the larva is very different in form from the adult. The larva can therefore be adapted to perform some particular function which the adult, due to its specialized form, cannot. These functions include:

1. Dispersal – Animals which do not move from place to place are said to be **sessile**. These organisms would become overcrowded and have to compete with each other for food and space if they did not disperse themselves effectively. To achieve efficient dispersal they often lay eggs which hatch into motile larvae. These disperse themselves before settling in some suitable habitat where they metamorphose into the sessile adult. Many marine organisms which live on rocky shores use this method. Mussels and barnacles, for example, have planktonic larvae. Slow moving organisms like starfish also use motile larvae to disperse themselves.

2. Feeding – Animals may use their larvae to exploit a different food source. Butterflies and moths, for example, feed on nectar as adults. This is only plentiful in the summer when flowers have developed, and even then the quantity in each flower is very small. If an animal is to complete its life cycle in one year, it needs to start as early as possible in the spring. At this time, flowers are likely to be in short supply and the total nectar available could not support the growth of offspring. In any case, nectar is almost exclusively carbohydrate and lacks the protein so essential to animal growth. The caterpillars (larvae) of butterflies and moths are adapted to feed on foliage. This is plentiful in spring and contains some protein. Growth occurs rapidly and the caterpillar can pupate in time for adults to emerge in summer, when nectar is abundant. In some species, e.g. mayflies, the larvae alone do the feeding, with the adults never doing so.

3. Overcoming adverse conditions – Some animals survive adverse conditions, such as cold and drought, by having larval forms which are physiologically adapted to withstand extremes.

4. Infecting hosts and building up numbers – Many platyhelminth parasites use larval stages to infect their hosts, or they may be specially adapted to withstand the conditions in a secondary host. The larvae of parasites like the liver fluke (*Fasciola hepatica*) are able to reproduce asexually and so rapidly build up their numbers.

What fundamentally distinguishes a larva from an adult is the absence of reproductive organs and its consequent inability to reproduce. Occasionally, however, larvae develop sexual organs and so attain sexual maturity. The process is called **neoteny**, and may explain sudden changes in evolution, such as the development of chordates. The axolotl is an example of an amphibian larva which has attained sexual maturity.

20.5 Growth and development in vertebrates

The development which follows fertilization in animals can be divided into three stages:

1. Cleavage – The mitotic division of the zygote to form a ball of identical cells.

2. Gastrulation – The arrangement of cells into definite layers.

3. Organogenesis – The differentiation of cells to form organs.

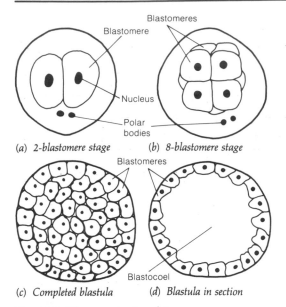

(a) 2-blastomere stage   (b) 8-blastomere stage

(c) Completed blastula   (d) Blastula in section

Fig. 20.8 Cleavage in Amphioxus

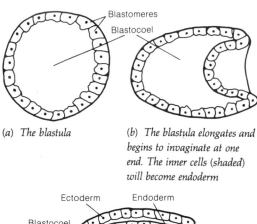

(a) The blastula

(b) The blastula elongates and begins to invaginate at one end. The inner cells (shaded) will become endoderm

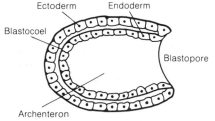

(c) The completed gastrula. A two-layered structure with a thin cavity of blastocoel which becomes obliterated in time

Fig. 20.9 Stages in gastrulation

20.5.1 Cleavage

After fertilization, the nucleus of the zygote divides mitotically, followed by cleavage of the cytoplasm. In this way a series of smaller and smaller cells called **blastomeres** are formed. These divisions continue, resulting in an embryonic structure called the **blastula**. The blastula has a cavity at the centre called a **blastocoel** (*blasto* – 'original'; *coel* – 'cavity'). The extent of cleavage depends on how much yolk is present, as yolk inhibits cleavage. Where an animal has little or no yolk in the egg, e.g. *Amphioxus*, cleavage involves the whole cell and the blastomeres are small and of equal size. When much yolk is present, as in the frog, it is normally concentrated at the vegetal pole, with much less, if any, at the animal pole (Fig. 19.2). Cleavage is inhibited at the vegetal pole and a few large cells called **macromeres** result. By contrast, cleavage occurs rapidly at the animal pole, producing many small cells called **micromeres**.

20.5.2 Gastrulation

Gastrulation involves the inpushing of one side of the ball of cells, to produce a cavity called the **archenteron**, which will later form the gut. The structure as a whole is called the **gastrula**. At this stage the cells of the structure have their final fates determined, i.e. it is possible to predict their functions in the developed organism. These **presumptive areas**, as they are called, can be mapped with accuracy. During gastrulation the cells of these presumptive areas migrate to their correct positions. The blastocoel becomes obliterated and the cells of the gastrula become arranged into **germ layers**. There are three germ layers:

1. The inner **endoderm**.
2. The central **mesoderm**.
3. The outer **ectoderm**.

TABLE 20.1 **The tissues and organs formed from each of the germ layers**

Germ layer	Tissue/organ formed during development
Ectoderm	Skin, scales, hair, feathers Jaws Nerves and central nervous system
Mesoderm	Striated muscle/smooth muscle Connective tissue (bone, cartilage, blood) Heart, blood system Kidney and excretory system Reproductive system Eyes
Endoderm	Alimentary canal Lining of gut, bladder and lungs Liver, pancreas and thyroid glands Germinative epithelium

20.5.3 Organogenesis

The cells of the gastrula continue to divide and become differentiated. The ectoderm folds inwards to become a **neural tube** which later develops into the nerve cord, the anterior part of which expands to form the brain. The embryo at this stage is called the **neurula**.

The mesoderm forms a **notochord** which in vertebrates is replaced by the vertebral column. The gut increases in length and becomes folded and gradually all major organs such as the heart develop.

20.6 Questions

1. (a) With the aid of annotated diagrams, compare and contrast the structure of the developing stem apex and the developing root apex in a typical flowering plant. (*10 marks*)
   (b) Discuss the factors that are important in initiating the germination of seeds. (*8 marks*)
   (*Total 18 marks*)

   Cambridge Board November 1983, Paper I, No. 5

2. (a) What is meant by the term 'growth'? (*4 marks*)
   (b) Explain how each of the following takes place.
   (i) Increase in girth of a flowering plant stem
   (ii) Increase in size of a mammal after birth (*12 marks*)
   (c) Comment on the importance of the larval stage in the life cycle of many insects. (*4 marks*)
   (*Total 20 marks*)

   London Board June 1988, Paper II, No. 5

3. (a) With reference to the seed of a named angiosperm, describe the structural and physiological changes that occur at germination.
   (b) Moistened deadnettle seeds were exposed to sunlight for varying periods of time. Subsequently, the percentage germination was noted. The results are recorded below.

Hours of sunlight	10	20	25	30	35	40	50	60
Percentage germination	0	10	12	31	50	52	53	52

   (i) Give a possible explanation for these results.
   (ii) Give possible ecological benefits to the plant, which derive from this phenomenon.

   Southern Universities Joint Board June 1983, Paper II, No. 6

4. Ten marks were drawn on the roots of a number of maize seedlings at intervals of 1 mm from the apex. After one hour, the distance *between the apex and each of the marks* was measured. The results (the mean from the marked seedlings) are shown below.

Apex to	1 mm	2	3	4	5	6	7	8	9	10
Distance (mm) after one hour	1.04	2.11	3.38	4.79	6.15	7.45	8.51	9.56	10.58	11.59

   (a) Fill in, on the chart, the length between successive marks, after one hour's growth. Express this increase as a percentage of the original length. (*4 marks*)
   (b) Comment on these results. (*4 marks*)

	Marked areas of root (mm)									
	to 1	1–2	2–3	3–4	4–5	5–6	6–7	7–8	8–9	9–10
Length after one hour (in mm)										
Increase as % of original length										

   (c) (i) Make a labelled diagram of a longitudinal section of a dicotyledonous stem from the apex to the area of differentiation of vascular tissues.
   (ii) Annotate your diagram to point out **three** differences between root and shoot apices. (*8 marks*)
   (*Total 16 marks*)

   Oxford and Cambridge Board July 1984, Paper I, No. 2

5. The effect of temperature on the growth rate of duckweed (*Lemna minor*) was investigated by placing 30 plants of *Lemna* in each of twelve containers of pond water.
   Six containers were placed by the side of the pond from which the duckweed had been collected and the other six containers were kept in a greenhouse at a temperature on average 10 °C higher than at the pond. Both sets of containers had their sides covered with black paper. Samples of equal volume were taken from each container every four days and the number of duckweed plants in each sample was counted. The results are tabulated below.

Time in days	Average number of duckweed plants per container	
	Greenhouse	Pond
0	30	30
4	440	100
8	780	180
12	1060	300
16	1140	520
20	1120	700
24	1020	780

   (a) Present the data in a suitable graphical form, using the graph paper on the opposite page. (*5 marks*)
   (b) Calculate the mean rate of growth during the first 12 days of the experiment for plants (i) in the greenhouse and (ii) by the pond. (Show the working of your calculations.) (*4 marks*)
   (c) Comment on the comparative growth rates in the first 12 days of this experiment. (*2 marks*)

(d) Suggest **three** reasons for the results obtained for the duckweed grown in the greenhouse containers during the last 12 days of the experiment. (3 marks)

(e) Comment on the following points in the experimental method.
  (i) A number of containers were used in each location.
  (ii) The sides of the containers were covered in black paper. (4 marks)

(f) State **two** additional precautions, not mentioned in the question, which should have been taken in carrying out this investigation. In each case, give **one** reason for your answer. (4 marks)

London Board June 1985, Paper I, No. 7   (Total 22 marks)

6.   (a) The diagram below shows a section through a wheat grain.
  (i) Identify the parts numbered 1 to 3. (3 marks)
  (ii) Describe the origin and function of endosperm. (3 marks)
  (iii) Name an important vitamin found in parts 1, 2 and 3. (1 mark)

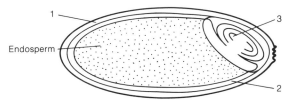

(b) The table shows the dry mass of endosperm and embryos from batches of 100 germinating wheat grains.

Days after germination	Dry mass from 100 germinating wheat grains (g)	
	Endosperm	Embryos
2	4.0	0.1
4	3.2	0.7
6	1.7	1.6
8	1.1	2.7
10	0.9	3.3

  (i) On one set of axes draw graphs of the dry mass of endosperm, the dry mass of embryos and their combined dry mass against time. (5 marks)
  (ii) Briefly describe the trend in each over the 10-day experimental period. Explain the reasons for these dry-mass changes. (8 marks)
  (Total 20 marks)

Associated Examining Board June 1983, Paper II, No. 1

7.   (a) Growth of an organism can be measured in different ways. Complete the table below by stating **one** disadvantage of measuring growth for each of the methods in the table.

Method	Disadvantage
Increase in length/height	
Increase in volume	
Increase in total mass	
Increase in dry mass	

(4 marks)

(b) Which of the above methods may be regarded as most suitable for measuring accurately the growth of (i) a tree and (ii) a rat? Give a reason for your choice in each case. (4 marks)

(c) Fill in the spaces in the following table which compares growth in flowering plants and mammals. Indicate a true statement by a tick (√) and an incorrect statement by a cross (×). The first statement has been completed for you.

Features of growth	Flowering plant	Mammal
Involves protein synthesis	√	√
Involves cell vacuolation		
Increase in complexity		
Growth is limited		
Constant production of new parts		
Light has no effect on growth		
Controlled by hormones		
Growth rate is uniform throughout the organism		

(7 marks)

London Board June 1986, Paper I, No. 14   (Total 15 marks)

8. Yeast is a unicellular fungus. The table shows the number of yeast cells counted through a microscope after the establishment at 0 hours of a yeast culture.

Hours	0	4	8	12	16	20	24	28	32	36	40	44	48
Number of yeast cells	2	5	8	15	29	52	86	116	134	142	143	142	130

(a) How would you establish a yeast culture in order to obtain such data? (3 marks)

(b) What method would you use to obtain reliable yeast-cell counts? (5 marks)

(c) Assuming yeast divides by binary fission once every four hours, construct a table similar to that above to show the *theoretical* population growth over a period of 36 hours starting with one cell at 0 hours. (*2 marks*)

(d) Draw on a graph, on a single set of axes, of the actual data in the table above and the theoretical data in your own table. (*5 marks*)

(e) Compare the actual and theoretical growth curves up to 36 hours. Explain the differences between them. (*3 marks*)

(f) How do you explain the shape of the actual growth curve after 36 hours? (*2 marks*)

(*Total 20 marks*)

Associated Examining Board June 1984, Paper II, No. 1

9. Sketch graphs to represent as clearly and accurately as you can the results you would expect to obtain in each of the given situations (*a*)–(*d*). Suggest suitable time scales on your graphs.

(a) The growth of a population of bacteria in an enclosed small bottle of nutrient broth. (Numbers against Time) (*2 marks*)

(b) The mean dry mass of a sample of seeds and resultant seedlings over a period of four weeks from the onset of germination. (Mass against Time) (*2 marks*)

(c) The growth of a named Arthropod, showing ecdysis, from hatching to maturity. (Length against Time) (*2 marks*)

(d) A small culture of yeast cells is added to an enclosed flask containing dilute sucrose solution on day 0. On day 5 a small culture of ciliate protozoans which will feed on the yeast cells, is added.

   On the same axes, plot:
   (i) the number of yeast cells from day 0 to day 10;
   (ii) the number of protozoan cells from day 5 to day 10. (Numbers against Time)

(*4 marks*)

(*Total 10 marks*)

Oxford Local June 1988, Paper I, No. 4

Powerstations produce energy in many different forms. Some of it is wasted in the form of steam (opposite)

Part III
Energetics

$\underline{21}$ Energy and organisms

Energy is defined as the 'capacity to do work'. It exists in a number of different forms: heat, light, electrical, magnetic, chemical, atomic, mechanical and sound. The laws which apply to energy conversions are the **laws of thermodynamics**.

21.1 First law of thermodynamics

This states: **energy cannot be created or destroyed but may be converted from one form into another**. Energy may also be stored. Water in a lake high up on a mountain is an example of stored energy. The energy it possesses is called **potential energy**. If the water is released from the lake it begins to flow downhill, and the energy of its motion is called **kinetic energy**. The stream of moving water may be used to drive a turbine which produces electricity (hydroelectric power). The kinetic energy is thus converted to electricity. This electricity may in turn be converted into light (light bulb), heat (electric fire/cooker), sound (record player/tape recorder) etc. During these changes not all the energy is converted into its intended form; some is lost as heat. By 'lost' we mean the energy is no longer available to do useful work because it is distributed evenly. Energy which is available to do work under conditions of constant temperature and pressure is called **free energy**. Reactions which liberate energy are termed **exogenic**, those which absorb free energy are termed **endogenic**.

21.2 Second law of thermodynamics

This states: **all natural processes tend to proceed in a direction which increases the randomness or disorder of a system**. The degree of randomness is called **entropy**. A highly ordered system has low entropy whereas a disordered one, with its high degree of randomness, has high entropy. We saw in Chapter 1 that entropy and free energy are inversely related. Systems with high entropy have little free energy, those with low entropy have more free energy. We also saw in Chapter 1 that the ability of living systems to maintain low entropy is what distinguishes them from non-living systems. The fact that living systems can decrease their entropy does not mean that they fail to obey the second law of thermodynamics. The reason that they are able to reduce their entropy is that they take in useful energy from their surroundings and release it in a less useful form. While the organism's entropy decreases, that of its surroundings increases to an even greater extent. The organism and its environment represent one system, the total entropy of which increases. The second law of thermodynamics is therefore not violated.

21.3 Energy and life

There are three stages to the flow of energy through living systems:

1. The conversion of the sun's light energy to chemical energy by plants during photosynthesis.

2. The conversion of the chemical energy from photosynthesis into ATP — the form in which cells can utilize it.

3. The utilization of ATP by cells in order to perform useful work.

The chemical reactions which occur within organisms are collectively known as **metabolism**. They are of two types:

1. The build-up of complex compounds from simple ones. These synthetic reactions are collectively known as **anabolism**.

2. The breakdown of complex compounds from simple ones. Such reactions are collectively known as **catabolism**.

A typical chemical reaction may be represented as:

$$A \rightarrow B + C$$

In this case A represents the **substrate** and B and C are the **products**. If the entropy of C and B is greater than A then the reaction will proceed naturally in the direction shown. A reaction which involves an increase in entropy is said to be **spontaneous**. The free energy of the products is less than that of the substrate. The word spontaneous could be misleading because the reaction is not instantaneous. Before any chemical reaction can proceed it must initially be activated, i.e. its energy must be increased. The energy required is called the **activation energy**. Once provided, the activation energy allows the products to be formed with a consequent loss of free energy and increase in entropy (Fig. 21.1). Chemical reactions are reversible and therefore C and B can be synthesized into A. Such a reaction is not, however, spontaneous and requires an external source of energy if it is to proceed. Most biological processes are in fact a cycle of reversible reactions. Photosynthesis and respiration, for example, are basically the same reaction going in opposite directions.

$$\text{Energy} + 6CO_2 + 6H_2O \underset{\text{respiration}}{\overset{\text{photosynthesis}}{\rightleftharpoons}} C_6H_{12}O_6 + 6O_2$$

As there is inevitably some loss of free energy in the form of heat each time the reaction is reversed, the process cannot continue without a substantial input of energy from outside the organisms. The ultimate source of this energy is the light radiation of the sun. The way in which organisms obtain their energy for metabolic and other processes is probably more important in determining their design than any other single factor. The fundamental differences between plants and animals are a result of their modes of nutrition.

Plants obtain their energy from the sun and use it to combine carbon dioxide and water in the synthesis of organic molecules. As the raw materials are readily available almost everywhere, there is no necessity for plants to move to obtain their nutrients. Indeed, in order to obtain sufficient light plants need to have a large surface area. They therefore need to be as large as possible in order to compete with other plants for light. For this reason many plants are

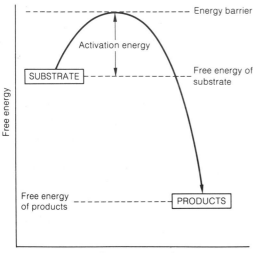

Fig. 21.1 Concept of activation energy

large. Locomotion for these plants would not only be difficult and slow, it would also be very energy-consuming. Plants therefore do not exhibit locomotion.

Animals obtain their energy from complex organic compounds. These occur in other organisms which must be sought. Most animals therefore exhibit locomotion in order to obtain their food. To help animals move from place to place they have developed a wide range of locomotory mechanisms. They are therefore more complex, and variable, in their design than plants. In carrying out locomotion, animals require a complex nervous system to coordinate their actions and a range of sense organs to help them manoeuvre and to locate food.

Being sessile, plants do not require nervous systems and sense organs. Having a large surface area to obtain light energy means that plants have no need for separate specialized surfaces for obtaining respiratory gas. Any additional oxygen needed, over and above that produced in photosynthesis, can be obtained by diffusion through the leaves and roots which already provide a substantial surface area. Animals, by contrast, being compact to assist locomotion, require specialized respiratory surfaces with a large surface area to compensate for their small external area.

Even in reproduction, the differences between plants and animals can be related to the means by which they obtain energy. As their method of nutrition favours being stationary, plants have to use an external agent to transfer the male gametes from one individual to another during sexual reproduction. Insects and wind are the main agents of pollination. Animals, being capable of locomotion, utilize it in finding a mate and therefore male gametes are either introduced directly into the female, as in terrestrial organisms, or released in the vicinity of the female, as in some aquatic ones. In either case, male and female are in close proximity.

In this section of the book, we shall look at how energy is obtained and utilized by organisms.

22 *Enzymes*

Enzymes may be described as **globular proteins with catalytic properties**. A catalyst is a substance which alters the rate of a chemical reaction without itself undergoing a permanent change. As they are not altered by the reactions they catalyze, enzymes can be used over and over again. They are therefore effective in very small amounts. Enzymes cannot cause reactions to occur, but only speed up ones which would otherwise take place extremely slowly. The word 'enzyme' means 'in yeast', and was used because they were first discovered by Eduard Buchner in an extract of yeast.

22.1 Enzyme structure and function

Enzymes are complex three-dimensional globular proteins, some of which have other associated molecules. While the enzyme molecule is normally larger than the substrate molecule it acts upon, only a small part of the enzyme molecule actually comes into contact with the substrate. This region is called the **active site**.

22.1.1 Enzymes and activation energy

In Chapter 21 we saw that before a reaction can take place it must overcome an energy barrier by exceeding its activation energy. Enzymes operate by lowering this activation energy and thus permit the reaction to occur more readily (Fig. 22.1). As heat is often the source of activation energy, enzymes often dispense with the need for this heat and so allow reactions to take place at lower temperatures. Many reactions which would not ordinarily occur at the temperature of an organism do so readily in the presence of enzymes.

22.1.2 Mechanism of enzyme action

Enzymes are thought to operate on a **lock and key mechanism**. In the same way that a key fits a lock very precisely, so the substrate fits accurately into the active site of the enzyme molecule. The two molecules form a temporary structure called the **enzyme-substrate complex**. The products have a different shape from the substrate and so, once formed, they escape from the active site, leaving it free to become attached to another substrate molecule. The sequence is summarized in Fig. 22.2.

More modern interpretations of the lock and key mechanism suggest that in the presence of the substrate the active site may change in order to suit the substrate's shape. The enzyme is flexible and moulds to fit the substrate molecule in the same way that clothing is flexible and can mould itself to fit the shape of the wearer.

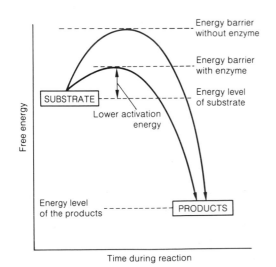

Fig. 22.1 How enzymes lower the activation energy

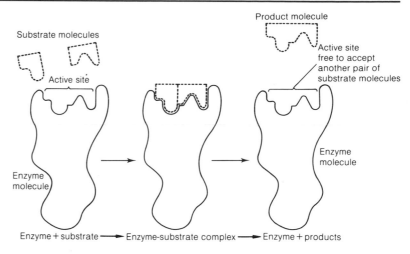

Fig. 22.2 Mechanism of enzyme action

22.2 Properties of enzymes

The properties of enzymes can be explained in relation to the lock and key mechanism of enzyme action.

22.2.1 Specificity

All enzymes operate only on specific substrates. Just as a key has a specific shape and therefore fits only complementary locks, so only substrates of a particular shape will fit the active site of an enzyme. Some locks are highly specific and can only be opened with a single key. Others are opened by a number of similar keys; yet others may be opened by many different keys. In the same way, some enzymes will act only on one particular isomer. Others act only on similar molecules; yet others will break a particular chemical linkage, wherever it occurs.

22.2.2 Reversibility

Chemical reactions are reversible, and equations are therefore often represented by two arrows to indicate this reversibility. (See opposite.)

At any one moment the reaction (shown left) may be proceeding predominantly in one direction. If, however, the conditions are changed, the direction may be reversed. It may be that the above reaction proceeds from left to right in acid conditions, but in alkaline conditions it goes from right to left. In time, reactions reach a point where the reactants and the product are in **equilibrium** with one another. Enzymes catalyze the forward and reverse reactions equally. They do not therefore alter the equilibrium itself, only the speed at which it is reached. Carbonic anhydrase is an enzyme which catalyzes a reaction in either direction depending on the conditions at the time. In respiring tissues where there is much carbon dioxide it converts carbon dioxide and water into carbonic acid. In the lungs, however, the removal of carbon dioxide by diffusion means a low concentration of carbon dioxide, and hence the carbonic acid breaks down into carbon dioxide and water. Both reactions are catalyzed by carbonic anhydrase, as shown opposite.

$$A + B \rightleftharpoons C + D$$

$$CO_2 + H_2O \xrightarrow{\text{carbonic anhydrase}} H_2CO_3 \quad \text{(in tissues)}$$

$$H_2CO_3 \xrightarrow{\text{carbonic anhydrase}} CO_2 + H_2O \quad \text{(in the lungs)}$$

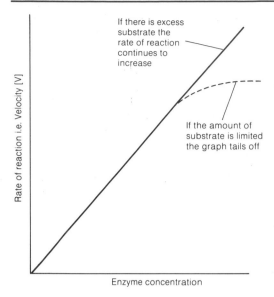

If there is excess substrate the rate of reaction continues to increase

If the amount of substrate is limited the graph tails off

Fig. 22.3 Graph to show the effect of enzyme concentration on the rate of an enzyme-controlled reaction

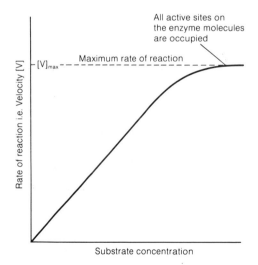

All active sites on the enzyme molecules are occupied

Maximum rate of reaction

$[V]_{max}$

Fig. 22.4 Graph to show the effect of substrate concentration on the rate of an enzyme-controlled reaction

22.2.3 Enzyme concentration

The active site of an enzyme may be used again and again. Enzymes therefore work efficiently at very low concentrations. The number of substrate molecules which an enzyme can act upon in a given time is called its **turnover number**. This varies from many millions of substrate molecules each minute, in the case of catalase, to a few hundred per minute for slow acting enzymes. Provided the temperature and other conditions are suitable for the reaction, and provided there are excess substrate molecules, the rate of a reaction is directly proportional to the enzyme concentration. If the amount of substrate is restricted it may limit the rate of reaction. The addition of further enzyme cannot increase the rate and the graph therefore tails off (Fig. 22.3).

22.2.4 Substrate concentration

For a given amount of enzyme, the rate of an enzyme-controlled reaction increases with an increase in substrate concentration — up to a point. At low substrate concentrations, the active sites of the enzyme molecules are not all used — there simply are not enough substrate molecules to occupy them all. As the substrate concentration is increased, more and more sites come into use. A point is reached, however, where all sites are being used; increasing the substrate concentration cannot therefore increase the rate of reaction, as the amount of enzyme is the limiting factor. At this point the graph tails off (Fig. 22.4).

22.2.5 Temperature

An increase in temperature affects the rate of an enzyme-controlled reaction in two ways:

1. As the temperature increases, the kinetic energy of the substrate and enzyme molecules increases and so they move faster. The faster these molecules move, the more often they collide with one another and the greater the rate of reaction.

2. As the temperature increases, the more the atoms which make up the enzyme molecules vibrate. This breaks the hydrogen bonds and other forces which hold the molecules in their precise shape. The three-dimensional shape of the enzyme molecules is altered to such an extent that their active sites no longer fit the substrate. The enzyme is said to be **denatured** and loses its catalytic properties.

The actual effect of temperature on the rate of reaction is the combined influence of these two factors and is illustrated in Fig. 22.5.
   The optimum temperature for an enzyme varies considerably. Many arctic and alpine plants have enzymes which function efficiently at temperatures around 10 °C, whereas those in algae inhabiting some hot springs continue to function at temperatures around 80 °C. For many enzymes the optimum temperature lies around 40 °C and denaturation occurs at about 60 °C.

22.2.6 pH

The precise three-dimensional molecular shape which is vital to the functioning of enzymes is partly the result of hydrogen bonding.

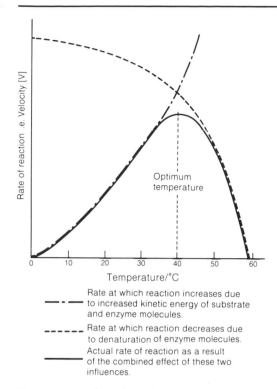

Fig. 22.5 Graph to show the effect of temperature on the rate of an enzyme-controlled reaction

Key to graph:
- — · — Rate at which reaction increases due to increased kinetic energy of substrate and enzyme molecules.
- — — — Rate at which reaction decreases due to denaturation of enzyme molecules.
- ——— Actual rate of reaction as a result of the combined effect of these two influences.

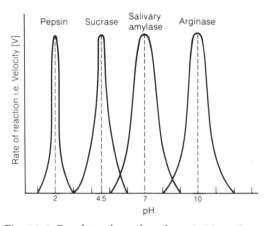

Fig. 22.6 Graph to show the effect of pH on the rate of reaction of four different enzymes

1. *Inhibitor absent* –
   Substrate is free to occupy activity site of enzyme. Reaction takes place at the normal rate.
2. *Inhibitor present* –
   Substrate competes with inhibitor for the active site of the enzyme. If the inhibitor occupies the site it prevents the substrate doing so. Rate of reaction is reduced.

These bonds may be broken by the concentration of hydrogen ions (H^+) present. pH is a measure of hydrogen ion concentration. It is measured on a scale of 1–14, with pH 7 being the neutral point. A pH less than 7 is acid, one greater than 7 is alkaline. The scale is logarithmic, however. A change of one pH point represents a tenfold change in the hydrogen ion concentration. Compared to water, which has a pH of 7, a solution with a pH of 6 is ten times more acid, one of pH 5 is 100 × more acid, one of pH 4 is 1000 × more acid and so on.

By breaking the hydrogen bonds which give enzyme molecules their shape, any change in pH can effectively denature enzymes. Each enzyme works best at a particular pH, and deviations from this optimum may result in denaturation. Fig. 22.6 illustrates the different pH optima of four enzymes.

22.2.7 Inhibition

The rate of enzyme-controlled reactions may be decreased by the presence of inhibitors. They are of two types: **reversible inhibitors** and **non-reversible inhibitors**.

Reversible inhibitors
The effect of this type of inhibitor is temporary and causes no permanent damage to the enzyme. Removal of the inhibitor therefore restores the activity of the enzyme to normal. There are two types: **competitive** and **non-competitive**.

Competitive inhibitors compete with the substrate for the active sites of enzyme molecules. The inhibitor may have a structure which permits it to combine with the active site. While it remains bound to the active site, it prevents substrate molecules occupying them and so reduces the rate of the reaction. The same quantity of product is formed, because the substrate continues to use any enzyme molecules which are unaffected by the inhibitor. It does, however, take longer to make the products. If the concentration of the substrate is increased, less inhibition occurs. This is because, as the substrate and inhibitor are in direct competition, the greater the proportion of substrate molecules the greater their chance of finding the active sites, leaving fewer to be occupied by the inhibitor.

Malonic acid is a competitive inhibitor. It competes with succinate for the active sites of succinic dehydrogenase, an important enzyme in the Krebs cycle (Section 26.3).

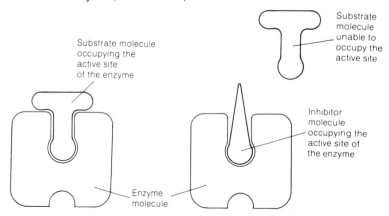

Fig. 22.7(a) Competitive inhibition

Non-competitive inhibitors do not attach themselves to the active site of the enzyme, but elsewhere on the enzyme molecule. They nevertheless alter the shape of the enzyme molecule in such a way that the active site can no longer properly accommodate the substrate. As the substrate and inhibitor molecules attach to different parts of the enzyme they are not competing for the same sites. An increase in substrate concentration will not therefore reduce the effect of the inhibitor.

Cyanide is a non-competitive inhibitor. It attaches itself to the copper prosthetic group of cytochrome oxidase, thereby inhibiting respiration (Section 26.4).

Non-reversible inhibitors
Non-reversible inhibitors leave the enzyme permanently damaged and so unable to carry out its catalytic function. Heavy metal ions such as mercury (Hg^{2+}) and silver (Ag^{+}) cause disulphide bonds to break. These bonds help to maintain the shape of the enzyme molecule. Once broken the enzyme molecule's structure becomes irreversibly altered with the permanent loss of its catalytic properties.

1. *Inhibitor absent –*
   The substrate attaches to the active site of the enzyme in the normal way. Reaction takes place as normal.
2. *Inhibitor present –*
   The inhibitor prevents the normal enzyme substrate complex being formed. The reaction rate is reduced.

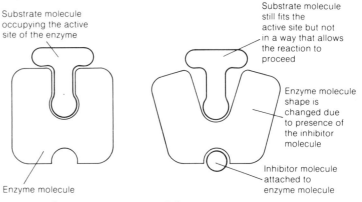

Fig. 22.7(b) Non-competitive inhibition

22.3 Enzyme cofactors

A **cofactor** is a non-protein substance which is essential for some enzymes to function efficiently. There are three types: **activators**, **coenzymes** and **prosthetic groups**.

22.3.1 Activators

Activators are substances which are necessary for the functioning of certain enzymes. The enzyme thrombokinase, which converts prothrombin into thrombin during blood clotting, is activated by calcium (Ca^{2+}) ions. In the same way salivary amylase requires the presence of chloride (Cl^{-}) ions before it will efficiently convert starch into maltose. It is possible that these activators assist in forming the enzyme-substrate complex by moulding either the enzyme or substrate molecule into a more suitable shape.

22.3.2 Coenzymes

Coenzymes are non-protein organic substances which are essential to the efficient functioning of some enzymes, but are not themselves bound to the enzyme. Many coenzymes are derived from vitamins, e.g. **nicotinamide adenine dinucleotide (NAD)** is derived from nicotinic acid, a member of the vitamin B complex. NAD acts as a coenzyme to dehydrogenases by acting as a hydrogen acceptor.

22.3.3 Prosthetic groups

Like coenzymes, prosthetic groups are organic molecules, but unlike them they are bound to the enzyme itself. Perhaps the best known prosthetic group is **haem**. It is a ring-shaped organic molecule with iron at its centre. Apart from its role as an oxygen carrier in haemoglobin, it is also the prosthetic group of the electron carrier cytochrome and of the enzyme catalase.

22.4 The range of enzyme activities

Enzymes are classified into six groups according to the type of reaction they catalyse. Table 22.1 summarizes this internationally accepted classification.

TABLE **22.1 The classification of enzymes**

Enzyme group	Type of reaction catalysed	Enzyme examples
1. Oxidoreductases	Transfer of O and H atoms between substances, i.e. all oxidation-reduction reactions	Dehydrogenases Oxidases
2. Transferases	Transfer of a chemical group from one substance to another	Transaminases Phosphorylases
3. Hydrolases	Hydrolysis reactions	Peptidases Lipases Phosphatases
4. Lyases	Addition or removal of a chemical group other than by hydrolysis	Decarboxylases
5. Isomerases	The rearrangement of groups within a molecule	Isomerases Mutases
6. Ligases	Formation of bonds between two molecules using energy derived from the breakdown of ATP	Synthetases

Each enzyme is given two names:

A **systematic** name, based on the six classification groups. These names are often long and complicated.

A **trivial** name which is shorter and easier to use.
The trivial names are derived by following three procedures:

1. Start with the name of the substrate upon which the enzyme acts, e.g. succinate.

2. Add the name of the type of reaction which it catalyses, e.g. dehydrogenation.

3. Convert the end of the last word to an -ase suffix, e.g. dehydrogenase.

The examples above gives succinic dehydrogenase. Another example would be DNA polymerase. This enzyme catalyses the formation (and breakdown) of the nucleic acid DNA by polymerization.
Some of the commercial uses of enzymes are considered in Sections 41.3.6 and 41.7.

22.5 Control of metabolic pathways

With many hundreds of reactions taking place in any single cell it is clear that a very structured system of control of metabolic pathways is essential. If the cell were merely a 'soup' of substrates, enzymes and products, the chances of particular reactants meeting would be small and the metabolic processes inefficient. In addition, different enzymes need different conditions, e.g. a particular pH, and it would be impossible to provide these in such an unstructured 'soup'. Cells contain organelles, and enzymes are often bound to these inner membranes in a precise order. This increases the chances of them coming into contact with their appropriate substrates, and leads to efficiency. The organelles may also have varying conditions to suit the specific enzymes they contain. By controlling these conditions, and the enzymes available, the cell can control the metabolic pathways within it.

Cells also make use of the enzyme's own properties to exercise control over metabolic pathways. The end-product of a pathway may inhibit the enzyme at the start.

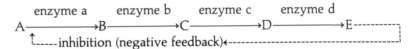

In the example above, the product E acts as an inhibitor to enzyme a. If the level of product E falls, this inhibition is reduced, and so more A is converted to B, and subsequently more E is produced. If the level of E rises above normal, inhibition of enzyme a increases and so the level of E is reduced. In this way homeostatic control of E is achieved, more details of which are given in Section 36.1. The mechanism is termed **negative feedback** because the information from the end of the pathway which is fed back to the start has a negative effect, i.e. a high concentration of E reduces its own production rate.

These forms of inhibition are, for obvious reasons, reversible, i.e. they do not permanently damage the enzymes. They frequently affect the nature of an enzyme's active site by binding with the enzyme at some point on the molecule remote from the active site. Such effects are termed **allosteric** and refer to the ability of the enzyme to have more than one shape. One shape renders the enzyme active, another renders it inactive.

22.6 Questions

1. (a) Explain what is meant by the terms:
    (i) enzyme;
    (ii) co-enzyme. *(6 marks)*
  (b) Discuss current ideas about the mode of action of enzymes. *(12 marks)*
  (c) **List** those factors which affect the rate of enzyme action. Briefly point out why the factors you mention will change the rate.
(12 marks)
(Total 30 marks)

Oxford Local June 1988, Paper II, No. 10

2. Discuss the role of enzymes in cell metabolism and comment on those factors which affect their activity.

London Board June 1986, Paper II, No. 8

3. (a) Give *one* named example of each of the following classes of enzyme, and describe the action of the enzyme in the metabolic pathway or body process where it occurs.
    (i) Hydrolases
    (ii) Dehydrogenases
    (iii) Polymerases *(12 marks)*
  (b) Explain how the following factors can inhibit enzyme action.
    (i) Temperature
    (ii) Hydrogen ion concentration (pH) *(8 marks)*
(Total 20 marks)

London Board June 1989, Paper II, No. 6

4. Explain the following terms related to enzyme activity:
  (a) lock and key theory;
  (b) active site;
  (c) competitive inhibition;
  (d) coenzyme;
  (e) optimum temperature *(10 marks)*

Southern Universities Joint Board June 1986, Paper I, No. 4

5. (a) What is meant by the active site of an enzyme? *(2 marks)*
  (b) Explain the part played by the active site in each of the following.
    (i) Enzyme specificity *(2 marks)*
    (ii) The catalytic action of enzymes *(2 marks)*
    (iii) Competitive inhibition *(2 marks)*
  (c) Using a named example, explain what is meant by non-competitive inhibition. *(3 marks)*
(Total 11 marks)

London Board June 1988, Paper I, No. 11

6. In a reaction in which protein is broken down to amino acids, the course of the reaction was observed by measuring the accumulation of amino acid. Only two factors, temperature of reaction and time of incubation of the reaction, were varied. The results given below were obtained:

Temperature in degrees Celsius		10	20	30	40	50	60
Milligram per litre of amino acid produced by the reaction	after 2.5 hours	50	110	150	270	240	100
	after 50 hours	650	820	900	690	500	130

  (a) Draw graphs of these data on one pair of axes, showing the rates of reaction. *(12 marks)*
  (b) Explain why the reactions probably involve a protease (an enzyme). *(2 marks)*
  (c) Explain the difference in rates of reaction and the shape of the curves after the two periods of incubation. *(4 marks)*
  (d) Calculate the Q_{10} between 10 °C and 20 °C for each of the two incubation periods and comment on the differences. *(6 marks)*
  (e) What other factors may influence the rate of enzymatic reactions? Give examples. *(6 marks)*
(Total 30 marks)

Oxford Local 1987, Specimen Paper II, No. 11

7. In an experiment on the effects of temperature on the action of sucrase (invertase) on sucrose, the following data were obtained:

Temp (°C)	Time (mins.) for complete hydrolysis of sucrose
0	50
10	15
20	8
30	4
40	6
50	28
60	110

(a) Using this information, plot a graph in the way that you feel most clearly shows how the **rate** of reaction varies with temperature. (*7 marks*)

(b) Comment on the result as seen between 10 °C and 30 °C. (*2 marks*)

(c) What does the term 'hydrolysis' mean? (*2 marks*)

(d) Write a chemical equation to summarise the effect of the enzyme on the sucrose. (*2 marks*)

(e) Name the products of this reaction. (*2 marks*)

(f) Explain briefly what is happening to the enzyme molecules between 40 °C and 60 °C. (*2 marks*)

(g) Comment on why the enzyme shows some activity at 0 °C when it is usually stated that enzymes would not be expected to work in the frozen state. (*2 marks*)

(h) Using the information shown on your graph, calculate the Q_{10} value for the rate of this reaction between 15 °C and 25 °C. (Show your working.) (*2 marks*)

(*Total 21 marks*)

Oxford Local June 1989, Paper I, No. 5

8. Equal volumes of four different concentrations of fat suspension were put into labelled test tubes (1 to 4). A fixed amount of a hydrolysing enzyme was added to each, the contents mixed and the time noted. The reaction was followed to completion using a pH probe, and the time noted again.

(a) (i) Suggest a name for the class of enzyme used.

(ii) How would it be known when the hydrolysis was complete?

(iii) If a pH probe was not available, and given that indicators might interfere with the reaction, describe how the reaction could nevertheless be followed using an indicator solution. (*3 marks*)

Table 1 shows some results from such an investigation.

Table 1

	Test tube number			
	1	2	3	4
Percentage fat concentration	0.05	0.1	0.2	0.4
Time for completion of reaction (minutes)	8	8	10	16

(b) Using the data in Table 1, complete Table 2. Note that, for example, in Tube 3 the fat concentration is four times that in Tube 1.

(c) Plot a graph of reaction rate against fat concentration. Ensure that the fat concentration axis includes a value of 0.6% so that the rest of this question may be answered. (*3 marks*)

Table 2

	Test tube number			
	1	2	3	4
Time equivalent for the breakdown of the amount of fat in tube 1	8		2.5	
Reaction rate $\left(\dfrac{1}{\text{time}}\right)$			0.4	

(d) Use your graph to predict how long it would take, under the same experimental conditions, to hydrolyse a tube containing a 0.6% fat suspension. (*2 marks*)

(*Total 11 marks*)

Joint Matriculation Board June 1989, Paper IA, No. 7

9. The results of an investigation into the effect of increasing substrate concentration on the rate of an enzyme reaction are shown in the graph.

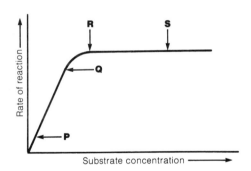

(a) (i) Name the factor which determines the rate of the reaction between points P and Q.

(ii) Name **two** factors which could account for the shape of the curve between R and S. (*3 marks*)

(b) (i) State **two** conditions which should be kept constant in this investigation.

(ii) What should be measured in order to determine the rate of an enzyme reaction? (*4 marks*)

(c) The investigation was repeated with the addition of a competitive inhibitor. The same amount of inhibitor was added to the substrate at **each** concentration.

(i) Draw the expected curve on the graph below.

(ii) Briefly explain how a competitive inhibitor would bring about this effect. (*5 marks*)

(d) Name **three** compounds which might have been used in the investigation.

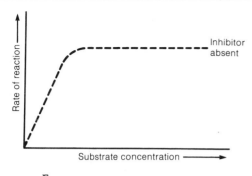

Test-tube number	Percentage concentration of ethanol added to each test-tube	Time required for complete starch hydrolysis/s
1	0	100
2	10	80
3	20	90
4	30	100
5	40	130
6	50	190
7	60	240
8	70	300

Enzyme
Substrate
Competitive inhibitor   (3 marks)
                                  (Total 15 marks)

Welsh Joint Education Committee June 1988, Paper A2,
                                           No. 5

10. An experiment was carried out to investigate the effect of alcohol on the activity of pancreatic amylase. Dilutions of ethanol in water were prepared in order to give a range of percentage concentrations from 0% to 70%.

To each of eight test-tubes was added 1 cm³ of the appropriate concentration of ethanol, followed by 2 cm³ of a 1% starch solution (a solution containing 1 g of starch per 100 cm³ of solution). 2 cm³ of pancreatic extract were added to each tube and the time, in seconds, required for starch hydrolysis was recorded. The results are summarized in the table below.

(a) Calculate the initial mass, in mg, of starch per
    test-tube.   (1 mark)
(b) Calculate
    (i) the percentage of ethanol in the reaction
         mixture for each of tubes 1 to 8 and
    (ii) the rate of starch hydrolysis in each test
         tube in milligrams per minute (mg min⁻¹).
         Present your figures in the form of a table
    with appropriate headings.   (4 marks)

(c) Draw a graph to show the relationship between
    ethanol concentration and rate of starch
    hydrolysis.   (5 marks)
(d) Organic solvents such as ethanol are known to
    alter the 3-dimensional structure of globular
    proteins.
    (i) Explain, in terms of enzyme structure and
         function, the effect of concentrations of
         ethanol greater than 6% in the reaction
         mixture.
    (ii) Suggest a hypothesis to account for the
         effect of concentrations of ethanol less than
         6%.   (5 marks)
(e) Ethanol is known to have adverse effects on
    mammals.
    (i) Which mammalian organ has the function
         of detoxifying ethanol? State *five* other
         important functions of this organ.
    (ii) State *two* short term and *two* long term
         adverse physiological effects of excess
         ethanol on the human body.   (5 marks)
                                  (Total 20 marks)

Northern Ireland Board June 1989, Paper II, No. 5X

23 *Autotrophic nutrition (photosynthesis)*

In Chapter 21 we saw that living systems differ from non-living ones in their ability to replace lost energy from the environment and so maintain themselves in an ordered condition (low entropy). Photosynthesis is the means by which this energy is initially obtained by living systems. All life is directly or indirectly dependent on this most fundamental process in living organisms.

Autotrophic (*auto* − 'self'; *trophic* − 'feeding') organisms use an inorganic form of carbon, such as carbon dioxide, to make up complex organic compounds. These complex compounds are more ordered and so possess more energy. In autotrophs this energy is provided from two sources: light and chemicals. The processes involved are termed photosynthesis and chemosynthesis respectively.

AUTOTROPHIC NUTRITION

light energy   chemical energy

PHOTOSYNTHESIS   CHEMOSYNTHESIS
All green plants   Certain bacteria

Photosynthesis is much the more common of the two processes. It is principally important because:

1. It is the means by which the sun's energy is captured by plants for use by all organisms.

2. It provides a source of complex organic molecules for heterotrophic organisms.

3. It releases oxygen for use by aerobic organisms.

23.1   Leaf structure

The leaf is the main photosynthetic structure of a plant, although stems, sepals and other parts may also photosynthesize. It is adapted to bring together the three raw materials, water, carbon dioxide and light, and to remove the products oxygen and glucose. The structure of the leaf is shown in Fig. 23.1, on page 316.

Considering that all leaves carry out the same process, it is perhaps surprising that they show such a wide range of form. This range of form is often the consequence of different environmental conditions which have nothing directly to do with photosynthesis. In dry areas, for example, leaves may be small in size with thick cuticles and sunken stomata to help reduce water loss. The presence of spines to deter grazing by herbivores is not uncommon. Other differences in leaf morphology are a result of the plant living in a sunny or shady situation.

The equation for photosynthesis may be summarized as:

$$6CO_2 + 6H_2O + sunlight \xrightarrow{\text{chlorophyll}} C_6H_{12}O_6 + 6O_2$$

$$carbon\ dioxide + water + sunlight \xrightarrow{\text{chlorophyll}} glucose + oxygen$$

$$gas + liquid + energy \xrightarrow{\text{chlorophyll}} liquid + gas\ (solution\ in\ water)$$

The adaptations of the leaf to photosynthesis are therefore:

1. To obtain energy (sunlight).

2. To obtain and remove gases (carbon dioxide and oxygen).

3. To obtain and remove liquids (water and sugar solution).

23.1.1 Adaptations for obtaining energy (sunlight)

As sunlight is the energy source which drives the photosynthetic process, it is often the factor which determines the rate of photosynthesis. To ensure its efficient absorption the leaf shows many adaptations:

1. Phototropism causes shoots to grow towards the light in order to allow the attached leaves to receive maximum illumination.

2. Etiolation causes rapid elongation of shoots which are in the dark, to ensure that the leaves are brought up into the light as soon as possible.

3. Leaves arrange themselves into a mosaic, i.e. they are arranged on the plant in a way that minimizes overlapping and so reduces the degree of shading of one leaf by another.

4. Leaves have a large surface area to capture as much sunlight as possible. They are held at an angle perpendicular to the sun during the day to expose the maximum area to the light. Some plants, e.g. the compass plant, actually 'track' the sun by moving their leaves so they constantly face it during the day.

5. Leaves are thin – If they were thicker, the upper layers would filter out all the light and the lower layers would not then photosynthesize.

6. The cuticle and epidermis are transparent to allow light through to the photosynthetic mesophyll beneath.

7. The palisade mesophyll cells are packed with chloroplasts and arranged with their long axes perpendicular to the surface. Although there are some air spaces between them, they still form a continuous layer which traps most of the incoming light. In some plants this layer is more than one cell thick.

8. The chloroplasts within the mesophyll cells can move – This allows them to arrange themselves into the best positions within a cell for the efficient absorption of light.

9. The chloroplasts hold chlorophyll in a structured way – The chlorophyll within a chloroplast is contained within the grana, where it is arranged on the sides of a series of unit membranes. The ordered arrangement not only presents the maximum amount of chlorophyll to the light but also brings it in close proximity to other pigments and substances which are necessary for its functioning. The structure of a chloroplast is shown in Fig. 4.8, Section 4.3.5.

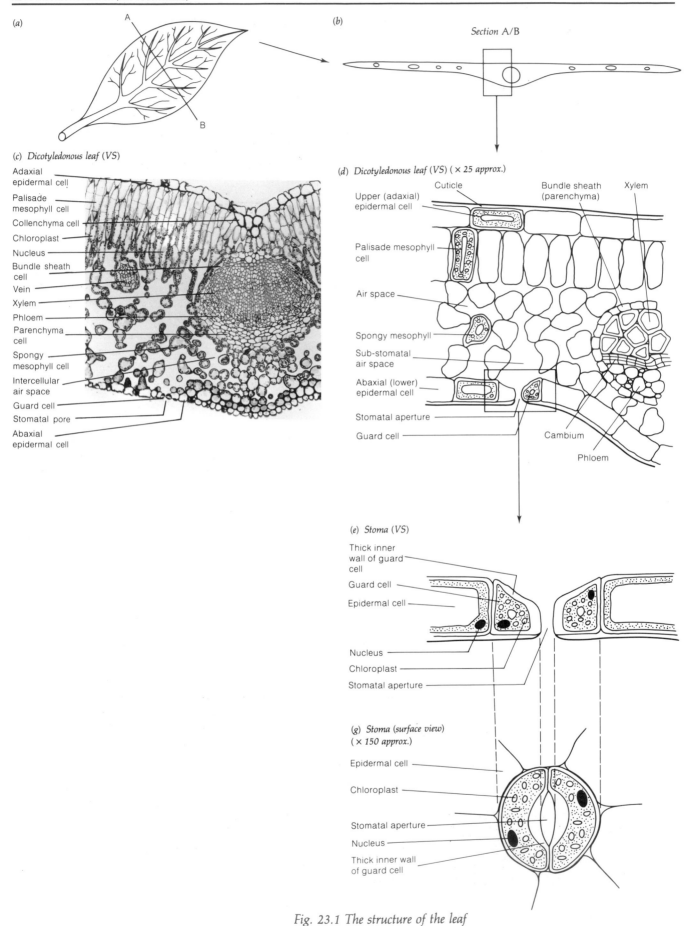

Fig. 23.1 The structure of the leaf

23.1.2 Adaptations for obtaining and removing gases

As gases diffuse relatively rapidly, the leaf has no special transport mechanism for carbon dioxide and oxygen. The leaf does, however, show a number of adaptations which ensure rapid diffusion of these gases:

1. Numerous stomata are present in the epidermis of leaves. There may be tens of thousands per cm² of leaf surface, which itself represents a very considerable area. Stomata are minute pores in the epidermis which, when open, permit unrestricted diffusion into and out of the leaf.

2. Stomata can be opened and closed – Plants need to be relatively impermeable to gases in order to prevent water loss, and yet they need the free entry of carbon dioxide for photosynthesis. To overcome this problem, they have stomatal pores which are bounded by two **guard cells**. Alterations in the turgidity of these cells opens and closes the stomatal pore, thus controlling the uptake of carbon dioxide and the loss of water. The detail of stomatal control is given in Section 33.3.1. Stomata open in conditions which favour photosynthesis and at this time some water loss is unavoidable. When photosynthesis cannot take place, e.g. at night, they close, thus reducing considerably the loss of water. At times of considerable water loss, the stomata may close anyway, regardless of the demands for carbon dioxide.

3. Spongy mesophyll possesses many airspaces – The mesophyll layer on the underside of the leaf has many air spaces. These communicate with the palisade layer and the stomatal pores. There is hence an uninterrupted diffusion of gases between the atmosphere and the palisade mesophyll. During photosynthesis carbon dioxide diffuses in and oxygen out of this layer. The air spaces avoid the need for these gases to diffuse through the cells themselves, a process which would be much slower. The palisade mesophyll also possesses air spaces to permit rapid diffusion around the cells of which it is made.

23.1.3 Adaptations for obtaining and removing liquids

As water is a liquid raw material for photosynthesis and as the sugar produced is carried away in solution, the leaf has to be adapted for the efficient transport of liquids.

1. A large central midrib is possessed by most dicotyledonous leaves. This contains a large vascular bundle comprising xylem and phloem tissue. The xylem permits water and mineral salts to enter the leaf and the phloem carries away sugar solution, usually in the form of sucrose.

2. A network of small veins is found throughout the leaf. These ensure that no cell is ever far from a xylem vessel or phloem sieve tube, and hence all cells have a constant supply of water for photosynthesis and a means of removing the sugars they produce. The xylem, and any sclerenchyma associated with the vascular bundle, also provide a framework of support for the leaf, helping it to present maximum surface area to the light.

(f) Stoma (VS)

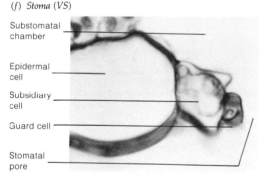

Substomatal chamber
Epidermal cell
Subsidiary cell
Guard cell
Stomatal pore

(h) Surface view of stomata

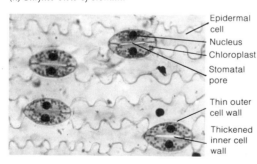

Epidermal cell
Nucleus
Chloroplast
Stomatal pore
Thin outer cell wall
Thickened inner cell wall

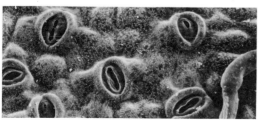

Stomata in surface view (scanning EM) (× 600 approx.)

23.2 Mechanism of light absorption

There are three features of light which make it biologically important:

1. Spectral quality (colour).

2. Intensity (brightness).

3. Duration (time).

To be of use as an energy source for organisms, light must first be converted to chemical energy. Radiant energy comes in discrete packets called **quanta** (Planck's quantum theory). A single quantum of light is called a **photon**. Light also has a **wave** nature and so forms a part of the electromagnetic spectrum. Visible light represents that part of this spectrum which has a wavelength between 400 nm (violet) and 700 nm (red).

The wavelength of light is the distance between successive peaks along a wave (Fig. 23.2a) and this is inversely proportional to its frequency, i.e. the smaller the wavelength, the greater the frequency. The amount of energy is inversely proportional to the wavelength, i.e. light with a short wavelength has more energy than light with longer wavelengths. In other words, a photon of blue light (wavelength 400 nm) has more energy than a photon of red light (wavelength 700 nm).

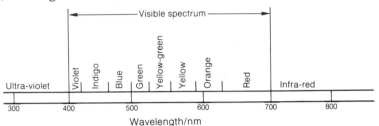

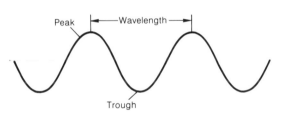

Fig. 23.2(a) Wave function of light

Fig. 23.2(b) The visible section of the electromagnetic spectrum

23.2.1 The photosynthetic pigments

A number of pigments are involved in photosynthesis of which **chlorophyll** is by far the most important. There are a number of different chlorophylls with chlorophylls a and b being the most common. Chlorophylls absorb light in the blue-violet and the red regions of the visible spectrum. The remaining light, in the green region of the spectrum, is reflected and gives chlorophyll its characteristic colour. All chlorophylls comprise a complex ring system called a **porphyrin ring** and a long hydrocarbon 'tail' (Fig. 23.3a). This 'tail' is lipid soluble (hydrophobic) and is therefore embedded in the thylakoid membrane. The porphyrin ring is hydrophilic and lies on the membrane surface.

There is a second group of pigments involved in photosynthesis – the **carotenoids**. There are many types but their basic structure comprises two small rings linked by a long hydrocarbon chain (Fig. 23.3b). The colour, which ranges from pale yellow through orange to red, depends upon the number of double bonds in the chain. The greater the number of double bonds, the deeper the colour.

The colour of carotenoids is normally masked in photosynthetic tissues by chlorophyll. Their colour does, however, become apparent when chlorophyll breaks down prior to leaf fall. The characteristic red, orange and yellow colour of leaves in the autumn is attributable to carotenoids as are many flower and fruit colours. They absorb

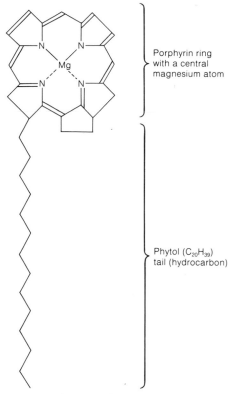

Fig. 23.3(a) *The general shape of the chlorophyll molecule*

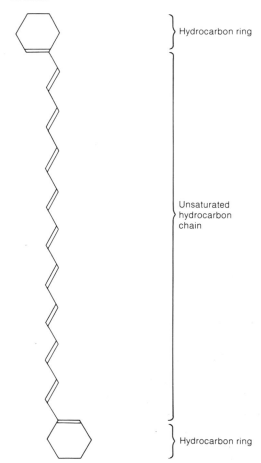

Fig. 23.3(b) *The general shape of a carotenoid molecule*

light in the blue-violet range of the spectrum. There are two main types of carotenoids: the **carotenes** and the **xanthophylls**. A common example of a carotene is β-carotene which gives carrots their familiar orange colour. It is easily formed into two molecules of vitamin A, making carrots a useful source of this vitamin to animals.

23.2.2 Absorption and action spectra

If a pigment such as chlorophyll is subjected to different wavelengths of light, it absorbs some more than others. If the degree of absorption at each wavelength is plotted, an **absorption spectrum** of that pigment is obtained. The absorption spectra for chlorophylls a and b are given in Fig. 23.4.

An **action spectrum** plots the biological effect of different wavelengths of light – in this case the effectiveness of different wavelengths of light in bringing about photosynthesis. As Fig. 23.4 shows, the action spectrum for photosynthesis is closely correlated to the absorption spectra for chlorophylls a and b and carotenoids. This suggests these pigments are those responsible for absorbing the light used in photosynthesis.

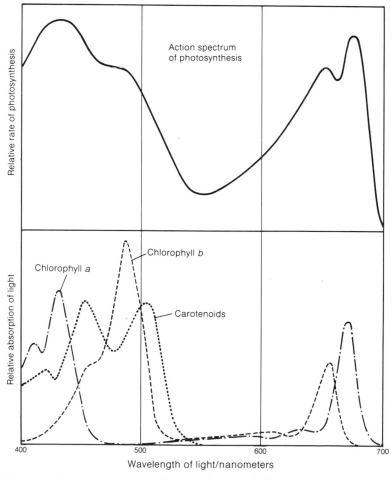

Fig. 23.4 *Action spectrum for photosynthesis and absorption spectra for common plant pigments*

23.3 Mechanism of photosynthesis

The overall equation for photosynthesis is:

$$6CO_2 + 6H_2O \xrightarrow[\text{chlorophyll}]{\text{sunlight}} C_6H_{12}O_6 + 6O_2$$

$$\text{carbon dioxide} + \text{water} \longrightarrow \text{glucose} + \text{oxygen}$$

Experiments show that the rate of photosynthesis is affected not only by light intensity but also by temperature (Section 23.4.4). As temperature does not affect processes such as the action of light on chlorophyll, this suggests that a second, purely chemical stage is involved, which is temperature-sensitive. By subjecting plants to short flashes of light separated by dark periods, it can be shown that the amount of photosynthesis is actually greater than if continuous light is provided. This also suggests that there are two parts to photosynthesis. The **light stage** involves light and chlorophyll which produces a particular product. This product is used in the second part of the process, the temperature-sensitive **dark stage**. In continuous light, the product from the light stage accumulates and so slows down that process. Periods of darkness allow the slower dark stage time to use up the product and so speed up the overall process. This explains why alternate flashes of light yield more photosynthetic product than continuous light. It must, however, be pointed out that the dark stage, despite its name, can occur in light or dark conditions. Its name simply indicates that light is *not* essential.

23.3.1 Light stage (photolysis)

The light stage of photosynthesis occurs in the grana of the chloroplasts and involves the splitting of water by light – **photolysis of water**. In the process, ADP is converted to ATP. This addition of phosphorus is termed **phosphorylation** and as light is involved it is called **photophosphorylation**. These processes are brought about by two photochemical systems which are summarized in Fig. 23.5.

1. Light energy is trapped in pigment system II and boosts electrons to a higher energy level.
2. The electrons are received by an electron acceptor.
3. The electrons are passed from the electron acceptor along a series of electron carriers to pigment system I which is at a lower energy level. The energy lost by the electrons is captured by converting ADP to ATP. Light energy has thereby been converted to chemical energy.
4. Light energy absorbed by pigment system I boosts the electrons to an even higher energy level.
5. The electrons are received by another electron acceptor.
6. The electrons which have been removed from the chlorophyll are replaced by pulling in other electrons from a water molecule.
7. The loss of electrons from the water molecule causes it to dissociate into protons and oxygen gas.
8. The protons from the water molecule combine with the electrons from the second electron acceptor and these reduce **nicotinamide adenine dinucleotide phosphate**.
9. Some electrons from the second acceptor may pass back to the chlorophyll molecule by the electron carrier system, yielding ATP as they do so. This process is called **cyclic photophosphorylation**.

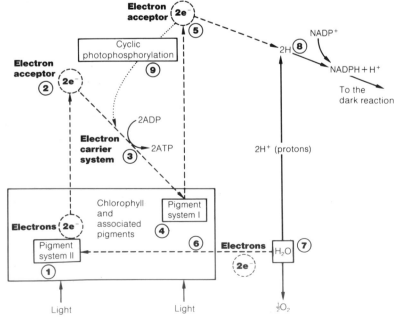

Fig. 23.5 Summary of the light stage of photosynthesis

1. *Carbon dioxide diffuses into the leaf through the stomata and dissolves in the moisture on the walls of the palisade cells. It diffuses through the cell membrane, cytoplasm and chloroplast membrane into the stroma of the chloroplast.*

2. *The carbon dioxide combines with a 5-carbon compound called **ribulose bisphosphate** to form an unstable 6-carbon intermediate.*

3. *The 6-carbon intermediate breaks down into two molecules of the 3-carbon **glycerate 3-phosphate** (**GP**).*

4. *Some of the ATP produced during the light stage is used to help convert GP into **triose phosphate** (glyceraldehyde 3-phosphate – GALP).*

5. *The reduced nicotinamide adenine dinucleotide phosphate ($NADPH + H^+$) from the light reaction is necessary for the reduction of the GP to triose phosphate. $NADP^+$ is regenerated and this returns to the light stage to accept more hydrogen.*

6. *Pairs of triose phosphate molecules are combined to produce an intermediate hexose sugar.*

7. *The hexose sugar is polymerized to form starch which is stored by the plant.*

8. *Not all triose phosphate is combined to form starch. A portion of it is used to regenerate the original carbon dioxide acceptor, ribulose bisphosphate. Five molecules of the 3-carbon triose phosphate can regenerate three molecules of the 5-carbon ribulose bisphosphate. More of the ATP from the light reaction is needed to provide the energy for this conversion.*

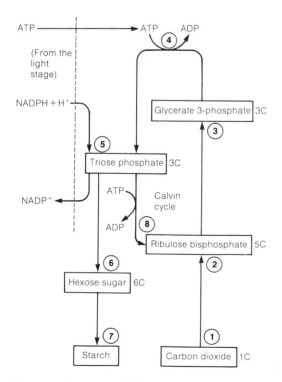

Fig. 23.6 Summary of the dark stage of photosynthesis

In the process summarized in Fig. 23.5, electrons from chlorophyll are passed into the dark reaction via $NADPH + H^+$. They are replaced by electrons from another source – the water molecule. The same electrons are *not* recycled back into the chlorophyll. This method of ATP production is thus called **non-cyclic photophosphorylation**.

There is a second method by which ATP can be generated. The electrons from the pigment system may return to the chlorophyll directly, via the electron carrier system, forming ATP in the process. Such electrons are recycled, harnessing energy from light and generating ATP. This is called **cyclic photophosphorylation**.

23.3.2 The dark stage

The dark stage of photosynthesis occurs in the stroma of the chloroplasts. It is referred to as the **light independent** stage as it takes place whether or not light is present. The details of the dark stage were analysed by Melvin Calvin and his co-workers and the process is often called the **Calvin cycle** (Figs. 23.6 and 23.7). It is basically the reduction of carbon dioxide using the reduced nicotinamide adenine dinucleotide phosphate ($NADPH + H^+$) and ATP from the light reaction.

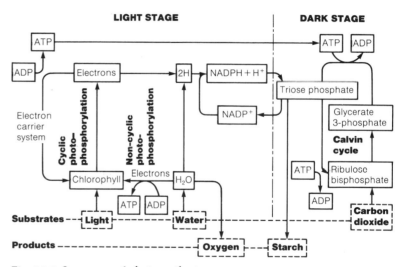

Fig. 23.7 Summary of photosynthesis

23.3.3 Alternative ways of reducing carbon dioxide – the C_4 plants

When plants are exposed to radioactively labelled carbon dioxide ($^{14}CO_2$) and allowed to photosynthesize, the first radioactively labelled organic product is normally glycerate 3-phosphate (GP). This is a 3-carbon compound. A few plants, notably sugar cane and maize, produce instead a radioactively labelled 4-carbon compound, **oxaloacetate**. These plants are therefore known as **C_4 plants** whereas those producing the more normal 3-carbon glycerate 3-phosphate (GP) are called **C_3 plants**.

C_4 plants use a different carbon dioxide acceptor. Instead of ribulose bisphosphate they use **phosphoenol pyruvate (PEP)**. The enzyme they use to combine the carbon dioxide with PEP is called **PEP carboxylase**. The main advantage of this enzyme is its great affinity for carbon dioxide. This allows it to 'trap' carbon dioxide efficiently even at very low concentrations. The system is summarized in Fig. 23.8, over the page.

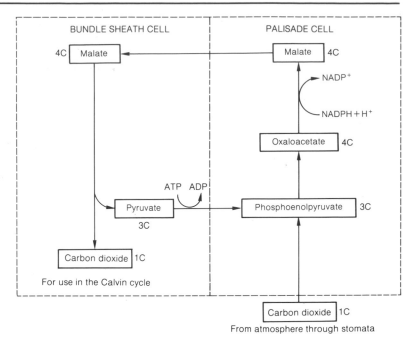

Fig. 23.8 Summary of carbon dioxide fixation in C_4 plants
(the Hatch–Slack pathway)

The C_4 system operates in addition to, and not instead of, the C_3 system. It has a number of advantages:

1. Photosynthesis can proceed at very low carbon dioxide concentrations – The great affinity of PEP carboxylase for carbon dioxide means that photosynthesis can still take place even when carbon dioxide is in short supply, e.g. in a tropical forest during the day when the demand for carbon dioxide by the plants leads to a temporary fall in its concentration. The C_4 pathway is more efficient at higher temperatures, making it especially useful to tropical plants.

2. Carbon dioxide can be temporarily stored for later use – The 4-carbon oxaloacetate produced may be converted to malate. This can be stored by the plant and later broken down into pyruvate releasing the carbon dioxide for use in the Calvin cycle in the normal way. This process allows plants to accumulate a store of carbon dioxide when it is in relative abundance and store it for later release when external supplies are reduced. This is particularly useful to plants living in situations where water is scarce at certain times. During these periods of drought, stomata are kept closed in the day to prevent excessive water loss. To permit photosynthesis to take place at these times, carbon dioxide is absorbed when stomata are open, stored as malate and released when the stomata are closed. The pyruvate produced is converted to PEP, ready to accept more carbon dioxide the following night.

3. Photorespiration is avoided – Photorespiration (Section 23.5) is a wasteful process which occurs in C_3 plants and reduces their photosynthetic efficiency. The biochemical pathways of C_4 plants prevent photorespiration occurring, making it more efficient at photosynthesis.

The leaf anatomy of C_4 plants is modified to suit it to its different method of carbon dioxide fixation (Fig. 23.9). Palisade cells are arranged around the bundle sheath of the vascular bundles. The bundle sheath cells are larger than normal and unusual in having chloroplasts. The fixation of carbon dioxide by PEP takes place in the palisade cells and the resulting C_4 compounds (e.g. malate) are passed to the bundle sheath cells. It is here that the carbon dioxide is released and used by the chloroplasts which carry out the Calvin cycle.

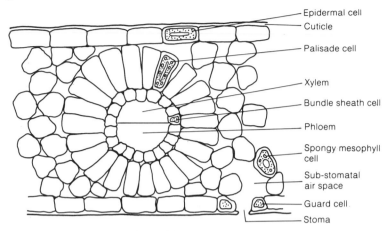

Fig. 23.9 *Arrangement of leaf cells in a C_4 plant*

23.3.4 Fate of photosynthetic products

From the products of photosynthesis a totally autotrophic plant must synthesize all organic materials necessary for its survival. Many of the raw materials for the synthesis of these organic molecules come from intermediates of the Calvin cycle.

Synthesis of other carbohydrates
The triose phosphate of the Calvin cycle can be synthesized into hexose sugars such as glucose and fructose by a reversal of the stages of glycolysis which occur during respiration (Section 26.2). The monosaccharides glucose and fructose may be combined to give the disaccharide **sucrose** which is the main form in which carbohydrate is transported throughout the plant in the phloem. The glucose may, on the other hand, be polymerized into **starch** for storage. In some plants, e.g. *Dahlia*, it is the fructose which is polymerized, in which case the storage polysaccharide formed is **inulin**. The glucose may alternatively be polymerized into another polysaccharide, **cellulose**. This makes up over 50% of plant cell walls. The chemical structure of polysaccharides is given in Section 3.5.

Synthesis of lipids
As we saw in Section 3.7, all lipids are esters of fatty acids, with glycerol as the most common alcohol found. Glycerate 3-phosphate (GP) may be converted to acetyl coenzyme A which in turn is used to synthesize a variety of fatty acids in chloroplasts as well as cytoplasm. Triose phosphate is easily converted into glycerol. The lipids are formed by combination of appropriate fatty acids with glycerol. As well as being important storage substances, especially in seeds, lipids are a major constituent of cell membranes and their waxy derivatives make up the waterproofing cuticle. Fatty acids provide some flower scents which are used to attract insects for pollination.

Synthesis of proteins
Conversion of the glycerate 3-phosphate of the Calvin cycle into acetyl coenzyme A is the starting point for amino acid synthesis. The acetyl CoA enters the Krebs cycle and from its intermediates a

wide variety of amino acids can be made by transamination reactions. The nitrogen necessary for these reactions is derived from the nitrates absorbed by plant roots and subsequently reduced. The amino acids are polymerized into proteins. Details of all these processes are given in Section 12.6.

Proteins are essential for growth and development and make up a major structural component of the cell, especially the cell membrane. All enzymes are proteins and they may also be used as storage material.

23.4 Factors affecting photosynthesis

The rate of photosynthesis is affected by a number of factors, the level of which determine the yield of material by a plant. Before reviewing these factors it is necessary to understand the principle of limiting factors.

23.4.1 Concept of limiting factors

In 1905, F. F. Blackman, a British plant physiologist, measured the rate of photosynthesis under varying conditions of light and carbon dioxide supply. As a result of his work he formulated the **principle of limiting factors**. It states: **At any given moment, the rate of a physiological process is limited by the one factor which is in shortest supply, and by that factor alone**.

In other words, it is the factor which is nearest its minimum value which determines the rate of a reaction. Any change in the level of this factor, called the **limiting factor**, will affect the rate of the reaction. Changes in the level of other factors have no effect. To take an extreme example, photosynthesis cannot proceed in the dark because the absence of light limits the process. The supply of light will alter the rate of photosynthesis – more light, more photosynthesis. If, however, more carbon dioxide or a higher

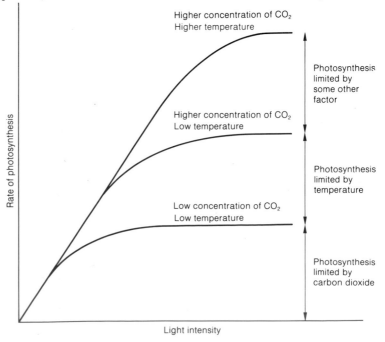

Fig. 23.10 *The concept of limiting factors as illustrated by the levels of different conditions on the rate of photosynthesis*

temperature is supplied to a plant in the dark, there will be no change in the rate of photosynthesis. Light is the limiting factor, therefore only a change in its level can affect the rate.

If the amount of light given to a plant is increased, the rate of photosynthesis increases up to a point and then tails off. At this point some other factor, such as the concentration of carbon dioxide, is in short supply and so limits the rate. An increase in carbon dioxide concentration again increases the amount of photosynthesis until some further factor, e.g. temperature, limits the process. These changes are illustrated in Fig. 23.10.

23.4.2 Effect of light intensity on the rate of photosynthesis

The rate of photosynthesis is often measured by the amount of carbon dioxide absorbed or oxygen evolved by a plant. These forms of measurement do not, however, give an absolute measure of photosynthesis because oxygen is absorbed and carbon dioxide is evolved as a result of cellular respiration. As light intensity is increased, photosynthesis begins, and some carbon dioxide from respiration is utilized in photosynthesis and so less is evolved. With a continuing increase in light intensity a point is reached where carbon dioxide is neither evolved nor absorbed. At this point the carbon dioxide produced in respiration exactly balances that being used in photosynthesis. This is the **compensation point**. Further increases in light intensity result in a proportional increase in the rate of photosynthesis until **light saturation** is reached. Beyond this point further increases in light intensity have no effect on the rate of photosynthesis. If, however, more carbon dioxide is made available

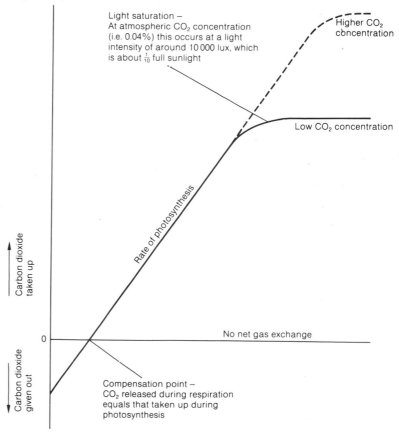

Fig. 23.11 Graph showing the effect of light intensity on the rate of photosynthesis, as measured by the amount of CO_2 exchanged

(c) Explain how **one named** radioactive isotope may be used to investigate the cycle you have described above. (3 marks)

(d) Name **two** types of molecule, other than carbohydrates, which are used for storage in plants. (2 marks)

(Total 19 marks)

Oxford and Cambridge Board June 1988, Paper I, No. 2

5. Apparatus used in an investigation to establish the sequence of biochemical changes in photosynthesis is illustrated below.

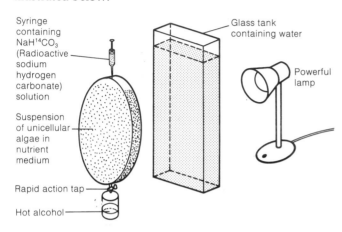

Syringe containing NaH^{14}CO$_3$ (Radioactive sodium hydrogen carbonate) solution

Glass tank containing water

Powerful lamp

Suspension of unicellular algae in nutrient medium

Rapid action tap

Hot alcohol

(a) Suggest **one** reason for each of the following.
(i) the flattened shape of the glass vessel containing the algae
(ii) the water tank
(iii) the hot alcohol
(iv) the NaH^{14}CO$_3$ solution (4 marks)

(b) Using this apparatus, explain how you would obtain a sample of algae that could be used to investigate the first products of photosynthesis. (2 marks)

(c) State what techniques would be used for
(i) the separation and identification of the photosynthetic products.
(ii) the estimation of the relative ^{14}C content of these products. (4 marks)

The results of this investigation are shown in the table below, which gives the ^{14}C content (in μ moles cm^{-3}) of four organic compounds (**A** to **D**) after five different periods of photosynthesis.

Compound	Time, in seconds, allowed for photosynthesis				
	5	15	60	180	600
A	0.3	2.5	6.2	10.3	7.9
B	1.0	2.0	3.1	3.2	3.2
C	0.05	0.11	0.16	1.0	1.0
D	0.01	0.02	0.08	0.17	1.7

(d) (i) Use the data in the table to place the compounds **A** to **D** in the order in which they would be formed.

$\longrightarrow$ $\longrightarrow$ $\longrightarrow$

(ii) Using your knowledge of photosynthesis, suggest a reason why the level of compound **B** remained steady in later samples.

(3 marks)

(Total 13 marks)

Joint Matriculation Board June 1989, Paper IA, No. 6

6. (a) The equation for photosynthesis is often cited as

$$6CO_2 + 6H_2O \longrightarrow C_6H_{12}O_6 + 6HO_2$$

Give **four** reasons for arguing that this equation is an over-simplification of the process.

(8 marks)

(b) Giving a reason in each case for your answer, state the expected effect of a 10°C rise in temperature on the rate of photosynthesis in
(i) very low light intensities
(ii) high light intensities. (4 marks)

(c) (i) Define the term 'compensation point'.
(ii) Plants show different compensation points. Suggest **one** reason for this. (4 marks)

(d) The assimilation number (rate of photosynthesis mg^{-1} chlorophyll) was calculated for two varieties of elm; a so-called 'green' and a 'yellow' variety. The results are given below.

Variety	mg chlorophyll in 10 g fresh mass of leaves	Assimilation number
green	16.2	6.9
yellow	1.2	82.0

Comment on the significance of these results.

(3 marks)

(Total 19 marks)

Oxford and Cambridge Board July 1984, Paper I, No. 6

7. Investigations were carried out using two strains of the same species of unicellular alga, one of which was a mutant that could not survive long periods of intense illumination. Light of known wavelength was passed through a tube containing the alga and measurements were taken both of the oxygen produced and of the light transmitted. The experimental arrangement is represented in the following diagram:

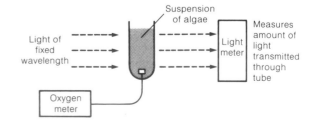

Suspension of algae

Light of fixed wavelength

Light meter

Measures amount of light transmitted through tube

Oxygen meter

The results obtained were used to plot the absorption and action spectra for each strain of alga. These are shown below:

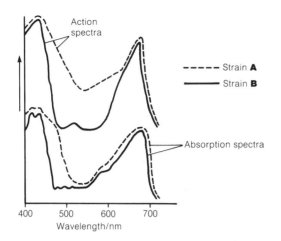

(a) (i) What is meant by the term *action spectrum*?
   (ii) What information from the experiment would have been used to plot the action spectra? *(2 marks)*

(b) (i) The amount of light **transmitted** through tubes without algae in them was 100%. Suggest how the figures plotted for the *absorption spectra* were derived from the results obtained with the light meter.
   (ii) Apart from temperature and pH (which have little effect), state **two** factors which should be standardised when using the apparatus shown above to measure the absorption spectra. *(3 marks)*

Information about the photosynthetic pigments found in these unicellular algae is given in the table below.

| Pigment | Absorption maxima (nm) | Rf values | |
		solvent I	solvent II
P	620	0.20	0.89
Q	545 and 575	0.60	0.29
R	420 and 660	0.65	0.11
S	490	0.91	0.19
T	430 and 645	0.82	0.92

The pigments from each strain were extracted and separated using two-dimensional paper chromatography, to give the chromatograms shown.

(c) Explain clearly the advantage of using two-dimensional rather than one-dimensional chromatography in separating these pigments. *(1 mark)*

(d) One of the strains of alga lacks one of the pigments.
   (i) Draw a box around this pigment on the appropriate chromatogram.
   (ii) Give the letter of the pigment concerned.

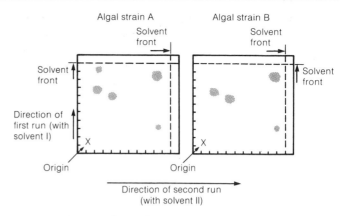

(iii) Explain what other evidence is available to confirm the absence of this pigment from one of the strains. *(5 marks)*

(e) From the information provided in the question, suggest the physiological rôle of the pigment in this species of alga. *(2 marks)*
(Total 13 marks)

Joint Matriculation Board June 1988, Paper IA, No. 6

8. (a) The graphs shown below illustrate the results obtained in experiments performed to find out what effects increased carbon dioxide levels could have on the rate of photosynthesis in one type of annual crop plant. All experiments were carried out in controlled conditions at 16 °C. The amount of artificial light given to the plants could be varied.

Rate of photosynthesis (as measured by evolution of oxygen) plotted against light intensity for three groups of plants supplied with different amounts of carbon dioxide.

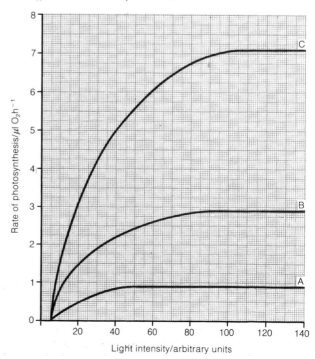

Group A: with 0.01% (by vol.) carbon dioxide.
Group B: with 0.03% (by vol.) carbon dioxide.
Group C: with 0.15% (by vol.) carbon dioxide.

(i) At what light intensity do those plants given 0.03% CO_2 reach their maximum rate of photosynthesis under these conditions? *(1 mark)*

(ii) Suggest a reason for the form of graph A beyond 54 units of light. *(1 mark)*

(iii) Suggest a reason for the form of graph C beyond 54 units of light. *(1 mark)*

(iv) Which graph best indicates the rate of photosynthesis one might expect to obtain from this species of plant in air at 16 °C? *(1 mark)*

(v) Suggest why the graphs do not commence at the origin. *(1 mark)*

(vi) State two possible ways (other than amount of evolved oxygen) by which the rate of photosynthesis of the plants might have been measured. *(2 marks)*

(vii) Suppose that, with Group B plants, the temperature had been increased by 10 °C at 30 light units and above. Draw on the graph, with a dotted line, any difference you might expect to the given curve. *(2 marks)*

(b) If the light intensity is reduced to a very low level the plants will be below the *compensation point*. Explain what is meant by this term.
(2 marks)
(Total 11 marks)

Oxford Local June 1988, Paper I, No. 8

9. Indigo carmine is a blue water-soluble dye. It can be decolourized by adding a few drops of a 10 per cent solution of sodium dithionite. The blue colour can be restored by gently blowing into the solution using a straw.

$$\text{indigo carmine (blue)} \underset{\text{oxygen}}{\overset{\text{sodium dithionite}}{\rightleftarrows}} \text{indigo carmine (colourless)}$$

(a) Draw and label a simple experimental arrangement by which you could demonstrate that this particular colour change is **not** caused by carbon dioxide present in the breath.
(3 marks)

An experiment using the indigo carmine indicator system investigated the effect of hydrogencarbonate concentration on the rate of photosynthesis at two different light intensities. An aquatic plant was used in the experiment.

The diagram shows the contents of a typical tube used in the experiment.

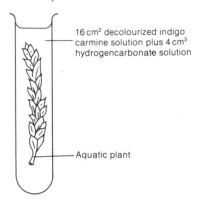

16 cm² decolourized indigo carmine solution plus 4 cm³ hydrogencarbonate solution

Aquatic plant

(b) (i) Suggest **one** problem you would need to overcome *when setting up* this experiment to ensure the contents of each tube are comparable.

(ii) Describe briefly how you would attempt to overcome the problem you suggest.

(iii) Assuming that all tubes *were* comparable at the start of the experiment, identify **one** possible source of inaccuracy in obtaining results.

(iv) Describe **one** simple modification to the experiment which might increase its accuracy. *(4 marks)*

The results of the experiment are shown in the following table.

Distance of tube from light source (cm)	20				10			
Concentration of hydrogencarbonate solution added (per cent)	0.25	0.5	0.75	1.0	0.25	0.5	0.75	1.0
Time taken for tube contents to turn a standard blue (minutes)	10.0	5.75	5.0	5.0	8.0	2.5	1.25	1.25
Rate of photosynthesis (relative to the slowest time)	1							

(c) (i) Calculate the missing values for the rate of photosynthesis (relative to the slowest time) and write them in the table.

(ii) Give any **two** conclusions which can be drawn from the results. *(6 marks)*
(Total 13 marks)

Joint Matriculation Board June 1985, Paper IA, No. 2

24 *Heterotrophic nutrition 1 – holozoic (feeding and digestion)*

24.1 Forms of heterotrophic nutrition

Heterotrophic organisms consume complex organic food material. This food originates with autotrophic organisms which synthesize it from simple inorganic raw materials. There are three main forms of heterotrophic nutrition:

1. Holozoic nutrition – Involves the consumption of complex food which is broken down inside the organism into simple molecules which are then absorbed. Most animals feed in this way, utilizing a specialized digestive system. Insectivorous plants are partly holozoic.

2. Saprophytic nutrition – Involves the consumption of complex organic food from the bodies of dead organisms. The food is either already in a soluble form or it is digested externally into simple molecules which then diffuse into the saprophyte. There is no digestive system. Some bacteria and fungi feed in this way.

3. Parasitic nutrition – Involves feeding on complex organic food derived from other living organisms. It is usually obtained in soluble form and so if a digestive system is present it is very simple. A few parasites ingest solid food and therefore possess digestive systems.

Saprophytic and parasitic nutrition are dealt with in Chapter 25.

Holozoic nutrition
Holozoic organisms obtain their energy from the consumption of complex organic food which is digested within their bodies. Their nutrition involves most, if not all, of the following stages:

1. Obtaining the food – May involve movement of the organism to the food source.

2. Ingestion – Organisms use a variety of feeding mechanisms which depend upon the size and nature of the food.

3. Physical (mechanical) digestion – By means of a variety of structures including teeth, radula and gizzard.

4. Chemical digestion – A process largely carried out by enzymes.

5. Absorption – Useful soluble materials must be absorbed from the digestive system into the body tissues.

6. Assimilation – The materials absorbed must enter individual cells and be incorporated into them.

7. Elimination (egestion) – Unwanted material which has been ingested must be removed from the body.

Holozoic organisms can be classified according to the type of food ingested. Organisms which feed on plant material are called **herbivores**, while those feeding on other animals are called **carnivores**. Organisms that feed on a diet combining plant and animal material are called **omnivores**. Some holozoic organisms consume liquid material and are known as **fluid feeders**, but the majority take in particles of solid food and are known as **phagotrophs**. Phagotrophs may take in very small particles, usually by a filtering mechanism, in which case they are called **microphagous** feeders. Alternatively, the particles ingested are relatively large in which case the organisms are called **macrophagous** feeders.

24.2 Diet

All organisms require a constant supply of essential nutrients. What these nutrients are and the amounts of each required by an organism varies from species to species. In mammals **carbohydrates** and **fats** are needed in relatively large quantities as sources of energy, and **proteins** are needed in large amounts for growth and repair. **Vitamins** and **minerals** are required in much smaller quantities for a variety of specific functions. **Water** is a vital constituent of the diet and **roughage** is necessary for efficient digestion.

24.2.1 Carbohydrates and fats (energy requirements)

Details of the chemistry of carbohydrates and fats is given in Chapter 3. The main function of both is to provide energy. The amount of energy in food is expressed in **joules**. (It was previously measured in calories, a term still commonly used in books and magazines concerned with diet. One calorie is equal to 4.18 joules.) To measure the amount of energy in different foods, a given mass is burned in

TABLE 24.1 **Recommended daily intake of energy according to age, activity and sex**

Age/years	Average body weight/kg	Degree of activity/ circumstances	Energy requirement/kJ	
			Male	Female
1	7	Average	3200	3200
5	20	Average	7500	7500
10	30	Average	9500	9500
15	45	Average Sedentary	11 500 11 300	11 500 9000
25	65 (male) 55 (female)	Moderately active Very active Sedentary	12 500 15 000 11 000	9500 10 500 9000
50	65 (male) 55 (female)	Moderately active Very active	12 000 15 000	9500 10 500
75	63 (male) 53 (female)	Sedentary	9000	8000
Any	—	During pregnancy	—	10 000
Any	—	Breast feeding	—	11 500

oxygen in a piece of apparatus called a **bomb calorimeter**. The total heat generated gives a measure of the food's energy content (also called its **calorific value**).

The energy required by an organism varies with sex, size, age and activity. Table 24.1 provides examples of the recommended daily energy intake for humans of various ages. Ideally two thirds of this should be derived from carbohydrates and the remainder from fats.

Much attention has been focused recently on the correlation between a high fat intake in the diet and heart disease. It is always difficult to draw relationships directly between one type of food and the incidence of a specific disease because foods contain a wide variety of substances. In addition, factors such as exercise, stress and smoking affect an individual's health and the way food is utilized. It does, however, seem that a high intake of fats, especially saturated fats (see Section 3.7.1), is a contributory factor in causing heart disease.

TABLE 24.2 **Vitamins required in the human diet**

Vitamin/name	Fat/water soluble	Major food sources	Function	Deficiency symptoms
A_1 Retinol	Fat soluble	Liver, vegetables, fruits, dairy foods	Maintains normal epithelial structure. Needed to form visual pigments	Dry skin. Poor night vision
B_1 Thiamin	Water soluble	Liver, legumes, yeast, wheat and rice germ.	Coenzyme in cellular respiration	Nervous disorder called beri-beri. Neuritis and mental disturbances. Heart failure
B_2 Riboflavin	Water soluble	Liver, yeast, dairy produce	Coenzymes (flavo-proteins) in cellular respiration	Soreness of the tongue and corners of the mouth
B_3 (pp factor) Niacin	Water soluble	Liver, yeast, wholemeal bread	Coenzyme (NAD, NADP) in cellular metabolism	Skin lesions known as pellagra. Diarrhoea
B_5 Pantothenic acid	Water soluble	Liver, yeast, eggs	Forms part of acetyl coenzyme A in cellular respiration	Neuromotor disorders, fatigue and muscle cramps
B_6 Pyridoxine	Water soluble	Liver, kidney, fish	Coenzymes in amino acid metabolism	Dermatitis. Nervous disorders
B_{12} Cyanocobalamine	Water soluble	Meat, eggs, dairy food	Nucleoprotein (RNA) synthesis. Needed in red blood cell formation	Pernicious anaemia. Malformation of red blood cells
Biotin	Water soluble	Liver, yeast. Synthesized by intestinal bacteria	Coenzymes in carboxylation reactions	Dermatitis and muscle pains
Folic acid	Water soluble	Liver, vegetables, fish	Nucleoprotein synthesis. Red blood cell synthesis	Anaemia
C Ascorbic acid	Water soluble	Citrus fruits, tomatoes, potatoes	Formation of connective tissues, especially collagen fibres	Non-formation of connective tissues. Bleeding gums – scurvy
D Calciferol	Fat soluble	Liver, fish oils, dairy produce. Action of sunlight on skin	Absorption and metabolism of calcium and phosphorus, therefore important in formation of teeth and bones	Defective bone formation known as rickets
E Tocopherol	Fat soluble	Liver, green vegetables	Function unclear in humans. In rats it prevents haemolysis of red blood cells	Anaemia
K Phylloquinone	Fat soluble	Green vegetables. Synthesized by intestinal bacteria	Blood clotting	Failure of blood to clot

24.2.2 Proteins

The chemistry of proteins is given in Section 3.9. As a last resort, the body may respire proteins to provide energy, but their main function is as a source of amino acids which are used to synthesize new proteins. These proteins are used in metabolism, growth and repair. Plants are able to synthesize all their own amino acids but animals are more limited. Humans, for example, require nine amino acids, called **essential amino acids**, in the diet. Although plant food contains proportionately fewer proteins, a properly balanced vegetable diet can nevertheless provide all the essential amino acids. It is only where there is a dependence on just one or two plant foods as sources of proteins that malnutrition results.

24.2.3 Vitamins

Vitamins are a group of essential organic compounds which are needed in small amounts for normal growth and metabolism. If the diet lacks a particular vitamin, a disorder called a **deficiency disease** results. The vitamins required vary from species to species. Table 24.2 lists those needed in a human diet and the roles they play. Vitamins are normally classified as **water soluble** (vitamins C and the B complex) or **fat soluble** (vitamins A, D, E and K). Whereas

TABLE 24.3 **Some essential minerals required in the human diet**

Mineral	Major food source	Function
Macronutrients Calcium (Ca^{2+})	Dairy foods, eggs, green vegetables	Constituent of bones and teeth, needed in blood clotting and muscle contraction. Enzyme activator
Chlorine (Cl^-)	Table salt	Maintenance of anion/cation balance. Formation of hydrochloric acid
Magnesium (Mg^{2+})	Meat, green vegetables	Component of bones and teeth. Enzyme activator
Phosphate (PO_4^{3-})	Dairy foods, eggs, meat, vegetables	Constituent of nucleic acids, ATP, phospholipids (in cell membranes), bones and teeth
Potassium (K^+)	Meat, fruit and vegetables	Needed for nerve and muscle action and in protein synthesis
Sodium (Na^+)	Table salt, dairy foods, meat, eggs, vegetables	Needed for nerve and muscle action. Maintenance of anion/cation balance
Sulphate (SO_4^{2-})	Meat, eggs, dairy foods	Component of proteins and coenzymes
Micronutrients (trace elements) Cobalt (Co^{2+})	Meat	Component for vitamin B_{12} and needed for the formation of red blood cells
Copper (Cu^{2+})	Liver, meat, fish	Constituent of many enzymes. Needed for bone and haemoglobin formation
Fluorine (F^-)	Many water supplies	Improves resistance to tooth decay
Iodine (I^-)	Fish, shellfish, iodized salt	Component of the growth hormone, thyroxine
Iron (Fe^{2+})	Liver, meat, green vegetables	Constituent of many enzymes, electron carriers, haemoglobin and myoglobin
Manganese (Mn^{2+})	Liver, kidney, tea and coffee	Enzyme activator and growth factor in bone development
Molybdenum (Mo^{4+})	Liver, kidney, green vegetables	Required by some enzymes
Zinc (Zn^{2+})	Liver, fish, shellfish	Enzyme activator, involved in the physiology of insulin

excess water-soluble vitamins are simply excreted in urine, fat-soluble vitamins tend to accumulate in fatty tissues of the body, and may even build up to lethal concentrations if taken in excess.

24.2.4 Minerals

The principal mineral ions required by plants and animals and their functions are listed in Chapter 3 in Table 3.1. The principal minerals required in a human diet, and their sources, are further summarized in Table 24.3.

24.2.5 Water

Water makes up about 70% of the total body weight of mammals and serves a wide variety of important functions which are discussed more fully in Section 33.1.8. Table 24.4 gives the daily water balance in a human not engaged in active work, i.e. there is no excessive sweating.

24.2.6 Roughage

Roughage is indigestible material which passes through the alimentary canal almost unchanged. As it does not cross an epithelial lining of the gut it never actually enters the body. It is not a metabolic product and its removal from the body is therefore called **egestion** or **elimination** and not excretion. Although it does not have a metabolic function, roughage is essential to the efficient functioning of the alimentary canal. It gives bulk to the material within the intestines, absorbing water and making the contents much more solid. In this form it stimulates peristalsis and is easier to move along the intestines. It thus helps prevent constipation and other intestinal disorders. Roughage, or **dietary fibre** as it is sometimes called, consists mostly of the cellulose cell walls of plants. In man, the removal of much roughage from processed food has led to an increase in intestinal disorders. This has shown the value of roughage and led to an emphasis on high-fibre diets as an aid to healthy living.

24.2.7 Milk

As milk is the only food received by mammals in the period after birth, it follows that it must provide all essential materials for growth and development. In this sense it is a balanced diet in itself. It cannot, however, sustain healthy development indefinitely for these reasons:
1. It contains little if any iron – This is no problem to a new-born baby as it accumulates iron from its mother before birth. This store cannot last indefinitely and alternative sources of iron are necessary in later life.
2. It contains no roughage – We saw in Section 24.2.6 the necessity of roughage and the problems associated with its long-term absence from the diet.
3. It contains a high proportion of fat – For a young, actively growing organism this is ideal, but as it grows the energy demand is reduced. This could lead to an increase in weight due to storage of the excess fat and a consequent increased risk of heart disease. For the early years, and as a supplement to the human diet in later life, milk nevertheless plays an invaluable rôle.

TABLE 24.4 **Human daily water balance**

Process	Water uptake /cm^3	Water output /cm^3
Drinking	1450	–
In food	800	–
From respiration	350	–
In urine	–	1500
In sweat	–	600
Evaporation from lungs	–	400
In faeces	–	100
TOTAL	2600	2600

24.2.8 Food additives

We are all exposed to food additives which have been used in increasing numbers and volumes during the past few decades. Most of the permitted additives are regulated by the European Community (EC) and labelled with the prefix E and a number. The EC coding groups classes of additives with similar functions together.

During recent years people have grown wary of the proliferation of 'E numbers' appearing on ingredients lists of prepared foods. Some manufacturers have countered opposition by listing the full names of additives rather than using their 'E numbers', but many have begun to reduce the number of artificial additives used.

It must be realized that not all additives are harmful. In many cases foods would be more susceptible to bacterial infections if they were not used and the health of the general population might suffer. Many of us would certainly have to change our life styles if all additives were eliminated. There would be fewer convenience foods and low-fat spreads, more foods would have to be sold close to their point of production or manufacture and shopping trips would need to be more frequent. However, a balance should be reached and there is no substitute for hygiene and adequate cooking. Preservatives may be needed but does it really matter what colour our food is? Tartrazine (E102) is one of the commonest food colourings, used in such things as orange juice, sauces and fish fingers, but it is now known that it can trigger hyperactivity and people with asthma suffer adversely.

A brief summary of the major groups of food additives is given below:

Colourings (E100–E180) – These are purely cosmetic and rarely add nutritional value; in fact they may disguise poor quality foods. They are banned from baby foods.

Preservatives (E200–E297) – These, perhaps the most easily justified additives, only make up 1% of all additives used. They do not spoil the texture, appearance or flavour as many old-fashioned methods of preservation do. However, they do mean that food can be kept bacteriologically safe for longer, even though its nutritional value may decline.

Antioxidants (E300–E321) – Antioxidants prevent oils and fats becoming rancid on contact with air. The natural antioxidant, vitamin E, is destroyed in processing and is therefore replaced by synthetic antioxidants.

Texture enhancers (E322–E495) – These include thickeners as well as emulsifiers and stabilizers. Emulsifiers are used to bind together fat and water, and stabilizers prevent them from separating out again. Together, they are useful in making low-fat spreads.

Synthetic flavourings – These make up a high proportion of additives used and they are vital to modern food processing. There are between 3000 and 6000 in use today. They are not subject to any regulation and do not have to be identified in detail on labels. They are used to replace the natural flavour lost during processing, and their use means that flavour is no longer an infallible guide to the food's quality.

Flavour enhancers and sweeteners (E620–E637) – Flavour enhancers have no flavour of their own but they make the flavours

TABLE 24.5 **Classification of animal feeding mechanisms (based on the classification by C. M. Yonge)**

A. Mechanisms for dealing with small particles
1. Pseudopodial   4. Tentacles
2. Flagella   5. Mucus
3. Cilia   6. Setae

B. Mechanisms for dealing with large particles
1. Swallowing inactive food
2. Scraping and boring
3. Seizing prey
  (a) Seizing and swallowing only
  (b) Seizing and masticating before swallowing
  (c) Seizing and externally digesting, before swallowing

C. Mechanisms for dealing with fluids and soft tissues
1. Piercing and sucking
2. Sucking only
3. Absorbing through the general body surface

of other food stronger. The best known is **monosodium glutamate**, which has been used by the Chinese for centuries. Sweeteners are used in nearly all processed foods, both sweet and savoury, and about 60% of the sugar consumed in the UK is in processed foods.

24.3 Feeding mechanisms

In attempting to classify the feeding mechanisms of holozoic organisms, there is little point in using the usual taxonomic groupings since the methods of feeding vary considerably within any group. What determines the feeding mechanism is the type of food (i.e. plant or animal) and its relative size. Members of the same class can feed in different ways and yet unrelated groups may use similar mechanisms to each other. For example, within the gastropod molluscs there are carnivores, herbivores and filter feeders. On the other hand, animals as diverse as sponges, annelids, molluscs,

TABLE 24.6 **Feeding mechanisms of small particle (microphagous) feeders**

Mechanism	Examples	Type of food eaten	Notes
Pseudopodia	Rhizopod protozoa e.g. *Amoeba*	Bacteria, unicellular algae, rotifers, ciliates, flagellates and organic detritus	Often associated with locomotion. Pseudopodia enclose food in a vacuole. Lysosomes discharge enzymes into the vacuole which undergoes an acid and alkaline phase of digestion. Once the food is broken down, soluble products are absorbed into the cytoplasm and undigested material is egested by **exocytosis**
Flagella	*Euglena*; sponges, e.g. *Spongilla*	Bacteria, unicellular algae, rotifers and organic detritus	The flagellum (or flagella) directs the food particles to a specialized region for their ingestion. In *Euglena* the flagellum is also used for locomotion
Cilia	Protozoans, e.g. *Paramecium*. Polychaete annelids, e.g. *Sabella* (fan worm) Pelycopod molluscs, e.g. *Mytilus* (mussel). Chordates, e.g. *Amphioxus*	Bacteria, unicellular algae, protozoans, invertebrate larvae and organic detritus	Cilia are used to create currents of water which carry the food to the mouth, or special feeding organ. The food is trapped by mucus which, when laden, is ingested. There is little selection of food except by size and density. The process is continuous and is very common in marine organisms owing to the large amounts of microscopic organic material in the sea
Tentacles	Echinodermata, e.g. *Holothuria* (sea cucumber)	Organic detritus	Tentacles covered in mucus are swept slowly over the sea bed, picking up organic particles. The tentacles are periodically inserted into the mouth one at a time and the food-laden mucus stripped off and swallowed
Mucus	Gastropod mollusc, e.g. *Vermetes*	Unicellular algae, protozoans, invertebrate larvae and organic detritus	Mucus plays an important rôle in feeding by cilia and tentacles, but *Vermetes* uses mucus alone. It forms a veil of mucus from the pedal gland which drifts in the water, trapping food on its sticky surface. The veil is then ingested and a new one formed
Setae	Crustaceans, e.g. *Daphnia* (water flea), *Balanus* (barnacle). Insects, e.g. *Culex* (mosquito) larvae	Unicellular algae, protozoans, bacteria, invertebrate larvae	Chitinous setae on appendages are swept through the water. They may possess hair-like cirri which help to trap food particles which are passed to the mouth and swallowed
Others (Chordate examples)	*Clupea* (herring)	Invertebrate larvae, crustaceans	Gill rakers filter out food from the water as it passes over the gills
	Cetorhinus (basking shark)	Planktonic organisms	The gills have special comb-like structures which filter out suspended particles as the water passes over them
	Flamingo	Blue-green algae	Algae are filtered out by the beak from the water and drawn into the mouth by the piston-like action of the tongue.
	Whalebone whales	Planktonic shrimp-like crustaceans called 'krill'	Food is collected by a row of keratinous plates in the mouth called the baleen. The filtered 'krill' is then swallowed

crustaceans, mammals and birds use similar filtering mechanisms. A useful classification of feeding types, based upon the nature and size of the ingested food, was devised by C. M. Yonge, and forms the basis of the following accounts. The categories are outlined in Table 24.5.

24.3.1 Small-particle feeders (microphagous feeders)

The food obtained in this way is usually microscopic in size, and typically consists of bacteria, unicellular algae or small invertebrate larvae. These are removed from the surrounding water by some form

TABLE 24.7 **Feeding mechanisms of large-particle (macrophagous) feeders**

Mechanism	Examples	Type of food eaten	Notes
Swallowing inactive food	*Arenicola* (lugworm) *Lumbricus* (earthworm)	Organic material, e.g. plant and animal remains and faeces, fungi, bacteria and protozoans	Non-selective swallowing of mud or soil takes place. It passes along the gut where the organic material is broken down by enzymes and the soluble products are absorbed: Earthworms may, in addition, deliberately ingest leaves
Scraping and boring	*Helix* (garden snail)	Terrestrial green plants	A ribbon-like structure, called the **radula**, bears a series of rows of horny teeth. It is pushed through the mouth and worked backwards and forwards on the vegetation as the mollusc crawls over it. The particles of plant material rasped off are swallowed
	Littorina (periwinkle)	Seaweeds	
	Teredo (shipworm)	Wood, e.g. from boats, piers etc.	Each of the shells of this bivalve mollusc bears strong ridges. The valves are opened and closed and the shells rotated by 90°. The particles of wood rasped off are passed to the mouth for ingestion
	Many insects, including caterpillars, termites, grasshoppers and locusts	Wood and leaves of terrestrial plants	Heavily chitinized jaws called **mandibles** rasp off minute pieces of vegetation which are swallowed. There are powerful muscles associated with the mandibles
Seizing prey **(i)** Seizing and swallowing only	Cnidarians, e.g. *Hydra, Obelia*	Small crustaceans	Prey is seized using tentacles which force it into the mouth. The prey may first be paralysed using nematocysts
	Nereis (ragworm)	Crustaceans and other annelids	Possess a muscular pharynx armed with chitinous teeth which can be everted through the mouth in order to capture the prey
	Many chordates	Varies according to the species	Many chordates simply swallow prey without any other treatment. The prey may first be paralysed, as in some snakes and mucus is used to assist swallowing
(ii) Seizing and masticating before swallowing	*Sepia* (squid)	Small fish	The prey is held against the mouth by the tentacles and horny, beak-like jaws fragment the food before swallowing
	Carcinus (crab)	Scavenge food such as fish	First pair of thoracic appendages are modified to form pincer-like **chelae** which seize the food, passing it to the mouth where the mandibles shred it prior to swallowing
	Most mammals	Varies according to the species	Teeth are used to masticate food prior to swallowing
(iii) Seizing and externally digesting before swallowing	Spiders	Insects	Insect is trapped in a web of silk and killed by poison injected from the spider's **chelicerae**. Powerful proteolytic enzymes are injected from the spider's **pedipalps** which soften the internal parts of the prey. The fluid mass is then sucked up by the spider
	Asterias (starfish)	Pelycopod molluscs	Starfish attaches its arms to the shells (valves) of the prey, slowly pulling them apart. It then extrudes its stomach and secretes enzymes into the shells, thus digesting the organic material which is drawn up as the stomach is pulled back into the starfish

of filtration mechanism. For this reason, organisms adopting this method are called **filter feeders**. Filter feeders are invariably aquatic and examples are found in all invertebrate phyla. They are often sedentary or slow moving, although the chordate examples, some whales and basking sharks, are free-swimming. Small-particle feeders use a variety of mechanisms for obtaining and ingesting food and these are summarized in Table 24.6, on p. 339.

24.3.2 Large-particle feeders (macrophagous feeders)

Feeding mechanisms involving ingestion of particles which are relatively large in comparison to the feeding organism occur throughout the animal kingdom. They are not restricted to any particular environment but are found in all habitats. The various methods are summarized in Table 24.7 and in addition the adaptations of a herbivorous and a carnivorous mammal to their respective diets are detailed in Section 24.6.

24.3.3 Fluid and soft-tissue feeders

Mechanisms for taking in fluids or soft tissues occur mostly among invertebrate animals. Chordate examples are rare. Most types are parasites, living their lives in close association with their host. Some, however, while feeding off other living organisms, do not live in association with them and rarely take two consecutive meals from the same individual. These are not true parasites.

TABLE 24.8 **Feeding mechanisms of organisms feeding on fluids and soft tissues**

Mechanism	Examples	Type of food eaten	Notes
Piercing and sucking	Leeches	Blood of fish, frogs and other chordates	Attaches itself to the host using muscular suckers, and pierces the skin with armoured jaws. Blood is sucked up from the wound and is prevented from clotting by an anticoagulant in the leech saliva
	Mosquitoes	Blood of mammals – females. Nectar and other plant juices – males	Food is sucked up through tubular mouth parts. Where blood is the food, anticoagulants from the salivary glands are first injected
	Aphids	Plant sap from the phloem	The specialized mouth parts are inserted into phloem sieve tubes in stems or leaves. A muscular pharynx is used to draw up the sap
	Vampire bats	Blood of mammals, especially cattle	Small, sharply pointed incisors pierce the skin. The blood so released is sucked or licked up. The saliva contains a proteolytic enzyme which destroys the fibrin basis of any clot which might form
Sucking only	Butterflies, moths	Nectar of flowering plants	A long proboscis is used to reach into the flower to obtain nectar, which is then sucked up into the gut
	Houseflies	Semi-fluid faeces, animal external secretions, decaying plant material	Liquids are drawn up through the proboscis. More solid material may first be made fluid by the secretion of enzymes, along the proboscis, which help to break down the food
Absorption through the general body surface	*Trypanosoma* (protozoan causing sleeping sickness)	Amino acids, sugars etc. from the blood plasma of its host	Lives in the blood-stream of mammals and therefore materials are absorbed by diffusion as required
	Taenia (tapeworm)	Soluble breakdown products of digestion from the intestines of its host	Lives in the intestine of mammals, surrounded by the products of digestion which it absorbs by diffusion. Its flattened shape give it a larger surface area to facilitate absorption

In their simplest form, fluid feeders may just absorb material over the whole body surface, in which case they live surrounded by their food. These organisms are sometimes termed **wallowers**. Other types simply suck up food like nectar which is secreted by another organism. The most complex types have a mechanism to pierce their 'host' in some way and then use a specialized suction apparatus for drawing food into their bodies, often through some tubular device. These various methods are summarized in Table 24.8.

24.4 Teeth and dentition in man

24.4.1 The structure of the tooth

Teeth are embedded in the bones of the upper and lower jaw. The visible part is referred to as the **crown** whereas that in the jaw bone is called the **root**. The outer layer of the crown is composed of **enamel**, the hardest substance in the body. Beneath this lies a softer material, although still harder than bone, called **dentine** which contains living cells and therefore needs to be supplied with oxygen, glucose etc. To satisfy this need, the centre of the tooth forms a hollow known as the **pulp cavity**. Within this cavity run nerves and blood vessels to deliver the necessary nutrients and remove waste products. The tooth is firmly anchored to the jaw by **peridontal fibres** and **cement**. The structure of a typical tooth is shown in Fig. 24.1.

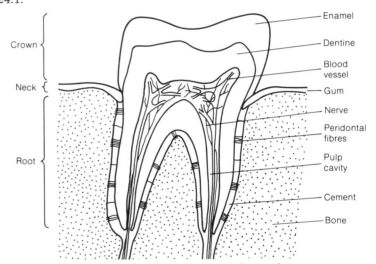

Fig. 24.1 Vertical section through a human molar tooth

24.4.2 Types of teeth

To accommodate the increase in size of the jaw during development, man is **diphyodont**, i.e. has two sets of teeth. The first set, called **milk** or **deciduous teeth**, begin to develop around the end of the first year of life and a complete set is usually present by the age of three. Between the ages of 5 and 13, they are progressively replaced by the **permanent teeth**. In addition to replacement of the 20 milk teeth, 12 molar teeth also develop giving a total of 32 in the permanent set. Where present, the teeth of most chordates are similar in shape and vary only in size. These animals are said to be **homodont** (*homo* – 'same'; *dont* – 'teeth'). Mammals, however, are **heterodont** (*hetero* – 'different'; *dont* – 'teeth') because four distinct types of teeth are recognized.

The **incisors** are found at the front of the mouth and have sharp, chisel-shaped edges for biting food. **Canines**, which occur further back in the mouth, are sharply pointed and used in killing prey and tearing flesh. They are well developed in carnivores and may be entirely absent in some herbivores. Behind the canines are the **premolars** which have two cusps and either a single or double root. They are flattened on top but the cusps give them a ridged structure. They are used for grinding food. At the back of the mouth are the **molars**, which are similar to premolars except they are larger, having 4 or 5 cusps and 2 or 3 roots. They crush and grind food. The **dental formula** is a shorthand method of expressing a mammal's dentition. As each mammal has the same type and number of teeth in each side of the jaw, the formula only takes account of one side of the mouth. Each type of tooth is identified by its initial letter(s) and is followed by the number of that type in the upper jaw over the number in the lower jaw.

An adult human dental formula is therefore:

$$I\frac{2}{2}C\frac{1}{1}Pm\frac{2}{2}M\frac{3}{3}$$

Section 24.6 illustrates the different types of teeth in a carnivorous and an herbivorous mammal.

24.4.3 Dental decay

Examinations of the skeletons of humans who lived many centuries ago show their teeth to have little or no dental decay. The teeth of Eskimos and some other groups likewise show little evidence of decay. It therefore seems that it is aspects of modern diet which account for the epidemic levels of dental disease in Western societies. In particular, an increased consumption of sugary foods has been shown to cause dental disease. The sugars cling to the surface of teeth and gums and provide a ready food source for bacteria. These bacteria may accumulate along with constituents of saliva and form soft deposits known as **plaque**. In time this, along with salts in saliva, can form a hard deposit called **calculus**. Whereas regular brushing can remove plaque, calculus can only be removed by scraping, a task which needs to be performed by a dentist or dental hygienist.

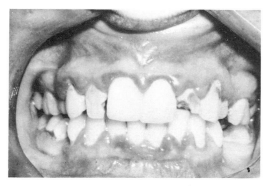

Badly damaged tooth

Plaque and calculus act rather like sponges in absorbing any sugary solutions in the mouth and thus act as reservoirs of food leading to the development of a large bacterial community. These bacteria may initially cause inflammation of the gums and in time attack and weaken the peridontal fibres which hold the tooth in its socket. This so-called **periodontal disease** results in receding gums and the eventual loosening of the tooth. The bacteria on teeth respire the sugar and produce acidic waste products. While enamel is immensely hard and resistant to wear, it is vulnerable to attack from acids. Being largely calcium phosphate, it is converted by acids into soluble substances and so the tooth is literally 'eaten away'. Holes and cavities are formed as a result of **dental caries**. These become filled with plaque which is difficult to remove by brushing and so further decay results. Once through the enamel, the dentine, being softer, is less resistant and dissolves more readily. At this stage the nerves in the pulp cavity may be exposed and so any mechanical disturbance or change in temperature, may stimulate them, causing toothache.

While cavities may be drilled to remove all the rotten portion of the tooth and then filled with an amalgam, it is obviously preferable to prevent dental disease. Reducing the intake of sugary food is helpful, but in any case regular brushing of teeth, especially after meals, will help to prevent the build-up of plaque. Dental check-ups twice a year allow early treatment of any disease and enable calculus to be removed. Use of fluoride, either as tablets, in toothpaste or by addition to drinking water, increases resistance to dental decay. It is most effective when administered to children while the permanent teeth are developing.

24.5 Digestion in man

The food obtained by an animal is not immediately available for use but must first be broken down into its basic constituents ready for absorption.

Holozoic organisms have a variety of alimentary systems designed to carry out this process. The food vacuoles of protozoans undergoing, as they do, a series of changes, represent the simplest system. While this has the advantage of allowing the organism to achieve an optimum concentration of enzymes in the small space within the vacuole, it suffers three disadvantages:

1. The organism is restricted to food small enough to be ingested by phagocytosis.
2. All enzymes must operate together within a small space and so specialization of certain regions to permit more efficient breakdown is not possible.
3. Acid and alkaline phases of digestion must take place within the same vacuole. The two phases must therefore be separated in time.

The evolution of other alimentary systems took place in response to selection pressures to overcome these difficulties. Firstly, a form of mechanical digestion, e.g. teeth or radula, is common, in order to break down larger pieces of food to a manageable size. Secondly, a variety of organs and glands, e.g. pancreas and salivary glands, arose, each responsible for particular aspects of digestion. Thirdly, the gut became elongated in order that acid and alkaline phases of digestion could occur simultaneously, separated not in time, but in space. Let us now look at digestion in the alimentary system of man.

24.5.1 Digestion in the mouth

Mechanical breakdown of food begins in the mouth or **buccal cavity**. Man is an omnivore and hence has an unspecialized diet of mixed animal and plant origin. His teeth reflect this lack of specialization, all types being present and developed to a similar extent. Apart from assisting speech in man, the tongue also manipulates the food during chewing and so ensures it is well mixed with **saliva** produced from three pairs of **salivary glands** (Fig. 24.2). Around 1.0–1.5 dm³ of saliva are produced daily. Saliva contains:

1. Water – Over 99% of saliva is water.
2. Salivary amylase – A digestive enzyme which hydrolyses starch to maltose.
3. Mineral salts (e.g. sodium hydrogen carbonate) – This helps to maintain a pH of around 6.5 which is the optimum for the action of salivary amylase.
4. Mucin – A sticky material which helps to bind food particles together and lubricate it to assist swallowing.

Taste buds on the tongue allow food to be selected — unpleasant tasting food being rejected. The thoroughly chewed food is rolled into a **bolus** and passed to the back of the mouth for swallowing.

24.5.2 Swallowing and peristalsis

The bolus is pushed by the tongue to the back of the mouth and then into the **pharynx**. In this region the **oesophagus**, which leads to the stomach, and the trachea, leading to the lungs, meet. A variety of reflexes ensure that food when swallowed passes down the oesophagus and not the trachea. One such reflex is the closure of the opening into the larynx (which leads to the trachea). This opening, called the **glottis**, is covered by a structure known as the **epiglottis** when food is passed to the back of the mouth. The opening to the nasal cavity is closed by the **soft palate**. In this way, which is illustrated in Fig. 24.2, the bolus enters the oesophagus, a muscular tube lined with stratified epithelium and mucus glands. Lubricated by

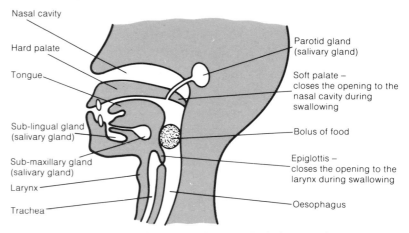

Fig. 24.2 Vertical section through the human skull showing the process of swallowing

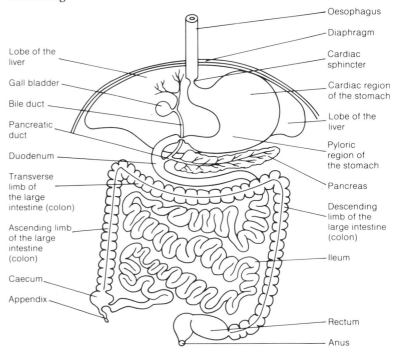

Fig. 24.3 Human digestive system

the mucus secreted by these glands, the bolus passes to the stomach by means of a wave of muscular contraction which causes constriction of the oesophagus behind the bolus. As this constriction passes along the oesophagus it pushes the bolus before it, down to the stomach. This process, which continues throughout the alimentary canal, is called **peristalsis**.

24.5.3 Digestion in the stomach

The stomach is roughly J-shaped, situated below the diaphragm. It is a muscular sac with a folded inner layer called the **gastric mucosa**. Embedded in this is a series of **gastric pits** which are lined with secretory cells (Fig. 24.4). These produce **gastric juice** which contains:

1. Water – The bulk of the secretion is water in which are dissolved the other constituents.

2. Hydrochloric acid – This is produced by **oxyntic cells** and with the water forms a dilute solution giving gastric juice its pH of around 2.0. It helps to kill bacteria brought in with the food and activates the enzymes pepsinogen and prorennin. It also initiates the hydrolysis of sucrose and nucleoproteins.

3. Pepsinogen – This is produced by the **zymogen** or **chief cells** in an inactive form to prevent it from hydrolysing the proteins of the cells producing it. Once in the stomach it is activated to **pepsin** by hydrochloric acid. Pepsin hydrolyses protein into polypeptides.

4. Prorennin – This too is produced by zymogen cells and is an inactive form of **rennin**, an enzyme which coagulates milk by converting the soluble **caseinogen** into the insoluble **casein**. It is therefore especially important in young mammals. Prorennin, too, is activated by hydrochloric acid.

5. Mucus – This is produced by **goblet cells** and forms a protective layer on the stomach wall, thus preventing pepsin and hydrochloric acid from breaking down the gastric mucosa (i.e. prevents autolysis). If the protection is not effective and the gastric juice attacks the mucosa, an ulcer results. Mucus also helps lubricate movement of food within the stomach.

During its stay in the stomach, food is thoroughly churned and mixed with gastric juice by periodic contractions of the stomach wall. In this way a creamy fluid called **chyme** is produced. Relaxation of the pyloric sphincter and contraction of the stomach allow the chyme to enter the duodenum. The chyme from any one meal is released gradually over a period of 3–4 hours. This enables the small intestine to work on a little material at a time and provides a continuous supply of food for absorption throughout the period between meals.

24.5.4 Digestion in the small intestine

In man the small intestine is over 6 m in length and its coils fill much of the lower abdominal cavity. It consists of two main parts: the much shorter **duodenum** where most digestion occurs and the longer **ileum** which is largely concerned with absorption. The walls of the small intestine are folded and possess finger-like projections called **villi**. The villi contain fibres of smooth muscle and regularly contract

(a) Entire stomach

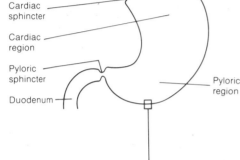

(b) Part of the stomach wall (VS)

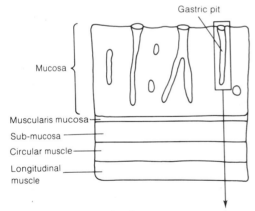

(c) Detail of gastric gland

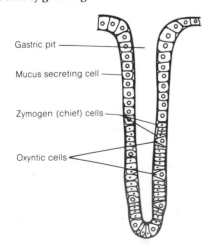

Fig. 24.4 Structure of the human stomach

and relax. This helps to mix the food with the enzyme secretions and keep fresh supplies in contact with the villi, for absorption. The digestive juices which operate in the small intestine come from three sources: the liver, the pancreas and the intestinal wall.

Bile juice

Bile juice is a complex green fluid produced by the liver. It contains no enzymes but possesses two other substances important to digestion.

1. Mineral salts (e.g. sodium hydrogen carbonate) – These help to neutralize the acid chyme from the stomach and so create a more neutral pH for the enzymes of the small intestine to work in.

2. Bile salts – sodium and potassium glycocholate and taurocholate – They **emulsify** lipids, breaking them down into minute droplets. This is a physical, not a chemical change, which provides a greater surface area for pancreatic lipase to work on.

The liver performs other functions, some associated with digestion, and these are detailed in Section 36.7.2.

Pancreatic juice

The pancreas is situated below the stomach and is unusual in that it produces both an exocrine secretion, the pancreatic juice, and an endocrine secretion, the hormone insulin. The endocrine function is not directly concerned with digestion and is described in Section 36.3. Pancreatic juice, in addition to water, contains:

1. Mineral salts (e.g. sodium hydrogen carbonate) – Help to neutralize acid chyme from the stomach and so provide a more neutral pH in which the intestinal enzymes can operate.

2. Proteases – These include **trypsinogen** which, when activated by enterokinase from the intestinal wall, breaks down proteins into peptides. The newly activated trypsin so produced activates another protease in the secretion, chymotrypsinogen into **chymotrypsin**; this too converts proteins into peptides. Also present is **carboxypeptidase** which converts peptides into smaller peptides and some amino acids.

3. Pancreatic amylase – Completes the hydrolysis of starch to maltose which began in the mouth.

4. Lipase – Breaks down fats into fatty acids and glycerol by hydrolysis.

5. Nuclease – Converts nucleic acids into their constituent nucleotides.

Intestinal juice (succus entericus)

Intestinal juice is produced by tubular glands called **crypts of Lieberkuhn** in the wall of the small intestine. In addition to water it contains:

1. Mucus – Produced in the duodenum by the coiled **Brunner's glands**, it helps to lubricate the intestinal walls and prevent autolysis.

2. Mineral salts (e.g. sodium hydrogen carbonate) – Produced by the Brunner's glands in order to neutralize the acid chyme from the stomach and so provide a more suitable pH for the action of enzymes in the intestine.

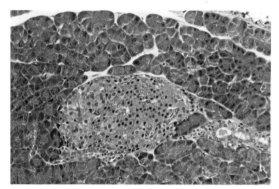

Cellular structure of pancreas showing Islets of Langerhans

3. Proteases (erepsin) – These include **aminopeptidase**, which converts peptides into smaller peptides and amino acids, and **dipeptidase**, which hydrolyses dipeptides into amino acids.

4. Enterokinase – A non-digestive enzyme which activates the trypsinogen produced by the pancreas.

5. Nucleotidase – Converts nucleotides into pentose sugars, phosphoric acid and organic bases.

6. Carbohydrases – These include **amylase**, which helps complete the hydrolysis of starch to maltose, **maltase**, which hydrolyses maltose to glucose, **lactase**, which hydrolyses the milk sugar lactose into glucose and galactose, and **sucrase**, which hydrolyses sucrose into glucose and fructose.

24.5.5 Absorption and assimilation

Digestion results in the formation of relatively small, soluble molecules which, provided there is a concentration gradient, could be absorbed into the body through the intestinal wall by diffusion. This, however, would be slow and wasteful and in any case, if the epithelial lining were permeable to molecules such as glucose, it could just as easily result in it diffusing out of the body when the concentration in the intestines was too low. For these reasons most substances are absorbed by **active transport** (Section 4.4.3) which only allows inward movement. Efficient uptake is often dependent on the presence of other factors. For example, glucose and amino acid absorption appear to be linked to the movement of sodium ions across the membranes of epithelial cells; calcium ion absorption requires the presence of Vitamin D.

Efficient absorption is also dependent on a large surface area being available. The wall of the ileum achieves this in four ways:

1. It is very long – almost 6 m in humans and up to 45 m in cattle.

2. Its walls are folded (**folds of Kerkring**) to provide large internal projections.

3. The folds themselves have numerous tiny finger-like projections called **villi** (Fig. 24.5).

4. The epithelial cells lining the villi are covered with minute projections about 0.6 μm in length, called **microvilli** (not to be confused with cilia). These collectively form a **brush border**.

Sugars, amino acids and other water-soluble materials such as minerals enter the blood capillaries of the villi. From here they enter arterioles which later merge to form the hepatic portal vein which carries blood to the liver. In general, the level of different absorbed foods in the hepatic portal vein varies, depending on the type of food eaten and the interval since ingestion. It is the main rôle of the liver to regulate these variations by storing excess where the level of a substance is above normal and by releasing its store when its level in the hepatic portal vein is low. For this reason, blood from the intestines is sent to the liver for homeostatic regulation before it passes to other organs where fluctuations in blood composition could be damaging. The liver is also able to break down any harmful substances absorbed, a process called **detoxification**. The fatty acids and glycerol from lipid digestion enter the epithelial cells lining the villi where they recombine into lipids. These then enter the lacteals

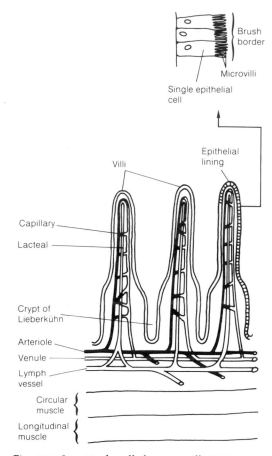

Fig. 24.5 *Intestinal wall showing villi (LS)*

Brush border
Microvilli
Single epithelial cell
Epithelial lining
Villi
Capillary
Lacteal
Crypt of Lieberkühn
Arteriole
Venule
Lymph vessel
Circular muscle
Longitudinal muscle

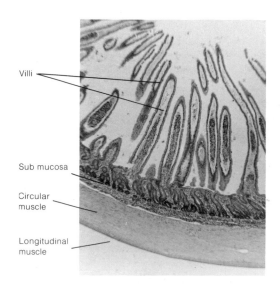

Villi
Sub mucosa
Circular muscle
Longitudinal muscle

Section of small nutrestine (× 35 approx.)

rather than the blood capillaries. From here they are transported in the lymph vessels before later joining the venous system of the blood near the heart.

The passage of food from the ileum into the large intestine or colon is controlled by the **ileo-caecal valve**. The caecum in man is little more than a slight expansion between the small and large intestine and the appendix is a small blind-ending sac leading from the caecum. In man neither structure performs any important digestive function but, as we shall see in Section 24.6, they are of considerable importance to herbivorous mammals.

24.5.6 Water reabsorption in the large intestine

Most of the water drunk by man is absorbed by the stomach. The large intestine or **colon** is mostly responsible for reabsorbing the water from digestive secretions. With the gastric and intestinal juices each producing up to $3\,dm^3$ (litre) of secretion every day and the saliva, pancreatic and bile juices each adding a further $1.5\,dm^3$ the total volume of digestive secretions may exceed $10\,dm^3$. As most of this volume is water, it follows that the body cannot afford to allow it simply to pass out with the faeces. While most water is absorbed in the ileum, the large intestine plays an important rôle in reabsorbing the remainder. In doing so it changes the consistency of the faeces from liquid to semi-solid.

Within the large intestine live a huge population of bacteria, such as *Escherichia coli*, which in man synthesize a number of vitamins including biotin and vitamin K. Deficiency of these vitamins is therefore rare, although orally administered antibiotics may destroy most of the bacteria and so create a temporary shortage. The vitamins produced are absorbed by the wall of the large intestine with water and some mineral salts. This wall is folded to increase the surface area available for absorption. Excess calcium and iron salts are actively transported from the blood into the large intestine for removal with the faeces.

24.5.7 Elimination (egestion)

The semi-solid faeces consist of a small quantity of indigestible food (roughage) but mostly comprise the residual material from the bile juice and other secretions, cells sloughed off the intestinal wall, a little water and immense numbers of bacteria. The wall of the large intestine produces mucus which, in addition to lubricating the movement of the faeces, helps to bind them together. After 24–36 hours in the large intestine the faeces pass to the rectum for temporary storage before they are removed through the anus, a process known as **defaecation**. Control of this removal is by two sphincters around the **anus**, the opening of the rectum to the outside. This control is not instinctive in man, and relaxation of the sphincter in a baby is a reflex response to the rectum becoming full. Control of the sphincters can, however, be learnt, allowing defaecation to become a voluntary action.

As much of the material making up the faeces is not the result of metabolic reactions within the body, it is said to be eliminated or egested rather than excreted. However, cholesterol and bile pigments from the breakdown of haemoglobin are metabolic products and are therefore excretory.

24.6   Adaptations to particular diets

Being an omnivore, man's dentition and alimentary canal are less specialized than those of organisms who restrict their diet to either plant or animal food. Those consuming plant food are called **herbivores** whereas those whose main diet is meat are called **carnivores**. Both types have specialized modifications of their digestive systems to adapt them to their respective diets.

24.6.1 Herbivorous adaptations of mammals

The principal material ingested by all herbivores is cellulose and yet few are able to produce cellulase, the enzyme responsible for its breakdown. Most herbivores, including all herbivorous mammals, rely on bacteria and protozoans, which inhabit the gut to carry out cellulose digestion. Plant food is a relatively tough material and its breakdown begins in the mouth where the dentition is modified to ensure that it is thoroughly ground up before swallowing. Adaptations of a typical herbivore like a sheep include:

1. A horny pad replaces the upper incisors and canines.

2. There is a pronounced gap, the **diastema**, between the incisors and premolars on both upper and lower jaws. This provides a space near the front of the mouth in which newly nibbled food can be kept separate from that undergoing mastication at the back of the mouth. The dental formula of a sheep is:

$$I\,\tfrac{0}{3}\,C\,\tfrac{0}{1}\,Pm\,\tfrac{3}{2}\,M\,\tfrac{3}{3}$$

3. The molars and premolars, collectively called the **cheek teeth**, are similar to each other with flat, ridged surfaces. The ridges are the result of the enamel of these teeth being unevenly distributed. Where the enamel is thin it quickly wears away, as does the now exposed softer dentine beneath. This forms depressions between the ridges of thicker enamel. The effect is not dissimilar to the flat, but abrasive, surface of a file.

4. The jaws are free to move laterally as well as vertically. This side to side movement allows the cheek to grind plant material between their two rows of ridged surfaces.

5. The teeth have **open roots**, which allow a constant supply of nutrients to them throughout the animal's life. They therefore grow continuously – a necessity if they are not to be completely worn away by constant grinding.

6. The stomach is divided into a number of chambers, some of which possess vast populations of microorganisms which digest cellulose. Food is regurgitated into the mouth before passing into the remaining stomach compartments. This process is detailed below.

7. The alimentary canal is relatively long, because digestion of plant material is difficult.

24.6.2 Digestion of cellulose by micro-organisms

Lacking the ability to produce their own cellulase, herbivorous mammals are reliant upon bacteria and protozoans to carry out cellulose breakdown for them. The herbivore must provide a region

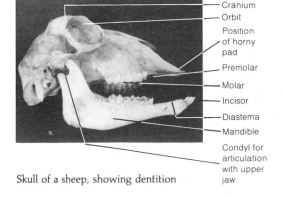

Cranium
Orbit
Position of horny pad
Premolar
Molar
Incisor
Diastema
Mandible
Condyl for articulation with upper jaw

Skull of a sheep, showing dentition

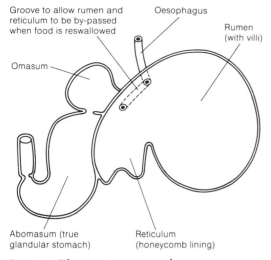

Groove to allow rumen and
reticulum to be by-passed
when food is reswallowed

Oesophagus

Rumen
(with villi)

Omasum

Abomasum (true
glandular stomach)

Reticulum
(honeycomb lining)

Fig. 24.6 The ruminant stomach

of the alimentary canal for these microorganisms to inhabit. This region must be separate from the main canal in order that food can be kept there long enough for the microorganisms to carry out the breakdown. Being separate, this compartment can also be kept free of the mammal's own digestive enzymes, which might otherwise destroy the microorganisms, and at a suitable pH for their activity. The accommodation for these cellulose-digesting bacteria and protozoans takes two main forms in mammals.

In **ruminants**, e.g. cattle, sheep and deer, a complex four-chambered stomach is present. When swallowed, the food enters the first two chambers, the **rumen** and **reticulum**. It is here that the microorganism carry out extracellular digestion of the cellulose by secreting cellulase. The products of this digestion are either absorbed by the walls of the rumen and reticulum which have villi or honey-combed ridges for this purpose, or are absorbed by the micro-organisms which are later digested.

The relationship between the mammal and microorganisms is symbiotic (Section 25.4.1), in which the mammal acquires the products of cellulose breakdown which it could not obtain alone, and the microorganisms receive a constant supply of food and a warm, sheltered environment in which to live. After some hours, and usually in the relative safety of some sheltered position, the herbivore regurgitates the food into the mouth, where it thoroughly chews it – 'chewing the cud'. On being re-swallowed, the food enters the final two chambers of the stomach, the **omasum** and the **abomasum** (true stomach) where the usual process of protein digestion in acid conditions takes place.

In rabbits and horses, the **caecum** and **appendix** are much enlarged and accommodate the microorganisms. Some absorption of the products of this digestion takes place through the walls of the caecum. In rabbits the yield is improved by the re-swallowing of the material

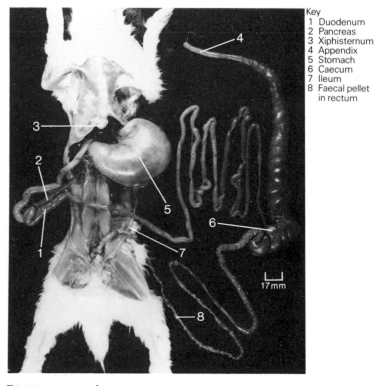

Key
1 Duodenum
2 Pancreas
3 Xiphisternum
4 Appendix
5 Stomach
6 Caecum
7 Ileum
8 Faecal pellet
   in rectum

17mm

Digestive system of a rat

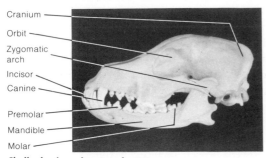

Skull of a dog, showing dentition

Cranium
Orbit
Zygomatic arch
Incisor
Canine
Premolar
Mandible
Molar

TABLE 24.9 **Comparison of herbivorous and carnivorous adaptations in mammals**

Herbivorous adaptations	Carnivorous adaptations
Sharp chisel-shaped incisors for cutting or gnawing	Incisors sharp, sometimes pointed, for nipping and biting
Upper incisors sometimes absent, e.g. in cattle and sheep	Upper incisors never absent
Canines, if present, small and incisor-like	Canines long and pointed for piercing and tearing
Diastema present	Diastema never present
Molars and premolars flattened with ridges of enamel for grinding food	Molars and premolars have pointed cusps for shearing food
Carnassial teeth absent	Last upper premolars and first molars form carnassial teeth
Open pulp cavity means teeth can grow continuously to compensate for wear	Closed pulp cavity means teeth do not continue to grow once full size
Teeth of upper jaws meet those of lower jaw end on to allow grinding of food	Teeth of upper jaw slide past the outside of those in the lower jaw to allow shearing of food
Lateral movement of lower jaw aids grinding of food	Absence of lateral movement of lower jaw helps prevent dislocation when capturing prey
Temporal and masseter muscles relatively small	Temporal and masseter muscles relatively large
Less well developed processes for muscle attachment	Well developed processes for muscle attachment
Specialized stomach or caecum and appendix to accommodate symbiotic cellulose digesting micro-organisms	No specialized adaptation of stomach; caecum and appendix small
Relatively long alimentary canal	Relatively short alimentary canal

from the caecum after it has left the anus – a process known as **coprophagy** or **refection**.

24.6.3 Carnivorous adaptations of mammals

Carnivores eat meat, which is mainly the muscle of another animal. This is rich in nutrients and therefore a much more concentrated source of food than plant material. This advantage may be offset to some degree by the time and energy expended in obtaining its prey. Once captured and ingested the digestion of meat presents little problem. The adaptations to this form of diet largely concern modifications to the jaw and its dentition. Adaptations of a typical carnivorous mammal such as a dog include:

1. The incisor teeth are sharp and used for nipping and biting.

2. The canines are long and pointed. They are used for piercing and killing the prey and tearing flesh from the body.

3. The molars and premolars have a number of sharp pointed cusps. The last upper premolar and first lower molar on each side of the mouth are particularly large and known as **carnassial teeth**.

4. The teeth of the upper jaw tend to overlap those of the lower jaw. The carnassial teeth therefore slide past one another rather than meeting each other end on. In this way they act like two blades of a pair of scissors and easily slice meat into manageable pieces.

5. The temporal and masseter muscles of the jaw are well developed and powerful. This enables carnivores to grip the prey firmly during the kill and helps in crushing bone.

6. There is no lateral jaw movement as in herbivores. Such movement leads to easier dislocation of the jaw, a distinct disadvantage when trying to grip struggling prey.

7. Vertical movement of the jaw is less restricted, allowing a wide gap for capturing and killing prey.

8. The alimentary canal is short, reflecting the relative ease with which meat can be chemically digested.

24.7 Nervous and hormonal control of secretions

The production of a digestive secretion must be timed to coincide with the presence of food in the appropriate region of the gut. In mammals the production of digestive secretions is under both nervous and hormonal control.

Nervous stimulation occurs even before the food reaches the mouth. The sight, smell or even the mere thought of food is sufficient to cause the salivary glands to produce saliva. This response is a conditioned reflex and is explained more fully in Section 38.7.3. Once in the mouth, contact of food with the tongue causes it to transmit nervous impulses to the brain. The brain in turn sends impulses which stimulate the salivary glands to secrete saliva. This is an unconditioned reflex response. At the same time the brain stimulates the stomach wall to secrete gastric juice, a response reinforced by nervous impulses transmitted as the food is swallowed.

The stomach is thus prepared to digest the food even before it reaches it. Once initiated, the response will continue for up to an hour. The stretching of the stomach due to the presence of food within it stimulates production of gastric juice after this time.

Hormonal control of secretions begins with the presence of food in the stomach. This stimulates the stomach wall to produce a

TABLE 24.10 **Summary of digestion**

Organ/ secretion	Production induced by	Site of action	pH of secretion	Contents	Effect
Salivary glands produce saliva	Visual or olfactory expectation and reflex stimulation	Mouth	Neutral or slightly alkaline	Salivary amylase	Amylose (starch)→maltose
				Mineral salts	Produce optimum pH for amylase action
				Mucin	Binds food particles into a bolus
Gastric glands in stomach wall produce gastric juice	Presence of food in mouth and swallowing. Presence of food in stomach. Hormones–gastrin and enterogasterone from stomach wall	Stomach	Very acid	Pepsin(ogen)	Proteins→peptides
				(Pro)rennin	Caseinogen→casein
				Hydrochloric acid	Activates pepsinogen and prorennin. Produces optimum pH for action of these enzymes
				Mucus	Lubrication and prevention of autolysis
Liver produces bile juice	Secretin stimulates production of bile and cholecystokinin causes it to be released	Duodenum	Neutral	Bile salts	Emulsify fats
				Mineral salts	Neutralize acid chyme
				Bile pigments	Excretory products from breakdown of haemoglobin
				Cholesterol	Excretory product
Pancreas produces pancreatic juice	Secretin stimulates production of mineral salts and pancreozymin production of enzymes	Duodenum	Neutral	Trypsin (ogen)	Protein→peptides + amino acids activates chymotrypsinogen
				Chymotrypsin (ogen)	Peptides→smaller peptides + amino acids
				Carboxypeptidase	Peptides→smaller peptides + amino acids
				Amylase	Amylose (starch)→maltose
				Lipase	Fats→fatty acids + glycerol
				Nuclease	Nucleic acids→nucleotides
				Mineral salts	Neutralize acid chyme
Wall of small intestine produces intestinal juice (succus entericus)	Presence of food stimulates the intestinal lining	Duodenum and ileum	Alkaline	Aminopeptidase	Peptides→amino acids
				Dipeptidase	Dipeptides→amino acids
				Enterokinase	Activates trypsinogen
				Nucleotidase	Nucleotides→organic base + pentose + sugar + phosphate
				Maltase	Maltose→glucose
				Lactase	Lactose→glucose + galactose
				Sucrase	Sucrose→glucose + fructose
				Mineral salts	Neutralize acid chyme

hormone called **gastrin** which passes into the bloodstream. Gastrin continues to stimulate the production of gastric juice for up to four hours. Because fat digestion takes longer and requires less acidic conditions, its presence in the stomach initiates the production of **enterogasterone** from the stomach wall. This hormone reduces the churning motions of the stomach and decreases the flow of the acid gastric juice. As stomach ulcers are irritated by gastric juice, sufferers are often urged to drink milk. Being rich in fat, it reduces the production of gastric juice.

When food leaves the stomach and enters the duodenum, it stimulates the production of two hormones from the duodenal wall. **Secretin**, via the blood stream, travels to the liver where it causes the production of bile and to the pancreas, where it stimulates the secretion of mineral salts. **Cholecystokinin-pancreozymin** causes the gall bladder to contract (releasing the bile juice into the duodenum) and stimulates the pancreas to secrete its enzymes.

24.8 Food poisoning – *Salmonella*

Food poisoning may be caused by a number of different bacterial species, but those belonging to the genus *Salmonella* are the most common, accounting for over half the reported cases in Britain. Most people ingest *Salmonella* frequently but as it takes around ten million live bacteria at one time to cause food poisoning, their occurrence usually goes unnoticed. The very young, the elderly and those weakened by other ailments (e.g. those in hospital) are most at risk and may be affected by doses which are harmless to other groups.

The bacteria are normally confined to the alimentary canal of an animal; it is after slaughter that meat is sometimes contaminated with *Salmonella*. Freezing or other means of preservation usually prevent a further build up of their numbers and proper cooking ensures they are killed before consumption. The problem arises if meat is not properly preserved and/or thoroughly cooked. Meat that has not been completely defrosted may not reach a sufficiently high temperature at its centre to kill all bacteria, including *Salmonella*. It is not the only potential source of *Salmonella*: eggs, unpasteurized milk and its products may also harbour the bacteria.

The symptoms of food poisoning, which can arise within a few hours of infection, include diarrhoea, abdominal pain, sickness and fever. The fluid loss caused by these symptoms may even prove fatal to vulnerable groups. The main treatment is to care for the patient, in particular to replenishing fluid loss; antibiotics have limited effectiveness. Prevention is the main weapon used against the disease and includes strict hygiene precautions on farms, at abattoirs and in the handling of meat, especially the pre-cooked variety both in shops and at home. Thorough cooking of meat and eggs is especially important.

24.9 Carnivorous plants

No plant is truly holozoic in the sense that it relies entirely for its survival on the digestion of complex organic material from other organisms. All carnivorous plants are autotrophic and obtain the majority of their nutrients by photosynthesis. Why then do these plants capture and digest insects? The object is to procure some

Sticky
secretion

Tentacle

Leaf

Sundew

nutrient which is lacking and cannot otherwise be acquired. Most often the missing nutrient is nitrogen, needed in plants for protein synthesis, and hence growth. Insectivorous plants like bladderwort live in ponds and ditches where nitrate concentrations are very low. Others, like sundew, live in wet, cold areas such as moorland. Here the cold makes decomposition of humus very slow with consequent slow release of nitrates. When they are released, the high level of rainfall causes them to be rapidly leached from the soil. The plant has thin 'tentacles' on its leaves which secrete a sticky fluid that glistens in the light – hence the name sundew. This glistening fluid attracts insects which become trapped on the sticky leaves. The 'tentacles' then secrete enzymes which kill and digest the insect. The amino acid products of this digestion are absorbed by the leaf and provide it with its main source of nitrogen. As sundew lives where water is plentiful and easy to absorb, and as minerals, including nitrogen, are obtained from the digestion of insects and not the soil, it is hardly surprising that the sundew has poorly developed roots.

24.10 Questions

1. (a) Describe the composition of a balanced diet.
        (4 marks)
    (b) Outline the mechanical and the chemical digestion of protein in mammals.   *(10 marks)*
    (c) Outline a method for determining the energy requirements of a mammal.   *(4 marks)*
        (Total 18 marks)

    Cambridge Board June 1988, Paper I, No. 3

2. (a) Describe the adaptations which enable the following organisms to obtain and ingest their food: a herbivorous mammal; a carnivorous mammal; a fluke.   *(6, 6, 3 marks)*
    (b) Explain the importance of the following in the human diet: a named vitamin; a named metal ion; a named non-metal element.   *(15 marks)*
        (Total 30 marks)

    Oxford Local 1987, Specimen Paper II, No. 8

3. Either

Give an account of the physiological rôles of vitamin A and vitamin D. Comment briefly on the importance of a deficiency of each for the human individual and in the community.

or

Discuss the statement that 'The diet of modern man contains too much fat, too much sugar and too little fibre'.   *(20 marks)*

Oxford and Cambridge Board June 1988, Paper II, No. 8

4. Recent reports on the British diet have made strong recommendations. Explain why salt and sugar should be decreased and why fibre should be increased.   *(6 marks)*

Southern Universities Joint Board June 1986, Paper I, No. 9

5. (a) What are *vitamins*?   *(4 marks)*
    (b) In experiments to understand the effects of lack of vitamins in the diet of small mammals, the data shown in the table below were obtained.
        Vitamin levels in cells, plasma, and urine were monitored over a period of 50 days. For 5 days the healthy animals were observed and given a full diet with plenty of vitamins. Then vitamins were withheld for 25 days, after which they were reintroduced into the diet.
        Results were as follows:

	Vitamin levels/arbitrary units		
Day	Cells	Plasma	Urine
0	60.0	55.0	50.0
5	60.0	55.0	50.0
8	60.0	50.5	20.0
10	59.5	44.0	3.0
15	58.5	8.0	**
20	54.0	**	**
25	40.0	**	**
29	2.0	**	**
32	13.0	8.5	**
35	44.5	22.0	**
40	56.0	47.5	12.0
42	58.5	51.0	37.0
45	60.0	55.0	50.0
50	60.0	55.0	50.0

** = negligible amounts, i.e. too low for measurement.

(i) Present the above data in graph form as suitable curves on the same axes.

(ii) Comment on the significance of the curves obtained and suggest explanations.

(14 marks)

(c) Choose any **three** vitamins necessary in the human diet. Explain their rôles in the tissues and the ensuing problems which arise when each is absent from the diet. *(12 marks)*

(Total 30 marks)

Oxford Local June 1988, Paper II, No. 6

6. The table below refers to the alimentary canals of a sheep and a dog.

If the statement is correct for that animal place a tick ($\sqrt{}$) in the appropriate box and if the statement is incorrect place a cross ($\times$).

Statement	Sheep	Dog
Crowns of teeth entirely covered with enamel		
Diastema present		
Stomach region of adult expanded into pouches containing symbiotic microorganisms		
Rennin present in gastric juice in young		
Alimentary canal connected by mesentery to body wall		

(Total 5 marks)

London Board June 1989, Paper I, No. 6

7. The efficiency of food absorption was investigated in two desert herbivores of similar size, one a reptile the other a mammal. The table below shows some of the findings.

Aspect studied	Reptile	Mammal
Average daily food intake (g dry mass)	0.55	9.7
Retention time of food in gut (hours)	125	4
Length of small intestine (cm)	20	42.
Gross internal area of small intestine (cm^2)	18	40
Microscopic area of small intestine (cm^2)	72	506
Extraction efficiency (%)	47	52

Note: Extraction efficiency is derived from:

$$\frac{\text{food dry mass} - \text{faeces dry mass}}{\text{food dry mass}} \times 100$$

(a) Explain briefly any **two** practical precautions you would take to ensure meaningful comparisons in such an investigation. *(2 marks)*

(b) Suggest briefly how retention time could have been determined. *(2 marks)*

(c) (i) Explain why the investigation was concerned especially with the small intestine.

(ii) Suggest a reason for the difference between the gross and microscopic internal area of the small intestine. *(2 marks)*

(d) Account for the difference in average daily food intake between the two animals given that they both eat the same sort of food. *(2 marks)*

(e) (i) State **one** problem that you would expect to cause inaccuracy when estimating the extraction efficiency.

(ii) Do you find it surprising that the extraction efficiencies of the two animals are comparable? Give reasons for your answer. *(3 marks)*

(Total 11 marks)

Joint Matriculation Board June 1989, Paper IA, No. 3

8. (a) The diagram below show three different types of epithelial tissue.

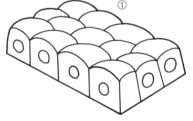
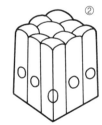
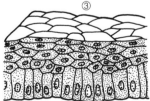

Name the tissues and indicate where they are located in the alimentary canal or associated organs.

	Name	Location
1		
2		
3		

(6 marks)

(b) The next diagrams show three different forms of exocrine glands found in the alimentary canal.

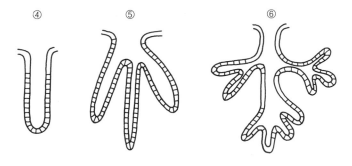

④ ⑤ ⑥

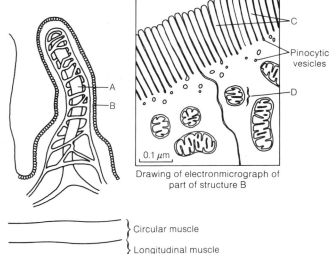

C

Pinocytic vesicles

D

A
B

0.1 μm

Drawing of electronmicrograph of
part of structure B

Name the glands and indicate where they are
located in the alimentary canal.

	Name	Location
4		
5		
6		

} Circular muscle

} Longitudinal muscle

(6 marks)

(c) The glands drawn above are involved in the
production of enzymes and mucus. State **two**
functions of mucus. *(2 marks)*
Give **two** examples of the ways in which the
activities of gut enzymes are controlled chemically.
(2 marks)

(d) In some recently published work it was stated
that people eating a high-fibre diet produce
300–500 grammes of faeces daily and the food
takes about 30 hours to pass through the gut,
whereas the majority of inhabitants in advanced
Western countries produce 80–120 grammes
of faeces daily and the food takes 70 hours to
pass through the gut.
Comment on these figures. *(4 marks)*
(Total 20 marks)

Oxford and Cambridge Board June 1988, Paper I, No. 1

9. (a) What are the main features of
(i) *autotrophic* nutrition; *(2 marks)*
(ii) *heterotrophic* nutrition? *(1 mark)*
The diagram shows part of a transverse section through
a mammalian ileum.
(b) Name the parts labelled **A** to **D** on the diagram.
(4 marks)
(c) Briefly describe how **three** features, shown in
the diagram, enable the ileum to carry out its
function of absorption. *(6 marks)*
(d) (i) Of what type of muscle do the layers of
circular and longitudinal muscle consist?
(1 mark)
(ii) What is the function of this muscle in the
ileum? How is this function achieved?
(2 marks)

(e) The duodenum and ileum receive secretions
from the pancreas and liver.
(i) List the components of the secretion
derived from the pancreas. *(2 marks)*
(ii) List the components of the secretion
derived from the liver. *(2 marks)*
(iii) Describe how the flow of the pancreatic
secretion is controlled. *(4 marks)*
(Total 24 marks)
Associated Examining Board June 1989, Paper II, No. 1

10. The table shows the energy and some of the nutrients
required each day by different individuals.

	Body mass in kg	Energy in kJ	Protein in g	Vit. D in μg	Iron in mg
1 year-old child	7.3	3200	20	10	6
18 year-old girl	56.0	9200	58	2.5	12
18 year-old boy	60.0	12 000	80	2.5	10

(a) Suggest **one** reason, in each case, why the 18
year-old boy requires
(i) only one quarter of the vitamin D
requirement of the child,
(ii) almost twice the iron requirement of the
child,
(iii) less iron than the 18 year-old girl.
(6 marks)
(b) (i) Calculate (to the nearest whole number)
the value of the protein requirement per
kilogram of body mass for the
1. 1 year-old child
2. 18 year-old boy.
(ii) Offer an explanation for this difference in
protein demand.

(iii) Explain why protein requirement is best met by eating meat or soya beans rather than most single types of vegetable food.

(4 marks)

(c) (i) Name **two** end products, other than carbon dioxide, formed by the total combustion of a piece of bacon in a bomb calorimeter. (Assume that bacon consists of protein and fat only.) (2 marks)

(ii) Explain why proteins and fats subjected to total combustion in a bomb calorimeter liberate more heat energy than when the same mass of each food substance is metabolised in the body. (2 marks)

(iii) Why does 1 gram of fatty acid produce more than twice as much heat energy as 1 gram of glucose? (2 marks)

(Total 16 marks)

Welsh Joint Education Committee June 1989, Paper A2, No. 6

25 *Heterotrophic nutrition 2 –*
parasites, saprophytes and symbionts

25.1   Introduction

Associations between organisms are never static and do not fall into clearly defined groups in spite of man's attempts to make them do so. The term **symbiont** has been used to describe a variety of associations in which both organisms may benefit or in which one may be severely damaged. For the purposes of this book the term **symbiosis** will be confined to those relationships in which both partners benefit to some degree and in which neither is damaged.

The terms commensalism and mutualism are sometimes regarded as special forms of symbiosis but will not be used in this book.

A **parasite** is an organism showing some degree of metabolic dependence on another organism, the **host**. In this relationship the host may suffer some harm, and if the population of parasites is high, may be killed. **Saprophytes** are organisms which are unable to use carbon dioxide as their sole source of carbon. They require organic compounds which have been synthesized by other plants and obtain these by absorption from dead organisms.

It must be realized that some organisms fall into different nutritional categories at various times during their lives. For example, the honey fungus *Armillaria mellea* is a parasite on the roots of forest trees. This parasite eventually kills its host and then lives as a saprophyte on the plant remains. The nematode hookworm *Ancylostoma duodenale* spends part of its life as a parasite absorbing human blood, part as a free-living worm feeding on soil bacteria and part as a non-feeding, free-living stage which can abruptly become parasitic.

25.2   Parasites

Parasitism is an association between two organisms in which one, the parasite, is metabolically dependent on the other, the host. This is invariably a nutritional dependence, the parasite absorbing either host tissues and fluids, or the contents of the host's intestine. In this relationship the host is harmed in some way. It is often difficult to distinguish between parasitism and predation or scavenging. However, most parasitologists feel that a parasitic relationship is one in which the parasite spends a significant length of time feeding on the host. A biting fly would not be considered a parasite by most biologists, but a leech would.

Another difference between a parasite and a predator is that the host can produce an immune response to a parasite but not to a predator. A parasite's success may be measured by its ability to resist this immune reaction.

There are two main categories of parasites: **endoparasites** which live inside the body of the host and **ectoparasites** which live on the outside. In both cases the parasite needs to be able to maintain its position in or on the host. Any organism must be able to reproduce and the offspring must be able to find a suitable habitat in which to develop. This is a particular problem for parasites which often need to produce large numbers of eggs or spores in order to ensure the success of a few. Parasitic life cycles may include elaborate mechanisms for successful transmission, often with several larval stages.

Plant parasites are sometimes divided into two groups according to the way they obtain their energy from cells. **Biotrophs** (living feeders) obtain their energy only from living host cells, whereas **necrotrophs** (dead feeders) can obtain their energy from cells which they have killed.

In this chapter, five parasites will be examined. In general, these examples will serve to illustrate some or all of the following features of parasites:

1. They have agents for penetration of the host.

2. They have a means of attachment to the host.

3. They have protection against the host's immune responses.

4. They show degeneration of unnecessary organ systems.

5. They produce many eggs, seeds or spores.

6. They have a vector or intermediate host.

7. They produce resistant stages to overcome the period spent away from the host.

25.2.1 *Phytophthora infestans*

Phytophthora infestans is an oomycete protoctist (Section 7.4.1) causing potato blight. This disease was partly responsible for the Irish famine in the mid-nineteenth century and is still a cause of serious potato crop losses in the British Isles.

Phytophthora cannot photosynthesize and obtains its nutrients from the living cells of potato plants. Spores which land on the leaves germinate to produce hyphae which enter the leaf via the stomata. The tips of these hyphae produce cellulases which enable them to penetrate the cell walls. Once inside the cells, the hyphae swell to form **haustoria** which produce digestive enzymes, and through which the soluble products are absorbed into the fungus.

Any parasite needs to produce large numbers of offspring if any are to establish themselves successfully on a new host. *Phytophthora* produces specialized hyphae called sporangiophores which grow through the stomata and give rise to large numbers of spores. These spores are produced within one week of the initial infection; they are light and readily dispersed by the wind. This parasite causes such widespread damage to the cells that the potato plant may be killed. When the sporangia are washed down from the leaves, tubers may also be infected and these rot during storage.

Potatoes are such an important crop that a great deal of research has been carried out on *Phytophthora infestans*. Varieties of potato have been bred which are resistant to attack, and fungicides can also prevent the parasite causing epidemics.

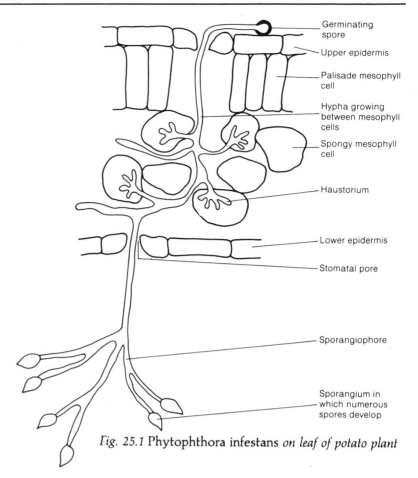

Germinating spore
Upper epidermis
Palisade mesophyll cell
Hypha growing between mesophyll cells
Spongy mesophyll cell
Haustorium
Lower epidermis
Stomatal pore
Sporangiophore
Sporangium in which numerous spores develop

Fig. 25.1 Phytophthora infestans *on leaf of potato plant*

Dodder

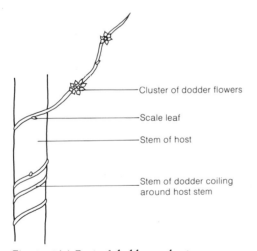

Cluster of dodder flowers
Scale leaf
Stem of host
Stem of dodder coiling around host stem

Fig. 25.2(a) Part of dodder on host

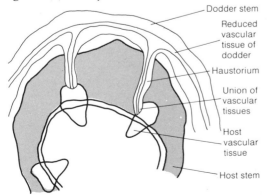

Dodder stem
Reduced vascular tissue of dodder
Haustorium
Union of vascular tissues
Host vascular tissue
Host stem

Fig. 25.2(b) Host stem with dodder attached (TS)

25.2.2 *Cuscuta* (dodder)

There are very few parasitic angiosperms and none are common. Dodder is an annual parasite of plants such as clover and heather. It is unable to photosynthesize because it lacks chlorophyll. It comprises a long, yellowish stem bearing small, widely separated scale leaves and clusters of pink flowers which produce numerous tiny seeds. The seedling is capable of a brief independent existence as its stem grows and swings about in order to contact a support. If it fails to touch a suitable host the seedling quickly dies. Mature dodder has no roots, has severely reduced xylem and is attached to the stem of its host. Flattened haustoria develop as outgrowths of the parasite and the production of cellulases enables them to penetrate the host's stem to eventually enter the vascular tissue. The haustoria develop tracheids where they contact the host xylem and phloem-like elements where they contact the phloem. This connection allows the photosynthetic products of the host to enter the parasite. The marked reduction of roots and vascular tissues allows the energy of dodder to be directed towards the production of vast quantities of tiny seeds and thus ensure the establishment of the next generation.

25.2.3 *Plasmodium* (malarial parasite)

Plasmodium is a protozoan of the phylum Apicomplexa. Four species are parasitic in man, all causing forms of malaria. Malaria is a debilitating fever and *Plasmodium falciparum* probably causes more human deaths in the tropics than any other organism. The malarial parasite has a very complex life cycle involving an asexual stage in the liver and red blood cells of man and a sexual stage which begins in man and continues in mosquitoes of the genus *Anopheles*.

Plasmodium, in common with almost all parasites, has very rapid phases of multiplication to produce vast numbers of offspring. Three stages, the sporozoite, merozoite and zygote, must also be able to penetrate the cells of their host. It also shows an additional parasitic feature, the use of a **vector**. A vector is a secondary host which, because it feeds on the primary host, ensures the transmission of the parasite. The life cycle of *Plasmodium vivax* will be described briefly in order to illustrate these parasitic features but a detailed consideration is beyond the scope of this book.

Following the bite of an infected mosquito, sickle-shaped forms of the parasite, called **sporozoites**, enter the blood of man. Within half an hour of infection these enter the liver cells where they undergo a prolonged period of rapid division. Numerous **merozoites** are produced which infect other liver cells and which may also enter red blood corpuscles. Within the erythrocytes, the multiplication continues and further merozoites are released into the blood. Eventually asexual reproduction is replaced by sexual reproduction and some of the merozoites develop into gametes. These remain dormant in the blood unless they are taken up by a mosquito. Female *Anopheles* mosquitoes need a meal of blood before ovulation and if they bite an infected man the gametes which are ingested become active. Fertilization takes place in the stomach of the mosquito and the zygote burrows into the stomach wall, encysts and meiosis takes place. The zygote then divides asexually to produce large numbers of sporozoites. These are released into the body cavity and many eventually reach the salivary glands. The life cycle continues when the mosquito bites man and releases these sporozoites into his blood.

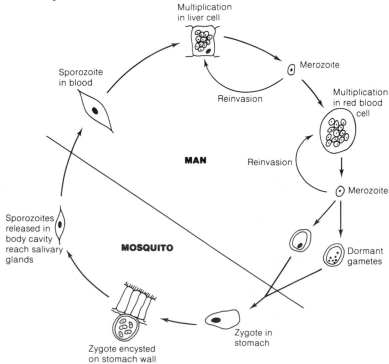

Fig. 25.3 Plasmodium vivax — *simplified life cycle*

Although large-scale eradication programmes have been in action for many years, malaria is still the most important transmissable human disease. Control of the secondary host, the mosquito, is vital if malaria is to be contained. Within the human host the parasite is most susceptible to drugs during the sporozoite and gamete stages but merozoites in the liver form a reservoir of the disease, causing relapses when resistance is low. People who suffer from sickle cell anaemia have a higher resistance to malaria than those with normal erythrocytes. This is discussed fully in Section 15.4.3.

25.2.4 Parasitic flatworms

Two groups of Platyhelminthes (flatworm) are parasitic: the Cestoda, or tapeworms, and the Trematoda, or flukes. The parasitic features of *Fasciola* (liver fluke) and *Taenia* (tapeworm) will be considered in this section.

Both have complex life cycles, that of *Fasciola* involving a vertebrate primary host and an invertebrate secondary host. The many eggs which are produced inside the primary host give rise to a series of larval stages, one of which enters the secondary host where its development continues. *Fasciola hepatica* is found in sheep and cattle world-wide and may cause epidemics of 'liver rot'. The life cycle of *Taenia* involves two vertebrate hosts.

Fasciola hepatica
Fasciola adults are leaf-shaped and capable of limited movement. They live in the liver and bile ducts to which they may attach by means of two suckers. They feed on blood and liver cells. A considerable part of the body is occupied by the hermaphrodite reproductive organs. Numerous eggs are produced and these pass down the bile duct to the intestine and leave the primary host in the faeces. From the egg hatches a free-swimming **miracidium** larva which produces enzymes capable of penetrating the foot of a snail, the secondary host. The parasite undergoes further larval stages within the secondary host before leaving it as a **cercaria** larva. This larva encysts on vegetation and only develops further when eaten by sheep or cattle. In the gut of the primary host, the cyst bursts and larvae bore through to the liver and gain their adult form. *Fasciola* is especially prevalent in regions of high rainfall and poor drainage.

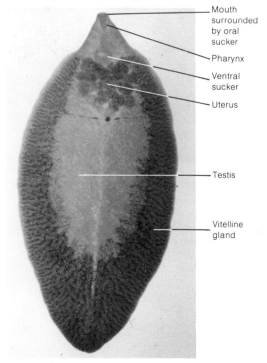

Fasciola

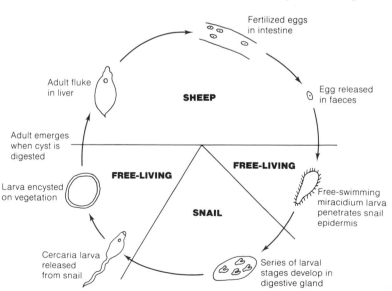

Fig. 25.4 Fasciola hepatica – *simplified life cycle*

Taenia
The adults of *Taenia solium*, the pork tapeworm, live attached to the intestinal mucosa of man. The intermediate host is a pig. Tapeworms

Scolex of *Taenia*

consist of a **scolex** followed by a series of segments known as **proglottids**. The scolex bears four suckers at the sides and a crown of hooks at the top. Although it is small it provides a firm attachment to the wall of the intestine and prevents the worm from being dislodged by the host's peristaltic movements.

Behind the scolex the narrow neck region gives rise to proglottids by a continuous process of budding. As the individual segments grow and mature they are pushed back from the scolex. The adult worm may be over three metres long.

Tapeworms are highly specialised endoparasites and neither scolex nor proglottids contain a mouth or alimentary canal. Predigested food can be absorbed over the entire body surface, facilitated by the worm's large surface area to volume ratio; each proglottid is no more than 1 mm thick. The worms have a thick cuticle and produce inhibitory substances to prevent their digestion by the host's enzymes. Simple nerve fibres and a pair of excretory canals run the length of the worm but most of its anatomy is concerned with reproduction. Each mature proglottid contains both male and female reproductive organs and, following fertilization, eggs receive shells and yolk before passing to the uterus where they accumulate. The sex organs degenerate leaving an egg-packed uterus which fills the proglottid. At intervals these segments, known as **gravid proglottids**, break off the chain and are expelled with the host's faeces. Each segment may contain up to 40 000 eggs which survive until eaten by the secondary host because they have resistant shells. Further development only takes place when the eggs are consumed by pigs. Embryos then emerge and move into the animal's muscles where they remain dormant until the 'meat' is eaten by man and the worm once again becomes active. Tapeworms are contracted by eating undercooked, infected meat.

The adult worms cause little discomfort in man, who is usually aware of the parasite only by the presence of creamy-white segments in the faeces. However, pork tapeworm has a unique feature in its life history which makes its eradication essential. If the eggs are eaten by man they develop in just the same way as if they were eaten by a pig. This causes human **cysticercosis** with the dormant embryos encysting in various organs and damaging the surrounding tissue. The adults can be eradicated from man by the use of appropriate drugs and this should reduce the frequency of

TABLE 25.1 **Parasitic diseases**

Parasite	Major group	Primary host	Disease caused	Secondary host, if any
Phytophthora infestans	Fungi	Potato	Potato blight	—
Puccinia graminis	Fungi	Wheat	Black stem rust	—
Eimeria	Apicomplexa	Poultry	Coccidiosis	—
Plasmodium	Apicomplexa	Man	Malaria	*Anopheles* mosquito
Schistosoma	Platyhelminthes	Man	Bilharzia (schistosomiasis)	Fresh-water snail
Fasciola	Platyhelminthes	Sheep	Liver rot	Snail
Taenia solium	Platyhelminthes	Man	Occasionally cysticercosis	Pig
Wucheria bancrofti	Nematoda	Man	Elephantiasis	Mosquito
Oncocerca volvulus	Nematoda	Man	River blindness	Blackfly

cysticercosis. Treatment must also be accompanied by thorough meat inspection and public health measures. There must be adequate sewage-treatment plants and the prohibition of the discharge of raw sewage into inland waters or the sea.

Beef tapeworm (*T. saginata*) has a similar life cycle to *T. solium* but cattle provide the intermediate host. *T. saginata* is more widespread world-wide but it is less serious because human cysticercosis does not arise in infections with this species.

Both *Fasciola* and *Taenia* have structural modifications for their parasitic mode of life. They have suckers for attachment and a body covering which helps protect them from the host's immune responses or digestive enzymes. *Fasciola* has a relatively well developed gut, and feeds on cells predominantly by extracellular digestion. *Taenia* having no gut, absorbs predigested food. Most parasitic Platyhelminthes are hermaphrodite and all have systems capable of producing large numbers of eggs. A combination of sexual and asexual stages in order to multiply rapidly is ideal for a parasite.

25.2.5 Economic importance of parasites

Many parasites cause disease in man, his crops and domesticated animals. Table 25.1 includes several fungal parasites and representative parasites from a number of animal groups. Diseases caused by bacteria and viruses are summarized in Chapter 7.

TABLE 25.2 **Economic importance of some saprophytes**

Saprophyte	Major group	Carbon source	Effect
Cladosporium resinae	Fungi	Jet fuel	Blocks filters and damages tank linings
Lentinus and others	Fungi	Lignin and cellulose	Causes white rot of wood, e.g. in power station cooling towers
Bacillus subtilis	Bacteria	Proteins	Breaks down protein of wool to amino acids, reducing its commercial value
Serpula lacrymans	Fungi	Cellulose	Dry rot
Botrytis cinerea	Fungi	Hydrocarbons	Degrades and recycles small amounts of hydrocarbons, including oil
Pseudomonas	Bacteria	Hydrocarbons	
Aerobacter	Bacteria	DDT	Their joint action breaks down pesticide residues in soil
Hydrogenomonas			
Clostridium and others	Bacteria	Cellulose and similar polysaccharides	Important component of sewage microflora, breaking down cellulose
Lactobacillus bulgaricus	Bacteria	Lactose	Yoghurt production
Saccharomyces (yeast)	Fungi	Sucrose	Fermentation products include CO_2 and ethanol used for baking and brewing
Aspergillus niger	Fungi	Sucrose	Used in citric acid production
Candida utilis	Fungi	Molasses or potato starch	Used for large-scale production of vitamin B using cheap organic carbon sources

25.3 Saprophytes

Heterotrophs are unable to use carbon dioxide as their sole source of organic materials. Saprophytes are heterotrophic organisms which obtain carbon by absorption from dead organisms or organic wastes. As the word saprophyte implies a plant, and many organisms in this category are not true plants, the word **saprobiont** is often preferred. True saprophytes do not invade living tissues but some parasitic fungi, e.g. *Ceratocystis ulmi* which causes Dutch elm disease, are able to feed saprophytically when they have killed their host.

The majority of saprophytes are bacteria or fungi and they absorb organic materials through their cell walls (see *Mucor*, Fig. 7.6). Monosaccharides may be absorbed directly but larger molecules must first be broken down by enzymes. These enzymes are produced by the saprophyte and released onto the food material where extracellular digestion occurs. The soluble products of digestion are then absorbed.

25.3.1 Economic importance of saprophytes

If nutrients are not to become exhausted, carbon, nitrogen and other elements contained in dead organisms must be made available to living plants and animals. Saprophytic bacteria and fungi play an essential rôle in recycling these chemicals as they break down dead organic material (Section 27.2).

Similar organisms cause spoilage of food as they utilize the organic compounds in such things as stored fruit and grain. Both recycling and food spoilage involve a large number of different saprophytes; some specific ones are mentioned in Table 25.2.

As saprophytes break down organic materials to obtain carbon, they form a number of by-products many of which can be used by man as the basis for various industrial processes, such as brewing, baking and cheese-making (Section 41.3).

25.4 Symbionts

Symbiosis is an association between two different organisms in which neither is harmed and both may benefit in some way. Examples of symbiosis include:

1. Lichen – This is an association between an alga, usually *Trebouxia*, and a fungus, usually an ascomycete. The benefit to the alga is not clear but the fungus receives and uses the products of algal photosynthesis.

2. Hydra-Chlorella symbiosis – *Chlorella* are unicellular green algae found in the endoderm of the cnidarian *Hydra*. They are able to photosynthesize and supply *Hydra* with maltose.

3. Mycorrhizas – These are structures formed by the association of roots of plants with fungi. They are common in many higher plants, including pine, oak, beech and birch. The fungus is dependent on symbiosis for carbon nutrition and the higher plant for inorganic nutrients.

Further examples of symbiosis will be considered in a little more detail in the following sections.

25.4.1 Gut symbionts

Symbiotic relationships are of great importance in the nutrition of some animal species in connection with digestive processes and vitamin requirements. Protozoans and bacteria living in the gut receive food and protection. They synthesize their own vitamins, especially vitamin K and vitamins of the B group, and any excess is made available to the host animal.

Symbionts are particularly important to animals with specialized diets, and in some invertebrates their presence is essential for the survival of the host. In termites the posterior region of the gut is expanded to form a pouch which houses a variety of flagellate protozoans such as *Trichonympha*. These symbionts produce the enzymes essential for the breakdown of the termites' main source of food – wood. Without them the insect would die. Many blood-sucking insects rely on symbiotic fungi and bacteria for essential vitamins, especially B, which they are not able to synthesize and which are not present in their diet.

Symbiotic bacteria, fungi and protozoans are essential for the digestion of cellulose in herbivores. Many herbivorous mammals have an enlarged caecum and appendix to hold large masses of vegetable matter while it is digested by the cellulases released by these symbionts. The most efficient herbivorous mammals, in terms of nutrients extracted from cellulose, are the ruminants. The rumen houses a greater variety of symbionts than the caecum and they are able to achieve a more complete breakdown of cellulose. Another reason why ruminant digestion is more effective is that the rumen is anterior to the main region of enzyme production and so, when the symbionts die, they pass through the digestive system with the food and form an important source of nutrients, especially protein. The caecum and appendix are beyond the main regions of enzymic digestion and so the materials incorporated into these symbionts will be lost to the host when they die. Further details of digestion in herbivorous mammals are given in Section 24.6.1.

25.4.2 Symbiosis and the nitrogen cycle

Many angiosperms and a few gymnosperms form swellings called **nodules** on their stems, roots or leaves. These nodules contain micro-organisms which are capable of **nitrogen fixation**. The best known examples are the root nodules formed in leguminous plants such as peas, beans and clover. In this case the nitrogen-fixing micro-organism is the bacterium *Rhizobium*. This association provides the bacteria with a carbon source and the legume with a source of nitrates, independent of their abundance in the soil. The development of a nodule is similar to the development of a lateral root except that at an early stage the central cells are filled with bacterial cells enclosed in a membrane. *Rhizobium* thus remains extracellular.

It has been estimated that symbiotic organisms fix about 100 million tonnes of nitrogen per year. Without this recycling of atmospheric nitrogen the level of soil nitrates would be far too low to support the present vegetation cover. Leguminous crops improve soil fertility and, in terms of efficiency of fixation, the biological process compares favourably with the commercial manufacture of nitrogenous fertilizer.

The exact site of nitrogen fixation has been the subject of much research. Neither the legume nor *Rhizobium* alone are capable of nitrogen fixation. Fixation depends on the symbiotic relationship, and a possible site for it is the membrane which separates the two organisms. The nitrogen cycle is considered in more detail in Section 27.2.2.

25.5 Questions

1. Distinguish between parasitism, predation and saprophytism. *(12 marks)*
Describe, using suitable examples, other types of close interspecific relationships between species. *(13 marks)*
(Total 25 marks)

Southern Universities Joint Board June 1986, Paper II, No. 7

2. Show how a knowledge of the ecology and life history of a disease-causing organism can facilitate the control of the disease.

Southern Universities Joint Board June 1984, Paper II, No. 22

3. (a) With reference to suitable **named** examples, discuss adaptations to a parasitic mode of life. *(14 marks)*

(b) Suggest reasons why some groups of organisms possess many parasitic species, while other groups possess few or none. *(4 marks)*
(Total 18 marks)

Cambridge Board June 1984, Paper I, No. 6

4. (a) Describe the essential features of the following relationships:
  (i) predator-prey;
  (ii) parasite-host;
  (iii) symbiosis (mutualism). *(6 marks)*

(b) How may **each** type of relationship influence the population sizes of the organisms involved? *(8 marks)*

(c) Referring to **one named** parasite and its host, discuss how the rate at which the parasite spreads throughout the host population may be related to its generation time, reproductive output, method of dispersal and density of the host. *(6 marks)*
(Total 20 marks)

Joint Matriculation Board June 1985, Paper II, No. 1

5. The life cycle of the fungus *Phytophthora infestans*, responsible for potato blight, is given at the top of the next column.
(a) *Phytophthora infestans* is an **obligate parasite**. Define this term. *(3 marks)*

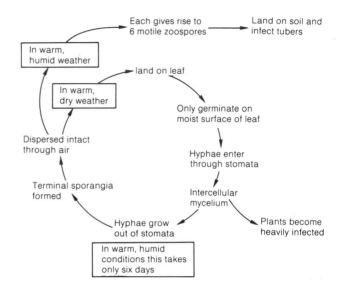

(b) How does the fungus obtain nutrients from the host? *(2 marks)*
(c) What weather conditions would restrict the spread of the disease? Give reasons for your answer. *(4 marks)*
(d) State **two** stages in the life cycle where preventative measures would be recommended and suggest what these might be. *(4 marks)*
(e) *Phytophthora infestans* was inoculated onto a growth medium with added fungicide (concentration 0.2 g cm^{-3}). A second sample was inoculated onto a growth medium without fungicide.
  After three months' growth, samples of each were transferred to fresh growth media with the same fungicide added in *varying* concentrations.
  The results are tabulated below.

Growth of mycelium /mm day	Conc. of fungicide/g cm^{-3}			
	None	0.1	0.2	0.5
Previously grown with fungicide	13.0	10.4	9.8	9.8
Previously grown without fungicide	12.3	1.5	1.0	0.5

(i) Compare the effects of increasing fungicide concentration as shown in the table. (*2 marks*)

(ii) Suggest reasons for the differences in the two cultures. (*4 marks*)

(*Total 19 marks*)

Oxford and Cambridge Board July 1984, Paper I, No. 5

6. (*a*) Distinguish between the terms *mutualism* (*symbiosis*) and *parasitism*. (*4 marks*)

(*b*) Describe how

(i) gut bacteria contribute to the well-being of herbivores,

(ii) nitrogen-fixing bacteria contribute to the well-being of legumes (*10 marks*)

(*c*) Explain the key rôles of soil microorganisms in the recycling of two inorganic nutrients.

(*6 marks*)

(*Total 20 marks*)

Oxford and Cambridge Board June 1989, Paper II, No. 3

7. The drawing shows a parasite which may be found in large numbers in the large intestine of the African elephant.

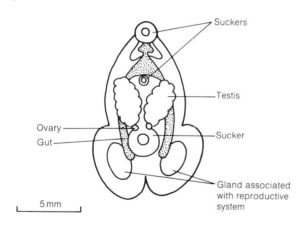

Using evidence from the drawing, give **two** features that are adaptations to a parasitic mode of life. In each case describe how the feature that you have mentioned represents an adaptation to parasitism. (*4 marks*)

Associated Examining Board June 1989, Paper I, No. 4

8. Some pea plants were dug up from a field and washed thoroughly to expose any nodules that were present on their roots. These nodules were removed, surface sterilized and transferred aseptically to a sterile, liquid culture medium. After two weeks' incubation, small samples of culture medium were removed and used to inoculate trays each containing a batch of pea plants growing in an inert medium.

Each batch of plants was 'watered' regularly with a nutrient solution containing a particular concentration of sodium nitrate. At the end of four weeks the mean number of root nodules and biomass were determined for each batch.

(*a*) (i) Suggest a suitable method for surface sterilising the root nodules.

(ii) Explain why surface sterilization and aseptic transfer were necessary in this particular investigation.

(iii) Which organism was most likely to have been isolated and grown in the liquid culture medium? (*4 marks*)

The results obtained from this investigation are shown on the graph.

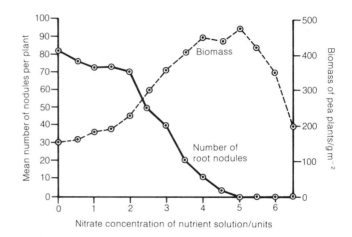

(*b*) *Using information from the graph*, state in what soil conditions you would expect pea plants to produce large numbers of nodules. (*1 mark*)

(*c*) Account as fully as possible for the results obtained for biomass in this investigation with applied nitrate concentrations of:

(i) 0 to 1.5 units;

(ii) 2.0 to 5.0 units;

(iii) 5.5 to 6.5 units. (*3 marks*)

(*d*) Draw clearly on the graph the plot for biomass you would expect to have obtained if the experiment had been repeated but with peas that had not been inoculated with culture medium. (*1 mark*)

(*Total 9 marks*)

Joint Matriculation Board June 1988, Paper IA, No. 2

9. The diagram overleaf shows a transverse section through the stem of the parasitic angiosperm dodder (*Cuscuta*) and part of a host plant.

(*a*) (i) Identify the structure labelled X and state its function. (*2 marks*)

(ii) Identify the host tissue labelled Y, with which structure X is in contact. (*1 mark*)

(*b*) State *two* ways in which extensive parasitism by dodder may damage the host plant.

(*2 marks*)

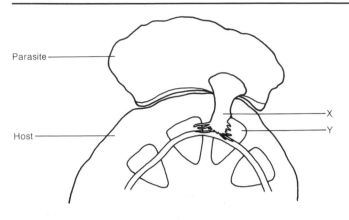

Parasite

Host

X

Y

(c) Parasites commonly show a reduction or loss of some of the features possessed by closely related organisms that are not parasitic.

 (i) State *three* such features reduced or lost in adult dodder plants. (*3 marks*)

 (ii) Briefly explain how such reduction or loss may have arisen in parasitic organisms.

(*2 marks*)

(*Total 10 marks*)

London Board January 1989, Paper I, No. 6

26 *Cellular respiration*

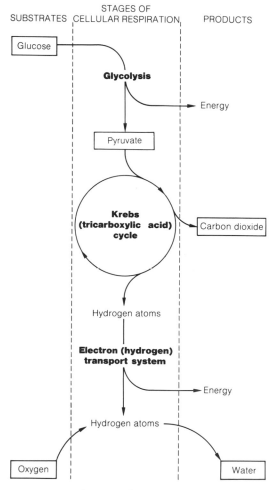

STAGES OF
SUBSTRATES CELLULAR RESPIRATION PRODUCTS

Glucose

Glycolysis

→ Energy

Pyruvate

**Krebs
(tricarboxylic acid)
cycle**

Carbon dioxide

Hydrogen atoms

**Electron (hydrogen)
transport system**

→ Energy

Hydrogen atoms

Oxygen   Water

Fig. 26.1 Outline of cellular respiration

In Chapter 21 we saw that living systems require a constant supply of energy to maintain low entropy and so ensure their survival. This energy initially comes from the sun and is captured in chemical form by autotrophic organisms during the process of photosynthesis (Chapter 23). While the carbohydrates, fats and proteins so produced are useful for storage and other purposes, they cannot be directly used by cells to provide the required energy. The conversion of these chemicals into forms like adenosine triphosphate, which can be utilized by cells, occurs during respiration.

Whatever form the food of an organism initially takes, it is converted into carbohydrate, usually the hexose sugar glucose, before being respired. Most respiration is the oxidation of this glucose to carbon dioxide and water with the release of energy, and the process can be conveniently divided into two parts:

1. Cellular (internal or tissue) respiration – the metabolic processes within cells which release the energy from glucose.

2. Gaseous exchange (external respiration) – the processes involved in obtaining the oxygen for respiration and the removal of gaseous wastes.

Gaseous exchange is dealt with in Chapter 31. Cellular respiration is the subject of this chapter and can be divided into three stages:

1. Glycolysis

2. Krebs (tricarboxylic acid) cycle

3. Electron (hydrogen) transport system

The relationship of these stages in cellular respiration is outlined in Fig. 26.1.

26.1 Adenosine triphosphate (ATP)

Adenosine triphosphate (ATP) is the short-term energy store of all cells. It is easily transported and is therefore the universal energy carrier.

26.1.1 Structure of ATP

ATP is formed from the nucleotide adenosine monophosphate (Fig. 12.13) by the addition of two further phosphate molecules. Its structure is shown in Fig. 26.2, over the page.

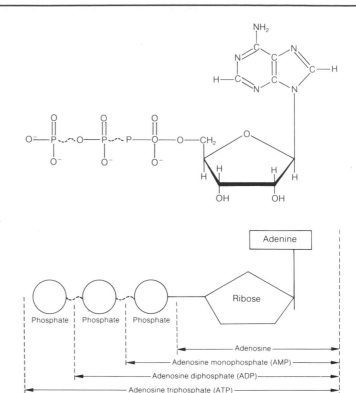

Fig. 26.2 Structure of adenosine triphosphate

26.1.2 Importance of ATP

The hydrolysis of ATP to ADP may be catalysed by a number of enzymes and the removal of the terminal phosphate yields 30.6 kJ mol^{-1} of free energy. Further hydrolysis of ADP to AMP yields a similar quantity of energy, but the removal of the last phosphate produces less than half this quantity. For this reason the last two phosphate bonds are often termed high energy bonds on account of the relatively large quantity of energy they yield on hydrolysis. This is misleading in that it implies that all the energy is stored in these bonds. The energy is in fact stored in the molecule as a whole, although the breaking of the bonds initiates its release.

AMP and ADP may be reconverted to ATP by the addition of phosphate molecules in a process called **phosphorylation**, of which there are two main forms:

1. Photosynthetic phosphorylation – occurs during photosynthesis in chlorophyll-containing cells (Chapter 23).

2. Oxidative phosphorylation – occurs during cellular respiration in all aerobic cells.

The addition of each phosphate molecule requires 30.6 kJ of energy. If the energy released from any reaction is less than this, it cannot be stored as ATP and is lost as heat. The importance of ATP is therefore as a means of transferring free energy from energy-rich compounds to cellular reactions requiring it. While not the only substance to transfer energy in this way, it is by far the most abundant and hence the most important. A metabolically active cell may require up to two million ATP molecules every second.

26.2 Glycolysis

Glycolysis (*glyco* – 'sugar'; *lyso* – 'breakdown') is the breakdown of a hexose sugar, usually glucose, into two molecules of the three-carbon compound **pyruvate (pyruvic acid)**. It occurs in all cells; in anaerobic organisms it is the only stage of respiration. Initially the glucose is insufficiently reactive and so it is phosphorylated prior to being split into two triose sugar molecules. These molecules yield some hydrogen atoms which may be used to give energy (ATP) before being converted into pyruvate. During its formation, the ATP used in phosphorylating the glucose is regenerated. Glycolysis takes place in the cytoplasm of the cell and its main stages are outlined below:

Stages of glycolysis

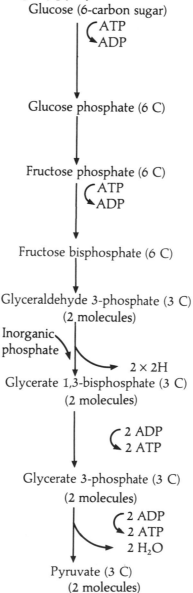

Stage 1
The glucose molecule is phosphorylated to make it more reactive. The phosphate molecule comes from the conversion of ATP to ADP.

Stage 2
The glucose molecule is reorganized into its isomer, fructose phosphate

Stage 3
Further phosphorylation takes place, by the donation of another phosphate molecule from ATP to make the sugar yet more reactive

Stage 4
The six-carbon sugar is split into two three-carbon sugars

Stage 5
More phosphorylation occurs but in this case the source of the phosphate is inorganic and not ATP. Two pairs of hydrogen atoms are removed

Stage 6
A phosphate molecule is lost from both molecules of glycerate 1,3-bisphosphate, thus yielding two molecules of ATP from ADP

Stage 7
A further pair of phosphates are removed forming two more ATPs. Each glycerate 3-phosphate molecule also has a water molecule removed.

Each glucose molecule produces two molecules of glycerate 3-phosphate and there is therefore a pair of every subsequent molecule for each glucose molecule. The energy yield is a net gain of two molecules of ATP (Stage 7). The two pairs of hydrogen atoms produced (Stage 5) may yield a further six ATPs (see Section 26.4), giving an overall total of eight ATPs.

26.3 Krebs (tricarboxylic acid) cycle

Although glycolysis releases a little of the energy from the glucose molecule, the majority still remains 'locked-up' in the pyruvate. These molecules enter the mitochondria and, in the presence of oxygen, are broken down to carbon dioxide and hydrogen atoms. The process is called the **Krebs cycle**, after its discoverer Hans Krebs. There are a number of alternative names, notably the **tricarboxylic acid cycle (TCA cycle)** and **citric acid cycle**. While the carbon dioxide produced is removed as a waste product, the hydrogen atoms are oxidized to water in order to yield a substantial amount of free energy. Before pyruvate enters the Krebs cycle it combines with a compound called coenzyme A to form **acetyl coenzyme A**. In the process, a molecule of carbon dioxide and a pair of hydrogen atoms are removed. The 2-carbon acetyl coenzyme A now enters the Krebs cycle by combining with the 4-carbon oxaloacetate (oxaloacetic acid) to give the 6-carbon citrate (citric acid). Coenzyme A is reformed and may be used to combine with a further pyruvate molecule. The citrate is degraded to the 5-carbon α-ketoglutarate (α-ketoglutaric acid) and then the 4-carbon oxaloacetate by the progressive loss of two carbon dioxide molecules, thus completing the cycle. For each turn of the cycle, a total of four pairs of hydrogen atoms are also formed. Of these, three pairs are combined with the hydrogen carrier **nicotinamide adenine dinucleotide (NAD)** and yield three ATPs for each pair of hydrogen atoms. The remaining pair combines with a different hydrogen carrier, **flavine adenine dinucleotide (FAD)** and yields only two ATPs. In addition, each turn of the cycle produces sufficient energy to form a single molecule of ATP. It must be remembered that all these products are formed from a single pyruvate molecule of which two are produced from each glucose molecule. The total yields from a single glucose molecule are thus double those stated. The significance of this will become apparent when considering the total quantity of energy released (Section 26.6).

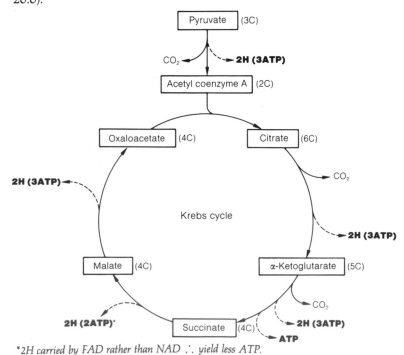

*2H carried by FAD rather than NAD ∴ yield less ATP.

Fig. 26.3 Summary of Krebs cycle

26.3.1 Importance of Krebs cycle

The Krebs cycle plays an important rôle in the biochemistry of a cell for three main reasons:

1. It brings about the degradation of macromolecules – The 3-carbon pyruvate is broken down to carbon dioxide.

2. It provides the reducing power for the electron (hydrogen) transport system – It produces pairs of hydrogen atoms which are ultimately the source of metabolic energy for the cell.

3. It is an interconversion centre – It is a valuable source of intermediate compounds used in the manufacture of other substances, e.g. fatty acids, amino acids, chlorophyll (Section 26.7).

26.4 Electron transport system

The electron transport system is the means by which the energy, in the form of hydrogen atoms, from the Krebs cycle, is converted to ATP. The hydrogen atoms attached to the hydrogen carriers NAD and FAD are transferred to a chain of other carriers at progressively lower energy levels. As the hydrogens pass from one carrier to the next, the energy released is harnessed to produce ATP. The series of carriers is termed the **respiratory chain**. The carriers in the chain include an **FAD**, **coenzyme Q** and iron-containing proteins called **cytochromes**. Initially hydrogen atoms are passed along the chain, but these later split into their protons and electrons, and only the electrons pass from carrier to carrier. For this reason, the pathway can be called the electron, or hydrogen, transport system. At the end of the chain the protons and electrons recombine, and the

NAD = nicotinamide adenine
dinucleotide
FAD = flavine adenine dinucleotide
CoQ = coenzyme Q
ATP = adenosine triphosphate

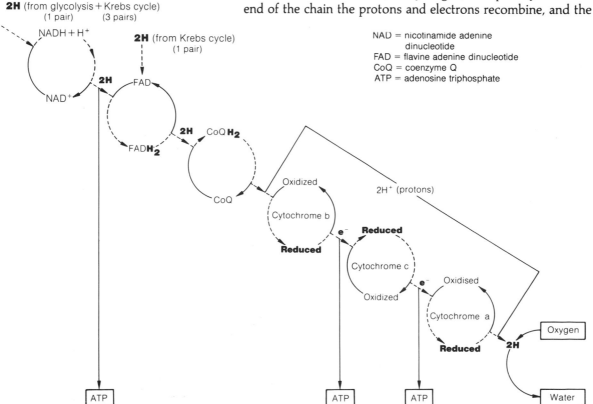

Fig. 26.4 Summary of the electron transport system

hydrogen atoms created link with oxygen to form water. This formation of ATP through the oxidation of the hydrogen atoms is called **oxidative phosphorylation**. It occurs in the mitochondria.

The rôle of oxygen is to act as the final acceptor of the hydrogen atoms. While it only performs this function at the end of the many stages in respiration, it is nevertheless vital as it drives the whole process. In its absence, only the anaerobic glycolysis stage can continue. The transfer of hydrogen atoms to oxygen is catalyzed by the enzyme **cytochrome oxidase**. This enzyme is inhibited by cyanide, so preventing the removal of hydrogen atoms at the end of the respiratory chain. In these circumstances the hydrogen atoms accumulate and aerobic respiration ceases, making cyanide a most effective respiratory inhibitor.

26.4.1 Mitochondria and oxidative phosphorylation

Mitochondria are present in all eukaryotic cells where they are the main sites of respiratory activity. Highly active cells, requiring much energy, characteristically have numerous large mitochondria packed with cristae. Such cells include:

Liver cells – Energy is required to drive the large and varied number of biochemical reactions taking place there.

Striated muscle cells – Energy is needed for muscle contraction, especially where this is rapid, e.g. flight muscle of insects.

Sperm tails – These provide energy to propel the sperm.

Nerve cells – Mitochondria are especially numerous adjacent to synapses where they provide the energy needed for the production and release of transmitter substances.

Intestinal epithelial cells – Many mitochondria occur beneath the microvilli on these cells to provide energy for the absorption of digested food by active transport.

We saw in Section 4.3.6 that mitochondria have an inner membrane which is folded to form cristae in order to increase its surface area. The cristae are lined with stalked particles (Fig. 4.9c). These particles are associated with oxidative phosphorylation and it is thought that the electron carriers of the electron transport system and their associated enzymes are attached to these particles in a precise sequence. Within the inner mitochondrial membrane there appears to be a mechanism which actively transports protons (H^+) from the matrix into the space between the inner and outer membranes of the organelle. This creates an electrochemical gradient of hydrogen ions across the inner membrane. According to the **chemi-osmotic theory** put forward by the British biochemist Peter Mitchell, in 1961, it is the energy of this 'charged' membrane which is used to synthesize ATP. In addition to carrying out oxidative phosphorylation, the mitochondria perform the reactions of the Krebs cycle. The enzymes for these reactions are mostly found within the matrix, with a few, like succinic dehydrogenase, attached to the inner mitochondrial membrane.

26.5 Anaerobic respiration (anaerobiosis)

Present thinking suggests that life originated in an atmosphere without oxygen and the first forms of life were therefore anaerobic. Many organisms today are also anaerobic; indeed, some find oxygen toxic. These forms are termed **obligate anaerobes**. Most anaerobic organisms will, however, respire aerobically in the presence of oxygen, only resorting to anaerobiosis in its absence. These forms are **facultative anaerobes**. The cells of almost all organisms are capable of carrying out anaerobic respiration, for a short time at least. From what we have so far learnt it is clear that, in the absence of oxygen, the Krebs cycle and electron transport system cannot operate. Only glycolysis can take place. This yields a little ATP directly (two molecules for each glucose molecule) and a total of two pairs of hydrogen ions. In the previous section we saw that these hydrogen ions possess much free energy. In the absence of oxygen, however, this energy cannot be released. Nevertheless, these hydrogen ions must be removed if glycolysis is to continue. They are accepted by the pyruvate formed at the end of glycolysis, to give either ethanol (alcohol) or lactate, in a process called **fermentation**. Neither process yields any additional energy, but both are merely mechanisms for 'mopping-up' the hydrogen ions.

26.5.1 Alcoholic fermentation

In alcoholic fermentation the pyruvate from glycolysis is first converted to ethanal (acetaldehyde) through the removal of a carbon dioxide molecule.

$$CH_3COCOOH \longrightarrow CH_3CHO + CO_2$$
pyruvate   ethanal   carbon dioxide

The ethanal then combines with the hydrogen ions, which are transported by the hydrogen carrier NAD, to form the alcohol, ethanol.

$$CH_3CHO \xrightarrow[\quad]{NADH+H^+ \quad NAD^+} CH_3CH_2OH$$
ethanal   ethanol

This form of fermentation occurs in yeast, where the alcohol produced may accumulate in the medium around the cells until its concentration rises to a level which prevents further fermentation, and so kills the yeast. The ethanol cannot be further broken down to yield additional energy.

The overall equation is:

$$C_6H_{12}O_6 \longrightarrow 2CH_3CH_2OH + 2CO_2$$
glucose   ethanol   carbon dioxide

Under anaerobic conditions, e.g. waterlogging of plant roots, the cells of higher plants may temporarily undergo this form of fermentation. Alcoholic fermentation is of considerable economic importance to man. It is the basis of the brewing industry, where the ethanol is the important product, and of the baking industry, where the carbon dioxide is of greater value (Section 41.3).

26.5.2 Lactate fermentation

In lactate fermentation the pyruvate from glycolysis accepts the hydrogen atoms from $NADH+H^+$ directly.

An industrial fermenter

$$\text{CH}_3\text{COCOOH} \xrightarrow{\quad \text{NADH} + \text{H}^+ \quad \text{NAD}^+ \quad} \text{CH}_3\text{CHOHCOOH}$$

pyruvate   lactate
(2-hydroxypropanoic acid)

Unlike alcoholic fermentation, the lactate can be further broken down, should oxygen be made available again, thus releasing its remaining energy. Alternatively it may be resynthesized into carbohydrate, or excreted.

This form of fermentation is common in animals. Clearly any mechanism which allows an animal to withstand short periods without oxygen (anoxia) has great survival value. Animals living in environments of fluctuating oxygen levels, such as a pond or river, may benefit from the temporary use of it, as might a baby in the period during and immediately following birth. A more common occurrence of lactate fermentation is in a muscle during strenuous exercise. During this period, the circulatory system may be incapable of supplying the muscle with its oxygen requirements. Lactate fermentation not only yields a little energy, but removes the pyruvate which would otherwise accumulate. Instead lactate accumulates, and while this in time will cause cramp and so prevent the muscle operating, tissues have a relatively high tolerance to it.

In the process of lactate fermentation, the organism accumulates an **oxygen debt**. This is repaid as soon as possible after the activity, by continued deep and rapid breathing following the exertion. The oxygen absorbed is used to oxidize the lactate to carbon dioxide and water, thereby removing it, and at the same time replenishing the depleted stores of ATP and oxygen in the tissue. In some organisms such as parasitic worms, where the food supply is abundant, the lactate is simply excreted, obviating the need to repay an oxygen debt.

26.6 Comparison of energy yields

Let us now compare the total quantity of ATP produced by the aerobic and anaerobic pathways.

Aerobic respiration
The ATP is derived from two sources: directly by phosphorylation of ADP and indirectly by oxidative phosphorylation using the hydrogen ions generated during glucose breakdown.

TABLE 26.1 **ATP yield during aerobic respiration of one molecule of glucose**

Respiratory process	Number of reduced hydrogen carrier molecules formed	Number of ATP molecules formed from reduced hydrogen carriers	Number of ATP molecules formed directly	Total number of ATP molecules
Glycolysis (glucose→pyruvate	$2 \times (\text{NADH} + \text{H}^+)$	$2 \times 3 = 6$	2	8
pyruvate→acetyl CoA	$1 \times (\text{NADH} + \text{H}^+) (\times 2)$	$2 \times 3 = 6$	0	6
Krebs (TCA) cycle	$3 \times (\text{NADH} + \text{H}^+) (\times 2)$ $1 \times \text{FADH}_2 (\times 2)$	$6 \times 3 = 18$ $2 \times 2 = 4$	$1 (\times 2)$	24
		Total ATP =		38

The figures given represent the yield for each pyruvate molecule which subsequently enters the Krebs (TCA) cycle. As there are two pyruvate molecules formed for each glucose molecule (Section 26.3), all these figures must be doubled (× 2) to give the quantities formed per glucose molecule.

The energy yield for each molecule of NADH + H$^+$ is three ATPs whereas for FADH$_2$ it is only two ATPs (Section 26.4).

The total of thirty-eight ATPs produced represents the maximum possible yield; the actual yield may be different depending upon the conditions in any one cell at the time. For example, the two NADH + H$^+$ may enter the mitochondria in two different, indirect ways.

Depending on the route taken, they may yield only four ATPs, rather than the six ATPs as shown in Table 26.1.

Each ATP molecule will yield around 30.6 kJ of energy. The total energy available from aerobic respiration is $38 \times 30.6 = 1162.8$ kJ. Compared to the total energy available from the complete oxidation of glucose of 2880 kJ, this represents an efficiency of slightly over 40%. This may not appear very remarkable, but it compares very favourably with man-made machines – the efficiency of a car engine is around 25%.

Anaerobic respiration

We have seen in the previous section that only glycolysis occurs during anaerobiosis and that the NADH + H$^+$ it yields is not available for oxidative phosphorylation. The total energy released is therefore restricted to the two ATPs formed directly. With each providing 30.6 kJ of energy, the total yield is a mere 61.2 kJ. Compared to the 2880 kJ potentially available from a molecule of glucose, the process is a little over 2% efficient. It must, however, be borne in mind that in lactate fermentation all is not lost, and the lactate may be reconverted to pyruvate by the liver, and so enter the Krebs cycle, thus releasing its remaining energy.

26.7 Alternative respiratory substrates

Sugars are not the only material which can be oxidized by cells to release energy. Both fats and protein may, in certain circumstances, be used as respiratory substrates, without first being converted to carbohydrate. The alternative pathways are shown on Fig. 26.5.

26.7.1 Respiration of fat

The oxidation of fat is preceded by its hydrolysis to glycerol and fatty acids. The glycerol may then be phosphorylated and converted into the triose phosphate glyceraldehyde 3-phosphate. This can then be incorporated into the glycolysis pathway and subsequently the Krebs cycle. The fatty acid component is progressively broken down in the matrix of the mitochondria into 2-carbon fragments which are converted to acetyl coenzyme A. This then enters the Krebs cycle with consequent release of its energy. The oxidation of fats has the advantage of producing a large quantity of hydrogen ions. These can be transported by hydrogen carriers and used to produce ATP in the electron (hydrogen) transport system. For this reason, fats liberate more than double the energy of the same quantity of carbohydrate.

379

26.7.2 Respiration of protein

Protein is another potential source of energy but is only used in cases of starvation. It must first be hydrolysed to its constituent amino acids which then have their amino (NH_2) group(s) removed in the liver – a process called **deamination**. The remaining portions of the amino acids then enter the respiratory pathway at a number of points depending on their carbon content: 5-carbon amino acids (e.g. glutamic acid) and 4-carbon amino acids (e.g. aspartic acid) are converted into the Krebs cycle intermediates, α-ketoglutarate and oxaloacetate respectively; 3-carbon amino acids like alanine are converted to pyruvate ready for conversion to acetyl coenzyme A. Other amino acids with larger quantities of carbon undergo transamination reactions to convert them into 3, 4 or 5-carbon amino acids.

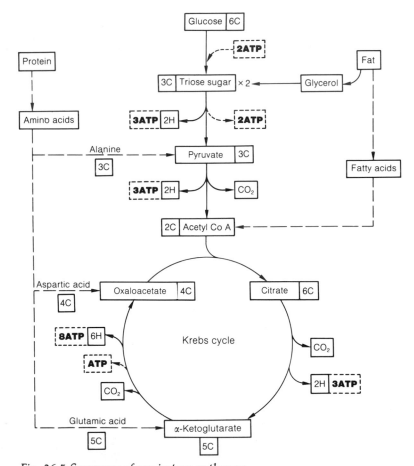

Fig. 26.5 Summary of respiratory pathways

26.8 Respiratory quotients

The **respiratory quotient (RQ)** is a measure of the ratio of carbon dioxide evolved by an organism to the oxygen consumed, over a certain period.

$$RQ = \frac{CO_2 \text{ evolved}}{O_2 \text{ consumed}}$$

For a hexose sugar like glucose, the equation for its complete oxidation is:

$$C_6H_{12}O_6 + 6O_2 \longrightarrow 6CO_2 + 6H_2O$$

The RQ is hence: $\dfrac{6CO_2}{6O_2} = 1.0$

In fats, the ratio of oxygen to carbon is far smaller than in a carbohydrate. A fat therefore requires a greater quantity of oxygen for its complete oxidation and thus has a RQ less than one.

$$C_{18}H_{36}O_2 + 26O_2 \longrightarrow 18CO_2 + 18H_2O$$
stearic acid

$$RQ = \frac{18CO_2}{26O_2} = 0.7$$

The composition of proteins is too varied for them to give the same RQ, but most have values around 0.9.

Organisms rarely, if ever, respire a single food substance, nor are substances always completely oxidized. Experimental RQ values therefore do not give the exact nature of the material being respired. Most resting animals have RQs between 0.8 and 0.9. With protein only respired during starvation, this must be taken to indicate a mixture of fat and carbohydrate as the respiratory substrates.

26.9 Questions

1. (a) Explain briefly how cellulose and lignin contribute to the structure of terrestrial flowering plants. *(6 marks)*
   (b) Cellulose is digested to glucose by certain bacteria found in the stomach of sheep. Some of this glucose is then metabolised anaerobically by these bacteria to form 2-carbon fragments of acetate (acetyl CoA). Sheep can absorb acetate and use it as a respiratory substrate.

   Outline the biochemical processes which occur in
   (i) the digestion of cellulose to glucose by the bacteria,
   (ii) the anaerobic metabolism of glucose to acetyl CoA via pyruvate by the bacteria,
   (iii) the aerobic respiration of acetyl CoA by the mammal. *(14 marks)*
   (Total 20 marks)
   Joint Matriculation Board June 1988, Paper IIB, No. 1

2. (a) Discuss the parts played by the following in aerobic respiration:
   (i) phosphorylation of hexose; *(3 marks)*
   (ii) acetyl coenzyme A; *(3 marks)*
   (iii) oxidative decarboxylation of pyruvate; *(4 marks)*
   (iv) electron transfer system. *(4 marks)*
   (b) Why does anaerobic respiration produce much less energy than aerobic respiration? *(4 marks)*
   (Total 18 marks)

   Cambridge Board November 1983, Paper I, No. 4

3. When an athlete is running in a 200-metre sprint, his rate of oxygen consumption rises above the resting level. After the race his oxygen consumption does not fall to the resting level for some time. A blood sample taken at the end of the race shows a significant increase in lactic acid. The diagram summarizes the metabolic processes involved in the release of energy in the muscles during and after exercise.

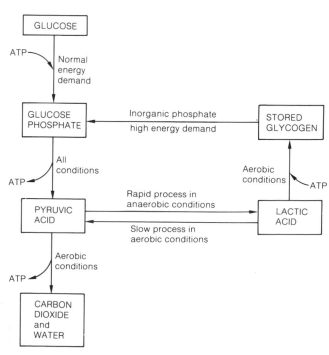

(a) (i) Explain why, although oxygen consumption rises during the race, more lactic acid is present in the blood at the end of the race. *(4 marks)*

(ii) Why is the athlete's oxygen consumption still higher than normal some time after the race? *(1 mark)*

(iii) After the race ATP is used to convert most lactic acid to glycogen. A smaller amount of lactic acid is fully oxidized with the production of ATP. With the help of information in the diagram, explain why this is an efficient use of lactic acid. *(4 marks)*

(iv) For every molecule of glucose 38 molecules of ATP are produced during aerobic respiration but only 2 molecules of ATP are produced during anaerobic respiration. Explain how this difference occurs. *(3 marks)*

(v) With the help of information in the diagram, explain why it is advantageous to respire glycogen rather than glucose during strenuous exercise. *(2 marks)*

(b) Athletes devote much time to training. Suggest **three** ways in which exercise may improve an athlete's subsequent performance. *(6 marks)*

(Total 20 marks)

Associated Examining Board June 1984, Paper II, No. 2

---

4. (a) (i) What are the three chemical groupings within the ATP molecule?

(ii) Which part of the molecule is important for its function in metabolism?

(iii) ATP is described as occupying a central place in metabolism; why is this so? *(5 marks)*

(b) What is the function of the electron transport chain (sometimes called the hydrogen transport chain) in the following processes?

(i) The oxidation of intermediates of glycolysis and the Krebs cycle (TCA cycle).

(ii) The production of ATP. *(6 marks)*

(c) How do the energy yields from the following processes differ?

(i) Anaerobic respiration of glucose in animals.

(ii) Aerobic respiration of glucose in animals.

(iii) Aerobic respiration of fats in animals. *(3 marks)*

(d) Outline the chemical changes which proteins and fats undergo when they are used as an energy source. Indicate where the products of these changes link up with carbohydrate respiration *(6 marks)*

(Total 20 marks)

Northern Ireland Board June 1984, Paper II, No. 4

---

5. (a) Explain the meaning of the term 'cellular respiration'. *(3 marks)*

(b) Describe the part played by each of following in cellular respiration.

(i) Glycogen

(ii) Fermentation

(iii) Oxidative phosphorylation *(12 marks)*

(c) What are the roles of electron transport systems in cell metabolism? *(5 marks)*

(Total 20 marks)

London Board June 1988, Paper II, No. 1

---

6. The diagram shows an outline of the metabolic pathways by which simple foodstuffs are broken down to release energy.

(a) State the name given to

(i) the chain of chemical reactions 1–4;

(ii) the reactions 5–8.

(b) State the site within the cell where reactions 1–4 occur.

(c) Use the letters on the diagram to identify

(i) a 6-carbon molecule;

(ii) a 4-carbon molecule;

(iii) a 3-carbon molecule.

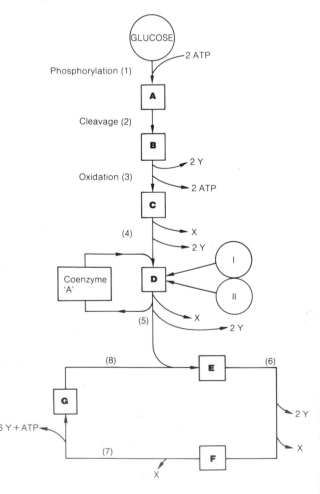

(d) Name the compounds represented by the letters X and D, and the end product represented by the letter Y.

(e) (i) Name the series of reactions involving the product Y.

(ii) State where in the cell these reactions occur.

(iii) List the two end-products of this series of reactions.

(f) Name the two energy sources (I, II) which might enter the pathway at D. (15 marks)

Welsh Joint Education Committee June 1983, Paper AI, No. 8

7. Give an outline of the process of aerobic respiration and discuss the significance of this process in maintaining the composition of the atmosphere. (20 marks)

London Board January 1989, Paper II, No. 7(b)

8. (a) Write a full description of the steps by which the energy of a glucose molecule is made available for living processes. (Diagrams alone are **not** sufficient.) (16 marks)

(b) Compare the yield of energy under aerobic and anaerobic conditions. (4 marks)

(Total 20 marks)

Welsh Joint Education Committee June 1989, Paper A1, No. 3

9. The figure gives some of the details of the conversion of a molecule of glucose to two molecules of pyruvate in glycolysis in a cell. The reactions numbered **1–7** are catalysed by different enzymes.

(a) What is meant by the word *glycolysis*?

(b) Where in the cell does glycolysis occur?

(c) Name the chemical processes that are occurring at **1** and **3**, at **4**, and at **5**.

(d) Name the class of enzymes that catalyses the reaction at **5**.

(e) Explain why inorganic phosphate is incorporated at **5**.

(f) In anaerobic respiration in muscles, the pyruvate produced in glycolysis is converted to lactate. Explain the significance of this reaction.

(g) (i) State what happens in aerobic respiration to a pair of hydrogen atoms after they are removed from glyceraldehyde 3-phosphate.

(ii) Name this particular process.

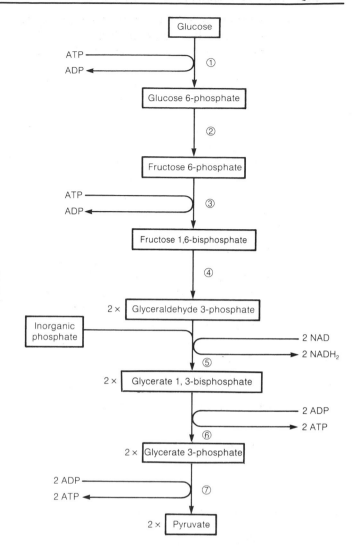

(iii) State exactly where in the cell this process occurs.

(iv) What is the significance of this process to the cell?

(Total 12 marks)

Cambridge Board June 1988, Paper III, No. 2

10. (a) In text books, the equation for aerobic respiration is often shown as
$$C_6H_{12}O_6 + 6O_2 = 6\ CO_2 + 6\ H_2O.$$
Give **four** different reasons why this equation does not fully represent the respiratory process. (4 marks)

(b) (i) The respiration rate in wheat leaves has been quoted as $310\,mg\ CO_2\ kg^{-1}$ fresh mass h^{-1} at 20 °C. Write out in full what the units kg^{-1} *fresh mass* h^{-1} mean. (2 marks)

(ii) Do you consider that the respiration rate of leaves would be better expressed in terms of leaf area rather than on a mass basis? Give a reason for your answer.

(*2 marks*)

(iii) State **two** other ways (apart from carbon dioxide output) by which the respiration rate of plant tissues may be measured.

(*2 marks*)

(c) The respiratory quotient (RQ) is defined as the ratio of the volume of carbon dioxide produced divided by the volume of oxygen uptake over the same period of time.

The table shows the changes in RQ as a soaked seed germinates.

	Treatment	R.Q.
1	4h soaking in water	6.0
2	4h soaking + 4h in air	1.8
3	4h soaking + 24h in air	1.0

(i) What would be the value of the respiratory quotient for the alcoholic fermentation of glucose by yeast cells? (*1 mark*)

(ii) Suggest a reason for the high RQ value obtained in treatment 1; (*2 marks*)
the fall in the RQ value in treatment 2; (*2 marks*)
the RQ value obtained in treatment 3. (*2 marks*)

(*Total 17 marks*)

Oxford and Cambridge Board June 1989, Paper I, No. 7

27 Energy and the ecosystem

Organisms live within a relatively narrow sphere over the earth's surface; it is less than 20 km thick, extending about 8 km above sea level and 10 km below it. The total volume of this thin film of land, water and air around the earth's surface is called the **biosphere**. It consists of two major divisions, the aquatic and terrestrial environments, with the aquatic environment being subdivided into freshwater, marine and estuarine. The terrestrial portion of the biosphere is subdivided into **biomes** which are determined by the dominant plants found there. They include tropical rain forest, temperate deciduous forest, coniferous forest, grasslands, tundra and desert. It is, of course, largely climatic conditions which determine the dominant plant type of a region, and hence the biome. It is equally possible to subdivide the terrestrial biosphere into **geographical zones**, e.g. Africa, Australia, North America, South America etc. In this case the divisions are made by barriers like oceans or mountain ranges.

A biome can be further divided into **zones** which consist of a series of small areas called **habitats**. Examples of habitats include a rocky shore, freshwater pond and a beech wood.

Within each habitat there are **populations** of individuals which collectively form a **community** (see Chapter 10). An individual member of the community is usually confined to a particular region of the habitat, called the **microhabitat**. The position any species occupies within its habitat is referred to as its **ecological niche**. It represents more than a physical area within the habitat as it includes an organism's behaviour and interactions with its living and non-living environment. As we saw in Section 10.2.3, no two species can occupy the same ecological niche.

The inter-relationship of the living (**biotic**) and non-living (**abiotic**) elements in any biological system is called the **ecosystem**. There are two major factors within an ecosystem:

1. The flow of energy through the system.

2. The cycling of matter within the system.

It is feasible to consider the biosphere as a single ecosystem because, in theory at least, energy flows through it and nutrients may be recycled within it. However, in practice there are much smaller units which are more or less self-contained in terms of energy and matter. A freshwater pond, for example, has its own community of plants to capture the solar energy necessary to supply all organisms within the habitat, and matter such as nitrogen and phosphorus are recycled within the pond with little or no loss or gain between it and other habitats. It is often easier to consider these smaller units as single ecosystems.

single food. Most animals feed on many different types. In the same way, an individual is normally a potential meal for many different species. The idea of a food chain as a sequence of species which feed exclusively off the individual below it in the series is clearly oversimplified. Individual food chains interconnect in an intricate and complex way. A simple species may form part of many different chains, not always occupying the same trophic level in each chain. Fig. 27.2 gives an example of a simplified food web for a woodland habitat.

27.1.3 Primary producers

It is the rôle of the photosynthetic organisms which make up the primary producers to manufacture organic substances using light, water and carbon dioxide. The rate at which they produce this organic food per unit area, per unit time, is called **gross primary productivity**. Not all this food is stored; around 20% is utilized by the plant, mainly during respiration. The remainder is called **net primary productivity**. It is this food which is available to the next link in the food chain, namely the primary consumers (herbivores). The net primary productivity depends upon climatic and other factors which affect photosynthesis. It is reduced in certain conditions such as cold, drought, absence of essential minerals, low light intensity etc. The type of primary producer varies from habitat to habitat and some examples are given in Table 27.1. Chemosynthetic bacteria (Section 23.6) must be considered as primary producers as they do provide energy for ecosystems, although from inorganic chemicals rather than light. Their overall contribution to the provision of the energy in a given ecosystem is very small compared to that provided by photosynthetic organisms.

27.1.4 Consumers

Those consumers which feed on the primary producers are called **primary consumers** or **herbivores**. Of the energy absorbed by a herbivore, only around 30% is actually used by the organism, the remainder being lost as urine and faeces. Most of this 30% is lost as heat, leaving less than 1% of the net productivity of the primary producer to be incorporated into the herbivore and so made available to the next animal in the food chain – the **secondary consumer**. The secondary consumers are the **carnivores** and they show almost double the efficiency of herbivores in incorporating available energy into themselves. This is largely because their protein-rich diet is much more easily digested. Not all secondary and tertiary consumers are predators; parasites and scavengers may also fall into these categories depending on the nature of their food.

27.1.5 Decomposers and detritivores

A dead organism contains not only a potential source of energy but also many valuable minerals. Decomposers (**lysotrophs**) are saprophytic microorganisms which exploit this energy source by breaking down the organic compounds of which the organism is made. In so doing they release valuable nutrients like carbon, nitrogen and phosphorus which may then be recycled (Section 27.2). Apart from dead organisms they also decompose the organic chemicals in urine, faeces and other wastes.

Detritus is the organic debris from decomposing plants and animals and is normally in the form of small fragments. It forms the diet of a group of animals called **detritivores**. They usually differ from decomposers in being larger and in digesting food internally rather than externally. Examples of detritivores include earthworms, woodlice, maggots, dog whelks and sea cucumbers.

27.1.6 Ecological pyramids

Pyramids of numbers

If a bar diagram is drawn to indicate the relative numbers of individuals at each trophic level in a food chain, a diagram similar to that shown in Fig. 27.3 (a) is produced. The length of each bar gives a measure of the relative numbers of each organism. The overall shape is roughly that of a pyramid, with primary producers outnumbering the primary consumers which in turn outnumber secondary consumers. Accepting that there is inevitably some loss when energy is transferred from one trophic level to the next in a food chain, it follows that to support an individual at one level requires more energy from the individual at the level below to compensate for this loss. In most instances this can only be achieved by having more individuals at the lower level (Fig. 27.3 (a)).

The use of pyramids of numbers has drawbacks, however:

1. All organisms are equated, regardless of their size. An oak tree is counted as one individual in the same way as an aphid.

2. No account is made for juveniles and other immature forms of a species whose diet and energy requirements may differ from the adult.

3. The numbers of some individuals are so great that it is impossible to represent them accurately on the same scale as other species in the food chain. For example, millions of blackfly may feed on a single rose-bush and this relationship cannot be effectively drawn to scale on a pyramid of numbers.

These problems may create some different shaped 'pyramids'. Take for example the food chain:

oak tree→aphid→ladybird

The pyramid of numbers produced is illustrated in Fig. 27.3 (b). It clearly bulges in the middle.

The food chain:

sycamore→caterpillar→protozoan parasites (of the caterpillar)

produces a complete inversion of the pyramid as illustrated in Fig. 27.3 (c).

These difficulties are partly overcome by the use of a pyramid of biomass, instead of one of numbers.

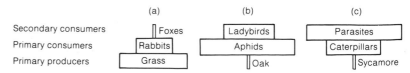

Fig. 27.3 Pyramids of numbers

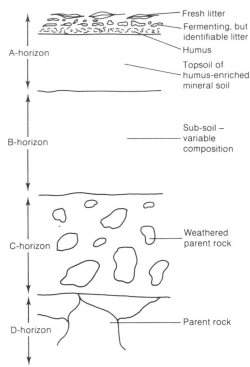

Fig. 27.7 A generalized soil profile

TABLE 27.2 **Classification of soil particles according to size**

Particle size (diameter in mm)	Particle type
2.00 − 0.200	Coarse sand
0.20 − 0.020	Fine sand
0.02 − 0.002	Silt
< 0.002	Clay

TABLE 27.3 **A comparison of clay and sandy soils**

Clay soil	Sandy soil
Particle size is less than 0.002 mm (2 μm)	Particle size from 0.02 mm to 2.0 mm
Small air spaces between particles giving poor aeration	Large air spaces between particles giving good aeration
Poor drainage; soil easily compacted	Good drainage; soil not compacted
Good water retention leading to possible waterlogging	Poor water retention and no waterlogging
Being a wet soil, evaporation of water causes it to be cold	Less water evaporation and therefore warmer
Particles attract many mineral ions and so nutrient content is high	Minerals are easily leached and so mineral content is low
Particles aggregate together to form clods, making the soil heavy and difficult to work	Particles remain separate, making the soil light and easy to work

2. Chemical weathering – the chemical breakdown as a result of water, acids, alkalis and minerals attacking certain rock types.

The general appearance of a soil seen in vertical section and called a **soil profile** is given in Fig. 27.7.

We have seen that the nature of any ecosystem is dependent upon the type of primary producer and its productivity. Both these factors are, in turn, largely determined by the properties of the soil on which the producer grows. The factors which determine a soil's properties are briefly described below.

Particle size and nature
The size of the constituent mineral particles of soil probably affects its properties, and hence the type of plant which grows on it, more than any other single factor. Soils are classified according to the size of their particles as shown in Table 27.2.

The **texture** of a soil is determined by the relative proportions of sand, silt and clay particles and this affects the agricultural potential of a soil. A clay soil, with its many tiny particles, has the advantage over a sandy soil, with its coarse particles, in holding water more readily and being less likely to have its minerals leached. On the other hand, it may easily become compacted, reducing its air content; it is slower to drain, colder and more difficult to work, especially when wet. A fuller comparison of clay and sandy soils is given in Table 27.3.

The nature of the particles as well as their size affects soil properties. Sand and silt are mainly silica (SiO_2) which is inert. Clay particles, however, have negative charges and these react with minerals in the soil, especially cations. This helps to prevent these nutrient minerals from being leached.

Organic (humus) content
This includes all dead plant and animal material as well as some animal waste products. Dead animals, leaves, twigs, roots and faeces are broken down by the decomposers and detritivores into a black, amorphous material called **humus**. It has a complicated and variable chemical make-up and is often acidic. It acts rather like a sponge in retaining water and in this way improves the structure of sandy soils. It is equally beneficial to a clay soil where it helps to lighten it by breaking up the clods and thereby improving aeration and drainage. Its slow breakdown releases valuable minerals in both types of soil. This breakdown is carried out by aerobic decomposers and thereby ceases in waterlogged conditions due to the lack of oxygen. In these circumstances the partly decomposed detritus accumulates as **peat**.

Water content
The water content of any well-drained soil varies markedly. Any freely drained soil which holds as much water as possible is said to be at **field capacity**. The addition of more water which cannot drain away leads to waterlogging and anaerobic conditions. Plants able to tolerate these conditions include the rushes (*Juncus* sp), sedges (*Carex* sp) and rice. They have air spaces among the root tissues which allow some diffusion of oxygen from the aerial parts to help supply the roots.

Air content
The space between soil particles is filled with air, from which the roots obtain their respiratory oxygen by direct diffusion. It is equally

essential to the aerobic micro-organisms in the soil which decompose the humus. They make heavy demands upon the available oxygen and may create anaerobic conditions.

Mineral content
As shown in Chapter 3 (Table 3.1), a wide variety of minerals is necessary to support healthy plant growth. Different species make different mineral demands and therefore the distribution of plants depends to some extent on the mineral balance of a particular soil. Some plants have particular nutrient requirements; the desert shrub, *Atriplex*, for example, requires sodium, a mineral not essential to most species.

Biotic content
Soils contain vast numbers of living organisms. They include bacteria, fungi and algae as well as animals like protozoans, nematodes, earthworms, insects and burrowing mammals. Bacteria and fungi carry out decomposition, while burrowing animals such as earthworms improve drainage and aeration by forming air passages in the soil. Earthworms also improve fertility by their thorough mixing of the soil which helps to bring leached minerals from lower layers within reach of plant roots. They may improve the humus content through their practice of pulling leaves into their burrows. By passing soil through their bodies they may make its texture finer.

pH
The pH of a soil influences its physical properties and the availability of certain minerals to plants. Plants such as heathers, azaleas and camelias grow best in acid soils, while dog's mercury and stonewort prefer alkaline ones. Most plants, however, grow best in an optimum pH close to neutral.

Temperature
All chemical and biological activities of a soil are influenced by temperature. The temperature of a soil may be different from that of the air above it. Evaporation of water from a soil may cool it to below that of the air whereas solar radiation may raise it above air temperature. Germination and growth depend on suitable temperatures and the optimum varies from species to species. The activity of soil organisms is likewise affected by changes in temperature, earthworms becoming dormant at low temperatures, for example.

Topography
Three features of topography may influence the distribution of organisms:

1. Aspect (slope) — South-facing slopes receive more sunlight, and are therefore warmer than north-facing ones (in the northern hemisphere).

2. Inclination (steepness) — Water drains more readily from steep slopes and these therefore dry more quickly than ones with a shallower gradient.

3. Altitude (height) — At higher altitudes the temperature is lower, the wind speed is greater and there is more rainfall.

Tropical rain forest

Tundra

27.3.2 Climatic factors

The world's major biomes are largely differentiated on the basis of climate. From the warm, humid tropical rain forest to the cold arctic tundra it is the prevailing weather conditions which determine the predominant flora and hence its attendant fauna. Each climatic zone has its own community of plants and animals which are suited to the conditions. The adaptations of plants and animals to these conditions are dealt with elsewhere in this book and what follows is merely a general review of the major climatic variables within ecosystems.

Light

As the ultimate source of energy for ecosystems, light is a fundamental necessity. Light is not only needed for photosynthesis, however. It plays a rôle in such photoperiodic behaviour as flowering in plants, and reproduction, hibernation and migration in animals. Phototaxis and phototropism in plants as well as visual perception in animals require light. There are three aspects to light – its wavelength, its intensity and its duration. The influence of these on photosynthesis is dealt with in Chapter 23, and on photoperiodism in Chapter 40.

Temperature

Just as the sun is the only source of light for an ecosystem, so it is the main source of heat. The temperature range within which life exists is relatively small. At low temperatures ice crystals may form within cells, causing physical disruption, and at high temperatures enzymes are denatured. Fluctuation in environmental temperature is more extreme in terrestrial habitats than aquatic ones because the high heat capacity of water effectively buffers the temperature changes in aquatic habitats. The actual temperature of any habitat may differ in time according to the season and time of day, and in space according to latitude, slope, degree of shading or exposure etc.

Water

Water is essential to all life and its availability determines the distribution of terrestrial organisms. The adaptations of terrestrial organisms to conserve water are discussed in Chapters 33 and 34. Even aquatic organisms do not escape problems of water shortage. In saline conditions water may be withdrawn from organisms osmotically, thus necessitating adaptations to conserve it. The salinity of water is a major factor in determining the distribution of aquatic organisms. Some fish, e.g. roach and perch, live exclusively in fresh water; others, like cod and herring, are entirely marine. A few fish, like salmon and eels, are capable of tolerating both extremes during their life.

Air and water currents

Air movements may affect organisms indirectly, for example by evaporative cooling or by a change in humidity. They may also affect them directly by determining their shape; the development of branches and roots of trees in exposed situations is an example. Wind is an important mechanism for dispersing seeds and spores. In the same way that the air currents determine the distribution of certain species in terrestrial habitats, so too do water currents in aquatic ones.

Humidity

Humidity has a major bearing on the rate of transpiration in plants and so affects their distribution. Although to a lesser degree, it affects the distribution of some animals by affecting the rate of evaporation from their bodies.

27.3.3 Biotic factors

Relationships between organisms are obviously varied and complex and are detailed throughout the book, but a brief outline of a few major biotic factors which affect organisms' distribution is given below.

Competition

Organisms compete with each other for food, water, light, minerals, shelter and a mate. They compete not only with members of other species – **interspecific competition** – but also with members of their own species – **intraspecific competition**. Where two species occupy the same ecological niche, the interspecific competition leads to the extinction of one or the other – the competitive exclusion principle (Section 10.2.3).

Predation

The distribution of a species is determined by the presence or absence of its prey and/or predators. The predator-prey relationship is an important aspect in determining population size (Section 10.2.2).

Antibiosis

Organisms sometimes produce chemicals which repel other organisms. These may be directed against members of their own species. Many mammals, for example, use chemicals to mark their territories, with the intention of deterring other members of the species from entering. Some ants produce a type of external hormone called a **pheromone** when they are in danger and, in sufficient concentrations, this warns off other members of the species. The chemicals may also be directed against different species. Many fungi, e.g. *Penicillium*, produce **antibiotics** to prevent bacterial growth in their vicinity.

Dispersal

Many organisms depend upon another species to disperse them. Plants in particular use a wide variety of animal species to disperse their seeds. Details are given in Section 18.6.2.

Pollination

Angiosperms utilize insects to transfer their pollen from one member of a species to another, and a highly complex form of interdependence between these two groups has developed. This is discussed in Section 18.2.

Mimicry

Many organisms, for a variety of reasons, seek to resemble other living organisms. Warning mimicry is used by certain flies which resemble wasps. Potential predators are warned off the harmless flies, fearing they may be stung.

Human influence

Man influences the distribution of other organisms more than any other single species. As a hunter, fisher, farmer, developer and polluter, to name a few activities, he dictates which organisms grow where. Some aspects of these influences are considered in Chapter 29.

Hoverfly mimicking a wasp

1. (a) Define the terms 'habitat' and 'niche'. (*4 marks*)

(b) With reference to a population of a **named** organism in a **named** habitat, describe the factors that control population size. (*10 marks*)

(c) How may this population vary in size with time? (*4 marks*)

(*Total 18 marks*)

Cambridge Board June 1984, Paper I, No. 9

2. Discuss how the activities of humans and other organisms affect the way in which nitrogen is cycled in an ecosystem. (*20 marks*)

London Board January 1989, Paper II, No. 7(a)

3. (a) Describe what you understand by the following.
   (i) A pyramid of biomass
   (ii) An ecological climax (*8 marks*)

(b) What part do the following play in an ecosystem?
   (i) Saprophytic fungi
   (ii) Herbivores (*8 marks*)

(c) Why is it considered important to conserve ecosystems? (*4 marks*)

(*Total 20 marks*)

London Board June 1988, Paper II, No. 6

4. The addition of artificial fertilisers to two neighbouring fields. **A** and **B**, was stopped in 1981.

(a) Explain in detail the biological principles and processes involved that could account for the following data:

Yields From Two Neighbouring Fields On The Same Farm					
Field **A**			Field **B**		
YEAR	CROP	YIELD per hectare	YEAR	CROP	YIELD per hectare
1982	Wheat	5.2 tonnes	1982	Wheat	5.1 tonnes
1983	Wheat	4.8 tonnes	1983	Clover (a legum- inous crop for hay)	6.2 tonnes
1984	Wheat	4.2 tonnes	1984	Wheat	5.4 tonnes

(*26 marks*)

(b) If field **B** had become water-logged in 1984, why might it be expected that the wheat yield would be less than 5.4 tonnes? (*4 marks*)

(*Total 30 marks*)

Oxford Local Board June 1988, Paper II, No. 7

5. (a) (i) What is meant by 'net primary production' and what is its ecological significance?

(ii) For a **named** ecosystem or habitat, outline how net primary production could be estimated. (*8 marks*)

(b) (i) What factors can limit the size of populations? Explain how they can do so.

(ii) For a given population, explain briefly how its size and geographical distribution may have evolutionary significance. (*12 marks*)

(*Total 20 marks*)

Joint Matriculation Board June 1989, Paper IIB, No. 5

6. Use the following terms to fill in the blanks in the way which seems to you to be the most accurate. (There are more terms than you will need.)

a mode; its niche; variance; abundant; dominant; its frequency; the density; the standard deviation; its habitat; an ecosystem; the mean; the average; a community; a species list; error; bias; the cover index; quadrats; significance; value.

The particular place where an organism usually lives is called .... The living organisms in such a place are collectively called .... The living organisms and the environment together make up .... Its structure is analysed by several methods. A convenient starting point is to draw up .... of the plants; then a subjective assessment of their relative frequency is made, designating each species as either ...., frequent, occasional or rare. .... are thrown at random, and the number of times a species occurs in one hundred throws gives .... However, this can give a misleading impression of the importance of a given species, as it may be small, widely spaced and numerous enough to give a high figure while its true importance is small. The .... avoids the undue bias of size and estimates the percentage of ground any one species occupies. The number of organisms per unit area is called ...., and is separate from either of the previous assessments.

Any random sampling system is prone to ....; these are measurable and a set of numbers from say fifty quadrats can be expressed as .... and .... about the mean. If a further fifty samples are taken from a different area and we wish to test the .... of the difference we use a t-test.

(*12 marks*)

Oxford Local 1987, Specimen Paper I, No. 15

7. Table 1 gives data from a simple food chain.

TABLE 1

	Number of individuals per unit area	Mean dry mass/ g per individual	Energy content/ kJ g^{-1} dry mass
Lettuces	10 m^{-2}	52	18.9
Slugs	24 m^{-2}	0.63	12.6
Thrushes	20 ha^{-1}*	60	23.5

* ha = hectare (10 000 m²)

(*a*) (i) To which trophic level do slugs belong?
   (ii) Outline **briefly** how each of the following might have been obtained.
   A The number of slugs.
   B The number of thrushes.
   C The energy content of the lettuces.

(*7 marks*)

(*b*) The relationship between the various trophic levels can be investigated in three ways: by analysing the numbers, biomass and energy content per unit area.
   (i) Using the information in Table 1, calculate the missing figures in Table 2 below. These relate the data to one square metre of the lettuce field. Write the figures in the spaces provided in the table.

TABLE 2

	Number per square metre	Biomass/g per square metre	Energy content/kJ per square metre
Lettuces	10		
Slugs	24		
Thrushes	0.002		

The values you have calculated above in Table 2 can be used to construct ecological pyramids. The pyramids of numbers, biomass and energy constructed from these figures are of different shapes or proportions. Account for the differences in shape or proportion between
   (ii) the pyramid of numbers and the pyramid of biomass;
   (iii) the pyramid of biomass and the pyramid of energy.

(*5 marks*)
(*Total 12 marks*)

Joint Matriculation Board June 1984, Paper IA, No. 1

8. Give the meaning of the following ecological terms:
   (*a*) community
   (*b*) edaphic factor
   (*c*) habitat   (*3 marks*)

Associated Examining Board June 1988, Paper I, No. 19

9. (*a*) State what is meant by the following terms.
      (i) Trophic level   (*2 marks*)
      (ii) Pyramid of biomass   (*2 marks*)
   (*b*) (i) The flow of energy through an ecosystem is described as linear.
         Explain what this means.   (*2 marks*)
      (ii) Explain why there are rarely more than four levels in a food chain.   (*2 marks*)
   (*c*) The following pyramid of biomass was obtained for plant plankton and animal plankton in a marine habitat. The figures represent dry mass in g per m³.

Comment on this pyramid.   (*2 marks*)
(*Total 10 marks*)

London Board January 1989, Paper I, No. 12

10. In an investigation into net primary production, several samples of heather were cut off at ground level, bagged and taken to the laboratory, where their dry mass was determined.
   (*a*) Describe how you would determine the dry mass of a sample of heather.
   (*b*) (i) What is *net primary production*?
      (ii) Suggest **two** components of net primary production which would not be present if this collecting technique was used.

(*Total 7 marks*)

Associated Examining Board June 1988, Paper I, No. 12

28 Ecological techniques

As the study of the inter-relationships of organisms with each other and their environment, ecology covers all aspects of biology. The study of this vast topic is fraught with difficulties, not least because it is not restricted to biology, but includes elements of physical science, geography and social science to name a few. There are two basic approaches to its study: the study of a community in relation to its environment – **synecology** – and the study of an individual species in relation to its environment – **autecology**. It is also possible to study ecology in other ways, by looking at a specific habitat, for example, or the evolution of a particular ecosystem. For practical reasons, any study at this level is often restricted to a particular habitat and even then can only be effectively undertaken by a team of individuals. Habitats vary in size and complexity and the techniques used depend upon the precise habitat being studied. Marine habitats include a rocky shore, sandy shore or sand dune, whereas freshwater habitats suitable for study are a pond or part of a river or lake. An estuary would combine aspects of both freshwater and marine ecology. Terrestrial habitats might include a grassland or an area of woodland.

It is clearly impossible in just a chapter to cover in detail the techniques adopted for all habitats which are suitable for study at this level. It is therefore intended to give only a general account of some ecological techniques used in a range of different habitats. It is assumed that the student will have the opportunity at some time during the A-level course to make a detailed study of at least one habitat and so learn at first hand much more about the particular techniques involved.

28.1 Synecology

Synecology is the study of a natural community and its surroundings. Any synecological investigation may be conveniently divided into the study of the **biotic** (living) and the **abiotic** (non-living) factors. A typical approach might be:

1. Make a general map of the area for study, and detailed maps or profiles of specific parts as they are investigated.

2. Identify the different species living within the study area with the aid of systematic keys (Section 6.1.2).

3. Make a quantitative study to determine the numbers and distribution of these species within the area, by a variety of sampling methods (Section 28.4).

4. Measure and analyse the abiotic variables within the area (Section 28.3).

5. Carry out experiments and observations on the organisms present, in an attempt to correlate their distribution with either the biotic or abiotic conditions. This is never easy and there are rarely simple answers, but it is good scientific method to isolate a problem, suggest a hypothesis to explain the answer and then devise experiments to test the validity of that hypothesis. Whatever the outcome, the practical experience will be invaluable.

6. If possible, a food web to show the energy relationships of some of the organisms within the community should be constructed.

28.2 Autecology

Autecology is the study of an individual species in relation to other species and its surroundings. An autecological investigation endeavours to find out as much as possible about a species' ecological niche. A typical approach might be to:

1. Identify the species being studied and discover as much as possible about its external features and the group to which it belongs.

2. Discover the type of habitat it lives in and its distribution within that habitat.

3. Investigate the biotic and abiotic factors which determine its distribution.

4. Determine its energy relationships with other organisms in the habitat, i.e. its position in the food web.

5. Determine other aspects of its physiology and behaviour, e.g. how it respires, excretes, moves from place to place, withstands adverse conditions etc. Relate these to its particular habitat.

6. Discover its life history, in particular when and how it reproduces and disperses its offspring. How long is its life span and what stages are there in its life cycle?

This approach involves more than observation in the field as some answers are better discovered by laboratory experiments or by reading. The facts about the organism overlap and so need not be investigated in any particular order. The overall idea is to build up a picture of every biological aspect of the species.

28.3 Measurement of environmental parameters

Abiotic factors, such as soil and climatic conditions, are important in determining both the distribution of organisms and their physical and physiological adaptations. It is therefore essential in any ecological study to be able to measure variations in environmental conditions both in time and space. Historically the measurement of some of these parameters was done by laboratory investigation and analysis of samples collected in the field. For example, the pH and oxygen content of a pond were calculated in this way. Modern techniques are much more likely to involve technologically sophisticated pieces of equipment which are used in the field and give an immediate reading. The same pond today would most probably have its pH and oxygen content determined in a matter of minutes using a pH and oxygen meter. With the trend to more electronic equipment bound to continue, no attempt has been made in this section to describe the more analytical methods of measuring environmental parameters.

Temperature

In ecological studies the precise temperature at any one moment is of little value. Of much greater significance are the **diurnal (daily)** and **seasonal** temperature variations. These variations can be measured using a mercury thermometer by taking readings at regular intervals. The best method of determining the highest and lowest temperature over a period of time is to use a maximum–minimum thermometer which leaves a marker at the highest temperature recorded and another at the lowest. These readings can therefore be taken at the end of a specific period, and the marker reset. For relatively inaccessible positions, e.g. under small stones, at different depths in a lake, a miniaturized **thermistor** may be used. This electrical instrument measures resistance, which changes with temperature. Using calibration scales it is possible to determine the actual temperature which corresponds to a given resistance.

pH

pH is a measure of the acidity or alkalinity of a solution. It is most easily determined by use of **universal indicator**, either as a liquid or impregnated on test paper. The addition of a few drops of universal indicator to a few cm³ of the solution under test will produce a coloured solution which, when compared with a colour chart, gives the corresponding pH. Alternatively, pH paper can be used. More complex, but more accurate, is the use of a **pH meter**. The probe of the instrument is rinsed in distilled water (pH 7 – neutral) before being placed in the test solution. The reading is noted. A calibration graph or scale must be made by taking readings of a series of buffer solutions of known pH. From this calibration graph the pH corresponding to any reading can be determined. To determine the pH of soil, about 1 cm³ of barium sulphate must be added to about 1 cm³ of the soil and both should be mixed with 10 cm³ of distilled water and 5 cm³ of universal indicator solution. After shaking, the contents should be allowed to settle and the pH determined as before, from the colour of the supernatant liquid. The barium sulphate will not affect the pH of the soil sample but is essential to flocculate any clay present.

Light

Two aspects of light, its duration and intensity, are generally important to ecological studies. The duration of daylight hours can be determined astronomically and is predictable for any location. The values are available for any given day, from most diaries. The intensity of light is most easily measured by use of an ordinary photographic light meter.

Humidity

Atmospheric humidity is normally expressed as **relative humidity**, i.e. the water content of a given volume of air relative to the same volume of fully saturated air. It is measured using a **whirling hygrometer**. This has two thermometers, the bulb of one being kept

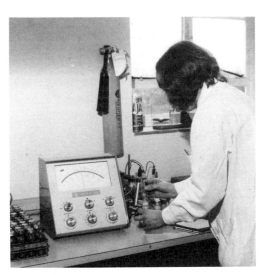

pH meter in use

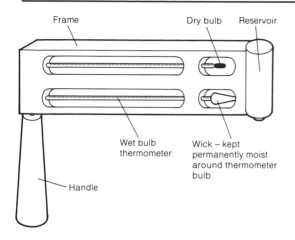

Fig. 28.1 Whirling hygrometer

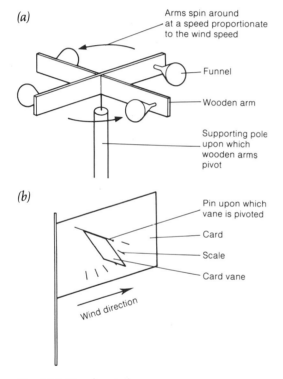

Fig. 28.2 Simple wind gauges

dry while the other is permanently wet. Both are mounted in a frame (Fig. 28.1). The hygrometer is rotated in the air until both thermometers give a constant reading. The wet bulb thermometer will always give a lower reading than the dry bulb one, due to the cooling effect of the evaporating water – the less humid the air, the more evaporation and the greater the temperature difference between the two thermometers. The actual humidity is determined by reading off the temperatures on a special scale.

Wind and water speed
Wind speed is measured by an instrument called an **anemometer**. Simple, but effective, wind gauges can be improvised and two examples are given in Fig. 28.2. These may not provide a direct measure of wind speed but will give comparative readings, e.g. inside and outside a wood, or at various vertical heights.

The easiest method of determining water speed is to time the movement of a floating object over a measured distance. To avoid inaccuracies caused by the wind blowing the portion floating above the water, it should be weighted so that it hardly breaks the surface. A specimen tube partly filled with water works admirably.

Salinity
The salinity of a water sample is best measured using a **conductivity meter**. This instrument measures the conductivity between two probes; the greater the salinity the greater the conductivity. It is also possible to determine the concentration of a particular ion. For example, chloride concentration can be determined by titration against silver nitrate solution, using potassium chromate as an indicator.

Oxygen level
An instrument known as an **oxygen meter** can be used to give a measurement of the oxygen concentration in a water sample. This has now largely replaced the chemical technique involving titration, known as the Winkler method.

28.4 Sampling methods

It is virtually impossible to identify and count every organism in a habitat. For this reason only small sections of the habitat are usually studied in detail. Provided these are representative of the area as a whole, any conclusions drawn from the findings will be valid. There are four basic sampling techniques.

28.4.1 Quadrats

A quadrat (Fig. 28.3) is a sturdily built wooden fame, often designed so it can be folded to make it more compact for storage and transport. It is placed on the ground and the species present within the frame are identified and their abundance recorded. Where the species are small and/or densely packed, one or more of the smaller squares within the frame may be used rather than the quadrat as a whole.

Sampling with a quadrat may be random or systematic. **Random sampling** can be as simple as throwing a quadrat over one's shoulder and counting the species within it wherever it falls. Even with the

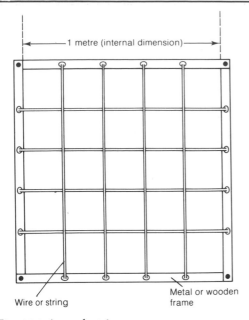

Fig. 28.3 A quadrat frame

best of intentions it is difficult not to introduce an element of personal bias using this method. A better form of random sampling is to lay out two long tape measures at right angles to each other, along two sides of the study area. Using random numbers generated on a computer or certain calculators, a series of coordinates can be obtained. The quadrat is placed at the intersection of each pair of coordinates and the species within it recorded. **Systematic sampling** involves placing the quadrat at regular intervals, for example, along a transect. It is sometimes necessary to sample the same area over many years in order to investigate seasonal changes or monitor ecological succession. In these circumstances a rectangular area of ground may be marked out by boundary stakes which are connected by rope. This is known as a **permanent quadrat**.

28.4.2 Point frames

A point frame, or point quadrat (Fig. 28.4), consists of vertical legs across which is fixed a horizontal bar with small holes along it. A long metal pin, resembling a knitting needle, is placed in each of the holes in turn. Each time the pin touches a species, it is recorded. The point frame is especially useful where there is dense vegetation as it can sample at many different levels.

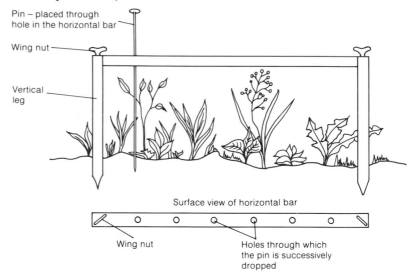

Fig. 28.4 Point frame (point quadrat)

28.4.3 Line transect

A line transect is used so that systematic sampling of an area can be carried out. A string or tape is stretched out along the ground in a straight line. A record is made of the organisms touching or covering the line all along its length, or at regular intervals. This technique is particularly useful where there is a transition of flora and/or fauna across an area, down a sea shore for example. If there is any appreciable height change along the transect, it is advisable to construct a profile of the transect to indicate the changes in level. This is especially important where vertical height is a major factor in determining the distribution of species. On a sea shore, for example, the height above the sea affects the duration of time any point is submerged by the tide. This has a considerable bearing on the species that can survive at that level. So, the distribution of species is related to the vertical height on the shore rather than the horizontal distance along it. This form of transect is called a **profile transect**.

TABLE 28.1 **Collecting apparatus of general use**

Apparatus	Purpose
Specimen tube (small)	With tight-fitting caps and of varying sizes, they can be used to contain small organisms while they are identified, or used to transport them
Screw-topped jars (large)	To collect and transport smaller organisms. Also useful for aquatic organisms
Polythene bags (various sizes)	To collect plant material, soil samples and other non-living material
Forceps	For transferring plants and hard-bodied animals as well as non-living materials, e.g. stones
Paint brush	For transferring small, delicate or soft-bodied organisms, e.g. aphids
Bulb pipette	For transferring small aquatic organisms
Pooter (aspirator) (Fig. 28.5)	For collecting and transferring small animals, e.g. insects and spiders
Widger	A useful all-purpose tool, similar to a spatula, used for digging, levering and transferring material
Sieve	For sifting sand, soil, mud and pond-water for small organisms
Hand lens	To magnify features of organisms to help in their identification
Enamel dish	For sorting specimens

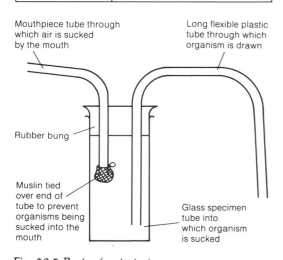

Mouthpiece tube through which air is sucked by the mouth

Long flexible plastic tube through which organism is drawn

Rubber bung

Muslin tied over end of tube to prevent organisms being sucked into the mouth

Glass specimen tube into which organism is sucked

Fig. 28.5 Pooter (aspirator)

28.4.4 Belt transect

A belt transect is a strip, usually a metre wide, marked by putting a second line transect parallel to the other. The species between the lines are carefully recorded, working a metre at a time. Another method is to use a frame quadrat in conjunction with a single line transect. In this case the quadrat is laid down alongside the line transect and the species within it recorded. It is then moved its own length along the line and the process repeated. This gives a record of species in a continuous belt, but the quadrat may also be used at regular intervals, e.g. every 5 m, along the line.

28.5   Collecting methods

Reliable methods of collecting organisms are an essential part of ecology. Because they photosynthesize, plants always occur in the light and are hence visible and usually large and stationary. All these factors make them easy to find and collect. Animals by contrast may live underground, in crevices, or simply be camouflaged; this makes them less conspicuous. Even when seen, the animal's ability to move from place to place presents problems of capture. Collecting all organisms within a habitat is normally impractical and therefore small areas are selected. Within these areas it is important to make thorough collections. Searches should be made in crevices, on other living organisms and beneath stones. It is extremely important to return all material to its original position, even to the extent of replacing stones the same way up. It is easy to completely destroy micro-habitats and so endanger the species which occupy them. Organisms should be identified on site, but if removing them is unavoidable, as few as possible should be taken and for as short a time as possible and always returned to the same location. When collecting specimens, as much information as possible should be recorded at the time. This should include details of the time, date, location, substrate, climate and any other relevant data.

Before describing some of the collecting methods commonly used for particular organisms in particular habitats, it is worth listing some collecting apparatus which is useful in most situations (Table 28.1).

In addition to collecting specimens by hand or with the use of a pooter, there are other pieces of apparatus of varying degrees of complexity which may be used to lure and trap animals.

Beating tray
This is a fabric sheet on a collapsible frame. It is held under a part of a bush or a tree which is then shaken or disturbed with a stick. The organisms which are dislodged are collected by hand or with the aid of a pooter (Fig. 28.5). It is used to collect small non-flying terrestrial organisms, e.g. beetles, spiders, caterpillars.

Light traps
Any light source will attract certain nocturnal flying insects. A very simple light trap may be made by placing a vertical sheet at the side of a light source, and a horizontal one beneath it. This is necessary as some insects prefer to rest vertically and others horizontally. More effective traps involve **mercury vapour lamps**. These emit much ultraviolet light which is particularly attractive to nocturnal insects such as moths. They fall into glass plates which slope towards a

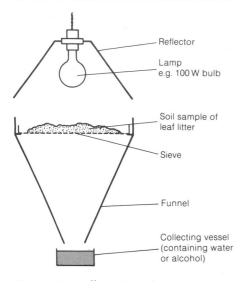

Fig. 28.6(a) Tullgren funnel

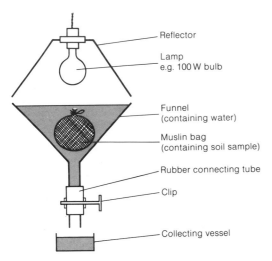

Fig. 28.6(b) Baermann funnel

narrow opening at the base of the lamp. Beneath the opening is a closed box with crumpled paper in which the insects become trapped. They are later anaesthetized before being removed.

Tullgren funnel

This is used to extract small animals from a sample of soil or leaf litter. The soil sample is placed on a coarse sieve and light and moderate heat are used to drive the animals downwards through the sieve. (Fig. 28.6a). They fall into a funnel which directs them into a collecting vessel. If the vessel contains water the animals will be unable to escape; alternatively, 70% alcohol can be used to preserve them directly. If required alive, damp blotting paper may be used, but predators such as centipedes need to be swiftly removed before they consume the other animals.

Baermann funnel

This is again used to extract soil animals and is particularly effective for worms, especially nematodes. The soil sample is contained within a muslin bag which is then submerged in water in a funnel (Fig. 28.6b). A tungsten bulb may be used as a source of heat which, along with the water, induces the organisms to leave the sample. They collect in the neck of the funnel from where they can be periodically removed.

Mammal traps

The best live trap is the **Longworth trap**. It is placed in situations which small mammals such as mice and voles frequent, like a runway. It comprises a metal box with a single entrance which closes firmly behind the mammal when it enters. The box is baited with bedding and the appropriate food to entice the animal to enter. It is normal to leave the trap in position for some time with the entrance locked open, so that the mammals become accustomed to it before it is used for sampling. The behaviour of many small mammals creates problems when this trap is used to assess population sizes, because some individuals, called 'trap-shy', never enter the trap, while others, called 'trap-happy', actively seek them out for the meal and bed they provide.

Pitfall traps

A jam-jar or similar vessel is sunk into the ground with its rim level with the soil. It is baited with the appropriate food, e.g. decaying meat to attract scavenging insects such as beetles, or honey to attract ants. Having fallen in, the insects are unable to climb the smooth walls of the jar to escape. To prevent flooding of the jar, a flat piece of wood or stone may be placed over the entrance, but raised above it on three or four small stones. In this way insects can still enter the jar beneath the cover from around the edges. For the same reason it should be placed in ground higher than that immediately around it, to prevent rain-water running off into the jar.

Netting

This is a popular method of capture and takes many forms. Hand-held nets with short handles give greater precision for catching insects in flight. With some insects it is better to stalk them until they settle before netting them. These nets are called **kite nets**. A more robust form, called the **sweep net**, is used to collect insects from foliage. It is swept along grass or through bushes, dislodging insects which fall into the net. This net may also be used to collect aquatic animals by sweeping it through streams or ponds. A **plankton net** is made of bolting silk because its fine mesh, while allowing water through, traps even microscopic organisms. It has a wide mouth held open by a circular metal frame and narrows down to a small collecting jar at the other end in which the plankton accumulates. The net is towed slowly through the water, usually behind a small boat.

28.6 Estimating population size

To count accurately every individual of any species within a habitat is clearly impractical, and yet much applied ecology requires information on the size of animal and plant populations. It is necessary therefore to use sampling techniques in particular ways in order to make estimates of the size of any population. The exact methods used depend not only on the nature of the habitat but also on the organism involved. Whereas, for example, it may be useful to know the number of individuals in an animal population, this may be misleading for a plant species, where the percentage cover may be more relevant.

28.6.1 Using quadrats

By sampling an area using quadrats and counting the number of individuals within each quadrat, it is possible to estimate the total number of individuals within the area. If, for example, an area of 1000 m² is studied and 100 quadrats, each 1 m², are sampled, it follows that a total of 100 m² of the area has been sampled. This represents one tenth of the total. The total number of individuals of a species in all 100 quadrats must therefore be multiplied by ten to give an estimate of the total population of that species in the area. The use of quadrats in estimating population size is largely confined to plants and sessile, or very slow-moving animals. Faster-moving animals would simply disperse upon being disturbed.

28.6.2 Capture–recapture techniques

The capture–recapture method of estimating the size of a population is useful for mobile animals which can be tagged or in some other way marked. A known number of animals are caught, clearly marked and then released into the population again. Some time later, a given number of individuals are collected randomly and the number of marked individuals recorded. The size of the population is calculated on the assumption that the proportion of marked to unmarked individuals in this second sample is the same as the proportion of marked to unmarked individuals in the population as a whole. This, of course, assumes that the marked individuals released from the first sample distribute themselves evenly among the remainder of the population and have sufficient time to do so. This may not be the case, due to deaths, migrations and other factors. Another problem is that while the tag or label may not itself be toxic, it often renders the individual more conspicuous and so more liable to predation. In this case the number of marked individuals surviving long enough to be recaptured is reduced and the size of the population will consequently be over-estimated. 'Trap-shy' and 'trap-happy'

Quadrat in use

individuals (see Section 28.5) will also adversely influence the estimate. The population size can be estimated using the calculation below:

$$\text{Estimated size of population} = \frac{\text{Total number of individuals in the first sample} \times \text{Total number of individuals in the second sample}}{\text{Number of marked individuals recaptured}}$$

The estimate calculated is called the **Lincoln index**.

The method can be used on a variety of animals; arthropods may be marked on their backs with non-toxic dabs of paint, fish can have tags attached to their opercula, mammals may have tags clipped to their ears and birds can have their legs ringed.

TABLE 28.2 **Abundance scales for two rocky shore species as devised by Crisp and Southward (1958)**

Abundance group	Symbol	Scale used for the limpet (*Patella*)	Scale used for barnacles, e.g. *Chthamalus stellatus*
Abundant	A	Over $50\,m^{-2}$	Over $1\,cm^{-2}$ (rocks well covered)
Common	C	$10–50\,m^{-2}$	$10–100\,dm^{-2}$ (up to one third of rock covered)
Frequent	F	$1–10\,m^{-2}$	$1–10\,dm^{-2}$ (never less than 10 cm apart)
Occasional	O	Less than $1\,m^{-2}$	$10–100\,m^{-2}$ (few less than 10 cm apart)
Rare	R	Only a few found in 30 minutes' searching	Less than $1\,m^{-2}$ (only a few found in 30 minutes' searching)

28.6.3 Abundance scales

The population size may be fairly accurately determined by making some form of **frequency assessment**. These methods are subjective and involve an experimenter making some estimate of the number of individuals in a given area, or the percentage cover of a particular species. This is especially useful where individuals are very numerous, e.g. barnacles on a rocky shore, or where it is difficult to distinguish individuals, e.g. grass plants in a meadow. The assessments are usually made on an abundance scale of five categories. These are given, with two examples of how they are used for different species, in Table 28.2.

1. For a given ecological problem which you have investigated either by yourself or as part of a team, give the objects of your investigation, how it was carried out and what conclusions you were able to draw.

Discuss the limitations of the methods you were able to use. *(20 marks)*

Oxford and Cambridge Board July 1984, Paper II, No. 9

2. For a named ecosystem you have studied, discuss **three** of the following:

(*a*) spatial zonation;
(*b*) seasonal succession;
(*c*) effects of weather;
(*d*) effects of climate;
(*e*) the rôle of decomposers;
(*f*) effects of man;
(*g*) rôle of predators;
(*h*) diurnal changes;
(*i*) effects of tides.

Oxford Local June 1983, Paper II, No. 7

3. For a named area where you have carried out ecological studies explain how you investigated **two** of the following:

(*a*) zonation;
(*b*) the density of any one population;
(*c*) distribution related to a change in the environment;
(*d*) the adaptations of one organism. *(8 marks)*

Southern Universities Joint Board June 1986, Paper I, No. 2

4. Longworth traps capture small mammals alive and are equipped with food and bedding. The diagram shows the distribution of Longworth traps during a 10-day investigation into the small-mammal population of an area of parkland. The traps were reset at noon each day, any captured animals being noted and immediately released unharmed. Above each trapping point in the diagram is the number of wood mice (*Apodemus sylvaticus*) trapped there and below it is the number of bank voles (*Clethrionomys glareolus*) trapped there. No other small mammals were trapped. The bracken in the parkland was dense; the woodland had a sparse herb-layer.

(*a*) (i) Construct a table to show the numbers of mice and voles caught in the rough grass, woodland and bracken. *(2 marks)*
    (ii) Calculate the percentage of mice and voles in the total catch. *(1 mark)*
    (iii) Comment on the relation that appears to exist between the animals caught and type of vegetation. *(9 marks)*
(*b*) The traps in this investigation were reset at noon each day. The usual procedure is to reset them at sunset and again at sunrise. Comment on the advantages and disadvantages of the two procedures. *(3 marks)*
(*c*) The release of the captured small mammals could have made a reliable estimate of the population impossible. Why is this? Suggest a procedure that would have eliminated this weakness without harming the animals. *(2 marks)*
(*d*) Describe **three** visible features that identify living mice and voles as mammals. *(3 marks)*

(Total 20 marks)

Associated Examining Board June 1984, Paper II, No. 6

5. Table 1 shows the height of samples of the plant *Achillea lanulosa* taken at different sites in a transect across the Sierra Nevada mountain range in the USA. For each site, the average height of plants is given, together with the elevation (height above sea level) of the site. The map on the next page shows the transect, and the location of the sites in relation to the geographical features.

TABLE 1

Site	Elevation /m	Mean height of plants/cm
Mather	1250	75
Aspen Valley	1800	50
Yosemite Creek	2000	49
Tenaya Lake	2250	32
Tuolumne Meadows	2425	21
Bighorn Lake	3050	15
Timberline	2750	20
Conway Summit	2325	25
Leevining	1950	43

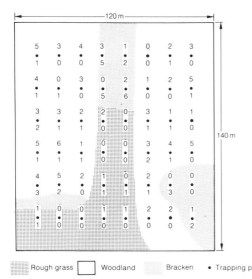

	120 m							
5	3	4	3	1	0	2	3
1	0	0	5	2	0	1	0
4	0	3	0	2	1	2	5
1	1	0	5	6	0	0	1
3	3	2	2	0	3	1	1
2	1	1	0	0	1	1	0
5	6	1	0	0	3	4	5
1	1	1	0	0	2	1	0
4	5	2	1	0	2	0	0
3	2	0	1	1	1	3	0
1	0	0	1	1	2	2	1
1	0	0	0	0	0	0	2

140 m

▓ Rough grass　□ Woodland　▒ Bracken　• Trapping point

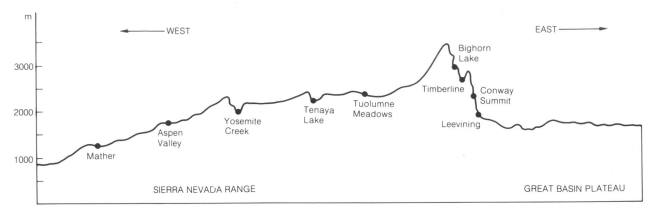

(a) Plot a graph of height of plants against elevation. (*4 marks*)

(b) What conclusions can you draw from your graph about the relationship between height of plants and elevation? (*2 marks*)

(c) Does the geographic distribution of the sites suggest any other factors which might influence the height of the plants? (*3 marks*)

(d) Fifty seeds from each location were collected, and planted under identical conditions in an experimental garden at Stanford University, California. The average height of the plants grown from the seeds collected in each location is shown in Table 2.

TABLE 2

Site	Average height of plants grown from seeds/cm
Mather	77
Aspen Valley	51
Yosemite Creek	49
Tenaya Lake	34
Tuolumne Meadows	23
Bighorn Lake	17
Timberline	23
Conway Summit	26
Leevining	45

(i) In view of the data in both tables, what conclusions can be drawn with respect to the cause of variation between these samples?

(ii) Discuss the implication of these figures for the emergence of new species of the genus *Achillea*. (*5 marks*)

(e) (i) What heights would you expect to find for plants sampled at intermediate sites along the transect?

(ii) What does this imply for the genetic composition of the plant populations along the length of the transects? (*4 marks*)

(f) Discuss how the trends shown in Table 1 could have been established. (*2 marks*)

(*Total 20 marks*)

Northern Ireland Board June 1985, Paper II, No. 8

6. (a) For one specific habitat,
   (i) discuss the effects of any **three** distinct environmental factors on the distribution of organisms;
   (ii) explain what measurements of these three environmental factors it would be appropriate to take in determining their effects on the distribution of organisms. (*12 marks*)

(b) Describe how you would determine the frequency in a community of
   (i) a *named* sessile (fixed) organism, and
   (ii) a *named* motile organism. (*8 marks*)

(*Total 20 marks*)

Joint Matriculation Board June 1988, Paper IIB, No. 8

7. Answer **one** of the following:

Either: (a) (i) Briefly discuss the causes and consequences of tides on the physical environment of the sea-shore. (*8 marks*)
   (ii) Describe how tides affect and control the distribution, structure and physiology of organisms which live on the sea-shore. (*22 marks*)

Or: (b) (i) Explain how you have investigated the relationships between the various trophic levels found in a fresh-water stream, pond or lake. (*18 marks*)
   (ii) With the aid of any suitable charts or diagrams, summarise and comment on your results and conclusions. (*12 marks*)

Or: (c) (i) Choose a terrestrial habitat such as woodland, moorland or sand-dunes that you have studied. State its location as precisely as possible. Explain how you carried out a scientific study of the plants and/or animals in that habitat. (*16 marks*)
   (ii) Name and briefly describe some of

the adaptations for survival shown
by the organisms in the habitat.

(*14 marks*)

(*Total 30 marks*)

Oxford Local Board June 1989, Paper II, No. 1

8. (*a*) Define, with examples, the terms *community* and
succession. (*5 marks*)

(*b*) Describe, in detail, how you would carry out
an ecological investigation to compare the
vegetation and the species of animals in **two**
different areas, such as a grazed grassland and
a grassland invaded by shrubs. (*13 marks*)

(*Total 18 marks*)

Cambridge Board November 1988, Paper I, No. 9

29 *Human activities and the ecosystem*

The effect of human activities on the environment is proportional to the size of the human population. As we saw in Section 10.1.5, the size of the human population has been rising exponentially and is presently increasing at the rate of one and a half million people each week, equivalent to 150 people per minute. The reasons for this increase are many but include more intensive forms of food production and better medical care. The latter has given many individuals a greater life expectancy and, more importantly, reduced child mortality. Reducing the mortality of individuals who have passed child-bearing age has little significance on the population, but reducing mortality among children means more people reach sexual maturity and so are able to produce offspring. The effect of this on the population is significant.

29.1 The impact of pre-industrial man on the environment

29.1.1 Man the hunter

As his population was small, pre-industrial man did not have a great impact on his environment. Early man hunted, fished and removed trees to make fires and shelters but, being nomadic, he did not remain in one place long enough to have a significant effect. When he moved on, the natural environment rapidly recovered. Early man did use fire and this may have accidentally got out of hand and burned down large areas of forest. Equally it may have been used deliberately to flush out prey to enable it to be captured. Man may thereby have created some grasslands at the expense of the forest, but his total impact was small.

29.1.2 Man the shepherd

In time man domesticated animals such as sheep, cattle, goats, llamas and alpacas. These herbivorous species required large areas of grassland on which to graze. To extend the grasslands man deliberately burnt large areas of forest. Domesticated species like cattle became symbols of wealth and power, as well as suppliers of milk and meat. This led to the build-up of large herds and subsequent overgrazing with resultant loss of soil fertility leading to erosion. These activities, which began in the Mediterranean and Near East, may have contributed to the development of many of the desert regions of these areas. Animal domestication also led to the extinction of their wild ancestors, probably through competition. Aurochs and European bison may have suffered this fate.

29.1.3 Man the farmer

The advent of agriculture marked the most significant event of pre-industrial man's impact on the environment. It probably originated in the Near East and involved the deliberate sowing of seeds to produce a crop. If man was to enjoy the fruits of this labour, he needed to harvest the crop at some later date. He therefore had to remain in one place or risk losing the crop to other animals. For the first time man formed permanent settlements. He built shelters for himself and his animals, and barns to store his crops. This required much wood and led to further destruction of the forests. More importantly, man cleared much forest to provide a greater area for sowing his crops. With no knowledge of minerals, he continued to grow the same crop on the same piece of land for many years, thus depleting it of essential nutrients. The soil could no longer support life, and this was a major factor in the formation of desert areas.

The impact of man upon the forests of the world is clearly illustrated in the United States of America. When first settled by Europeans in the early seventeenth century, there were an estimated 170 million hectares of forest. Now there are eight million. Most of this clearance was carried out to permit the cultivation of corn and wheat in the north, and tobacco and cotton in the south.

29.2 Exploitation of natural resources

Prior to the industrial revolution, the energy expended in the production of food by man and his beasts of burden came from the food itself. Much of the crop therefore went into producing the next one. With the industrial revolution came machines which carried out ploughing, sowing, harvesting etc. Instead of food, these machines used fossil fuels like coal, and more recently oil, as energy sources. A very much smaller proportion of the crop harvested was therefore needed to produce the next one. More food was left and a larger population could be supported. This partly explains the exponential rise in the size of the human population since the industrial revolution. The use of fertilizers, pesticides and better crops are other significant factors.

Man is dependent for his survival on the earth's resources, and these take two forms: renewable and non-renewable.

29.2.1 Renewable resources

Renewable resources, as the name suggests, can be replaced. They are things which grow, and are materials based on plants or animals, e.g. trees and fish. They are not, however, produced in limitless quantities and their supply is ultimately exhausted if the rate at which they are removed exceeds that at which they have been produced. Renewable resources have a **sustainable yield**. This means that the amount removed (yield) is equal to, or less than, the rate of production. If the trees in a forest take 100 years to mature, then one hundredth of the forest may be felled each year without the forest becoming smaller. A sustainable yield can be taken indefinitely.

Whilst wood is a renewable resource, its production is not without ecological problems. Trees grow relatively slowly and so give a small yield for a given area of land. For this reason, it is not economic to use fertile farmland for their cultivation. Instead, poorer quality land typical of upland areas is often used. As conifers grow more rapidly, these softwood species are more often cultivated than endemic hardwoods such as elm, oak, ash and beech. Large areas in Scotland, Wales and the Lake District have become **afforested**. The trees are often grown in rows and many square miles are covered by the same species. Not only does this arrangement have an unnatural appearance, but the density of the trees permits little, if anything, to grow beneath them and the forest floor is a barren place. There is little diversity of animal life within these forests. The demand for wood, not only for construction but also for paper, necessitates this intensive form of wood production.

Another renewable resource is fish. Unlike many of man's renewable resources, fish are not generally farmed. For the most part man removed them from the seas with no attempt to replace stocks by breeding. The replacement is left to nature. As the seas are considered a common resource for all, no previous attempt has been made to control the amount of fish removed by each country. While fishing was carried out by small boats, working locally, its impact on stocks was negligible because a sustainable yield was removed. Modern fishing methods involve large factory ships, capable of travelling thousands of miles and catching huge hauls of fish, which can be processed and frozen on board. Sonar equipment, echo sounders and even helicopters may be used in locating shoals. These methods have led to **over-fishing**, because sustainable yields have been exceeded and stocks depleted. It takes many years for such stocks to recover. The total weight of fish caught increased from 20 million tonnes in 1940 to 70 million tonnes in 1970. Some controls now exist and international agreement has been reached on quotas of fish which each country can take. These quotas are often bitterly disputed and the difficulty of enforcement has led to many being ignored. There are regulations concerning the mesh size of nets. If the mesh is sufficiently large, younger, and therefore smaller, fish escape capture. These survive to grow larger and, more importantly, are able to reach sexual maturity. These fish can then spawn thus ensuring some replenishment of the stock. One species to have suffered from over-fishing is the North Sea herring. Depletion of its stocks have made fishing it practically uneconomic in recent years.

29.2.2 Non-renewable resources

These are resources which, for all practical purposes, are not replaced as they are used. Minerals such as iron and fuels like coal and oil are non-renewable. There is a fixed quantity of these resources on the planet and in time they will be exhausted. Oil and natural gas supplies are unlikely to last more than 50 years, although much depends upon the rate at which they are burned.

Mineral and ore extraction have been carried out for a considerable time with important metals such as iron, copper, lead, tin and aluminium being mined. In theory these metals can be recycled, but in practice this is often difficult or impossible for various reasons:

1. The metal may be oxidized or otherwise converted into a form unsuitable for recycling. Iron for example rusts.

2. The quantities of the metal within a material may be so small that it is not worthwhile recovering it. The thin layer of tin on most metal cans is not economically worth recovering.

3. The metal is often combined with many other materials, including other metals, which make it difficult to separate. A motor car may contain small amounts of many metals including zinc, lead, tin, copper and aluminium. The cost in labour of separating these metals makes it uneconomic to carry out.

The supply of many ores is becoming more scarce and the belief that supply would satisfy demand for the foreseeable future is being questioned. The lead, tin, copper, gold and silver mines of Wales, Cornwall and the Lake District have almost entirely ceased their activities. As the supply is reduced, the price increases and it could be that sources previously considered uneconomic may prove worthwhile exploiting again.

Fossil fuels are continually being formed, but the process is so slow compared to their rate of consumption that for all practical purposes they may be considered as a non-renewable resource. Over 80% of the world's consumption of fossil fuels occurs in developed countries, where only 25% of its population lives. The burning of fossil fuels produces a range of pollutants and even their extraction is not without its hazards. As the supply of these fuels is becoming rapidly depleted, man has sought alternative energy sources. Nuclear power is a potentially long-term supplier of energy, but it has inherent dangers as the accident at Chernobyl in Russia in April 1986 illustrated. It is therefore treated by the public with some suspicion. Attempts continue to be made to harness wind, wave and solar energy effectively. In the end it could be **biological fuels** that man may have to look to to supply his growing energy needs. The energy content of the organic matter produced annually by photosynthesis exceeds man's annual energy consumption by 200 times. The main end-product of this photosynthesis is cellulose, most of which is unused by man. Some of it can be burnt as wood or straw to provide heat or electricity. Much can be converted to other fuels like methane (CH_4), methanol (CH_3OH), ethanol (C_2H_5OH) and other gases. These processes need not use valuable food resources; the energy may be obtained from plants with no food value or from the discarded parts of food plants, estimated to total 20 million tonnes dry mass year^{-1} in the UK alone. The gasohol programme in Brazil, where sugar cane wastes are used to produce a motor vehicle fuel, is an example of this (Section 41.5.1). Wastes such as animal manure (45 million tonnes dry mass year^{-1} in the UK), human sewage (6 million tonnes dry mass year^{-1} in the UK) and other domestic and industrial wastes (30 million tonnes dry mass year^{-1} in the UK) could be converted to useful fuels like methane by biogas digesters (Section 41.5.2). These conversions can be carried out by bacteria, often as part of fermentation reactions. The day may not be far away when large industrial plants convert these wastes into useful fuels and energy forms and where crops cultivated entirely for conversion to fuels are commonplace.

Part of the Chernobyl nuclear reactor after the explosion in April 1986

29.3 Pollution

Pollution is a difficult term to define. It has its origins in the Latin word *polluere* which means 'contamination of any feature of the environment'. Any definition of pollution should take account of the fact that:

1. It is not merely the addition of a substance to the environment but its addition at a rate faster than the environment can accommodate it. There are natural levels of chemicals such as arsenic and mercury in the environment, but only if these levels exceed certain critical values can they be considered pollutants.

2. Pollutants are not only chemicals; forms of energy like heat, sound, α-particles, β-particles and X-rays may also be pollutants.

3. To be a pollutant, a material has to be potentially harmful to life. In other words, some harmful effect must be recognized.

Using the above criteria, it is arguable that there is such a thing as natural pollution. We know for example that sulphur dioxide, one product of the combustion of fossil fuels, is a pollutant, and yet 70% of the world's sulphur dioxide is the result of volcanic activity. To avoid 'natural pollution' some scientists like to add a fourth criterion, namely that pollution is only the result of human activities.

29.3.1 Air pollution

The layer of air which supports life extends about 8 km above the earth's surface and is known as the **troposphere**. While there may be small localized variations in the levels of gases in air, its composition overall remains remarkably constant. Almost all air pollutants are gases added to this mixture. It has existed since man first used fire and in 1307 a resident of London was executed for causing air pollution. It is only since the industrial revolution in the eighteenth century that its effects have become significant. Almost all air pollution is the result of burning fossil fuels, either in the home, by industry or in the internal combustion engine.

Smoke

Smoke is tiny particles of soot (carbon) suspended in the air, which are produced as a result of burning fossil fuels, particularly coal and oil. It has a number of harmful effects:

1. When breathed in, smoke may blacken the alveoli, causing damage to their delicate epithelial linings. It also aggravates respiratory ailments, e.g. bronchitis.

2. While it remains suspended in the air, it can reduce the light intensity at ground level. This may lower the overall rate of photosynthesis.

3. Deposits of smoke, or more particularly soot and ash, may coat plant leaves, reducing photosynthesis by preventing the light penetrating or by blocking stomata.

4. Smoke, soot and ash become deposited on clothes, cars and buildings. These are costly to clean.

Sulphur dioxide

Fossil fuels contain between 1 and 4% sulphur and as a result around 30 million tonnes of sulphur dioxide is emitted from the chimneys of Europe each year. Much of this combines readily with other chemicals like water and ammonia and is quickly deposited. It may increase soil fertility in areas where sulphates are deficient, or even help to control diseases such as blackspot of roses by acting as a fungicide. Nevertheless its effects, especially in high concentrations, are largely harmful:

Smoke gives rise to some very serious problems of pollution

TABLE 29.1 **Tolerance of moss and lichen species to sulphur dioxide**

Annual average sulphur dioxide concentration in $\mu\text{g m}^{-3}$	Species tolerant and therefore able to survive
Greater than 60	*Lecanora conizaeoides* (lichen) *Lecanora dispersa* (lichen) *Ceratodon purpureus* (moss) *Funaria hygrometrica* (moss)
Less than 60	*Parmelia saxatilis* (lichen) *Parmelia fulginosa* (lichen)
Less than 45	*Grimma pulvinata* (lichen) *Hypnum cupressiforme* (moss)

1. It causes irritation of the respiratory system and damage to the epithelial lining of the alveoli. It can also irritate the conjunctiva of the eye.

2. It reduces the growth of many plants, e.g. barley, wheat, lettuce, while others such as lichens may be killed.

The tolerance of lichen and moss species to sulphur dioxide is very variable and makes them useful **indicator species** for measuring sulphur dioxide pollution. Table 29.1 shows the tolerance of some mosses and lichen species to sulphur dioxide.

As one moves from the centre of a major industrial city like Newcastle upon Tyne, the concentration of sulphur dioxide falls rapidly. At the same time the number of species of lichen and moss increases. In the centre only the most tolerant species are found, whereas on the outskirts less tolerant ones also occur (Fig. 29.1). Using Table 29.1, we can see that if an area of a city possesses *Lecanora dispersa* and *Funaria hygrometrica* but none of the other species, then the levels of sulphur dioxide must exceed $60\ \mu\text{g m}^{-3}$.

If all species in the table are present, the sulphur dioxide level must be less than $45\ \mu\text{g m}^{-3}$.

Sulphur dioxide is soluble in water. The sulphur dioxide in the atmosphere therefore will be dissolved in rain-water as it falls. The sulphur dioxide and water combine to form sulphurous and sulphuric acids. The rain therefore has a low pH and is known as **acid rain**. Due to the prevailing winds, much of the sulphur dioxide from Europe, including that from Britain, is carried over Scandinavia. It is here that acid rain causes the greatest problems. Coniferous trees are particularly vulnerable and considerable damage has been caused to some forests. Lakes in the region are extremely acid and many species within them have been killed, although it is not clear whether this is the direct result of the acidity or the accumulation of aluminium leached from soils as a result of acid rain.

Carbon dioxide
Carbon dioxide is formed during the respiration of organisms, and by the burning of fossil fuels. That produced as a result of respiration is taken up by plants during photosynthesis, ensuring it does not accumulate. The additional carbon dioxide produced in the burning of fossil fuels has caused a rise in atmospheric carbon dioxide concentration. Scientists believe that this change in air composition prevents more of the sun's heat escaping from the earth, much in the way the glass in a greenhouse does. They argue that the rise in temperature that this so-called **greenhouse effect** produces will cause the gradual melting of the polar ice caps and consequent rise in sea level. This would in turn cause flooding of low-lying land, upon which, as it happens, many of the world's capital cities lie. These beliefs can only be supposition; indeed, it could be argued that smoke emissions reduce the amount of heat reaching the earth and so have the opposite effect.

Carbon monoxide
Carbon monoxide occurs in exhaust emissions from cars and other vehicles. It is poisonous on account of having an affinity for haemoglobin some 250 times greater than that of oxygen. Upon combining with haemoglobin, it forms a stable compound which is not released and prevents oxygen combining with it. Continued inhalation leads to death as all haemoglobin becomes combined with

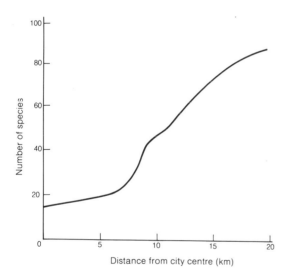

Fig. 29.1 Number of lichen species as one moves from the centre of Newcastle upon Tyne

carbon monoxide, leaving none to transport oxygen. In small concentrations it may cause dizziness and headache. Even on busy roads levels of carbon monoxide rarely exceed 4%, and it does not accumulate due to the action of certain bacteria and algae which break it down, according to the equation:

$$4CO + 4H_2O \longrightarrow 4CO_2 + 8H^+ + 8e^-$$

$$\underset{\text{carbon monoxide}}{} \quad \underset{\text{water}}{} \quad \underset{\text{carbon dioxide}}{} \quad \underbrace{\text{protons} \quad \text{electrons}}_{\text{hydrogen}}$$

Cigarette smoking is known to increase the carbon monoxide concentration of the blood; up to 10% of a smoker's haemoglobin may be combined with carbon monoxide at any one time.

Nitrogen oxides

Nitrogen oxides, like nitrogen dioxide, are produced by the burning of fuel in car engines and emitted as exhaust. In themselves they are poisonous, but more importantly they contribute to the formation of **photochemical smog**. Under certain climatic conditions pollutants become trapped close to the ground. The action of sunlight on the nitrogen oxides in these pollutants causes them to be converted to **peroxacyl nitrates (PAN)**. These compounds are much more dangerous, causing damage to vegetation, and eye and lung irritation in man.

Lead

The toxicity of lead has been known for some time. It has long been used in making water pipes and water obtained through these may be contaminated with it. As lead is not easily absorbed from the intestines this does not present a major health hazard. Much more dangerous is the lead absorbed from the air by the lungs. Most lead in the air is emitted from car exhausts. **Tetraethyl lead (TEL)** is added to petrol as an **anti-knock** agent to help it burn more evenly in car engines. Each year in Britain alone, around 50 000 tonnes of lead are added to the atmosphere in this way. While much of this is deposited close to roads, that which remains in the atmosphere and is absorbed by the lungs could have the following adverse effects:

1. Digestive problems, e.g. intestinal colic.
2. Impairing the functioning of the kidney.
3. Nervous problems, including convulsions.
4. Brain damage and mental retardation in children.

Anti-knock agents which do not contain lead exist and in some countries legislation permits only this type. They are, however, more expensive and cause an increase in the actual cost of petrol. However, governments can, and sometimes do, offset these costs by reducing the tax on unleaded fuel, thus making it more attractive.

Control of air pollution

On 9 December 1952, foggy conditions developed over London. Being very cold, most houses kept fires burning, with coal as the major fuel. The smoke from these fires mixed with the fog and was unable to disperse, resulting in a smog which persisted for four days. During this period some 4000 more people died than would be expected at this time of the year. Most of these additional deaths were due to respiratory disorders. These alarming consequences of smog prompted the government to seek ways of controlling smoke emissions from chimneys. This led ultimately to the **Clean Air Act**

Photochemical smog in Los Angeles

of 1956. Among other things this created smokeless zones, in which only smoke-free fuels could be burned. Grants were made available to assist with the cost of having fires converted to take these smokeless fuels. For many years now most cities have been smokeless and the smogs, once a common feature of winter, no longer occur.

Other methods of controlling air pollution include the use of non-lead anti-knock agents and the removal of pollutants such as sulphur dioxide before smoke is emitted from chimneys. The latter is achieved by passing the smoke through a spray of water in which much of the sulphur dioxide dissolves. The use of electric cars is a further means of limiting air pollution.

Ozone depletion and the greenhouse effect

Between 15 and 40 kilometres above the earth is a layer of ozone which is formed by the effect of ultra-violet radiation on oxygen molecules. In this way, a large amount of the potentially harmful ultra-violet radiation is absorbed and so prevented from reaching the earth's surface. There is evidence that this beneficial ozone layer is being damaged by atmospheric pollution to the point where a hole in it has appeared over the Antarctic and possibly the Arctic too.

A number of pollutants can affect the ozone layer, the **chlorofluorocarbons (CFC's)** being the best known. CFC's are used as propellants in aerosol sprays, in refrigerators, and make up the bubbles in many plastic foams, e.g. expanded polystyrene. They are remarkably inert and therefore reach the upper stratosphere unchanged. Along with other ozone depleting gases such as **nitrous oxide** (NO), CFC's are contributing to global warming – the so-called 'Greenhouse Effect' mentioned earlier in this section. In addition, the ultra-violet radiation causes skin cancer: an increase in the incidence of this disease is already evident.

29.3.2 Water pollution

Pure water rarely, if ever, exists naturally. Rainwater picks up additives as it passes through the air, not least sulphur dioxide (Section 29.3.1). Even where there is little air pollution, chlorides and other substances are found in rainwater. As water flows from tributaries into rivers it increasingly picks up minerals, organic matter and silt. If not the result of man's activities, these may be considered as natural additives and therefore not pollutants. Before we consider water pollution, it is worth looking at the uses to which man puts water.

1. First and foremost it is used for drinking – not just for man himself but also for his domesticated livestock.

2. It is used for bathing, and washing clothes, utensils, cars etc.

3. Sewage is removed by water.

4. It is used to irrigate crops.

5. It has many industrial uses, for example as a coolant.

6. It is a source of food in that fish live and breed in water.

7. It is a source of power in hydro-electric schemes.

8. It has a recreational use; fishing, sailing, diving etc.

For domestic use alone, each individual in Britain uses an average of 150 litres of water each day.

Sewage and its disposal

Sewage is quite simply anything which passes down sewers. It has two main origins: from industry and from the home (domestic). Domestic effluent is 95–99% water, the remainder being organic matter. In itself, the organic material is harmless, but it acts as a food source for many saprophytic organisms, especially bacteria. Where oxygen is available, aerobic saprophytes decompose the organic material – a process called **putrefaction** – and in so doing use up oxygen. This creates a **biochemical oxygen demand (BOD)**.

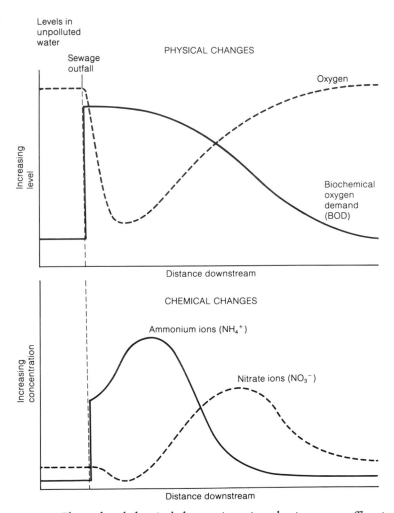

Fig. 29.2 *Physical and chemical changes in a river due to sewage effluent*

Where sewage is deposited untreated into relatively small volumes of water, i.e. rivers and lakes rather than the oceans, the BOD may be great enough to remove entirely the dissolved oxygen. This causes the death of all aerobic species, including fish, leaving only anaerobic ones. The BOD is offset by new oxygen being dissolved, and in fast-moving, shallow, turbulent streams this is sufficient to prevent anaerobic conditions. Unfortunately many centres of population are situated near river estuaries where the waters are slower, deeper and less turbulent. The amount of oxygen dissolving is much less and so any untreated sewage added to these waters quickly results in them becoming anaerobic. With only around 5 cm^3 of dissolved oxygen in each dm^3 (litre) of fresh water, every individual produces enough organic matter each day to remove the oxygen from 9000 dm^3 of water. Where untreated sewage enters a river it creates a BOD which gradually decreases further down stream as organic material is decomposed. Part of this sewage is combined nitrogen; each human produces 8 g of this daily, mostly in the form of urea and uric acid. This combined nitrogen is converted to ammonia by bacteria. While the ammonia may be toxic, its effects are temporary, as nitrifying bacteria rapidly oxidize it to nitrates. These relationships are illustrated in Fig. 29.2.

The chemical and physical changes brought about by sewage are accompanied by changes in the fauna and flora of the water. Where the level of organic material is high, saprophytic bacteria concentrations, including filamentous bacteria known as **sewage fungus**, increase as they feed on the sewage. The algal levels initially fall, possibly due to the sewage reducing the amount of light which penetrates the water. Further downstream the algal levels rise above normal because the bacterial breakdown of the sewage releases many minerals, including nitrates. These minerals, which previously limited algal growth, now allow it to flourish. As the minerals are used up, algal population levels return to normal.

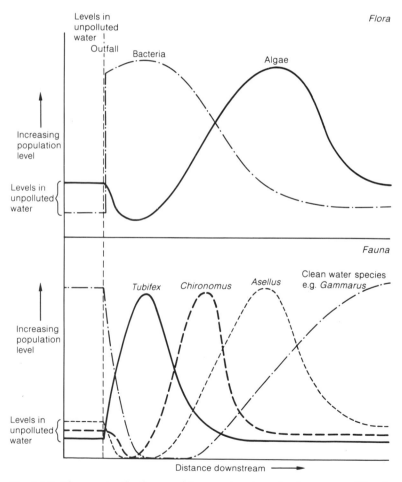

Fig. 29.3 Changes in the flora and fauna of a river due to sewage effluent

The population levels of animal species vary according to the level of oxygen in the water. Most tolerant of low oxygen levels are worms of the genus *Tubifex* whose haemoglobin has a particularly high affinity for oxygen which it obtains even at very low concentrations. These worms can therefore survive close to a sewage outfall; indeed, as other species cannot survive there, *Tubifex* are free from competitors and predators and so their numbers increase greatly.

Further downstream, as oxygen levels rise, other species such as the larvae of the midge *Chironomus* are also able to tolerate low oxygen levels. These compete with *Tubifex* for the small amount of available oxygen, and the worm population is reduced as a consequence. A continuing rise in oxygen level further from the outfall results in the appearance of species like the water louse, *Asellus*. Its presence adds to the competition, causing reduction in the populations of *Tubifex* and *Chironomus*. Finally, as the sewage is completely decomposed, oxygen levels in the water return to normal and clean-water species, like the fresh-water shrimp, *Gammarus*, are present again. The ecological equilibrium is restored and population levels return to those found above the outfall. These changes in fauna and flora are illustrated in Fig. 29.3.

These organisms act as indicator species for polluted water. Where repeated additions of sewage occur at different points along the river, the water may be anaerobic for much of its length. In addition to the death of aerobic species, these conditions can result in the build-up of ammonia and hydrogen sulphide from anaerobic decomposition of sewage. These chemicals are toxic and result in an almost lifeless river. This was the situation with most large British rivers until the introduction of **sewage treatment works**. These works not only remove organic material but also potentially dangerous pathogenic organisms such as those causing cholera and typhoid.

The process of sewage treatment is outlined in Fig. 29.4. It consists of a series of stages:

1. Screening – Large pieces of debris are filtered off to prevent them blocking pipes and equipment in the treatment works. This filtering is performed by a screen of metal rods, about 2 cm apart. The debris which is trapped on the screen is periodically scraped off and either buried or broken up into smaller pieces ready to undergo normal sewage treatment. Alternatively, the sewage enters a machine called a comminuter which reduces all the sewage into pieces small enough to enter the treatment works without risk of blockage.

2. Detritus removal – The sewage enters a tank or channel in which the rate of flow is reduced sufficiently to allow heavy inorganic material such as grit to deposit out. The lighter organic matter is, however, carried along in the water flow. The material that settles out is called **detritus** and can be dumped without further treatment.

3. Primary sedimentation – The sewage flows into large tanks which have a conical shaped base with a central exit pipe. The flow across these tanks is very slow, and may take several days. Fine silt and sand along with any organic material settle out and become deposited at the bottom of these tanks. The addition of ferric chloride, which causes flocculation, assists sedimentation in these tanks. The material which settles out is called **sludge** and is periodically pumped from the bottom of the tank to sludge digestion tanks. The sewage which has had most solid material removed is now known as **effluent**. It is removed from the top of the sedimentation tanks and either enters **activated sludge tanks** or passes through **percolating filters**.

4a. Activated sludge method – The effluent is inoculated with aerobic microorganisms which break down dissolved organic material. It then flows into long channels through which air is blown in a fine stream of bubbles from the bottom. This provides oxygen for aerobic microorganisms rapidly to decompose the organic matter into carbon dioxide and some nitrogen oxides. One problem with this method is that detergents in the sewage can cause foaming.

4b. Percolating filter method – The alternative to the activated sludge method is to spray the effluent onto beds of sand, clinker and stones in which live a large variety of aerobic organisms, especially bacteria.

In both the above processes microorganisms oxidize the various dissolved substances. Urea for example may be decomposed as follows:

(i)
$$CO(NH_2)_2 + H_2O \xrightarrow{\text{bacteria producing urease}} 2NH_3 + CO_2$$

urea　　water　　　　　　　　　　ammonia　　carbon dioxide

(ii)
$$2NH_3 + 3O_2 \xrightarrow{\text{Nitrosomonas}} 2NO_2^- + 2H^+ + H_2O$$

ammonia　oxygen　　　　　　　　nitrite　hydrogen　water
　　　　　　　　　　　　　　　　　　　　ions

(iii)
$$NO_2^- + O_2 \xrightarrow{\text{Nitrobacter}} 2NO_3^-$$

nitrite　oxygen　　　　　　　　　nitrate

5. Final (humus) sedimentation – The effluent from the sludge tanks or percolating filters contains a large number of microorganisms. It therefore passes into further sedimentation tanks to allow these organisms to settle out. The sediment, known as **humus**, is then passed into the sludge treatment tanks.

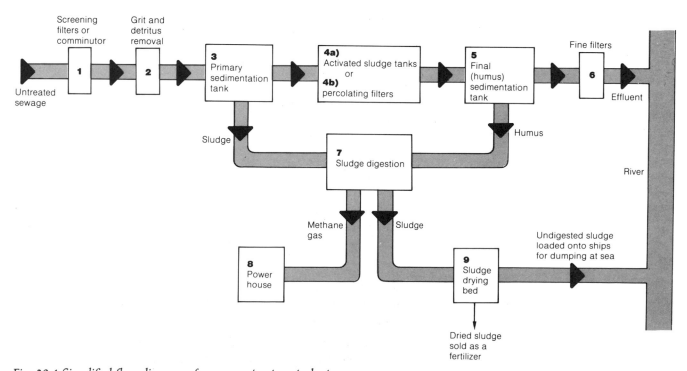

Fig. 29.4 Simplified flow diagram of a sewage treatment plant

6. Fine filters – These remove any suspended particles in the effluent which may then be safely discharged into rivers.

7. Sludge digestion – The sludge and humus are pumped into large covered tanks where they are hydrolysed into simpler compounds, leaving gases such as methane, and a digested sludge.

8. Use of methane for generating power – The methane produced during sludge digestion is usually collected and used as a fuel to drive turbines in a power-house. The electricity generated can be used to power the equipment and lights at the sewage works, making them, in some cases, self-sufficient in energy.

9. Sludge drying beds – The sludge is led off into large tanks where its water content is reduced by air-drying. The resultant semi-solid material may either be loaded onto ships for dumping at sea or sold as fertilizer.

The removal of solid material during sewage treatment is highly efficient, being reduced from $400 \, \text{mg dm}^{-3}$ in untreated sewage to $10 \, \text{mg dm}^{-3}$ once treated. Similarly, the amount of organic carbon is reduced from $250 \, \text{mg dm}^{-3}$ to $20 \, \text{mg dm}^{-3}$. Ninety-nine per cent of complex chemicals like the pesticide DDT are also removed. Potential pollutants such as zinc, copper and phosphorus may only be 50% removed. The phosphorus is a particular problem as it is widely used as a water softener in detergents and is therefore present in high concentrations in sewage. Most pathogenic organisms are removed by sewage treatment although *Salmonella paratyphi* (causes paratyphoid) and *Enteramoeba histolytica* (causes dysentery) may survive in small numbers. The eggs of worm parasites like *Ascaris* and *Taenia* have also been found in sewage works effluent.

Toxic chemicals
There is a large variety of toxic chemicals released into rivers and seas around the world. These include copper, zinc, lead, mercury and cyanide. Fish are killed by fairly low concentrations of copper but algae are even more susceptible and die at concentrations as low as one part in two million.

Mercury is a particularly hazardous chemical as it forms strong complexes with the -SH groups in proteins, and so causes disruption of cell membranes and denaturation of enzymes. The kidney, liver and brain are most affected, resulting in loss of sensation, paralysis and death. An estimated 250 tonnes of mercury enters the world's oceans each year from the natural weathering of rocks. A further 3000 tonnes is added as the result of environmental pollution. One graphic illustration of the effect of mercury occurred during the 1950s in the Japanese fishing village of Minamata where mercury, used as a catalyst in a nearby plastics factory, was discharged into the sea. In itself not especially toxic, the mercury accumulated in mud on the sea-bed. Here anaerobic methane-producing bacteria converted it to dimethyl mercury, a neurotoxin. The bacteria were consumed by filter-feeding shellfish which formed a major food source for the inhabitants of the village. Numbness, locomotory disorders, convulsions and blindness were experienced by the villagers, forty-six of whom died as a result of mercury poisoning. One hundred and twenty people died in similar circumstances at the nearby village of Niigata. Even babies born to mothers from the village some three years later showed mental retardation as a result of the poisoning.

Eutrophication by sewage and fertilizer

Eutrophication is a natural process during which the concentration of salts builds up in bodies of water. It occurs largely in lakes and the lower reaches of rivers, and the salts normally accumulate until an equilibrium is reached where they are exactly counterbalanced by the rate at which they are removed. Lakes and rivers with low salt concentrations are termed **oligotrophic** and the salts are frequently the factor limiting plant growth. Waters with high concentrations are termed **eutrophic** and here there is much less limitation on growth. **Algal blooms** occur where the waters become densely populated with species of blue-green bacteria in particular. The density of these blooms increases to a point where light is unable to penetrate to any depth. The algae in the deeper regions of the lake are therefore unable to photosynthesize, and die. Decomposition of these dead organisms by saprophytic bacteria creates a considerable biochemical oxygen demand (BOD) resulting in deoxygenation of all but the very upper layers of the water. As a consequence all aerobic life in the lower regions dies.

Salts necessary for eutrophication of lakes and rivers come from three sources:

1. Leaching from the surrounding land – This natural process is slow and is offset by the removal of salts as water drains from lakes or rivers.

2. Sewage – Even when treated, sewage effluent contains much phosphate as a result of the decomposition of detergents and washing powders (Section 29.4.1).

3. Fertilizers – An increasing quantity of inorganic fertilizer is now applied to farmland to increase crop yield. A major constituent of these fertilizers is nitrate. As this is highly soluble it is readily leached and quickly runs off into lakes and rivers.

Oil

The effects of oil pollution are localized, but nonetheless serious. Oil is readily broken down by bacteria, especially when thoroughly dispersed. Most oil pollution is either the result of illegal washing at sea of storage tanks of oil tankers or accidental spillage. The first major oil pollution incident in Great Britain occurred in 1967 when the Torrey Canyon went aground off Lands End. It released 60 000 tonnes of crude oil which was washed up on many Cornish beaches. Sea birds are particularly at risk because the oil coats their feathers, preventing them from flying; it also reduces their insulatory properties, causing death by hypothermia. The Torrey Canyon incident alone is estimated to have killed 100 000 birds. On shores, the oil coats seaweed, preventing photosynthesis, and covers the gills of shellfish, interfering with feeding and respiration. The effects are, however, temporary and shores commonly recover within two years. Detergents, used to disperse oil, can increase the ecological damage as they are toxic. With larger 'super-tankers', the potential danger from oil pollution is increased. The wrecking of the Amoco Cadiz off the Brittany coast in 1978 with the release of 200 000 tonnes of crude oil made the Torrey Canyon incident appear small by comparison. As a result of the Amoco Cadiz disaster, considerable damage was caused to commercial shellfish beds on the Brittany coast.

Oil polluted sea bird

Thermal pollution

All organisms live within a relatively narrow range of temperature. Wide fluctuations in temperature occur more often in terrestrial environments as the high specific heat of water buffers temperature changes. For this reason aquatic organisms are less tolerant of temperature fluctuations. Most thermal pollution of water is the result of electricity generation in power-stations. The steam used to drive the turbines in these stations is condensed back to water in large cooling towers. The water used in the cooling process is consequently warmed, being discharged at a temperature some 10–15°C higher than when removed from the river. Although warmer water normally contains less dissolved oxygen, the spraying of water in cooling towers increases its surface area and thereby actually increases its oxygen content. The main effect of thermal pollution is to alter the ecological balance of a river by favouring warm-water species at the expense of cold-water ones. Coarse fish such as roach and perch may, for example, replace salmon and trout.

29.3.3 Terrestrial pollution

Pollution of land may be separated into two parts:

1. The dumping of wastes and deposits.
2. The use of pesticides.

The dumping of wastes and deposits

Many commercial activities result in the production of large volumes of solid waste material. For the most part this waste is dumped in pits or heaps. **Spoil heaps** consist of waste material from various mining activities, like gravel digging. While they may be unsightly they rarely present a direct hazard. **Slag heaps** are wastes from ore-digging and metal refining activities, and particularly from the mining of coal. The heaps themselves, especially when they result from mining metal ores, are toxic owing to the high concentration of heavy metal ions. Vegetation is difficult to grow on them and only varieties carefully selected for their tolerance of a particular metal will survive. Some 400 million tonnes of these wastes are piled up each year. They are unsightly and, in the absence of vegetation, may be unstable. This instability brought tragedy to the Welsh village of Aberfan when, on Friday 21 October 1966, a slag heap above the town, destabilized by heavy rain, began to slip. It engulfed the primary school and a number of houses, killing 116 children and 28 adults.

Domestic rubbish once contained a high proportion of ash, cinders and other solid, non-combustible material. It was frequently dumped in old quarries and pits and on low-lying land where it formed a relatively stable base upon which top soil could be placed and the land used for recreational or agricultural purposes. Present-day rubbish contains little ash or cinder and comprises mostly combustible material like paper and plastics. Incineration is often the best way to dispose of this rubbish, with the heat produced possibly being used to generate electricity. An incinerator at Edmonton in North London supplies electricity to the National Grid and so makes a sizeable profit from the disposal of its waste. Even when rubbish is dumped, the high content of organic material leads to decomposition by bacteria. These produce methane and other gases and there are plans at one large tip in the Midlands to collect the gas and burn it

This land was considered for the shallow burial of low-level radioactive waste

to produce electricity. It is hoped that enough will be generated to supply up to 10 000 homes. More details about the production of methane and alcohol from waste are given in Section 41.5.

Pesticides

It is difficult to define what exactly is a 'pest', but it is generally accepted to be an organism which is in competition with man for food or soil space, or is potentially hazardous to health. It may even be an organism which is simply a nuisance and so causes annoyance. Pesticides are poisonous chemicals which kill pests, and they are named after the pests they destroy; hence insecticides kill insects, fungicides kill moulds and other fungi, rodenticides kill rodents such as rats and mice, and herbicides kill weeds. Unlike other pollutants, where their poisonous nature is an unfortunate and unwanted property, pesticides are quite deliberately produced and dispensed in order to exploit their toxicity.

An ideal pesticide should have the following properties:

1. It should be **specific**, in that it is toxic only to the organisms at which it is directed and harmless to all others.

2. It should **not persist** but be unstable enough to break down into harmless substances. It is therefore temporary and has no long-term effect.

3. It should **not accumulate** either in specific parts of an organism or as it passes along food chains.

TABLE 29.2 **Some major pesticides**

Name of pesticide	Type of pesticide	Additional information
Inorganic pesticides Calomel (mercuric chloride)	Fungicide	Used for dusting seeds to control transmission of fungal diseases
Copper compounds (e.g. copper sulphate)	Fungicide and algicide	One of the first pesticides ever used was Bordeaux mixture (copper sulphate + lime)
Sodium chlorate	Herbicide	Used to clear paths of weeds. Persistent, although not very poisonous
Organic pesticides Organo-phosphorus compounds (e.g. malathion and parathion)	Insecticides	Although very toxic they are not persistent and therefore not harmful to other animals if used responsibly. May kill useful insects such as bees, however
Organo-chlorine compounds (e.g. DDT, BHC, dieldrin, aldrin)	Insecticides	DDT is fairly persistent and accumulates in fatty tissue as well as along food chains. Aldrin may persist for more than 10 years. Resistance to them is now common. Most kill by inhibiting the action of choline-esterase
Hormones (e.g. 2,4-D, 2,4,5-T)	Herbicides	Selective weedkillers which kill broad-leaved species. Stimulate auxin production and so disrupt plant growth. May contain a dangerous impurity – dioxin

Pesticides have been used for some time. A mixture of copper sulphate and lime, called Bordeaux mixture, was used over 100 years ago to control fungal diseases of vines. The problem is that in an attempt to produce food more economically and control human disease, pesticides have been used in large amounts in most regions of the world. A summary of some major pesticides is given in Table 29.2.

Most pesticides are not persistent. Warfarin, for example, readily kills any rodent which eats it, but as it is quickly broken down inside the rodent's body, it is harmless to anything which eats the corpse, e.g. maggots. Some pesticides, dichlorodiphenyltrichlorethane (DDT), for example, are unfortunately persistent. First synthesized in 1874, its insecticidal properties were not appreciated until 1939. It was used extensively during the Second World War, in which it played a vital rôle in controlling lice, fleas and other carriers of disease. It was subsequently used in killing mosquitoes and so helped control malaria. Not only is DDT persistent, it also accumulates along food chains. If, for example, garden plants are sprayed with it in order to control greenfly, some of the flies will survive despite absorbing the DDT. These may then be eaten by tits who further concentrate the chemical in their bodies, especially in the fat tissues where it accumulates. If a number of tits, each containing DDT, are consumed by a predator, e.g. a sparrowhawk, the DDT builds up in high enough concentrations to kill the bird. Even where the concentrations are not sufficient to kill, they may still cause harm. It is known that DDT can alter the behaviour of birds, sometimes preventing them building proper nests. It may cause them to become infertile and can result in the egg shells being so thin that they break when the parent bird sits on them during incubation. In Britain these effects led to a marked decline in the 1950s and 60s of populations of peregrine falcons, sparrow hawks, golden eagles and other predatory birds. As a consequence, Britain, along with many other countries, restricted the use of DDT with the result that populations of these birds have now recovered.

Owing to the persistence of DDT, it remains in the environment, despite the death of the organism containing it. With over one million tonnes of the chemical having already been used it now occurs in all parts of the globe and is found in almost all animals. Indeed, many humans contain more DDT than is permitted by many countries in food for human consumption.

With such widespread use of DDT, it is not surprising that selection pressure has resulted in insect varieties which are able to break it down and so render it useless. The development of **resistance** is now common among insect disease vectors like mosquitoes, and has set back prospects of eradicating malaria.

Herbicides make up 40% of the world's total pesticide production, and in developed countries the figure exceeds 60%. Some herbicides like paraquat kill all vegetation. While paraquat is highly poisonous it is rapidly broken down by bacteria and rendered harmless. Other weedkillers are selective, destroying broad-leaved plants (mostly dicotyledons) but not narrow-leaved ones (mostly monocotyledons). As most of man's cereal crops are narrow-leaved and the weeds that compete with them are broad-leaved, such selective weedkillers are extensively used. They are similar to the plant's natural hormones, auxins, and as such are quickly broken down and rendered harmless. The two best known examples are 2, 4-dichlorophenoxyacetic acid (2, 4-D) and 2, 4, 5-trichlorophenoxyacetic acid (2, 4, 5-T). In the production of 2, 4, 5-T an impurity called **dioxin** is formed. Dioxin is one of the most toxic compounds known to man, a single gram being sufficient to kill in excess of 5000 humans. Even in minute quantities it may cause cancer, a skin disorder called chloracne and abnormalities in unborn babies. The chemical gained notoriety when used as a defoliant by America during the Vietnam war in the 1970s. It was a constituent of 'Agent Orange', 50 million dm³ (litres) of which were sprayed over jungle areas to cause the leaves to drop so that enemy camps could be revealed. The dioxin produced physical and mental defect in children born in the area, as well as in those born to American servicemen working in the region. In 1976, an accident at a factory in Seveso, Italy, resulted in the release of dioxin into the atmosphere. Despite evacuation of the area, thousands of people suffered with chloracne, miscarriages, cancer and foetal abnormalities.

29.3.4 Radioactive pollution

There are two main forms of radiation: electromagnetic waves and sub-atomic particles.

Electromagnetic waves are not all potentially dangerous; only those of short wavelength, and therefore high energy content, such as X-rays and γ-rays, are normally hazardous.

Sub-atomic particles include **α-particles** which are positively charged. They are identical with the nucleus of a helium atom, consisting of two protons and two neutrons. **β-particles** are negatively charged, consisting of high-speed electrons. **Neutrons** have no charge. In isolation they are radioactive as they decay.

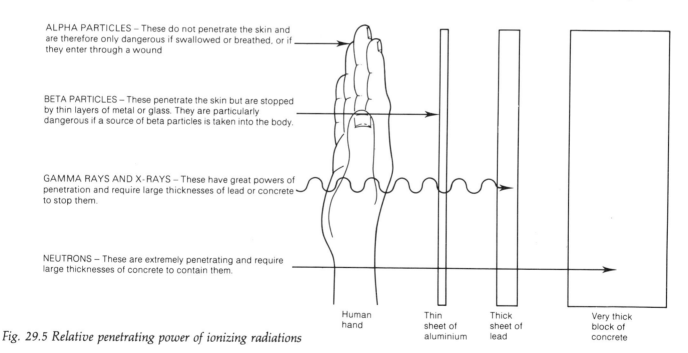

ALPHA PARTICLES – These do not penetrate the skin and are therefore only dangerous if swallowed or breathed, or if they enter through a wound

BETA PARTICLES – These penetrate the skin but are stopped by thin layers of metal or glass. They are particularly dangerous if a source of beta particles is taken into the body.

GAMMA RAYS AND X-RAYS – These have great powers of penetration and require large thicknesses of lead or concrete to stop them.

NEUTRONS – These are extremely penetrating and require large thicknesses of concrete to contain them.

Human hand

Thin sheet of aluminium

Thick sheet of lead

Very thick block of concrete

Fig. 29.5 Relative penetrating power of ionizing radiations

Types of ionizing radiation
Sub-atomic particles and the more energetic electromagnetic waves produce ions when they strike materials. They are therefore referred to as **ionizing radiation**. In living tissues, this ionization frequently results in chemical changes which cause injury. The penetrating powers of various forms of ionizing radiation are shown in Fig. 29.5.

Some radioactive sources emit their radiation for short periods, others for considerable lengths of time. The characteristic time taken

TABLE 29.3 **Half-lives of some biologically important elements**

Element	Half-life
Carbon – ^{14}C	5760 years
Sodium – ^{24}Na	15 hours
Manganese – ^{54}Mn	300 days
Strontium – ^{90}St	28 years
Iodine – ^{129}I	16 000 000 years·
^{131}I	8 days
Caesium – ^{137}Cs	30 years
Plutonium – ^{239}Pu	240 000 years

TABLE 29.4 **Sources of radiation in the UK**

Source	Radiation dose in microsieverts $/\mu Sv$
Cosmic rays	310
Rocks and minerals	380
Food and drink	370
Medical	500
Background radiation in the air	800
Miscellaneous, e.g. watches, TV sets	8
Nuclear fallout	10
Nuclear waste	3

for the activity of a particular radioactive substance to decay to half its value – i.e. for half the atoms present to disintegrate – is called the **half-life**. Half-lives vary from less than a millionth of a second to millions of years; the half-life of some biologically important elements is given in Table 29.3.

The sources of radiation are mostly natural and little can be done to alter these. In any case, organisms have evolved to tolerate these levels. What is of concern is the additional radiation that is produced as the consequence of man's activities. The two major sources of man-made radiation are radioactive fall-out, from the testing of nuclear weapons, and wastes from the nuclear power industry. As Table 29.4 shows, neither at present contributes substantially to the total background radiation.

Biological effects of radiation
It is probably fair to assume that all ionizing radiation is dangerous and there is therefore no absolutely safe level, but the risk at low levels of radiation is small. It is not only the intensity of the radiation but also the duration of the exposure which is important. The product of these two is called the **dose**. Radiation damage is proportional to the size of the dose.

Some sources of radiation are particularly hazardous for a number of reasons. Strontium 90, for example, is a particular risk because:

1. Its half-life of 28 years means it continues to cause damage for most of an individual's life.

2. It is a waste product of nuclear weapons and the nuclear power industry.

3. It is absorbed into grass from the air, and cattle eating the grass then concentrate it into their milk which is consumed by humans.

4. Once absorbed by humans it is concentrated further in bone tissue and so is adjacent to bone marrow which produces blood cells and as such is one of the most rapidly dividing tissues in the body.

5. Dividing tissues are more vulnerable to radiation damage than other tissues. In this instance, strontium 90 can cause leukaemia.

The major danger of ionizing radiation is that it causes the breaking of hydrogen bonds in the DNA molecule; the result can be the loss of a nitrogen base, i.e. gene mutation. These mutations can of course be inherited. Intense radiation doses may also cause chromosome mutations. As any mutation in the gonads may lead to production of defective gametes and hence deformed offspring, the gonads are especially at risk. All dividing tissue is particularly vulnerable and for this reason pregnant women are never X-rayed for risk of damaging the foetus, in which most cells are rapidly dividing.

The somatic rather than genetic effects occur in two phases. **Early effects** include damage to the gut, blood cells and bone tissue, as well as skin burns, loss of hair and infertility. These effects are referred to as **radiation sickness**. **Delayed effects** include an increased risk of cancer as well as the possibility of hereditary defects.

In the same way that pesticides may be concentrated as they pass along food chains, so too may radioactive materials. Levels of radioactive phosphorus (^{32}P) found in a North American river were 1000 times greater in the algae living in it. The aquatic birds feeding off these algae had a level 7500 times greater than that in the river, and their eggs, whose shells contain calcium phosphate, had levels

TABLE 29.5 **Sound levels**

Sound	Level dB	Effect
Moon rocket at 300 m	200	Skin burned, possibly lethal
Jet take off at 30 m	140	Temporary deafness
Inside discotheque	120	Ear drums ring
Screaming in someone's ear	110	Unpleasant
Noisy machine in a factory	100	Uncomfortable
Pneumatic drill at 10 m	90	Hearing damaged if continuous for 8 hours
Telephone at 1 m	80	Annoying
Inside small car	70	Normal conversation difficult
Busy office	60	Normal conversation
Average living room	50	Everyday noise
Average bedroom	40	Quiet conversation
Ideal library	30	Very quiet
Empty broadcasting studio	20	Silence
Totally soundproof room	10	Heartbeat audible
Total absence of sound	0	May cause hallucinations due to sensory deprivation

some 200 000 times greater. Once again the potential danger is that these concentrated levels occur close to the rapidly dividing cells of the embryonic bird inside the shell.

29.3.5 Noise pollution

Noise is not normally lethal, although in extreme cases it may cause hearing damage. It is, however, often irritating and stressful. Loud and sudden noise causes adrenalin to be released into the blood, causing increased blood pressure among other changes. Sound is measured in decibels (dB) and Table 29.5 illustrates the level of various sounds. It should be noted that the decibel scale is logarithmic and therefore an increase of 10 dB represents a doubling of the noise level. In other words, 40 dB is *twice* as loud as 30 dB, and 90 dB is *twice* as loud as 80 dB and some sixty-four times louder than 30 dB.

29.4 Conservation

There has been a growing interest in conservation as a result of increasing pressures placed upon the natural environment, the widespread loss of natural habitats and the growing numbers of extinct and endangered species. As early as 1872, the Yellowstone National Park in the USA was established in order to protect a particularly valuable natural environment. Australia (1886) and New Zealand (1894) established national parks soon after. It was not until 1949 that the first national park in Britain was established, but prior to that many societies such as the Royal Society for the Protection of Birds (1889) and the National Trust (1895) had been set up to promote conservation. There are now a large number of agencies responsible for conservation in one form or another. These include international groups like the World Wide Fund for Nature, large national bodies such as the Department of the Environment (DoE), Nature Conservancy Council (NCC), and the National Trust (NT); commercial organizations like the water authorities and the Forestry Commission; charitable groups like the Royal Society for the Protection of Birds (RSPB) as well as County Trusts for Nature Conservation and Farming and Wildlife Advisory Groups. The main impetus for conservation has come as a result of the pressures created by an ever-increasing human population – likely to be 6000 million before the end of the century. Many species have become **extinct**, i.e. they have not been definitely located in the wild during the past 50 years. Others are **endangered**, i.e. they are likely to become extinct if the factors causing their numbers to decline continue to operate. At least 25 000 plant species are considered to be endangered.

There are a number of reasons why organisms become endangered:

1. Natural selection – It is, and always has been, part of the normal process of evolution that organisms which are genetically better adapted replace ones less well adapted.

2. Habitat destruction – Man exploits many natural habitats, destroying them in the process. Timber cutting destroys forests and endangers species like the orang-utan. Industrial and agricultural development threaten many plant species of the Amazon forest. Clearing of river banks destroys the natural habitat of the otter, and modern farming methods remove hedgerows and drain wetlands, endangering the species which live and breed there.

3. Competition from man and his animals – Where a species is restricted to a small area, e.g. the giant tortoises in the Galapagos Islands, they are often unable to compete with the influx of man and his animals. Because their habitat is restricted, in this case by water, they cannot escape.

4. Hunting and collecting – Man hunts tigers for sport, crocodiles for their skins, oryx as trophies, elephants for ivory, whales for oil and rhinoceros for their horn. Other organisms are collected for the pet trade, e.g. tamarins and parrots; and for research purposes, e.g. frogs. These are in addition to the numerous species hunted purely as food.

5. Destroyed by man as being a health risk – Many species are persecuted because they carry diseases of man's domesticated species, e.g. badgers (tuberculosis of cattle), and eland (various cattle diseases).

6. Pollution – Oil pollution threatens some rare species of sea birds. The build-up of certain insecticides along food chains endangers predatory birds like the peregrine falcon and the golden eagle (Section 29.3.3).

To combat these pressures a number of conservation techniques are used:

1. Development of national parks and nature reserves – These are habitats legally safeguarded and patrolled by wardens. They may preserve a vulnerable food source, e.g. in China areas of bamboo forest are protected to help conserve the giant panda.

2. Planned land use – On a smaller scale, specific areas of land may be set aside for a designated use. The types of activities permitted on the land are carefully controlled by legislation. Such areas include Green Belts, Areas of Outstanding Natural Beauty, Sites of Special Scientific Interest, and country parks.

3. Legal protection for endangered species – It is illegal to collect or kill certain species, e.g. the koala in Australia. In Britain, the Wildlife and Countryside Act gives legal protection to many plants and animals. Even legislation such as the Clean Air Act, may indirectly protect some species from extinction. Despite stiff penalties, such laws are violated because of the difficulty of enforcing them.

4. Commercial farming – The development of farms which produce sough-after goods, e.g. mink farming, deer farming, may produce enough material to satisfy the market and so remove the necessity to kill these animals in the wild.

5. Breeding by zoos – Endangered species may be bred in the protect environment of a zoo. When numbers have been sufficiently increased they may be reintroduced into the wild.

6. Removal of animals from threatened areas – Organisms in habitats threatened by man, or natural disasters such as floods, may be removed and resettled in more secure habitats.

7. Control of introduced species – Organisms introduced into a country by man often require strict control if they are not to out-compete the endemic species. Feral animals (domesticated individuals which escape into the wild) must be similarly controlled.

8. Ecological study of threatened habitats – Careful analysis of all natural habitats is essential if they are to be managed in a way that permits conservation of a maximum number of species.

9. Education – It is of paramount importance to educate the population in ways of preventing habitat destruction and encouraging the conservation of organisms.

29.5 Questions

1. Discuss the advantages and disadvantages of the use of pesticides, indicating how the disadvantages may be overcome. *(20 marks)*

Welsh Joint Education Committee June 1984, Paper AII, No. 17

2. Write an essay on conservation. *(20 marks)*

London Board June 1989, Paper II, No. 8(b)

3. Discuss the effects on the environment of fertilizers, pesticides and fuels.

Associated Examining Board June 1985, Paper III, No. 2

4. During the past thirty years many soft-water lakes in Sorlandet (Southern Norway) have lost their fish populations and have become more acid in their surface waters. At the same time there has been a marked increase in the combustion of fossil fuels in industrial Europe.

(*a*) What are the possible links between these observations? *(4 marks)*

(*b*) The diagrams above right show the relationship between fish stocks and (i) pH, and (ii) excess sulphate concentration in the lake water.

  (i) The relationship between pH and fish stocks of lakes

  (ii) The relationship between excess sulphate concentration and fish stocks of lakes

Which of these two factors, pH or excess sulphate concentration, had the greater influence on fish stocks? Give evidence from the diagrams in support of your answer. *(3 marks)*

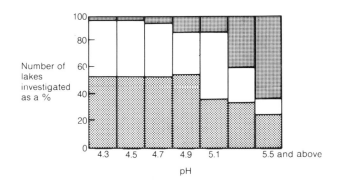

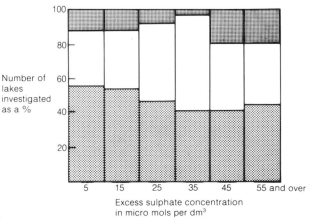

Key

Lakes with good fish population

Lakes with a sparse fish population

Lakes with no fish

(*c*) What would be the likely effect on fish stocks of increasing by 0.2 units the pH of lakes which currently have a pH range of (i) 4.3–4.5 and (ii) 5.1–5.3? *(3 marks)*

(Total 10 marks)

London Board June 1986, Paper I, No. 5

5. What evidence is there that air pollution could damage the environment? Describe how such pollution can be monitored.

Southern Universities Joint Board June 1986, Paper II, No. 12

6. (a) What is meant by sewage? (*4 marks*)
(b) Describe the sequence of stages used in the treatment of the sewage produced by an urban community. (*10 marks*)
(c) What are the problems associated with the disposal of large amounts of sewage? (*6 marks*)
(*Total 20 marks*)

London Board June 1986, Paper II, No. 4

7. The diagram below shows some of the effects of discharging sewage and copper waste into a river.
(a) What is meant by the term *biochemical oxygen demand* (BOD)? Explain the changes in BOD shown in the diagram. (*3 marks*)
(b) Explain the changes in nitrate level shown in the diagram. (*4 marks*)

(c) Compare and comment on the curves for the sewage fungus and the algae in the diagram.
(*5 marks*)

(d) Using evidence from the diagram, suggest a method by which an organism might be used as a pollution indicator. Your answer should include practical details of your method.
(*6 marks*)

(e) Suppose that the chemical works also discharged thermal pollution. Suggest **one** possible effect on the river's chemical content and **one** possible effect on its biological content. (*2 marks*)
(*Total 20 marks*)

Associated Examining Board June 1984, Paper II, No. 4

8. Study the following passage and answer the questions which follow.

In the years after 1945 when the powerful synthetic organic insecticides appeared, insect suppression was made so easy and effective that virtually every other method of pest control was dropped in favour of these miracle chemicals. But in neglecting the investigation and development of alternative controls we rendered

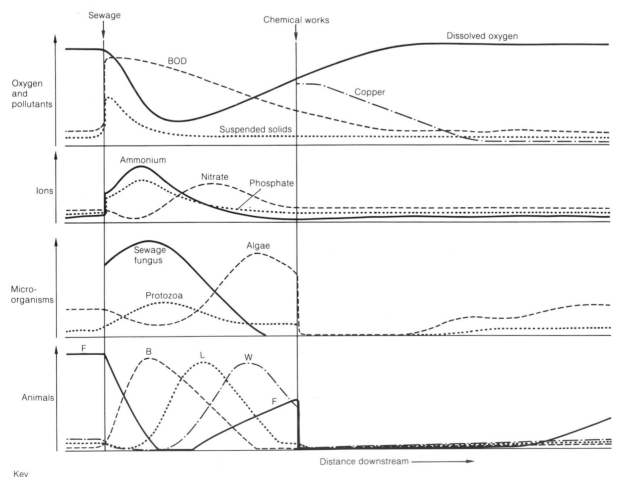

Key
BOD = biochemical oxygen demand
B = burrowing worms
L = bottom-living insect larvae
W = water lice on plants and stones
F = clean-water fish   (*modified after Mellanby*)

ourselves vulnerable to insect depredation in the event that the chemicals failed. Very few imagined that this would ever come about. But today it is happening and major disaster threatens in many areas. The signs are manifold: in Northeastern Mexico, the cotton industry has been destroyed by the tobacco budworm, an insect that cannot be killed by insecticides; in Central America, the human population, virtually malaria-free for more than a decade, is now threatened with an epidemic as the vector mosquito *Anopheles albimanus Wiedemann*, verges on total resistance to available insecticides; the cotton bollworm responds to 'control' insecticides by returning in even greater numbers to devastate cotton crops; the world over, in the wake of insecticide usage, spider mites have leapt into the forefront as crop pests and threaten economic disaster in a variety of commodities. Today there are more insect species of pest status than ever before.

A number of factors have contributed to this situation, but insecticide disruption of naturally occurring biological control lies at the heart of the matter. The basic flaw in the modern insecticide is its broad toxicity. The materials kill indiscriminately, and when applied in the field their broadly toxic action can virtually strip the treated areas of arthropod life. Thus where they are used, the ecosystem web is often shattered overnight, and a biotic vacuum created in which violent reactions are almost inevitable. Backlashes occur literally everywhere that broad spectrum insecticides have been used. The agriculturalist placed himself on an insecticidal treadmill because of three ecologically based phenomena: (1) target pest resurgence, (2) secondary pest outbreaks, and (3) pesticide resistance. All these are directly related to the disruption of biological control and lead increasingly to a 'pesticide addiction' from which it is difficult to withdraw. Target pest resurgence occurs when the insecticide not only kills a high percentage of the pest population but also a large proportion of its natural enemies. Secondary pest outbreaks occur where insecticides applied to control noxious insects destroy the natural enemies of comparatively uncommon species occupying the same habitat. Target pest resurgence and secondary pest outbreaks in turn contribute directly to insecticide resistance because they necessitate heavier and more frequent treatments which speed genetic selection for resistance.

Adapted from R. van den Bosch and P. S. Messenger (1973) *Biological Control* Intertext Books.

(a) Give a concise account of the following terms used in the passage above, giving specific examples and indicating their biological basis:
    (i) target pest resurgence   *(3 marks)*
    (ii) secondary pest outbreaks   *(3 marks)*
    (iii) pesticide resistance.   *(3 marks)*

(b) What is meant by 'pesticide addiction' in this passage?   *(2 marks)*
(c) A field of cotton was given a single treatment on the 8th July with a mixture of DDT and toxaphene against *Lygus hesperus*, a serious cotton pest. The table below gives the numbers of beet armyworm larvae (*Spodoptera exigua*) in a field of cotton (measured as the numbers of larvae per 1600 sweeps with a sweep net).

Date	8 July	15 July	29 July	5 Aug	12 Aug
Treated plot	0	30	43	121	162
Untreated plot	4	2	7	6	10

    (i) Present the data above in a suitable graphical form.   *(4 marks)*
    (ii) Discuss these results in the light of the passage above and suggest reasons for the differences between the treated and untreated plots. Why was it important to measure the population of armyworms in an untreated plot?   *(5 marks)*
    (Total 20 marks)
Northern Ireland Board June 1989, Paper II, No. 6X

9. Identical plots of land were cleared of weeds prior to sowing pea seeds. The plots were kept weed-free for different periods immediately after sowing. After a period of nine weeks, all the plants were harvested and the pea plants and weeds in each plot were weighed. The results are shown in the graph below.

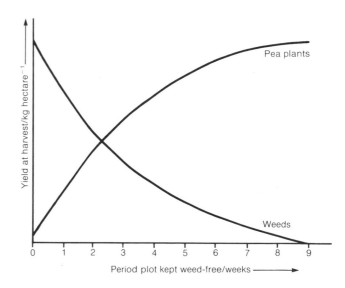

(a) (i) What conclusions can be drawn about the competition between the pea plants and the weeds?
    (ii) Describe how you might estimate the total mass of weeds growing in a large field.
    (5 marks)

(b) Competition can also occur between plants of the same species. Design an experiment to test the hypothesis that maize plants grown at a high density produce fewer grains per plant than maize plants grown at a low density.

(*6 marks*)

(c) Assuming the hypothesis in (*b*) was supported by experimental evidence and that maize plants respond to competition with weeds in a similar way to pea plants, state the advice you would give to a farmer who wishes to obtain the maximum yield of maize from a field. (*2 marks*)

(*Total 13 marks*)

Welsh Joint Education Committee June 1988, Paper A2,
No. 10

Trachea of insect (opposite)

▬Part IV▬
Transport and exchange mechanisms

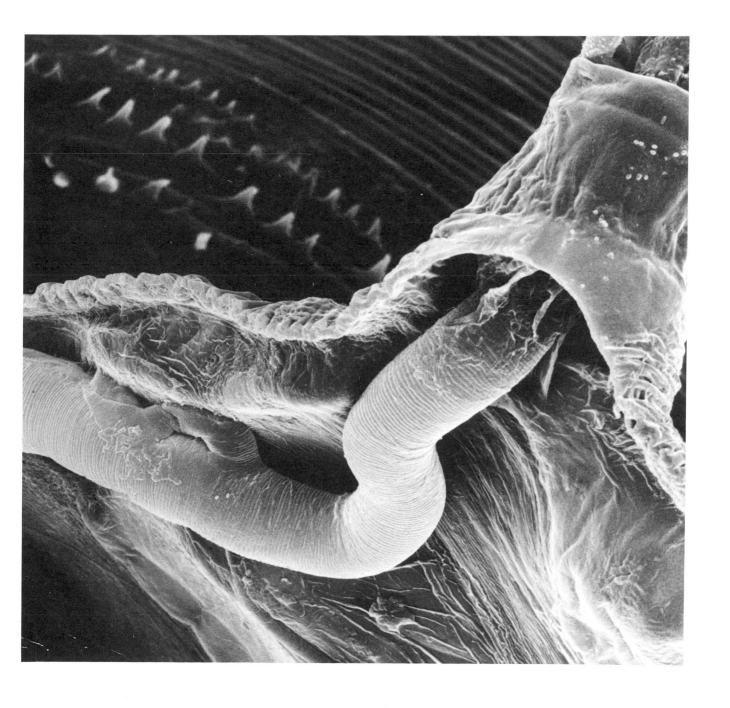

30 *Why organisms need transport and exchange mechanisms*

All organisms need to exchange materials between themselves and their environment. Respiratory gases and the raw materials for growth must pass into an organism and waste products must be removed. This exchange is carried out passively by diffusion and osmosis and actively by active transport, pinocytosis and phagocytosis (Section 4.4). To be efficient, exchange mechanisms require the surface area over which transfer occurs to be large when compared to the volume of the organism. Where diffusion is involved the exchange surface needs to be moist and the distance across which diffusion occurs must be as small as possible. In small organisms such as protozoans and unicellular algae their surface area is sufficiently large compared with their volume to allow efficient exchange of most materials over the whole surface of their bodies.

When organisms became multicellular and so grew in size, they could only meet their exchange demands by simple diffusion if their requirements were very modest, for example if they had a very low metabolic rate. An increase in size inevitably meant an increased distance from the surface to the centre of the organism. Even if sufficient exchange occurred at the surface, the centre could still be starved of raw materials, because the rate of delivery was inadequate to supply the demand, if it was dependent on diffusion alone. One means of overcoming this problem is to become flattened in shape, so ensuring that no part of the body is far from the surface which supplies its nutrients. This explains the shape of the flatworms (Platyhelminthes). A further solution is to leave the central region of the organism hollow, or fill it with non-metabolizing material. Where hollow, it is an advantage to arrange for the external medium to enter the space. This allows exchange to occur across both inner and outer surfaces. Cnidarians utilize this method, allowing them to be relatively large and yet compact, without the need for specialized exchange or transport systems.

Further increases in size and/or metabolic rate necessitated the development of specialized exchange surfaces to compensate for a smaller surface area/volume ratio and/or an increased oxygen demand. In insects, flight necessitated a high metabolic rate and hence a more efficient delivery of oxygen to the tissues and subsequent removal of carbon dioxide. To achieve this they developed tubular ingrowths, the tracheae, which carried the air directly to the respiring tissues. These have the advantage of allowing oxygen and carbon dioxide to diffuse through a gaseous medium, rather than through the aqueous medium of cells – a much slower process. In addition, mass flow of the air is possible and this too speeds the movement of gases. With tracheae there is no need for a circulatory system to carry respiratory gases. While a blood system is present in insects, it possesses no respiratory pigments and its purpose is to carry nutrients, wastes and phagocytic cells.

Where large size is combined with a high metabolic rate, both specialized exchange surfaces and an efficient means of transport become essential. In water the exchange surfaces for respiratory gases take the form of gills. In their simplest forms these are branched external outpushings of the body wall, as in amphibians like the axolotl. A similar mechanism is used by crustaceans, but here they are covered, and hence protected, by a shell or carapace. When covered, gills require a means of ventilation to supply them with fresh respiratory medium. This ventilation may be carried out by muscular action as in most crustaceans, or by cilia, as in bivalve molluscs. In fish the gills are more elaborate. They form highly branched, blood-filled extensions around gill slits which lead from the pharynx. A regular current of water is pumped over these internal gills. While gills are in principle adequate exchange surfaces for terrestrial organisms, they suffer the disadvantage of lacking support in the less dense medium of air. They therefore collapse, reducing their surface area and making them inefficient. An additional problem is that respiratory surfaces need to be kept moist to allow efficient diffusion. Gills could be kept moist, but due to their positions they would lose intolerably large quantities of water through evaporation. The solution for large, highly metabolizing, terrestrial organisms was to develop lungs. These comprise tiny elastic sacs, the alveoli, which are supported by connective tissue. The tubes, called bronchi, leading to these sacs are supported by cartilagenous rings to prevent collapse and the lungs as a whole are supported and protected by a bony cage of ribs. Being located deep within the body and communicating to the outside only by means of a narrow tube, the trachea, evaporative losses are kept to a minimum. The linings of the alveoli are thin and well supplied with blood. Muscular action ensures constant ventilation of the lungs.

With increasing size and specialization of organisms, tissues and organs became increasingly dependent upon one another. Materials needed to be exchanged not only between organs and the environment, but also between different organs. To this end, animals developed circulatory systems. These comprise a fluid which either flows freely over all cells (open system) or is confined to special vessels which communicate within diffusing distance of cells (closed system). The fluid is circulated by cilia, body muscle or a specialized pump (the heart) or some combination of these mechanisms. Closed blood systems are used by the larger, more highly evolved animals as they allow greater control of the distribution of the blood, making them more efficient at meeting changes in the demands of different tissues. Control of the heart beat assists this process.

The blood itself must be adapted to transport a wide variety of substances. Many are simply dissolved in a watery solution (the plasma), but others like oxygen are carried by special chemicals (respiratory pigments); these may be contained in specialized cells (red blood cells). Being distributed to all parts of the body, the blood is ideally situated to convey the body's defence and immune system (the white blood cells). The liquid nature of blood, so necessary for rapid transport around the body, suffers the disadvantage that it leaks away when damage is caused to the cavities or vessels containing it. Consequently a mechanism has evolved to ensure rapid clotting in these circumstances.

In plants their method of nutrition, necessitating as it does the need to capture light, means that they have an exceedingly large surface area for this purpose. This same surface therefore serves for the exchange of gases. As plants do not carry out locomotion, their metabolic rate is relatively low compared to most animals and therefore diffusion suffices. In addition, all respiring tissues are near to the surface of the plant. Large trees, for example, possess dead xylem tissue at the centre of their trunks and large branches. Respiratory gases therefore need only be conveyed very short distances and there is no specialized system for their transport — diffusion suffices. The same is not true of water and photosynthetic products. These often need to be transported the total length of the plant, 100 m or more in some cases.

In an attempt to gain a competitive advantage in the struggle for light, many plants have evolved to be very tall. The water required for photosynthesis is, however, obtained from the roots, which are firmly anchored in the soil. A transport system is therefore necessary to convey water from the roots to the leaves. At the same time, the sugars manufactured in the leaves must be transported in the opposite direction to sustain their respiration.

Plants, unlike animals, do not possess contractile cells like muscles. They therefore depend largely upon passive rather than active mechanisms for transporting materials. The evaporation of water from stomata creates an osmotic gradient across the leaf which draws in water from the xylem. Xylem forms a continuous unimpeded column of narrow tubes from the roots to the leaves. Owing to its cohesive properties, removal of water at the top of this column pulls up water from the bottom in a continuous stream. An osmotic gradient is responsible for the movement of water from the soil into the roots and across the cortex to the xylem. Only in the process of getting water into the xylem in the root is energy expended by the plant. The flow of sugars from the leaves to the roots is less clearly understood. Theories like 'mass flow' involve a passive movement; others, such as the 'transcellular strand theory', suggest an active mechanism.

31 *Gaseous exchange*

31.1 Respiratory surfaces

All aerobic organisms must obtain regular supplies of oxygen from their environment and return to it the waste gas carbon dioxide. The movement of these gases between the organism and its environment is called **gaseous exchange**. Gaseous exchange always occurs by **diffusion** over part or all of the body surface. This is called **a respiratory surface** and in order to maintain the maximum possible rate of diffusion respiratory surfaces have a number of characteristics:

1. Large surface area to volume ratio – This may be the body surface in small organisms or infoldings of the surface such as lungs and gills in larger organisms.

2. Permeable

3. Thin – Diffusion is only efficient over distances up to 1 mm since the rate of diffusion is inversely proportional to the square of the distance between the concentrations on the two sides of the respiratory surface.

4. Moist – since oxygen and carbon dioxide diffuse in solution.

5. Efficient transport system – This is necessary to maintain a diffusion gradient and may involve a vascular system.

Organisms can obtain their gases from the air or from water. The oxygen content of a given volume of water is lower than that of air, therefore an aquatic organism must pass a greater volume of the medium over its respiratory surface in order to obtain enough oxygen.

TABLE 31.1 **Water and air as respiratory media**

Property	Water	Air
Oxygen content	Less than 1%	21%
Oxygen diffusion rate	Low	High
Density	Relative density of water about 1000 times greater than that of air at the same temperature	
Viscosity	Water much greater, about 1000 times that of air	

31.2 Mechanisms of gaseous exchange

As animals increase in size most of their cells are some distance from the surface and cannot receive adequate oxygen. Many larger animals also have an increased metabolic rate which increases their oxygen demand. These organisms need to develop specialized respiratory surfaces such as gills or lungs. These surfaces allow the gases to enter and leave the body more rapidly. There remains the problem of transporting the gases between the respiring cells and the respiratory surface. Generally the gases are carried by the blood vascular system. The presence of respiratory pigments like haemoglobin increase the oxygen-carrying capacity of the blood (Section 32.1.1). The diffusion gradients may be further maintained by ventilation movements, e.g. breathing.

Gaseous exchange will be considered in detail for a number of different organisms. They will serve to demonstrate the differences between aquatic and terrestrial organisms and also show the problems associated with increased size.

31.2.1 Small organisms

Small organisms have a large surface area to volume ratio and do not require specialized structures for gaseous exchange. *Amoeba* are less than 1 mm in diameter and gases diffuse over their whole surface. Cnidarians, like the sea anemone *Actinia,* are hollow and have all their cells in contact with the water which surrounds them. Platyhelminthes (flatworms) also rely on diffusion over the whole body surface and this is facilitated by their flattened shape which considerably increases their surface area to volume ratio. All these organisms must live in water from which they obtain dissolved oxygen; they would rapidly desiccate in a terrestrial environment.

31.2.2 Flowering plants

Plants have a low metabolic rate, requiring less energy per unit volume than animals. Unicellular algae employ the whole surface of the cell for gaseous exchange. In the larger flowering plants this is not possible because their outer surfaces are waterproof to prevent the desiccation which results from living on land. Gases pass through small pores in the leaves and green stems. These pores are the **stomata** whose structure is illustrated in Section 23.1 and the mechanisms of which are described in Chapter 33. Woody plants still have stomata in the leaves but the stems have small areas of loosely packed bark cells called **lenticels**. Gases diffuse through the stomata and lenticels. Within the plant, oxygen moves through the intercellular air spaces to the respiring cells. Carbon dioxide moves in the reverse direction. Both gases move by diffusion through the spaces and then through the moist cell walls into the respiring cells themselves. Cells which contain chloroplasts have a further source of oxygen because it is released as a waste product of photosynthesis and immediately taken up by the mitochondria. Similarly, the carbon dioxide released from the mitochondria can be used by the chloroplasts for photosynthesis. The rate of photosynthesis is affected by the intensity of light and so the carbon dioxide used and oxygen released by this process vary considerably during the day. The balance of respiratory and photosynthetic gases therefore changes. This is referred to in Section 23.4.2 and also in Fig. 23.11.

31.2.3 Insects

For insects, diffusion of gases over the whole body surface is no longer possible. These predominantly terrestrial animals are covered by an exoskeleton and waterproof cuticle. Gases must enter and leave through pores called **spiracles**. Spiracles are usually found in pairs in ten of the body segments (second and third thoracic and first eight abdominal segments) and are not just simple holes. Each is surrounded by hairs which help to retain water vapour and may be closed by a system of valves operated by tiny muscles. Respiring cells inside the insect release carbon dioxide and as this accumulates it is detected by chemoreceptors and the spiracles open.

The spiracles open into a complex series of tubes running throughout the body. These tubes are called **tracheae**. They are supported by rings of chitin which prevent their collapse when the pressure inside them falls. The tracheae divide to form smaller **tracheoles** extending right into the tissues. Respiratory gases are carried in the **tracheal system** between the environment and the respiring cells; they are not transported by the blood. The tracheal system carries oxygen rapidly to the cells and allows the insects to develop high metabolic rates. The ends of the fine tracheoles are fluid-filled. At rest the tissue cells are hypotonic to the fluid in the tracheoles. As activity increases the muscles respire anaerobically and lactic acid accumulates. This raises the osmotic pressure of the cells so that they are hypertonic to the fluid in the tracheoles and water therefore moves out of them. As fluid is lost, air is drawn further into the tracheoles, making more oxygen available for cellular respiration. When activity ceases the metabolites are oxidized, the osmotic pressure is lowered and the fluid re-enters the tracheoles. The system is ventilated by contraction of the abdominal muscles of the insect flattening the body. This reduces the volume of the tracheal system. The volume increases again as the elastic nature of the body returns the insect and tracheal system to their original shape. Larger insects, such as locusts, have some of the tracheae expanded to form air-sacs which act as bellows.

The tracheal system provides an extremely efficient means of gaseous exchange, but it does have its limitations. Since it relies entirely on diffusion for the gases to move from the environment to the respiring cells, insects are not able to attain a large size. In addition, the chitinous linings of the tracheae must be moulted with the rest of the exoskeleton.

31.2.4 Cartilaginous fish

Cartilaginous fish, like sharks, dogfish and rays, exchange respiratory gases via a series of **gills** on either side of the pharynx. There are five pairs of gills, each situated in a **gill pouch**. Gills comprise a cartilagenous rod called a **branchial arch** which supports a series of

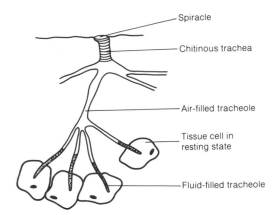

Fig. 31.1 Part of an insect tracheal system

Spiracle

Chitinous trachea

Air-filled tracheole

Tissue cell in resting state

Fluid-filled tracheole

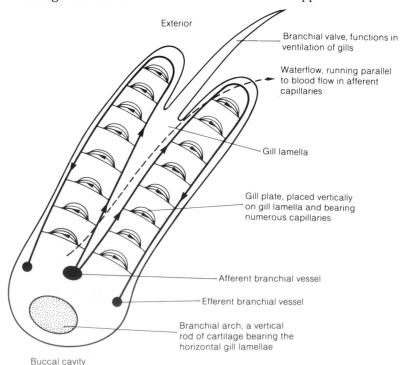

Exterior

Branchial valve, functions in ventilation of gills

Waterflow, running parallel to blood flow in afferent capillaries

Gill lamella

Gill plate, placed vertically on gill lamella and bearing numerous capillaries

Afferent branchial vessel

Efferent branchial vessel

Branchial arch, a vertical rod of cartilage bearing the horizontal gill lamellae

Buccal cavity

Fig. 31.2 Water flow over gill lamellae in a cartilaginous fish

horizontal septa or **lamellae**. The surface area of these lamellae is further increased by the presence of vertical **gill plates** on each of them. The free edge of the septum is extended to form a **branchial valve**.

In order to ventilate these gills water is drawn in through the mouth and spiracle and it leaves when the branchial valves open; there is no operculum. The ventilation mechanism is illustrated in Fig. 31.3.

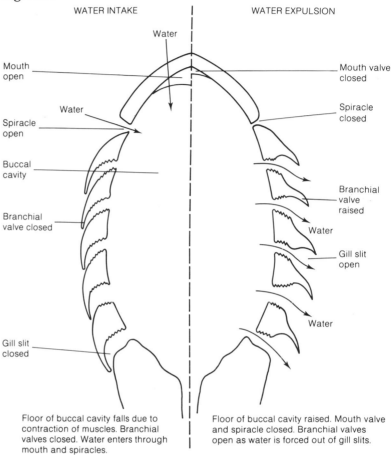

Floor of buccal cavity falls due to contraction of muscles. Branchial valves closed. Water enters through mouth and spiracles.

Floor of buccal cavity raised. Mouth valve and spiracle closed. Branchial valves open as water is forced out of gill slits.

Fig. 31.3 Ventilation of gills in a cartilaginous fish

Contraction of the hypobranchial muscles lowers the floor of the pharynx. This increases the volume of the buccal cavity, decreases the pressure within it and water enters through the mouth and spiracle. The fall in pressure in the pharynx and buccal cavity pulls the branchial valves tightly closed. Water flows around the gills and gaseous exchange takes place between the water and the blood. When the floor of the pharynx and buccal cavity are raised, the mouth and spiracle are closed. This forces water over the gill lamellae and out of the gill slits as the branchial valves are opened. The alternation of the buccal **pressure pump** and branchial **suction pump** ensures a more or less continuous flow of water over the gills.

Afferent branchial arteries carry deoxygenated blood from the ventral aorta to the gill capillaries where gaseous exchange occurs. Oxygenated blood leaves via the efferent branchial vessels.

Much of the water which flows over the gills is passing in the same direction as the blood. This is known as **parallel flow** and provides a relatively inefficient means of obtaining oxygen from the water since a diffusion gradient is not maintained and equilibrium is soon reached. (See Fig. 31.4.)

Water and blood flow side by side in the same direction. Oxygen is exchanged by diffusion. The oxygen concentration in the blood leaving the gills can never exceed that of the water leaving the gills. Equilibrium is soon reached.

(a)

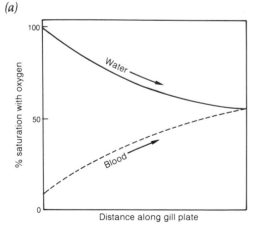

(b)

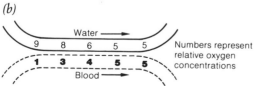

Numbers represent relative oxygen concentrations

Fig. 31.4 Parallel flow in the gills of a cartilaginous fish

31.2.5 Bony fish

Bony fish have four pairs of bony branchial arches supporting gill lamellae. These lamellae form a double row arranged in a V-shape. As with the cartilagenous fish, the lamellae bear gill plates at right angles to their surface. There are no branchial valves but the gill slits are covered by a bony flap called an **operculum**. This helps to protect the delicate gills but also plays a part in their ventilation.

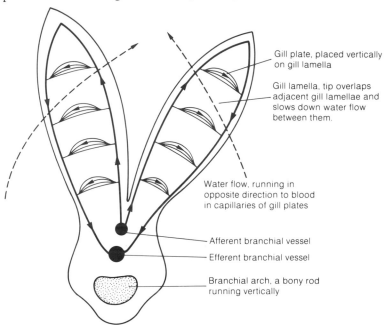

Fig. 31.5 Water flow over gill lamellae in a bony fish

Blood flows in the opposite direction to the water. At each point of contact, the water has a higher oxygen concentration than the blood. Therefore diffusion can occur over the whole region and much higher blood oxygen concentrations can be reached than with parallel flow.

(a)

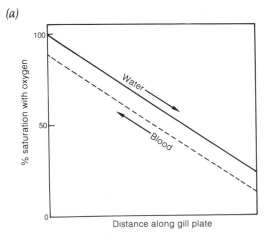

(b)

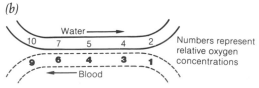

Fig. 31.6 Counter-flow in the gills of a bony fish

Deoxygenated blood enters the gill capillaries via the afferent branchial vessels. Oxygenated blood leaves in the efferent branchial artery to join the dorsal aorta along which blood passes to the rest of the body.

The gills of bony fish demonstrate extremely well the **counter current principle**. The essential feature of this is that the blood and water flow over the gill plates in opposite directions. This allows a fairly constant diffusion gradient to be maintained between the blood and the water, right across the gill. It ensures that blood which is already partly loaded with oxygen meets water which has had very little oxygen removed from it. Similarly, blood with very low oxygen saturation meets water which has already had much of its oxygen removed. (See Fig. 31.6.)

This mechanism allows bony fish to achieve 80% absorption of oxygen, compared to about 50% in the parallel flow system of a dogfish. The overlapping ends of the gill lamellae also slow down the passage of the water so that there is a greater time for diffusion to occur. Alternation of a buccal pressure pump and an opercular suction pump allows water to be drawn over and between the gills more or less continuously. To take in water, the floor of the buccal cavity is lowered, this increases its volume and the pressure within it decreases. Water enters through the mouth. At the same time the operculum is pressed close to the body and tiny opercular muscles increase the volume of the opercular cavity slightly. This causes some water to move out of the buccal cavity and into the opercular region, bathing the gills. Water is expelled by raising the floor of the buccal cavity. As pressure inside the buccal cavity increases, flaps of skin close the mouth and the water is forced over the gills and out of the body under the free edge of the operculum.

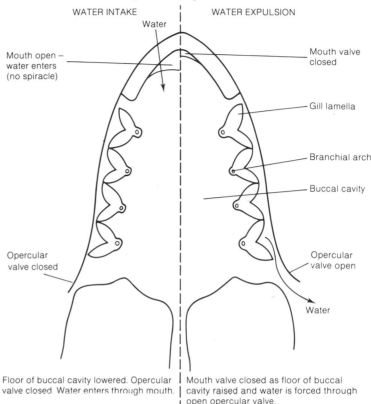

WATER INTAKE | WATER EXPULSION

Water

Mouth open – water enters (no spiracle)

Mouth valve closed

Gill lamella

Branchial arch

Buccal cavity

Opercular valve closed

Opercular valve open

Water

Floor of buccal cavity lowered. Opercular valve closed. Water enters through mouth. | Mouth valve closed as floor of buccal cavity raised and water is forced through open opercular valve.

Fig. 31.7 Ventilation of gills in a bony fish

31.2.6 Birds

Birds are endothermic and have a high metabolic rate. The lungs for gaseous exchange are surprisingly small but there is also an extensive series of air sacs in the body cavity and the long bones. These air sacs are not well supplied with blood vessels and so are unlikely to be sites of gaseous exchange. Exactly how they function is uncertain but it is thought that they help to maintain a unidirectional flow of air from the posterior sacs, through the lungs and the anterior sacs to the exterior. This one-way flow eliminates the dead-air found in mammalian lungs. In addition, blood flow in the pulmonary blood vessels is in the opposite direction to the air flow, creating an efficient countercurrent system.

The air movement is brought about by a complex series of ventilation movements involving the abdominal and intercostal muscles. When the bird is flying, contraction of the large pectoral muscles pulls on the sternum and increases the ventilation rate. This provides the extra gaseous exchange essential for prolonged flight.

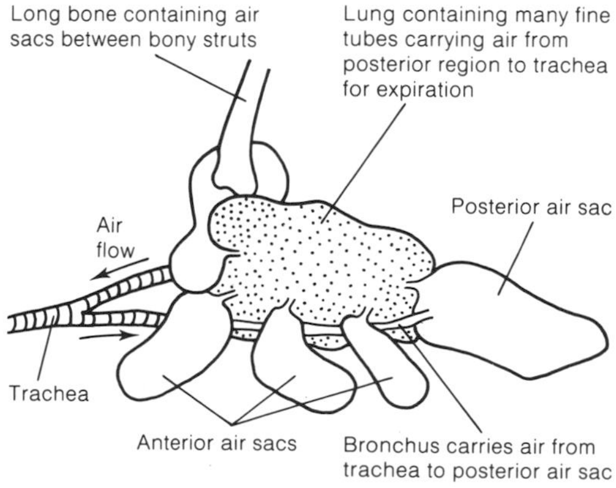

Long bone containing air sacs between bony struts

Lung containing many fine tubes carrying air from posterior region to trachea for expiration

Posterior air sac

Air flow

Trachea

Anterior air sacs

Bronchus carries air from trachea to posterior air sac

Fig. 31.8 Part of the respiratory system of a bird

31.2.7 Mammals

Lungs are the site of gaseous exchange in mammals. They are found deep inside the thorax of the body and so their efficient ventilation is essential. The lungs are delicate structures and, together with the heart, are enclosed in a protective bony case, the **rib cage**. There are twelve pairs of ribs in humans, all attached dorsally to the thoracic vertebrae. The anterior ten pairs are attached ventrally to the sternum.

The remaining ribs are said to be 'floating'. The ribs may be moved by a series of intercostal muscles. The thorax is separated from the abdomen by a muscular sheet, the **diaphragm**.

Air flow in mammals is **tidal**, air entering and leaving along the same route. It enters the nostrils and mouth and passes down the **trachea**. It enters the lungs via two **bronchi** which divide into smaller **bronchioles** and end in air-sacs called **alveoli**. These regions are illustrated in Fig. 31.9.

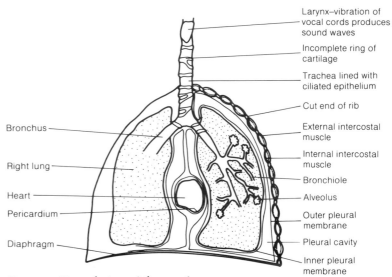

Labels, clockwise from top right:
- Larynx–vibration of vocal cords produces sound waves
- Incomplete ring of cartilage
- Trachea lined with ciliated epithelium
- Cut end of rib
- External intercostal muscle
- Internal intercostal muscle
- Bronchiole
- Alveolus
- Outer pleural membrane
- Pleural cavity
- Inner pleural membrane
- Diaphragm
- Pericardium
- Heart
- Right lung
- Bronchus

Fig. 31.9 *Ventral view of thorax of man*

Regions of the respiratory system

Within the nasal channels mucus is secreted by goblet cells in the ciliated epithelium. This mucus traps particles and the cilia move them to the back of the buccal cavity where they are swallowed. The mucus also serves to moisten the incoming air and it is warmed by superficial blood vessels. Within this region there are also olfactory cells which detect odours.

Air then passes through the pharynx and past the **epiglottis**, a flap of cartilage which prevents food entering the trachea. The **larynx**, or voice box, at the anterior end of the trachea is a box-like, cartilagenous structure with a number of ligaments, the **vocal cords**, stretched across it. Vibration of these cords when air is expired produces sound waves. The trachea is lined with ciliated epithelium and goblet cells. The mucus traps particles and the cilia move them to the back of the pharynx to be swallowed. The trachea is supported by incomplete rings of cartilage which prevent the tube collapsing when the pressure inside it falls.

The trachea divides into two **bronchi**, one entering each lung. The bronchi are also supported by cartilage. The **bronchioles** branch throughout the lung; as the tubes get finer cartilagenous support gradually ceases. These eventually end in **alveoli**. There are over 350 million alveoli in *each* lung of man and these form a *total* respiratory surface of about 90 m². Each lung is surrounded by an air-tight cavity called the **pleural cavity**. This is bounded by two membranes, or **pleura**, which secrete **pleural fluid** into the cavity. The fluid is a lubricant, preventing friction when the lungs expand at inspiration. Pressure in the pleural cavity is always about 500 Pa lower than in the lungs and this allows them to expand and fill the thorax.

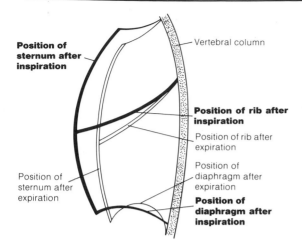

Position of sternum after inspiration

Vertebral column

Position of rib after inspiration

Position of rib after expiration

Position of diaphragm after expiration

Position of sternum after expiration

Position of diaphragm after inspiration

Fig. 31.10 Relative positions of ribs, diaphragm and sternum after breathing in and out

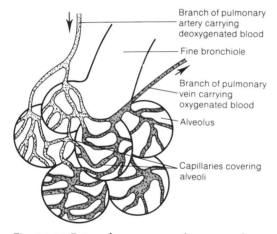

Branch of pulmonary artery carrying deoxygenated blood

Fine bronchiole

Branch of pulmonary vein carrying oxygenated blood

Alveolus

Capillaries covering alveoli

Fig. 31.11 External appearance of a group of alveoli

Breathing in (inspiration) in man

In order for air to enter the lungs from the exterior the pressure inside the lungs must be lower than atmospheric pressure. This lowering of pressure is brought about as follows.

When the external intercostal muscles contract and the internal intercostal muscles relax, the ribs move upwards and outwards (anteriorly and ventrally). The diaphragm muscle contracts and flattens. These two movements cause the volume of the thorax to increase and therefore the pressure inside it falls. The elastic lungs expand to fill the available space and so their volume increases and the pressure within them falls. This causes air to rush into the lungs from the exterior.

Breathing out (expiration) in man

Breathing in is an active process but breathing out is largely passive. The volume of the thorax is decreased as the diaphragm muscle relaxes and it resumes its dome-shape. The external intercostal muscles also relax, allowing the ribs to move downwards (posteriorly) and inwards (dorsally). This may be assisted by contraction of the internal intercostal muscles. As the volume of the thorax decreases, the pressure inside it increases and air is forced out of the lungs as their elastic walls recoil.

Exchange at the alveoli

Each minute alveolus (diameter 100 μm) comprises squamous epithelium and some elastic and collagen fibres. It is surrounded by a network of blood capillaries which come from the pulmonary artery and unite to form the pulmonary vein. These capillaries are extremely narrow and the red corpuscles (erythrocytes) are squeezed as they pass through. This not only slows down the passage of the blood, allowing more time for diffusion, but also results in a larger surface area of the red blood cell touching the endothelium and thus facilitating the diffusion of oxygen. The oxygen in the inspired air dissolves in the moisture of the alveolar epithelium and diffuses across this and the endothelium of the capillary into the erythrocyte. Inside the red blood cell, the oxygen combines with the respiratory pigment **haemoglobin** to form **oxyhaemoglobin** (Section 32.1.1). Carbon dioxide diffuses from the blood into the alveolus to leave the lungs in the expired air.

31.3 Control of ventilation in man

Ventilation of the respiratory system in man is primarily controlled by the **breathing centre** in a region of the hindbrain called the medulla oblongata. The ventral portion of this centre controls inspiratory movements and is called the **inspiratory centre**; the remainder controls breathing out and is called the **expiratory centre**. Control also relies on **chemoreceptors** in the carotid and aortic bodies of the blood system. These are sensitive to minute changes in the concentration of carbon dioxide in the blood. When this level rises, increased ventilation of the respiratory surfaces is required. Nerve impulses from these chemoreceptors stimulate the inspiratory centre in the medulla. Nerve impulses pass along the phrenic and thoracic nerves to the diaphragm and intercostal muscles. Their increased rate of contraction causes faster inspiration. As the lungs expand, **stretch receptors** in their walls are stimulated and impulses pass along the vagus nerve to the expiratory centre in the medulla. This automatically 'switches off' the inspiratory centre, the muscles relax and expiration takes place. The stretch receptors are no longer stimulated, the expiratory centre is 'switched off' and the inspiratory centre 'switched on'. Inspiration takes place again. This complex example of a feedback mechanism is illustrated in Fig. 31.12 and further examples of homeostatic control are considered in Chapter 36.

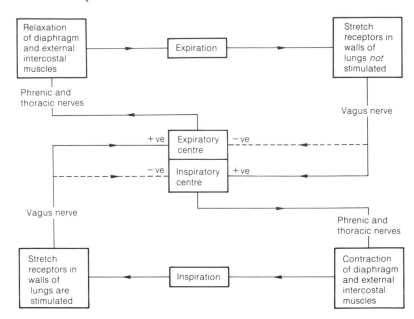

Fig. 31.12 Control of ventilation

The breathing centre may also be stimulated by impulses from the forebrain resulting in a conscious increase or decrease in breathing rate.

The main stimulus for ventilation is therefore the change in carbon dioxide concentration and stimulation of stretch receptors in the lungs; changes in oxygen concentration have relatively little effect. At high altitudes the reduced atmospheric pressure makes it more difficult to load the haemoglobin with oxygen. In an attempt to obtain sufficient oxygen a mountaineer takes very deep breaths. This forces more carbon dioxide out of the body and the level of carbon dioxide in the blood therefore falls. The inspiratory centre is no longer stimulated and breathing becomes increasingly laboured, causing great fatigue. Given time, man can adapt to these conditions by excreting more alkaline urine. This causes the pH of the blood to fall, the chemoreceptors are stimulated and so is the inspiratory centre.

31.4 Measurements of lung capacity

Human lungs have a volume of about 5 dm³ but in a normal breath only about 0.45 dm³ of this will be exchanged (**tidal volume**). During forced breathing the total exchanged may rise to 3.5 dm³ (**vital capacity**) which leaves a **residual volume** of about 1.5 dm³. These terms, and others associated with lung capacity, are illustrated in Fig. 31.13.

Air that reaches the lungs on inspiration mixes with the residual air so that it does not 'stagnate' but is gradually changed. This mixing

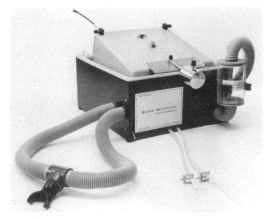

Spirometer

of relatively small volumes of fresh air with a much larger volume of residual air keeps the level of gases in the alveoli more or less constant.

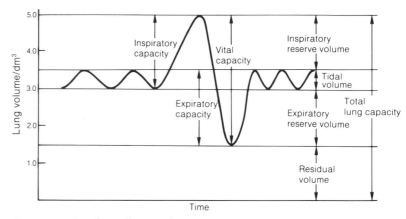

Fig. 31.13 Graph to illustrate lung capacities

TABLE 31.2 **Comparison of inspired, alveolar and expired air**

Gas	% Composition by volume		
	Inspired air	Alveolar air	Expired air
Oxygen	20.95	13.8	16.4
Carbon dioxide	0.04	5.5	4.0
Nitrogen	79.01	80.7	79.6

Measurements of respiratory activity may be made using a spirometer attached to a kymograph which records all its movements.

The ventilation rate is calculated as the number of breaths per minute × tidal volume.

31.5 Questions

1. (a) Distinguish between the processes of aerobic and anaerobic respiration. (*6 marks*)
   (b) Describe how oxygen from the air reaches
      (i) liver cells in a mammal;
      (ii) pith cells in the stem of a flowering plant. (*14 marks*)
      (*Total 20 marks*)

   London Board 1983, Paper II, No. 3

2. (a) State precisely the respiratory surface(s) employed by the following organisms:
      (i) a protozoan;
      (ii) a flatworm;
      (iii) a tadpole;
      (iv) an adult frog;
      (v) a fish;
      (vi) a lizard;
      (vii) a mammalian foetus. (*2 marks*)
   (b) What are the essential features of a good respiratory surface? (*2 marks*)
   (c) With the aid of a fully labelled diagram, describe the process of inspiration in a human being. (*5 marks*)

   (d) (i) What factors influence breathing?
      (ii) Which part of the brain controls breathing, and how is this control achieved? (*6 marks*)
      (*Total 15 marks*)

   Northern Ireland Board June 1983, Paper I, No. 2

3. Describe the important features of gaseous exchange surfaces, with reference to their function in the insect tracheal system, mammalian lungs and leaf palisade mesophyll. (*18 marks*)

   Cambridge Board June 1984, Paper I, No. 2

4. (a) Distinguish between *breathing, gas exchange* and *cellular respiration*. (*4 marks*)
   (b) Describe the mechanism of breathing in a mammal. (*9 marks*)
   (c) Explain how breathing is controlled by the nervous system. (*5 marks*)
      (*Total 18 marks*)

   Cambridge Board June 1989, Paper I, No. 3

5. (*a*) What are the properties of a respiratory surface?
(*4 marks*)

(*b*) Describe the mechanisms involved in the ventilation movements of the following animals.
(i) A terrestrial insect
(ii) A mammal (*10 marks*)

(*c*) Explain the rôles of the following structures in gaseous exchange in plants.
(i) Stomata
(ii) Lenticels (*6 marks*)
(*Total 20 marks*)

London Board June 1989, Paper II, No. 2

6. (*a*) Describe the principal structures and processes involved in gaseous exchange in a **named** fish.
(*15 marks*)

(*b*) Explain to what extent the 'counter-current mechanism' is important in this process.
(*5 marks*)

(*c*) Suppose that you are asked to investigate the rate of breathing (or ventilation) in fish. Design an experiment to show quantitatively how changing one named factor might affect breathing rate. Outline clearly the procedure you would adopt. (Note that principles of experimental design will be considered to be more important than fine detail.) (*10 marks*)
(*Total 30 marks*)

Oxford Local Board June 1989, Paper II, No. 5

7. (*a*) The diagram below shows part of the gas-exchange system of insects.
(i) Identify parts **A** to **E** and give the function of each. (*10 marks*)
(ii) Describe how oxygen passes from the air to the muscles at rest and during exercise.
(*5 marks*)

(*b*) Certain water bugs and water beetles can remain submerged for a time because, when they dive, they take with them a bubble of air which remains in contact with their gas-exchange system. This diagram shows the partial pressures (*p*) of oxygen and nitrogen in the air, in the water and in the gas within the bubble when the insect has been active just below the water surface for a short time.

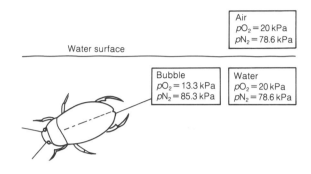

Water surface

Air
$pO_2 = 20\,kPa$
$pN_2 = 78.6\,kPa$

Bubble
$pO_2 = 13.3\,kPa$
$pN_2 = 85.3\,kPa$

Water
$pO_2 = 20\,kPa$
$pN_2 = 78.6\,kPa$

(i) Account for the partial pressures of oxygen and nitrogen in the air and in the water.
(*2 marks*)

(ii) Account for the differences between the partial pressures of oxygen and nitrogen in the bubble and in the water. (*3 marks*)
(*Total 20 marks*)

Associated Examining Board June 1985, Paper II, No. 3,
Alternative 1

8. The respirometer shown opposite can be used to measure the rate of oxygen uptake and carbon dioxide release by germinating peas. Tubes A and B are immersed in a water bath at 25 °C, with the manometer suspended outside.

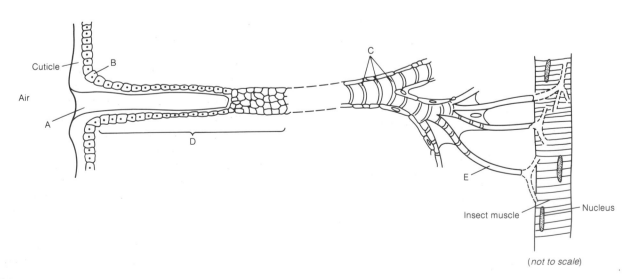

Cuticle

Air

Insect muscle

Nucleus

(*not to scale*)

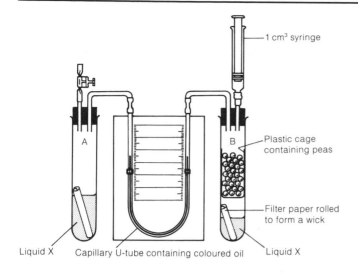

Liquid X — Capillary U-tube containing coloured oil — Liquid X

(a) (i) Name liquid X when the respirometer is used to measure oxygen uptake. (*1 mark*)
   (ii) Suggest why the volume of liquid X is greater in tube A than in tube B. (*1 mark*)
(b) Explain the function of the following parts of the apparatus:
   (i) side A (*1 mark*)
   (ii) the syringe (*1 mark*)
The table shows the results for oxygen uptake and carbon dioxide release obtained with pea seeds during the first 48 hours of germination, using the respirometer illustrated.

Time /h	Oxygen uptake /cm³ g⁻¹ h⁻¹	Carbon dioxide release /cm³ g⁻¹ h⁻¹
12	0.03	0.08
24	0.03	0.10
36	0.08	0.11
48	0.15	0.14

(c) What can be deduced from these results about the metabolism of the pea seeds during the first 48 hours of germination? (*2 marks*)
(*Total 6 marks*)

Joint Matriculation Board (Nuffield) June 1988, Paper IA, No. 7

9. In an investigation of gas exchange, the percentages by volume of certain gases in inspired air, expired air and alveolar air were determined in a resting human. The results are shown in the table below.

	Inspired air	Expired air	Alveolar air
Oxygen	20.90	15.3	13.9
Nitrogen	78.60	74.9	no data
Carbon dioxide	0.03	3.6	4.9
Water vapour	0.47	6.2	no data

(a) Explain why the percentage volume of oxygen in expired air is intermediate between the inspired and alveolar values. (*3 marks*)
(b) Nitrogen is not used physiologically, yet it forms a lower percentage volume in expired air than in inspired air. Suggest why this should be so. (*3 marks*)
(c) Suggest a simple way by which alveolar air might be sampled. (*2 marks*)
(d) In an experiment, the carbon dioxide concentration in inspired air was increased whilst the oxygen concentration was maintained at a constant level.
The effects of this change on the volume of air breathed in and out per minute, and on breathing rate, were investigated.
The results are shown in the graphs below.

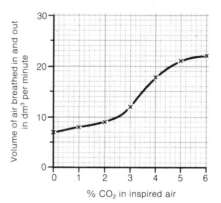

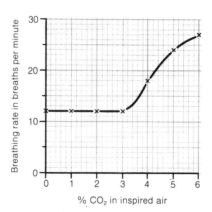

(i) Describe the effect of the increase in CO_2 concentration on the volume of air breathed in and out per minute. (*2 marks*)
(ii) Explain why the breathing rate does not change until air containing more than 3% CO_2 is inspired. (*2 marks*)
(iii) Calculate the average volume of a single breath when the CO_2 concentration in inspired air is 5%. Show your working. (*3 marks*)
(e) Explain why the concentration of oxygen in inspired air can be reduced to low levels without affecting breathing rate. (*1 mark*)

(f) At rest, an elephant breathes at about 5 breaths per minute, and a rat at about 150 breaths per minute. Suggest reasons for this difference in breathing rate. (4 marks)

(Total 20 marks)

London Board January 1989, Paper I, No. 13

10. Answer **BOTH** parts A **AND** B

PART A

In an experiment, Subject **A** was connected to a spirometer filled with oxygen and fitted with a cylinder of soda lime to absorb exhaled CO_2.

The spirometer trace is displayed below:

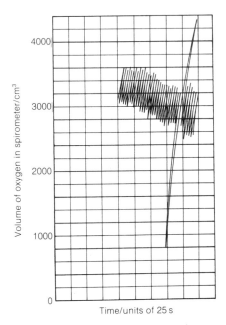

Time/units of 25 s

(a) From the spirometer trace determine:
    (i) the ventilation rate in breaths min^{-1} and the resting tidal volume in the first 50 s,
    (ii) inspiratory reserve volume and expiratory reserve volume,
    (iii) vital capacity. (5 marks)

(b) Calculate **A**'s rate of oxygen consumption in cm^3 min^{-1} over the whole period of the experiment. (2 marks)

(c) Explain why it would be highly dangerous to carry out the above experiment using air instead of oxygen. (3 marks)

PART B

In the second experiment Subject **B**, whose resting tidal volume was 400 cm^3, was asked to sit on an exercise bicycle and to commence pedalling at the time shown by the downward vertical arrow on the spirometer trace. The exercise lasted for 50 s. The volume of oxygen remaining in the spirometer at the end of the exercise is indicated by the horizontal arrow pointing to a line ruled across the spirometer trace.

The spirometer trace is displayed below.

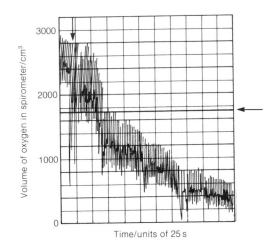

Time/units of 25 s

(a) Explain why **B**'s tidal volume before exercise was higher than normal, and state two important ways in which his pattern of breathing changed during the exercise.

(2 marks)

(b) Describe clearly the mechanism by which exercise caused the changes noted in (a).

(3 marks)

(c) If the total volume of CO_2 exhaled by B during the exercise period was 1000 cm^3, calculate B's respiratory quotient (RQ) for the exercise period.

(2 marks)

(d) What information is provided by a knowledge of RQ values? Illustrate your answer by reference to specific examples. (3 marks)

(Total 20 marks)

Northern Ireland Board June 1988, Paper II, No. 5Y

32 *Blood and circulation (transport in animals)*

As all cells are bathed in an aqueous medium, the delivery of materials to and from these cells is carried out largely in solution. The fluid in which the materials are dissolved or suspended is blood. The cellular components of mammalian blood are described in Section 5.3.3. While a number of ideas on blood were put forward by Greek and Roman scientists, it was the English physician William Harvey (1578–1657) who first showed that it was pumped into arteries by the heart, circulated around the body and returned via veins.

32.1 Functions of blood

Blood performs two distinct functions: the transport of materials (summarized in Table 32.1) and defence against disease.

TABLE 32.1 **Summary of the transport functions of blood**

Materials transported	Examples	Transported from	Transported to	Transported in
Respiratory gases	Oxygen	Lungs	Respiring tissues	Haemoglobin in red blood cells
	Carbon dioxide	Respiring tissues	Lungs	Haemoglobin in red blood cells. Hydrogen carbonate ions in plasma
Organic digestive products	Glucose	Intestines	Respiring tissues/liver	Plasma
	Amino acids	Intestines	Liver/body tissues	Plasma
	Vitamins	Intestines	Liver/body tissues	Plasma
Mineral salts	Calcium	Intestines	Bones/teeth	Plasma
	Iodine	Intestines	Thyroid gland	Plasma
	Iron	Intestines/liver	Bone marrow	Plasma
Excretory products	Urea	Liver	Kidney	Plasma
Hormones	Insulin	Pancreas	Liver	Plasma
	Anti-diuretic hormone	Pituitary gland	Kidney	Plasma
Heat	Metabolic heat	Liver and muscle	All parts of the body	All parts of the blood

32.1.1 Respiratory pigments

The solubility of oxygen is low, with only $0.58 \, cm^3$ dissolving in $100 \, cm^3$ of water at $25 \, °C$. At human body temperature ($37 \, °C$) the quantity is even less, just $0.46 \, cm^3$, because the solubility decreases as the temperature increases. A group of Antarctic fish are able to survive by carrying oxygen around their bodies in aqueous solution. This is possible because at the low water temperature at which these fish live the oxygen is more soluble and their metabolic rate, and hence their oxygen demand, is very low. However, most vertebrates, and many invertebrates, have evolved a group of coloured proteins capable of loosely combining with oxygen, in order to increase the oxygen-carrying capacity of the blood. These are known as **respiratory pigments**. With a few exceptions, the pigments with large relative molecular mass (RMM) are found in the plasma while those of smaller RMM occur within cells to prevent them being lost by ultrafiltration in the kidneys.

TABLE 32.2 **Distribution of respiratory pigments**

Pigment	Colour	Metallic ion present	Relative molecular mass	Site	Main groups it occurs in
Chlorocruorin	Green	Iron	3 000 000	Plasma	Polychaete annelids
Haemoerythrin	Red/brown	Iron	66 000–120 000	Cells	Annelids
Haemocyanin	Blue	Copper	400 000–7 000 000	Plasma	Gastropod molluscs Cephalopod molluscs Crustaceans
Haemoglobin	Red	Iron	16 000–3 000 000	Plasma	Annelids Molluscs
				Cells	Mammals Birds Reptiles Amphibians Fish

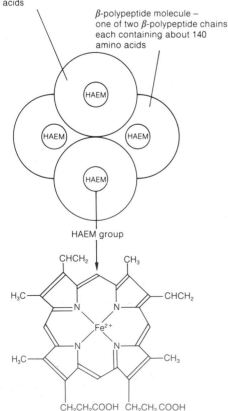

α-polypeptide molecule – one of two α-polypeptide chains each containing about 140 amino acids

β-polypeptide molecule – one of two β-polypeptide chains each containing about 140 amino acids

HAEM

HAEM

HAEM

HAEM

HAEM group

CHCH₂

CH₃

H₃C

CHCH₂

Fe^{2+}

H₃C

CH₃

CH₂CH₂COOH CH₂CH₂COOH

Fig. 32.1 The structure of haemoglobin

The important property of respiratory pigments is their ability to combine readily with oxygen where its concentration is high, i.e. at the respiratory surface, and to release it as readily where its concentration is low, i.e. in the tissues. All these pigments contain a metallic atom in their structure. Different pigments may occur in the same phylum or even within the same animal. A summary of the distribution of the respiratory pigments found in animals is given in Table 32.2, above.

The best known and most efficient respiratory pigment is heamoglobin. It occurs in most animal phyla, protozoans and even a few plants. The haemoglobin molecule is made up of an iron porphyrin compound – the **haem** group – and a protein – **globin**. While the globin group varies considerably from species to species, the haem group is always the same. Each haem group contains a ferrous iron atom, each of which is capable of carrying a single oxygen molecule. Different haemoglobins have a different number of haem groups and so vary in their ability to carry oxygen. A single molecule of human haemoglobin, for example, has a RMM of 68 000 and possesses four haem groups. It therefore is capable of carrying four molecules of oxygen. The arrangement of the haemoglobin molecule is given in Fig. 32.1.

32.1.2 Transport of oxygen

An efficient respiratory pigment readily picks up oxygen at the respiratory surface and releases it on arrival at tissues. This may appear contradictory as a substance with a high affinity for oxygen is unlikely to release it easily. Respiratory pigments overcome the problem by having a high affinity for oxygen when its concentration is high, but this is reduced when the oxygen concentration is low. Oxygen concentration is measured by partial pressure, otherwise called the **oxygen tension**. Normal atmospheric pressure is approximately 100 kiloPascals. As oxygen makes up around 21% of the atmosphere, the oxygen tension (partial pressure) of the atmosphere is around 21 kPa.

When a respiratory pigment such as haemoglobin is exposed to a gradual increase in oxygen tension it absorbs oxygen rapidly at first, but more slowly as the tension continues to rise. This relationship between the oxygen tension and the saturation of haemoglobin is called the **oxygen dissociation curve** and is illustrated below.

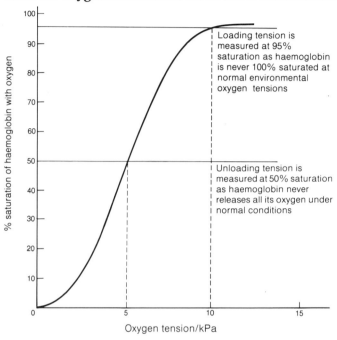

Fig. 32.2 Oxygen dissociation curve for adult human haemoglobin

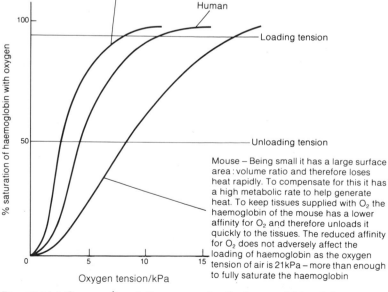

Fig. 32.3(a) Oxygen dissociation curves for the haemoglobin of three mammals

The different haemoglobins found in animals vary in their affinity for oxygen. The oxygen dissociation curves for a number of animals are given in Fig. 32.3, p. 457 and below. Before attempting to explain the significance of these differences, it should be noted that:

1. The more the dissociation curve of a particular pigment is displaced to the right, the less readily it picks up oxygen, but the more easily it releases it.

2. The more the dissociation curve of a particular pigment is displaced to the left, the more readily it picks up oxygen, but the less readily it releases it.

The release of oxygen from haemoglobin is facilitated by the presence of carbon dioxide – a phenomenon known as the **Bohr effect**. Where carbon dioxide concentration is high, i.e. in respiring tissues, oxygen is released readily; where carbon dioxide concentration is low, i.e. at the respiratory surface, oxygen is taken up readily. These effects are shown in Fig. 32.4.

The Bohr effect is not shown by the haemoglobin of animals like *Arenicola* which live in environments with low oxygen tensions. As the habitat of these animals frequently has a high level of carbon dioxide, the Bohr effect would cause the dissociation curve to shift to the right, reducing the haemoglobin's affinity for oxygen. As the environment contains little oxygen, this reduced affinity for it would prevent them taking it up in sufficient quantities for their survival.

Not only do respiratory pigments vary between species, there are often different types in the same species. In humans, for example, the foetus has a haemoglobin which differs in two of the four polypeptide chains from the haemoglobin of an adult. This gives the foetal haemoglobin a dissociation curve to the left of that of the adult and therefore a greater affinity for oxygen (Fig. 32.5a, on the next page). Only in this way can the foetal haemoglobin absorb oxygen from the maternal haemoglobin in the placenta. At birth the production of foetal haemoglobin gives way to that of the adult type.

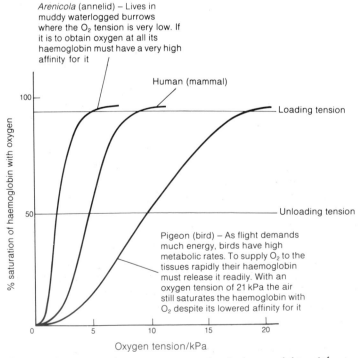

Fig. 32.3(b) Oxygen dissociation curves for the haemoglobin of three animals from different groups

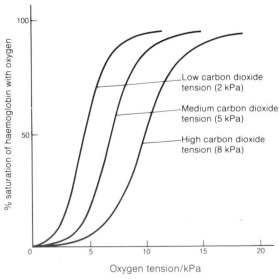

Fig. 32.4 Oxygen dissociation curve of human haemoglobin, illustrating the Bohr effect

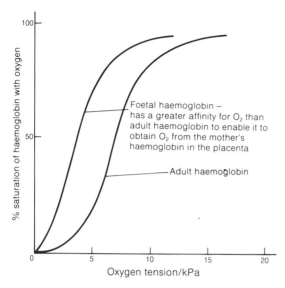

Fig. 32.5 (a) Comparison of the oxygen dissociation curves of adult and foetal haemoglobin

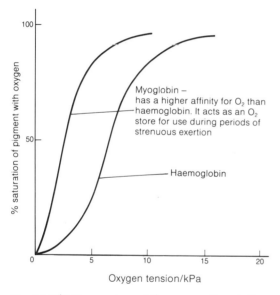

Fig. 32.5(b) Comparison of the oxygen dissociation curves of human haemoglobin and myoglobin

Another respiratory pigment in vertebrates is **myoglobin**. It consists of a single polypeptide chain and a single haem group, rather than the four found in haemoglobin. Like foetal haemoglobin, myoglobin has a dissociation curve displaced to the left of that of the adult and therefore has a greater affinity for oxygen (Fig. 32.5b). Myoglobin occurs in the muscles of all vertebrates, where it acts as a store of oxygen. In periods of extreme exertion, when the supply of oxygen by the blood is insufficient to keep pace with demand, the oxygen tension of muscle falls to a very low level. At these very low oxygen tensions, myoglobin releases its oxygen to keep the muscles working efficiently. Once exercise has ceased the myoglobin store is replenished from the haemoglobin in the blood. Being red, myoglobin is largely responsible for the red colour of meat. The breast meat of much poultry is white as it is made up of the muscles which operate the wings. As most poultry is non-flying these muscles are relatively inactive and so have no need of an oxygen-storing pigment like myoglobin. By contrast, the breast meat of flighted birds is very dark.

Haemoglobin has a greater affinity for carbon monoxide than it does for oxygen. When carbon monoxide is inhaled, even in small quantities, it combines with haemoglobin in preference to oxygen to form a stable compound, **carboxyhaemoglobin**. The carbon monoxide is not released at normal atmospheric oxygen tensions and the haemoglobin is therefore permanently prevented from transporting oxygen. Obviously if sufficient carbon monoxide is inhaled vital tissues become deprived of oxygen, resulting in death from **asphyxia**.

32.1.3 Air breathing animals living under water

A problem which has long fascinated scientists is exactly how certain air-breathing organisms can remain submerged in water for long periods of time without resurfacing for air. Some organisms, such as frogs, have permeable vascular skins through which oxygen diffuses, and this is adequate to supply their needs, provided they remain relatively inactive. Some insects store air in their tracheal systems. In the case of the gnat (*Culex*) larva, up to 1.5 mm³ of air may be held in this way, sufficient for it to remain submerged for up to ten minutes. Larger insects, like the water scorpion, with greater tracheal capacities can extend this period to thirty minutes. The water beetle (*Dytiscus*) traps atmospheric air beneath its elytra (wing covers). In addition to the air held in the tracheal system this allows the beetle to remain underwater for thirty-six hours. Some beetles even trap oxygen bubbles released by photosynthesizing aquatic plants and use this to supplement their oxygen supplies.

Of all air-breathing animals which live under water none have been so well investigated as the diving mammals like seals, whales and dolphins. The duration of a single dive in seals rarely exceeds twenty minutes, whereas that of a sperm whale may extend to seventy-five minutes; bottlenosed dolphins have been known to dive for up to two hours. The remarkable ability of these mammals to endure such long periods without replenishing their air supplies is a consequence of:

1. A larger total volume of blood – The blood accounts for 10–15% of the body weight compared to 7% in humans. This is accommodated in large sinuses and vena cavae.

2. Increased concentration of red blood cells – There is a greater concentration of red blood cells than in terrestrial mammals.

3. Greater haemoglobin concentration – The concentration of haemoglobin in each red blood cell is increased. Coupled with the increased number of red blood cells this allows diving mammals to hold 35 cm³ of oxygen/100 cm³ of blood – about double the capacity of humans.

4. Reduced sensitivity to blood pH – Most mammals respond to a decrease in blood pH (due to an increase in carbon dioxide concentration) by increasing the heart beat and respiratory rate. These effects would be harmful during a dive and so diving mammals fail to show these responses.

5. Muscles rich in myoglobin – The muscles are rich in myoglobin which acts as an oxygen store.

6. Reduction in cardiac output – During a dive the heart-beat is reduced (**bradycardia**), reducing the amount of blood circulated and thereby conserving oxygen. In seals the heart beat may drop from 150 beats per minute to ten beats per minute.

7. Restriction of blood supply to vital organs – Sphincters on the larger veins allow the blood supply to organs such as the kidney, stomach and muscles to be reduced, thereby conserving oxygen. These organs continue to operate by anaerobic respiration. The supply to organs like the brain is unaffected.

8. Tolerance of high lactate levels – Many organs, e.g. muscles, respire anaerobically throughout the dive. They show a high tolerance of lactate which would cause severe cramp in other mammals.

9. Reduced metabolic rate during a dive – There is evidence that the overall metabolic rate is reduced during a dive. As they normally have a higher metabolic rate this simply reduces to a level similar to that of their terrestrial counterparts.

10. Larger tidal volume – Although the overall lung capacity is much the same as that of other mammals, those that dive exchange far more of their air when breathing. Whereas the tidal volume in man is a mere 10% of the total lung capacity, in diving mammals it is around 80%.

11. Lungs may be almost entirely collapsed – To allow large exchanges of air, the lungs can be almost entirely collapsed. To facilitate this, fewer ribs are attached to the sternum. This also makes the rib cage more flexible, permitting it to collapse partially when under pressure during a deep dive.

12. Cartilaginous rings extend further into lungs – The rings extend down into the bronchioles to prevent these collapsing under pressures experienced during a deep dive.

13. Expulsion of air during the dive – This reduces the danger of excessive nitrogen becoming dissolved in the blood. This can have a narcotic effect or even cause bubbles of nitrogen to arise in the blood on resurfacing – a situation comparable to 'the bends' in human divers.

14. Closure of the nostrils – Anatomical modifications allow the nostrils to be closed during a dive to prevent entry of water into the lungs.

Because human divers have none of these adaptations they can only survive by taking with them a supply of air. The problem is that as one dives deeper the pressure increases by around 1 atmosphere (100 kPa) for each 10 metres. In order to allow air to pass from the storage tanks into the lungs, its pressure must be correspondingly increased. Under these conditions a greater concentration of oxygen and nitrogen enters the blood. The oxygen may be toxic and the nitrogen has a narcotic effect. A further problem arises if the diver surfaces rapidly. In these circumstances the nitrogen may come out of solution and form bubbles. This gives painful symptoms known as 'the bends'. To avoid this, divers need to spend specified periods of time at certain depths on their way to the surface. This gives time for the additional nitrogen to be expelled via the lungs. This is known as **decompression**. The use of an oxygen-helium mixture for breathing eliminates the twin problems of oxygen toxicity and nitrogen narcosis. It does, however, increase the loss of body heat and produces voice distortion resulting in unintelligible, high pitched speech. Heated diving suits and even helium speech unscramblers overcome these problems.

32.1.4 Living at high altitude

The amount of oxygen in the atmosphere is the same at high altitudes as it is at sea level, namely 21%. The respiratory problems associated with living at high altitude are a result of the reduced atmospheric pressure, not of oxygen concentration. The reduced pressure means that it is more difficult to load the haemoglobin with oxygen. Above about 6000 m the pressure is inadequate to load haemoglobin effectively. Some human settlements exist at these altitudes and the inhabitants have become **acclimatized**. Acclimatization involves:

1. Adjustment of blood pH – The reduced loading of haemoglobin leads to deeper breathing – **hyperventilation** – in an attempt to compensate for the lack of oxygen in the blood. This leads to excessive removal of carbon dioxide and a raised blood pH. Nervous responses are triggered causing reduced depth of breathing – undesirable in the circumstances. In acclimatized individuals the hydrogen carbonate ions are removed by the kidney, restoring blood pH to normal.

2. Increased oxygen uptake – More oxygen is absorbed by the lungs as a result of an improved capillary network in the lungs, and deeper breathing.

3. Improved transport of oxygen to the tissues – This is the result of:

(a) **increased red blood cell concentration** – this may rise from 45% to 60% of the total blood volume;
(b) **increased haemoglobin concentration** – this may rise from 15 g per 100 cm^3 of blood to 20 g per 100 cm^3.

4. Changes in haemoglobin affinity for oxygen – The oxygen dissociation curve is shifted to the right to facilitate release of oxygen to the tissues. Above 3500 m this advantage is offset by the reduced affinity of haemoglobin for oxygen in the lungs and so is not shown by those living at these altitudes, where a shift to the left favours survival.

5. Increased myoglobin levels in muscles – With its higher affinity for oxygen, this facilitates the exchange of oxygen from the blood to the tissues.

32.1.5 Transport of carbon dioxide

Carbon dioxide is more soluble in water than oxygen, but its transport in solution is still inadequate to meet the needs of most organisms. There are three methods of carrying carbon dioxide from the tissues to the respiratory surface:

1. In aqueous solution – A small amount, around 5%, of carbon dioxide is transported in physical solution in blood plasma.

2. In combination with haemoglobin – A little carbon dioxide, around 10%, will combine with the amino groups ($-NH_2$) in the four polypeptide chains which make up each haemoglobin molecule (Hb).

$$Hb-N{\Large\langle}^{H}_{H} + CO_2 \rightleftharpoons Hb-N{\Large\langle}^{H}_{COO^-} + H^+$$

haemoglobin + carbon dioxide   carbamino haemoglobin + hydrogen ions

3. In the form of hydrogen carbonate – The majority of the carbon dioxide (85%) produced by the tissues combines with water to form carbonic acid. This reaction is catalysed by the zinc-containing enzyme **carbonic anhydrase**. The carbonic acid dissociates into hydrogen and hydrogen carbonate ions.

$$H_2O + CO_2 \underset{\text{anhydrase}}{\overset{\text{carbonic}}{\rightleftharpoons}} H_2CO_3 \rightleftharpoons H^+ + HCO_3^-$$

water   carbon dioxide   carbonic acid   hydrogen ion   hydrogen carbonate ion

The above reactions take place in red blood cells. The hydrogen ions produced combine with haemoglobin which loses its oxygen. The oxygen so released diffuses out of the red blood cell, through the capillary wall and tissue fluid into a respiring tissue cell. The hydrogen carbonate ions diffuse out of the red blood cell into the plasma where they combine with sodium ions from the dissociation of sodium chloride to form sodium hydrogen carbonate. It is largely in this form that the carbon dioxide is carried to the respiratory surface where the processes are reversed, releasing carbon dioxide which diffuses out of the body. The loss of negatively charged

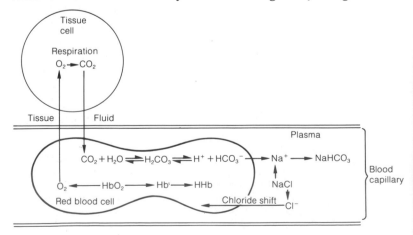

Fig. 32.6 The chloride shift

hydrogen carbonate ions from the red blood cells is balanced by the inward diffusion of negative chloride ions from the dissociation of the sodium chloride. In this way the electrochemical neutrality of the red blood cell is restored. This is known as the **chloride shift** and is illustrated in Fig. 32.6.

32.1.6 Transport of other materials

Apart from respiratory gases, the blood transports a wide variety of other materials from the points where they are absorbed or synthesized to various destinations where they are removed or utilized. Most of these materials are water-soluble and are carried in the plasma. Some examples are given in Table 32.1 (p. 455). The blood is also used to distribute hormones and heat around the body. Many important proteins are carried in suspension in the plasma. These include prothrombin and fibrinogen, used in the clotting process, and antitoxins and antibodies produced by white blood cells to defend against disease.

32.1.7 Clotting of the blood

If a blood vessel is ruptured it is important that the resultant loss of blood is quickly arrested. If not, the pressure of the blood in the circulatory system could fall dangerously low. At the same time it is important that clotting does not occur during the normal circulation of blood. If it does, the clot might lodge in some blood vessel, cutting off the blood supply to a vital organ and possibly resulting in death from **thrombosis**. For this reason the clotting process is very complex, involving a large number of stages. Only under the very specific conditions of injury are all stages completed and clotting occurs. In this way the chances of clotting taking place in other circumstances is reduced. The following account includes only the major stages, of what is a more complex process.

Cellular fragments in the blood called **platelets (thrombocytes)** are involved in the clotting or **coagulation** of the blood. At the site of a wound the damaged cells and ruptured platelets release **thromboplastins**. In the presence of calcium ions and vitamin K these cause the inactive plasma protein, **prothrombin**, to become converted to its active form, **thrombin**. This in turn converts another plasma protein, the soluble **fibrinogen**, to **fibrin**, its insoluble form. The fibrin forms a meshwork of threads in which red blood cells become trapped. These dry to form a clot beneath which repair of the wound takes place. The clot not only prevents further blood loss, it also prevents entry of bacteria which might otherwise cause infection. The clotting process is summarized in Fig. 32.7.

Clotting of blood is prevented by substances such as oxalic acid, which precipitate out the calcium ions as calcium oxalate, and heparin, which inhibits the conversion of prothrombin to thrombin. These substances are known as **anticoagulants**.

32.1.8 Defence against infection – phagocytosis

Two types of white blood cell, the neutrophils and monocytes, are capable of amoeboid movement. Both types carry out **phagocytosis**. This is the process by which large particles are taken up by cells via plasma membrane-derived vesicles. White cells carry out phagocytosis for two reasons: to protect the organism against

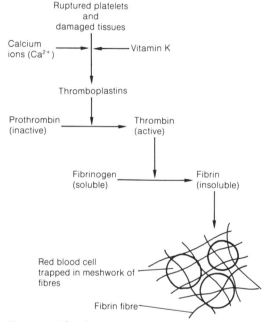

Fig. 32.7 The clotting process – a summary of the main stages

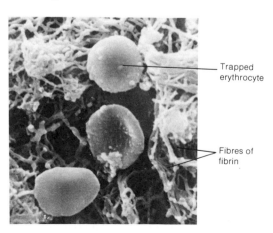

The clotting process (scanning EM) (× 2500 approx.)

1. *The neutrophil is attracted to the bacterium by chemoattractants. It moves towards the bacterium along a concentration gradient.*

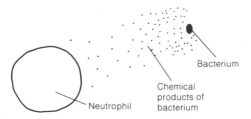

2. *The neutrophil binds to the bacterium.*

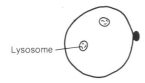

3. *Lysosomes within the neutrophil migrate towards the phagosome formed by pseudopodia engulfing the bacterium.*

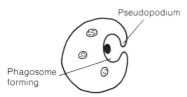

4. *The lysosomes release their lytic enzymes into the phagosome where they break down the bacterium.*

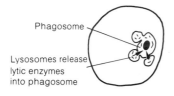

5. *The breakdown products of the bacterium are absorbed by the neutrophil.*

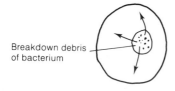

Fig. 32.8 Summary of phagocytosis of a bacterium by a neutrophil

pathogens and to dispose of dead, dying or damaged cells and cellular debris.

In protecting against infection the phagocyte is attracted to chemicals produced naturally by bacteria. The recognition is aided by the presence of **opsonins** – plasma proteins which attach themselves to the surface of the bacteria. The phagocytes have specific proteins on their surface that bind to these chemo-attractants. This causes the phagocyte to move towards the bacteria, possibly along a concentration gradient. The phagocyte strongly adheres to a bacterium on reaching it. This stimulates the formation of pseudopodia which envelop the bacterium, forming a vacuole called a **phagosome**. **Lysosomes** within the phagocyte migrate towards the phagosome into which they release lytic enzymes that break down the bacterium. The breakdown products are finally absorbed by the phagocyte. Fig. 32.8 summarizes the process.

Some phagocytic cells called **macrophages** are found throughout body tissues. They are part of the **reticulo-endothelial system** and are mostly concentrated in lymph nodes and in the liver.

Phagocytosis causes **inflammation** at the site of infection. The hot and swollen area contains many dead bacteria and phagocytes which are known as **pus**. Inflammation results when **histamine** is released as a result of injury or infection. This causes dilation of blood capillaries from which plasma, containing antibodies, escapes into the tissues. Neutrophils also pass through the capillary walls in a process called **diapedesis**.

32.2 The immune response

Immunity is the ability of an organism to resist disease. It involves the recognition of foreign material and the production of chemicals which help to destroy it. These chemicals, called antibodies, are produced by lymphocytes of which there are two types: **T-lymphocytes**, which are formed in bone marrow but mature in the thymus gland, and **B-lymphocytes**, which are formed and mature in the bone marrow.

32.2.1 Antigen–antibody reaction

Foreign material such as a bacterium has a protein or polysaccharide coat. This coat acts as an **antigen (immunogen)** which causes an animal to produce a chemical which is attracted to the antigen. This chemical is called an **antibody (immunoglobulin)**. The lymphocytes of an animal may produce thousands of antibodies many of which are specific to a single antigen. Antibodies combine with their appropriate antigens and neutralize their action, preventing them from causing harm. There are a number of different antibodies each performing a different function:

1. Agglutinins – antibodies which cause foreign particles to stick together in clumps. This makes the material more vulnerable to attack from other types of antibody.

2. Precipitins – antibodies which cause precipitation of the antigen-antibody complex.

3. Antitoxins – antibodies which neutralize toxins produced by micro-organisms so that they cease to take effect.

4. Lysins – antibodies which break down foreign material, causing it to disintegrate.

5. Opsonins – antibodies which stimulate macrophages to engulf foreign particles which possess the appropriate antigen.

Antibodies are not restricted to the blood; they occur throughout the body. They may act as the first line of defence when they are poured out onto the mucus surfaces of the respiratory organs and alimentary canal. They are also present in tears. The **thymus**, a gland in the thorax, appears to play an important rôle in immunity during embryonic life and for a period after birth. It secretes a hormone, **thymosin**, which alters the properties of lymphocytes so that they are able to react with antigens.

32.2.2 Types of immunity and immunization

There are two basic types of immunity, passive and active.

Passive immunity is the result of antibodies being passed into an individual in some way, rather than being produced by the individual itself. This passive immunity may occur naturally in mammals when, for example, antibodies pass across the placenta from a mother to her foetus or are passed to the newborn baby in the mother's milk. In both cases the young developing mammal is afforded some protection from disease until its own immune system is fully functional.

Alternatively, passive immunity may be acquired artificially by the injection of antibodies from another individual. This occurs in the treatment of tetanus and diphtheria in man, although the antibodies are acquired from other mammals, e.g. horses. In all cases, passive immunity is only temporary.

Active immunity occurs when an organism manufactures its own antibodies. Active immunity may be the natural result of an infection. Once the body has started to manufacture antibodies in response to a disease-causing agent, it may continue to do so for a long time after, sometimes permanently. It is for this reason that most people suffer diseases such as mumps and measles only once. It is possible to induce an individual to produce antibodies even without them suffering disease. To achieve this, the appropriate antigen must be injected in some way. This is the basis of **immunization (vaccination)** of which there are a number of different types depending on the form the antigen takes.

1. Living attenuated microorganisms – Living pathogens which have been treated, e.g. by heating, so that they multiply but are unable to cause the symptoms of the disease. They are therefore harmless but nonetheless induce the body to produce appropriate antibodies. Living attentuated microorganisms are used to immunize against measles, tuberculosis, poliomyelitis and rubella (German measles).

2. Dead microorganisms – Pathogens are killed by some means and then injected. Although harmless they again induce the body to produce antibodies in the same way it would had they been living. Typhoid, cholera, influenza and whooping cough are controlled by this means.

3. Toxoids – The toxins produced by some diseases, e.g. diphtheria and tetanus, are sufficient to induce antibody production by an individual. To avoid these toxins causing the symptoms of the disease they are first detoxified in some way, e.g. by treatment with formaldehyde, and then injected.

4. Extracted antigens – The chemicals with antigenic properties may be extracted from the pathogenic organisms and injected; influenza vaccine is produced in this way.

5. Artificial antigens – Through genetic engineering it is now possible to transfer the genes producing antigens from a pathogenic organism to a harmless one which can easily be grown in a laboratory. Mass production of the antigen is then possible in a fermenter ready for separation and purification before use.

32.2.3 Acquired Immune Deficiency Syndrome (AIDS)

Acquired immune deficiency syndrome or AIDS is caused by the **Human Immunodeficiency Virus (HIV)** – a retrovirus, more details of which are given in Section 7.1.2. HIV infects T-helper cells, a type of T-lymphocyte which helps both B-lymphocytes (which produce antibodies) and other T-lymphocytes (which kill cells infected by viruses) to carry out their functions (Section 5.3.3). Neither type of lymphocyte can therefore operate properly and so the body's immune system is rendered ineffective, not only against HIV but other infections. Hence AIDS victims are frequently killed, usually within two years of developing the disease, by opportunist organisms which take advantage of impaired resistance.

Once infected with HIV an individual is said to be **HIV positive**, a condition which persists throughout their life. As the virus remains dormant for about eight years, on average, an HIV positive person does not suffer any symptoms during this period but can act as a carrier, often unwittingly spreading the disease. The virus can be detected in virtually all body fluids of an HIV positive individual. However, since it is only in blood, semen or vaginal fluid that the concentration is high enough to infect others, it is spread through sexual intercourse or transfer of infected blood from one person to another – such as, when drug users share a hypodermic needle, or from mother to baby during childbirth. Faeces, urine, sweat, saliva and tears have such a low incidence of HIV in an infected person that contact with these presents practically no risk of contracting AIDS. In any case, the virus quickly dies outside the human body, and therefore even blood, semen and vaginal secretions must be transferred directly. Thus, the risk from contaminated clothing, etc. is negligible.

Preventative precautions such as restricting the number of sexual partners, using a condom during sexual intercourse, and intravenous drug users not sharing a needle, are at present the only means of containing the disease. While antibiotics may be used to help combat the opportunistic infections there is, as yet, no cure for HIV. A drug currently under test, **zydovudine (AZT)** holds out some hope. It prevents reverse transcriptase (Section 7.1.2) making DNA from RNA and so interferes with the virus's ability to replicate. AZT's effect on the host's cells' DNA, causing anaemia, is a drawback at present. Other research includes producing a form of the receptor on the host cell to which HIV binds (the CD4 antigen) and using it to 'coat' the virus so that it can no longer attach to the host cell.

32.2.4 Blood groups

Blood groups are an example of an antigen–antibody system. The membrane of red blood cells contains polysaccharides which act as antigens. They may induce the production of antibodies when

Blood group	Antigens present
A	A
B	B
AB	A and B
O	None

Blood group	Antigen	Antibodies
A	A	b
B	B	a
AB	A and B	None
O	None	a and b

introduced into another individual. While there are over twenty different blood grouping methods, the ABO system, first discovered by Landsteiner in 1900, is the best known. In this system there are just two antigens, A and B, which determine the blood group (see left). For each of these antigens there is an antibody, which is given the corresponding lower case letter. The presence of an antigen and its corresponding antibody together causes an immune response resulting in the clumping together of red cells (**agglutination**) and their ultimate breakdown (**haemolysis**). For this reason, an individual does not produce antibodies corresponding to the antigens present but produces all others as a matter of course. These antibodies are present in the plasma. The composition of each blood group is therefore as given on the left.

In transfusing blood from one person, the **donor**, to another, the **recipient**, it is necessary to avoid bringing together corresponding antigens and antibodies. However, if only a small quantity of blood is to be transfused, then it is possible to add the antibody to the antigen, because the donor's antibodies become so diluted in the recipient's plasma that they are ineffective. It is not, however, feasible to add small quantities of antigen to the corresponding antibody as even a tiny amount of antigen will cause an immune response. This, after all, is why small numbers of invading bacteria are immediately destroyed as part of the body's defence mechanism. It is therefore possible to safely add antibody a to antigen A and antibody b to antigen B, in small quantities, but not the reverse. For this reason, blood group O, with no antigens present, may be given in small amounts to individuals of all other blood groups. Group O is therefore referred to as the **universal donor**. Individuals of this group are, however, restricted to receiving blood from their own group. In the same way group AB, with no antibodies, may receive blood with either antigen. In other words, group AB can receive blood from all groups, and is therefore termed the **universal recipient**. They can, however, only donate to their own group. When agglutination occurs between two groups, they are said to be **incompatible**. Table 32.3 shows the compatibility of blood groups in the ABO system.

TABLE 32.3 **Compatibility of blood groups in the ABO system**

				Recipient's blood group			
	Group			A	B	AB	O
Group		Antigens		A	B	A and B	None
		Antigens	Antibodies	b	a	None	a and b
Donor's blood group	A	A	b	√	X	√	X
	B	B	a	X	√	√	X
	AB	A and B	None	X	X	√	X
	O	None	a and b	√	√	√	√

√ Compatible – bloods do not clot   X Incompatible – bloods clot

Despite this knowledge and careful matching of blood groups there continued to be inexplicable failures of transfusions up to 1940. It was then that Landsteiner discovered a new antigen (actually a system of antigens) in rhesus monkeys, which was also present in humans. This became known as the **rhesus system** and the antigen as **antigen D**. Where an individual possesses the antigen he or she is said to be **rhesus positive**; where it is absent he or she is **rhesus negative**. There is no naturally occurring antibody to antigen D, but if blood with the antigen is transfused into a person without it (rhesus negative), antibody d production is induced in line with the usual immune response. For this reason, before transfusion, blood is matched with respect to the rhesus factor as well as the ABO system.

One problem associated with the rhesus system arises in pregnancy. As blood groups are genetically determined (Section 14.5.2), it is possible for the foetus to inherit from the father a blood group different from that of the mother. The foetus may, for example, be rhesus positive while the mother is rhesus negative. Towards the end of pregnancy, and especially around birth, fragments of blood cells may cross from the foetus to the mother. The mother responds by producing the rhesus antibody (d) in response to the rhesus antigen (D) on the foetal red blood cells. These antibodies are able to cross the placenta. As the build-up and transfer of rhesus antibodies takes some time, and as the problem only arises during the latter stages of pregnancy, their concentration is rarely sufficient to have any effect on the first child. The production of rhesus antibodies by the mother continues for only a few months, but subsequent foetuses may again induce production and are therefore subject to a greater influx of rhesus antibodies. These break down the foetal red blood cells — a condition known as **haemolytic disease of the newborn.** It requires a number of foetal blood transfusions throughout the pregnancy if it is not to prove fatal. Knowledge that the mother is rhesus negative can, however, avert the danger. If this is the case, and the father is known to be rhesus positive, a potential problem exists. In this event rhesus antibodies (d) from blood donors, are injected into the mother immediately after the first birth. These destroy any foetal cell fragments with antigen D, which may have entered her blood, before they induce the mother to manufacture her own antibodies. The injected antibodies are soon broken down by the mother, and in the absence of new ones being produced subsequent foetuses are not at risk.

The proportion of different blood groups varies throughout the world. In the British population the proportions are O–46%, A–42%, B–9%, AB–3%, although there are variations between different areas. In England, for example, numbers with group A slightly exceed group O. Some South American tribes are exclusively group O whereas some North American Indian tribes are three quarters group A. Over a third of European gypsies have group B. The proportion of rhesus positive individuals in most groups is between 75% and 85%.

32.2.5 Tissue compatibility and rejection

We have seen that blood, if adequately matched, can be transfused from one person to another. It should therefore be equally feasible to transplant organs in the same way. The problem lies in the complexity of organs; they possess a far greater number of antigens and so perfect cross-matching can rarely be achieved. In the absence of perfect cross-matching the recipient treats the donated organ as foreign material and so an immune response is initiated. The organ is therefore rejected. Despite these difficulties there have been major advances in the grafting and transplanting of tissues. Clearly if a tissue is grafted from one part of an organism to another, there are no problems of rejection as all material is genetically identical and so compatible. Skin is frequently grafted by this means. Equally transplants between genetically identical individuals like identical twins do not present problems of rejection. Unfortunately, most humans do not have genetically identical brothers or sisters and so depend upon organs from others when the need for a transplant arises. To minimize the chances of rejection, careful matching takes place, to find tissues which are as nearly compatible as possible. This minimizes the extent of the immune response, reducing the risk of rejection. Such compatible tissues are often, but not always, found in close relatives. In addition the recipient is treated with **immunosuppressant drugs** which lower the activity of their natural immune response, so delaying rejection long enough for the transplanted tissue to be accepted. The problem with these drugs is that the recipient is vulnerable to other infections and even minor ones can prove fatal.

32.2.6 Chemotherapy and immunity

The use of chemicals to prevent and cure diseases and disorders is known as **chemotherapy**. The earliest examples used natural chemicals extracted from plants, but nowadays the chemicals are largely synthetically manufactured. Among the most widely used synthetic drugs are the **sulphonamides**. These were found to be effective against certain bacterial infections during the mid-1930s and were used extensively against urinary and bowel infections as well as pneumonia. The development of strains of bacteria resistant to sulphonamides and the wider use of antibiotics have reduced the importance of these drugs.

Antibiotics are chemicals produced by various fungi and bacteria, which suppress the growth of other micro-organisms. Since the discovery by Sir Alexander Fleming, in 1928, of **penicillin**, many other types of antibiotics have been marketed, e.g. ampicillin, streptomycin and chloramphenicol. All interfere with some stage of bacterial metabolism and so suppress their growth. Penicillin for example prevents the synthesis of certain components of the bacterial cell wall, while streptomycin affects the functioning of ribosomes. Because bacterial cells are prokaryotic, these effects are not experienced by the eukaryotic cells of the host and so antibiotics may be safely used throughout the body. Once again, the development of resistant species has reduced the effectiveness of some antibiotics (see Section 16.4.3).

32.3 The circulatory system

Only very small animals, where cells are never far from the outside, exist without a specialized transport system. The larger and more active an animal, the more extensive and efficient is its transport system. These systems frequently incorporate a pump, valves and an elaborate means of controlling distribution of the blood. Very simple animals have an **open blood system**. Here the blood moves freely over the tissues, through a series of spaces known collectively as the **haemocoel**. The blood is not confined to vessels. In its simplest form, e.g. in nematodes, the blood is transported haphazardly by the muscular movements of the body as it moves around. In most arthropods, including insects, there is some circulation of the blood

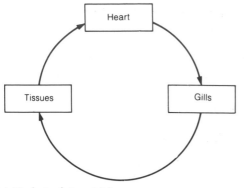

(a) *Single circulation of fish*

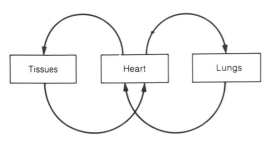

(b) *Double circulation of other vertebrates*

Fig. 32.9 *Single and double circulations*

by a tubular heart, which pumps it into the haemocoel. In all open blood systems the blood is moved at very low pressure and there is little control over its distribution.

To allow more rapid transport and greater control of distribution, larger and more active organisms have evolved **closed blood systems**. Here blood is confined to vessels. The pumping action of the heart sustains high pressure within these vessels and a combination of vasodilation, vasoconstriction and valves ensures a much more controlled distribution of blood. Closed systems occur in cephalopod molluscs, echinoderms, annelids and vertebrates.

Within vertebrates, there are two different systems of circulating the blood. In fish, the blood passes from the heart, over the gills and then to the rest of the body before returning to the heart. As the blood passes only once through the heart during a complete circulation of the body it is called a **single circulation**. The disadvantage of this system is that the resistance created by the fine network of blood capillaries in the gills causes the blood pressure to drop from around 11 kPa as it leaves the heart to 7 kPa as it leaves the gills. This blood then meets more resistance as it passes through the capillaries of the body tissues and its pressure is even further reduced. The flow is therefore rather sluggish. To overcome this problem, other vertebrates have developed a **double circulation**. Here the blood is returned to the heart after passing over the respiratory surface and before it is pumped over the body tissues. This helps to sustain a high blood pressure and so allow more rapid circulation. In mammals and birds the complete separation of the heart into two halves allows oxygenated and deoxygenated blood to be kept separate. This improves the efficiency of oxygen distribution, something which is essential to sustain the higher metabolic rate of these endothermic animals.

32.3.1 Blood vessels

In a closed circulation there are three types of vessel. **Arteries** carry blood away from the heart ('a' for 'artery' = 'a' for 'away' from the

TABLE 32.4 **A comparison of arteries, veins and capillaries**

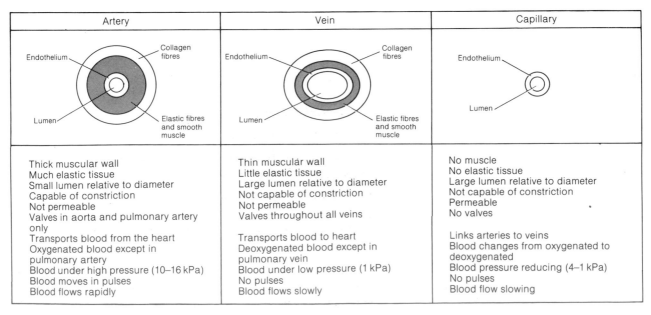

Artery	Vein	Capillary
Thick muscular wall Much elastic tissue Small lumen relative to diameter Capable of constriction Not permeable Valves in aorta and pulmonary artery only Transports blood from the heart Oxygenated blood except in pulmonary artery Blood under high pressure (10–16 kPa) Blood moves in pulses Blood flows rapidly	Thin muscular wall Little elastic tissue Large lumen relative to diameter Not capable of constriction Not permeable Valves throughout all veins Transports blood to heart Deoxygenated blood except in pulmonary vein Blood under low pressure (1 kPa) No pulses Blood flows slowly	No muscle No elastic tissue Large lumen relative to diameter Not capable of constriction Permeable No valves Links arteries to veins Blood changes from oxygenated to deoxygenated Blood pressure reducing (4–1 kPa) No pulses Blood flow slowing

heart), **veins** carry blood to the heart whereas the much smaller **capillaries** link arteries to veins. A comparison of the structure of these three vessels is given in Table 32.4.

The diameter of arteries and veins gradually diminishes as they get further from the heart. The smaller arteries are called **arterioles** and the smaller veins are called **venules**.

32.3.2 Mammalian circulatory system

The purpose of the mammalian circulatory system is to carry blood between various parts of the body. To this end, each organ has a major artery supplying it with blood from the heart and a major vein which returns it. These arteries and veins are usually named by preceding them with the adjective appropriate to that organ, e.g. each kidney has a renal artery and renal vein. A general plan of the mammalian circulation is given in Fig. 32.10.

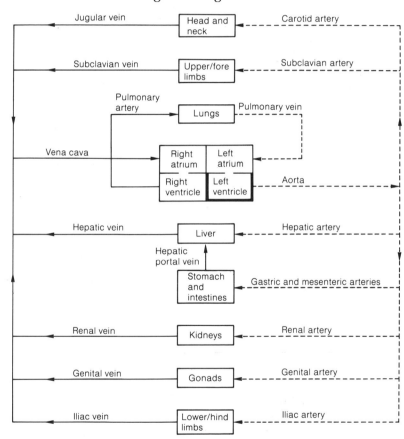

Fig. 32.10 General plan of the mammalian circulatory system

The flow of blood is maintained in three ways:

1. The pumping action of the heart – This forces blood through the arteries into the capillaries.

2. Contraction of skeletal muscle – The contraction of muscles during the normal movements of a mammal squeeze the thin-walled veins, increasing the pressure of blood within them. Pocket valves in the veins ensure that this pressure directs the blood back to the heart.

3. Inspiratory movements – When breathing in, the pressure in the thorax is reduced. This helps to draw blood towards the heart, which is within the thorax.

471

32.3.3 Foetal circulation

The circulatory system of a foetus is fundamentally similar to that of an adult, but modifications are necessary because the lungs are not functional. It is necessary to almost completely by-pass the lungs to avoid the wasted expenditure of energy needed to force blood over the pulmonary capillary network. This is achieved in two ways:

1. The foramen ovale – This is a hole in the wall between the right and left atria. Deoxygenated blood returning to the right atrium from the body passes through the foramen ovale into the left atrium instead of passing to the lungs via the right ventricle and pulmonary artery. From the left atrium this blood can pass via the left ventricle and aorta to the placenta where it is oxygenated.

2. The ductus arteriosus – This is a short vessel which links the pulmonary artery with the aorta. Deoxygenated blood from the pulmonary artery, which would otherwise have passed to the lungs, is directed along it to the aorta, from where it may pass to the placenta to pick up oxygen from the maternal blood. These modifications are summarized in Fig. 32.11.

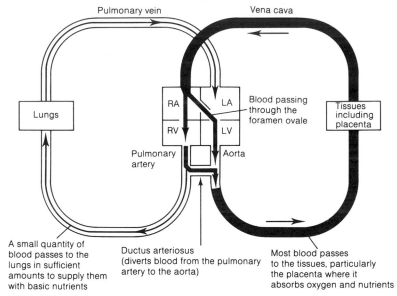

Fig. 32.11 Foetal circulation

Another difference in the foetal circulation includes the presence of a **ductus venosus**, which helps blood by-pass the liver. As the functions of the foetal liver are largely performed by that of the mother, forcing large amounts of blood through the foetal liver is unnecessary and wasteful of energy. The foetal circulatory system also includes umbilical arteries and veins which carry blood to and from the placenta. There are no equivalent vessels in the adult.

As a consequence of the first breath taken by the newly born baby, there is a sudden reduction in pressure in the lungs. This draws blood along the pulmonary artery and causes the closure of the foramen ovale, which soon becomes permanently sealed. Shortly after birth the ductus arteriosus also becomes sealed off. Very occasionally the foramen ovale fails to close completely. In this case some mixing of oxygenated and deoxygenated blood occurs across the inter-atrial wall. The presence of deoxygenated blood in the general circulation often gives the baby a slightly blue appearance and gives rise to the term 'blue baby', otherwise called a 'hole-in-the-heart baby'. The condition can be rectified by surgery.

32.4 Heart structure and action

A pump to circulate the blood is an essential feature of most circulatory systems. These pumps or hearts generally consist of a thin-walled collecting chamber – the **atrium** or **auricle** and a thick-walled pumping chamber – the **ventricle**. Between the two are valves to ensure the blood flows in one direction, namely, from the atrium to the ventricle. In fish, with their single circulation, the heart has two chambers only. With the evolution of a double circulation came the development of two atria, one to receive blood from the systemic (body) circulation and the other to receive it from the pulmonary (lungs) circulation. In amphibia and most reptiles there is a three-chambered heart – two atria and a single ventricle. This has the disadvantage of allowing oxygenated blood from the pulmonary system to mix with deoxygenated blood from the systemic system. Ridges within the ventricle minimize this mixing so that in frogs, for example, three quarters of the oxygenated blood from the pulmonary system is pumped into the systemic one. Being ectothermic, and therefore usually having a lower metabolic rate, allows this inefficiency to be tolerated. In the endothermic mammals and birds, however, the ventricle is completely partitioned into two, allowing complete separation of oxygenated and deoxygenated blood. This four-chambered heart is really two two-chambered hearts side by side.

32.4.1 Structure of the mammalian heart

The mammalian heart is made up of two thin-walled atria which are elastic and distend as blood enters them. The left atrium receives oxygenated blood from the pulmonary vein while the right atrium receives deoxygenated blood from the vena cava. When full, the atria contract together, forcing the remaining blood into their respective ventricles. The right ventricle then pumps blood to the lungs.

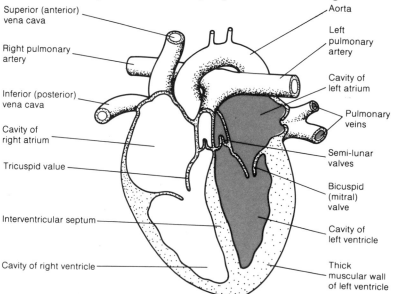

Fig. 32.12 *The structure of the mammalian heart as seen in vertical section from the ventral side*

Owing to the close proximity of the lungs to the heart, the right ventricle does not need to force blood far and is much less muscular than the left ventricle which has to pump blood to the extremities

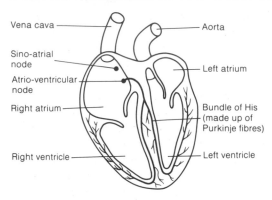

Vena cava

Aorta

Sino-atrial node

Left atrium

Atrio-ventricular node

Right atrium

Bundle of His (made up of Purkinje fibres)

Right ventricle

Left ventricle

Fig. 32.13(a) VS through mammalian heart to show position of sino-atrial node, atrio-ventricular node and bundle of His

1.
Blood enters atria and ventricles from pulmonary veins and venae cavae

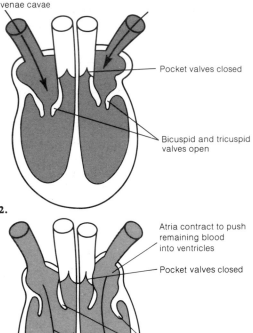

Pocket valves closed

Bicuspid and tricuspid valves open

2.

Atria contract to push remaining blood into ventricles

Pocket valves closed

Bicuspid and tricuspid valves open

Blood pumped from atria to ventricles

3.
Blood pumped into pulmonary arteries and the aorta

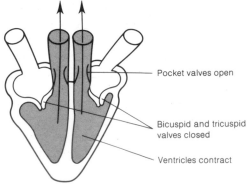

Pocket valves open

Bicuspid and tricuspid valves closed

Ventricles contract

Fig. 32.13(b) The cardiac cycle

of the body. To prevent backflow of blood into the atria when the ventricles contract, there are valves between the atria and ventricles. On the right side of the heart these comprise three cup-shaped flaps, the **tricuspid valves**. On the left side of the heart only two cup-shaped flaps are present; these are the **bicuspid** or **mitral valves**. To prevent these valves inverting under the pressure of blood, they are attached to papillary muscles of the ventricular wall by fibres known as the **chordae tendinae**. Fig. 32.12, on the previous page, illustrates the structure of the heart.

Blood leaving the ventricle is prevented from returning by pocket valves in the aorta and pulmonary artery. These close when the ventricles relax. The heart itself consists largely of cardiac muscle, whose structure is discussed in Section 5.4.3. The muscle is richly supplied with blood vessels and also contains connective tissue which gives strength and helps to prevent the muscle tearing.

32.4.2 Control of heart beat (cardiac cycle)

All vertebrate hearts are **myogenic**, that is, the heart beat is initiated from within the heart muscle itself rather than by a nervous impulse from outside it. Where it is initiated by nerves, as in insects, the heart is said to be **neurogenic**.

The initial stimulus for a heart beat originates in a group of histologically different cardiac muscle cells known as the **sino-atrial node (SA node)**. This is located in the wall of the right atrium near where the vena cavae enter it. The SA node determines the basic rate of heart beat and is therefore known as the **pacemaker**. In humans, this basic rate is 70 beats/minute but can be adjusted according to demand by stimulation from the autonomic nervous system. A wave of excitation spreads out from the SA node across both atria, causing them to contract more or less at the same time.

The wave of excitation reaches a similar group of cells known as the **atrio-ventricular node (AV node)** which lies between the two atria. To allow blood to be forced upwards into the arteries, the ventricles need to contract from the apex upwards. To achieve this the new wave of excitation from the AV node is conducted along **Purkinje fibres** which collectively make up the **bundle of His**. These fibres lead along the interventricular septum to the apex of the ventricles from where they radiate upwards. The wave of excitation travels along these fibres, only being released to effect muscle contraction at the apex. The ventricles contract simultaneously from the apex upwards. These events are known as the **cardiac cycle** and are summarized in Fig. 32.13(b).

32.4.3 Factors modifying heart beat

The human heart normally contracts 70 times a minute, but this can be varied from 50 to 200 times a minute. In the same way the volume of blood pumped at each beat can be varied. The volume

1. *Diastole*
   Atria are relaxed and fill with blood. Ventricles are also relaxed.

2. *Atrial systole*
   Atria contract pushing blood into the ventricles. Ventricles remain relaxed.

3. *Ventricular systole*
   Atria relax. Ventricles contract pushing blood away from heart through pulmonary arteries and the aorta.

pumped multiplied by the number of beats in a given time is called the **cardiac output**. Changes to the cardiac output are effected through the autonomic nervous system. Within the medulla oblongata of the brain are two centres. The **cardio-acceleratory centre** is linked by the sympathetic nervous system to the SA node. When stimulated these nerves cause an increase in cardiac output. **The cardio-inhibitory centre** is linked by parasympathetic fibres within the vagus nerve, to the SA node, AV node and bundle of His. Stimulation from these nerves decreases the cardiac output.

Which of these centres stimulates the heart depends on factors like the pH of the blood. This in turn depends upon its carbon dioxide concentration. Under conditions of strenuous exercise, the carbon dioxide concentration of the blood increases as a consequence of the greater respiratory rate. The pH of the blood is therefore lowered. Receptors in the carotid artery detect this change and send nervous messages to the cardio-acceleratory centre which increases the heart beat, thereby increasing the rate at which carbon dioxide is delivered to the lungs for removal. A fall in carbon dioxide level (rise in pH) of blood causes the carotid receptors to stimulate the cardio-inhibitory centre, thus reducing the heart beat.

Another means of control is by stretch receptors in the aorta, carotid artery and vena cava. When the receptors in the aorta and carotid artery are stimulated, it indicates that there is distention of these vessels as a result of increased blood flow in them. This causes the cardio-inhibitory centre to stimulate the heart to reduce cardiac output. Stimulation of receptors in the vena cava indicates increased blood in this vessel, probably as a result of muscular activity increasing the rate at which blood is returned from the tissues. Under these conditions the cardiac centres in the brain increase the cardiac output.

32.4.4 Maintenance and control of blood pressure

Changes in cardiac output will alter blood pressure, which must always be maintained at a sufficiently high level to permit blood to reach all tissues requiring it. Another important factor in controlling blood pressure is the diameter of the blood vessels. When narrowed – **vasoconstriction** – blood pressure rises; when widened – **vasodilation** – it falls. Vasoconstriction and vasodilation are also controlled by the medulla oblongata, this time by the **vasomotor centre**. From this centre nerves run to the smooth muscles of arterioles throughout the body. Pressure receptors, known as **baroreceptors**, in the carotid artery detect blood pressure changes and relay impulses to the vasomotor centre. If blood pressure falls, the vasomotor centre sends impulses along sympathetic nerves to the arterioles. The muscles in the arterioles contract, causing vasoconstriction and a consequent rise in blood pressure. A rise in blood pressure causes the vasomotor centre to send messages via the parasympathetic system to the arterioles, causing them to dilate and so reduce blood pressure.

A rise in blood carbon dioxide concentration also causes a rise in blood pressure. This increases the speed with which blood is delivered to the lungs and so helps remove the carbon dioxide more quickly. Hormones like adrenaline similarly raise blood pressure.

32.4.5 Heart disease

As the organ pumping blood around the body, any interruption to the heart's ceaseless beating can have serious, often fatal, consequences. There are many defects and disorders, some acquired, like atherosclerosis, others congenital, such as a hole-in-the-heart (Section 32.3.3). Some affect the pacemaker leading to an irregular heart rhythm, others the valves allowing blood to 'leak' back into the atria when the ventricles contract. By far the most common is **coronary heart disease** which affects the pair of blood vessels – the coronary arteries – which serve the heart muscle itself. There are three ways in which blood flow in these arteries may be impeded:

Coronary thrombosis – a blood clot which becomes lodged in a coronary vessel.

Atherosclerosis – narrowing of the arteries due to thickening of the arterial wall caused by fat, fibrous tissue and salts being deposited on it. The condition is sometimes referred to as hardening of the arteries.

Spasm – repeated contractions of the muscle in the coronary artery wall.

It is often a combination of these factors, rather than one in isolation, which results in a **heart attack**. If the main coronary artery is blocked, the whole of the heart muscle or **myocardium** may be deprived of blood, resulting in death. If only a branch vessel is affected the loss of blood supply affects only a portion of the myocardium and, after a period of severe chest pain and temporary incapacitation followed by a number of days of complete rest, recovery normally follows. **Angina** is the result of reduced blood flow in the coronary arteries due to atherosclerosis, but sometimes the result of thrombosis or spasm. Chest pain, and breathlessness often occur when an angina sufferer is undertaking strenuous physical effort.

Many factors are known to increase the risk of coronary heart disease. Smoking is a major contributor increasing the likelihood of both thrombosis and atherosclerosis. A raised level of fat, especially cholesterol, in the blood is a major cause of atherosclerosis. Saturated fat of the type found in most meat and animal products, such as milk, is particularly dangerous. High blood pressure or hypertensive disease, a high level of salt in the diet and diabetes are other factors which contribute to atherosclerosis and hence coronary heart disease. Stress is suspected of increasing the risk of heart disease, but scientific evidence is hard to come by as it is difficult to accurately measure stress levels. There appears to be an inherited factor with individuals with a family history of heart disease being more susceptible to heart attacks. Older people are more at risk than younger ones, males more at risk than females. One thing generally accepted is that exercise can reduce the risk of coronary heart disease.

32.5 Lymphatic system

The lymphatic system consists of widely distributed **lymph capillaries** which are found in all tissues of the body. These capillaries merge to form **lymph vessels** which possess valves and whose structure is similar to that of veins. The fluid within these vessels, the **lymph**, is therefore carried in one direction only, namely, away from the tissues. The lymph vessels from the right side of the head and thorax and the right arm combine to form the **right lymphatic duct** which drains into the right subclavian vein near the heart. The lymph vessels from the rest of the body form the **thoracic duct** which drains into the left subclavian vein.

Along the lymph vessels are series of **lymph nodes**. These contain a population of phagocytic cells, e.g. lymphocytes which remove bacteria and other foreign material from lymph. During infection these nodes frequently swell. Lymph nodes are the major sites of lymphocyte production.

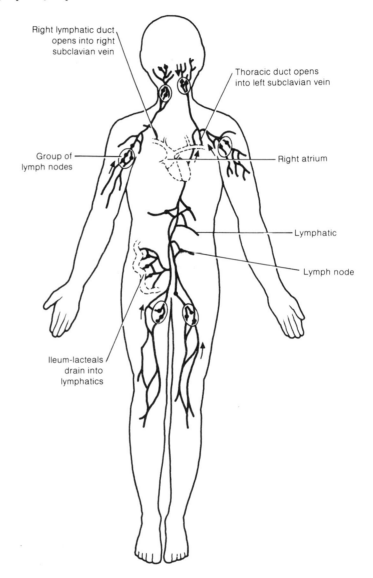

Fig. 32.14 The human lymphatic system

The movement of lymph through the lymphatic system is achieved in three ways:

1. Hydrostatic pressure – The pressure of tissue fluid leaving the arterioles helps push lymph along the lymph system.

2. Muscle contraction – The contraction of skeletal muscle compresses lymph vessels, exerting a pressure on the lymph within them. The valves in the vessels ensure that this pressure pushes the lymph in the direction of the heart.

3. Inspiratory movements – On breathing in, pressure in the thorax is decreased. This helps to draw lymph towards the vessels in the thorax.

Lymph is a milky liquid derived from tissue fluid. It contains lymphocytes and is rich in fats obtained from the lacteals of the small intestines. These fats might damage red blood cells and so are carried separately, until they are later added to the general circulation in safe quantities.

32.5.1 Tissue fluid and its formation

As blood passes from arterioles into the narrow capillaries, a hydrostatic pressure is created which helps fluid escape through the capillary walls. This fluid is called **tissue (intercellular) fluid** and it bathes all cells of the body. It contains glucose, amino acids, fatty acids, salts and oxygen which it supplies to the tissues, from which it obtains carbon dioxide and other excretory material. Tissue fluid is thus the means by which materials are exchanged between blood and tissues.

The majority of this tissue fluid passes back into the venules by osmosis. The plasma proteins, which did not leave the blood, exert an osmotic pressure which draws much of the tissue fluid back into the blood. The fluid which does not return by this means passes into the open-ended lymph capillaries, from which point it becomes known as lymph.

32.6 Questions

1. (a) Describe the mode of action of the mammalian heart. *(8 marks)*
   (b) Describe the mechanisms which regulate heart rate during and after vigorous exercise. *(8 marks)*
   (c) Explain the cause and likely effects of the condition known as 'hole in the heart'. *(4 marks)* *(Total 20 marks)*

   Joint Matriculation Board June 1989, Paper IIB, No. 6

2. Describe how contractions are initiated and propagated in the mammalian heart and explain how cardiac output is regulated so that the demands of tissues are met. *(20 marks)*

   Oxford and Cambridge Board June 1989, Paper II, No. 7

3. What is immunity? Explain the ways in which the human body gains immunity

   Southern Universities Joint Board June 1986, Paper II, No. 22

4. Fill in the blanks in the following account of mammalian blood.

   Blood consists of a liquid plasma in which are suspended solid elements. The three main solid elements are ...., ... and .... Water makes up 90% of the plasma and acts as a solvent for a variety of chemicals, including several inorganic ions. The three principal positive inorganic ions (cations) in the plasma are ...., ... and .... and the main negative ions (anions) are ...., ..., and .... About 8% of the plasma's weight consists of proteins, two abundant examples of which (apart from haemoglobin) are .... and .... They influence important physico-chemical properties of the blood including its .... and its ....
   Blood is therefore the major transport system of the body, and in addition to nutrients, it carries respiratory gases. Most of the oxygen is carried in combination with the protein haemoglobin, but most of the CO_2 is carried as .... The continued normal functioning of the body also depends on the transport of additional organic substances in the blood, for example .... and .... *(8 marks)*

   Welsh Joint Education Committee June 1983, Paper AI, No. 1

5. (a) The table shows the percentage saturation of haemoglobin (Hb) resulting from a range of oxygen partial pressures (pO_2) at two different carbon dioxide partial pressures (pCO_2):

pO_2/mm Hg	% saturation Hb	
	$pCO_2 = 40$ mm Hg	$pCO_2 = 80$ mm Hg
10	11	8
20	38	22
30	59	42
40	75	60
50	83	71
60	90	81
70	94	88
80	98	95
90	99	98
100	99	99

Plot these figures on a single pair of axes; this will give you the O_2 dissociation curve of Hb at two different pCO_2. (Remember that the independent variable must be on the horizontal axis.)

   (i) At pCO_2 of 80 mm Hg, what percentage saturation of Hb is produced by a pO_2 of:
      1. 25 mm Hg   2. 65 mm Hg
   (ii) At pO_2 of 45 mm Hg, what percentage saturation of Hb is obtained at pCO_2 of:
      1. 40 mm Hg   2. 80 mm Hg   *(6 marks)*

   (b) Given the following information:

	pCO_2 /mm Hg	pO_2 /mm Hg	% sat. Hb
Pulmonary capillaries	40	X	97
Muscle capillaries	80	19	Y

   (i) Determine, from your graphs, values for X and Y.
   (ii) What is the significance of these figures? *(3 marks)*

   (c) The table shows figures for the percentage saturation of the intramuscular oxygen carrier myoglobin (Mb) at different pO_2:

pO_2/mm Hg	% saturation Mb
1	5
2	30
4	70
8	90
15	96
30	98
100	99

Plot these figures on the same sheet as your first graph, using the same set of axes.

   (i) Using the information given already, determine the percentage saturation of Mb in the muscles.
   (ii) What are the implications of these figures? *(6 marks)*

(d) (i) Make a sketch graph showing the two O_2 dissociation curves for Hb you have already plotted (section (a)). To this add the O_2 dissociation curve you would expect for foetal Hb at pCO_2 of 80 mm Hg.

(ii) Explain the relationship between the curves you have drawn. (5 marks)

(Total 20 marks)

Northern Ireland Board June 1983, Paper II, No. 8

6. (a) Present, in diagrammatic form, a classification for the composition of human blood. Drawings of constituents are **not** required. (6 marks)

(b) List the functions of each cellular constituent. (7 marks)

(c) Explain how it is that arterial blood, which leaves the heart in spurts and under great pressure, flows evenly and under relatively low pressure in the capillary beds. (7 marks)

(Total 20 marks)

Oxford and Cambridge Board June 1988, Paper II, No. 1

7. The drawing is made from an electronmicrograph of part of a healthy human lung.

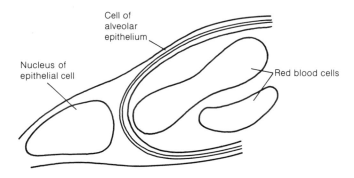

(a) Suggest an explanation for the different shapes of the red blood cells in this drawing. (1 mark)

(b) (i) By marking a line on the drawing, show the shortest path that a molecule of oxygen would follow in diffusing from an alveolus to a haemoglobin molecule in a red blood cell.

(ii) Given that the diameter of a human red blood cell is approximately 7.5 μm, calculate the length of this path. (2 marks)

(c) Suggest how each of the following adaptations of a red blood cell is related to its ability to transport large amounts of oxygen:

(i) the shape of the cell, (2 marks)

(ii) the absence of a nucleus. (2 marks)

(d) Suggest a possible advantage to a mammal in having haemoglobin confined to red blood cells rather than free in plasma. (1 mark)

(Total 6 marks)

Associated Examining Board June 1989, Paper I, No. 6

8. Below is a diagrammatic representation of gas exchange in respiring tissue.

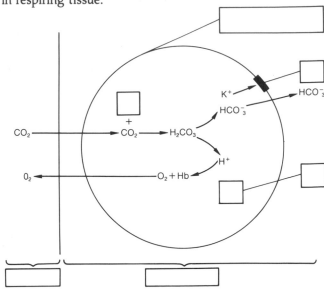

(a) In the diagram, write in each box the label, symbol or formula that you consider appropriate. (7 marks)

(b) Name one enzyme involved in the process represented above. (1 mark)

(c) What 'shift' is depicted in the diagram? (1 mark)

(d) Why is oxygen being transferred? (2 marks)

(Total 11 marks)

Oxford Local June 1984, Paper I, No. 9

9. An investigation was carried out to determine the effect of exercise on the flow of blood through various organs and tissues.

Varying degrees of exercise were performed by a person using an exercise bicycle. The work done by the person on the bicycle was measured in arbitrary units and ranged from 0 (rest) to 50 arbitrary units (severe exercise).

At each work rate, the investigators measured the volume of blood passing through certain organs. The results are given in the table below.

Work done in arbitrary units	Volume of blood in cm³ per min				
	Brain	Heart	Liver	Skeletal muscle	Other tissues
0	700	200	1650	750	1700
10	840	300	1500	1500	1860
20	1200	480	1600	3200	1520
30	1650	770	1650	5500	1430
40	2250	1200	1650	9000	900
50	3000	1800	1800	12400	1000

(a) (i) Plot the results for the heart, liver and skeletal muscles as a graph. (5 marks)

(ii) By reference to your graph, describe and comment on the changes in blood flow through these organs and tissues in response to exercise. (5 marks)

(b) The *cardiac output* is the volume of blood pumped per minute by the heart into the aorta.

(i) Calculate the increase in cardiac output between the conditions of rest and severe exercise. Show your working. (2 marks)

(ii) State *two* ways in which cardiac output may be increased. (2 marks)

(c) What volume of blood was pumped per minute into the pulmonary artery when the person was exercising at a rate of 50 arbitrary units? State your reasoning. (2 marks)

(d) Why does blood flow to the brain need to be increased during exercise? (2 marks)

(e) How is the circulatory system of a mammal able to vary the percentage total blood flow passing to different organs at different times? (2 marks)
(Total 20 marks)

London Board June 1988, Paper I, No. 14

10. Distinguish between each member of the following pairs:

(a) artery and vein; (4 marks)

(b) plasma and intercellular fluid; (2 marks)

(c) antigen and antibody; (2 marks)

(d) lymphocyte and neutrophil granulocyte (= polymorphonuclear leucocyte). (2 marks)
(Total 10 marks)

Oxford Local Board June 1988, Paper I, No. 5

33 *Uptake and transport in plants*

For plants there are certain advantages in large size. They can, for example, compete more readily for light. As a result, many trees are tall, some exceeding 100 m. The leaves, as the sites of photosynthesis, must be in these aerial parts to obtain light. The water so essential to photosynthesis is, however, collected by the roots which may be some considerable distance beneath the soil surface. An efficient means of transporting this water, and certain minerals, to the leaves is necessary. The sugars formed as a result of photosynthesis in the leaves must be transported in the opposite direction to sustain respiration in the roots. In the absence of muscle or other contractile cells, plants depend to a large extent on passive rather than active means of transport.

33.1 The water molecule

Water is the most abundant liquid on earth and is essential to all living organisms. It is, however, no ordinary molecule. It possesses some unusual properties as a result of the hydrogen bonds which readily form between its molecules. These properties make water an ideal constituent of living things.

33.1.1 Structure of the water molecule

The water molecule is made up of two atoms of hydrogen and one of oxygen. Its basic atomic structure is given in Chapter 2, Fig. 2.4. This is a simplified representation for in practice the two hydrogen atoms are closer together, as shown in Fig. 33.1. By weight, 99.76% of water molecules consist of $^1H_2{}^{16}O$; the remainder is made up of various isotopes such as ^2H and ^{18}O. The commonest isotope is deuterium (^2H) and when this is incorporated into a water molecule it is known as **heavy water**, as a result of its greater molecular mass. Heavy water may be harmful to living organisms.

33.1.2 Polarity and hydrogen bonding

The distribution of the charges on a water molecule is unequal. The charge on the hydrogen atoms is slightly positive while that on the oxygen atom is negative (Fig. 33.1). Molecules with unevenly distributed charges are said to be **polar**. Attraction of oppositely charged poles of water molecules causes them to group together. The attractive forces form **hydrogen bonds**. Although individual bonds are weak, they collectively form important forces which hold water molecules together. This makes water a much more stable substance than would otherwise be the case.

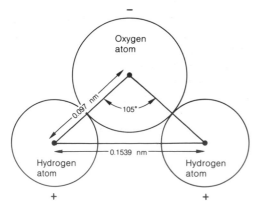

Fig. 33.1 Structure of water molecule

33.1.3 Thermal properties

Due to the hydrogen bonds, more energy is required to separate water molecules. It therefore requires more heat than expected to convert liquid water into vapour; thus evaporation of sweat is an effective means of cooling in mammals. For this reason, the boiling point of water is much higher than expected. In the same way, its **specific heat** is abnormally high. More heat is needed to raise the temperature of a given mass of water. Water is heated and cooled more slowly than expected – it in effect buffers sharp temperature changes.

At temperatures of 0 °C and below, water forms the crystalline substance ice. The arrangement of the water molecules in ice makes it less dense than liquid water. As a result, ice will form at the surface of a body of water such as a pond or lake. This insulates the water beneath and thus prevents the whole of the body of water freezing solid. Living organisms can therefore survive beneath the ice providing there is sufficient food and oxygen available. Water has its maximum density at 4 °C; above this temperature the additional heat breaks some hydrogen bonds and so the water molcules are less densely compacted.

33.1.4 Dissociation, pH and buffers

There is a natural tendency for water molecules to dissociate into ions.

$$H_2O \rightleftharpoons H^+ + OH^-$$
water   hydrogen ion   hydroxyl ion

The positively charged hydrogen ions become attached to the slightly negatively charged oxygen atom of another water molecule.

$$H^+ + H_2O \rightleftharpoons H_3O^+$$
hydrogen ion   water molecule   oxonium ion

The complete reaction may be summarized thus:

$$2H_2O \rightleftharpoons H_3O^+ + OH^-$$
water   oxonium ion   hydroxyl ion

At 25°C the concentration of oxonium ions (hydrogen ions) in pure water is $10^{-7}\,mol\,dm^{-3}$ and this is given the value of 7 on the **pH scale**. When the concentration of hydrogen ions in a solution is $10^{-3}\,mol\,dm^{-3}$ this represents a pH of 3. The pH scale ranges from 1 (very acid) to 14 (very alkaline). The pH scale is logarithmic and so a pH of 6 is ten times more acid than one of pH 7, pH 5 is one hundred times more acid than pH 7 and so on. The pH within most cells is in the range 6.5–8.0. It remains fairly constant because substances within the cells act as **buffers**. Buffer solutions do not appreciably change their pH, despite the addition of small amounts of acids or bases. This is important in cells, because fluctuations in pH could affect the efficiency with which their enzymes work. Apart from dissociating itself, water readily causes the dissociation of other substances. This makes it an excellent solvent.

33.1.5 Colloids

A solid placed in a liquid may dissolve. If not, it may float or sink.

In some cases an intermediate situation may arise whereby the solid becomes finely dispersed as particles throughout the liquid. These particles, normally 1–100 nm in diameter, are known as the **disperse phase** while the liquid around them is called the **dispersion medium**. Collectively they form a **colloid**. Many protein and polysaccharide molecules form colloids. Their most important feature is the large surface area of contact between the particles and the liquid. Cytoplasm is a colloid, made up largely of protein molecules dispersed in water. Cytoplasm is an example of a **hydrophilic colloid**. Here the particles attract water molecules around them and it is this which prevents them aggregating into large particles which could settle out. The attraction of water by hydrophilic colloids causes them to absorb water in a process called **imbibition**. It is by this process that dry seeds initially absorb water, often with remarkable force – exceeding 100 000 kPa in some cases.

33.1.6 Cohesion and surface tension

Cohesion is the tendency of molecules of a substance to attract one another. The magnitude of this attractive force depends upon the mass of the particles and their distance apart. Gases, with their smaller molecular masses, have small cohesive forces. In liquids, the cohesive forces are much greater. Unlike gases, liquids cannot be expanded or compressed to any degree. Hydrogen bonding increases the cohesive forces between water molecules. One effect of these large cohesive forces in water is that the molecules are pulled inwards towards each other, so forming spherical drops rather than spreading out in a layer. The inward pull of the water molecules creates a skin-like layer at the surface. This force is called **surface tension**.

The cohesive forces between molecules accounts for the upward pull of water in xylem when evaporation occurs at the leaves. It is surface tension which allows insects like pond-skaters to walk on the surface of the water.

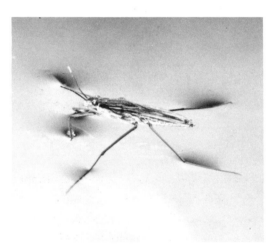

Pond skater walking on water

33.1.7 Adhesion and capillarity

Adhesion is the tendency of molecules to be attracted to ones of a different type. Considerable adhesive forces exist between the walls of xylem vessels and the water within them. The magnitude of these forces has been estimated at up to 100 000 kPa. **Capillarity** is also the result of intermolecular forces between various molecules. If one end of an open glass tube is held vertically beneath the surface of water, the liquid can be seen to rise up the tube. The smaller the diameter of the tube the higher it rises. Xylem vessels, with their diameters around 0.02 mm, have considerable capillarity forces which contribute to the movement of water up a plant. Capillarity also plays an important rôle in the upward movement of water in soil.

33.1.8 The importance of water to living organisms

Life arose in water and many organisms still live surrounded by it. Those that left water to colonize land nonetheless keep their cells bathed in it. Water is therefore the main constituent of all organisms – in jellyfish up to 98% and in most herbaceous plants 90%. Even mammals consist of around 65% water. The importances of water to organisms are many and there is only room here to list a few.

Metabolic role of water

1. Hydrolysis – Water is used to hydrolyse many substances, e.g. proteins to amino acids, fats to fatty acids and glycerol and polysaccharides to monosaccharides.

2. Medium for chemical reactions – All chemical reactions take place in an aqueous medium.

3. Diffusion and osmosis – Water is essential to the diffusion of materials across surfaces such as the lungs or alimentary canal.

4. Photosynthetic substrate – Water is a major raw material in photosynthesis.

Water as a solvent

Water readily dissolves other substances and therefore is used for:

1. Transport – Blood plasma, tissue fluid and lymph are all predominantly water and are used to dissolve a wide range of substances which can then be easily transported.

2. Removal of wastes – Metabolic wastes like ammonia and urea are removed from the body in solution in water.

3. Secretions – Most secretions comprise substances in aqueous solution. Most digestive juices have salts and enzymes in solution; tears consist largely of water and snake venoms have toxins in suspension in water.

Water as a lubricant

Water's properties, especially its viscosity, make it a useful lubricant. Lubricating fluids which are mostly water include:

1. Mucus – This is used externally to aid movement in animals, e.g. snail and earthworm; or internally in the vagina and gut wall.

2. Synovial fluid – This lubricates movement in many vertebrate joints.

3. Pleural fluid – This lubricates movement of the lungs during breathing.

4. Pericardial fluid – This lubricates movement of the heart.

5. Perivisceral fluid – This lubricates movement of internal organs, e.g. the peristaltic motions of the alimentary canal.

Supporting rôle of water

With its large cohesive forces water molecules lie close together. Water is therefore not easily compressed, making it a useful means of supporting organisms. Examples include:

1. Hydrostatic skeleton – Animals like the earthworm are supported by the pressure of the aqueous medium within them.

2. Turgor pressure – Herbaceous plants and the herbaceous parts of woody ones are supported by the osmotic influx of water into their cells.

3. Humours of the eye – The shape of the eye in vertebrates is maintained by the aqueous and vitreous humours within them. Both are largely water.

4. Amniotic fluid – This supports and protects the mammalian foetus during development.

5. Erection of the penis – The pressure of blood, a largely aqueous fluid, makes the penis erect so that it can be introduced into the vagina during copulation.

6. Medium in which to live – Water provides support to the organisms which live within it. Very large organisms, e.g. whales, returned to water as their sheer size made movement on land difficult.

Miscellaneous functions of water

1. Temperature control – Evaporation of water during sweating and panting is used to cool the body.

2. Medium for dispersal – Water may be used to disperse the larval stages of some terrestrial organisms. In mosses and ferns it is the medium in which sperm are transferred. The build-up of osmotic pressure helps to disperse the seeds of the squirting cucumber.

3. Hearing and balance – In the mammalian ear the watery endolymph and perilymph play a rôle in hearing and balance.

33.2 Water relations of a plant cell

For practical purposes the plant cell can be divided into three parts (Fig. 33.2):

1. The vacuole – This contains an aqueous solution of salts and sugars and organic acids.

2. The cytoplasm surrounded by membranes – The inner membrane is called the **tonoplast** and the outer one, the **plasma membrane**; both are partially permeable.

3. The cell wall – This is made of cellulose fibres and is completely permeable to even large molecules. It may be impregnated with substances like lignin, in which case it is impermeable to molecules.

We saw in Section 4.4.2 that the **water potential (Ψ)** of a system is the difference between the chemical potential of the water in the system and the chemical potential of pure water at the same temperature and atmospheric pressure. The water potential of pure water at standard temperature and pressure is zero.

The presence of solute molecules in the vacuole of a plant cell makes its water potential more negative (lower). The greater the concentration of solutes, the more negative is the water potential. This change in water potential as a consequence of the presence of solute molecules is called the **solute potential (Ψ_s)**. (This was previously called 'osmotic potential'.) As the solute molecules always lower the water potential, the value of the solute potential is always negative.

If a plant cell is surrounded by pure water, the water potential of the vacuole, containing solute molecules as it does, will be more negative than the surrounding medium. Water therefore enters by osmosis, thus creating a hydrostatic pressure which pushes outwards on the cell wall. This is known as the **pressure potential (Ψ_p)**. In this instance, as in most plant cells, the pressure potential is positive. In xylem vessels, however, where transpiration is pulling water up the plant, the pressure potential is negative.

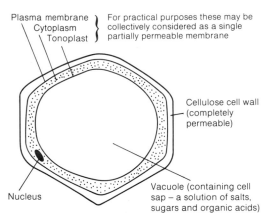

Plasma membrane ⎫ For practical purposes these may be
Cytoplasm   ⎬ collectively considered as a single
Tonoplast   ⎭ partially permeable membrane

Cellulose cell wall (completely permeable)

Nucleus

Vacuole (containing cell sap – a solution of salts, sugars and organic acids)

Fig. 33.2 Typical plant cell showing osmotically important structures

The relationship between water potential, solute potential and pressure potential is shown in the equation:

$$\Psi = \Psi_s + \Psi_p$$

| Ψ | = | Ψ_s | + | Ψ_p |
| water potential | | solute potential | | pressure potential |

If the same cell is placed in a solution which has a more negative solute potential than that of its cell sap, it will lose water by osmosis. In this case the external solution has a more negative water potential than the internal solution of the cell sap. As a result, water is drawn out of the cell by osmosis and the protoplast (the living part of the cell) shrinks. As it does so the pressure potential decreases. A point is reached where the protoplast no longer presses on the cell wall, and hence the pressure potential falls to zero. This point is termed **incipient plasmolysis**. Any further loss of water causes the protoplast to shrink more and so pull away from the cell wall. This condition is called **plasmolysis** and the cell is said to be **flaccid**. It is important to remember that water will always move from a less negative (higher) water potential to a more negative (lower) one. Pure water has a water potential of zero.

Fig. 33.3 Chart to show differences between cells placed in external solutions of different water potential

Water potential (Ψ) of external solution compared to internal solution	Less negative (higher)	Equal	More negative (lower)
Net movement of water	Enters cell	Neither enters nor leaves cell	Leaves cell
Protoplast	Swells	No change	Shrinks
Condition of cell	Turgid	Incipient plasmolysis	Plasmolysed (flaccid)

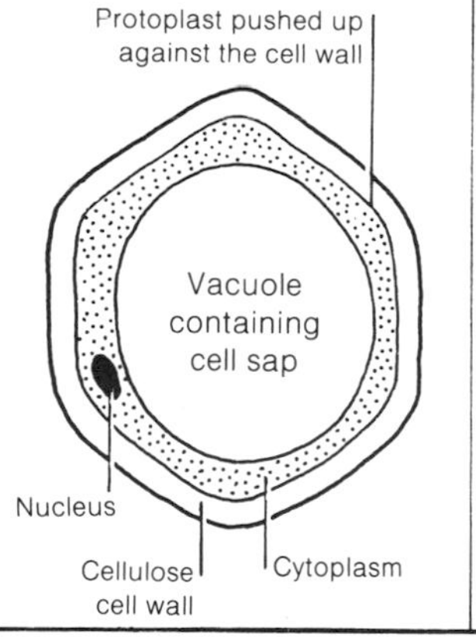

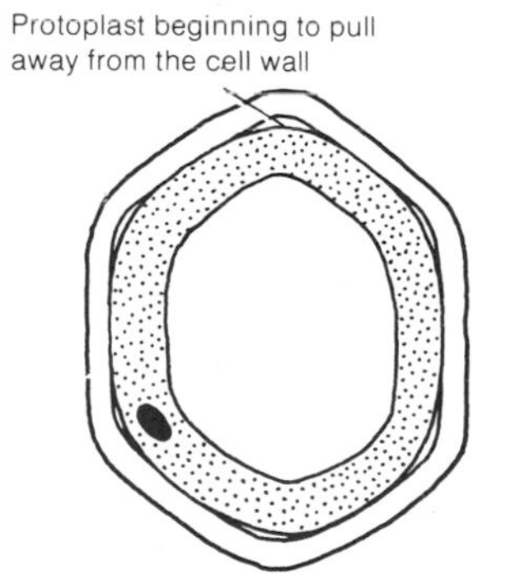

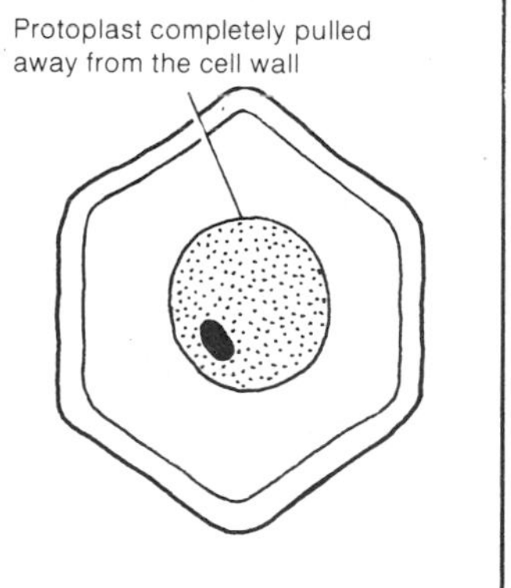

33.3 Transpiration

The evaporation of water from plants is called **transpiration**. This evaporation takes places at three sites:

1. Stomata – Most water loss, up to 90%, takes place through these minute pores which occur mostly, but not exclusively, on leaves. Some are found on herbaceous stems.

2. Cuticle – A little water, perhaps 10%, is lost through the cuticle, which is not completely impermeable to gases. The thicker the cuticle the smaller the water loss.

Tree bark showing lenticels

3. Lenticels – Woody stems have a superficial layer of cork which considerably reduces gas exchange over their surface. At intervals the cells of this layer are loosely packed, appearing externally as raised dots. These are the lenticels through which gaseous exchange, and hence water loss, may occur. The amount of transpiration through these is relatively insignificant.

Transpiration would appear to be the inevitable, if undesirable, consequence of having leaves punctured with stomata. The necessity of stomata for photosynthesis is obvious. It is unfortunate that any opening in a leaf which allows gases in will inevitably allow water out. Transpiration is not essential as a means of bringing water to the leaves. This could occur by purely osmotic means. As water was used for photosynthesis in a leaf mesophyll cell, its water potential would become more negative (lower). Water could then enter this cell from adjacent ones whose water potential would also become more negative (lower). They in turn would draw it from adjacent cells and so on down to the root hair cells of the plant, which would draw their water from the soil solution. While mineral salts are drawn up the plant in the transpiration stream, these could be diffused, or actively transported, in the absence of transpiration. Mineral uptake from the soil is largely independent of the rate of transpiration and could certainly occur in its absence. The cooling effect of transpiration would only be beneficial when environmental temperatures were high and even then its effects would be insignificant. The definition of transpiration as a 'necessary evil' is therefore not without some justification.

33.3.1 Stomatal mechanism
Stomata occur primarily on leaves where they are normally more abundant on the abaxial (lower) surface, where there are typically around 180 per mm². They may even be entirely absent on the adaxial (upper) surface. The structure of a stoma is given in Chapter 23, Fig. 23.1. It consists of a pair of specialized epidermal cells, the **guard cells** which surround a small pore a few microns wide known as the **stomatal aperture**. Guard cells are different from normal epidermal cells in being kidney-shaped. Unlike other epidermal cells, they possess chloroplasts and have denser cytoplasm with a more prominent nucleus. The inner walls of guard cells are thicker and less elastic than the outer ones. It is this which makes the inner wall less able to stretch and results in the typical kidney shape. Moreover, any increase in the volume of a guard cell, owing, for example, to the osmotic uptake of water, causes increased bowing of the cell owing to the greatest expansion occurring in the outer wall. When this occurs in the two guard cells of a stoma, the stomatal aperture enlarges (Fig. 33.4).

Even when all stomata on a leaf are fully open, the total area of the apertures rarely exceeds 2% of the leaf area.

It is possible to show that opening of a stoma is the result of pressure potential changes within the guard cells by puncturing turgid guard cells using a micro-needle. This releases the pressure potential and the cells collapse, closing the stoma.

What exactly causes changes in pressure potential within guard cells? While there are a number of factors which influence pressure potential, they all operate through changes in the water potential. It has long been observed that stomata open in the light and close in the dark. Initially it was thought that in the light the chloroplasts of

(a) Stoma closed – guard cells less turgid

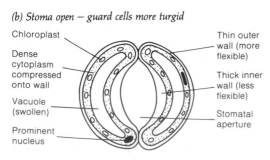

Chloroplast

Dense cytoplasm

Vacuole

Prominent nucleus

Thin outer wall (more flexible)

Thick inner wall (less flexible)

(b) Stoma open – guard cells more turgid

Chloroplast

Dense cytoplasm compressed onto wall

Vacuole (swollen)

Prominent nucleus

Thin outer wall (more flexible)

Thick inner wall (less flexible)

Stomatal aperture

Fig. 33.4 Surface view of stoma

the guard cells produced sugars by photosynthesis. These made the water potential of the guard cells more negative (lower) which therefore drew in water from adjacent cells, causing increased pressure potential and the opening of the stomata. When it was found that the opening occurred too rapidly to be accounted for in this way, a second theory was substituted. It was thought that changes in the carbon dioxide concentration of leaves was responsible for stomatal opening and closing. It could be observed that stomata opened when carbon dioxide concentrations were low and closed when it was high. At high light intensities photosynthesis uses up carbon dioxide. Being acidic in solution, its removal causes a rise in pH in the guard cells. This favours the conversion of starch to sugar by the enzyme starch phosphorylase. The sugar so produced makes the water potential of the guard cells more negative (lower) which therefore take in water from the surrounding cells. The resultant increase in pressure potential causes the stomata to open. The reverse situation occurs in the dark, when carbon dioxide produced in respiration causes a fall in pH. This theory cannot completely explain the process, however. In many plants the changes in sugar concentration are inadequate to produce the necessary alteration in pressure potential. Futhermore, in some species, like onion (*Allium cepa*), no starch at all is formed in guard cells.

Another theory suggests that it is the mineral salt concentration of the guard cells which effects opening and closing. In the dark, minerals such as potassium ions (K^+) leave guard cells and enter the adjacent subsidiary cells. In the light, it is known that ATP is rapidly produced in guard cells. This is thought to provide energy for the active transport of potassium ions back into the guard cells. These ions make the water potential of the guard cells more negative (lower), causing an influx of water by osmosis, increasing their pressure potential and effecting stomatal opening.

The process would appear to involve malate. In the light, starch grains in the chloroplasts of guard cells diminish in size. It is postulated that this starch is being converted to malic acid which dissociates into hydrogen and malate ions. The hydrogen ions are exchanged with potassium ions from the subsidiary cells. The combined effects of the potassium and malate ions in making the water potential of the guard cells more negative (lower), causes water to pass into them by osmosis. The resultant increase in pressure potential opens the stoma.

33.3.2 Movement of water across the leaf

Under normal circumstances the humidity of the atmosphere is less than that in the sub-stomatal air-space. Provided the stomata are open (see Section 33.3.1) the less negative (higher) water potential in the sub-stomatal air-space causes water vapour to diffuse out into the atmosphere through the stomatal aperture. Provided there is some air movement around the leaf, the water vapour is swept away once it leaves the stomata. The water lost from the sub-stomatal air-space is replaced by more evaporating from the spongy mesophyll cells surrounding the space. The water is brought to the spongy mesophyll cells from xylem in the leaf – in three ways:

1. The apoplast pathway – Most water travels from cell to cell via the cell wall, which is made up of cellulose fibres, between which are water-filled spaces. As the water evaporates into the sub-stomatal air-space from the wall of one cell, it creates a tension which pulls

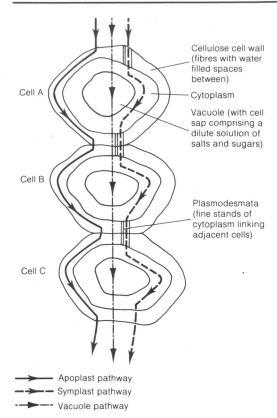

Cellulose cell wall (fibres with water filled spaces between)

Cytoplasm

Vacuole (with cell sap comprising a dilute solution of salts and sugars)

Plasmodesmata (fine stands of cytoplasm linking adjacent cells)

Cell A

Cell B

Cell C

———→ Apoplast pathway
– ▸ –▸– Symplast pathway
-·–▸–·- Vacuole pathway

Fig. 33.5 Alternative routes for water transport across cells

in water from the spaces in the walls of surrounding cells. The pull is a result of the cohesive forces between the water molecules which, due to hydrogen bonding, are particularly strong.

2. The symplast pathway — Some water is lost to the sub-stomatal air-space from the cytoplasm of cells surrounding it. The water potential of this cytoplasm is thereby made more negative (lower). Between adjacent cells are tiny strands of cytoplasm, known as **plasmodesmata**, which link the cytoplasm of one cell to that of the next. Water may pass along these plasmodesmata from adjacent cells with a less negative (higher) water potential. This loss of water makes the water potential of this second cell more negative (lower), which may in turn replace it with water from other cells with less negative (higher) water potentials. In this way a water potential gradient is established between the sub-stomatal space and the xylem vessels of the leaf. The symplast pathway carries less water than the apoplast pathway, but is of greater importance than the vacuolar pathway.

3. The vacuolar pathway — A little water passes by osmosis from the vacuole of one cell to the next, through the cell wall, membranes and cytoplasm of adjacent cells. In the same way as the symplast pathway, a water potential gradient between the xylem and the sub-stomatal air space exists. It is along this gradient that the water passes.

In Fig. 33.5, water leaving cell C causes its water potential to become more negative. Assuming all three cells were originally of equal water potential, then, compared to cell C, cell B now has a less negative water potential. Water therefore flows from cell B to cell C. In the same way, loss of water from B to C causes water to enter B from A. Remember that this mechanism applies only to the symplast and vacuolar pathways. The apoplast pathway is due to cohesion tension and is independent of a water potential gradient.

33.3.3 Movement of water up the stem

Water moves up the stem and into the leaves through xylem vessels and tracheids. Their structure is dealt with in Section 5.7.1. Xylem and phloem together form **vascular bundles** which are arranged mostly towards the outside of a stem. The reason for this is that the vascular bundles, along with associated sclerenchyma, give support to the stems of herbaceous plants. The main forces acting on stems are lateral ones, owing largely to the wind. These forces are best resisted by an outer cylinder of supporting tissue. Hence the vascular bundles are predominantly at the periphery of stems as shown in Fig. 33.6.

The evidence supporting the view that xylem carries water up the stem includes:

1. A leafy shoot is cut under water containing a dye, e.g. eosin. The cut end is kept in the solution of the dye and left for a few hours. The shoot is removed and cut at various levels up the stem. Only the xylem is found to be stained red, indicating that it alone transports water. If left long enough in the dye, the veins of the leaves also become stained.

2. A ring of tissue removed from the outside of a woody stem does not affect the flow of water up the stem, provided only the bark, including the phloem, are removed. If, however, the outer layers of xylem are also removed, upward transport ceases and the leaves wilt.

3. If the cut end of a shoot is placed in a solution of a metabolic poison, the uptake of the solution continues as normal. If the process were an active one, it would be expected that the poison would kill the cells in which it travelled and so prevent water transport. As the process continues, it must be assumed that it occurs passively. As xylem cells are dead, these would seem the most likely sites of this passive process.

4. Plants which are allowed to draw up fatty solutions soon wilt. It is found by staining and microscopic examination that these fats have blocked the lumina of the xylem cells. The wilting must therefore be the result of this blockage and hence xylem must be the route by which water rises up the stem.

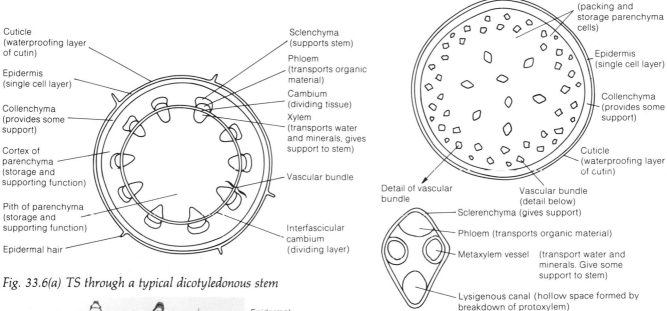

Fig. 33.6(a) TS through a typical dicotyledonous stem

Fig. 33.6(b) TS through a typical monocotyledonous stem

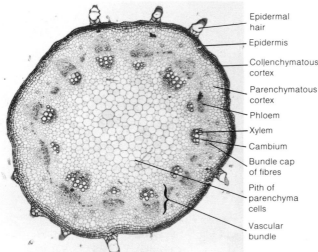

TS stem of *Helianthus* (× 20 approx.)

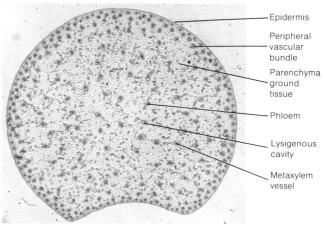

TS stem *Zea* (× 8 approx.)

The theory of the mechanism by which water moves up the xylem is known as the **cohesion-tension theory**. The transpiration of water from the leaves draws water across the leaf (Section 33.3.2). This water is replaced by that entering the mesophyll cells from the xylem by osmosis. As water molecules leave xylem cells in the leaf, they pull up other water molecules. This pulling effect, known as the **transpiration pull**, is possible because of the large cohesive forces between water molecules. The pull creates a tension in the xylem cells which, if cut, draw in air rather than exude water. Such is the cohesive force of this column of liquid that it is sufficient to raise water to heights in excess of 100 m, i.e. large enough to supply water to the top of the tallest known trees, the Californian redwoods.

Other forces may contribute to the movement of water up the stems of plants. These make only a small contribution to transport in large trees but may be significant in smaller herbaceous plants. **Adhesion forces** between the water molecules and the walls of xylem vessels help water to rise upwards in xylem – a phenomenon known as **capillarity**. This force can cause water to rise to a height of 3 m. **Root pressure**, which is discussed in Section 33.5.3, also contributes.

33.3.4 Xylem structure related to its rôle of water transport

The relationship between structure and function is always a close one and this is especially true of xylem vessels and tracheids. The structure of xylem is given in Section 5.7.1. Correlations between xylem structure and its function of water transport include the seven points below.

1. Both vessels and tracheids consist of long cells (up to 5 mm in length) joined end to end. This allows water to flow in a continuous column.

2. The end walls of xylem vessels have broken down to give an uninterrupted flow of water from the roots to the leaves. Even in tracheids where end walls are present, large bordered pits reduce the resistance to flow caused by the presence of these walls.

3. The walls are impregnated with lignin. This makes them impermeable to water and so prevents water escaping *en route* (see note **4** below).

4. There are pits at particular points in the lignified wall which permit lateral flow of water where this is necessary.

5. The lignified walls are especially rigid to prevent them collapsing under the large tension forces set up by the transpiration pull.

6. The impregnation of the cellulose walls with lignin increases the adhesion of water molecules and helps the water to rise by capillarity.

7. The narrowness of the lumina of vessels and tracheids (0.01–0.2 mm in diameter) increases the capillarity forces.

33.3.5 Measurement of transpiration

Transpiration in a cut leafy shoot can be measured using a potometer, a diagram of which is given in Fig. 33.7. To be strictly accurate, the instrument measures the rate of water uptake of a shoot, but in practice this is almost exactly the same as the rate of transpiration. A little of the water taken up may be used in photosynthesis and other metabolic processes, but the vast majority is transpired. The experiment is carried out as follows:

1. Cut a leafy shoot off a plant. As it is under tension, cutting the shoot will cause air to enter the xylem so, if possible, firstly hold the part where the cut is to be made under water. If this is not feasible, it will be necessary to trim back the cut shoot a few centimetres to remove the xylem containing air.

2. Submerging the potometer, fill it with water, using the syringe to help pump out any air bubbles. Fit the leafy shoot to the rubber tube, ensuring a tight fit.

3. Remove the apparatus from the water and allow excess water to drain off. Gently shake the shoot to remove as much water as possible.

4. Seal joints around the rubber tube with vaseline to keep the apparatus watertight.

5. Introduce an air bubble into the water column by using the syringe to push the water almost to the end of the capillary tube. Leave a small air-space. Place the open end of the capillary tube in a vessel of water and draw up more water behind the air-space.

6. When the shoot is dry, the syringe may be depressed with the tap open until the air bubble in the capillary tube is pushed back to the zero mark. The tap should then be turned to close off the syringe.

7. Measure the distance moved by the air bubble in the calibrated capillary tube in a given time. Repeat the procedure a number of times, using the syringe to return the air bubble to zero each time.

8. Calculate the water uptake in $mm^3\ min^{-1}$ using the average of the results obtained.

9. The experiment can be repeated under differing conditions, e.g. in light and dark, at different air temperatures and humidities, in still and moving air.

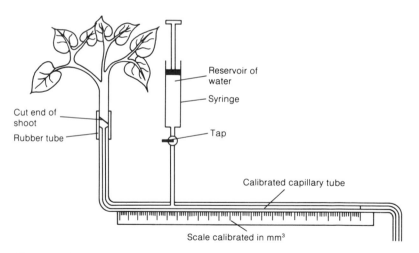

Fig. 33.7 A potometer

33.4 Factors affecting transpiration

A number of factors influence the rate of transpiration. These may be divided for convenience into external (environmental) factors and internal ones related to the structure of the plant itself.

33.4.1 External factors affecting transpiration

External factors include all aspects of the environment which alter the diffusion gradient between the transpiring surface and the atmosphere. Among these are:

1. Humidity – The humidity, or vapour pressure, of the air affects the water potential gradient between the atmosphere within the leaf and that outside. When the external air has a high humidity, the gradient is reduced and less water is transpired. Conversely, low humidities increase the transpiration rate.

2. Temperature – A change in temperature affects both the kinetic movement of water molecules and the relative humidity of air. A rise in temperature increases the kinetic energy of water molecules and so increases the rate of evaporation of water. At the same time it lowers the relative humidity of the air. Both changes increase the rate of transpiration. A fall in temperature has the reverse effect, namely, a reduction in the amount of water transpired.

3. Wind speed – In the absence of any air movement the water vapour which diffuses from stomata accumulates near the leaf surface. This reduces the water potential gradient between the moist atmosphere in the stomata and the drier air outside. The transpiration rate is thus reduced. Any movement of air tends to disperse the humid layer at the leaf surface, thus increasing the transpiration rate. The faster the wind speed, the more rapidly the moist air is removed and the greater the rate of transpiration.

4. Light – The stomata of most plants open in light and close in the dark. The suggested mechanisms by which these changes are brought about are given in Section 33.3.1. It follows that, up to a point, an increase in light intensity increases the transpiration rate and vice versa.

5. Water availability – A reduction in the availability of water to the plant, for example as the result of a dry soil, means there is a reduced water potential gradient between the soil and the leaf. The transpiration rate is reduced as a result.

33.4.2 Internal factors affecting transpiration

There are a number of anatomical and morphological features of plants which also influence the transpiration rate. Many specific adaptations of plants designed to reduce water loss are dealt with in Section 33.4.3. Only general features applicable to all plants are dealt with here.

1. Leaf area – As a proportion of water loss occurs through the cuticle, the greater the total leaf area of a plant, the greater the rate of transpiration regardless of the number of stomata present. In addition, any reduction in leaf area inevitably involves a reduction in the total number of stomata.

2. Cuticle – The cuticle is a waxy covering over the leaf surface which reduces water loss. The thicker this cuticle the lower the rate of cuticular transpiration.

3. Density of stomata – The greater the number of stomata for a given area the higher the transpiration rate. Stomatal density on the abaxial (lower) epidermis of plants may vary from around 2000 cm^2

in an oat (*Avena sativa*) leaf to 45 000 cm² on an oak (*Quercus* spp.) leaf.

4. Distribution of stomata − In most dicotyledonous plants the leaves are positioned with their adaxial (upper) surfaces towards the light. The upper surfaces are subject to greater temperature rises than the lower ones owing to the warming effect of the sun. Transpiration is therefore potentially greater from the upper surface. Many plants, like the oak (*Quercus* spp.) and the apple (*Malus* spp.) limit their stomata entirely to the abaxial (lower) surface to reduce their overall water loss.

33.4.3 Xerophytic adaptations

Xerophytes are plants which have adapted to conditions of unfavourable water balance, i.e. conditions where the rate of loss is potentially greater than the availability of water. Most plants live in areas where ample water is available. These are known as **mesophytes**. Xerophytic plants have evolved a wide range of features designed to reduce the rate of transpiration. These are known as **xeromorphic** features. These adaptations are not confined to xerophytic plants but may also occur in mesophytes. By no means all xeromorphic adaptations occur in plants of hot, dry desert regions. Many species in cold regions cannot supply adequate water to the leaves as the soil water is frozen, and so unobtainable, for much of the year. Other species with a reasonable supply of water may suffer excessive loss from the leaves as a result of living in exposed, windy situations. These too exhibit xeromorphic features. Other plants live in salt marshes where the concentration of salts in the soil make the obtaining of water difficult. These **halophytes**, as they are called, also exhibit xeromorphic features.

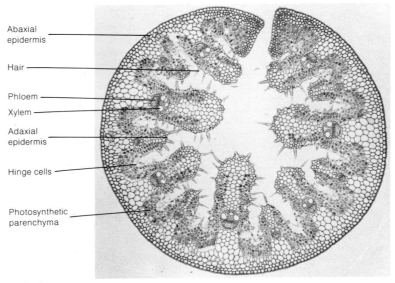

Abaxial epidermis

Hair

Phloem

Xylem

Adaxial epidermis

Hinge cells

Photosynthetic parenchyma

TS leaf of *Ammophila*, showing xerophytic features

Adaptations in xerophytes take a number of general forms:

1. Reduction in the transpiration rate − Clearly anything which lowers the rate at which the plant loses water helps to conserve it when in short supply.

2. Storage of water – In plants living where water supply is intermittent, e.g. in the desert, there is considerable advantage in rapidly absorbing the water when available and storing it for use during periods of drought. Plants which store water are termed **succulents**. The adaptations of these plants are not limited to specialized water storage tissue but include mechanisms for rapid water absorption, and reducing the rate at which it is lost.

3. Resistance to desiccation – Some species exhibit a remarkable tolerance to water loss and resistance to wilting.

The xeromorphic adaptations of plants are summarized in Table 33.1.

TABLE 33.1 **Xeromorphic adaptations of plants**

General form	Specific adaptation	Examples	Mechanism by which adaptation functions
Reduction of the transpiration rate	Thick cuticle	Evergreens, especially gymnosperms	Reduces cuticular transpiration by forming a waxy barrier preventing water loss
	Rolling of leaves	*Ammophila* (marram grass); *Calluna* (ling)	Moist air is trapped within the leaf, preventing water diffusing out through stomata which are confined to the inner surface
	Layer of protective hairs on leaf (pubescence)	*Ammophila* (marram grass); *Calluna* (ling)	Moist air is trapped in the hair layer, increasing the length of the diffusion path, so reducing transpiration
	Depression of stomata	*Pinus* (pine); *Ilex* (holly)	Lengthens the diffusion path by trapping still, moist air above the stomata, so reducing transpiration
	Reduction of surface area/volume ratio of leaves	*Pinus* (pine)	The leaves are small and circular in cross-section to reduce the transpiration area. The shape also gives structural rigidity to help prevent wilting
	Absence of leaves	Most cacti, e.g. *Opuntia* (prickly pear)	Dispensing with leaves altogether limits water loss to the stems which have considerably fewer stomata. These stems are flattened to provide an adequate area for photosynthesis
	Orientation of leaves	*Lactuca* (compass plant)	The positions of the leaves are constantly changed so that the sun strikes them obliquely. This reduces their temperature and hence the transpiration rate
	More negative (lower) water potential of cell sap	Many xerophytes and most halophytes, e.g. *Salicornia* (glasswort)	The cells accumulate salts which make their water potential more negative (lower). This makes it more difficult for water to be drawn from them
Succulence	Succulent leaves	E.g. *Bryophyllum*	Stores water
	Succulent stems	Most cacti, e.g. *Opuntia* (prickly pear)	Stores water
	Closing of stomata during daylight	Most cacti and other C_4 plants	The more efficient use of carbon dioxide by C_4 plants allows them to keep stomata closed during much of the day, so reducing transpiration
	Shallow but extensive root systems	Most succulents	Allows efficient absorption of water over a wide area when the upper layers of the soil are moistened by rain
Resistance to desiccation	Reduction of transpiration surface through loss or adaptation of leaves	*Berberis* (barberry); many cacti	Leaves reduced to spines to protect plant from grazing. Flattened stems perform photosynthesis
		Ruscus (butcher's broom)	Leaves are lost and photosynthesis is performed by a flattened stem known as a cladode
		Acacia	The lamina of the leaf is lost and the petiole becomes flattened to carry out photosynthesis
	Lignification of leaves	*Hakea*	Lignified tissue supports the leaf, preventing it wilting in times of drought and thereby allowing it to continue photosynthesis
	Reduction in cell size	Many xerophytes	The proportion of cell wall material is greater with many small cells. This gives additional support, making the plant less liable to wilt

33.4.4 Hydrophytic adaptations

Plants which live wholly or partly submerged in water are called **hydrophytes**. There are three types distinguished by their habitat:

1. Marsh plants – These live in soil which is always very wet and often waterlogged. Examples include the sedges (*Carex* spp.) and the rushes (*Juncus* spp.).

2. Swamp plants – These live in permanently waterlogged soil over which there is a shallow layer of standing water. Examples include the reeds (*Phragmites* spp.).

3. Aquatic plants – These live partially or completely submerged in water. Examples include water lilies (*Nymphaea* spp.) and duckweeds (*Lemna* spp.).

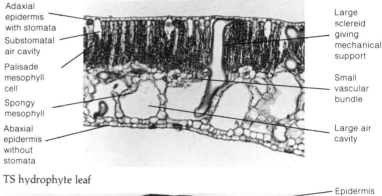

TS hydrophyte leaf

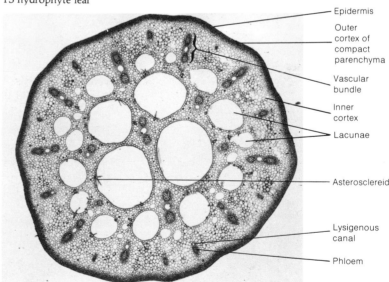

TS stem of *Nymphaea*, a hydrophyte

The greatest problem for hydrophytes is obtaining oxygen for respiration. Waterlogged soil has no air-spaces and the water in the soil spaces contains very little oxygen for two reasons. Firstly, oxygen is not very soluble in water, and secondly, the bacterial decay of organic material in the soil rapidly uses up that which is available. Most waterlogged soils are therefore anaerobic.

One common feature of hydrophytes is the presence of **aeration tissue**. This typically comprises large air-spaces, called **lacunae**, between the cells of the stems and leaves. These form an extensive communicating network throughout the submerged parts of the plant. For example, during the day, the oxygen produced by photosynthesis will largely remain in the spaces from where it can diffuse towards the roots for use in respiration. Aerating tissue also confers buoyancy, raising leaves to the surface where they can take maximum advantage of the light.

Hydrophytes, especially completely submerged ones, show a marked absence of supporting tissues like sclerenchyma. Not only is such tissue unnecessary as the water provides support, it would make the plant more rigid, rendering it liable to breakage by water currents. Xylem, another supporting tissue, is poorly developed not simply because support is unnecessary but because its function of water transport is less important in a plant largely surrounded by it. Where leaves are submerged, stomata are absent or non-functional. Where they float, as in water lilies (*Nymphaea* spp.), only the upper surface possesses them. Aerial leaves show normal stomatal distribution.

33.5 Uptake of water by roots

If a plant is to survive, the large quantities of water lost through transpiration must be replaced. Water absorption is largely carried out by the younger parts of roots which bear extensions of the epidermal cells, known as **root hairs**. These root hairs only remain functional for a few weeks, being replaced by others formed on the younger regions nearer the growing apex. As the root becomes older the epidermis is replaced by a layer known as the **exodermis** through which some water absorption still takes place.

33.5.1 Root structure

Beneath the epidermis of a young root is a broad cortex of parenchyma cells. The vascular tissue occurs as a central column rather than a peripheral cylinder as in stems. The reason is that roots are subject to pulling forces in a vertical direction rather than the lateral forces experienced by stems. Vertical forces are better resisted by a central column of supporting tisssue such as xylem. As roots, embedded as they are in soil, are subject to fewer stresses and strains than aerial parts of the plant, the other major supporting tissue, sclerenchyma, is reduced or absent.

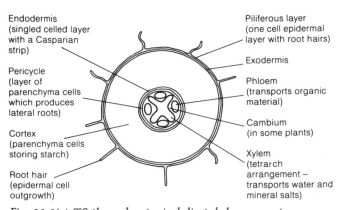

Endodermis (singled celled layer with a Casparian strip)

Pericycle (layer of parenchyma cells which produces lateral roots)

Cortex (parenchyma cells storing starch)

Root hair (epidermal cell outgrowth)

Piliferous layer (one cell epidermal layer with root hairs)

Exodermis

Phloem (transports organic material)

Cambium (in some plants)

Xylem (tetrarch arrangement – transports water and mineral salts)

Fig. 33.8(a) TS through a typical dicotyledonous root, as seen under a microscope (100 ×)

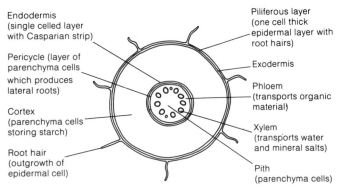

Endodermis (single celled layer with Casparian strip)

Pericycle (layer of parenchyma cells which produces lateral roots)

Cortex (parenchyma cells storing starch)

Root hair (outgrowth of epidermal cell)

Piliferous layer (one cell thick epidermal layer with root hairs)

Exodermis

Phloem (transports organic material)

Xylem (transports water and mineral salts)

Pith (parenchyma cells)

Fig. 33.8(b) TS through a typical monocotyledonous root, as seen under a microscope (100 ×)

Around the central vascular tissue is a single-celled ring known as the **endodermis**. These living cells are elongated vertically. Part of the wall is impregnated with **suberin**. This forms a distinctive band known as the **Casparian strip**. Inside the endodermis is the **pericycle**, a layer of parenchyma cells between the endodermis and the vascular tissue. It is from the pericycle that lateral roots originate. The internal structure of roots is shown in Fig. 33.8.

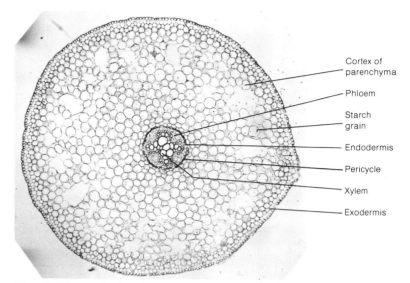

Cortex of parenchyma
Phloem
Starch grain
Endodermis
Pericycle
Xylem
Exodermis

TS root of *Ranunculus* (× 20)

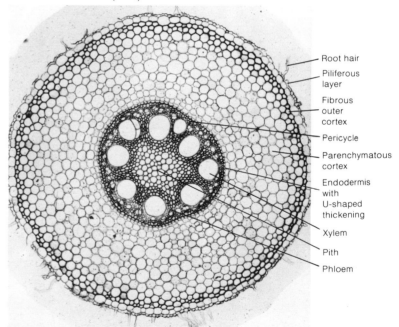

Root hair
Piliferous layer
Fibrous outer cortex
Pericycle
Parenchymatous cortex
Endodermis with U-shaped thickening
Xylem
Pith
Phloem

TS root of *Zea* (× 35 approx.)

33.5.2 Mechanisms of uptake

The same three pathways which are responsible for movement of water across the leaf also bring about its movement in the root:

1. The apoplast pathway – Compared to the leaf (Section 33.3.2) this pathway has one major difference in the root. Movement through the cell walls is prevented by the suberin of the Casparian strip in the endodermis. The water is thereby forced to enter the living protoplast of the endodermal cell, as the only available route to the xylem (Fig. 33.9).

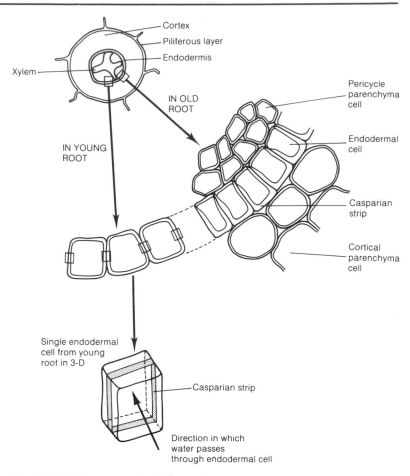

Fig. 33.9 Water transport in the root

The significance of this is that salts may then be actively secreted into the vascular tissue from the endodermal cells. This makes the water potential in the xylem more negative (lower), causing water to be drawn in from the endodermis. While the mechanism has yet to be proven scientifically, there is some circumstantial evidence supporting this view:

(a) There are numerous starch grains in endodermal cells which could act as an energy source for the process.

(b) Depriving roots of oxygen prevents water being exuded from cut stems. Lowering the temperature reduces the rate of exudation.

(c) Treating roots with metabolic poisons like cyanide also prevents water being exuded from cut stems.

Whatever the process, it is clearly an active one.

2. The symplast pathway – This operates in the same way as in the leaf (Section 33.3.2). Water leaving the pericycle cells to enter the xylem causes the water potential of the pericycle cells to become more negative (lower). These therefore draw in water from adjacent cells which in turn have a more negative (lower) water potential. In this way a water potential gradient is established across the root from the xylem to the root hair cells, which draws water across it.

3. The vacuolar pathway – This operates in the same way as in the leaf (Section 33.3.2), using the same water potential gradient described above.

A summary diagram of the passage of water throughout the plant is given in Fig. 33.10.

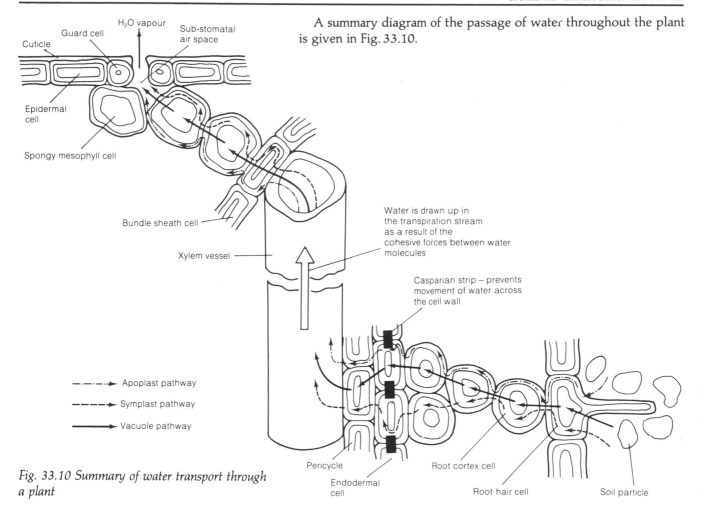

Fig. 33.10 Summary of water transport through a plant

33.5.3 Root pressure

The soil solution is normally very dilute and hence has a less negative (higher) water potential than the solution in the root hair cells. As a result, water enters the root hair cells by osmosis and their water potential becomes less negative (higher) as a consequence. Water therefore enters an adjacent cortical cell. This water potential gradient creates a force known as **root pressure**. If the stem of a plant is cut near to the roots, this root pressure causes water to be exuded from the cut stump. The process probably involves the pumping of salts into the xylem (see Section 33.5.2) as it is halted by the presence of metabolic poisons. In some plants, root pressure may be sufficiently large to force liquid out of pores on the leaves called **hydathodes**. The process is known as **guttation**. Root pressure may contribute to the movement of water up the stem, especially in herbaceous plants, but its contribution is far less than that of transpiration. An instrument for measuring root pressure is illustrated in Fig. 33.11.

33.6   Uptake and translocation of minerals

All plants require a supply of minerals, the functions of which are summarized in Chapter 3, Table 3.1. These minerals are largely absorbed by roots through root hairs, although leaves can also absorb them if sprayed with a suitable solution. Such sprays are called **foliar feeds**.

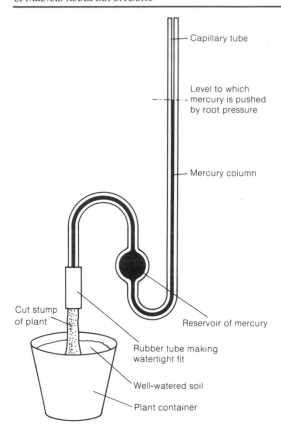

Fig. 33.11 Manometer for measuring root pressure

33.6.1 Mechanisms of mineral uptake

Minerals may be absorbed either passively or actively:

Passive absorption – If the concentration of a mineral in the soil solution is greater than its concentration in a root hair cell, the mineral may enter the root hair cell by **diffusion**.

Active absorption – If the concentration of a mineral in the soil solution is less than that in a root hair cell it may be absorbed by active transport, details of which are given in Section 4.4.3. Most minerals are absorbed in this way. The process is selective. Requiring energy as it does, the rate of absorption is dependent upon respiration.

Fig. 33.12(a) shows that when salts are added to plants growing in water, their respiration rate increases, presumably in order to provide the energy for their absorption. The addition of the respiratory inhibitor potassium cyanide prevents active mineral uptake, leaving only absorption by passive means (Fig. 33.12b). Increases in temperature increase the rate of respiration and hence the rate of mineral uptake (Fig. 33.12c).

Once absorbed, the mineral ions may move along the cell walls (apoplast pathway) by either diffusion or mass flow. In the latter case they are carried along in solution by the water being pulled up the plant in the transpiration stream. When these minerals reach the endodermis the Casparian strip prevents further movement along the cell walls. Instead, the ions enter the cytoplasm of the cell from where they diffuse or are actively transported into the xylem. Minerals may alternatively pass through the cytoplasm of cortical cells (symplast pathway) to the xylem into which they diffuse or are actively pumped.

33.6.2 Transport of minerals in the xylem

Analysis of the contents of xylem vessels reveals the presence of mineral salts and water, although sugars and amino acids may also be present. The evidence supporting the rôle of xylem in transporting minerals includes:

1. The presence of mineral ions in xylem sap.

2. A similarity between the rate of mineral transport and the rate of transpiration.

3. Evidence that other solutes, e.g. the dye eosin, are carried in the xylem (Section 33.3.3).

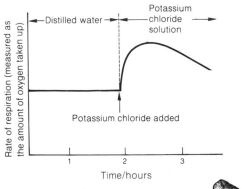

Fig. 33.12(a) Relationship between the rate of respiration and mineral uptake

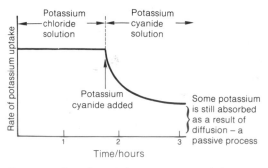

Fig. 33.12(b) Effect of the respiratory inhibitor cyanide on mineral uptake

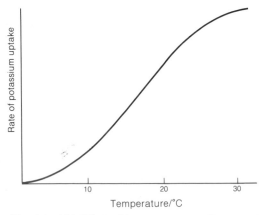

Fig. 33.12(c) Effect of temperature on the rate of potassium uptake in plants

4. Experiments using radio-active tracers (Fig. 33.13). The interpretation of the experiments is that where lateral transfer of minerals can take place, minerals pass from the xylem to phloem. Where it is prevented, the transport of minerals takes place almost exclusively in the xylem.

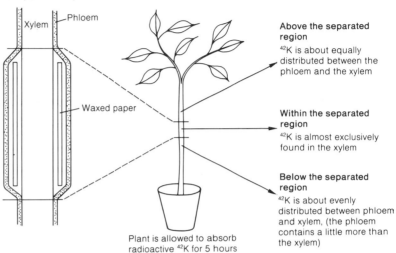

A 22.5 cm section of stem has its phloem and xylem separated by wax paper which is impervious to water and minerals. Lateral transport is thus prevented

By use of a fan, the transpiration rate of the plant is increased

After 5 hours absorbing radioactive ^{42}K, sections of the stem are tested for the amount of ^{42}K in the phloem and xylem

Xylem — Phloem

Waxed paper

Above the separated region
^{42}K is about equally distributed between the phloem and the xylem

Within the separated region
^{42}K is almost exclusively found in the xylem

Below the separated region
^{42}K is about evenly distributed between phloem and xylem, (the phloem contains a little more than the xylem)

Plant is allowed to absorb radioactive ^{42}K for 5 hours

Fig. 33.13 Summary of experiment of Stout and Hoagland (1939) to demonstrate that minerals are translocated up the plant in the xylem

Once in the xylem, minerals are carried up the plant by the mass flow of the transpiration stream. Once they reach the places where they will be utilized, called **sinks**, they either diffuse or are actively transported into the cells requiring them.

33.7 Translocation of organic molecules

The organic materials produced as a result of photosynthesis need to be transported to other regions of the plant where they are used for growth or stored. This movement takes place in the phloem.

33.7.1 Evidence for transport of organic material in the phloem

The evidence supporting the view that organic material formed as a result of photosynthesis is carried in the phloem includes:

1. When phloem is cut, the sap which exudes is rich in organic materials such as carbohydrates. The fact that sap is exuded suggests the contents of the phloem are under pressure.

2. The sugar content of phloem varies in relation to environmental conditions. Where conditions favour photosynthesis, the concentration of sugar in phloem increases. There is also a diurnal variation in the sugar content of phloem which reflects the diurnal variation in the rate of photosynthesis in relation to light intensity. Fig. 33.14 shows how the sucrose content of leaves increases to a maximum around 1500 hours, as a result of the high light intensity and temperature favouring photosynthesis at this time. This peak of sugar concentration is reflected in the phloem of the stem a little time later. Little variation of sucrose concentration in the xylem took place.

503

3. Removal of a complete ring of phloem from around a stem causes an accumulation of sugars above the ring, indicating that their downward progress has been interrupted.

4. If radio-active $^{14}CO_2$ is given to plants as a photosynthetic substrate, the sugars later found in the phloem contain ^{14}C. When phloem and xylem are separated by waxed paper, the ^{14}C is almost entirely found in the phloem.

5. Aphids have needle-like mouthparts with which they penetrate phloem in order to obtain sugars. If a feeding aphid is anaesthetized and then the mouthparts cut from the body, they remain as tiny tubes from which samples of the phloem contents exude. Analyses of these exudates confirm the presence of carbohydrates and amino acids, and a diurnal variation in their concentrations.

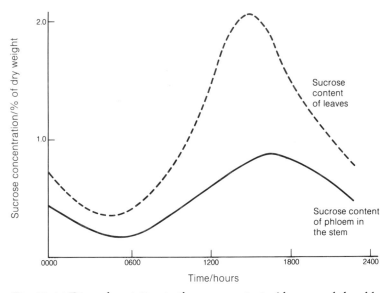

Fig. 33.14 Diurnal variation in the sugar content of leaves and the phloem in stems

33.7.2 Structure of phloem sieve tubes

Phloem tissue is living and comprises **sieve tubes** (or sieve elements), phloem parenchyma and phloem fibres. The basic structure of this tissue, as seen under a light microscope, is described in Section 5.7.2. In angiosperms, specialized parenchyma cells known as **companion cells** are always found associated with sieve tube cells.

The sieve tube cells are the only components of phloem obviously adapted for the longitudinal flow of material. They are elongated with a characteristic series of pores 2–6 μm in diameter in the end walls. These are lined with the polysaccharide **callose** and form a **sieve plate**. The sieve tube cells have a well defined plasma membrane and their cytoplasm contains numerous plastids and mitochondria. Within the lumen of the cells are longitudinal strands of cytoplasm 1–7 μm wide. They are made up of **phloem protein**. The strands are continuous from cell to cell through the pores of the sieve plate and are known as **transcellular strands**. There is some question as to the existence of these strands as some researchers believe them to be no more than artefacts. Mature sieve tube cells lack a nucleus and are called **sieve tube elements**.

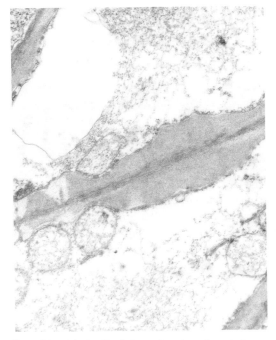

Sieve plate, at the junction between two sieve elements, in leaf of *Zinnia elegans* (EM) (× 21 000 approx.)

The companion cells have a thin cellulose cell wall and dense cytoplasm. Within the cytoplasm is a large nucleus, numerous mitochondria, plastids, small vacuoles and an extensive rough endoplasmic reticulum. Companion cells are metabolically active. They are closely associated with the sieve tube element with which they communicate by means of numerous plasmodesmata.

The structure of a sieve tube element and companion cell as seen under an electron microscope is given in Fig. 33.15.

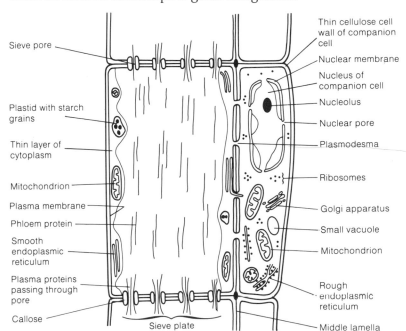

Fig. 33.15 *Structure of a longitudinal section of a sieve tube element and companion cell, as seen by an electron microscope*

33.7.3 Mechanisms of translocation in phloem

There is much controversy regarding the mechanism by which materials are translocated in phloem. One thing, however, is agreed: the observed rate of flow is much too fast for diffusion to be the cause. The theories put forward include the following hypotheses.

Mass flow (pressure flow) hypothesis
This theory, put forward by Munch in 1930, is still probably the most widely accepted. Photosynthesis forms soluble carbohydrates like sucrose.

Photosynthesizing cells in the leaf therefore have their water potentials made more negative (lower) by the accumulation of this sucrose. As a result, water which has been transported up the stem in the xylem enters these cells. This causes an increase in their pressure potential. At the other end of the plant, in the roots, sucrose is either being utilized as a respiratory substrate or is being converted to starch for storage. The sucrose content of these cells is therefore low, giving them a less negative (higher) water potential and a consequently lower pressure potential. There is therefore a gradient of pressure potential between *the source* of sucrose (the leaves) and its point of utilization – *the sink* (the roots and other tissues). The two are linked by the phloem and as a result liquid flows from the leaves to other tissues along the sieve tube elements. A simple physical model to illustrate this mechanism is given in Fig. 33.16.

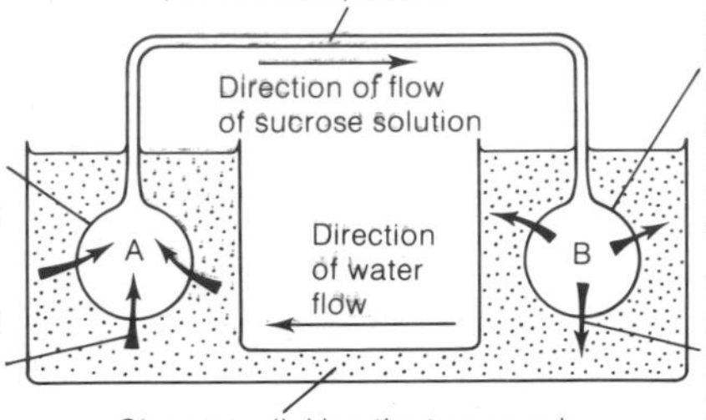

Glass tube linking cell A to cell B represents the phloem

Cell A – has membrane permeable only to water. It contains a solution with a high concentration of sucrose. It represents the leaf of a plant

Direction of flow of sucrose solution

Direction of water flow

Cell B – has a membrane permeable only to water. It contains a solution with little sucrose. It represents respiring or storage regions of the plant e.g. roots

Water enters cell A by osmosis because it has a lower water potential than the water surrounding it

Water leaves because it is forced out due to the high pressure potential created in A

Glass tube linking the two vessels – represents the xylem

Fig. 33.16 Physical model to illustrate the mass flow theory of translocation in phloem

Provided sucrose is continually produced in A (leaf) and continually removed at B (e.g. root) the mass flow of sucrose from A→B continues

Evidence supporting the mass flow theory includes:

1. There is a flow of solution from phloem when it is cut or punctured by the stylet of an aphid.

2. There is some evidence of concentration gradients of sucrose and other materials, with high concentrations in the leaves and lower concentrations in roots.

3. Some researchers have observed mass flow in microscopic sections of living sieve elements.

4. Viruses or growth chemicals applied to leaves are only translocated downwards to the roots when the leaf to which they are applied is well illuminated and therefore photosynthesizing. When applied to shaded leaves, no downward translocation occurred.

It is likely that the sucrose produced in mesophyll cells as a result of photosynthesis needs to be actively transported against a concentration gradient into the sieve elements. The energy from this process is provided by ATP.

One criticism of the mass flow theory is that it offers no explanation for the existence of sieve plates which act as a series of barriers impeding flow. Indeed, as the process is passive after the initial stage, there is no necessity for the phloem to be a living tissue at all. One suggested function of the sieve plates is a means of sealing off damaged sieve tube elements. As the material within the elements is under pressure, any damage could lead to wasteful loss of sugar solution. It has been observed that once an element is damaged the sieve plate is quickly sealed by deposition of callose across the pores.

Electro-osmosis hypothesis
Originally put forward by Spanner in 1958, and since modified on several occasions, the theory proposes that potassium ions are actively transported by companion cells, across the sieve plate. The movement of these ions draws polar water molecules across the plate. The movement is still one of mass flow, but the theory at least offers some function for the sieve plates and explains the high metabolic rate observed in companion cells. However, there is no direct evidence of a potential difference existing across sieve plates.

Transcellular strand hypothesis
Thaine in 1962 proposed that transcellular strands, which he believes to extend from cell to cell via pores in the sieve plate, carry out a form of cytoplasmic streaming. Some type of peristaltic movement

in the strands is considered to aid the movement of the solutes. The process, being active, accounts for the many mitochondria in both sieve tube elements and companion cells. It will, however, require more positive proof of the existence of the actual process before it becomes widely accepted.

33.8 Questions

1. (a) Describe briefly the various types of cell found in the phloem of a flowering plant. (*6 marks*)
   (b) Give an account of the experimental evidence which indicates that the phloem transports sucrose. (*6 marks*)
   (c) State **one** possible mechanism of sucrose transport in the phloem and discuss the evidence for, and against, it. (*6 marks*)
   (*Total 18 marks*)

   Cambridge Board November 1988, Paper I, No. 4

2. (a) Explain how water is taken up by a root and suggest how the absorbed water moves to the xylem. (*9 marks*)
   (b) Describe how the cohesion tension hypothesis attempts to explain the upward movement of water from the root xylem to the leaf. (*6 marks*)
   (c) Plants of salt marshes may be said to suffer from a 'physiological drought'.
      (i) What do you understand by the term 'physiological drought' in this context?
      (ii) What physiological and anatomical modifications might you expect to find in plants growing in such habitats? Explain how they would enable the plant to survive such conditions. (*5 marks*)
      (*Total 20 marks*)

   Joint Matriculation Board June 1985, Paper IIB, No. 5

3. (a) With the aid of labelled diagrams, describe the structures involved in the translocation of organic solutes between different parts of an angiosperm plant. (*10 marks*)
   (b) Describe the mechanisms that have been suggested to account for movement of solutes during translocation. Which do you consider the most likely explanation or explanations? (*14 marks*)
   (c) Explain one experiment which has been performed that has enhanced our knowledge of translocation in flowering plants. (*6 marks*)
   (*Total 30 marks*)

   Oxford Local Board June 1988, Paper II, No. 8

4. Cylinders of approximately equal size were cut from a large potato tuber using a cork borer. They were divided into two groups A and B. The mass of each of the cylinders was measured and recorded. Single cylinders from group A were each placed into a different sucrose solution. Cylinders from group B were treated similarly except that 0.1 g of gibberellic acid was added to each solution. After 4 hours, each cylinder was removed from its solution, reweighed, and the percentage change in mass calculated. The results are given in the table.

	Percentage change in mass	
	Group A	Group B
Sucrose molarity	Sucrose solution only	Sucrose solution + gibberellic acid
0	+ 7	+ 42
0.1	+ 6	+ 38
0.2	+ 5	+ 36
0.3	+ 1	+ 27
0.4	− 4	+ 6
0.5	− 8	+ 3
0.6	− 14	− 3
0.7	− 17	− 4
0.8	− 16	− 7

(a) How should the potato cylinders be treated immediately before weighing? (*1 mark*)
(b) (i) Plot the data for *group A only* on graph paper. (*3 marks*)
   (ii) Comment briefly on the information shown by the graph. (*1 mark*)
(c) A 1M sucrose solution has a water potential of − 3.40 MPa. Use this information, and the graph you have plotted, to determine the water potential of the potato tissue. Show how you arrived at your answer. (*2 marks*)

Gibberellic acid is produced naturally in potato tubers, where it stimulates the production of carbohydrase enzymes.

(d) Account for the different results for groups A and B of the potato cylinders. (*2 marks*)
   (*Total 9 marks*)

   Joint Matriculation Board (Nuffield) June 1988, Paper IA, No. 3

5. Explain how the structure of vascular tissues in a dicotyledonous plant is related to their rôle in transport.
(20 marks)

London Board June 1985, Paper II, No. 7

6. (a) (i) Define the terms *water potential* and *osmotic potential*. (6 marks)
(ii) Explain how these two potentials interact to bring about the movement of water from the soil across the root cortex. (8 marks)
(b) (i) Outline the likely mechanism of stomatal opening and closing. (8 marks)
(ii) Discuss to what extent you consider stomata control the rate of transpiration. (8 marks)
(Total 30 marks)

Oxford Local 1987, Specimen Paper II, No. 4

7. Porometers can be used to measure air flow through a leaf under different environmental conditions. The figure shows a simple porometer attached to the underside of a leaf.

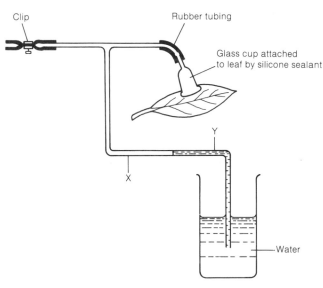

To obtain results, the time is taken for the meniscus to travel from X to Y.
(a) (i) Explain how the meniscus would be set at X at the beginning of the experiment.
(ii) Describe **two** different ways in which the apparatus might be modified to give a faster flow of fluid from X to Y. (3 marks)
The apparatus shown was used with detached leaves to obtain the results shown in the graph.
(b) Using your knowledge of leaves, interpret the changing shape of the graph between
(i) 0 and 1 hours
(ii) 1 and 5 hours
(iii) 5 and 9 hours (5 marks)

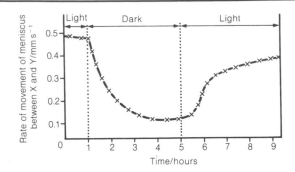

(c) (i) Suggest **one** objection to the use of this apparatus for gathering information about the response of leaves to periods of light and darkness.
(ii) Would similar results have been obtained if the porometer had been attached to the *upper* surface of the leaf. Explain your answer. (2 marks)
(d) Describe in outline how you would use this experimental method to investigate the effect of different wavelengths of light on the porosity of detached leaves. (4 marks)
(Total 14 marks)

Joint Matriculation Board June 1988, Paper IA, No. 5

8. How would you demonstrate incipient plasmolysis of EITHER onion epidermal cells OR cells of a small cube of beetroot?
What is water potential?
Calculate the water potential of a cell A when the solute potential (Ψ_s) of the cell sap is $-1.5\,MPa$ and the pressure potential (Ψ_p) is $+0.7\,MPa$.
What is the water potential of cell B? (Ψ_s is $-0.8\,MPa$ and Ψ_p is $+0.4\,MPa$.)
Cell A is next to cell B. Will water move from A to B or from B to A, and with what force? (10 marks)

Southern Universities Joint Board June 1986, Paper I, No. 11

9. For each of the following indicate whether it is true or false and, if false, state the correct version.

(a) In plants which exhibit secondary thickening the secondary phloem is laid down externally to the primary phloem and along the same radius.
(b) Mature sieve cells lack a nucleus but retain cytoplasm.
(c) There are more mitochondria in active sieve tube cells than in companion cells.
(d) Mature xylem tracheid cells are metabolically more active than mature phloem sieve cells.
(e) The phloem of a tree can transport organic substances upwards only. (5 marks)

Southern Universities Joint Board June 1986, Paper I, No. 5

10. (a) List **five** of the main functions of water in **plants**.
(5 marks)

(b) When strips of potato tuber are put in distilled water they increase in length because their cells become larger.

  (i) Using water potential terminology, explain how water is drawn into plant cells.
(5 marks)

  (ii) What ultimately limits this movement?
(2 marks)

(c) Make a labelled diagram to show the condition of a plant cell which has been placed in a $0.8 \, mol \, dm^{-3}$ sucrose solution for 30 minutes.
(3 marks)

(d) If the rate of water uptake and the rate of water loss in a herbaceous plant is measured over the same period, results such as those shown in the graph may be obtained.

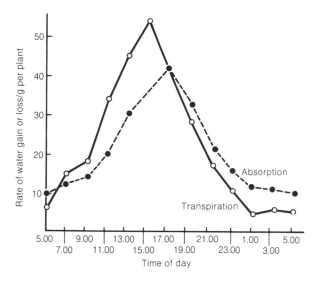

Explain what has happened to the water content of the plant over the period shown. (4 marks)

(e) Describe the likely appearance of the plant at 16.00 hours.
(1 mark)
(Total 20 marks)

Oxford and Cambridge Board June 1988, Paper I, No. 6

11. The graph shows the growth of a single leaf with its import (+) and export (−) of sucrose and phosphorus over a period of 40 days.

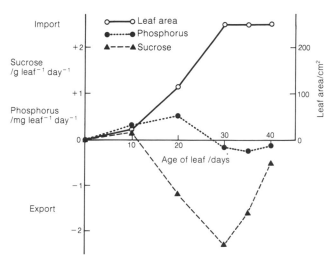

(a) Describe the pattern of growth in the leaf.
(3 marks)

(b) (i) How is sucrose transported to and from the leaf?
(4 marks)

  (ii) Describe and explain the movement of sucrose in the 40-day period.
(6 marks)

(c) (i) Suggest how phosphorus is likely to be transported to and from the leaf.
(3 marks)

  (ii) Describe and explain the movement of phosphorus in the 40-day period.
(4 marks)
(Total 20 marks)

Associated Examining Board June 1985, Paper II, No. 1

34 *Osmoregulation and excretion*

34.1 Introduction

Animals living under terrestrial conditions tend to lose water by evaporation. Those living in surroundings more concentrated (hypertonic) than their tissue fluids lose water by osmosis. Animals in a dilute (hypotonic) environment have to face the problem of water flooding into the body by osmosis. Both problems have led to structural and physiological adaptations in order to maintain the balance of water and solutes. These homeostatic processes are termed **osmoregulation**. Homeostasis is the subject of Chapter 36.

The complex chemical reactions which occur in all living cells produce a range of waste products which must be eliminated from the body in a process known as **excretion**. Most nitrogenous waste comes from the breakdown of excess proteins which cannot be stored in the body. The form of these excretory products is influenced partly by the availability of water for their excretion. Animals living under conditions of water shortage cannot afford to lose large volumes of water in order to remove their nitrogenous waste. If water is plentiful it may be used to facilitate excretion.

It is important not to confuse the terms excretion, secretion and elimination. **Excretion** is the expulsion from the body of the waste products of metabolism. **Secretion** is the production by the cells of substances useful to the body, such as digestive juices or hormones. **Elimination**, or egestion (see Section 24.5.7), is the removal of undigested food and other substances which have never been involved in the metabolic activities of cells.

34.2 Excretory products

All animals produce carbon dioxide as a waste product of aerobic respiration and the elimination of this is dealt with in Chapters 31 and 32. Other excretory products include bile pigments, water and mineral salts. However, in this section we shall concentrate on the variety of nitrogenous excretory products released by animals. There are three main waste products of nitrogenous metabolism: ammonia, urea and uric acid. No animal excretes one of these to the exclusion of the others but the predominance of one over the others is determined by three factors:

1. The production of enzymes necessary to convert ammonia into either urea or uric acid.

2. The availability of water in the habitat for the removal of the nitrogenous excretory material.

3. The animal's ability to control water loss or uptake by the body.

Many aquatic animals excrete mainly ammonia and are called

ammoniotelic. Other aquatic animals and some terrestrial forms excrete predominantly urea and are said to be **ureotelic**. The remaining terrestrial animals are **uricotelic**, excreting mainly uric acid.

34.2.1 Ammonia

Ammonia is derived from the breakdown of proteins and nucleic acids in the body. Ammonia is very toxic and is never allowed to accumulate within the body tissues or fluids. It is extremely soluble and diffuses readily across cell membranes. In spite of its toxicity it is the main excretory product of marine invertebrates and all freshwater animals. Marine invertebrates have body fluids which are isotonic with sea water, there is no water balance problem and ammonia rapidly diffuses out of the cells into the surrounding water. Freshwater animals have fluids which are hypertonic to their surroundings and have to cope with a constant influx of water by osmosis. This water must be steadily pumped out if the solute balance is to be maintained. Ammonia is removed dissolved in this water.

34.2.2 Urea

Urea, $CO(NH_2)_2$, comprises two molecules of ammonia linked by CO_2.

However, its synthesis in living tissues is much more complex than this simple equation suggests. An outline of the machinery involved was first presented by Hans Krebs in 1932 but modern biochemical and radioactive tracer techniques have served to show up additional stages. Urea is produced by a cyclic process known as the **urea** or **ornithine cycle**. This cycle is closely linked to the Krebs cycle (see Section 26.3) through oxaloacetate. The cycle has been especially studied in mammalian liver but it is a normal pathway for the metabolism of some amino acids and is thought to be widespread. In simplified terms the ornithine cycle is as follows.

The enzyme arginase is used to split the amino acid arginine into urea and ornithine. Ornithine is reconverted to arginine by the addition of two more nitrogen atoms in ammonia derived from the deamination of proteins. This addition of two nitrogen atoms occurs in two stages. The first molecule of ammonia enters the cycle in combination with carbon dioxide as carbamyl phosphate and joins with ornithine to form **citrulline**. The second molecule of ammonia enters the cycle as **aspartic acid**. Arginine is restored by the removal of fumaric acid. This fumaric acid can be used to regenerate aspartic acid, or it may enter the Krebs cycle via oxaloacetate.

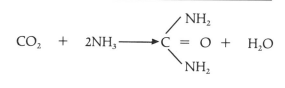

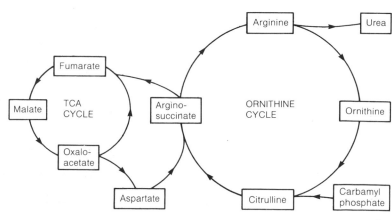

Fig. 34.1 Ornithine cycle and its relationship to the Krebs cycle

The steps in the ornithine cycle are catalysed by a series of enzymes, e.g. arginase, which catalyses the breakdown of arginine into ornithine and urea.

$$HN=C\begin{array}{l} NH_2 \\ NH \\ | \\ (CH_2)_3 + H_2O \\ | \\ CHNH_2 \\ | \\ COOH \end{array} \xrightarrow{\text{arginase}} \begin{array}{l} NH_2 \\ | \\ (CH_2)_3 \\ | \\ CHNH_2 \\ | \\ COOH \end{array} + \; C\begin{array}{l} NH_2 \\ =O \\ NH_2 \end{array}$$

arginine   ornithine   urea

The rest of the cycle is geared to the regeneration of arginine. Urea is much less toxic than ammonia and, although it is more soluble, less water is needed for its elimination because the tissues can tolerate higher concentrations of it. Indeed, as we shall see in Section 34.3.5, cartilaginous fish retain urea in the blood in order to raise their osmotic pressure above that of the sea water. Lungfish remain inactive (aestivate) during dry periods and at this time their tissues tolerate high levels of urea. Under normal conditions, with plentiful water, they excrete ammonia. This indicates that the excretion of urea is an adaptation to restricted water supplies and so it is not just aquatic organisms which excrete urea, it is also excreted by many terrestrial ones.

34.2.3 Uric acid

Uric acid is a more complex molecule than urea; it is a purine in the same group as adenine and guanine. Like urea, it involves the expenditure of quite considerable energy in its formation, but this is outweighed by the advantages it confers on an animal for which it is the main nitrogenous excretory product. Uric acid is virtually insoluble in water and is therefore non-toxic. It requires very little water for its removal from the body and it is therefore a suitable product for animals living in arid conditions e.g. terrestrial reptiles and insects. Containing little, if any, water, its storage within organisms does not greatly increase their mass. This is an advantage to flying organisms, e.g. birds and insects. It is removed as a solid pellet or thick paste.

The complex cycle by which uric acid is formed in the body will not be covered in this book; suffice it to say that it is part of a normal series of reactions for the production and interconversion of nucleotides.

34.2.4 Other nitrogenous excretory products

In addition to the three major nitrogenous excretory products – ammonia, urea and uric acid – there are many other substances which, especially in invertebrates, may form over 50% of the total nitrogenous excretion. Many of these have not been identified but three that have will be briefly mentioned here:

Guanine
This is even less soluble than uric acid and requires no water for its

elimination. It is the major excretory product of spiders.

Trimethylamine oxide
This may form up to one third of the excreted nitrogen of marine teleosts. It is soluble and non-toxic but the mechanism of its synthesis is not understood. It accumulates in the flesh of dead fish, giving them their characteristic odour.

Creatinine
Creatine, or its derivative creatinine, has been found in the urine of many invertebrates but only exceeds 10% of the total excreted nitrogen in some marine fishes.

34.3 Osmoregulation and excretion in animals

In this section we shall review osmoregulation and excretion in a number of different organisms. The animals considered live in a variety of habitats and so face different problems of water balance. Even where two animals live in the same habitat their abilities to cope with osmotic imbalance will vary, partly owing to their evolutionary history and behaviour and partly owing to their size and the complexity of their appropriate organ systems.

34.3.1 Protozoa

All protozoa are small unicells with a relatively large surface area to volume ratio. They are separated from their surroundings only by a plasma membrane through which carbon dioxide and the waste nitrogenous product, ammonia, can readily diffuse. Protozoa living in sea water are isotonic with their environment, have no net gain or loss of water and so do not have any osmoregulatory structures. Freshwater protozoa, on the other hand, are hypertonic to their surroundings and water constantly enters the cell by osmosis. If the protozoan cell is not to increase its volume this water must be removed by the constant filling and bursting at the surface of the **contractile vacuole**. These vacuoles are almost always found in freshwater protozoa, although their size, position and structure is extremely variable. The water which enters the cell by osmosis must be moved into the contractile vacuole across its bounding membrane. This process requires the expenditure of energy and it is significant that there are always large numbers of mitochondria around the contractile vacuoles. These mitochondria are the site of many of the respiratory enzymes and it has been found that treatment with cyanide, which prevents oxidative phosphorylation by inhibiting the cytochrome system, stops the activity of the contractile vacuole and leads to swelling of the cell. Ammonia may be found in the water expelled by the contractile vacuole but any excretory rôle is probably secondary since ammonia diffuses so readily through the cell membrane.

34.3.2 Insects

Arthropods are found in a wide range of habitats — terrestrial, freshwater and marine. They therefore show a corresponding range of excretory and osmoregulatory mechanisms. These adaptations invariably involve the development of specialized organs (e.g. antennal glands of many crustaceans; Malpighian tubules of insects)

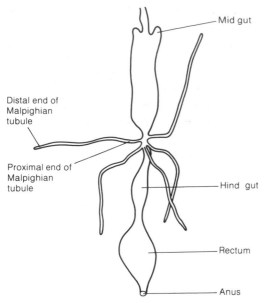

Fig. 34.2 Position of the Malpighian tubules

since the outer surface is covered by an impermeable cuticle and diffusion of excretory products through it is not possible.

Insects are among the most successful terrestrial animals. The problem of preventing water loss by evaporation is overcome by the waterproof waxy cuticle on the exoskeleton and a tracheal respiratory system (Chapter 31). The problem of water loss by excretion is overcome by the production of uric acid. Uric acid is concentrated in the **Malpighian tubules**. These form a bunch of blind-ending tubes which project into the blood-filled body cavity (haemocoel) from the junction of the mid gut and hind gut (see Fig. 34.2).

Insect cells release uric acid into the blood where it combines with potassium and sodium hydrogen carbonates and water to form potassium and sodium urate, carbon dioxide and water. Potassium and sodium urate are actively taken up by the Malpighian tubules and the water then enters by osmosis. The rate of uptake is increased by the writhing movements of the tubules due to the contraction of tiny muscles in their walls. As the soluble potassium and sodium urate pass down the tubules, they combine with carbon dioxide and water (from respiration) to form hydrogen carbonates and uric acid. At the proximal end of the Malpighian tubule the walls have many microvilli and it is here that hydrogen carbonates are actively reabsorbed into the haemocoel, lowering the osmotic pressure within the Malpighian tubule, so that water passes out by osmosis. There is a subsequent lowering of pH and concentration of the uric acid which precipitates out as crystals. These crystals pass into the rectum where they mix with the waste materials from the digestive system. Further water reabsorption takes place through the rectal epithelium, so that a very concentrated excretory product is eliminated from the body. These processes are illustrated in Fig. 34.3.

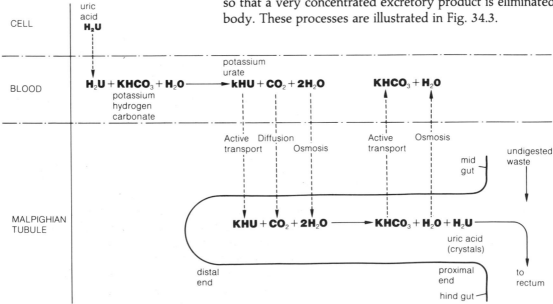

Fig. 34.3 Functioning of the Malpighian tubule

The production and concentration of uric acid by insects contributes greatly to the conservation of water and enables them to live in some of the hottest and driest parts of the world.

34.3.3 Freshwater teleosts

Freshwater teleosts have body fluids which are hypertonic to their surroundings and they are therefore subject to the osmotic uptake of water. Most of this water enters the fish through the highly permeable gills, the body itself being covered with impermeable scales and mucus. The excess water is removed from the body by the kidneys which have many large glomeruli. A large volume of glomerular filtrate is produced from which salts are selectively reabsorbed into the blood, resulting in the production of copious amounts of very dilute urine. There is some loss of ions in this urine but many more are lost across the highly permeable gill membranes. These must be replaced in the food and by active uptake from the environment by special chloride secreting cells in the gills. As well as being the site of gaseous exchange, water uptake and salt loss, the gills are also the main excretory organ of freshwater teleosts. Since water is plentiful, the nitrogenous excretory product is ammonia. Some of this ammonia is excreted by the kidneys, but as it is so soluble and diffuses readily, most of it is expelled by the gills.

34.3.4 Marine teleosts

Fish are thought to have evolved in the sea and then invaded fresh water. Present-day marine teleosts result from the secondary re-invasion of sea water. They therefore have body fluids which, instead of being isotonic with sea water like the marine invertebrates, are actually hypotonic to it. There is little movement of water through the scale-covered body but the fish is liable to water loss by osmosis across the highly permeable gills. In order to maintain sufficient water inside the body, marine teleosts drink sea water and secretory cells in the gut actively absorb the salts and transfer them to the blood. The chloride secretory cells in the gills remove the sodium and chloride ions from the blood and other ions like sulphate and magnesium are removed by the kidney.

The kidneys of marine teleosts are adapted to produce very small amounts of urine since the animal already has a problem of excessive water loss. There are only a few, small glomeruli and the excretory product is not ammonia but trimethylamine oxide which requires less water for its elimination.

Euryhaline fish
Most fish are only able to maintain their osmotic balance as long as the conditions around them remain more or less constant. However, there are some fish which are able to adapt their osmoregulatory mechanisms to allow them to move between fresh water and sea water. These are known as **euryhaline** fish (*eury* = 'broad'; *halo* = 'salt'). There are two groups of euryhaline fish. Those which, like the salmon, hatch in fresh water, mature in sea water and return to fresh water to spawn are called **anadromic** fish (*ana* = 'up'; *dromein* = 'to run'). **Catadromic** fish (*cata* = 'down'), like the eel, hatch in the sea, mature in fresh water and migrate to the sea again to spawn. When eels move into fresh water they start to take up water by osmosis but after several days an osmotic balance is achieved and salts are taken up by the gills. On returning to the sea they lose water rapidly and have to compensate for this by drinking sea water and expelling the excess salts via the gills. The chloride secretory cells of the gills therefore have an active transport mechanism which works in two directions: pumping in salts when in fresh water, and pumping them out when in sea water.

34.3.5 Marine elasmobranchs

Cartilaginous fish are thought to have evolved in fresh water and then moved into the sea. Although the concentration of mineral salts in the blood of elasmobranchs is similar to that in marine teleosts they are not hypotonic to sea water; instead, they are slightly hypertonic to it. They manage to raise the osmotic pressure of their blood by producing and retaining a certain level of urea and trimethylamine oxide. This is unusual since urea normally disrupts the hydrogen bonds between amino acids, denaturing proteins and inhibiting enzyme action. The presence of urea at a concentration often one hundred times that tolerated by other vertebrates actually appears to be necessary for many of the metabolic activities of elasmobranchs.

Excess water entering the body is excreted by the kidneys along with excess urea and trimethylamine oxide. However, the kidneys are not used for the elimination of excess sodium and chloride ions. Instead these are removed from the body fluids by active transport into the rectum. The high osmotic pressure of the body fluids is also maintained because the gills are relatively impermeable to nitrogenous waste, whose excretion can therefore be more accurately regulated by the kidney.

34.3.6 Birds

The main organ of nitrogenous excretion in birds, as in all vertebrates, is the kidney. Birds lose little water through their relatively impermeable skin which lacks sweat glands and is covered in feathers. However, their high metabolic rate means that a great deal of water is lost as a result of respiration and birds must conserve water as much as possible. The main excretory product is therefore uric acid, which requires little water for its elimination. The kidney has small glomeruli but long loops of Henle which provide a large surface area for the reabsorption of water from the glomerular filtrate into the blood. This results in urine which is very hypertonic to the body fluids. The urine passes into the cloaca where more water is reabsorbed from it and the faeces so that a semi-solid residue is expelled from the body. Marine birds such as gannets and penguins ingest large quantities of salt in sea water and the fish they eat. This is removed from the blood by special **salt glands** behind the orbit. These open into the nasal cavity and secrete a sodium chloride solution four times more concentrated than the body fluids.

34.3.7 Mammals

The majority of mammals are terrestrial and have urea as their main nitrogenous excretory product. This is expelled via kidneys which also regulate the water and salt balance of the body. Details of the structure and functions of a mammalian kidney are covered in Section 34.4.

34.4 The mammalian kidney

In all vertebrates the main organ of nitrogenous excretion is the kidney. Kidneys are composed of a number of basic units called **nephrons**. In mammals, these nephrons are particularly numerous, with long tubules for water reabsorption, and together they form a very compact kidney known as a **metanephric kidney**. This form is also found in reptiles and birds since it is the most adapted for the production of concentrated urine so necessary for terrestrial animals.

34.4.1 Structure of the nephron

The main regions of the mammalian nephron are shown in Fig. 34.4, with details of enlargements in Fig. 34.7. Basically it comprises a glomerulus and a long tubule with several clearly defined regions. The **glomerulus** is a mass of blood capillaries which are partially enclosed by the blind-ending region of the tubule called the **Bowman's capsule**. The blood supply to the glomerulus is from

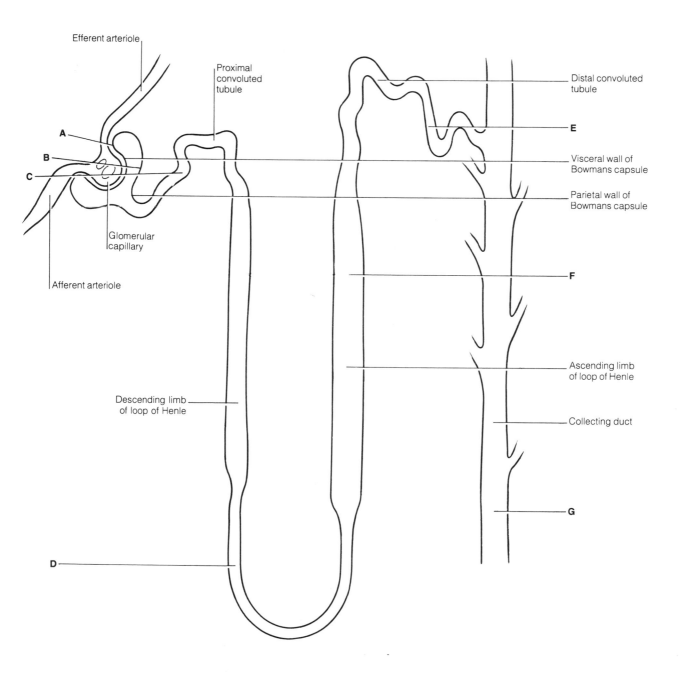

Fig. 34.4 Regions of the nephron

the afferent arteriole of the renal artery; blood leaves the glomerulus via the narrower efferent arteriole. The inner, or visceral, layer of the Bowman's capsule is made up of unusual cells called **podocytes** (see Fig. 34.7A) while the outer layer is unspecialized squamous epithelial cells. The remaining regions of the nephron are the proximal convoluted tubule, whose surface area is increased by the presence of microvilli, the descending and ascending limbs of the loop of Henle, which function as a counter-current multiplier, the distal convoluted tubule and the collecting duct.

Before the functions of these various regions are considered in detail, the gross structure and position of the kidneys will be considered.

34.4.2 Gross structure of the kidney

The paired kidneys are held in position in the abdominal cavity by a thin layer of tissue called the peritoneum and they are usually surrounded by fat. In man, each kidney is about 7–10 cm long and 2.5–4.0 cm wide, packed with blood vessels and an estimated one million nephrons. Each kidney is supplied with blood from the renal artery and drained by a renal vein. The urine which is produced by the kidney is removed by a ureter for temporary storage in the urinary bladder. A ring of muscle called a sphincter closes the exit from the bladder. Sense cells in the bladder wall are stimulated as the bladder fills, triggering a reflex action which results in relaxation of the bladder sphincter and simultaneous contraction of the smooth muscle in the bladder wall. The expulsion of urine from the body via the urethra is known as **micturition**. Although micturition is controlled by the autonomic nervous system, humans learn to control it by voluntary nervous activity.

Within each kidney there are a number of clearly defined regions. The outer region, or **cortex**, mainly comprises Bowman's capsules and convoluted tubules with their associated blood supply. This gives the cortex a different appearance from the inner **medulla** with its loops of Henle, collecting ducts and blood vessels. These structures in the medulla are in groups known as **renal pyramids** and they project into the **pelvis** which is the expanded portion of the ureter.

34.4.3 Functions of the nephron

Apart from being organs of nitrogenous excretion, the kidneys also play a major rôle in maintaining the composition of the body fluids in a more or less steady state, in spite of wide fluctuations in water and salt uptake. This dual rôle of excretion and osmoregulation is best studied by a detailed consideration of the functioning of one nephron the main regions of which are shown in Fig. 34.4 and in the eight detailed drawings of Fig. 34.7.

Ultrafiltration in the Bowman's capsule
The cup-shaped Bowman's capsule encloses a mass of capillaries, the **glomerulus**, originating from the afferent arteriole of the renal artery. The capillary walls are made up of a single layer of endothelial cells perforated by pores about 0.1 μm in diameter. The endothelium is closely pressed against the basement membrane which in places is the only membrane between the blood and the cavity of the Bowman's capsule (see Fig. 34.4). The blood pressure in the kidneys is higher than in other organs. This high pressure is maintained because in each Bowman's capsule the afferent arteriole has a larger

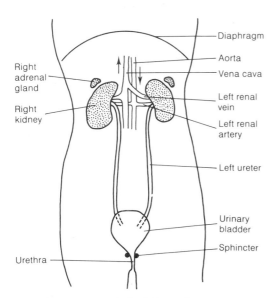

Fig. 34.5 Position of the kidneys in man

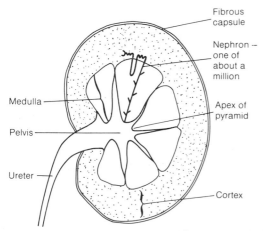

Fig. 34.6 Mammalian kidney to show position of a nephron (LS)

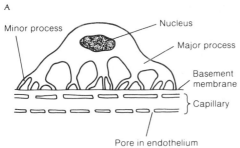

A

Minor process — — Nucleus

— Major process

— Basement membrane

— Capillary

Pore in endothelium

A Podocyte from visceral wall of Bowman's capsule

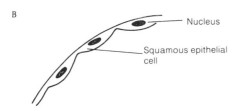

B

— Nucleus

— Squamous epithelial cell

B Parietal wall of Bowman's capsule

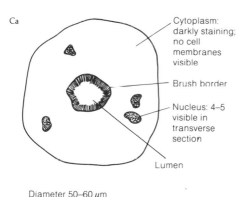

Ca

Cytoplasm: darkly staining; no cell membranes visible

Brush border

Nucleus: 4–5 visible in transverse section

Lumen

Diameter 50–60 μm

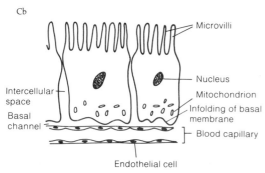

Cb

— Microvilli

— Nucleus

Intercellular space — Mitochondrion

Basal channel — Infolding of basal membrane

— Blood capillary

Endothelial cell

C (a) TS proximal convoluted tubule
(b) Detail of two cells from proximal convoluted tubule

Fig. 34.7 Detailed diagrams of the areas marked in Fig. 34.4 (cont.)

diameter than the efferent arteriole. As a result of this pressure, substances are forced through the endothelial pores of the capillary, across the basement membrane and into the Bowman's capsule by ultrafiltration. The glomerular filtrate contains substances with a relative molecular mass (RMM) less than 68 000, e.g. glucose, amino acids, vitamins, some hormones, urea, uric acid, creatinine, ions and water. Remaining in the blood, along with some water, are red and white corpuscles, platelets and plasma proteins which are too large to pass the filter provided by the basement membrane. Further constriction of the efferent arteriole in response to hormonal and nervous signals results in an increased hydrostatic pressure in the glomerulus and substances with an RMM greater than 68 000 may pass into the glomerular filtrate. This filtering process is extremely efficient. All the blood in the circulatory system passes through the kidneys every 4–5 minutes and as it does so, over one fifth of the volume of the plasma is filtered into the Bowman's capsule. The glomerular filtrate passes from the Bowman's capsule along the kidney tubule (nephron). As it does so, the fluid undergoes a number of changes, since the urine excreted has a very different composition from the glomerular filtrate. These differences are brought about primarily by selective reabsorption of substances useful to the body. The urine when compared with the glomerular filtrate will contain, for example, less glucose, amino acids and water and a relatively higher percentage of urea and other nitrogenous waste products.

Ultrafiltration is a passive process and selection of substances passing from the blood into the glomerular filtrate is made entirely according to relative molecular mass. Both passive and active processes are involved in the selective reabsorption of substances from the nephron. The composition is further altered by the active secretion of substances, such as creatinine, from the blood into the tubule. Without selective reabsorption man would produce about 180 dm³ of urine per day whereas the actual volume produced is approximately 1.5 dm³.

Proximal convoluted tubule

This is the longest region of the nephron. It comprises a single layer of epithelial cells, with numerous microvilli forming a brush border (see opposite). The base of each cell is convoluted where it is adjacent to a blood capillary and there are numerous intercellular spaces. Another notable feature of these cells is the presence of large numbers of mitochondria providing the ATP necessary for active transport. These cells are ideally adapted for reabsorption and over 80% of the glomerular filtrate is reabsorbed here, including all the food substances and most of the sodium chloride and water. Amino acids, glucose and ions diffuse into the cells of the proximal convoluted tubule and these are actively transported into the intercellular spaces from where they diffuse into the surrounding capillaries. The constant removal of these substances from the cells of the convoluted tubule causes others to enter from the lumen of the tubule by diffusion. The active uptake of sodium accompanied by appropriate anions, e.g. chloride, raises the osmotic pressure in the cells and water enters them by osmosis. About half the urea present in the tubular filtrate also returns to the blood by diffusion. Proteins of small molecular mass which may have been forced out of the blood in the Bowman's capsule are taken up at the base of the microvilli by pinocytosis. As a result of all this activity, the tubular filtrate is isotonic with blood in the surrounding capillaries.

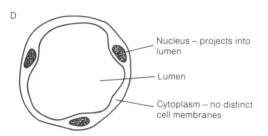

Nucleus – projects into lumen

Lumen

Cytoplasm – no distinct cell membranes

Diameter 15–20 μm

D *Loop of Henle – thin segment (TS)*

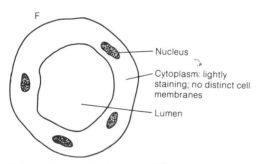

Nucleus

Cytoplasm: lightly staining; no distinct cell membranes

Lumen

F *Loop of Henle – thick segment (TS)*

Fig. 34.7 (cont.)

Loop of Henle

It is the presence of the loop of Henle which enables birds and mammals to produce urine which is hypertonic to the blood. The concentration of the urine is directly related to the length of the loop of Henle. It is short in semi-aquatic mammals which have a correspondingly narrow medulla, and extremely long in desert-dwelling mammals such as the desert rat *Dipodomys* which therefore has a wide medulla. *Dipodomys* produces a small volume of urine, ten times more concentrated than that produced in large volumes by a beaver. The loop of Henle is made up of two regions, the descending limb and the ascending limb. Throughout their length there runs a parallel blood capillary known as the vasa recta. Although the loop of Henle and the vasa recta are close, substances do not pass directly from one to the other. Instead they are also found in the interstitial region of the medulla which lies between them. It is important to remember this fact when considering the rôle of the loop of Henle.

The loop of Henle does not just directly reabsorb water from the filtrate into the blood stream. It causes a build-up of sodium chloride in the medulla and this results in the movement of water out of the collecting ducts by osmosis. In this section we shall consider how sodium chloride becomes concentrated in the medulla by a method known as a **countercurrent multiplier**.

The descending limb of the loop of Henle is narrow and its walls are permeable to water. The walls of the wider, ascending limb are thick and impermeable to water. Along the length of the ascending limb sodium and chloride ions are removed from the glomerular filtrate by active transport. The active transport of sodium is often referred to as a sodium pump and it involves the expenditure of energy. These ions pass into the spaces between the two limbs, making the interstitial fluid and the blood in the vasa recta very concentrated, especially towards the apex of the loop of Henle. Flow of blood in the vasa recta is slow and so the high salt concentration which builds up in this region of the medulla is maintained. This high concentration causes water to be drawn out of the descending limb of the loop of Henle by osmosis. The water moves straight into the vasa recta and so does not dilute the interstitial fluid of the medulla. The difference in the osmotic concentration between the ascending and descending limbs at any one level is small but over the whole length of the loop these have a cumulative effect. The concentration of the filtrate at the apex of the loop is therefore much greater than the concentration at either end of it. The longer the loop, the greater the difference in concentration. The fact that the fluid in the two limbs flows in opposite directions, and the effect is cumulative, gives rise to the term countercurrent multiplier.

A countercurrent mechanism also operates between the two limbs of the vasa recta whose walls are freely permeable to water, ions and urea. As well as helping to maintain a high concentration around the apex of the loop of Henle this mechanism also allows the osmotic concentration of the plasma leaving the kidney to remain more or less constant regardless of the osmotic concentration of the blood entering it. The way this works is that as the descending capillary enters the medulla it encounters an increasingly concentrated interstitial fluid. This causes water to leave the plasma by osmosis and sodium chloride and urea to enter it. As the ascending limb leaves the medulla the surroundings become gradually less concentrated, water re-enters and sodium chloride and urea leave the plasma. All these movements are passive, requiring no expenditure of energy.

It may seem strange that the concentration of the filtrate leaving the loop of Henle is lower than that entering it. It must be remembered that the build-up of sodium chloride in the region surrounding the apex of the loop will enable the filtrate to be further concentrated by reabsorption of water from it in the collecting duct.

Distal convoluted tubule

The cells in this region are very similar to those of the proximal convoluted tubule, having a brush border and numerous mitochondria. The permeability of their membranes is affected by hormones (see Section 34.5) and so precise control of the salt and water balance of the blood is possible. The distal convoluted tubule also controls the pH of the blood, maintaining it at 7.4 by excreting hydrogen ions and retaining hydrogen carbonate ions if the pH falls, and the reverse if it rises. As a result the pH of the urine may vary between 4.5 and 8.5.

Collecting duct

The permeability of the walls of the collecting duct, like those of the distal convoluted tubule, are affected by hormones. This hormonal effect, together with the hypertonic interstitial fluids built up by the loop of Henle in the medulla, determine whether hypotonic or hypertonic urine is released from the kidney.

If the walls of the collecting duct are water-permeable, water leaves the ducts to pass into the hyperosmotic surroundings and concentrated urine is produced. If the ducts are impermeable to water the final urine will be less concentrated. Hormonal control of the permeability of the walls of the collecting duct to water will be considered in the next section

The mechanism by which urine is concentrated is illustrated in Fig. 34.8.

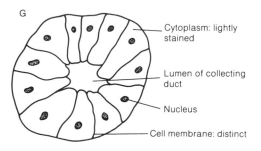

E Distal convoluted tubule (TS)

Diameter 20–50 μm

- Nucleus
- Lumen – wider than in proximal convoluted tubule
- Cytoplasm – lightly staining: no distinct cell membranes

G Collecting duct (TS)

Diameter 50–60 μm

- Cytoplasm: lightly stained
- Lumen of collecting duct
- Nucleus
- Cell membrane: distinct

Fig. 34.7 (cont.)

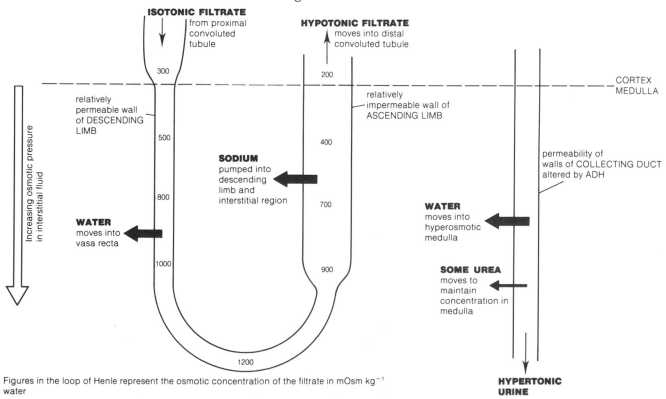

Figures in the loop of Henle represent the osmotic concentration of the filtrate in mOsm kg⁻¹ water

Fig. 34.8 Counter-current multiplier of the loop of Henle

34.5 Hormonal control of osmoregulation and excretion

If the kidney is to regulate the amount of water and salts present in the body, very precise monitoring systems are required. The two hormones ADH and aldosterone are particularly important in this respect.

34.5.1 Antidiuretic hormone (ADH)

Antidiuretic hormone (ADH) affects the permeability of the distal convoluted tubule and collecting duct.

A rise in blood osmotic pressure may be caused by any one of, or combination of, three factors:

1. Little water is ingested.

2. Much sweating occurs.

3. Large amounts of salt are ingested.

The rise in blood osmotic pressure is detected by **osmoreceptors** in the hypothalamus which results in nerve impulses passing to the posterior pituitary gland which releases ADH. ADH increases the permeability of the distal convoluted tubule and collecting duct to water. This water passes into the hyperosmotic medulla and a more concentrated (hypertonic) urine is released from the kidney.

ADH also increases the permeability of the collecting duct to urea which passes into the medulla, increasing the osmotic concentration and causing more water to be lost from the descending loop of Henle. If the osmotic pressure of the blood falls owing to

1. large volumes of water being ingested

2. little sweating

3. low salt intake

then ADH production is inhibited and the walls of the distal convoluted tubule and collecting duct remain impermeable to water and urea. As a result, less water is reabsorbed and hypotonic urine is released. Anyone who is unable to produce sufficient levels of ADH will produce large volumes of very dilute urine, whatever their diet. This condition is termed **diabetes insipidus**. ADH is also referred to in Section 37.2. Fig. 34.9 illustrates the regulation of ADH production.

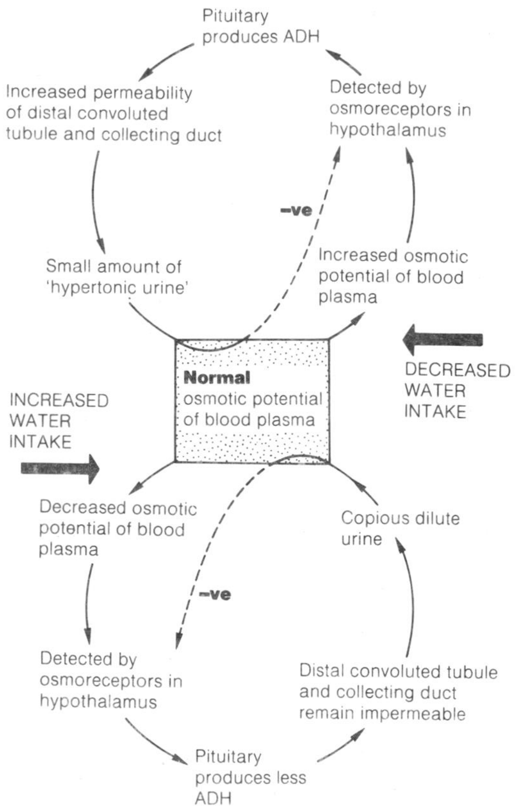

Fig. 34.9 Regulation of ADH production

34.5.2 Aldosterone

Aldosterone is the hormone responsible for maintaining a more or less constant sodium level in the plasma and it has a secondary effect on water reabsorption. The control of aldosterone production is very complex.

Any loss of sodium which causes a decrease in blood volume causes a group of secretory cells lying between the afferent arteriole and the distal convoluted tubule to release an enzyme. These cells are known as the **juxtaglomerular complex** and the enzyme they release is **renin**. Renin causes a plasma globulin produced by the liver to form the hormone **angiotensin**. It is this hormone which stimulates the release of aldosterone from the adrenal cortex.

Aldosterone causes sodium ions to be actively taken up from the glomerular filtrate into the capillaries which surround the tubule. This uptake will be accompanied by an osmotically equivalent volume of water, thus restoring the sodium level of the plasma and the volume of the blood.

Further details of aldosterone are given in Section 37.6.

34.6 Kidney failure – Dialysis and transplants

Kidneys may fail as a result of damage or infection. Upon the loss of one kidney the remaining one will adapt to undertake the work of its partner, but the loss of both is inevitably fatal. Survival depends upon either regular treatment on a kidney machine or a transplant. A kidney machine carries out **dialysis**, a process in which the patient's blood flows on one side of a thin membrane while a solution, called the dialysate, flows in the opposite direction on the other side. This counter-current flow ensures the most efficient exchange of material across the membrane (Section 31.2.5). As the membrane is permeable to small molecules such as urea, this waste product will diffuse from the blood where it is relatively highly concentrated to the dialysate where its concentration is lower. To prevent useful substances like glucose and salts, which are also highly concentrated in blood, diffusing out, the dialysate's composition is the same as that of normal blood. This means that any substance which is in excess, e.g. salts, will also diffuse out until they are in equilibrium with the dialysate. Large molecules such as blood proteins are too large to cross the membrane and there is therefore no risk of them being lost to the dialysate.

A patient's blood needs to pass through the kidney machine many times to ensure the removal of all wastes. Thus, it is necessary for dialysis to take place for up to ten hours every few days. While wastes accumulate in the blood when the patient is away from the machine, adherence to a strict diet ensures they do not build up to a dangerous level before the next treatment. Fig. 34.10 on the next page illustrates the mechanism of kidney dialysis.

While the kidney machine is an invaluable life-saver, it has many drawbacks. Not only does a considerable time have to be spent connected to the machine, a rigid diet has to be maintained and there is always the risk of anaemia, infection or bone disease. The cost of each treatment is such that there are insufficient machines for all who need them. The preferred solution for many is therefore a **kidney transplant**: a failed kidney must be replaced by a healthy one from a human donor. The donated kidney must be matched so that it is as similar as possible to the failed one since this reduces the chance of it being rejected by the recipient's immune system. A close relative is more likely to have a compatible kidney and therefore live donors are often used. Here a person donates a kidney, safe in the knowledge that he/she can live a normal life with the remaining one. The donor is, however, vulnerable since a disease in the remaining kidney could make dialysis necessary, and there is always a risk in any operation. Against this is the satisfaction of allowing someone to lead a near normal life, free from all the constraints dialysis imposes. However, less than 15% of transplants come from live donors.

The remaining transplants involve the use of healthy kidneys from people who die as a result of other causes. A road accident victim may, for example, provide two functional kidneys. The kidney must be removed within an hour of death, cooled to delay deterioration, and be transplanted within 12 hours. As permission is needed to remove any organ, and as such a request made of a distressed relative is difficult, many people carry donor cards. Their owners sign to say that upon their death they consent to the use of their kidneys (and often other organs) being used for transplants. The card must of course, be carried by the owner at all times to avoid any delay in removing organs.

There are around 2000 kidney transplants in the UK each year, with around twice that number of people awaiting a suitable donor.

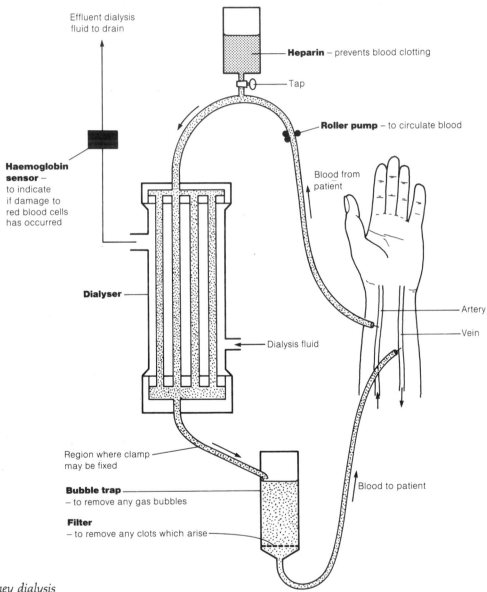

Fig. 34.10 The mechanism of kidney dialysis

34.7 Excretion in plants

Plants, being primary producers, synthesize organic compounds according to their requirements. There are no excess proteins and any that are broken down to amino acids are recycled as new proteins. Plants do not therefore have nitrogenous waste products in the same way as animals. The carbon dioxide and water released by respiration are used for photosynthesis and so these do not need to be excreted during daylight. The only gaseous waste product may be oxygen which on bright sunny days is produced by photosynthesis at a faster rate than it can be used in respiration. Non-nitrogenous organic wastes may be formed in plants but are never subject to a regular excretory process. Instead they are stored in non-living parts of the plant, e.g. heartwood, where they do not affect the living tissues. Some mineral ions which are taken up in excess may not be used, e.g. calcium. This may combine with oxalic acid or pectic acid to form crystals of calcium oxalate or calcium pectate which can be safely stored in plant cells. Other excess ions, such as manganese and iron and organic acids like tannic acid, may accumulate in leaves, petals, fruits and seeds. They will then be lost from the plant when the leaves and petals fall or when fruits and seeds are dispersed.

34.8 Questions

1. What is the importance of osmotic control in animals? Describe the methods by which the water and salt content of the body is regulated in mammals and fish.

(20 marks)

London Board June 1985, Paper II, No. 8

2. (a) Outline the ways in which water is lost and gained by terrestrial animals. *(4 marks)*
(b) Outline the ways in which these water losses are controlled in terrestrial animals by:
(i) structural, *(3 marks)*
(ii) physiological, *(7 marks)*
(iii) behavioural adaptations. *(4 marks)*
(c) To what extent do terrestrial plants have similar adaptations to animals for the prevention of excessive water loss? *(2 marks)*
(Total 20 marks)

Northern Ireland Board June 1988, Paper II, No. 4

3. (a) Describe the role of the mammalian kidney in the following processes.
(i) Osmoregulation
(ii) Excretion *(14 marks)*
(b) Explain the significance of the following as excretory products.
(i) Ethanol in yeast
(ii) Ammonia in freshwater teleosts *(6 marks)*
(Total 20 marks)

London Board January 1989, Paper II, No. 2

4. (a) Make a labelled diagram to show the form of a nephron, including its blood supply, in a mammalian kidney. *(3 marks)*
(b) Explain how ultrafiltration occurs within the kidney. *(7 marks)*
(c) Describe the role of the kidney as a homeostatic organ of
(i) salt balance,
(ii) acid-base balance. *(10 marks)*
(Total 20 marks)

Joint Matriculation Board June 1988, Paper IIB, No. 5

5. The diagram below shows the water balance of the desert seed-eating kangaroo rat at two different environmental humidities (R.H.) at 25 °C.

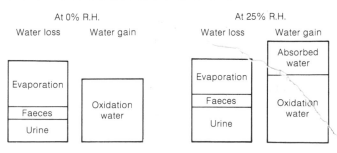

(a) Calculate the percentage of water loss by evaporation at 0% R.H. Show your working. *(2 marks)*
(b) Why was less water lost by evaporation at 25% R.H.? *(1 mark)*
(c) What is meant by *water gain (oxidation water)*? *(3 marks)*

(d) 'The reproduction rate of larger animals is lower than in smaller animals. Thus in regions where there is little or no danger of desiccation, smaller animals have a selective advantage. But in desert regions, larger animals have the advantage'. Comment on this statement.

(3 marks)

(e) 'A camel may wander around in the searing heat of the desert without water, losing a quarter of its body weight. The dehydrated camel can then drink 200 litres or so of water. Variations in the animal's water balance are potentially dangerous because of drastic changes in the concentration of the blood. The camel's kidneys are remarkable; they can produce a dark syrupy urine at one moment and a watery almost colourless urine within half an hour of drinking water' (Report in the *New Scientist* Dec. 1986)

(i) Why are drastic changes in the concentration of the blood potentially dangerous? *(3 marks)*

(ii) The concentration of urine in the camel is influenced by the activities of the pituitary gland and the anti-diuretic hormone (ADH). How do these activities account for the fluctuations in the concentration of the urine in the camel as reported in the passage above? *(4 marks)*

(iii) It is claimed that rehydration in the camel is associated with increased salt absorption from the alimentary canal. Why may these two events be associated? *(3 marks)*

(Total 19 marks)

Oxford and Cambridge Board June 1988, Paper I, No. 8

6. The following figures show the progress of a patient who developed acute kidney failure after emergency heart surgery. The operation was carried out on Thursday, 18th October. Blood dialysis began on the Saturday afterwards and was needed over the next 13 days. As a result of treatment the patient eventually recovered full health.

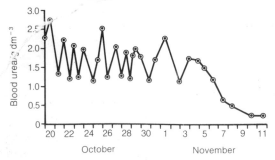

(a) (i) Using this information, state how many times the patient received dialysis.

(ii) Explain how you arrived at this answer.

(2 marks)

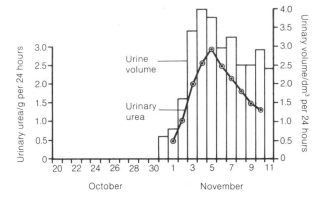

(b) On what date did urine production start again?

(1 mark)

(c) The patient's mass fell by 7 kg between 20 October and 11 November. Suggest **two** possible reasons for this change in mass.

(2 marks)

The efficiency of the working of the kidneys can be judged from renal : plasma ratios. For urea, for example, this is calculated as the concentration of urea in urine divided by the concentration of urea in the blood. The table shows how this was calculated for November 5th.

Date	Urea concentration in urine /g dm^{-3}	Blood urea /g dm^{-3}	Renal:plasma ratio for urea
November 5th	$\frac{2.9}{3.8} = 0.76$	1.5	0.5
November 10th			

(d) (i) Complete the table to show the missing values for November 10th.

(ii) On which of these two days were the kidneys working more efficiently? Explain your answer. *(4 marks)*

(Total 9 marks)

Joint Matriculation Board June 1989, Paper IA, No. 4

7. (a) Amino acid metabolism in animals leads to the formation of nitrogenous waste products. Explain briefly why nitrogenous waste does not normally occur in plants. *(2 marks)*

(b) Analysis of the glomerular filtrate and the urine of a mammal yielded the following mean daily values:

	Glomerular filtrate	Urine
Urea	60 g	35 g
Water	180 dm³	1.5 dm³

(i) 150 dm³ of water is reabsorbed by the proximal tubules. Calculate the percentage of water from the filtrate that is reabsorbed elsewhere.

(ii) Name **two** other regions of the tubules where this further reabsorption of water takes place. *(3 marks)*

(c) In mammalian kidneys, the relative length of the loops of Henle shows considerable variation from one species to another. Suggest, with reasons, the type of habitat in which you would expect to find species with extremely long loops of Henle. *(3 marks)*

(d) Nitrogenous waste in animals may occur as ammonia, urea or uric acid. Ammonia is very soluble and highly toxic; urea is soluble and mildly toxic; uric acid is insoluble and non-toxic. The table shows the percentages of these three compounds in the urine of four different animals.

	Ammonia	Urea	Uric acid
Freshwater fish	56	6	0
Seawater fish	7	81	0
Lizard	0	0	91
Bird	3	4	72

(i) Offer an explanation for the difference in the main excretory compound in freshwater and seawater fish. *(5 marks)*

(ii) Both lizards and birds are terrestrial, egg-laying animals. How do these characteristics relate to the nature of their main excretory products? *(2 marks)*

(Total 15 marks)

Welsh Joint Education Committee June 1985, Paper AI, No. 10

8. An investigation was carried out to determine the rates of flow and the composition of fluids in human kidneys. These were measured at positions A, B, C and D, shown on the diagram below, which represents a nephron and associated blood vessels.

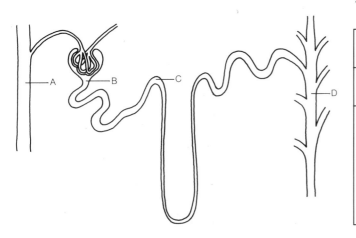

The results are given in the table below.

Position	Total flow rate in cm³ per min	Solute concentrations in g per 100 cm³		
		Protein	Glucose	Urea
A	1000	7.5	0.1	0.03
B	100	0.0	0.1	0.03
C	20	0.0	0.0	0.15
D	1	0.0	0.0	1.80

(a) Explain how the following are brought about.
  (i) The change in flow rate between B and D *(2 marks)*
  (ii) The change in protein concentration between A and B *(2 marks)*
  (iii) The change in glucose concentration between B and C *(2 marks)*
  (iv) The change in urea concentration between B and C *(2 marks)*
  (v) The change in urea concentration between C and D *(2 marks)*

(b) The smallest plasma protein molecules have molecular weights of about 69 000. Haemoglobin has a molecular weight of about 65 000. If red blood cells are damaged in the bloodstream, haemoglobin may appear in the urine.
Comment on the significance of these observations. *(2 marks)*

(Total 12 marks)

London Board June 1988, Paper I, No. 12

9. In an experiment, individuals of two different species A and B of *Amoeba* were transferred from their natural habitats to different dilutions of sea water. Each individual was given time to adjust to its new environment, and then the rate of contraction of its contractile vacuole was studied. The following results were obtained.

Concentration of sea water (Normal sea water = 100%)	Number of vacuolar contractions per hour	
	Species A	Species B
5%	82	20
10%	74	63
15%	65	64
20%	58	56
30%	34	31
40%	14	13
50%	0	6
60%	0	0

(a) Plot the results of the experiment as a graph.
(5 marks)

(b) (i) What is the function of a contractile vacuole?
(1 mark)

(ii) Explain how it carries out this function.
(4 marks)

(c) Explain, by reference to the data, the difference in vacuolar contraction in the two species of *Amoeba* when placed in the higher concentrations of sea water. (3 marks)

(d) What information may be deduced about the natural habitats of the two species from the rates of vacuolar contractions? (4 marks)

(Total 17 marks)

London Board June 1986, Paper I, No. 8

10. Read through the following account of kidney function and then write on the dotted lines the most appropriate word or words to complete the account.

Blood entering the kidney from the .... passes into an afferent arteriole which divides to form the .... inside the cup of a .... Much of the blood is forced into the tubule by the process of ultrafiltration. Only blood cells and large molecules, such as ...., and some fluid remain in the blood vessel. The filtrate is a watery fluid rich in food substances such as glucose and .... Normally all the glucose is reabsorbed by the blood vessels surrounding the proximal tubule though a low level of the hormone .... may cause some to be excreted in the urine. Most of the .... are also reabsorbed causing a passive movement of most of the water out of the tubule due to the higher osmotic potential (osmotic pressure) now exerted by the blood. Further reduction in the water content of the filtrate takes place when .... diffuse into the loop of Henle and are later pumped out, so producing a .... filtrate. Any increase in the .... of the blood stimulates receptor cells in the .... of the brain which results in the secretion of .... by the pituitary gland. This hormone .... the permeability of the wall of the tubule so that .... water leaves the tubule.

(Total 14 marks)

London Board June 1983, Paper I, No. 4

A multipolar neurone from the cerebellum (opposite)

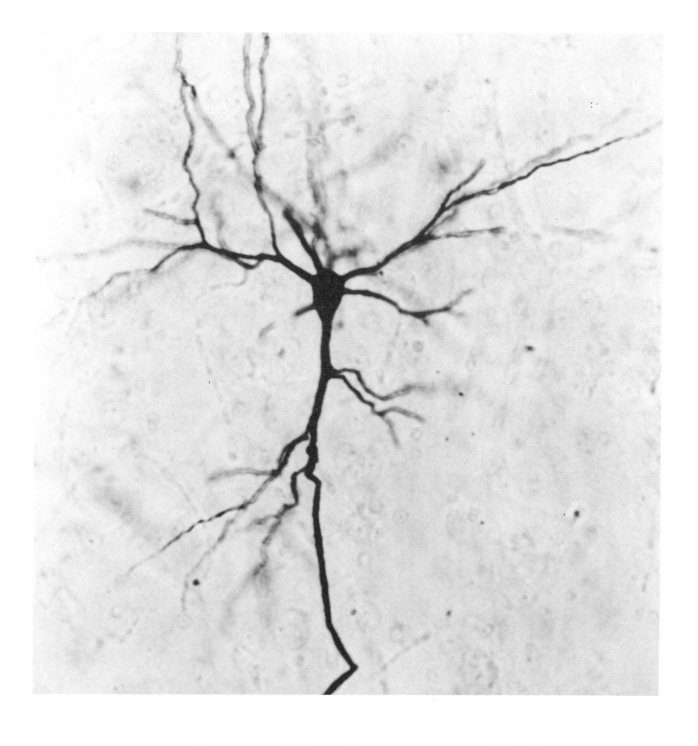

35 *How control systems developed*

The ability to respond to stimuli is a characteristic of all living organisms. While many stimuli originate from outside, it is also necessary to respond to internal changes. In a single celled organism, such responses are relatively simple. No part of it is far from the medium in which it lives and so it can respond directly to environmental changes. The inside of a single cell does not vary considerably from one part to another and so there are few internal differences to respond to.

With the development of multicellular organisms came the differentiation of cells which specialized in particular functions. With specialization in one function came the loss of the ability to perform others. This division of labour, whereby different groups of cells each carried out their own function, made the cells dependent upon one another. Cells specializing in reproduction, for example, depend on other cells to obtain oxygen for their respiration, yet others to provide glucose and others to remove waste products. These different functional systems must be coordinated if they are to perform efficiently. If, for example, an animal needs to exert itself in order to capture its food, the muscular activity involved must itself be coordinated. The locomotory organs need to operate smoothly and efficiently, with each muscle contracting at exactly the correct time. In addition, more oxygen and glucose will be required and an increased amount of carbon dioxide will need to be removed from the tissues. If breathing is increased so that the oxygen concentration of the blood rises, it is essential that the heart increases its output accordingly. Without coordination between the two systems an increased effort by one could be neutralized by the other. No bodily system can work in isolation, but all must be integrated in a coordinated fashion.

There are two forms of integration in most multicellular animals: nervous and hormonal. The nervous system permits rapid communication between one part of an organism and another, in much the same way as a telephone system does in human society. The hormonal system provides a slower form of communication and can be likened to the postal system. Both systems need to work together. A predator, for example, may be detected by the sense organs, which belong to the nervous system, but in turn cause the production of adrenaline, a hormone. While the nervous system coordinates the animal's locomotion as it makes its escape, the adrenalin ensures that an increased breathing rate and heart-beat supply adequate oxygen and glucose to allow the muscles to operate efficiently. The link between these two coordinating systems is achieved by the **hypothalamus**. It is here that the nervous and hormonal systems interact.

All organisms, plant and animal, must respond to environmental changes if they are to survive. To detect these changes requires sense organs. Those detecting external changes are located on the surface of the body and act as a vital link between the internal and external environments. Many other sense cells are located internally to provide information on a constantly changing internal environment. In responding to stimuli, an organism usually modifies some aspect of its functioning. It may need to produce enzymes in response to the presence of food, become sexually aroused in response to certain behaviour by a member of the opposite sex or move away from an unpleasant stimulus. In most cases the organ affecting the change is some distance away from the sense cell detecting the stimulus. A rapid means of communication between the sense cell and the effector organ is essential. In animals the nerves perform this function.

The stimuli received by many sense organs, e.g. the eyes, are very complex and require widely differing responses. The sight of a female of the same species may elicit a totally different response from the sight of a male of the same species. Each response involves different effector organs. The sense organs must therefore be connected by nerves to all effector organs, in much the same way as a telephone subscriber is connected to all other subscribers. One method is to have an individual nerve running from the sense organ to all effectors. Clearly this is only possible in very simple organisms where the sense organs respond to a limited number of stimuli and the number of effectors is small. The nerve nets of coelenterates work in this way.

Large, complex organisms require a different system, because the number of sense organs and effectors is so great that individual links between all of them is not feasible. Imagine having a separate telephone cable leading from a house to every other subscriber's house in Britain, let alone the world. Animals developed a **central nervous system** to which every effector and sense organ has at least one nerve connection. The central nervous system (brain and spinal cord) acts like a switchboard in connecting each incoming stimulus to the appropriate effector. It works in much the same way as a telephone exchange, where a single cable from a home allows a subscriber to be connected to any other simply by making the correct connections at various exchanges.

Where then should the brain be located? The development of locomotion in animals usually resulted in a particular part of the animal leading the way. This anterior region was much more likely to encounter environmental changes first, e.g. changes in light intensity, temperature, pH etc. It was obvious that most sense organs should be located on this anterior portion. There would be little point locating them in the posterior region, as a harmful substance would not be detected until it had already caused damage to the anterior of the animal. Most sensory information therefore originated at the front. To allow a rapid response, the 'brain' was located in this region. This led, in many animals, to the formation of a distinct head, **cephalization**, concerned primarily with detection and interpretation of stimuli. It was still essential for the brain to receive stimuli from the rest of the animal and to communicate with effector organs throughout the body. An elongated portion of the CNS therefore extends the length of most animals. In vertebrates this is the spinal cord.

With increasingly complex stimuli being received, the brain developed greater powers of interpretation. In particular it developed the ability to store information about previous experiences in order to assist it in deciding on the appropriate response to a future situation. With this ability to learn came the capacity to use previous experience and even to make responses to situations never previously encountered. Thus intelligence developed.

The hormone or endocrine system is concerned with longer-term changes, especially in response to the internal environment. It is an advantage to maintain a relatively constant internal environment. Not only can chemical reactions take place at a predictable rate, but the organism also acquires a degree of independence from the environment. It is no longer restricted to certain regions of the earth but can increase its geographical range. It does not have to restrict its activities to particular periods of the day, or seasons, when conditions are suitable. The maintenance of a constant internal environment is called **homeostasis** and is largely controlled by hormones. In the same way that the brain coordinates the nervous system, the activities of the endocrine system are controlled by the **pituitary gland**.

Responding to changes in the internal and external environments is no less important to survival in plants. Because they lack contractile tissue and do not move from place to place independently, they have no need for very rapid responses. There is therefore no nervous system or anything equivalent to it. Plant responses are hormonal. Their movements are as a result of growth, rather than contractions, and as a consequence are much slower than those of animals.

36 *Homeostasis*

We have seen that matter tends to assume its lowest energy state. It tends to change from an ordered state to a disordered one, i.e. tends towards high entropy. The survival of biological organisms depends on their ability to overcome this tendency to disorderliness. They must remain stable. This need for constancy was recognized in the nineteenth century by Claude Bernard. He contrasted the constancy of the fluid which surrounds all cells (*milieu interieur*) with the ever changing external environment (*milieu exterieur*). Bernard concluded: '*La fixité du milieu interieur est la condition de la vie libre.*' (The constancy of the internal environment is the condition of the free life.)

By bathing cells in a fluid, the tissue fluid, whose composition remains constant, the chemical reactions within these cells can take place at a predictable rate. Not only are the cells able to survive but they can also function efficiently. The whole organism thus becomes more independent of its environment.

The term **homeostasis** (*homoio* = 'same'; *stasis* = 'standing') was not coined until 1932. It is used to describe all the mechanisms by which a constant environment is maintained. Some examples of homeostatic control have already been discussed, for example osmoregulation in Chapter 34. Further examples are examined in this chapter and the following one.

36.1   Principles of homeostasis

Before examining the detailed operation of homeostatic systems, it is necessary to look at the fundamental principles common to them. Organisms are examples of open systems, since there is exchange of materials between themselves and the environment. Not least, they require a constant input of energy to maintain themselves in a stable condition against the natural tendency to disorder. The maintenance of this stability requires control systems capable of detecting any deviation from the usual and making the necessary adjustments to return it to its normal condition.

36.1.1 Control mechanisms and feedback

Cybernetics (*cybernos* = 'steersman') is the science of control systems, i.e. self-regulating systems which operate by means of feed-back mechanisms.

The essential components of a control system are:

1. Reference point – the set level at which the system operates.

2. Detector – signals the extent of any deviation from the reference point.

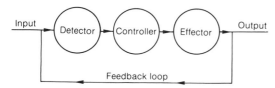

Fig. 36.1 Principal components of a typical control system

3. Controller – coordinates the information from various detectors and sends out instructions which will correct the deviation.

4. Effector – brings about the necessary change needed to return the system to the reference point.

5. Feedback loop – informs the detector of any change in the system as a result of action by the effector.

The relationship between these components is given in Fig. 36.1.

An everyday example of such a control system occurs in the regulation of a central heating system in a home, where the various components are:

1. Reference point – temperature determined by the occupant, e.g. 20 °C, and set on the thermostat.

2. Detector – the thermostat which constantly monitors the temperature of the room in which it is situated.

3. Controller – the programmer which can be set to turn the heating on and off at set times. It is connected to the boiler, hot water cylinder, circulation pump and thermostat.

4. Effector – the boiler, circulation pump, radiators and associated pipework.

5. Feedback loop – the movement of air within the room.

If the temperature of the room falls below 20 °C, the thermostat sends an electrical message to the programmer. The programmer coordinates this information with that in its own programme, i.e. whether the heating is set to operate at this particular time of day. If it is set to operate, it sends appropriate electrical messages which turn on the boiler and the circulation pump. Hot water flows around the central heating system to the radiator in the room. The heat from this radiator warms the air in the room which circulates until it reaches the thermostat. Once the temperature of this air reaches 20°C, the thermostat ceases to send information to the programmer which then turns off the circulation pump. As the feedback causes the system to be turned off, it is called **negative feedback**. It is possible to have positive feedback systems. Although these are rare in living organisms one example is described in Section 38.1.2.

A similar system operates to control body temperature in birds and mammals. Temperature detectors in the skin provide information on changes in the external temperature which is conveyed to the hypothalamus of the brain, which acts as the controller. This initiates appropriate corrective responses in effectors, such as the skin and blood vessels, in order to maintain the body temperature constant. Details of these processes are given in the following section.

36.2 Temperature control

The temperature of environments inhabited by living organisms ranges from 90 °C in hot springs to −40 °C in the Arctic. Most organisms, however, live in a narrow range of temperature from 10–30 °C. To survive, most animals need to exert some control over their body temperature. This regulation of body temperature is called **thermoregulation**. In all organisms heat may be gained in the two main ways.

1. Metabolism of food.

2. Absorption of solar energy – This may be absorbed directly or indirectly from

(a) heat reflected from objects;
(b) heat convected from the warming of the ground;
(c) heat conducted from the ground.

Heat may be lost in four main ways:

1. Evaporation of water, e.g. during sweating.

2. Conduction from the body to the ground or other objects.

3. Convection from the body to the air or water.

4. Radiation from the body to the air, water or ground.

36.2.1 Ectothermy and endothermy

The majority of animals obtain most of their heat from sources outside the body. These are termed **ectotherms**. The body temperature of these animals frequently fluctuates in line with environmental temperature. Animals whose temperature varies in this way are called **poikilotherms** (*Poikilos* = 'various'; *thermo* = 'heat'). While most ectotherms have a body temperature approaching that of the environment, it is rarely equal to it. During exercise, for example, the metabolic heat produced may raise the body temperature above that of the environment. Moist-bodied ectotherms frequently have temperatures a little below that of the environment owing to the cooling effect of evaporation. Ectotherms do attempt to regulate their temperatures within broad limits. The methods used are largely behavioural.

Mammals and birds maintain constant body temperatures irrespective of the environmental temperature. As their heat is derived internally, by metabolic activities, they are called **endotherms**. As the temperature of the body remains the same, they are sometimes called **homoiotherms** (*homoio* = 'same'; *thermo* = 'heat'). The body temperature of endotherms is usually in the range 35–44 °C.

The higher the body temperature the higher the metabolic rate of the animal. This is especially important for birds where the energy demands of flight make a high metabolic rate an advantage. Most birds therefore have body temperatures in the range 40–44 °C. The problem with a high body temperature is that the environment is usually cooler, and heat is therefore continually lost to it. The higher the body temperature, the greater the gradient between internal and external temperatures and so the more heat is lost. Mammals, with less dependence on a very high metabolic rate than birds, therefore have body temperatures in the range 35–40 °C to minimize their total heat loss. The body temperatures of all endotherms are something of a compromise, balancing the advantage of a high temperature, in increasing metabolic activity, and the disadvantage of increased heat loss due to a greater temperature gradient between the internal and external environments. The evolutionary advantage of being endothermic is that it gives much more environmental independence. It is no coincidence that the most successful animals in the extremes of temperatures found in deserts and at the poles are mammals and birds. The independence that a constant high body temperature brings allows these groups to extend their geographical range considerably.

36.2.2 Structure of the skin

Most heat exchange occurs through the skin, as it is the barrier between the internal and external environments. It is in mammals that the skin plays the most important rôle in thermoregulation. Mammalian skin has two main layers, an outer epidermis and an inner dermis.

The epidermis

This comprises three regions:

1. The Malpighian layer (germinative layer) — The deepest layer made up of actively dividing cells. The pigment **melanin**, which determines the skin colour, is produced here. It absorbs ultra-violet light and so helps to protect the tissues beneath from its damaging effects. The Malpighian layer has numerous infoldings which extend deep into the dermis, producing sweat glands, sebaceous glands and hair follicles. There are no blood vessels in the epidermis and so the cells of this layer obtain food and oxygen by diffusion and active transport from capillaries in the dermis.

2. Stratum granulosum (granular layer) — This is made up of living cells which have been produced by the Malpighian layer. As they are pushed towards the skin surface by new cells produced beneath, they accumulate the fibrous protein, **keratin**, lose their nuclei and die.

3. Stratum corneum (cornified layer) — This is the surface layer of the skin and comprises flattened, dead cells impregnated with keratin. It forms a tough, resistant, waterproof layer which is constantly replaced as it is worn away. It is thickest where there is greatest wear, i.e. on the palms of the hand and the soles of the feet. Through this layer extend sweat ducts and hair.

The dermis

The dermis is largely made up of connective tissue consisting of collagen and elastic fibres. It possesses:

1. Blood capillaries — These supply both the epidermis and dermis with food and oxygen. Special networks supply the sweat glands and hair follicles. They play an important rôle in thermoregulation.

2. Hair follicles — These are formed by inpushings of the Malpighian layer. Cells at the base multiply to produce a long cylindrical hair, the cells of which become impregnated with keratin and die. The more melanin in the hair, the darker its colour. Attached to it is a small bundle of smooth muscle, contraction of which causes the hair to become erect.

3. Sebaceous glands — Situated at the side of the hair follicle these produce an oily secretion called **sebum** which waterproofs the hair and epidermis. It also keeps the epidermis supple and protects against bacteria.

4. Sweat glands — These are coiled tubes made up of cells which absorb fluid from surrounding capillaries and secrete it into the tube, from where it passes to the skin surface via the sweat duct. Sweat has a variable composition, consisting mainly of water, dissolved in which are mineral salts and urea. Evaporation of sweat from the skin surface helps to cool the body.

5. Sensory nerve endings – There is a variety of different sensory cells concerned with providing information on the external environment. These include:

(a) touch receptors (Meissner's corpuscles);
(b) pressure receptors (Pacinian corpuscles);
(c) pain receptors;
(d) temperature receptors.

6. Subcutaneous fat – Beneath the dermis is a layer of fat (adipose) tissue. This acts both as a long-term food reserve and as an insulating layer.

The structure of human skin is shown in Fig. 36.2.

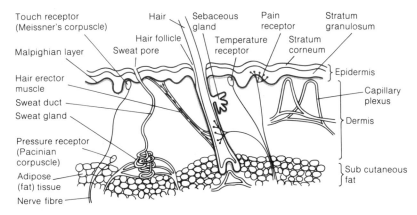

Fig. 36.2 VS through human skin

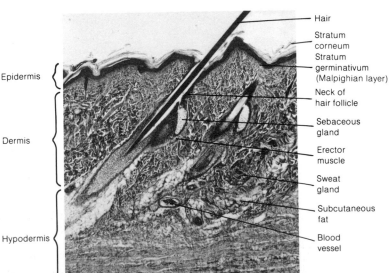

VS Human skin showing hair (× 25 approx.)

36.2.3 Maintenance of a constant body temperature in warm environments

Endothermic organisms which live permanently in warm climates have developed a range of adaptations to help them maintain a constant body temperature. These adaptations may be anatomical, physiological or behavioural, and include the following:

1. Vasodilation – Blood in the network of capillaries in the skin may take three alternative routes. It can pass through capillaries close to the skin surface, through others deeper in the dermis or it may pass beneath the layer of subcutaneous fat. In warm climates, superficial arterioles dilate in order to allow blood close to the skin surface. Heat from this blood is rapidly conducted through the epidermis to the skin surface from where it is radiated away from the body (Fig. 36.3).

2. Sweating – The evaporation of each gram of water requires 2.5 kJ of energy. Being furless, man has sweat glands over the whole body, making him efficient at cooling by this means. Animals with fur generally have sweat glands confined to areas of the skin where fur is absent, e.g. pads of the feet in dogs. Sweating beneath a covering of thick fur is inefficient as it prevents air movements which would otherwise evaporate the sweat. Birds lack sweat glands altogether as their covering of feathers makes its evaporation almost impossible. Humans, on the other hand, may produce up to 1000 cm^3 hr^{-1} of sweat.

3. Panting and licking – Where animals have few or no sweat glands, cooling by evaporation of water nonetheless takes place from the mouth and nose. Panting in dogs may result in the breathing rate increasing from 30 to 300 breaths min^{-1}. This results in excessive removal of carbon dioxide from the blood which is partly offset by a reduction in the depth of breathing. Even so, dogs are able to tolerate a depletion of carbon dioxide which would prove fatal to other organisms. Panting is common in birds. Licking, while not as effective as sweating, may help cool the body. It has been reported in kangaroos, cats and rabbits.

4. Insulation – A layer of fur or fat may help prevent heat gain when external temperatures exceed those of the body. In warm climates the fur is usually light in colour to help reflect the sun's radiation. At high environmental temperatures the hair erector muscles are relaxed and the elasticity of the skin causes the fur to lie closer to its surface. The thickness of insulatory warm air trapped is thus reduced (Fig. 36.4) and body heat is more readily dissipated.

5. Large surface area to volume ratio – Animals in warm climates frequently have large extremities, such as ears, when compared to related species from cold climates. African foxes have much longer ears than their European counterparts which in turn have longer ears than the Arctic fox. Being well supplied with blood vessels and covered with relatively short hair, ears make especially good radiators of heat.

6. Variation in body temperature – Some desert animals allow their body temperatures to fluctuate within a specific range. In camels this range is 34–41 °C. By allowing their body temperature to rise during the day, they reduce the temperature gradient between their body and the environment and so reduce heat gain. In addition, it delays the onset of sweating and so helps to conserve water.

7. Behavioural mechanisms – Many desert animals avoid the period of greatest heat stress by being nocturnal. Some hibernate during the hottest months. This summer hibernation is known as **aestivation**. Other animals avoid the sun by sheltering under rocks or burrowing beneath the surface.

Arctic fox (*top*) and desert fox (*bottom*) showing differences in length of ears

36.2.4 Maintenance of a constant body temperature in cold environments

Endothermic animals living in cold environments show adaptations to the climate. These include:

1. Vasoconstriction – In cold conditions, the superficial arterioles contract, so reducing the quantity of blood reaching the skin surface. Blood largely passes beneath the insulating layer of subcutaneous fat and so loses little heat to the outside. Both vasodilation and vasoconstriction are illustrated in Fig. 36.3.

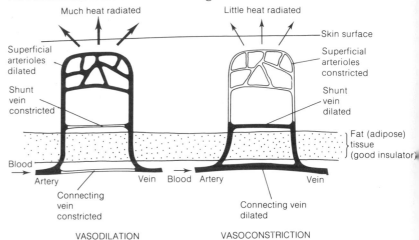

Fig. 36.3 Vasodilation and vasoconstriction

2. Shivering – At low environmental temperatures, the skeletal muscle of the body may undergo rhythmic, involuntary contractions which produce metabolic heat. This shivering may be preceded by asynchronous twitching of groups of muscle.

3. Insulation – Insulation is an effective means of reducing heat loss from the body. It may be achieved by an external covering of fur or feathers and/or an internal layer of subcutaneous fat. The thickness of the fur is related to the environmental temperature, with animals in cold regions having denser and thicker fur. One problem with effective insulation is that it prevents the rapid heat loss necessary during strenuous exercise. For this reason the fur on the underside of the body may be thinner to facilitate heat loss. In birds, specialized down feathers provide particularly efficient insulation. Both fur and feathers function by trapping warm air next to the body. In cold conditions, the hair erector muscles contract to pull up the hairs and so increase the thickness of the layer of air trapped, improving insulation (Fig. 36.4).

Subcutaneous fat alone is only half as effective as fur but makes a useful additional contribution. In aquatic mammals (whales, dolphins, seals) fur would be ineffective and has therefore all but disappeared. To compensate for this loss the subcutaneous fat is extremely thick and forms an effective insulating layer, known as **blubber**.

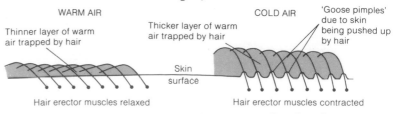

Fig. 36.4 Elevation and depression of hair in controlling heat loss

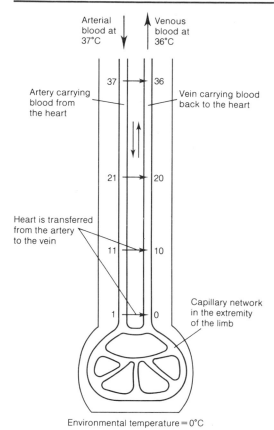

Arterial blood at 37°C

Venous blood at 36°C

Artery carrying blood from the heart

Vein carrying blood back to the heart

Heart is transferred from the artery to the vein

Capillary network in the extremity of the limb

Environmental temperature = 0°C

Fig. 36.5 Counter-current heat exchange system (rete mirabilis)

4. Small surface area to volume ratio – Animals in colder climates have a tendency to be more compact, with smaller extremities, than related species in warm climates. In this way heat loss is reduced.

5. Variations between superficial and core temperature – The extremities of animals in cold regions are maintained at lower temperatures than the core body temperature. This reduces the temperature gradient between them and the environment. This is especially important in order to reduce heat loss from the feet which are in contact with the cold ground. The reduction of heat loss from these extremities is achieved by **countercurrent heat exchangers** found in the limbs of certain birds and mammals. Blood in veins returning from the limbs passes alongside blood in arteries. Heat from the warm blood entering the limb is transferred to cold blood in the vein returning from it. The limb is thereby kept at a lower temperature and cold blood is prevented from entering the core of the body. This system is illustrated in Fig. 36.5.

6. Increased metabolic rate – In addition to an increase in heat produced by muscles during shivering, the liver may also increase its metabolic rate during cold conditions. Low temperatures induce increased activity of the adrenal, thyroid and pituitary glands. All these produce hormones which help to increase the body's metabolic rate and so produce additional heat. This requires increased consumption of food, arctic animals consuming more food per gram of body weight than their tropical relatives. Rats kept at 3 °C take in 50% more food than rats kept at 20 °C.

7. Behavioural mechanisms – In cold regions, animals are usually active during the day (diurnal). Huddling of groups of individuals is a common way of reducing heat loss. Reproductive behaviour is adapted to allow the young to be born at a time when food is available by the time they are weaned. Delayed fertilization and implantation of the embryo may occur (Sections 19.7 and 19.8).

8. Hibernation – One special behavioural mechanism utilized by endothermic animals in cold climates is hibernation. During times of greatest cold and hence shortest supply of food, mammals like squirrels and dormice may undergo a period of long sleep. During this time the metabolic rate is reduced 20–100 times below that of normal with a consequent reduction in food and oxygen consumption. The hibernation may last several months, during which time fat reserves accumulated during the summer are used. These reserves take the form of **brown adipose tissue** which is easily metabolized at low temperatures. Breathing and heartbeat become slow and irregular. The body temperature falls close to that of the environment, being 1–4 °C higher. To prevent the body freezing at temperatures of − 4 °C and below, well insulated and sheltered nests are essential, often underground. For this reason hibernation is virtually impossible in areas of permafrost.

36.2.5 Rôle of the hypothalamus in the control of body temperature

Control of body temperature is effected by the **hypothalamus**, a small body at the base of the brain. Within the hypothalamus is the thermoregulatory centre which has two parts: a heat gain and a heat loss centre. The hypothalamus monitors the temperature of blood passing through it and in addition receives nervous information from

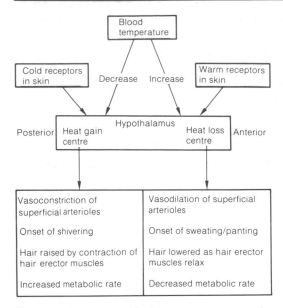

Fig. 36.6 Summary of body temperature control by the hypothalamus

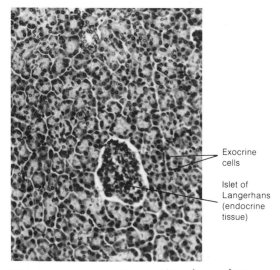

Histology of pancreas, showing Islets of Langerhans ($\times 65$)

receptors in the skin about external temperature changes. Any reduction in blood temperature will bring about changes which conserve heat. A rise in blood temperature has the opposite effect. These effects are summarized in Fig. 36.6.

36.3   Control of blood sugar

All metabolizing cells require a supply of glucose in order to continue functioning. The nervous system is especially sensitive to any reduction in the normal glucose level of 90 mg glucose in 100 cm^3 blood. A rise in blood sugar level can be equally dangerous. The supply of carbohydrate in mammals fluctuates because they do not eat continuously throughout the day and the quantity of carbohydrate varies from meal to meal. There may be long periods when no carbohydrate is absorbed from the intestines. Cells, however, metabolize continuously and need a constant supply of glucose to sustain them. A system which maintains a constant glucose level in the blood, despite intermittent supplies from the intestine, is essential. The liver plays a key rôle in glucose homeostasis. It can add glucose to the blood in two ways:

(i) by the breakdown of glycogen (**glycogenolysis**);

(ii) by converting protein into glucose (**gluconeogenesis**).

It can remove glucose from the blood by converting it into glycogen (**glycogenesis**) which it then stores. A normal liver stores around 75 g of glycogen, sufficient to maintain the body's supply of glucose for about twelve hours. The interconversion of glucose and glycogen is largely under the control of two hormones, produced by the pancreas. In addition to being an exocrine gland producing pancreatic juice, the pancreas is also an endocrine gland. Throughout the pancreas are groups of histologically different cells known as the **islets of Langerhans**. The cells within them are of two types: α cells, which produce the hormone **glucagon**, and β-cells, which produce the hormone **insulin**. Both hormones are discharged directly into the blood. Some hours after a meal, the glucose formed as a result of carbohydrate breakdown is absorbed by the intestines. The blood capillaries from the intestine unite to form the hepatic portal vein which carries this glucose-rich blood to the liver. Insulin from the pancreas causes excess glucose to become converted to glucose-6-phosphate and ultimately glycogen which the liver stores. The same process can occur in many body cells, especially muscle. Some time later, when the level of glucose in the hepatic portal vein has fallen below normal, the liver reconverts some of its stored glycogen to glucose, to help maintain the glucose level of the blood. This change involves a phosphorylase enzyme in the liver which is activated by the pancreatic hormone glucagon.

Should the glycogen supply in the liver become exhausted, glucose may be formed by other means. Once a low level of blood glucose is detected by the hypothalamus it stimulates the pituitary gland to produce adrenocorticotrophic hormones (ACTH) which cause the adrenal glands to release the glucocorticoid hormones, e.g. cortisol. These cause the liver to convert amino acids and glycerol into glucose. In times of stress, another hormone from the adrenal glands, adrenaline, causes the breakdown of glycogen in the liver and so helps to raise the blood sugar level. Further details of these effects are given in Section 37.6.

The control of blood sugar level is summarized in Fig. 36.7.

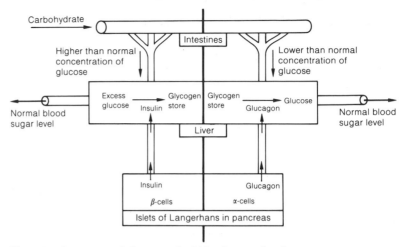

Fig. 36.7 Summary of the control of blood sugar level

36.4 Control of respiratory gases

Metabolizing cells all require a continuous supply of oxygen sufficient to satisfy their respiratory needs. The homeostatic control of both oxygen and carbon dioxide levels in the blood is achieved by the breathing (respiratory) centres in the medulla oblongata of the brain. Details of the mechanisms of control are given in Section 31.3.

36.5 Control of blood pressure

Having achieved homeostatic control of the composition of respiratory gases, sugar and other metabolites in the blood, it is clearly essential for a mammal to control the distribution of this blood. It is essential that blood pressure be kept above a certain minimum level if the supply of essential materials to cells is to be maintained. This is achieved by controlling the rate at which the heart pumps blood and by the vasoconstriction or vasodilation of blood vessels. The mechanisms by which these are controlled are detailed in Sections 32.4.4 and 32.4.5 respectively.

36.6 Cellular homeostasis

To some extent each cell of the body is an independent unit capable of selecting which materials enter and leave it. Each cell exerts homeostatic control of the level of each metabolite within it. As the synthesis of substances within a cell is carried out by enzymes, it follows that this control is achieved by controlling enzyme production as follows:

1. If the level of a particular metabolite falls, it stimulates the relevant operon whose operator gene switches on its corresponding structural genes (Section 12.7).

2. Each structural gene forms mRNA by transcription.

3. The various mRNAs are then translated into the particular proteins (enzymes) by the ribosomes in conjunction with tRNA.

4. These enzymes enter the cell's metabolic centre where they synthesize the formation of the specific metabolite.

5. The metabolite enters the cell's metabolic pool where its concentration rises.

6. When the metabolite's level returns to normal it causes the operator gene to switch off (negative feedback).

7. mRNA production ceases as in due course does the production of the metabolite.

These events are summarized in Fig. 36.8.

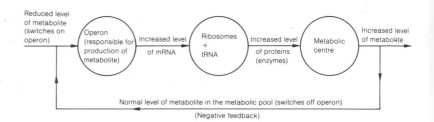

Fig. 36.8 Summary of cellular homeostasis

36.7 The liver

The liver makes up 3–5% of the body weight. It probably originated as a digestive organ but its functions are now much more diverse, many being concerned with homeostasis.

36.7.1 Structure of the liver

In an adult human, the liver is typically 28 cm × 16 cm × 9 cm although its exact size varies considerably according to the quantity of blood stored within it. It is found immediately below the diaphragm to which it is attached. Blood is supplied to the liver by two vessels: the hepatic artery brings oxygenated blood from the aorta, whereas the hepatic portal vein supplies blood rich in soluble digested food from the intestines. A single vessel, the hepatic vein, drains blood from the liver. In addition, the bile duct carries bile produced in the liver to the duodenum. The relationships of these structures is given in Fig. 36.9.

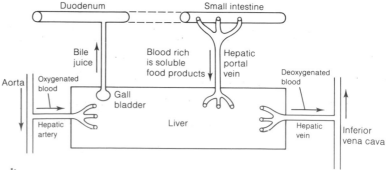

Fig. 36.9 Blood system associated with the mammalian liver

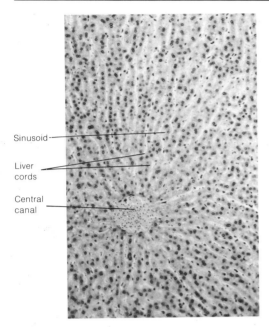

Liver lobule (× 60)

The branches of the hepatic artery and those of the hepatic portal vein combine within the liver to form common venules which lead into a series of channels called **sinusoids**. These are lined with liver cells or **hepatocytes**. The sinusoids eventually drain into a branch of the hepatic vein called the **central vein**. Between the hepatocytes are fine tubes called **canaliculi** in which bile is secreted. The canaliculi combine to form bile ducts which drain into the gall bladder where the bile is stored before being periodically released into the duodenum. The structure of the liver is shown in Fig. 36.10.

(a) Liver, as seen under a light microscope (TS)

(b) Single liver lobule

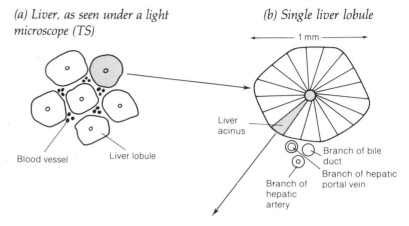

(c) Single acinus

(d) Detail of part of acinus

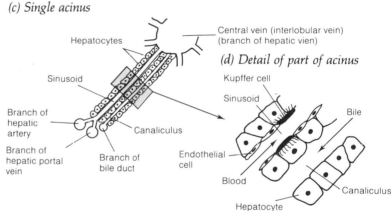

Fig. 36.10 Structure of the mammalian liver

The functional unit of the liver is the **acinus**. As blood from the hepatic portal vein and hepatic artery mixes it passes along the sinusoids which are lined with hepatocytes. Materials are exchanged between these cells and the blood. To facilitate this exchange, the hepatocytes have microvilli to increase their surface area. They also possess a large nucleus, many mitochondria, lysosomes and glycogen granules — all indicate a highly metabolic rôle for these cells. The canaliculi are also lined with microvilli and these appear to remove bile from the hepatocytes by active transport.

The sinusoids are lined with flattened **endothelial cells**. Their structure is similar to that in many other organs except for the presence of pores up to 10 nm in diameter. In addition, there are specialized cells lining the sinusoid. These are **Kupffer** cells. They are highly phagocytic and form part of the **reticulo-endothelial system**. They are ideally situated for ingesting any foreign organisms or particulate matter which enter the body from the intestines. They also engulf damaged and worn out blood cells, producing the bile pigment bilirubin as a by-product of this process. The bilirubin is passed into the canaliculi for excretion in the bile.

36.7.2 Functions of the liver

The liver is the body's chemical workshop and has an estimated 500 individual functions. Some of these have been grouped under the following twelve headings:

1. Carbohydrate metabolism – The liver's major rôle in the metabolism of carbohydrates is in converting excess glucose absorbed from the intestine into glycogen. This stored glycogen can later be reconverted to glucose when the blood sugar level falls. This interconversion is under the control of the hormones insulin and glucagon produced by the islets of Langerhans in the pancreas (Section 36.3).

2. Lipid metabolism – Lipids entering the liver may either be broken down or modified for transport to storage areas elsewhere in the body. Once the glycogen store in the liver is full, excess carbohydrate will be converted to fat by the liver. Excess cholesterol in the blood is excreted into the bile by the liver, which conversely can synthesize cholesterol when that absorbed by the intestines is inadequate for the body's need. The removal of excess cholesterol is essential as its accumulation may cause atherosclerosis (narrowing of the arteries) leading to thrombosis. If in considerable excess its presence in bile may lead to the formation of **gall stones** which can block the bile duct.

3. Protein metabolism – Proteins are not stored by the body and so excess amino acids are broken down in the liver by a process called **deamination**. As the name suggests, this is the removal of the amino group ($-NH_2$) to form ammonia (NH_3) which in mammals is converted to the less toxic urea ($CO(NH_2)_2$). This occurs in the ornithine cycle, the main stages of which are shown in Fig. 34.1, Section 34.2.2.

Transamination reactions whereby one amino acid is converted to another are also performed by the liver. All non-essential amino acids may be synthesized in this way, should they be temporarily deficient in the diet.

4. Synthesis of plasma proteins – The liver is responsible for the production of vital proteins found in blood plasma. These include albumins and globulins as well as the clotting factors prothrombin and fibrinogen.

5. Production of bile – The liver produces bile salts and adds to them the bile pigment bilirubin from the breakdown of red blood cells. With sodium chloride and sodium hydrogen carbonate, cholesterol and water this forms the green-yellow fluid known as bile. Up to $1\,dm^3$ of bile may be produced daily. It is temporarily stored in the gall bladder before being discharged into the duodenum. The bile pigments are purely excretory. The remaining contents have digestive functions and are described in Section 24.5.4.

6. Storage of vitamins – The liver will store a number of vitamins which can later be released if deficient in the diet. It stores mainly the fat-soluble vitamins A, D, E and K, although the water-soluble vitamins B and C are also stored. The functions of these vitamins are given in Table 24.2, Section 24.2.1.

7. Storage of minerals – The liver stores minerals, e.g. iron, potassium, copper and zinc, the functions of which are dealt with in Table 3.1, Section 3.1. It is the liver's stores of these minerals, along with vitamins, which makes it such a nutritious food.

8. Formation and breakdown of red blood cells – The foetus relies solely on the liver for the production of red blood cells. In an adult this rôle is transferred to the bone marrow. The adult liver, however, continues to break down red blood cells at the end of their 120-day life span. The Kupffer cells lining the sinusoids carry out this breakdown, producing the bile pigment bilirubin which is excreted in the bile. The iron is either stored in the liver or used in the formation of new red blood cells by the bone marrow. The liver produces **haematinic principle**, a substance needed in the formation of red blood cells. Vitamin B_{12} is necessary for the production of this principle, and its deficiency results in pernicious anaemia.

9. Storage of blood – The liver, with its vast complex of blood vessels, forms a large store of blood with a capacity of up to $1500 \, cm^3$. In the event of haemorrhage, constriction of these vessels forces blood into the general circulation to replace that lost and so helps to maintain blood pressure. In stressful situations adrenaline also causes constriction of these vessels, creating a rise in blood pressure.

10. Hormone breakdown – To varying degrees, the liver breaks down all hormones. Some, such as testosterone, are rapidly broken down whereas others, like insulin, are destroyed more slowly.

11. Detoxification – The liver is ideally situated to remove or render harmless, toxic material absorbed by the intestines. Foreign organisms or material are ingested by the Kupffer cells while toxic chemicals are made safe by chemical conversions within hepatocytes. Alcohol and nicotine are two substances dealt with in this way.

12. Production of heat – The liver, with its considerable metabolic activity, can be used to produce heat in order to combat a fall in body temperature. This reaction, triggered by the hypothalamus, is in response to adrenaline, thyroxine and nervous stimulation. Whether the liver's activities produce excess heat under ordinary circumstances is a matter of some debate.

36.8 Questions

1. Explain how homeostasis is achieved in a mammal by reference to water balance, temperature control, and blood sugar level. *(10,10,10 marks)*

Oxford Local 1987, Specimen Paper II, No. 6

2. (a) What is homeostasis and why is it important? *(2 marks)*

(b) Outline in **general terms** how negative feedback mechanisms assist in enabling homeostasis to occur. *(3 marks)*

(c) Describe how negative feedback operates to control **three** of the following in mammals:
  (i) blood thyroxine levels;
  (ii) body temperature;
  (iii) blood osmotic potential (osmotic pressure);
  (iv) blood glucose levels. *(15 marks)*
  (Total 20 marks)

Joint Matriculation Board June 1984, Paper IIB, No. 2

3. Claude Bernard, a 19th century physiologist, observed

547

that glucose was always present in the liver and bloodstream of a carnivorous mammal despite the fact that its diet contained no sugar.

(a) Give the modern explanation for this observation. *(1 mark)*

(b) The concentration of glucose in the bloodstream remains fairly constant in normal life processes. If, however, the pancreas is surgically removed, there is a drastic increase in the general level of glucose in the blood.
   (i) Indicate a change that you would expect this operation to cause in the liver.
   (ii) What do these changes in the blood and liver suggest about the function of the pancreas? *(3 marks)*

(c) Humans suffering from diabetes mellitus can be treated by regular injections of insulin.
   (i) Explain why insulin is usually injected rather than taken by mouth.
   (ii) Describe the effect of an overdose of insulin on the concentration of glucose in the blood.
   (iii) Suggest how this effect could be countered. *(3 marks)*

(d) Explain briefly how the regulation of blood sugar level illustrates the principle of negative feedback. *(2 marks)*

(e) In Bernard's time, it was believed that the liver acted in some way on the blood passing through it, so that sugar was formed directly by the blood itself.
   In an experiment to test this hypothesis, the liver was removed from a freshly-killed mammal and a current of water passed through it. When all the blood had been flushed out, a sample of the water leaving the organ was tested for sugar.
   (i) Name the structure to which you would attach the water supply in order to pass the current of water through the liver.
   (ii) State, with reasons, whether or not the results would support the 19th century belief, if
      A sugar was present in the water;
      B sugar was absent from the water.
   (iii) Suggest **one** reason why the conclusions from this experiment might be invalid. *(6 marks)*
   (Total 15 marks)

Welsh Joint Education Committee June 1984, Paper AI, No. 10

4. (a) What is meant by homeostasis? *(1 mark)*
   (b) Explain how the following adaptations might assist in homeostasis:
      (i) an elongated loop of Henle in a desert mammal;
      (ii) the thick fur pelt in an arctic mammal;
      (iii) the subcutaneous fat in a marine mammal. *(6 marks)*
   (c) State **three** major processes by which water may be lost from a mammal and in each case give the reason for this loss. *(6 marks)*
   (d) Giving **two** examples in each case, describe how organisms (other than mammals) adapt to (i) daily and (ii) seasonal changes in temperature. *(8 marks)*
   (Total 21 marks)

London Board June 1984, Paper I, No. 7

5. (a) Make a labelled diagram to show the structure of mammalian skin. *(6 marks)*
   (b) Name **three** structures of the skin derived from ectoderm and **three** derived from mesoderm. *(3 marks)*
   (c) Discuss how internal temperature is controlled in mammals. *(7 marks)*
   (d) How are plants adapted to respond to changes in environmental temperature? *(4 marks)*
   (Total 20 marks)

Oxford and Cambridge Board June 1988, Paper II, No. 3

6. (a) Describe briefly the structure of the mammalian liver. *(5 marks)*
   (b) Give an account of the range of functions of the mammalian liver. *(13 marks)*
   (Total 18 marks)

Cambridge Board June 1989, Paper I, No. 4

7. (a) (i) Distinguish between ectothermy as shown by reptiles and endothermy as shown by mammals. *(2 marks)*
      (ii) Give *one* advantage and *one* disadvantage of ectothermy. *(2 marks)*
      (iii) Give *one* advantage and *one* disadvantage of endothermy. *(2 marks)*
   (b) Drawings A, B, C and D below show the heads of four species of hare inhabiting different latitudes of North America, living in

A   B   C   D

progressively colder habitats. Species A is the most southerly species and inhabits hot deserts. Species D is the most northerly, and lives in the Arctic.

(i) Select *two* anatomical features shown in the drawings, excluding colour, which are related to temperature regulation and which change progressively with latitude from species A to species D. Describe the variations shown in these two features and indicate their significance in temperature regulation. (*4 marks*)

(ii) How may the difference in colour between species A and species D be related to temperature regulation? (*2 marks*)

(*Total 12 marks*)

London Board January 1989, Paper I, No. 11

8. The diagram shows a mechanism by which mammals maintain a constant body temperature despite fluctuations in the external temperature.

(*a*) (i) What is the general name given to this kind of regulatory mechanism?

(ii) Name the form of feedback provided by this mechanism. (*2 marks*)

(*b*) P, Q, R and S indicate four different responses. Give an example of **each**. (*4 marks*)

(*c*) Complete the table by placing a tick($\sqrt{}$) in the column or columns which describe the method of co-ordination involved between the sites described. (*4 marks*)

Sites	Autonomic nervous system	Sensory neurone	Motor neurone	Hormone
Skin receptors and heat regulation centres in the hypothalamus.				
Heat regulation centres in the hypothalamus and an effector organ.				
An effector organ and a site of metabolic activity				

(*d*) What is the significance of **each** of the following in temperature control?

(i) Mammals such as the polar bear and the whale, living in cold regions, tend to be large.

(ii) Foxes living in Africa have far larger ears than those living in Northern regions.

(iii) A species of Australian grasshopper is dark coloured at low temperatures but becomes a lighter colour at higher temperatures.

(iv) Some greenhouse plants wilt at very high temperatures even when the water supply is adequate. (*4 marks*)

(*Total 14 marks*)

Welsh Joint Education Committee June 1988, Paper A2, No. 13

9. The diagram summarizes the mean skin temperatures (in °C) recorded over the course of a day in a study of temperature regulation in an elephant.

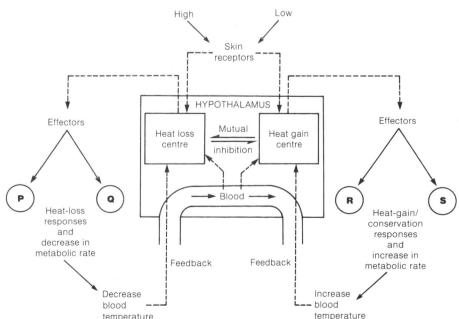

(a) In the course of a day, the mean rectal temperature was 36.4 °C and it varied between 36.0 °C and 36.8 °C. Suggest a reason for this variation in rectal temperature.   (*1 mark*)

(b) Explain the difference between the intermuscular temperature and the temperature of the skin on the front leg.   (*2 marks*)

(c) It has been suggested that temperature control is a particular problem for tropical mammals as large as an elephant. Why should this be so?
          (*2 marks*)

(d) (i) Suggest a function for the large ears of the elephant in temperature control.   (*2 marks*)

   (ii) Give **one** observation or measurement which could be made that might help to confirm your hypothesis.   (*2 marks*)

          (*Total 7 marks*)

Associated Examining Board June 1989, Paper I, No. 8

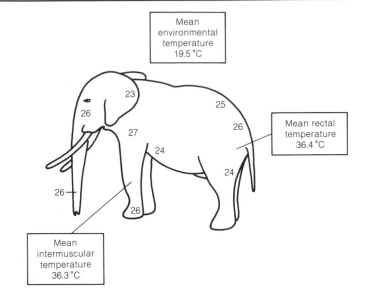

37 *The endocrine system*

Animals possess two principal coordinating systems, the **nervous system** and the **endocrine system**. The nervous system gives rapid control and details of its functioning are provided in Chapter 38. The endocrine system on the other hand regulates long-term changes. The two systems interact in a dynamic way in order to maintain the constancy of the animal's internal environment, while permitting changes in response to a varying external environment. Both systems secrete chemicals, the nervous system as a transmitter between neurones and the endocrine system as its sole means of communication between various organs and tissues in the body. It is worth noting that adrenaline may act both as a hormone and a nervous transmitter.

This chapter discusses the principles of the endocrine system, the nature of hormones and the activities of specific endocrine glands. Because of their close association with particular organ systems, the activities of certain endocrine glands are dealt with elsewhere in this book. Reproductive hormones are described in Section 19.3 and digestive hormones in Section 24.5.

37.1 Principles of endocrine control

In animals, two types of gland are recognized: **exocrine glands**, which convey their secretions to the site of action by special ducts, and **endocrine glands**, which lack ducts and transport their secretions instead by the blood. For this reason, the term **ductless glands** is often applied to endocrine glands. The glands may be discrete organs or cells within other organs.

The secretions of these glands are called **hormones**. Derived from the Greek word *hormon*, which means 'to excite', hormones often inhibit actions as well as excite them. All hormones are effective in small quantities. Most act on specific organs, called **target organs**, although some have diffuse effects on all body cells.

Most, but not all, endocrine glands work under the influence of a single master gland, the **pituitary**. In this way the actions of individual glands can be coordinated. Such coordination is essential as hormones work, not in isolation, but interacting with each other. Most organs are influenced by a number of different hormones. If the pituitary is considered to be the master of the endocrine system then the **hypothalamus** can be thought of as the manager. It not only assists in directing the activities of endocrine glands, it also acts as the all-important link between the endocrine and nervous systems.

The positions of the major endocrine organs in a human are shown in Fig. 37.1.

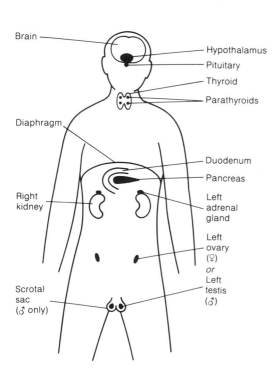

Fig. 37.1 Location of major endocrine glands in humans

TABLE 37.1 **The chemical nature of commonly occurring hormones**

Chemical group	Hormones
Polypeptides (less than 100 amino acids)	Oxytocin Vasopressin Insulin Glucagon
Protein	Prolactin Follicle stimulating hormone Luteinizing hormone Thyroid stimulating hormone Adrenocorticotrophic hormone Growth hormone
Amines (derivatives of amino acids)	Adrenaline Noradrenaline Thyroxine
Steroids (derivatives of lipids)	Oestrogen Progesterone Testosterone Cortisone Aldosterone

37.1.1 Chemistry of hormones

A hormone is a chemical messenger produced by an endocrine gland. It is discharged into the blood stream which carries it around the body to its target organs. Its effects are unrelated to its energy content. Hormones do not belong to any one chemical group. They are generally molecules of medium size, being large enough to have their own specific composition but sufficiently small for them to circulate freely in the body. Some hormones are amines, whereas others are polypeptides and proteins. A few are steroids which are derived from lipids. A summary of the chemical nature of commonly-occurring hormones is given in Table 37.1. The largest hormones are proteins which may possess up to 300 amino acids. The smallest possess less than ten amino acids. The same hormone differs little from one species to another, although its effects may vary in different animals.

37.1.2 Nature of hormone action

Hormones exert their influence by acting on molecular reactions in cells. They achieve this by one or more of the following cell processes:

1. Transcription of genetic information (e.g. oestrogen).

2. Protein synthesis (e.g. growth hormone).

3. Enzyme activity (e.g. adrenaline).

4. Exchange of materials across the cell membrane (e.g. insulin).

While a hormone is transported to all cells by the blood, it only affects specific ones. The explanation is that only target cells possess special chemicals called **receptor molecules** on their surface. These receptors are specific to certain hormones. Both the receptor and hormone have complementary molecular shapes which fit one another in a 'lock and key' manner, much in the way that enzymes and substrates combine (Section 22.1.2).

1. *Hormone approaches receptor site.*

2. *Hormone fuses to receptor site, and in doing so activates the second messenger inside the membrane.*

3. *The activated second messenger moves away from the membrane and initiates a specific chemical change within the cell.*

(a) Use of second messenger, e.g. protein and polypeptide hormones

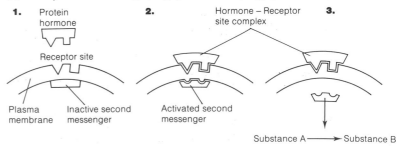

1. *Hormone approaches receptor site.*

2. *Hormone combines with the receptor to form a complex.*

3. *Hormone-receptor complex passes across the membrane into the cell where it influences some activity within the cell.*

(b) Steroid hormone mechanism of action

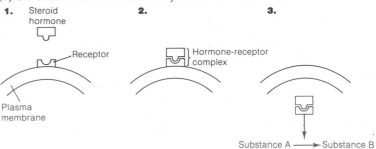

Fig. 37.2 Mechanism of hormone action

There appears to be an alternative means by which this receptor-hormone complex influences the cell. The complex may induce the production of a second messenger which has a specific effect within the cell. The polypeptide and protein hormones act in this way. Steroid hormones, being lipid derivatives, can pass through the cell membrane. The whole complex enters the cell where it exerts its influence. These mechanisms are illustrated in Fig. 37.2, on the previous page. In some cases the complex simply alters the permeability of the cell membrane. Insulin operates in this manner by increasing the permeability of cell membranes to glucose.

37.2 The pituitary gland

Situated at the base of the brain and immediately above the roof of the mouth, the pituitary produces a large number of hormones, many of which influence the activity of other endocrine glands. In its turn, the pituitary depends upon information received from the hypothalamus, to determine which hormone it should secrete, and when. The pituitary gland itself has two distinct portions. The **anterior pituitary** is a region of glandular tissue which communicates with the hypothalamus by means of tiny blood vessels. The **posterior pituitary** is of nervous origin and is in effect an outgrowth of the hypothalamus. Communication with the hypothalamus is by nerves rather than blood vessels. The structure is shown in Fig. 37.3.

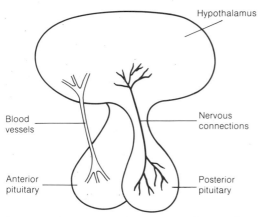

Fig. 37.3 *Structure of the pituitary gland and its relationship to the hypothalamus*

TABLE 37.2 **Functions of hormones secreted by the pituitary**

37.2.1 The anterior pituitary

This portion of the pituitary gland produces six hormones. Most have other endocrine glands as their target organs. These hormones, called **trophic hormones**, stimulate the activity of their respective

Hormone	Abbreviation	Function
Anterior pituitary Thyroid stimulating hormone (thyrotrophic hormone, thyrotrophin)	TSH	1. Stimulates the growth of the thyroid gland 2. Stimulates the thyroid gland to produce its hormones, e.g. thyroxine
Adrenocorticotrophic hormone	ACTH	1. Regulates growth of the adrenal cortex 2. Stimulates the adrenal cortex to produce its hormones, e.g. cortisone
Follicle stimulating hormone	FSH	1. Initiates cyclic changes in the ovaries, e.g. development of the Graafian follicles 2. Initiates sperm formation in the testes
Luteinizing hormone (interstitial cell stimulating hormone)	LH (ICSH)	1. Causes release of the ovum from the ovary and consequent development of the follicle into the corpus luteum 2. Stimulates secretion of testosterone from interstitial cells in the testes
Prolactin (luteotrophic hormone, luteotrophin)	LTH	1. Maintains progesterone production from the corpus luteum 2. Induces milk production in pregnant females
Growth hormone	GH	1. Promotes growth of skeleton and muscles 2. Controls protein synthesis and general body metabolism
Posterior pituitary Anti-diuretic hormone (vasopressin)	ADH	1. Reduces the quantity of water lost from the kidney as urine 2. Raises blood pressure by constricting arterioles
Oxytocin		1. Induces parturition (birth) by causing uterine contractions 2. Induces lactation (secretion of milk from the nipple)

endocrine glands. The only non-trophic hormone is growth hormone which, rather than influencing other endocrine glands, affects body tissues in general. The production of all these hormones is determined by small peptide molecules produced by the hypothalamus and passed to the pituitary via small connecting blood vessels. The functions of the hormones produced are given in Table 37.2.

37.2.2 The posterior pituitary

This portion of the pituitary gland stores two hormones: **anti-diuretic hormone (ADH)** or **vasopressin** and **oxytocin**. Both have remarkably similar chemical structures, differing in just one of their nine amino acids. Despite this they exert very different influences as detailed in Table 37.2. The production of these hormones appears to be by neurosecretory cells in the hypothalamus from where they pass to the posterior pituitary for storage. The release of the hormones is initiated by nervous impulses from the hypothalamus.

37.3   The hypothalamus

Lying at the base of the brain to which it is attached by numerous nerves, the hypothalamus in humans weighs a mere 4 g. Despite its small size it performs many vital functions.

1. It regulates activities such as thirst, sleep and temperature control.

2. It monitors the level of hormones and other chemicals in the blood passing through it.

3. It controls the functioning of the anterior pituitary gland.

4. It produces anti-diuretic hormone and oxytocin which are stored in the posterior pituitary gland.

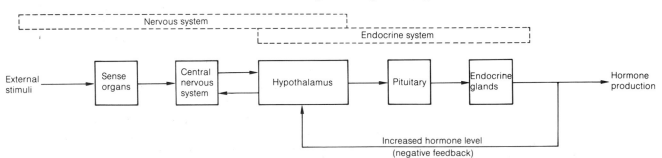

Fig. 37.4 Role of the hypothalamus as the link between nervous and endocrine systems

The hypothalamus is the link between the nervous and endocrine systems as illustrated in Fig. 37.4. By monitoring the level of hormones in the blood, the hypothalamus is able to exercise homeostatic control of them. For example, the control of thyroxine production by the thyroid gland is achieved by this means:

1. The hypothalamus produces **thyrotrophin releasing factor (TRF)** which passes to the pituitary along blood vessels.

2. TRF stimulates the anterior pituitary gland to produce **thyroid stimulating hormone (TSH).**

3. TSH stimulates the thyroid gland to produce thyroxine.

4. As the level of thyroxine builds up in the blood it suppresses TRF production from the hypothalamus and TSH production by the anterior pituitary gland. By this form of negative feedback the level of thyroxine in the blood is maintained at a constant level.

Fig. 37.5 summarizes these effects.

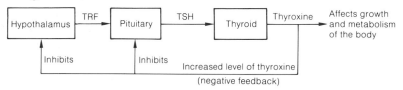

Fig. 37.5 Homeostatic control of thyroxine production

37.4 The thyroid gland

Found in the neck close to the larynx, the thyroid gland, which weighs around 25 g and whose structure is shown in Fig. 37.6, produces three hormones: **triiodothyronine (T$_3$), thyroxine (T$_4$)** and **calcitonin**.

Triiodothyronine and thyroxine are very similar chemically and functionally. They regulate the growth and development of cells. In this respect they are especially important in young mammals. In addition, these hormones increase the rate at which glucose is oxidized by cells. One consequence of this is the production of heat and these hormones are therefore produced when an organism is exposed to severe cold. Emotional stress and hunger may elicit a similar production of these hormones. The overall effect is to control the metabolic rate of cells and in so doing the hormones work in close conjunction with insulin, adrenaline and cortisone.

Both triiodothyronine and thyroxine are derivatives of the amino acid tyrosine and both contain iodine. Thyroxine possesses four iodine molecules while triiodothyronine has only three. In times of iodine shortage the latter is produced in preference to the former in order to make maximum use of limited iodine. If the iodine supply is severely reduced, the thyroid is unable to make adequate supplies of these hormones and underactivity of the thyroid results.

Abnormalities of the thyroid

The two main thyroid abnormalities are underactivity **(hypothyroidism)** and overactivity **(hyperthyroidism)**.

Underactivity (hypothyroidism) has a more marked effect on immature mammals as mental and physical retardation occur in addition to general sluggishness. The condition is known as **cretinism**. In adults, where mental, physical and sexual development is already complete, underactivity of the thyroid causes mental and physical sluggishness as well as a reduced metabolic rate. The latter leads to a reduced heart and ventilation rate, a lowered body temperature and obesity. The condition is known as **myxoedema**. As a result of underactivity, a swelling called a **goitre** may arise in the throat. The cause of underactivity is often the result of an insufficient supply of thyroid stimulating hormone (TSH). The symptoms of underactivity can be eliminated by taking thyroxine orally.

Overactivity (hyperthyroidism) leads to an increased metabolic rate. This results in an increased heart and ventilation rate and a raised body temperature. Nervousness, restlessness and irritability are other symptoms. A goitre may also be apparent. Extreme cases may result in such overactivity that heart failure occurs. This

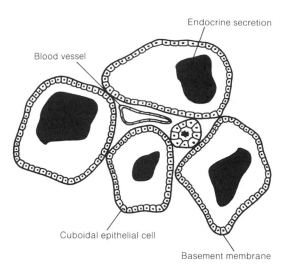

Fig. 37.6 TS thyroid gland of monkey

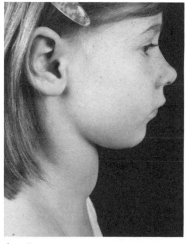

A goitre

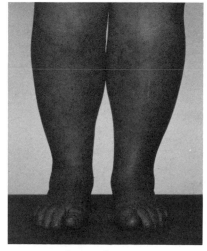

Symptoms of myxoedema

condition is called **thyrotoxicosis**. One main cause of overactivity is a blood protein which stimulates the thyroid to increase its production of triiodothyronine and thyroxine. Controlling overactivity is achieved by the surgical removal of part of the thyroid gland or the destruction of part of the gland by some means, e.g. administration of radioactive iodine.

Calcitonin

This hormone is concerned with calcium metabolism. Calcium in addition to being a major constituent of bones and teeth, is essential for blood clotting and the normal functioning of muscles and nerves. In conjunction with parathormone from the parathyroid gland, calcitonin controls the level of calcium ions (Ca^{2+}) in the blood. A peptide of 32 amino acids, calcitonin is produced in response to high levels of Ca^{2+} in the blood and it causes a reduction in the Ca^{2+} concentration.

37.5 The parathyroid glands

Following the surgical removal of part of the thyroid gland in an attempt to control overproduction of thyroxine, it was noticed that patients suffered disruption to their calcium metabolism. Examination of the excised part of the thyroid led to the discovery of four tiny glands embedded in it. These are the parathyroid glands. They produce a single hormone, **parathormone**, which maintains the level of blood calcium at a sufficiently high level to permit normal muscle and nervous activity. Parathormone raises the level of calcium ions (Ca^{2+}) in the blood in three ways:

1. It increases the rate of calcium reabsorption by the kidney, at the expense of phosphate ions.

2. It increases the rate of calcium absorption from the gut.

3. It causes the release of calcium reserves from bones.

Parathormone is a peptide of 84 amino acids and works antagonistically to calcitonin, the former raising the level of Ca in the blood while the latter reduces it. These events are summarized in Fig. 37.7.

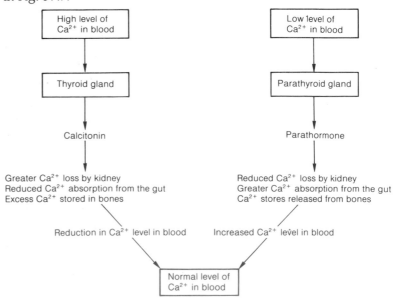

Fig. 37.7 Summary of the control of calcium in the blood

Over production of parathormone leads to excessive removal of calcium from bones which thus become brittle and liable to fracture. Excessive calcium is removed by the kidneys causing the formation of kidney stones. Under production of parathormone results in a low level of blood calcium, leading to nervous disorders and the uncontrollable contraction of muscle known as **tetany**.

37.6 The adrenal glands

Situated above each kidney in humans is a collection of cells weighing about 5 g. These are the adrenal glands. They have two separate and independent parts:

1. The adrenal cortex – This consists of the outer region of the glands.

2. The adrenal medulla – This consists of the inner region of the glands.

37.6.1 The adrenal cortex

Making up around 80% of the adrenal gland, the cortex produces a number of hormones which have relatively slow, long-lasting effects on body metabolism, kidney function, salt balance and blood pressure. All the hormones produced are **steroids** formed from **cholesterol**. Being lipid-soluble they are able to pass across cell membranes along with their receptor molecule (Section 37.1.2). Hormones from the adrenal cortex are collectively called **corticoids** and fall into two groups:

1. Glucocorticoids which are concerned with glucose metabolism.

2. Mineralocorticoids which are concerned with mineral metabolism.

Glucocorticoid hormones
This group of hormones includes **cortisol** which is produced in response to stress. In stressful situations like shock, pain, emotional distress, extreme cold or infection, the hypothalamus induces the anterior pituitary gland to produce adrenocorticotrophic hormone (ACTH). This in turn causes the adrenal cortex to increase its production of glucocorticoids, including cortisol. Where stress is prolonged, the size of the adrenal glands increases. The glucocorticoid hormones combat stress in a number of ways:

1. Raise the blood sugar level, partly by inhibiting insulin and partly by the formation of glucose from fats and proteins.

2. Increase the rate of glycogen formation in the liver.

3. Increase the uptake of amino acids by the liver. These may either be deaminated to form more glucose or used in enzyme synthesis.

Where the production of glucocorticoids is deficient, a condition known as **Addison's disease** occurs. Symptoms include a low blood sugar level, reduced blood pressure and fatigue. The condition of the body deteriorates when stresses such as extreme temperatures and infection are experienced. Over production of glucocorticoids causes **Cushing's syndrome** where there is high blood sugar level mainly

557

TABLE 37.3 **Effects of the hormones adrenaline and noradrenaline on the body and the purpose of these responses**

Effect	Purpose
Bronchioles dilated	Air is more easily inhaled into the lungs. More oxygen is therefore made available for the production of energy by glucose oxidation
Smooth muscle of the gut relaxed	The diaphragm can be lowered further, increasing the amount of air inhaled at each breath, making more oxygen available for the oxidation of glucose
Glycogen in the liver converted to glucose	Increases blood sugar level, making more glucose available for oxidation
Heart rate increased	Increase the rate at which oxygen and glucose are distributed to the tissues
Volume of blood pumped at each beat increased	
Blood pressure increased	
Blood diverted from digestive and reproductive systems to muscles, lungs and liver	Blood rich in glucose and oxygen is diverted from tissues which have less urgent need of it to those more immediately involved in producing energy
Peristalsis and digestion inhibited	Reduction of these processes allows blood to be diverted to muscle and other tissues directly involved in exertion
Sensory perception increased	Heightened sensitivity produces a more rapid reaction to external stimuli
Mental awareness increased	Allows more rapid response to stimuli received
Pupils of the eyes dilated	Increases range of vision and allows increased perception of visual stimuli
Hair erector muscles contract	Hair stands upright. In many mammals this gives the impression of increased size and may be sufficient to frighten away an enemy

due to excessive breakdown of protein. This breakdown causes wasting of tissues, especially muscle. There is a high blood pressure and symptoms of diabetes.

Mineralocorticoid hormones

This group of hormones includes **aldosterone** which regulates water retention by controlling the distribution of sodium and other minerals in the tissues. Aldosterone cannot increase the total sodium in the body, but it can conserve that already present. This it achieves by increasing the reabsorption of sodium (Na^+) and chloride (Cl^-) ions by the kidney, at the expense of potassium ions which are lost in urine. Control of aldosterone production is complex. In response to a low level of sodium ions in the blood, or a reduction in the total volume of blood, special cells in the kidney produce **renin** which in turn activates a plasma protein called **angiotensin**. It is angiotensin which stimulates production of aldosterone from the adrenal cortex. This causes the kidney to conserve both water and sodium ions. Angiotensin also affects centres in the brain creating a sensation of thirst, in response to which the organism seeks and drinks water, thus helping to restore the blood volume to normal.

Over production of aldosterone, often as a result of a tumour, leads to excessive sodium retention by tissues; high blood pressure and headaches then arise. The retention of sodium leads to a consequent fall in potassium levels leading to muscular weakness. Under production of aldosterone leads to a fall in level of sodium in the tissues. In extreme cases this is fatal.

37.6.2 The adrenal medulla

The central portion of the adrenal gland is called the **adrenal medulla**. It produces two hormones, **adrenaline** (epinephrine) and **noradrenaline** (norepinephrine). Both are important in preparing the body for action. The cells producing them are modified neurones, and noradrenaline is produced by the neurones of the sympathetic nervous system. These hormones therefore link the nervous and endocrine systems. They are sometimes called the 'flight or fight hormones' as they prepare an organism to either flee from or face an enemy or stressful situation. The effects of both hormones are to prepare the body for exertion and to heighten its responses to stimuli. These effects and their purposes are summarized in Table 37.3.

In one respect adrenaline and noradrenaline differ. Whereas adrenaline dilates blood vessels, noradrenaline constricts them. This difference explains the constriction of blood vessels around the gut while those supplying muscles, lungs and liver are dilated. It appears that receptors on some blood vessels are sensitive to noradrenaline and so constrict while others are sensitive to adrenaline and so dilate.

37.7 The pancreas

The structure of the pancreas and its rôle as an exocrine gland are considered in Section 24.5.4. At intervals within the exocrine cells are the **islets of Langerhans** which are part of the endocrine system. Cells known as alpha cells produce the hormone **glucagon** whereas beta cells secrete the hormone **insulin**. The two operate antagonistically, with glucagon stimulating the breakdown of glycogen to glucose while insulin initiates the conversion of glucose to glycogen. The mechanism by which these two control the level of sugar in blood is described in Section 36.3.

Insulin is a polypeptide of 51 amino acids and any deficiency in its production leads to a disorder called **diabetes mellitus**. As a result, the blood sugar increases to a potentially dangerous level, possibly causing blindness and kidney failure. The kidney is unable to reabsorb all the glucose passing through it and so a classic symptom of this disorder is the presence of glucose in the urine. Treatment involves the administration of insulin. Originally this was extracted from the pancreases of animals, a treatment first used by Banting and Best in 1921. With the elucidation of the structure of insulin by Sanger in 1950 came the synthetic production of the hormone.

Glucagon is a smaller peptide comprising 29 amino acids. The symptoms of abnormal glucagon production are not well defined.

A summary of the major endocrine glands and the function of the hormones they produce is given in Table 37.4.

TABLE 37.4 **The major endocrine glands and the effects of the hormones they produce**

Endocrine gland	Hormone produced	Effect
Pituitary	The pituitary hormones and their various effects are given in Table 37.2	
Thyroid	Triiodothyronine (T_3) Thyroxine (T_4)	Regulate growth and development of cells by affecting metabolism
	Calcitonin	Lowers calcium level of blood
Parathyroid	Parathormone	Raises blood calcium level while lowering that of phosphate
Adrenal cortex	Glucocorticoid hormones, e.g. cortisol	Helps body resist stress by raising blood sugar level and blood pressure
	Mineralocorticoid hormones, e.g. aldosterone	Increases reabsorption of sodium by kidney tubules
Adrenal medulla	Adrenaline Noradrenaline	Prepare body for activity in emergency or stressful situations
Pancreas (islets of Langerhans)	Insulin (from β cells)	Lowers blood sugar level by stimulating conversion of glucose to glycogen
	Glucagon (from α cells)	Raises blood sugar level by stimulating conversion of glycogen to glucose
Stomach wall	Gastrin	Initiates secretion of gastric juice
Duodenum	Secretin	Stimulates production of bile by the liver and mineral salts by the pancreas
	Cholecystokinin-pancreozymin	Causes contraction of the gall bladder and stimulates the pancreas to produce enzymes
Kidney	Renin	Activates the plasma protein angiotensin
Testis	Testosterone	Produces male secondary sex characteristics
Ovary	Oestrogen (from follicle cells)	Produces female secondary sex characteristics
	Progesterone (from corpus luteum)	Inhibits ovulation and generally maintains pregnancy
Placenta	Chorionic gonadotrophin	Maintains the presence of the corpus luteum in the ovary

37.8 Insect hormones

Most invertebrate hormones are neurosecretory substances. In insects they are important in the control of **ecdysis** (moulting). The process involves two main hormones, **ecdysone** (moulting hormone) and **neotinin** (juvenile hormone). In a holometabolous insect (complete metamorphosis) all moults require ecdysone. If, however, a high concentration of neotinin is also present, the larval moult produces another larval stage. As growth proceeds, the level of neotinin diminishes. At low concentrations of neotinin, ecdysone causes the larva to moult into a pupa. In the absence of neotinin, ecdysone causes the pupa to moult into an imago (adult).

The production of ecdysone and neotinin is controlled by the brain of the insect as follows:

1. The brain produces **brain hormone (prothoracicotrophic hormone)** which passes to a pair of bodies called the **corpora cardiaca** which lie next to the brain. Here the hormone is stored.

2. In response to external stimuli, e.g. day length, temperature, food supply, the brain sends nerve impulses to the corpora cardiaca stimulating them to release their stored brain hormone.

3. The brain hormone released passes to the **prothoracic glands** which are stimulated to produce **ecdysone** (moulting hormone).

4. Neotinin (juvenile hormone) is produced by a region behind the brain called the **corpus allatum**. The production of neotinin diminishes during the insect's development but is resumed in the adult.

The control of insect metamorphosis is summarized in Section 20.4.2, Fig. 20.7.

37.9 Other hormone-like substances

There exist in organisms a number of substances which function in a very similar way to hormones.

37.9.1. Pheromones

Pheromones also form part of the chemical coordination system of certain organisms. Unlike hormones, they operate not within an individual but between members of a species. For this reason they have been called 'social hormones'.

The female silk moth (*Bombyx mori*) produces a scent which attracts male silk moths. So sensitive is the male moth to the pheromone that it can be attracted to a female who is many kilometres away. Commercial variants of such pheromones have been used to control moth populations by luring individuals to their death.

Ants, termites and bees all produce chemicals which aid others in their social groups to locate a food source. In the fire-ant, a volatile liquid is spotted on the ground at intervals when an individual returns to the nest having found a food supply. Other ants in the group follow the scent to the food and they too deposit pheromone on their return, thus reinforcing the trail. Ants also produce an 'alarm pheromone' which warns of danger. In low concentrations it attracts other ants to its aid. These ants also release pheromone, thus increasing its concentration. At very high levels other ants become repelled. This is presumably a mechanism to avoid the whole colony being killed when the danger is beyond their control.

In locusts, the first male to emerge from an egg is thought to produce a pheromone which speeds the maturation of the remaining eggs. This helps to coordinate development.

Worker bees lick a pheromone produced by the mandibular glands of the queen which she spreads over her body. The workers transfer the chemical among themselves. It prevents maturation of their ovaries, maintaining their sterility, and so controlling the size of the colony.

37.9.2 Prostaglandins

A group of specialized lipids which act as chemical messengers are the prostaglandins. They differ from hormones in that they are not produced at a discrete site nor do they act at positions far from their origins. Originally discovered in semen, it now appears they are produced throughout the body. The mechanism of prostaglandin action is not known but it is thought that they may operate as an intermediate stage between a hormone binding to its receptor on a target cell and the activation of the second messenger. Not surprisingly, prostaglandins have diverse effects. These include:

1. Initiation of uterine contractions (this perhaps explains their presence in semen. They would presumably cause the uterus to contract helping to carry the sperm into the oviducts).

2. Regulation of smooth muscle tone.

3. Assistance in blood clotting.

4. Creation of inflammation in response to injury and infection.

5. Control of the secretion of certain hormones.

37.9.3 Endorphins (natural opiates)

It has been known for some time that many drugs, including **morphine** which is derived from the opium poppy, relieve pain by becoming bound to specific receptor sites on cells in the human brain. The question was therefore posed, why should the brain possess such sites? Do they exist in order to bind natural chemicals rather than externally administered ones? In due course natural opiates were discovered and named endorphins ('inside morphine').

Endorphins operate in ways similar to both hormones and neurotransmitters. Their effects include:

1. Relief of pain.

2. Reduction in thyroxine activity.

3. Lowered ventilation and cardiac rate.

4. Water conservation.

5. Influence on hibernation.

Endorphins are peptides which include some amino acid sequences found in several hormones. Further work is necessary to determine the exact rôle of these substances.

1. Explain the rôle of the endocrine system in:

    (a) the production of milk by the mammary gland;
    (b) ovulation;
    (c) the level of sugar in circulating blood. (*8 marks*)

Southern Universities Joint Board June 1984, Paper I, No. 10

2. The diagram shows the major organs that respond to a decrease in the temperature of the skin of a mammal. The routes by which the organs are stimulated are shown by arrows.

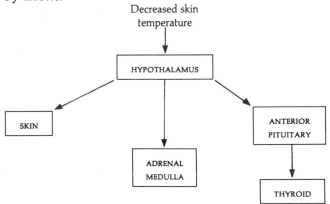

    (a) Describe how each of the five organs is stimulated. (*10 marks*)
    (b) Describe **one** way in which
      (i) the skin reduces heat loss from the body; (*2 marks*)
      (ii) the adrenal medulla increases the temperature of the body; (*3 marks*)
      (iii) the thyroid increases the temperature of the body. (*3 marks*)

    (c) How is 'negative feedback' involved in the control of body temperature? (*2 marks*)
    (*Total 20 marks*)

Associated Examining Board June 1984, Paper II, No. 3, Alternative 1

3. Read through the following account of how a constant level of glucose is maintained in the blood and then write on the dotted lines the most appropriate word or words to complete the account.

Carbohydrate metabolism involves the maintenance of a steady level of glucose in the blood. This level is kept to about .... mg per 100 cm³ regardless of dietary intake. The level is maintained by the interconversion of glucose and ...., the latter being stored mainly in the .... and in the .... A rise in the level of blood glucose results in increased production of the hormone .... from the .... cells of the .... situated in the organ known as the

.... This hormone has the effect of .... the conversion of .... to .... so that the blood glucose level ....

During a period of strenuous exercise, the blood sugar level .... and is returned to normal by the production of .... (*Total 14 marks*)

London Board June 1986, Paper 1, No. 4

4.   (a) In the table below, name (i) a hormone which in a mammal raises the blood sugar level and (ii) another hormone which lowers the blood sugar level. For each hormone specify the site at which it is produced.

	Name	Site of production
(i)	....	....
(ii)	....	....

    (*4 marks*)

  (b) In the table below, name (i) a hormone which in a mammal is responsible for the onset of the oestrous cycle and (ii) another hormone which in a mammal is responsible for the suspension of the oestrous cycle. For each hormone, specify the site at which it is produced.

	Name	Site of production
(i)	....	....
(ii)	....	....

    (*4 marks*)

  (c) Describe briefly the inter-relationships of the activities of the pairs of hormones in each event. (*4 marks*)
    (*Total 12 marks*)

London Board June 1980, Paper I, No. 2

5.   (a) Describe the action of anti-diuretic hormone in a mammal and show how its production and release is regulated. (*7 marks*)
  (b) State the factors involved in the initiation of a nerve impulse and describe how the impulse is propagated along the neurone fibre. (*7 marks*)
  (c) Compare the action of the endocrine system with that of the autonomic nervous system. (*6 marks*)
    (*Total 20 marks*)

Joint Matriculation Board June 1989, Paper IIB, No. 7

6. Write concisely on the following:
   (*a*) thyroxine,
   (*b*) insulin *or* adrenaline (epinephrine).   (*12 marks*)
What do you understand by the term *'cascade' effect?*
(*2 marks*)
What is its significance?   (*2 marks*)
Give a brief description of **one** cascade effect.
(*4 marks*)
(*Total 20 marks*)

Oxford and Cambridge Board June 1989, Paper II, No. 4

7. Read through the following account of the rôles played by hormones in the menstrual cycle, and then add the most appropriate word or words to complete the account.

The development of a Graafian follicle in the ovary is controlled by the two hormones .... and ...., both secreted by the .... gland. As the follicle matures, it releases increasing amounts of .... which stimulates the initial development of the uterine mucosa, progressively inhibits the secretion of ...., and causes the secretion of .... hormone. This hormone has several effects, two of which are the stimulation of .... and the conversion of the burst follicle into a .... This structure now secretes a hormone, ..., which promotes further development of the uterine mucosa. If .... fails to occur the secretion of this hormone eventually ceases, its level in the blood .... and a new cycle starts.

(*Total 11 marks*)

London Board June 1988, Paper I, No. 8

8. The following actions may be stimulated by hormones in a mammal.
Write the name of *one* appropriate hormone beside each of the actions described.

Action	Hormone
Ejection of milk from mammary glands	
Hydrolysis of glycogen in liver	
Reabsorption of sodium ions in kidney	
Ovulation	
Acceleration of heart rate	
Development of male secondary sexual characteristics	
Contraction of smooth muscle in uterus wall	
Increased glucose uptake by muscle cells	

(*Total 8 marks*)
London Board June 1989, Paper I, No. 10

38 *The nervous system*

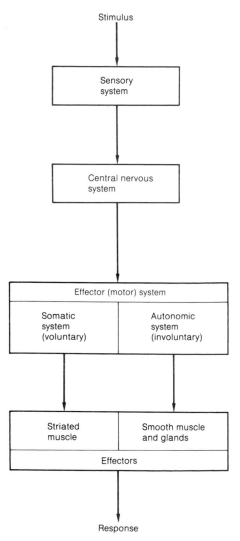

Stimulus

Sensory system

Central nervous system

Effector (motor) system

Somatic system (voluntary)	Autonomic system (involuntary)

Striated muscle	Smooth muscle and glands

Effectors

Response

Fig. 38.1 Inter-relationships of the various components of the nervous system

The ability to respond to stimuli is a fundamental characteristic of living organisms. While all cells of multicellular organisms are able to perceive stimuli, those of the nervous system are specifically adapted to this purpose.

The nervous system performs three functions:

1. To collect information about the internal and external environment.

2. To process and integrate the information, often in relation to previous experience.

3. To act upon the information, usually by coordinating the organism's activities.

One remarkable feature of the way in which these functions are performed is the speed by which the information is transmitted from one part of the body to another. In contrast to the endocrine system (Chapter 37), the nervous system responds virtually instantaneously to a stimulus. The cells which transmit nerve impulses are called **neurones**. They form the structural and functional basis of the nervous system and are described in Section 5.5.

The nervous system may be sub-divided into a number of parts. The collecting of information from the internal and external environment is carried out by **receptors**. Along with the neurones which transmit this information, the receptors form the **sensory system**. The processing and integration of this information is performed by the **central nervous system (CNS)**. The final function whereby information is transmitted to **effectors**, which act upon it, is carried out by the **effector (motor) system**, which has two parts. The portion which activates involuntary responses is known as the **autonomic nervous system** whereas that activating voluntary responses is termed the **somatic system**. The sensory and effector (motor) neurones are sometimes collectively called the **peripheral nervous system (PNS)**.

38.1 The nerve impulse

Nerve impulses are transmitted along cells called **neurones**, the basic structure of which is illustrated in Fig. 5.13(b) in Section 5.5. There are, however, several forms of neurone, each with its own specific shape and some of these are illustrated in Fig. 38.2.

The neurone consists of a **cell body** which contains the nucleus and other cell organelles. Arising from it is a long process which carries the impulse away from the cell body. This is the **axon**. There is a second set of branching processes, called **dendrites**, which carry the impulse towards the cell body. All living cells maintain a potential difference, called the **membrane potential**, across their membranes.

(a) Unipolar neurone
e.g. arthropod motor neurone

(b) Bipolar neurone
e.g. from mammalian retina

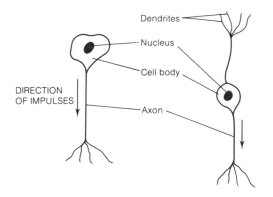

Dendrites

Nucleus

Cell body

DIRECTION
OF IMPULSES

Axon

(c) Branched unipolar neurone
e.g. vertebrate sensory neurone

(d) Multipolar neurone
e.g. from the mammalian spinal cord

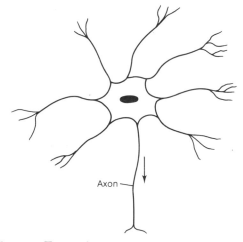

Axon

Fig. 38.2 *Types of neurone*

This remains constant. What makes neurones different is their ability to alter this potential difference.

38.1.1 Resting potential

In its normal state, the membrane of a neurone is negatively charged internally with respect to the outside. The potential difference varies somewhat depending on the neurone but lies in the range 50–90 mV, most usually around 70 mV. This is known as the **resting potential** and in this condition the membrane is said to be **polarized**. The resting potential is the result of the distribution of four ions: potassium (K^+), sodium (Na^+), chloride (Cl^-) and proteins (COO^-). Initially the concentration of potassium (K^+) and protein (COO^-) is higher inside the neurone, while the concentration of sodium (Na^+) and chloride (Cl^-) is higher outside. The membrane, however, is considerably more permeable to potassium ions (K^+) than any of the others. As the potassium ion (K^+) concentration inside the neurone is twenty times greater than that outside, potassium ions (K^+) rapidly diffuse out. This outward movement of positive ions means that the inside becomes slightly negative relative to the outside. As more potassium ions move out, the less able they are to do so. In time an equilibrium is reached whereby the rate at which they leave is exactly balanced by the rate of entry. It is therefore the electrochemical gradient of potassium ions which largely creates the resting potential.

The differences in concentration of ions across the membrane are maintained by the active transport of the ions against the concentration gradients. The mechanisms by which these ions are transported are called pumps. As sodium ions (Na^+) are moved in this way they are often referred to as **sodium pumps**. However, as potassium ions are also actively transported they are more accurately **cation pumps**. These cation pumps exchange sodium and potassium ions by actively transporting in potassium ions and removing sodium ions. Being active, this transport requires ATP.

38.1.2 Action potential

By appropriate stimulation, the charge on a neurone can be reversed. As a result, the negative charge inside the membrane of − 70 mV changes to a positive charge of around + 40 mV. This is known as the **action potential** and in this condition the membrane is said to be **depolarized**. Within about 2 milliseconds (two thousandths of a second) the same portion of the membrane returns to resting potential (− 70 mV inside). This is known as **repolarization**. These changes are illustrated graphically in Fig. 38.3.
Provided the stimulus exceeds a certain value, called the **threshold value**, an action potential results. Above the threshold value the size of the action potential remains constant, regardless of the size of the stimulus. In other words, the action potential is either generated, in which case it is always the same, or it is not. This is called an **all or nothing** response. The size of the action potential does not decrease as it is transmitted along the neurone but always remains the same.

The action potential is the result of a sudden increase in the permeability of the membrane to sodium. This allows a sudden influx of sodium ions because there is a high concentration outside which has been maintained by the sodium pump. The influx of sodium ions begins to depolarize the membrane and this depolarization in turn

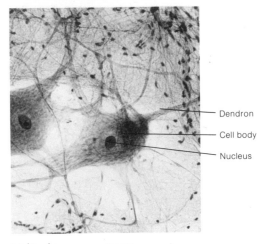

Multipolar neurones (× 17 approx.)

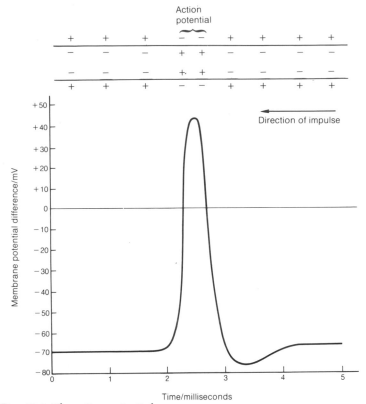

Fig. 38.3 *The action potential*

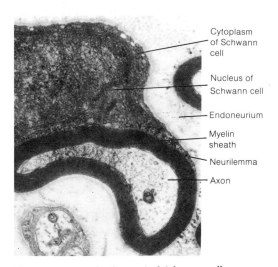

Electronmicrograph of axon and Schwann cell

increases the membrane's permeability to sodium, leading to greater influx and further depolarization. This runaway influx of sodium ions is an example of **positive feedback**. When sufficient sodium ions have entered to create a positive charge inside the membrane, the permeability of the membrane to sodium starts to decrease.

At the same time as the sodium ions begin to move inward, so potassium ions start to move in the opposite direction along a

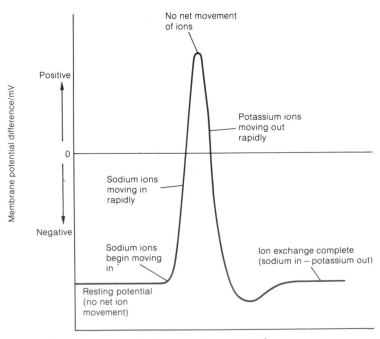

Fig. 38.4 *Ion movements during an action potential*

diffusion gradient. This outward movement of potassium is, however, much less rapid than the inward movement of sodium. It nevertheless continues until the membrane is repolarized. The changes are summarized in Fig. 38.4, on the previous page.

38.1.3 Refractory period

Following an action potential, the outward movement of potassium ions quickly restores the resting potential. However, for about one millisecond after an action potential the inward movement of sodium is prevented in that region of the neurone. This means that a further action potential cannot be generated for at least one millisecond. This is called the **refractory period**.

The refractory period is important for two reasons:

1. It means the action potential can only be propagated in the region which is not refractory, i.e. in a forward direction. The action potential is thus prevented from spreading out in both directions until it occupies the whole neurone.

2. By the end of the refractory period the action potential has passed further down the nerve. A second action potential will thus be separated from the first one by the refractory period which therefore sets an upper limit to the frequency of impulses along a neurone.

The refractory period can be divided into two portions:

1. The **absolute refractory period** which lasts around 1 ms during which no new impulses can be propagated however intense the stimulus.

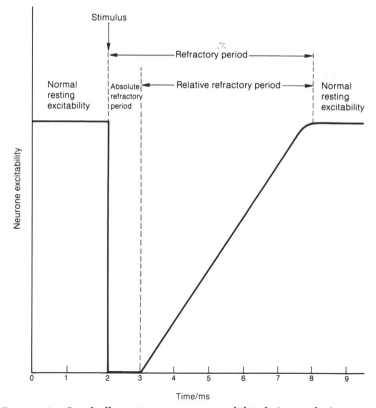

Fig. 38.5(a) Graph illustrating neurone excitability before and after a nerve impulse

Where the stimulus is at the threshold value the excitability of the neurone must return to normal before a new action potential can be formed. In the time interval shown, this allows just 2 action potentials to pass i.e. a low frequency of impulses. Where the stimulus exceeds the threshold value a new action potential can be created before neurone excitability returns to normal. In the time interval shown this allows six action potentials to pass, i.e. a high frequency of impulses.

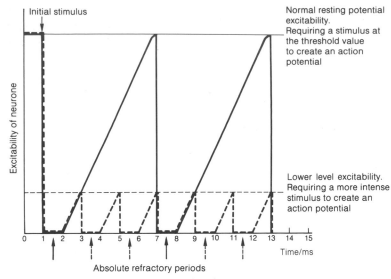

Fig. 38.5(b) Determination of impulse frequency

2. The **relative refractory period** which lasts around 5 ms during which new impulses can only be propagated if the stimulus is more intense than the normal threshold level.

Fig. 38.5(a) illustrates the refractory period in graph form, while Fig. 38.5(b) demonstrates how the refractory period determines the frequency of impulses along a neurone.

38.1.4 Transmission of the nerve impulse

Once an action potential has been set up, it moves rapidly from one end of the neurone to the other. This is the nerve impulse and is described in Fig. 38.6.

According to the precise nature of a neurone, transmission speeds vary from 0.5 metres msec^{-1} to over 100 metres msec^{-1}. Two factors are important in determining the speed of conduction:

(a) The diameter of the axon; the greater the diameter the faster the speed of transmission.
(b) The myelin sheath; myelinated neurones conduct impulses faster than non-myelinated ones.

The myelin sheath, which is produced by the **Schwann cells**, is not continuous along the axon, but is absent at points called **nodes of Ranvier** which arise every millimetre or so along the neurone's length. As the fatty myelin acts as an electrical insulator, an action

1.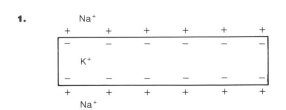

2.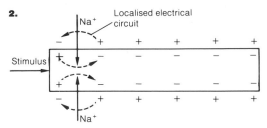

3.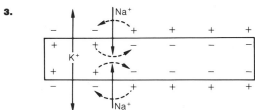

4.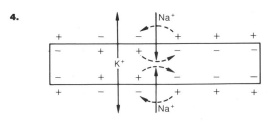

5.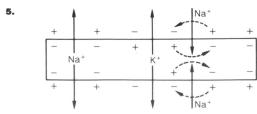

Fig. 38.6 Transmission of an impulse along an unmyelinated neurone

1. *At resting potential there is a high concentration of sodium ions outside and a high concentration of potassium ions inside it.*

2. *When the neurone is stimulated sodium ions rush into the axon along a concentration gradient. This causes depolarization of the membrane.*

3. *Localized electrical circuits are established which cause further influx of sodium ions and so progression of the impulse. Behind the impulse, potassium ions begin to leave the axon along a concentration gradient.*

4. *As the impulse progresses, the outflux of potassium ions causes the neurone to become repolarized behind the impulse.*

5. *After the impulse has passed and the neurone is repolarized sodium is once again actively expelled in order to increase the external concentration and so allow the passage of another impulse.*

potential cannot form in the part of the axon covered with myelin. They can, however, form at the nodes. The action potentials therefore jump from node to node, increasing the speed with which they are transmitted (Fig. 38.7).

The insulating myelin causes ion exchange to occur at the nodes of Ranvier. The impulse therefore jumps from node to node.

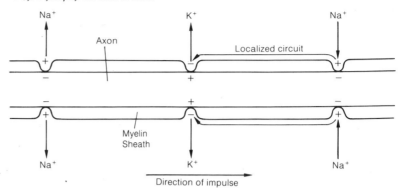

Fig. 38.7 Transmission of an impulse along a myelinated neurone

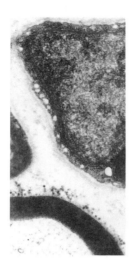

EM of neurone (TS) (× 1600 approx.)

38.2   The synapse

The word synapse (*syn* = 'with'; *apsis* = 'knot') means 'to clasp'. It is the point where the axon of one neurone clasps or joins the dendrite or cell body of another. The gap between the two is around 20 nm in width. The synapse must in some way pass information across itself from one neurone to the next. This is achieved in the vast majority of synapses by **chemical** transmission, although at some synapses the transmission is **electrical**.

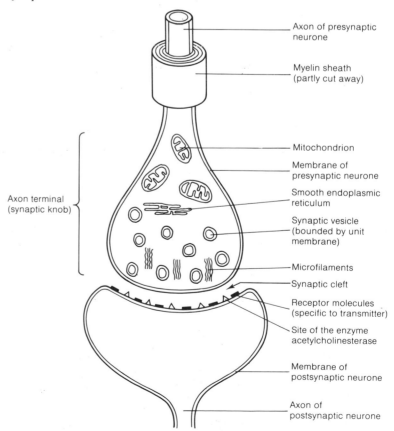

Fig. 38.8 Structure of a chemical synapse

569

1. *The arrival of the impulses at the synaptic knob alters its permeability allowing calcium ions to enter.*

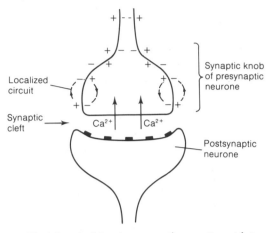

2. *The influx of calcium ions causes the synaptic vesicle to fuse with the presynaptic membrane so releasing acetylcholine into the synaptic cleft.*

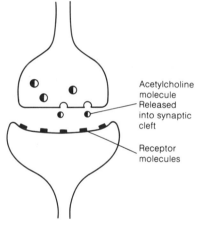

3. *Acetylcholine fuses with receptor molecules on the postsynaptic membrane. This alters its permeability allowing sodium ions to rush in.*

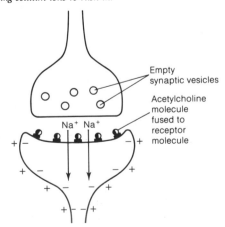

Fig. 38.9 *Sequence of diagrams to illustrate synaptic transmission (only relevant detail is included in each drawing) (cont.)*

38.2.1 Structure of the synapse

As it is the most frequent type, we shall deal here with the structure of the chemical synapse.

At the synapse the nerve axon is expanded to form a bulbous ending called the **axon terminal (bouton terminale)** or **synaptic knob**. This contains many mitochondria, microfilaments and structures called **synaptic vesicles**. The vesicles contain a transmitter substance such as acetylcholine or noradrenaline. The neurone immediately before the synapse is known as the **presynaptic neurone** and is bounded by the presynaptic membrane. The neurone after the synapse is the **postsynaptic neurone**, bounded by the postsynaptic membrane. Between the two is a narrow gap, 20 nm wide, called the **synaptic cleft**. The postsynaptic membrane possesses a number of large protein molecules known as **receptor molecules**. The structure of the synapse is illustrated in Fig. 38.8.

38.2.2 Synaptic transmission

When a nerve impulse arrives at the synaptic knob it alters the permeability of the presynaptic membrane to calcium, which therefore enters. This causes the synaptic vesicles to fuse with the membrane and discharge their transmitter substance which, for the purposes of this account, will be taken to be acetylcholine. The empty vesicles move back into the cytoplasm where they are later refilled with acetylcholine.

The acetylcholine diffuses across the synaptic cleft, a process which takes 0.5 ms. Upon reaching the postsynaptic membrane it fuses with the receptor molecules. In **excitatory synapses** this alters the permeability of the postsynaptic membrane, allowing sodium ions to enter and thus creating a new potential known as the **excitatory postsynaptic potential** in the postsynaptic neurone. These events are detailed in Fig. 38.9.

Once acetylcholine has depolarized the postsynaptic neurone, it is hydrolysed by the enzyme **acetylcholinesterase** which is found on the postsynaptic membrane. This breakdown of acetylcholine is essential to prevent successive impulses merging at the synapse. The resulting choline and ethanoic acid (acetyl) diffuse across the synaptic cleft and are actively transported into the synaptic knob of the presynaptic neurone into which they diffuse. Here they are coupled together again and stored inside synaptic vesicles ready for further use. This recoupling requires energy which is provided by the numerous mitochondria found in the synaptic knob.

The excitatory postsynaptic potentials build up as more transmitter substance arrives until sufficient depolarization occurs to exceed the threshold value and so generate an action potential in the postsynaptic neurone. This additive effect is known as **summation**. All events so far described relate to an **excitatory synapse**, but not all synapses operate in this way. Some, known as **inhibitory synapses**, cause the postsynaptic membrane to become more negative internally. It is thus more difficult for the threshold value to be exceeded and therefore less likely that a new action potential will be created.

4. *The influx of sodium ions generates a new impulse in the postsynaptic neurone.*

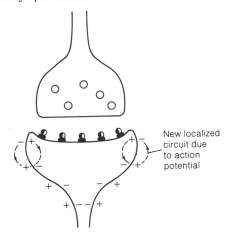

New localized circuit due to action potential

5. *Acetylcholinesterase on the postsynaptic membrane hydrolyses acetylcholine into choline and ethanoic acid (acetyl). These two components then diffuse back across the synaptic cleft into the presynaptic neurone.*

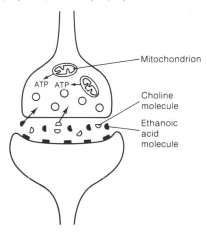

Mitochondrion

Choline molecule

Ethanoic acid molecule

6. *ATP released by the mitochondria is used to recombine choline and ethanoic acid (acetyl) molecules to form acetylcholine. This is stored in synaptic vesicles for future use.*

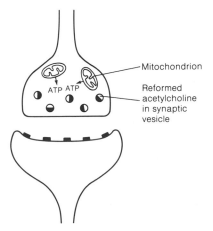

Mitochondrion

Reformed acetylcholine in synaptic vesicle

Fig. 38.9 (cont.) Sequence of diagrams to illustrate synaptic transmission (only relevant detail is included in each drawing)

38.2.3 Functions of synapses

Synapses have a number of functions:

1. Transmit information between neurones – The main function of synapses is to convey information between neurones. It is from this basic function that the others arise.

2. Pass impulses in one direction only – As the transmitter substance can only be released from one side of a synapse, it ensures that nerve impulses only pass in one direction along a given pathway.

3. Act as junctions – Neurones may converge at a synapse. In this way a number of impulses passing along different neurones may between them release sufficient transmitter to generate a new action potential in a single postsynaptic neurone. In this way responses to a single stimulus may be coordinated.

4. Filter out low level stimuli – Background stimuli at a constantly low level, e.g. the drone of machinery, produce a low frequency of impulses and so cause the release of only small amounts of transmitter at the synapse. This is insufficient to create a new impulse in the postsynaptic neurone and so these impulses are carried no further than the synapse. Such low level stimuli are of little importance and the absence of a response to them is rarely, if ever, harmful. Any change in the level of the stimulus will be responded to in the usual way.

5. Allow adaptation to intense stimulation – In response to a powerful stimulus, the high frequency of impulses in the presynaptic neurone causes considerable release of transmitter into the synaptic cleft. Continued high-level stimulation may result in the rate of release of transmitter exceeding the rate at which it can be reformed. In these circumstances the release of transmitter ceases and hence also any response to the stimulus. The synapse is said to be **fatigued**. The purpose of such a response is to prevent overstimulation which might otherwise damage an effector.

38.2.3 Effects of drugs on synaptic transmission

There are many chemicals which are known to influence synaptic transmission. They fall into two categories according to the way in which they affect transmission.

Excitatory drugs
These amplify the process of synaptic transmission in one or more of the following ways:

1. By acting on the receptor molecules of the post-synaptic neurone in exactly the same way as the natural transmitter does, i.e. they mimic the transmitter.

2. By stimulating the release of more of the natural transmitter.

3. By slowing down or even preventing the normal breakdown of the transmitter thus leaving it to continue to stimulate the post synaptic neurone.

Examples include **amphetamines** which mimic the action of noradrenaline by causing its release from nerve endings in the brain, and **caffeine** which accelerates cell metabolism leading to the release of more natural transmitter. **Nicotine**, the drug found in tobacco, also mimics natural neurotransmitters.

Inhibitory drugs
These decrease the process of synaptic transmission in one of the following ways:

1. By preventing release of the synaptic transmitter.

2. By blocking the action of the transmitter at the receptor molecules on the post-synaptic neurone.

Examples include propanolol, one of a group of substances known as β-**blockers**, and atropine, the drug found in deadly nightshade (*Atropa belladonna*).

It must be remembered that some natural transmitters are inhibitory and, therefore, the use of a drug which decreases their influence, actually increases the outward effect. For example, glycine acts as an inhibitory transmitter in the spinal cord. **Strychnine** inhibits glycine, thereby stimulating increased activity of the central nervous system leading to convulsions.

38.3 The reflex arc

A **reflex** is an automatic response which follows a sensory stimulus. It is not under conscious control and is therefore involuntary. The pathway of neurones involved in a reflex action is known as a **reflex arc**. The simplest forms of reflex in vertebrates include those concerned with muscle tone. An example of which is the **knee jerk reflex** which may be separated into six parts:

1. Stimulus – A blow to the tendon situated below the patella (knee cap). This tendon is connected to the muscles that extend the leg, and hitting it causes these muscles to become stretched.

2. Receptor – Specialized sensory structures, called **muscle spindles**, situated in the muscle detect the stretching and produce a nervous signal.

3. Sensory neurone – The signal from the muscle spindles is conveyed as a nervous impulse along a sensory neurone to the spinal cord.

4. Effector (motor) neurone – The sensory neurone forms a synapse inside the spinal cord with a second neurone called an effector neurone. This effector neurone conveys a nervous impulse back to the muscle responsible for extending the leg.

5. Effector – This is the muscle responsible for extending the leg. When the impulse from the effector neurone is received, the muscle contracts.

6. Response – The lower leg jerks upwards as a consequence of the muscle contraction.

This reflex arc has only one synapse, that between the sensory and effector neurone in the spinal cord. Such reflex arcs are therefore termed **monosynaptic**. These reflexes do not involve any neurones connected to the brain which therefore plays no role in the response. As the reaction is routine and predictable, not requiring any analysis, it would be wasteful of the brain's capacity to burden it with the millions of such responses that are required each day. Any reflex arc which is localized within the spinal cord and does not involve the brain is called a **spinal reflex**.

Reflexes involving two or more synapses are termed **polysynaptic**. Typical polysynaptic spinal reflexes include the withdrawal of parts of the body from painful stimuli, e.g. removal of the hand or foot from a hot or sharp object. Due to the response involved, such an action is called a **withdrawal reflex**. Fig. 38.10 illustrates a withdrawal reflex where the hand is placed on a sharp object. This reflex involves an additional stage, namely the **connector (intermediate, internuncial,** or **relay) neurone**, within the spinal cord.

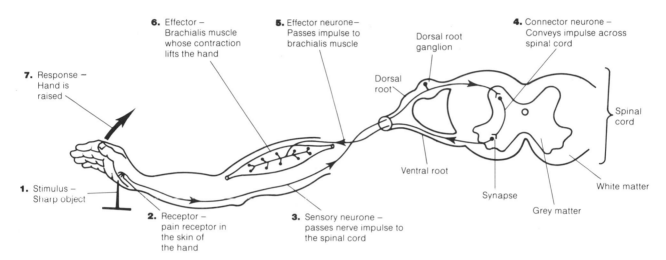

Fig. 38.10 Reflex arc involved in withdrawal from an unpleasant stimulus

These simple reflexes are important in making involuntary responses to various changes in both the internal and external environment. In this way homeostatic control of things like body posture may be maintained. Control of breathing, blood pressure and other systems are likewise effected through a series of reflex responses. Another example is the reflex constriction or dilation of the iris diaphragm of the eye in response to changes in light intensity. Details of this are given in Section 38.6.4.

Brain reflexes have neurone connections with the brain and are usually far more complex, involving multiple responses to a stimulus. While reflexes are themselves involuntary, they may be modified in the light of previous experience. These are called **conditioned reflexes** and are discussed in Section 38.7.3.

38.4 The autonomic nervous system

The autonomic (*auto* = 'self'; *nomo* = 'govern') nervous system controls the involuntary activities of smooth muscle and certain glands. It forms a part of the peripheral nervous system and can be sub-divided into two parts: the **sympathetic nervous system** and the **parasympathetic nervous system**. Both systems comprise effector neurones, which connect the central nervous system to their effector organs. Each pathway consists of a **preganglionic neurone** and a **postganglionic neurone**. In the sympathetic system the synapses between the two are located near the spinal cord whereas in the parasympathetic system they are found near to, or within, the effector organ. This, and other differences, are illustrated in Fig. 38.11.

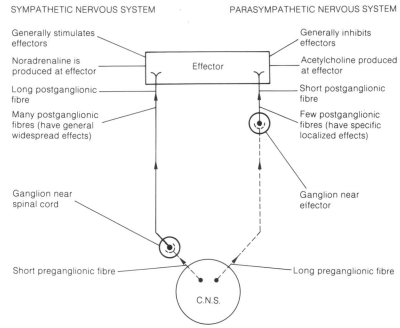

Fig. 38.11 Comparison of the sympathetic and parasympathetic nervous systems

TABLE 38.1 **Comparison of some effects of sympathetic and parasympathetic nervous systems**

Sympathetic nervous system	Parasympathetic nervous system
Increases cardiac output	Decreases cardiac output
Increases blood pressure	Decreases blood pressure
Dilates bronchioles	Constricts bronchioles
Increases ventilation rate	Decreases ventilation rate
Dilates pupils of the eyes	Constricts pupils of the eyes
Contracts anal and bladder sphincters	Relaxes anal and bladder sphincters
Contracts erector pili muscles, so raising hair	No comparable effect
Increases sweat production	No comparable effect
No comparable effect	Increases secretion of tears

The effects of the sympathetic and parasympathetic nervous systems normally oppose one another, i.e. they are **antagonistic**. If one system contracts a muscle, the other usually relaxes it. The balance between the two systems accurately regulates the involuntary activities of glands and organs. It is possible to control consciously certain activities of the autonomic nervous system through training. Control of the anal and bladder sphincters are examples of this.

Table 38.1 lists some of the effects of the sympathetic and parasympathetic nervous systems.

38.5 The central nervous system

The central nervous system (CNS) acts as the coordinator of the nervous system. It comprises a long, approximately cylindrical structure – the **spinal cord**, and its anterior expansion – the **brain**.

38.5.1 The spinal cord

The spinal cord is a dorsal cylinder of nervous tissue running within the vertebrae which therefore protect it. It possesses a thick membranous wall and has a small canal, the **spinal canal**, running through the centre. The central area is made up of nerve cell bodies, synapses and unmyelinated connector neurones. This is called **grey matter** on account of its appearance. Around the grey matter is a region largely composed of longitudinal axons which connect different parts of the body. The myelin sheath around these axons give this region a lighter appearance, hence it is called **white matter**.

At intervals along the length of the spinal cord there extend spinal nerves. There are thirty-one pairs of these nerves in humans. They separate into two close to the spinal cord. The uppermost (dorsal) of these is called the **dorsal root**, while the lower (ventral) one is called the **ventral root**. The dorsal root carries only sensory neurones while the ventral root possesses only effector ones; a sort of spinal nerve one-way system. The cell bodies of the sensory neurones occur within the dorsal root, forming a swelling called the **dorsal root ganglion**. The structure of the spinal cord is illustrated in Fig. 38.12.

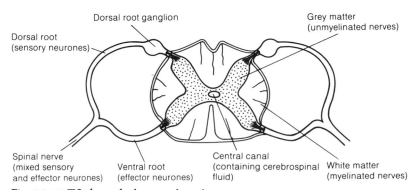

Fig. 38.12 TS through the spinal cord

38.5.2 Structure of the brain

As an elaboration of the anterior region of the spinal cord, the brain has a basically similar structure. Both grey and white matter are present, as is the spinal canal although it is expanded to form larger cavities called **ventricles**. Broadly speaking, the brain has three regions: the **forebrain**, **midbrain** and **hindbrain**.

In common with the entire central nervous system, the brain is surrounded by protective membranes called **meninges**. There are three in all and the space between the inner two is filled with **cerebro-spinal fluid**, which also fills the ventricles referred to above. The cerebro-spinal fluid supplies the neurones in the brain with respiratory gases and nutrients and removes wastes. To achieve this, it must first exchange these materials with the blood. This it does within the ventricles which are richly supplied with capillaries. Having exchanged materials, the fluid must be circulated throughout the CNS in order that it may be distributed to the neurones. This function is

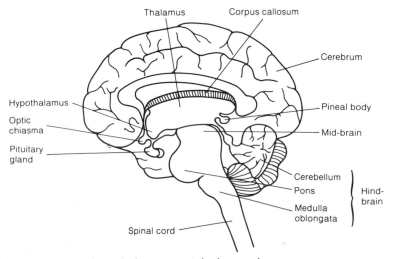

Fig. 38.13 VS through the centre of the human brain

performed by cilia found on the epithelial lining of the ventricles and central canal of the spinal cord. The structure of the brain is illustrated in Fig. 38.13, at the bottom of the previous page.

38.5.3 Functions of the brain

The hindbrain

Medulla oblongata

This region of the brain contains many important centres of the autonomic nervous system. These centres control reflex activities, like ventilation rate (Section 31.3), heart rate (Section 32.4.3) and blood pressure (Section 32.4.4). Other activities controlled by the medulla are swallowing, coughing and the production of saliva.

Cerebellum

This is a large and complex association area concerned with the control of muscular movement and body posture. It receives sensory information relating to the tone of muscles and tendons as well as from the organs of balance in the ears. Its rôle is not to initiate movement but to coordinate it. Any damage to the cerebellum not surprisingly results in jerky and uncoordinated movement.

The midbrain

The midbrain acts as an important link between the hindbrain and the forebrain. In addition it houses both visual and auditory reflex centres. The reflexes they control include the movement of the head to fix on an object or locate a sound.

The forebrain

The thalamus

Lying as it does at the middle of the brain, the thalamus forms an important relay centre, connecting other regions of the brain. It assists in the integration of sensory information. Much of the sensory input received by the brain must be compared to previously stored information before it can be made sense of. It is the thalamus which conveys the information received to the appropriate areas of the cerebrum. Pain and pleasure appear to be perceived by the thalamus.

The hypothalamus

This is the main controlling region for the autonomic nervous system. It has two centres, one for the sympathetic nervous system and the other for the parasympathetic nervous system. At the same time it controls such complex patterns of behaviour as feeding, sleeping and aggression. As another of its rôles is to monitor the composition of the blood, it not unexpectedly possesses a very rich supply of blood vessels. It is also an endocrine gland and details of this function are given in Section 37.3.

The cerebrum

In vertebrates, the size of the cerebrum relative to the body increases from fish, through amphibians and reptiles, to mammals. Even with mammals there is a graded increase in this relative size, with humans having far and away the largest. In addition, it is highly convoluted in humans, considerably increasing its surface area and hence its capacity for complex activity.

The cerebrum is divided into left and right halves known as **cerebral hemispheres**. The two halves are joined by the **corpus callosum**. In general terms, the cerebrum performs the functions of receiving sensory information, interpreting it with respect to that

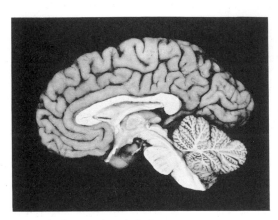

Longitudinal section through the brain

stored from previous experiences and transmitting impulses along motor neurones to allow effectors to make appropriate responses. In this way, the cerebrum coordinates all the body's voluntary activities as well as some involuntary ones. In addition, it carries out complex activities like learning, reasoning and memory.

The outer 3 mm of the cerebral hemispheres is known as the **cerebral cortex** and in humans this covers an especially large area. Within this area the functions are localized, a fact verified in two ways. Firstly, if an electrode is used to stimulate a particular region of the cortex, the patient's response indicates the part of the body controlled by that region. For example, if a sensation in the hand is felt, then it is assumed that the area receives sensory information from the hand. Equally, if the hand moves, this must be the effector centre for the hand. The second method involves patients who have suffered brain damage by accidental means. If the injured person is unable to move his arm, then the damaged portion is assumed to be the effector centre for that arm.

The association areas of the cerebral cortex help an individual to interpret the information received in the light of previous experience. The **visual association area**, for example, allows objects to be recognized, and the **auditory association area** performs the same function for sounds. In humans, there are similar areas which permit understanding of speech and the written word. Yet another centre, the **speech effector centre**, coordinates the movement of the lips and tongue as well as breathing, in order to allow a person to speak coherently.

Despite the methods outlined above that are used to investigate the functions of the brain, certain areas at the front of the cerebral cortex produce neither sensation nor response when stimulated. These are aptly termed **silent areas**. It is possible that they determine certain aspects of personality as their surgical removal has been known to relieve anxiety. The patients, while tranquil, become rather irresponsible and careless, however. The major localized regions of the cerebral cortex are outlined in Fig. 38.14.

Diffusely situated throughout the brain stem is a system called the **reticular activating system**. It is used to stimulate the cerebral

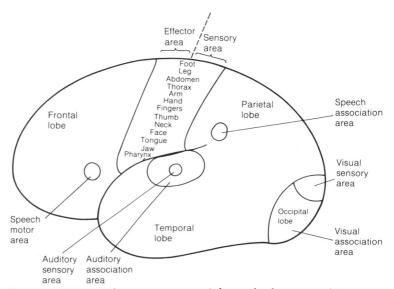

Fig. 38.14 *Map of the major regions of the cerebral cortex and their functions*

TABLE 38.2 **Comparison of endocrine and nervous systems**

Endocrine system	Nervous system
Communication is by chemical messengers – hormones	Communication is by nervous impulses
Transmission is by the blood system	Transmission is by nerve fibres
Target organ receives message	Effector (muscle or gland) receives message
Transmission is relatively slow	Transmission is very rapid
Effects are widespread	Effects are localized
Response is slow	Response is rapid
Response is often long-lasting	Response is short-lived
Effect may be permanent and irreversible	Effect is temporary and reversible

cortex and so rouse the body from sleep. The system is therefore responsible for maintaining wakefulness. The reticular activating system also appears to monitor impulses reaching and leaving the brain. It stimulates some and inhibits others. By so doing it is likely that the system concentrates the brain's activity upon the issues of most importance at any one time. For example, if searching avidly for a lost contact lens, the visual sense may be enhanced. On the other hand, if straining to hear a distant voice, the system may shift the emphasis to increase auditory awareness.

38.5.4 Comparison of endocrine and nervous systems

Both endocrine and nervous systems are concerned with coordination and in performing this function they inevitably operate together. At the same time the systems operate independently and therefore display differences. A comparison of the two systems is given in Table 38.2.

38.6 Sensory perception

All organisms experience changes in both their internal and external environments. Their survival depends upon responding in an appropriate way to these changes, and they have therefore developed elaborate means of detecting stimuli. To some degree, all cells are sensitive to stimuli, but some have become highly specialized to detect a particular form of energy. These are **receptor** cells. In general terms, these receptors convert whichever form of energy it is that they respond to into a nervous impulse, i.e. they act as **biological transducers.**

In its simplest form, a sensory receptor comprises a single neurone in which a single dendrite receives the stimulus and creates an action potential, which it then conveys along its axon to the remainder of the nervous system. This is called a **primary sense cell**, of which the cones of the vertebrate retina are an example. Sometimes the function of receiving the stimulus is performed by a cell outside the nervous system which then passes a chemical or electrical message to a neurone which creates an action potential. This is called a **secondary sense cell**, of which the taste cells on a human tongue are an example.

Whichever form a sensory cell takes, it is often found in groups, usually in conjunction with other tissues, and together they form **sense organs**. At one time it was usual to classify receptors according to their positions. Hence **exterioceptors** collected information from the external environment, **interoceptors** collected it from the internal environment, and **proprioceptors** provided information on the relative position and movements of muscles. It is now more usual to base classification upon the form of stimulus energy. This gives five categories:

1. **Mechanoreceptors** – detect movements, pressures and tensions, e.g. sound.

2. **Chemoreceptors** – detect chemical stimuli, e.g. taste and smell.

3. **Thermoreceptors** – detect temperature changes.

4. **Electroreceptors** – detect electrical fields (mainly in fish).

5. Photoreceptors – detect light and some other forms of electromagnetic radiation.

38.6.1 Mechanoreception

Mechanoreceptors respond to mechanical stimuli which include gravity, vibrations and pressure changes. They are considered as primitive receptors and normally respond to low frequency stimuli. In mammals, touch and pressure receptors are found near the skin surface where they respond to movements of the skin caused by external mechanical stimuli. The proprioceptors found in muscle are a form of mechanoreceptor. These respond to the contraction and stretching of muscle and so give constant information to the CNS on its precise position. One sense organ incorporating numerous mechanoreceptors is the ear. Its structure and function are dealt with later, in Section 38.6.3.

38.6.2 Chemoreception

Chemoreceptors detect chemicals. In mammals, they are primarily involved in taste or **gustation** and smell or **olfaction**.

The taste receptors are located on the tongue and elsewhere in the mouth. The receptor cells are grouped together in taste buds. Hair-like extensions of the receptor cells are exposed at the tongue surface and it is these which detect the chemicals present. There are four basic tastes, sweet, sour, bitter and salt, each detected by a different receptor. The distribution of these receptors is varied, with salt being detected all over the tongue and sweet chemicals at the tip. Sour and bitter are detected at the sides and back respectively.

Most of what we call 'taste' is the result of vapours passing from the mouth into the nasal passages, and is therefore actually smell. The olfactory receptors occur at the back of the nose. They are very densely packed in animals with acute senses of smell. A dog, for example, may have a density of 40 million cm^{-3}. Unlike most other stimuli it is difficult to categorize smell, or to provide some scale of intensity.

Chemoreception serves a number of functions in animals including:

1. Locating food and its acceptance or rejection.

2. Locating a mate.

3. Detecting predators.

4. Detecting dangers, e.g. fire.

5. Locating and marking territories.

38.6.3 The mammalian ear

The mammalian ear performs both as an organ of hearing and one of balance. It is broadly divided into three regions: the **outer ear**, **middle ear** and **inner ear**.

The outer ear comprises an external flap of skin-covered elastic cartilage, known as the **pinna**. The pinna collects and focuses sound waves and directs them along the **ear canal (external auditory meatus)**. In some mammals the pinna may be rotated to help locate the source of a sound without moving the head. This has the advantage of avoiding sudden movements of the head which might reveal the individual's location, making it vulnerable to predators. Across the inner end of the ear canal is stretched the **tympanic membrane (eardrum)** which separates the outer ear from the middle ear.

Within the middle ear are three connected bones called the **ear ossicles** which are held in position by muscles. The bones are the **malleus** (hammer), **incus** (anvil) and **stapes** (stirrup). The middle ear is air-filled and so it is important that the pressure within it is kept equal to that of the atmosphere. Failure to do so causes the tympanic membrane to be stretched, so reducing the amplitude of its vibration and dulling the sense of hearing. The equalization of pressure within the middle ear with that of the atmosphere is achieved through a narrow tube called the **Eustachian tube** which connects the middle ear to the pharynx. It is usually during swallowing that air enters or leaves the middle ear.

The inner ear comprises a complex of fluid-filled tubes. Certain of these tubes form a coil known as the **cochlea** which is concerned with hearing, while the remainder form the **semi-circular canals** which are organs of balance. The full structure of the ear is illustrated in Fig. 38.15.

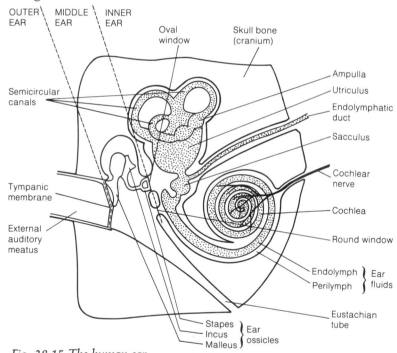

Fig. 38.15 *The human ear*

Mechanism of hearing

Sound travels as waves and the distance between identical points on these waves is known as the wavelength (Section 23.2, Fig. 23.2). The longer the wavelength, the lower the frequency; the shorter the wavelength, the higher the frequency. The human ear can detect wavelengths which range from 40 to 16 000 Hz (cycles sec^{-1}) but is most sensitive to the range 800–8500 Hz. The frequency of sound waves is known as **pitch** whereas its loudness is referred to as **intensity**.

Sound waves are collected by the pinna and focused into the ear canal down which they travel until they meet the tympanic membrane (eardrum). They cause the tympanic membrane to vibrate and these vibrations are transmitted to the oval window by the ear ossicles. As the area of the oval window is only about one twentieth of that of the tympanic membrane, the vibrations are amplified over twenty times. In other words, relatively small movements of the tympanic membrane produce relatively large displacements of the oval window.

The oval window lies at one end of a long, hair-pin canal filled with perilymph (Fig. 38.16b). Between the upper and lower portions of the hair-pin is another canal containing endolymph which communicates with the semi-circular canals. The hair-pin is 35 mm long in humans, but to economize on space it is coiled (Fig. 38.16a). As the last of the ear ossicles, the stapes, vibrates, it pushes the oval window in and out in a piston-like manner. Being a liquid, the perilymph behind the oval window cannot be compressed or expanded and so movements of the oval window cause similar movements of the perilymph. These displacements of perilymph cause similar displacements of the round window, so as the oval window moves inwards the perilymph displaced causes the round window to bulge outwards into the middle ear (Fig. 38.16b). This is possible because the middle ear is air-filled and the displacement of the round window simply compresses the air.

(a) *Side view to illustrate spiral shape of the cochlea*

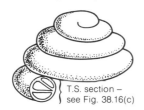

T.S. section –
see Fig. 38.16(c)

(c) *Transverse section through the cochlea*

Upper chamber (vestibular canal)
Reissner's membrane
Perilymph
Middle chamber (median canal)
Endolymph
Tectorial membrane
Sensory hair cell
Perilymph
Branch of auditory nerve
Lower chamber (tympanic canal)
Basilar membrane

Fig. 38.16 *Structure and function of the cochlea*

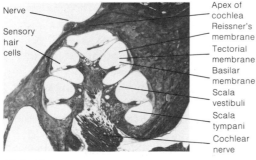

Nerve
Sensory hair cells
Apex of cochlea
Reissner's membrane
Tectorial membrane
Basilar membrane
Scala vestibuli
Scala tympani
Cochlear nerve

TS Cochlea (× 17 approx.)

(b) *Diagrammatic section through an 'unwound' cochlea showing the relationship between displacements of the round and oval windows*

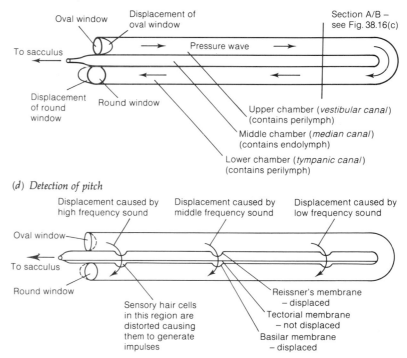

Oval window
Displacement of oval window
Section A/B – see Fig. 38.16(c)
To sacculus
Pressure wave
Displacement of round window
Round window
Upper chamber (*vestibular canal*) (contains perilymph)
Middle chamber (*median canal*) (contains endolymph)
Lower chamber (*tympanic canal*) (contains perilymph)

(d) *Detection of pitch*

Displacement caused by high frequency sound
Displacement caused by middle frequency sound
Displacement caused by low frequency sound
Oval window
To sacculus
Round window
Sensory hair cells in this region are distorted causing them to generate impulses
Reissner's membrane – displaced
Tectorial membrane – not displaced
Basilar membrane – displaced

How then does this arrangement detect sound? It is thought that the pressure waves set up as a result of the piston-like action of the stapes on the oval window lead to displacements of **Reissner's membrane** (between the upper and middle chambers) which in turn displace endolymph in the middle chamber. Being incompressible, the endolymph displaces the basilar membrane. As the basilar membrane is quite elastic but the tectorial membrane is more rigid this leads to movements of the basilar membrane while the tectorial membrane remains relatively fixed. Strung between these two membranes are sensory hair cells (Fig. 38.16c). As the basilar membrane is displaced relative to the tectorial membrane the sensory hair cells become distorted. This distortion sets up an action potential which is conveyed to the CNS along a branch of the auditory nerve. In this way the brain is made aware that a sound stimulus has reached the ear. The tectorial and basilar membranes along with the sensory hair cells make up a special structure called the **organ of Corti**.

How is the pitch of the sound determined? The basilar membrane broadens and thickens the further away it is from the round window. Microscope observations have shown that the basilar membrane nearest to the round window vibrates more when a sound is of high frequency, while the region of the basilar membrane closest to the apex of the cochlea vibrates most when stimulated by low frequency sound. It seems therefore that different frequency sound waves stimulate different regions of the cochlea. By determining which region of the cochlea is sending impulses, the brain can interpret the pitch of sound entering the ear. These events are summarized in Fig. 38.16(d).

Finally, how is the intensity of sound determined? It seems likely that at any point along the basilar membrane there are a number of different sensory hair cells each with a different threshold at which it is stimulated. The louder the sound at any one frequency, the greater the number of sensory hair cells which will be stimulated at any one point on the basilar membrane.

Maintenance of balance

While the ear plays an important rôle in helping to maintain balance, it must not be forgotten that proprioceptors in the joints and muscles of the body also provide the brain with a wealth of essential information.

The parts of the ear concerned with balance are the **semi-circular canals**. These are three curved canals containing endolymph which communicate with the middle chamber of the cochlea via the **utriculus** and **sacculus**. Each of the three canals is arranged in a plane at right angles to the other two (Fig. 38.17a). A movement in any one plane will result in movement of the canals in the same direction as the head. However, the inertia of the endolymph within the canals means that the endolymph remains more or less stationary. There is therefore relative movement between the canals and endolymph within them in the same way that there is between a bottle and the liquid within it when the bottle is shaken. The movement of the endolymph is much greater in the canal which is in the same plane as the plane of movement.

Each of the three canals possesses a swollen portion, the **ampulla**, within which there is a flat gelatinous plate, the **cupula**. The movement of endolymph displaces this cupula in the opposite direction to that of the head movement (Fig. 38.17b). Sensory hairs at the base of the cupula detect the displacement and send impulses to the brain via the **vestibular nerve**. The brain can then initiate motor impulses to various muscles to correct the imbalance.

The utriculus and sacculus also aid balance by providing information on the position of the body relative to gravity as well as changes in position due to acceleration and deceleration. The information is provided by chalk granules known as **otoliths** which are embedded in a jelly-like material. Various movements of the head cause these otoliths to displace sensory hair cells on regions of the walls of the utriculus and sacculus which respond to vertical and lateral movements respectively. The sensory hair cells thus send appropriate sensory impulses to the brain.

Fig. 38.17(a) The two vertical and one horizontal planes, each at right angles to the others, in which the three semi-circular canals are arranged

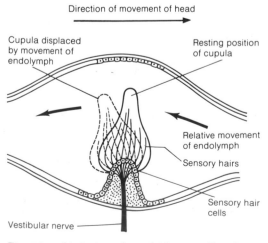

Fig. 38.17(b) Section through the ampulla of a semi-circular canal

38.6.4 The mammalian eye

That part of the electromagnetic spectrum which can be detected by the mammalian eye lies in the range 400–700 nm. The eye acts like a television camera in producing an ever-changing image of the visual field at which it is directed.

Each eye is a spherical structure located in a bony socket of the skull called the **orbit**. It may be rotated within its orbit by **rectus muscles** which attach it to the skull. The external covering of the eye is the **sclera**. It contains many collagen fibres and helps to maintain the shape of the eyeball. The sclera is transparent over the anterior portion of the eyeball where it is called the **cornea**. It is the cornea that carries out most refraction of light entering the eye. A thin transparent layer of living cells, the **conjunctiva** overlies and protects much of the cornea. The conjunctiva is an extension of the epithelium of the eyelid. Tears from **lachrymal glands** both lubricate and nourish the conjunctiva and cornea.

Inside the sclera lies a layer of pigmented cells, the **choroid**, which prevents internal reflection of light. It is rich in blood capillaries which supply the innermost layer, the **retina**. This contains the light-sensitive **rods and cones** which convert the light waves they receive into nerve impulses which pass along neurones to the **optic nerve** and hence to the brain. There is an especially light-sensitive spot on the retina which contains only cones. This is the **fovea centralis**. The amount of light entering the eye is controlled by the **iris**. This is a heavily pigmented diaphragm of circular and radial muscle whose contractions alter the diameter of the aperture at its centre, called the **pupil**, through which light enters. Just behind the pupil lies the transparent, biconvex **lens**. It controls the final focusing of light onto the retina. It is flexible and elastic, capable of having its shape altered by the **ciliary muscles** which surround it. These are arranged circularly and radially and work antagonistically to focus incoming light on the retina by altering the lens' shape and hence its focal length. The region in front of the lens is called the **anterior chamber** and contains a transparent liquid called **aqueous humour**. Behind the lens is the much larger **posterior chamber** which contains the transparent jelly-like **vitreous humour** which helps to maintain the eyeball's shape. The structure of the eye is illustrated in Fig. 38.18.

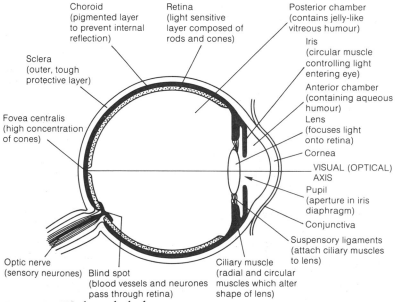

Fig. 38.18 VS through the human eye

Control of the amount of light entering the eye

Controlling the amount of light entering the eye is of importance because if too little light reaches the retina the cones may not be stimulated at all. Alternatively, if the quantity of light is too great the retinal cells may be overstimulated, causing dazzling. Control is exercised by the iris diaphragm as outlined in Fig. 38.19.

	BRIGHT LIGHT	DIM LIGHT
1.	More photoreceptor cells in the retina are stimulated by an increase in light intensity	Fewer photoreceptor cells are stimulated due to decrease in light intensity
2.	Greater number of impulses pass along sensory neurones to the brain	Fewer impulses pass along sensory neurones to the brain
3.	Brain sends impulses along parasympathetic nervous system to the iris diaphragm	Brain sends impulses along the sympathetic nervous system to the iris diaphragm
4.	In the iris diaphragm, circular muscle contacts and radial muscle relaxes	In the iris diaphragm, circular muscle relaxes and radial muscle contracts
5.	Pupil constricts	Pupil dilates
6.	Less light enters the eye	More light enters the eye
	Anterior view of iris and pupil — Radial muscle relaxed, Pupil constricted, Circular muscle contracted	Anterior view of iris and pupil — Radial muscle contracted, Pupil dilated, Circular muscle relaxed

Fig. 38.19 Mechanism of control of light entering the eye

Focusing of light rays onto the retina

Light rays entering the eye must be **refracted** (bent) in order to focus them onto the retina and so give a clear image. Most refraction is achieved by the cornea. However, the degree of refraction needed to focus light rays onto the retina varies according to the distance from the eye of the object being viewed. Light rays from objects close to the eye need more refraction to focus them on the retina than do more distant ones. The cornea is unable to make these adjustments and so the lens has become adapted to this purpose. Being elastic, it can be made to change shape by the ciliary muscle which encircles it. The muscle fibres are arranged circularly and the lens is supported by **suspensory ligaments** (Fig. 38.18). When the circular ciliary muscle contracts, the tension on the suspensory ligaments is reduced and the natural elasticity of the lens causes it to assume a fatter (more convex) shape. In this position it increases the degree of refraction of light. When the circular ciliary muscle is relaxed, the suspensory ligaments are stretched taut, thus pulling the lens outwards and making it thinner (less convex). In this position it decreases the degree of light refraction. By changing its shape in this manner the lens can focus light rays from near and distant objects on the retina. The process is called **accommodation**. How the eye accommodates for distant and near objects is shown in Figs. 38.20(a) and (b).

The retina

The retina possesses the **photoreceptor** cells. These are of two types, **rods** and **cones**. Both act as transducers in that they convert light energy into the electrical energy of a nerve impulse. Both cell types are partly embedded in the pigmented epithelial cells of the choroid. The microscopic structure of the retina is shown in Figs. 38.21(a) and (b). In cats and some other nocturnal mammals, there exists a

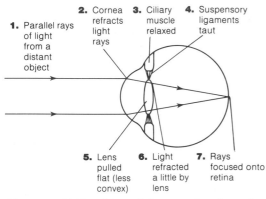

1. Parallel rays of light from a distant object
2. Cornea refracts light rays
3. Ciliary muscle relaxed
4. Suspensory ligaments taut
5. Lens pulled flat (less convex)
6. Light refracted a little by lens
7. Rays focused onto retina

Fig. 38.20(a) Condition of the eye when focused on a distant object

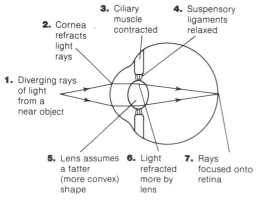

1. Diverging rays of light from a near object
2. Cornea refracts light rays
3. Ciliary muscle contracted
4. Suspensory ligaments relaxed
5. Lens assumes a fatter (more convex) shape
6. Light refracted more by lens
7. Rays focused onto retina

Fig. 38.20(b) Condition of the eye when focused on a near object

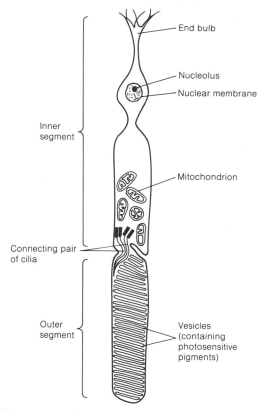

Fig. 38.21(a) Structure of a single rod cell

reflecting layer, called the **tapetum**, behind the retina. This reflects light back into the eye and so affords further opportunities for rod cells to absorb it. This greatly improves vision in dim light and is why cats' eyes give bright reflections when light shines into them — it is the reflected light from the tapetum which is seen.

As Fig. 38.21 shows, the basic structure of rods and cones is similar. However, there are both structural and functional differences and these are detailed in Table 38.3.

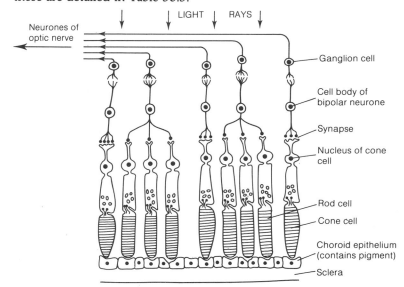

Fig. 38.21(b) Microscopic structure of the retina

TABLE 38.3 **Differences between rods and cones**

Rods	Cones
Outer segment is rod-shaped	Outer segment is cone-shaped
Occur in greater numbers in the retina – being 20 times more common than cones	Fewer are found in the retina – being one twentieth as common as rods
Distributed more or less evenly over the retina	Much more concentrated in and around the fovea centralis
None found at the fovea centralis	Greatest concentration occurs at the fovea centralis
Give poor visual acuity because many rods share a single neurone connection to the brain	Give good visual acuity because each cone has its own neurone connection to the brain
Sensitive to low-intensity light, therefore mostly used for night vision	Sensitive to high-intensity light, therefore mostly used for day vision
Do not discriminate between light of different wavelengths, i.e. not sensitive to colour	Discriminate between light of different wavelengths, i.e. sensitive to colour
Contain the visual pigment rhodopsin which has a single form	Contain the visual pigment iodopsin which occurs in three forms

Each rod possesses up to a thousand vesicles in its outer segment. These contain the photosensitive pigment **rhodopsin** or **visual purple**. Rhodopsin is made up of the protein **opsin** and a derivative of vitamin A, **retinal**. Retinal normally exists in its *cis* isomer form, but light causes it to become converted to its *trans* isomer form. This change initiates reactions which lead to the splitting of rhodopsin into opsin and retinal – a process known as **bleaching**. This splitting in turn leads to the creation of a generator potential in the rod cell which, if sufficiently large, generates an action potential along the neurones leading from the cell to the brain.

Before the rod cell can be activated again in the same way, the opsin and retinal must first be resynthesized into rhodopsin. This resynthesis is carried out by the mitochondria found in the inner segment of the rod cell, which provide ATP for the process. Resynthesis takes longer than the splitting of rhodopsin but is more rapid the lower the light intensity. A similar process occurs in cone cells except that the pigment here is **iodopsin**. This is less sensitive to light and so a greater intensity is required to cause its breakdown and so initiate a nerve impulse.

Colour vision

It is thought that there are three forms of iodopsin, each responding to light of a different wavelength. Each form of iodopsin occurs in a different cone and the relative stimulation of each type is interpreted by the brain as a particular colour. This system is known as the **trichromatic theory** because there are three distinct types of cone, each responding to three different colours of light, blue, green and red. Other colours are perceived by combined stimulation of these three. Equal stimulation of red and green cones, for example, is perceived as yellow. Fig. 38.22 shows the extent of stimulation of each type of cone at different wavelengths of light.

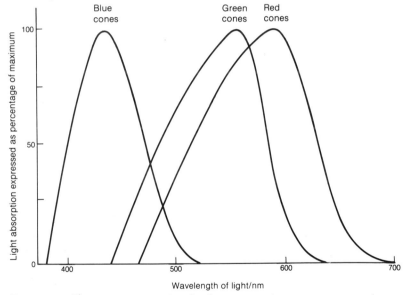

Fig. 38.22 Absorption spectra for the three types of cone occurring in the retina as proposed in the trichromatic theory of colour vision

Colour blindness can be explained in terms of some deficiency in one or more cone type. Deficiencies in the red and green cones, for example, give rise to the relatively common **red-green colour blindness**. The condition is due to a defect on a gene which is linked to the X chromosome (Section 14.4.4). Alternative theories to the trichromatic theory have been put forward in recent years but these have yet to achieve general acclaim.

38.7 Behaviour

In order to survive, organisms must respond appropriately to changes in their environment. Broadly speaking, behaviour is the response of an organism to these changes and it involves both endocrine and nervous systems. Behaviour has a genetic basis and is unique to each species. It is often adapted, however, in the light of previous experience. The study of behaviour is called **ethology**.

Although both plants and animals exhibit behaviour, that of animals is considerably more complex. This section deals only with animal behaviour; plant behaviour is covered in Chapter 40.

38.7.1 Reflexes, kineses and taxes

Reflexes are the simplest form of behavioural response. Section 38.3 describes a simple reflex response to a stimulus. These reflexes are involuntary responses which follow an inherited pattern of behaviour. How then do these responses improve an animal's chance of survival? The **withdrawal reflex** illustrates their importance. If the hand is placed on a hot object, the reflex response causes it to be immediately withdrawn. In this way damage is avoided. Invertebrate animals have a similar **escape reflex**. An earthworm, for example, withdraws down its burrow in response to vibrations of the ground. By behaving in this way the earthworm is less likely to be captured by a predator and so improves its chances of survival.

A more complex form of behaviour is a **kinesis**. This is a form of orientation behaviour. The response is non-directional, i.e. the animal does not move towards or away from the stimulus. Instead it simply moves faster, and changes direction more frequently when subjected to an unpleasant stimulus. The greater the intensity of the stimulus the faster it moves and the more often it changes direction. This is known as **orthokinesis**. It is exhibited by woodlice which prefer damp situations. If a woodlouse finds itself in a dry area, it simply moves faster and keeps changing its direction. It thereby increases the speed with which it is likely to move out of the dry area. If this brings it into a more moist situation its movement slows, indeed it may cease altogether. It therefore spends a much longer period of time in the more humid conditions. Naturally it cannot remain stationary indefinitely, and factors such as hunger will ultimately cause it to move in search of food.

A **taxis** involves the movement of a whole organism in response to a stimulus, where the direction of the movement is related to the direction of the stimulus, usually towards it (positive +), or away from it (negative −). Tactic responses are classified according to the nature of the stimulus, e.g. light − **phototaxis**; chemicals − **chemotaxis**. The flatworm, *Planaria*, shows a positively chemotactic response to the presence of food. By moving its head from side to side it can detect the relative intensity of chemical stimulation experienced on each side of the head. This allows it to locate the source of food. Other examples of taxes are given in Section 40.1.2.

38.7.2 Innate behaviour

Innate or **instinctive** behaviour is inherited and is highly specific. It is normally an inborn pattern of behaviour which cannot be altered. In practice almost all instincts can be modified to some degree in response to experiences. However, innate behaviour is relatively inflexible when compared to learning. Much instinctive behaviour is highly complex and consists of a chain of actions, the completion of each stage in the chain acting as the stimulus for the commencement of the next stage.

This pattern of events is illustrated by a species of digger wasp. As its parents die long before each wasp hatches, there is no opportunity for it to learn from its parents and it must rely on inherited patterns of behaviour. When the time comes to lay eggs, a female digger wasp digs a nest hole and in it constructs small cells, each of which it provisions with a paralysed caterpillar to act as a food source for its offspring. Having done so, it lays a single egg on the roof of each cell and seals it. If upon sealing the cell a caterpillar is placed where it is visible to the wasp, the wasp opens the cell, adds the caterpillar, lays another egg and seals the cell. It will repeat this many times, to the point where the cell is so crammed with caterpillars that there is insufficient air or space for the eggs to develop. It is obvious that the wasp has no appreciation of the purpose of its actions. If it had it would surely not pursue its pattern of behaviour to the point of jeopardizing the survival of its eggs. Instead the wasp reacts automatically, responding to the sight of a caterpillar by carrying out the next stage in its inherited pattern of events. Removal of the roof of the cell where the egg is normally laid causes the wasp some agitation. However, it continues to lay the egg as if the cell roof was still intact, rather than adapt its behaviour to lay it elsewhere. This illustrates the relative inflexibility of innate behaviour.

The characteristics of innate behaviour are:

1. It is inherited and not acquired, although some modifications may result from experience.

2. It is similar among all members of a species and there are no individual differences other than those between males and females of the species.

3. It is unintelligent and often accompanied by no appreciation of the purposes it serves.

4. It often comprises a chain of reflexes. Completion of each link in the chain provides the stimulus for the commencement of the next link.

38.7.3 Learned behaviour

Learned behaviour is behaviour which is acquired and modified in response to experience. As such it takes time to refine and so is of greatest benefit to animals with relatively long life spans. The chief advantage of learning over innate behaviour is its adaptability; learned behaviour can be modified to meet changing circumstances. The simplest form of learned behaviour is **habituation**. This involves learning to ignore stimuli because they are followed by neither reward nor punishment. A snail crawling across a board can be made to withdraw into its shell by hitting the board firmly. Repetition of this action results in the snail taking no notice of the stimulus, i.e. it has learnt to ignore the stimulus as it is neither beneficial nor harmful to it.

Associative learning involves the association of two or more stimuli. One form of associative learning is the **conditioned reflex** exemplified by the classic experiments performed on dogs by the Russian physiologist I. P. Pavlov:

1. He allowed dogs to hear the ticking of a metronome and observed no change in the quantity of saliva produced.

2. He presented the dogs with the taste of powdered meat and measured the quantity of saliva produced.

3. He presented the powdered meat and the noise of the metronome simultaneously on 5 to 6 occasions.

4. He presented the noise of a ticking metronome *only* and observed that the dogs salivated in response to it whereas previously they had not done so (Stage **1**).

5. Repetition of Stage **4** leads to a reduction in the quantity of saliva produced until the stimulus fails to produce any response.

This association of one stimulus with another is the basis upon which birds reject certain caterpillars. If a caterpillar is distasteful it is in its interests to advertise itself by being brightly patterned. Having associated a particular pattern with the unpleasant memory of eating such a caterpillar, the bird ignores it as a potential source of food. This often leads to mimicry by otherwise edible insects, which thereby avoid being eaten.

The features of a conditioned reflex are:

1. It is the association of two stimuli presented together.

2. It is a temporary condition.

3. The response in involuntary.

4. It is reinforced by repetition.

5. Removal of the cerebral cortex causes loss of the response.

A second form of associative learning is **operant conditioning (trial and error learning)**. This form of learning, studied by Skinner, differs from the conditioned reflex in the way it becomes established. Animals learn by trial and error. If mistakes are followed by an unpleasant stimulus while correct responses are followed by a pleasant one, the animal learns a particular pattern of behaviour.

The features of operant learning are:

1. The associative stimulus *follows* the action, i.e. it does not need to be simultaneous with it.

2. Repetition improves the response.

3. The action is involuntary.

4. While temporary, the association is less easily removed than in a conditioned reflex.

5. Removal of the cerebral cortex does *not* cause loss of the response.

Latent learning arises when an animal stores information while exploring its environment, and uses it at some later time. A rat placed in a maze with no reward as a stimulus will later complete the maze, when a reward is present, more rapidly than a rat which has never been in the maze.

The highest form of learning is **insight** or **intelligent behaviour**. It involves the recall of previous experiences and their adaptation to help solve a new problem. The rapidity with which a solution is achieved excludes any possibility of trial and error. Chimpanzees will acquire bananas fixed to the roof of their cage by piling up boxes upon which they climb to reach them. In the same way sticks may be joined together to form a long pole which is used to obtain bananas which are out of reach outside the cage. Chimpanzees may even chew the ends of the sticks so they can be made to fit one another.

Imprinting is a simple but specialized form of learning. Unlike other forms of learning, imprinted behaviour is fixed and not easily

Pigeon choosing in a Skinner box

adapted. Newly hatched geese will follow the first thing they see. Ordinarily this would be their mother and the significance of this behaviour is therefore obvious. However, they will follow humans or other objects should these be seen first. This principle is often used in training circus animals. If a trainer becomes imprinted in an animal's mind, it becomes much easier to train.

38.7.4 Social behaviour

Even individuals of solitary species interact, if only to bring their gametes together. At the opposite end of the spectrum, the individuals of some species are completely dependent on one another for their survival. It is vital that these social groupings adapt their behaviour so that it is directed towards the interests of the group rather than the individual.

The advantages of a social group include greater opportunities for locating food and better protection against predators. To achieve this there needs to be efficient communication between individuals and an appreciation of each individual's rôle within the community. Frequently there is a **social hierarchy** or **pecking order**, with each individual having its own fixed status within the group in much the same way that there is a dominance hierarchy in human organizations and institutions.

Social behaviour may be illustrated by a honey-bee colony. Here there is a **caste system** where there are distinct types each with a specified rôle in the group. The **queen** is the single fertile female; the remaining females, the **workers,** are sterile. All males, called **drones**, are fertile. It is particularly in foraging for food that the honey-bees demonstrate complex social cooperation. This form of social behaviour was studied in detail by the German zoologist, Karl von Frisch. He discovered that worker bees returning from foraging missions reveal the location of food sources to other workers using a special **bee dance**. These dances communicate the direction of the food source from the hive and its distance away.

There are two forms of the dance: if the food source is within 100 m of the hive, a **round dance** is performed; if greater than 100 m, a **waggle dance** is carried out. In both dances, the speed at which it is performed is inversely proportional to the distance away from the hive that the food lies. The round dance does not indicate the direction of the food from the hive, but as it lies within 100 m the other workers easily locate it after a brief search.

The waggle dance is more complex. The worker bee moves in a figure-of-eight pattern, waggling its abdomen as it does so. The waggle part of the dance occurs as it moves along the line between the two loops of the figure-of-eight. The number of waggles gives some indication of the quantity of food discovered. The angle of this waggle relative to the vertical is the same as the angle relative to the sun which the other bees should take on leaving the hive, if they are to locate the food. Examples to illustrate this are shown in Fig. 38.23.

The obvious question arises – how do bees locate food sources on cloudy days? They seem able to determine the plane of polarized light, which penetrates the cloud cover, and so are able to locate the sun. To navigate accurately the bees must be able to make allowances for movements in the position of the sun according to the time of day. Honey-bees achieve this with the aid of an internal biological clock, which permits them to have a continually changing picture of

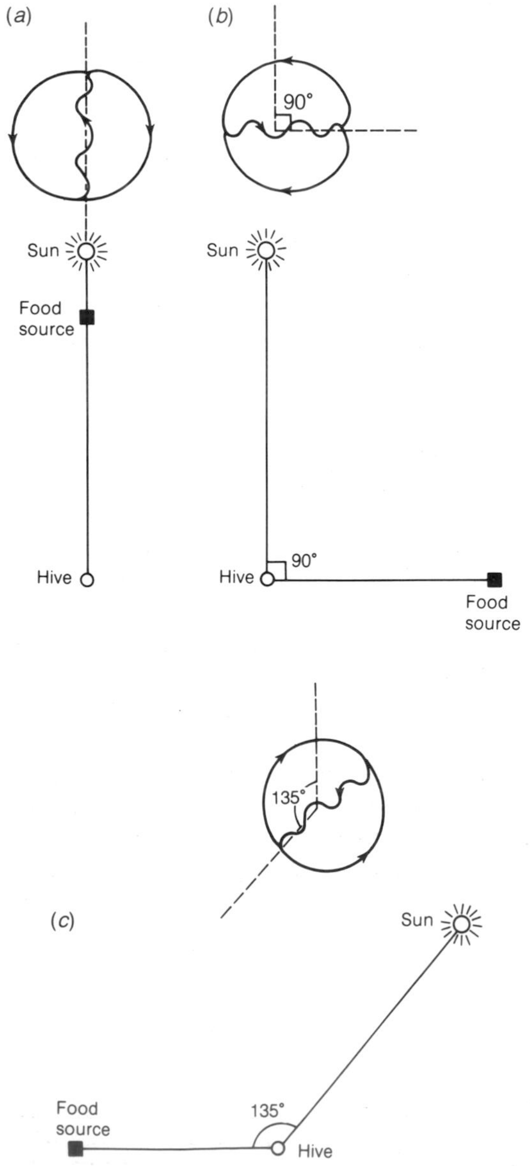

Fig. 38.23 Waggle dance of the honey-bee

the sun's movement during the day. Further information on the nature of the food source may be provided by samples of the nectar and/or pollen brought back to the hive by the foraging worker bee.

These complex bee dances serve to illustrate the importance of social behaviour in communicating valuable information quickly and accurately between individuals in the colony.

38.8 Questions

1. (a) Describe how sound waves in air within the external auditory meatus are transduced into impulses along the auditory nerve. (*8 marks*)
   (b) How is the ear able to distinguish between sounds of different pitch and different loudness? (*4 marks*)
   (c) A loud sound may stimulate preparation for emergency. Describe how this is brought about through co-ordination of the sympathetic and adrenal systems. (*8 marks*)
   (*Total 20 marks*)

   Oxford and Cambridge Board June 1988, Paper II, No. 4

2. (a) Describe a reflex arc. (*6 marks*)
   (b) Explain how information is transmitted across a synapse. (*8 marks*)
   (c) Describe the functions of synapses in the operation of the nervous system. (*4 marks*)
   (*Total 18 marks*)

   Cambridge Board November 1988, Paper I, No. 5

3. The diagram below shows part of two nerve fibres and a synapse. The figures indicate the value in mV of the potential across the membrane between the cytoplasm of the fibres and the extracellular fluid at intervals along each fibre.

   (a) (i) Draw a circle round *one* region of the diagram where an action potential exists. Explain your choice. (*2 marks*)
   (ii) By means of an arrow on the diagram, indicate the direction in which action potentials would normally travel along these fibres.
   Explain your choice of direction. (*2 marks*)
   (b) Identify structures X and Y, and state how each is involved in the transmission of nerve impulses. (*4 marks*)
   (c) (i) What is the major chemical constituent of structure Z? (*1 mark*)
   (ii) State *two* effects of structure Z on the transmission of action potentials. (*2 marks*)
   (*Total 11 marks*)

   London Board January 1989, Paper I, No. 8

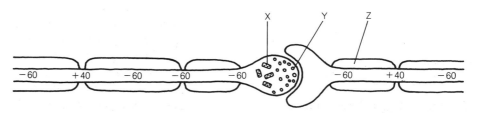

4. The diagram below represents a recording of the changes in potential difference across a nerve fibre membrane during the passage of an action potential.

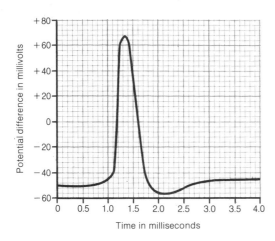

Time in milliseconds

(a) At approximately what times during the recording did the fibre membrane become more than usually permeable to
   (i) sodium ions
   (ii) potassium ions? *(2 marks)*

(b) Give
   (i) the direction
   (ii) the cause
   (iii) the effect of the net movement of sodium ions across the membrane during the period of increased sodium permeability. *(3 marks)*

(c) Explain why sodium and potassium ion concentrations stay more or less constant on both sides of the nerve fibre membrane, no matter how many action potentials are conducted along it. *(2 marks)*

(d) Indicate briefly how, during nerve impulse transmission, depolarisation at one point along a fibre may be passed on to the adjacent region. *(2 marks)*

(Total 9 marks)

London Board June 1989, Paper I, No. 5

5. (a) Describe, with the aid of examples, how the brain is involved in:
   (i) voluntary muscle actions;
   (ii) involuntary actions;
   (iii) sensory perception. *(15 marks)*

(b) Distinguish between instinctive and learned behaviour. *(5 marks)*

(Total 20 marks)

London Board June 1982, Paper II, No. 2

6. The diagram below represents part of the retina of the human eye.

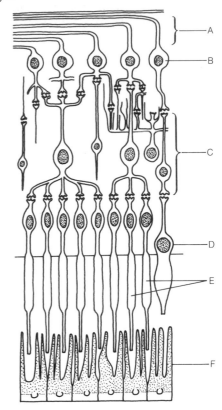

(a) Name the structures labelled **A** to **F**. *(6 marks)*

(b) Indicate by an arrow on the diagram the direction in which light rays enter the eye. *(1 mark)*

(c) What is the significance of the fact that
   (i) several cells of type labelled **E** connect to one cell in layer **C**? *(2 marks)*
   (ii) one cell of type labelled **D** connects to one cell in layer **C**? *(2 marks)*

(d) Comment on the distribution of cell types labelled **D** and **E**
   (i) in the retina generally, *(2 marks)*
   (ii) at the fovea (yellow spot). *(2 marks)*
   (iii) at the blind spot. *(1 mark)*

(e) How is the light energy received by the cells of the retina converted into a nervous impulse? *(4 marks)*

(Total 20 marks)

Oxford and Cambridge Board June 1989, Paper I, No. 3

7. (a) Contrast the modes of action of the nervous and endocrine systems in mammals. *(4 marks)*

(b) Give an illustrated account of the ways by which impulses are transmitted
   (i) along neurones and
   (ii) across synapses in the mammalian nervous system. *(16 marks)*

(Total 20 marks)

Welsh Joint Education Committee June 1989, Paper A1, No. 4

8. *Fig. 1* shows a simplified representation of apparatus used to investigate the response of a muscle to stimuli. A fresh muscle and nerve preparation from an amphibian was used. This was bathed in physiological saline throughout the experiment.

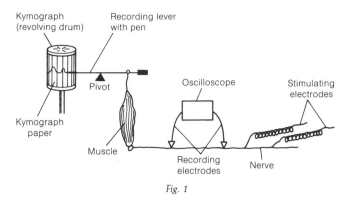

Fig. 1

Fig. 2 shows traces that were recorded from the **oscilloscope** when different voltages (referred to as low, medium and high) were applied to the nerve via the stimulating electrodes.

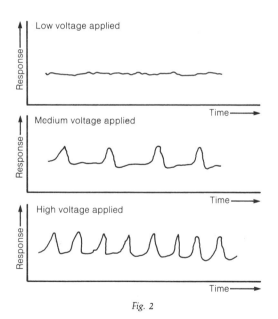

Fig. 2

(a) Describe **two** features of the nerve impulse which can be inferred from these traces. Explain your reasoning in each case. (*4 marks*)

The apparatus was then used to determine the speed of a nerve impulse. With the kymograph drum revolving at full speed, stimuli were applied along the nerve at different distances from the muscle. The results produced on the kymograph papers are shown in *Fig. 3*. (Note that each vertical line is separated by a time interval of 0.002 seconds)

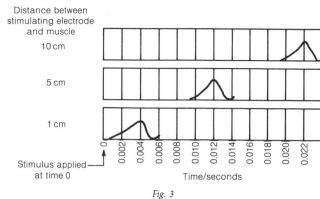

Fig. 3

(b) (i) Suggest why stimuli were applied at different distances from the muscle.
(ii) Using these results, calculate the speed of the nerve impulse in cm s^{-1}. (*3 marks*)

The experiment was repeated with the nerve poison *curare* added to the solution bathing the nerve-muscle preparation. No response on the kymograph paper was observed when a medium voltage was applied at the stimulating electrodes, although direct physical stimulation of the muscle did produce a response.

(c) (i) From this information suggest how curare acts on nerves.
(ii) Describe and explain a simple experiment you could use to confirm your suggestion.
(*5 marks*)
(*Total 12 marks*)

Joint Matriculation Board June 1988, Paper IA, No. 4

9. (a) The diagram shows a cross-section through the human eye.

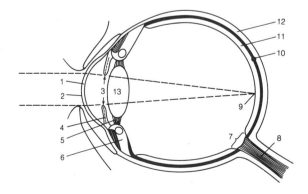

Identify the features labelled 1 to 13. (*2 marks*)
(b) Describe how the eye is able to accommodate to both near and distant objects. (*2 marks*)
(c) Ophthalmologists often place a few drops of atropine (an antagonist of acetylcholine) in the eyes of their patients before performing a retinal examination.

(i) What do you deduce is the effect of atropine on the pupil?

(ii) What is the mechanism of this action?

(4 marks)

(d) In the human retina there are approximately 150 million sensory cells (7 million cones, the rest being rods). There are only 1 million fibres in an optic nerve.

(i) What is the implication of these figures?

(ii) With the aid of a simple labelled diagram of the human retina, explain the structural basis of visual acuity. (Visual acuity is a measure of the finest detail which the unaided eye can distinguish.)

(7 marks)

(e) (i) Give a brief account of the trichromatic theory of colour vision.

(ii) Outline the experimental evidence for this theory.

(5 marks)

(Total 20 marks)

Northern Ireland Board June 1984, Paper II, No. 3

39 *Movement and support in animals*

Movement is a displacement from one point to another, and occurs at all levels of organization:

Atomic – the raising of an electron to a higher energy level during the light stage of photosynthesis.

Molecular – osmotic movement of water molecules, diffusion of gases, nutrients and wastes into and out of an organism, cytoplasmic streaming.

Cellular – transport of red and white blood cells around the body, locomotion of unicellular organisms like *Amoeba*, swimming of gametes.

Organ – beating of the heart.

Organism – locomotion of multicellular individuals by walking, swimming, flying etc.

Population – migrations of lemmings, swallows, salmon etc.

Locomotion is the movement of a whole organism from one place to another. While both plants and animals exhibit movement, only animals carry out locomotion. There are a number of reasons why animals move from place to place:

1. To obtain food – Most animals need to obtain food outside their immediate vicinity in order to satisfy their nutritional requirements. Herbivores exhaust the plant supply in their area and need to move to new pastures; carnivores must chase and capture their prey.

2. To escape predators – The survival of many animal species is dependent upon their ability to avoid capture by predators.

3. To find a mate – Where fertilization is internal two animals must come together before gametes may be transferred.

4. To distribute offspring – If overcrowding is to be avoided, and the genetic variety produced in sexual reproduction is to be exploited, animals must move into new areas. Scattered populations are less vulnerable to epidemic disease.

5. To reduce competition – The movement of individuals away from centres of population prevents overcrowding and reduces intraspecific competition.

6. To avoid danger – Survival may depend on an organism being able to move away from dangers – both biotic, e.g. predators, and abiotic, e.g. fire.

7. To maintain position – In the absence of swim bladders, sharks must swim horizontally simply to maintain a fixed vertical position.

8. To avoid waste products – As waste products are often toxic and may carry disease, it is an advantage for animals to move away from them.

Running with the hare

It is clearly an advantage for any organism which moves from place to place to keep its size relatively small. In this way it expends less energy during locomotion and, if it is terrestrial, reduces the problems of support. Plants, however, need to be relatively large as an increased surface area is essential for the efficient absorption of light, carbon dioxide and water. So on the whole, the nutrition of plants is incompatible with locomotion, while in animals locomotion is often an essential part of obtaining food. Some animals, however, do not exhibit locomotion, i.e. they are sessile. A stationary mode of life presents some problems. Food must be brought to the organism and this is often achieved by creating through the animal a current of the medium in which it lives. From this the animal filters or captures food. As air has little suspended material in it, but water frequently has much, sessile animals are almost exclusively aquatic. A second problem for these animals is the transfer of gametes. Internal fertilization is impractical and so of necessity it is external. Once again, in animals this is only feasible in an aquatic medium because gametes cannot swim through air.

Movement, whether of an animal from one position to another or of material through the animal, is achieved in three ways: by use of pseudopodia, cilia (or flagella) or muscle.

39.1 Pseudopodia

Pseudopodia are temporary projections of a cell. The ability to form them is restricted to a few cells, notably rhizopod protozoans like *Amoeba*, and macrophage cells such as white blood cells. The plasma membrane of these cells is slightly adhesive to enable it to stick to its substrate. Beneath the membrane the cytoplasm is divided into two regions. The outer **ectoplasm** consists of a more rigid form of cytoplasm known as **plasmagel**. The inner layer, **endoplasm**, has a more liquid form of cytoplasm known as **plasmasol**. During the formation of a pseudopodium the central core of endoplasm is observed to move forwards, pushing out the tip of the pseudopodium. It then forms the more viscous ectoplasm and flows backwards along the outer portion of the cell (Fig. 39.1).

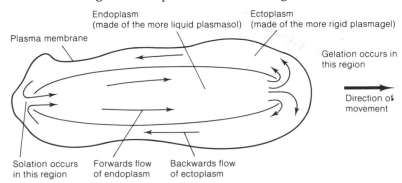

Fig. 39.1 Cytoplasmic movements during the formation of a pseudopodium in Amoeba

At the anterior end, the plasmasol becomes stiffer and forms the more rigid plasmagel in a process called **gelation**. At the posterior end, the plasmagel forms the more liquid plasmasol in a process called **solation**.

The exact mechanism by which movement is brought about in these cells is uncertain and the recent discovery of the proteins actin and myosin in these, and indeed in all eukaryotic cells, has led to speculation that the process may be much the same as that of muscle contraction. There have been two other theories put forward. In the first, and more generally accepted one, it is postulated that protein molecules in the cytoplasm can exist in a contracted or extended state. In the forward-flowing endoplasm the proteins are extended and so form the more liquid plasmasol. At the tip of the pseudopodium the protein molecules contract by coiling, pulling forward more plasmasol from behind. This contraction forms the more rigid plasmagel which flows backwards at the periphery of the cell. At the rear the molecules extend by uncoiling and then flow forwards as plasmasol in the centre of the cell.

The second theory proposes that the contractile forces moving the cell occur not at the anterior but in the posterior region. This causes a force at the rear which pushes the central column of plasmasol forwards, forming a pseudopodium.

39.2 Cilia and flagella

Cilia and flagella were considered in Section 4.3.15 and their structure is illustrated in Fig. 4.12. While cilia are found in many animals, it is only in very small organisms, especially protozoans, where they serve as the sole means of locomotion, although they contribute to locomotion in some creeping flatworms and molluscs. Flagella, however, are largely locomotory and move not only unicellular organisms but also individual cells, e.g. sperm, of multicellular ones. As we saw in Chapter 4, flagella and cilia are fundamentally similar, the additional length of a flagellum being its main distinguishing feature. Their methods of propelling cells are, however, very different. A flagellum produces a wave-like motion usually in one plane, although it may occasionally be spiral. Commonly the flagellum is at the rear of a cell from where it propels it forwards. The mode of action of a flagellum is illustrated in Fig. 39.2 (a).

In a cilium, there are two distinct phases to its movement. During the **effective stroke**, the cilium is fairly rigid and extends rapidly in a complete arc. It is this action which presents the greatest resistance on the medium and propels the cell forwards. During the **recovery stroke**, the cilium is more flexible and returns slowly in a folded position. This action creates far less resistance against the medium. The mode of action of a cilium is illustrated in Fig. 39.2 (b).

The structures of cilia and flagella are remarkably similar in all organisms. The usual arrangement of filaments is shown in the photo on the next page and Fig. 4.12 (b) and is known as the **9 + 2 arrangement**. The main departure from this system is the absence of the two central filaments, in which case the structure is non-motile. Each of the peripheral filaments is made up to two microtubules, A and B, and a hook-like extension made of a protein called **dynein**.

Bending results when the microtubules of a pair slide past one another, a process involving the use of ATP. It seems likely that the five pairs on one side act together in this way during the effective stroke, followed by the four on the opposite side during the recovery stroke. The rôle of the central filaments may be to send the messages from the basal body which initiates sliding. On the other hand, they may act as temporary anchorage points for the outer filaments. Cilia do not function independently as do flagella, but rather beat in coordinated waves. The wave of effective strokes begins at some point on the membrane and spreads along its length in a single

(a) Stages in the movement of a flagellum

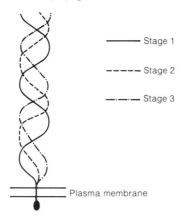

——— Stage 1

----- Stage 2

—·—·— Stage 3

Plasma membrane

(b) Stages in the movement of a cilium

Direction of rapid forward stroke

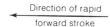

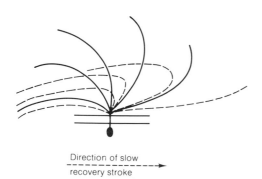

Direction of slow recovery stroke

Fig. 39.2 Mode of action of cilia and flagella

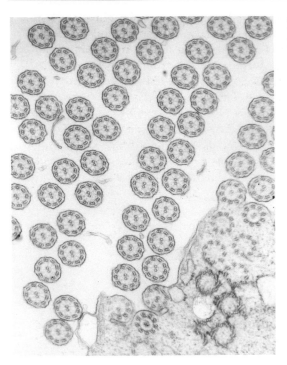

TS cilia (EM) (× 26 500 approx.)

direction. This synchronized motion is called **metachronal rhythm**. It is not yet clear how such coordinated movement is controlled. Both cilia and flagella play a variety of rôles in living organisms.

Functions of flagella

1. Locomotion – Flagellate protozoans like *Trypanosoma* use a flagellum as an organ of locomotion.

2. Feeding – In sponges, flagella create a flow of water through the organism from which food is removed.

3. Reproduction – The sperm of vertebrates are propelled to the ovum by means of their flagellate tails. Gametes in other organisms are transported in the same way.

Functions of cilia

1. Locomotion – Protozoans such as *Paramecium* depend on cilia as their sole means of propulsion.

2. Feeding – Cilia create a water flow in many filter-feeding organisms, e.g. *Mytilus* (mussel).

3. Reproduction – Cilia may contribute to the movement of ova down the oviduct in mammals.

4. Gaseous exchange – During gaseous exchange in mammals, dust and dirt are trapped by mucus lining the nasal passages, trachea and bronchi. This dirt-laden mucus is removed by the action of cilia. In some aquatic invertebrates, cilia create water flow which brings in oxygen and removes carbon dioxide.

5. Transport of nutrients – The central canal and ventricles of the central nervous system are lined with cilia which circulate the cerebro-spinal fluid that carries nutrients to the nerve cells and removes their wastes which are exchanged with the blood.

39.3 Muscular movement

In mammals there are three types of muscle: cardiac, smooth and skeletal. **Cardiac muscle** occurs exclusively in the heart where it functions to circulate blood around the body. Its structure is described in Section 5.4.3 and its function in Section 32.4. **Smooth muscle** makes up the walls of most tubular structures, such as the alimentary canal, blood vessels and ducts of the urino-genital system. Its structure is described in Section 5.4.2. While both types of muscle have an effector nerve supply from the central nervous system, they are not under voluntary control and will continue to contract and relax rhythmically in the absence of the effector nerve supply. By contrast, the third type of muscle, **skeletal muscle**, is under voluntary control and will only contract when innervated by a motor nerve. Its basic structure is described in Section 5.4.1. As skeletal muscle is responsible for the majority of locomotion in animals, a more detailed account of its structure and functions is given below.

39.3.1 Structure of skeletal muscle

An individual muscle is made up of hundreds of **muscle fibres**. These fibres are cylindrical in shape with a diameter of around 50 μm. They

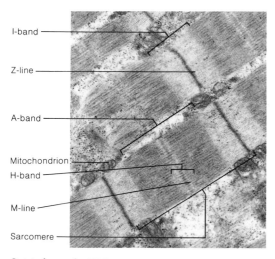

Striated muscle (EM)

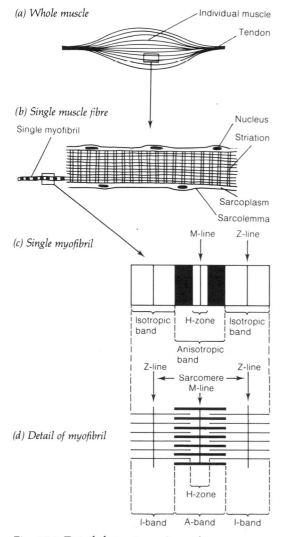

Fig. 39.3 Detailed structure of muscle

vary in length from a few millimetres to several centimetres. Each fibre has many nuclei and a distinctive pattern of bands or cross striations. It is bounded by a membrane – the **sarcolemma**. The fibres are composed of numerous **myofibrils** arranged parallel to one another. Each repeating unit of cross striations is called a **sarcomere** and in mammals has a length of 2.5–3.0 μm. The cytoplasm of the myofibril is known as **sarcoplasm** and possesses a system of membranes called the **sarcoplasmic reticulum**.

The myofibril has alternating dark and light bands known as the **anisotropic** and **isotropic** bands respectively. Confusion between their names can be avoided by reference to the following:

dArk band
↓
Anisotropic band

lIght band
↓
Isotropic band

Each isotropic (light) band possesses a central line called the **Z line** and the distance between adjacent Z lines is a **sarcomere**. Each anisotropic (dark) band has at its centre a lighter region called the **H zone**, which may itself have a central dark line – the **M line**. This pattern of bands is the result of the arrangement of the two types of protein found in a myofibril. **Myosin** is made up of thick filaments and **actin** of thin ones. Where the two types overlap, the appearance of the muscle fibre is much darker. Anisotropic bands are therefore made up of both actin and myosin filaments whereas the isotropic band is made up solely of actin filaments. These arrangements and the overall structure of skeletal muscle are shown in Fig. 39.3.

Myosin filaments are approximately 10 nm in diameter and 2.5 μm long. They consist of a long rod-shaped fibre and a bulbous head which projects to the side of the fibre. These heads are of major significance in the contraction of muscle (Section 39.3.4). Actin filaments are thinner and slightly shorter than those of myosin being approximately 5 nm in diameter and 2.0 μm long. The filaments comprise two different strands of actin molecules twisted around one another. Associated with these filaments are two other proteins: **tropomysin**, which forms a fibrous strand around the actin filament, and **troponin**, a globular protein vital to contraction of muscle fibre.

39.3.2 The neuromuscular junction

Skeletal muscle will not contract of its own accord but must be stimulated to do so by an impulse from an effector nerve. The point where the effector nerve meets a skeletal muscle is called the **neuromuscular junction** or **end plate**. If there was only one junction of this type it would take time for a wave of contraction to travel across the muscle and so not all the fibres would contract simultaneously and the movement would be slow. As rapid contraction is frequently essential for survival, animals have evolved a system whereby there are many end plates spread throughout a muscle. These simultaneously stimulate a group of fibres known as an **effector (motor) unit**; contraction of the muscle is thus rapid and powerful. This arrangement also gives control over the force generated by a muscle as not all the units need be stimulated at one time. If only slight force is needed only a few units will be stimulated. The structure of an end plate is shown in Fig. 39.4.

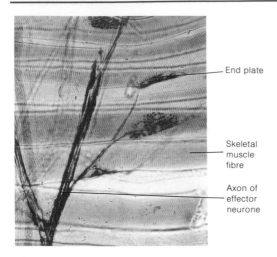

End plate (× 350 approx.)

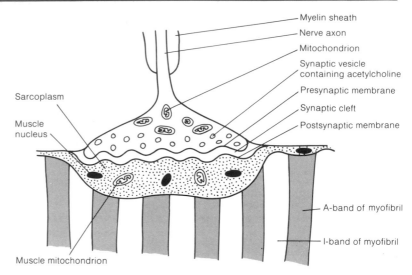

Fig. 39.4 Neuromuscular junction – the end plate

When a nerve impulse is received at the end plate, synaptic vesicles fuse with the end plate membrane and release their acetylcholine. The transmitter diffuses across to the sarcolemma where it alters its permeability to sodium ions which now rapidly enter, depolarizing the membrane. Provided the threshold value is exceeded, an action potential is fired in the muscle fibre and the effector (motor) unit served by the end plate contracts. Breakdown of the acetylcholine by acetylcholinesterase ensures that the muscle is not overstimulated and the sarcolemma becomes repolarized. This sequence of events is much the same as the mechanism of synaptic transmission (Section 38.2.2).

39.3.3 Muscular contraction

Muscle cells have the ability to contract when stimulated and are therefore able to exert a force in one direction. Before a muscle can be contracted a second time, it must relax and be extended by the action of another muscle. This means that muscles operate in pairs with each member of the pair acting in the opposite direction to the other. These muscle pairs are termed **antagonistic**. Muscles may be classified according to the type of movement they bring about. For example, a **flexor** muscle bends a limb, whereas an **extensor** straightens it. Flexors and extensors therefore form an antagonistic pair. These and other types of skeletal muscle are listed in Table 39.1.

Skeletal muscle, as the name suggests, is attached to bone. This attachment is by means of **tendons** which are connective tissue made up largely of **collagen fibres**. Collagen is relatively inelastic and extremely tough. When a muscle is contracted, the tendons do not stretch and so the force is entirely transmitted to the bone. Each muscle is attached to a bone at both ends. One attachment, called the **origin**, is fixed to a rigid part of the skeleton, while the other, the **insertion**, is attached to a moveable part. There may be a number of points of insertion and/or origin. For example, two antagonistic muscles which bend the arm about the elbow are the **biceps** (flexor) and **triceps** (extensor). The biceps has two origins both on the scapula (shoulder blade) and a single insertion on the radius. The triceps has three origins, two on the humerus and one on the scapula and a single insertion on the ulna. These arrangements are illustrated in Fig. 39.5.

TABLE 39.1 **Antagonistic pairs of muscles and the movements they perform**

Muscle action	Opposing muscle action
Flexor – bends a limb	**Extensor** – straightens a limb
Adductor – moves a limb laterally away from the body	**Abductor** – moves a limb from a lateral position in towards the body
Protractor – moves a limb forwards	**Retractor** – moves a limb backwards

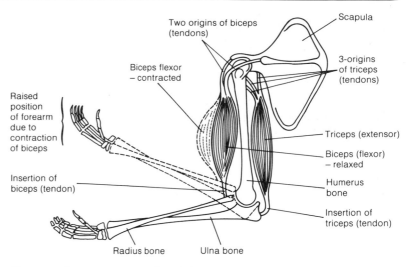

Fig. 39.5 Antagonistic muscles of the forearm

39.3.4 Mechanism of muscular contraction – the sliding filament theory

Much of our present knowledge about the mechanism of muscular contraction has its origins in the work of H. E. Huxley and J. Hanson in 1954. They compared the appearance of striated muscle when contracted and relaxed, observing that the length of the anisotropic band (A band) remained unaltered. They concluded that the filaments of actin and myosin must in some way slide past one another – the **sliding filament theory**. It appears that the actin filaments, and hence the Z lines to which they are attached, are pulled towards each other, sliding as they do over the myosin filaments. No shortening of either type of filament occurs. These changes are illustrated in Fig. 39.6. From this diagram it should be clear that upon contraction the following can be observed in the appearance of a muscle fibre:

1. The isotropic band (I band) becomes shorter.

2. The anisotropic band (A band) does not change in length.

3. The Z lines become closer together, i.e. the sarcomere shortens.

4. The H zone shortens.

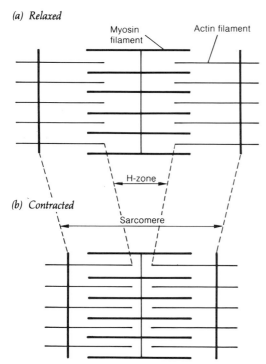

(a) Relaxed

(b) Contracted

The H-zone, sarcomere and I-band all shorten. The A-band is unaltered

Fig. 39.6 Changes in appearance of a sarcomere during muscle contraction

How exactly do the actin and myosin filaments slide past one another? The explanation seems to be related to cross bridges between the two types of filament, which can be observed in photoelectronmicrographs of muscle fibres. The bulbous heads along the myosin filaments form these bridges, and they appear to carry out a type of 'rowing' action along the actin filaments. The sequence of events involved in these movements is shown in Fig. 39.7.

Each myosin filament has a number of these bulbous heads and each progressively moves the actin filament along, as it becomes attached and reattached. This process is similar to the way in which a ratchet operates and for this reason it is often termed a **ratchet mechanism**. The result of this process is the contraction of a muscle or **twitch**. Having contracted, the muscle then relaxes. In this condition the myosin heads are drawn back towards the myosin filament and the fibrous tropomyosin blocks the attachment sites of the actin filament (Fig. 39.8). This prevents linking of myosin to actin and so stops further muscle contraction. As the separation of the

(a) *The head of the myosin molecule is 'cocked' ready to attach to the actin filament*

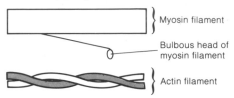

Myosin filament

Bulbous head of myosin filament

Actin filament

(b) *Myosin head attaches to a monomer unit on the actin molecule*

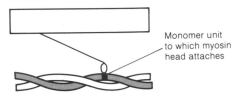

Monomer unit to which myosin head attaches

(c) *The myosin changes position in order to attain a lower energy state. In doing so it slides the actin filament past the stationary myosin filament*

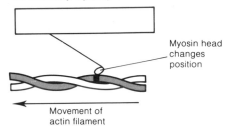

Myosin head changes position

Movement of actin filament

(d) *The myosin head detaches from the actin filament as a result of an ATP molecule fixing to the myosin head*

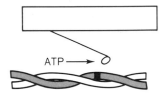

ATP →

(e) *The ATP provides the energy to cause the myosin head to be 'cocked' again. The hydrolysis of the ATP gives rise to ADP + P*

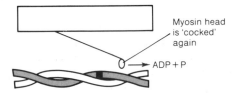

Myosin head is 'cocked' again

→ ADP + P

(f) *The 'cocked' head of the myosin filament reattaches further along the actin filament and the cycle of events is repeated*

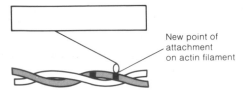

New point of attachment on actin filament

Fig. 39.7 Sliding filament mechanism of muscle contraction

myosin and actin requires the binding of ATP to the myosin head, and as this can only be produced in a living organism, the muscles at death remain contracted. This results in a stiffening of the body known as **rigor mortis**.

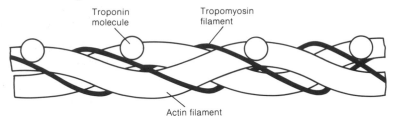

Troponin molecule

Tropomyosin filament

Actin filament

Fig. 39.8 Relationship of tropomyosin and troponin to the actin filament

Before a relaxed muscle may be contracted again, the actin filament must be unblocked by somehow moving the tropomyosin so that the myosin heads may once again bind with it. When a muscle is stimulated the wave of depolarization created spreads, not only along the sarcolemma but also throughout a series of small tubes known as the **T-system**. The T-system is in contact with the sarcoplasmic reticulum, both of which are rich in calcium ions (Ca^{2+}). On depolarization they both release these ions. The calcium so released binds to part of a protein molecule called **troponin**. This causes the troponin to change shape and in so doing move the tropomyosin molecule away from the actin filament. This unblocks the actin and so allows the myosin head to become attached to it, causing the muscle to contract. The calcium ions are then actively pumped back into the sarcoplasmic reticulum and T-system ready for use in initiating further muscle contractions.

The energy for muscle contraction is provided by ATP which is formed by oxidative phosphorylation during the respiratory breakdown of glucose. The supply of glucose is provided from the store of glycogen found in muscles. The resynthesis of ATP after its hydrolysis requires a substance called **phosphocreatine**.

39.3.5 Summary of muscle contraction

The events described fit the observed facts of muscle contraction. Further research may reveal additional detail, but the basic mechanism is unlikely to be modified. In view of the complexity of the process, a summary of the main stages is given below:

1. Impulse reaches the neuromuscular junction (end plate).

2. Synaptic vesicles fuse with the end-plate membrane and release a transmitter (e.g. acetylcholine).

3. Acetylcholine depolarizes the sarcolemma.

4. Acetylcholine is hydrolysed by acetylcholinesterase.

5. Provided the threshold value is exceeded, an action potential (wave of depolarization) is created in the muscle fibre.

6. Calcium ions (Ca^{2+}) are released from the T-system and sarcoplasmic reticulum.

7. Calcium ions bind to troponin, changing its shape.

8. Troponin displaces tropomyosin which has been blocking the actin filament.

9. The myosin heads now become attached to the actin filament.

10. The myosin head changes position, causing the actin filaments to slide past the stationary myosin ones.

11. An ATP molecule becomes fixed to the myosin head, causing it to become detached from the actin.

12. Hydrolysis of ATP provides energy for the myosin head to be 'cocked'.

13. The myosin head becomes reattached further along the actin filament.

14. The muscle contracts by means of this ratchet mechanism.

15. The following changes in the muscle fibre occur:

   (a) I band shortens;
   (b) Z lines move closer together (i.e. sarcomere shortens);
   (c) H zone shortens.

16. Calcium ions are actively absorbed back into the T-system.

17. Troponin reverts to its original shape, allowing tropomyosin to again block the actin filament.

18. Phosphocreatine is used to regenerate ATP.

39.4 Skeletons

Organisms originally evolved in water which gave support. Nevertheless some skeletal system was still necessary for most aquatic organisms, either as a rigid framework for the attachment of muscles or for protection. When organisms colonized land, where air provides little support, a skeleton was also necessary to support them against the pull of gravity. Skeletons therefore fulfil three main functions: support, locomotion and protection.

39.4.1 Types of skeleton

Skeletons take a wide variety of forms but they are generally classified into three main types: **hydrostatic, exoskeleton** and **endoskeleton**.

Hydrostatic skeleton
As liquids are incompressible they can form a resilient structure against which muscles are able to contract. This type of skeleton is typical of soft-bodied organisms such as annelids where liquid is secreted and trapped within body cavities. Around the liquid, muscles are arranged segmentally. They are not attached to any part of the body but simply contract against one another. The antagonistic arrangement involves circular muscles, whose contraction makes the body longer and thinner, and longitudinal muscles, whose contraction makes it shorter and thicker. In an earthworm (*Lumbricus terrestris*), chaetae on each segment anchor it to the substrate, so that alternate contractions of the two types of muscle bring about movement. Not all circular or longitudinal muscles contract together; instead, waves of contraction pass along the body.

Exoskeleton
An exoskeleton is a more or less complete external covering which

provides protection for internal organs and a rigid attachment for muscles. It is characteristic of arthropods where it comprises a three-layered cuticle secreted by epidermal cells beneath it. The outer layer, the **epicuticle**, is thin and waxy and so forms a waterproof covering. Beneath this is a rigid layer of **chitin** (Section 3.5.4) impregnated with tanned proteins. This is the **exocuticle**. The inner layer, or **endocuticle**, is a more flexible layer of chitin. The exoskeleton may be impregnated with salts, e.g. calcium carbonate, which give it additional strength. This is frequently the case in crustacean exoskeletons.

To permit uninhibited movement, the inflexible parts of the exoskeleton are separated by flexible regions where the rigid exocuticle is absent. Openings to glands and the digestive, respiratory and reproductive systems puncture the exoskeleton as do sensory hairs. Since the exoskeleton cannot expand, it imposes a limit on growth. To overcome this, it is periodically shed by a process known as moulting or **ecdysis** (Section 20.4).

Endoskeleton
An endoskeleton forms an internal framework within an organism. Endoskeletons occur in certain protozoans, molluscs (cephalopods) and vertebrates. In the vertebrates, the skeleton is cellular although the bulk of it is a non-cellular matrix secreted by the cells. In the Chondrichthyes (sharks and rays), cartilage forms the entire skeleton. This has the advantage of combining rigidity with a degree of flexibility. Most vertebrates have a skeleton made up of bone which provides a strong, rigid framework. This, however, needs special articulating points known as **joints** if movement is to be possible.

39.4.2 Structure of bone

Bone is essentially calcareous material in a complex crystalline form deposited in a matrix of collagen fibres. It is a connective tissue and details of its histology are given in Section 5.3.2. As bones are usually observed in their dried state, they often give the impression of dead, immutable structures. In fact they are living tissue which is very plastic; capable of moulding itself to meet the mechanical requirements demanded of it.

When a bone, e.g. the femur, is examined in detail, it is found to have a complex internal structure. There is a hollow shaft, the **diaphysis**, which contains **marrow**, a tissue producing various kinds of blood cell. At each end is an expanded head, the **epiphysis**, which articulates with other bones or to which tendons are attached. While the diaphysis and epiphysis are composed of **hard (compact) bone**, the remainder of the structure is made up of **spongy (cancellous) bone**. This has a honeycomb appearance and provides strength with a minimum of additional mass. A tough, fibrous membrane, the **periosteum**, surrounds the bone. The structure of the femur is shown in Fig. 39.9.

Bone has a number of functions:

1. Provides a framework for supporting the body.

2. Provides a means of attachment for muscles which then operate the bones as a system of levers for locomotion.

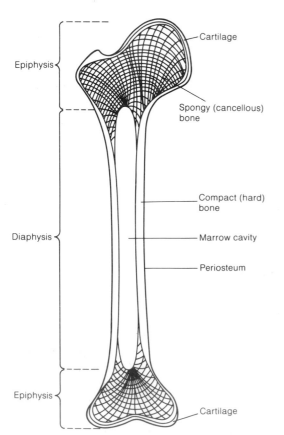

Epiphysis

Diaphysis

Epiphysis

Cartilage

Spongy (cancellous) bone

Compact (hard) bone

Marrow cavity

Periosteum

Cartilage

Fig. 39.9 Vertical section through the femur

3. Protects delicate parts of the body, e.g. the rib cage protects the heart and lungs, the cranium protects the brain.

4. Acts as a reservoir for calcium and phosphorus salts, helping to maintain a constant level in the blood stream (Section 37.5).

5. Produces red blood cells and certain white blood cells, e.g. granulocytes.

39.4.3 Joints

The skeleton has to fulfil two conflicting functions. On the one hand, it needs to be rigid in order to provide support and attachment for muscles; on the other hand, it needs to be flexible in order to permit movement. In the case of bony endoskeletons this paradox is overcome by having a series of flexible joints between the individual bones of the skeleton. The various types of joint found in a mammalian skeleton can be classified into three groups according to the degree of movement possible:

1. Immovable (suture) joints – No movement is possible between the bones.

2. Partly movable (gliding) joints – Only a little movement is possible between individual bones.

3. Movable (synovial) joints – There is considerable freedom of movement between bones; the actual amount depends upon the precise nature of the joint.

The different joints found in mammals are described in Table 39.2.

In movable joints, the ends of the bones are covered with a layer of **cartilage**. This prevents damage to the articulating surfaces of bones as a result of friction between them. The joint is surrounded by a fibrous covering called the **synovial capsule**. The inner lining of this capsule is known as the **synovial membrane**. It secretes a mucus-containing lubricant fluid called **synovial fluid** which also provides nutrients for the cartilage at the ends of the bones. The structure of a typical synovial joint is illustrated in Fig. 39.10.

The bones of a joint are held together by means of strong, but elastic, **ligaments**.

TABLE 39.2 **Joints of the mammalian skeleton**

Name of joint	Type of joint	Example
Suture	Immovable	Between the bones of the cranium Between the sacrum and ilia of the pelvic girdle
Gliding	Partly movable	Between adjacent vertebrae In the wrist and ankle
Pivot	Partly movable	Between the axis and atlas vertebrae
Hinge	Synovial	In the elbow and knee In the fingers and toes
Ball and socket	Synovial	At the shoulder and hip

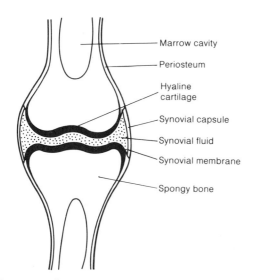

Fig. 39.10 Structure of a typical synovial joint

- Marrow cavity
- Periosteum
- Hyaline cartilage
- Synovial capsule
- Synovial fluid
- Synovial membrane
- Spongy bone

39.5 Locomotion

Locomotion is the movement of an organism from one place to another. The reasons why organisms carry out locomotion are listed in the introduction to this chapter.

39.5.1 Locomotion in fish

Water has a greater relative density than air and so, at the same time as lending support to organisms within it, it also offers considerable resistance to their movement. This resistance makes streamlining essential if rapid locomotion is to be achieved. It does, however, provide a sufficiently dense medium against which the organs of locomotion can push.

The propulsive force in a fish is provided by blocks of muscle segmentally arranged on either side of the vertebral column. These

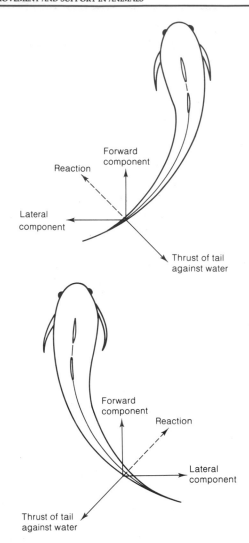

Fig. 39.11 Swimming action in fish

(a) *Heterocercal tail – Found in Chondrichthyes (cartilaginous fish)*

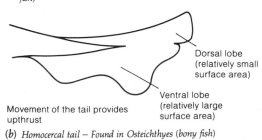

Movement of the tail provides upthrust

Dorsal lobe (relatively small surface area)

Ventral lobe (relatively large surface area)

(b) *Homocercal tail – Found in Osteichthyes (bony fish)*

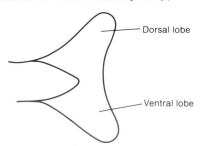

Dorsal lobe

Ventral lobe

Dorsal and ventral lobes have equal surface areas. Movement of the tail does *not* provide upthrust.

Fig. 39.12 Homocercal and heterocercal tails

blocks are called **myotomes**. The myotomes contract and relax alternately on each side of the flexible vertebral column. The fish is thereby bent into a series of waves which create a side-to-side lashing of the tail. This creates a forward and sideways thrust, which is greatly increased in magnitude by the broad caudal fin which presents a large surface area with which to push against the water. As the lateral thrust alternates between left and right, the forces cancel one another out. This leaves an overall forward thrust, with little sideways motion. These events are summarized in Fig. 39.11.

In the Osteichthyes (bony fish), buoyancy is maintained by use of a **swim bladder**. This is a gas-filled compartment in the fish which reduces the overall density of the organism close to that of the surrounding water. By increasing and decreasing the amount of gas in the bladder, the fish is able to float higher or sink deeper in the water. In some fish the bladder contains air which may be gulped into the pharynx at the water surface and passed to the swim bladder via a connecting duct. The air may be expelled at any time by reversing the process. A more sophisticated arrangement in many fish involves the exchange of oxygen between the swim bladder and the blood, using gas glands in its lining.

The Chondrichthyes (cartilaginous fish) lack a swim bladder and so are heavier than the surrounding water and tend to sink. To avoid this the ventral lobe of the caudal fin is larger than the dorsal lobe, making the tail asymmetric – a **heterocercal tail** (Fig. 39.12). This gives an upthrust and so, provided the fish keeps swimming, it is able to maintain its vertical position in the water.

The control of direction and the stability of fish is maintained by the fins. These prevent instability in three major directions, as illustrated in Fig. 39.13.

(a) *Yawing (Lateral change of direction – controlled by median fins)*

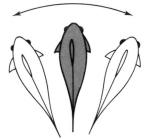

(b) *Pitching (Vertical change of direction – controlled by paired fins)*

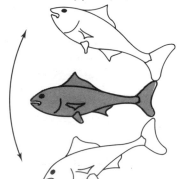

(c) *Rolling (Rotational change of direction – controlled by median fins)*

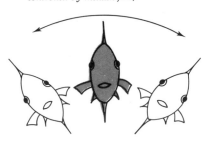

Fig. 39.13 Yawing, pitching and rolling

1. Yawing is the lateral movement of fish. This is controlled by vertical, median fins (dorsal and ventral fins) as well as the general lateral flattening of the body.

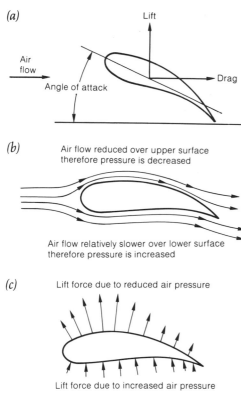

Fig. 39.14 Creation of lift by a bird's wing

2. Rolling is the rotation of the body around its longitudinal axis and this too is controlled by the vertical, median fins (dorsal and ventral fins).

3. Pitching is the vertical up-and-down movement of the head of a fish. It is controlled by the paired fins (pectoral and pelvic fins).

39.5.2 Locomotion in birds

To raise the body of an organism off the ground against the pull of gravity requires immense muscular exertion and yet birds perform this feat seemingly effortlessly. Flight in birds is achieved by use of wings – modified limbs – whose movements through the air create forces which counteract the pull of gravity while at the same time moving the body forwards.

In its simplest form, a wing is a flat plate projecting sideways from the body of the bird. If the wing is held edgewise in a current of air at a slight upward angle – **the angle of attack** (Fig. 39.14a) – the wing will tend to lift. This **lifting force** is due to a greater air pressure being exerted beneath the wing than on top of it. Anything which increases the pressure beneath the wing, or decreases the pressure above it, will increase the lifting force. The wing is shaped so that air travelling over the upper surface takes a longer route than that travelling beneath it. This means air on the upper surface is accelerated compared to that on the lower surface. There is therefore a lower pressure exerted above the wing than below it, and the lifting force is increased (Fig. 39.14).

As air passes over the wing it creates a backwards force called **drag**. Efficient flight requires the lifting force to be large relative to the drag force.

Birds use two main methods of flight. In **gliding flight**, birds hold out their wings to create as much air resistance as possible. The bird therefore descends slowly, often covering large horizontal distances as it does so. Gliding birds typically have large wing spans – in the albatross this may exceed 3.0 m. To attain height initially, the birds utilize rising currents of air. These may occur as a result of air being forced upward as it passes over obstructions, e.g. mountains and cliffs, or as a result of rising currents of warm air, called **thermals**.

In **flapping flight**, the wings are moved up and down rhythmically. If the bird is to gain height it must create a greater resistance against the air on the downward stroke than it does on the upward one. This it achieves by presenting the full surface of the wing to the air on the downstroke, but twisting the wing so that only the front edge is presented to it on the upstroke. At the same time, the feathers are arranged so that no air passes through them on the downstroke while it is free to do so on the upstroke. The downstroke is brought about by the contraction of the large and powerful **pectoralis major muscles**. The smaller **pectoralis minor muscles** are responsible for the upstroke.

39.5.3 Adaptations to flight in birds

Much of the anatomy and physiology of birds is modified to make flight possible. Most adaptations are concerned either with permitting the rapid release of large amounts of energy or in conserving mass. There are adaptations to every major system of birds, and these include:

Movement of a bird's wing

Feathers

1. Contour feathers combine being airproof with very little mass.

2. Contour feathers can be rotated to allow air to pass between them on the upstroke.

3. Down feathers provide excellent insulation which is necessary to help maintain the bird's high body temperature.

Skeleton and muscles

4. The fore limbs are modified into wings with an elongated second digit.

5. The hind limbs are lengthened to prevent wings touching the ground during take off.

6. The aerofoil shape of the wing provides uplift.

7. The first digit is modified to form the **alula** (bastard wing) which helps to prevent drag and turbulence.

8. Some of the bones in the skeleton are fused, e.g. metacarpals and carpals. This provides a more rigid wing with which to push against the air.

9. The total number of bones is reduced to decrease mass.

10. The bones are hollow to reduce mass. A frigate bird with a wing span of 2 m may have a skeleton weighing as little as 110 g.

11. The sternum possesses a large keel for attachment of the large pectoralis muscles.

12. The pectoralis muscles which move the wings are especially large and powerful, making 20 – 40% of the total mass of the bird.

13. There are no teeth and so mass is reduced.

Respiratory system

14. Air sacs extend from the lungs into body spaces including the bones. These allow the bird to absorb oxygen more efficiently.

Blood and circulatory systems

15. A high body temperature (39–42 °C) permits a greater metabolic rate and so a faster release of energy.

16. Separate pulmonary and systemic systems provide more efficient oxygen transport.

17. The innominate arteries which serve the pectoralis muscles are especially large.

18. The erythrocytes are nucleated and so can divide while in circulation. This removes the need for blood-producing bone marrow and so saves mass.

19. The blood sugar level is high in order to supply adequate respiratory substrate to meet the bird's high energy demands.

20. The blood pressure is high to supply oxygen and glucose more rapidly to muscles.

21. The heart beat is very rapid, up to 1000 beats min^{-1} in extreme cases – again to increase the blood supply to the muscles.

Digestive system

22. Concentrated digestive juices reduce the need for water and saves mass.

23. A gizzard is present to carry out physical digestion instead of teeth which would have a greater mass.

24. Digestion is rapid to provide adequate respiratory substrate to sustain the high metabolic rate.

Reproductive system

25. The gonads become considerably reduced in size during the non-breeding season, so conserving mass. In a starling for example they may be 1500 times less heavy at this time.

Nervous system

26. The cerebellum is well developed in order to carry out the complex task of coordinating feathers and muscles during flight.

27. The eyes are well developed with a high concentration of cones in the retina. This permits acute vision, allowing them to spot food from the air at a distance.

28. The eye has a transparent third eyelid - the **nictitating membrane** — which protects the cornea from damage by dust and dirt in the air.

Excretory system

29. Uric acid rather than urine is excreted, conserving water and reducing mass by avoiding the need to store dilute urine in a bladder.

General adaptations

30. External ears are absent, so reducing projections, and the body has an overall streamlined shape to reduce drag.

39.5.4 Locomotion in mammals

Much of the earlier work in this chapter has been concerned with locomotion in mammals and it only remains to review some of the principles behind the methods they use.

The limbs of the body act as a series of levers; in quadrupeds like a dog all four limbs are involved, whereas in man, with his bipedal gait, only the hind limbs make a major contribution. It is the retractor and extensor muscles which play a major rôle in walking and running. The feet are placed on the ground with the limb in a flexed position. Extension of the limb thrusts the animal forwards and slightly upwards. During walking in quadrupeds, only one limb is raised at a time leaving the remaining three in contact with the ground to form a stable tripod of support. As the speed of movement increases the number of limbs in contact with the ground is reduced until, when running at high speed, all four may be airborne at the same time. Mammals have become adapted to a wide range of locomotory methods. Whales and dolphins, for example, have become aquatic and therefore show similar adaptations to those of fish. Likewise, bats have developed flight and have certain adaptations shown by birds.

1. Describe, with the aid of diagrams, how structure and function in the bird are specialized for the performance of flight. *(20 marks)*

Welsh Joint Education Committee June 1985, Paper AII, No. 9

2. Compare the ways in which movement occurs in plants and animals. *(20 marks)*

London Board June 1989, Paper II, No. 7(a)

3. (a) Describe the parts played in movement by the following structures.
  (i) The exoskeleton of an insect limb
  (ii) The long bones of the forelimb of a mammal *(12 marks)*
  (b) What are the advantages and disadvantages of an exoskeleton to a terrestrial organism? *(8 marks)*
  (Total 20 marks)

London Board January 1989, Paper II, No. 6

4. Review the structure of striated voluntary muscle tissue, as observed by the light- and the electron-microscope.
  Describe what happens in a muscle fibre when it contracts. Explain how the contraction is initiated by the nervous system.

Southern Universities Joint Board June 1983, Paper II, No. 9

5. A skeleton serves a number of functions in animals. Bearing this in mind, discuss the relative advantages and disadvantages of exoskeletons and endoskeletons. *(20 marks)*

Oxford and Cambridge Board June 1984, Paper II, No. 8

6. Describe how the following are involved in the sequence of events by which contraction of a muscle fibre is brought about.
  (a) Acetylcholine *(3 marks)*
  (b) Calcium ions *(3 marks)*
  (c) Actin and myosin filaments *(3 marks)*
  (d) ATP *(3 marks)*
  (Total 12 marks)

London Board June 1988, Paper I, No. 9

7. (a) Describe the mechanism of locomotion in (i) a ciliate protozoan, and (ii) *Amoeba* *(5, 5 marks)*
  (b) Discuss why these mechanisms are restricted to small organisms, and explain why large animals, such as fish, use muscles for locomotion. *(8 marks)*
  (Total 18 marks)

Cambridge Board June 1985, Paper I, No. 9

8. (a) (i) There are three types of animal skeleton: hydrostatic skeleton, exoskeleton and endoskeleton. For each type of skeleton give an example of an animal which possesses it.
  (ii) For each of your examples, describe the means by which muscle action produces locomotion. *(12 marks)*
  (b) Describe the mechanisms of:
   (i) amoeboid movement, and
   (ii) the action of cilia and flagella. *(4 marks)*
  (c) Movements in higher plants are normally much slower than in animals, as they are mainly growth movements. However, some quite rapid movements are possible in plants, owing to changes in the water content of cells. Describe two examples of such movements in named plants. *(4 marks)*
  (Total 20 marks)

Northern Ireland Board June 1985, Paper II, No. 5

9. The figure is an electron micrograph of muscle tissue.

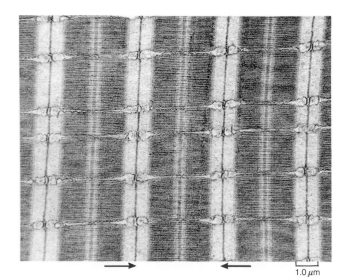

1.0 µm

  (a) Name the type of muscle tissue shown in the figure and explain why it is so called.
  (b) (i) Name the region between the two black lines indicated by the arrows.
   (ii) Draw a simplified diagram of this region and label the structures and bands shown.

(iii) Indicate how contraction occurs according to the sliding filament theory of muscle contraction.

(iv) State what happens to the length of the **A**, **H** and **I** bands as the muscle contracts.

(v) Giving your reasons, decide whether the muscle in the figure is fully contracted, about half contracted or fully relaxed.

(Total 15 marks)

Cambridge Board November 1989, Paper III, No. 1

40 *Control systems in plants*

The ordered growth and development of plants shows that, like animals, they are capable of coordinating their activities. Unlike animals, they possess no nervous system and so plant coordination is achieved almost entirely by hormones. These hormones are similar to those of animals in being organic substances which in low concentrations cause changes in other parts of the organism. Unlike animals, plant hormones almost always affect some aspect of growth. This growth may lead to movements of plant parts, although these are relatively slow responses compared to those of animals.

40.1 Plant responses

Plants do not possess any contractile tissue with which to move. Their survival may, however, depend on their ability to move towards certain stimuli such as light and water. These movements are usually performed by growth responses.

40.1.1 Tropisms

A tropism is a growth movement of part of a plant in response to a directional stimulus. The direction of the response is related to that of the stimulus and in almost all cases the plant part moves towards or away from it. Each response is named according to the nature of the stimulus, e.g. a response to light is termed phototropism. The direction of the response is described as positive, if movement is towards the stimulus, or negative if it is away from it. Some examples of tropisms are given in Table 40.1.

TABLE 40.1 **Examples of tropic responses**

Stimulus	Name of response	Examples
Light	Phototropism	In almost all plants, shoots bend towards a directional light source (i.e. are positively phototropic), some roots bend away (i.e. are negatively phototropic) while leaves position themselves at right angles (i.e. are diaphototropic)
Gravity	Geotropism	In almost all plants, shoots bend away from gravity (i.e. are negatively geotropic) and roots bend towards it (i.e. are positively geotropic). The leaves of dicotyledonous plants position themselves at right angles (i.e. are diageotropic)
Water	Hydrotropism	Almost all plant roots bend towards moisture (i.e. are positively hydrotropic) while stems and leaves show no response
Chemicals	Chemotropism	Some fungal hyphae grow away from the products of their metabolism (i.e. are negatively chemotropic). Pollen tubes grow towards chemicals produced at the micropyle (i.e. are positively chemotropic)
Touch	Thigmotropism	The tendrils of peas (*Pisum*) twine around supports. The shoots of beans (*Phaseolus*) spiral around supports
Air	Aerotropism	Pollen tubes grow away from air (i.e. are negatively aerotropic)

40.1.2 Taxes

A taxis is the movement of a freely motile organism, or a freely motile part of an organism, in response to a directional stimulus. The direction of the response is related to that of the stimulus, being towards it (positive) or away from it (negative). They occur in both plants and animals. As with tropisms, the type of stimulus determines the name of a tactic response. Examples are given in Table 40.2.

TABLE 40.2 **Examples of tactic responses**

Stimulus	Name of response	Examples	
		Plants	**Animals**
Light	Phototaxis	*Euglena* swims towards light provided it is not too intense (i.e. is positively phototactic)	Earthworms (*Lumbricus*) and woodlice (*Oniscus*) move away from light (i.e. are negatively phototactic)
Temperature	Thermotaxis	The green alga, *Chlamydomonas*, will swim to regions of optimum temperature. Motile bacteria behave in a similar way	Blowfly larvae and many other small animals move away from extremes of temperature
Chemicals	Chemotaxis	Antherozooids (sperm) of mosses, liverworts and ferns, are attracted to chemicals produced by the archegonium (i.e. they are positively chemotactic)	Many show negative chemotactic responses to specific chemicals – a fact exploited in the use of insect repellents

40.1.3 Nasties

A **nasty** is the movement of part of a plant in response to a stimulus. The direction of the response is *not* necessarily related to the direction of the stimulus. While some nastic responses are the result of growth, others are due to changes in turgor pressure. Examples of nastic responses are given in Table 40.3.

TABLE 40.3 **Examples of nastic responses**

Stimulus	Name of response	Examples
Light	Photonasty	Many leguminous plants (e.g. French bean – *Phaseolus*) lower their leaves in the dark and raise them in light. The flowers of the wood sorrel, *Oxalis*, open in daylight and close at night
Temperature	Thermonasty	Petals of the crocus and tulip open at temperatures around 16°C and close at ones below this
Touch (shock)	Thigmonasty (seismonasty)	Venus' fly trap (*Dionaea*) closes its leaves rapidly when touched. The leaves of the sensitive plant (*Mimosa pudica*) collapse when touched

40.2 Plant hormones

Plant hormones, or **plant growth substances** as they are often called, are chemical substances produced in plants which accelerate, inhibit or otherwise modify growth. In Section 20.3.3 we saw that growth at the apices of plants occurs in three stages: cell division, cell elongation and cell differentiation. Plant hormones may affect any, or all, of these processes. There are five groups of growth substances generally recognized: **auxins, gibberellins, cytokinins, abscissic acid (inhibitor)** and **ethene** (ethylene).

40.2.1 Auxins

Auxins are a group of chemical substances of which **indoleacetic acid (IAA)** is the most common. They have been isolated from a large number of plants. Charles Darwin was one of the earliest to investigate the response of plant shoots to light (phototropism) which eventually led to the discovery and isolation of auxins. The historical record of these developments is traced in Fig. 40.1.

The auxin indoleacetic acid, a derivative of the amino acid tryptophane, is largely produced at the apices of shoots and roots. Fig. 40.2, on page 616, outlines its chemical structure.

The transport of auxin occurs in one direction, namely, away from the tip, i.e. its movement is **polar**. Short distance movement from cell to cell occurs by diffusion, but long distance transport is possible via phloem.

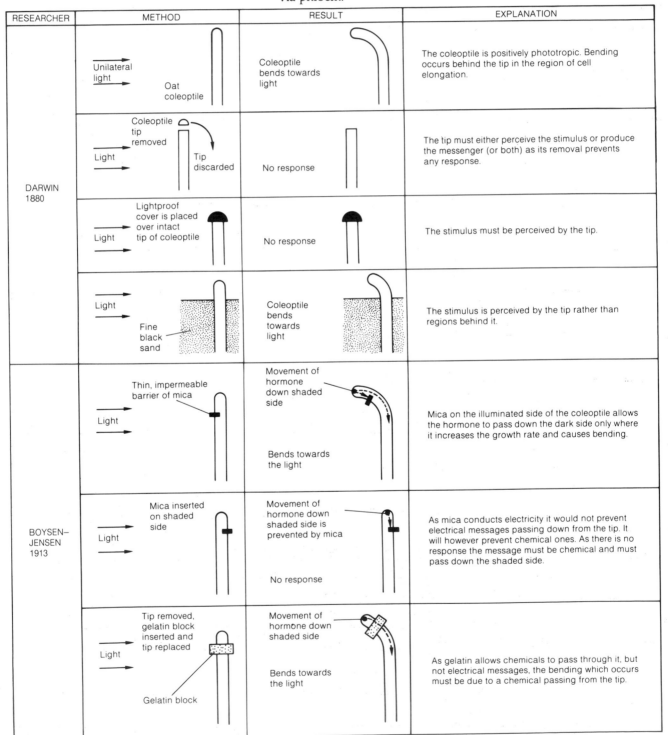

Fig. 40.1 Diagrammatic summary of the historical events leading to the discovery of auxin and an understanding of its mechanism of action (cont.)

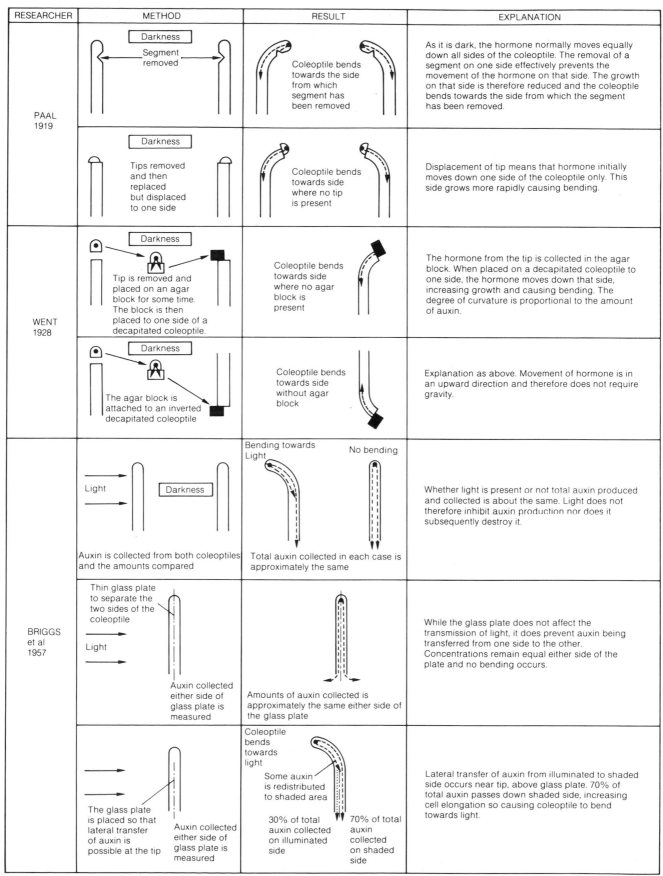

RESEARCHER	METHOD	RESULT	EXPLANATION
PAAL 1919	Darkness — Segment removed	Coleoptile bends towards the side from which segment has been removed	As it is dark, the hormone normally moves equally down all sides of the coleoptile. The removal of a segment on one side effectively prevents the movement of the hormone on that side. The growth on that side is therefore reduced and the coleoptile bends towards the side from which the segment has been removed.
	Darkness — Tips removed and then replaced but displaced to one side	Coleoptile bends towards side where no tip is present	Displacement of tip means that hormone initially moves down one side of the coleoptile only. This side grows more rapidly causing bending.
WENT 1928	Darkness — Tip is removed and placed on an agar block for some time. The block is then placed to one side of a decapitated coleoptile.	Coleoptile bends towards side where no agar block is present	The hormone from the tip is collected in the agar block. When placed on a decapitated coleoptile to one side, the hormone moves down that side, increasing growth and causing bending. The degree of curvature is proportional to the amount of auxin.
	Darkness — The agar block is attached to an inverted decapitated coleoptile	Coleoptile bends towards side without agar block	Explanation as above. Movement of hormone is in an upward direction and therefore does not require gravity.
BRIGGS et al 1957	Light — Darkness — Auxin is collected from both coleoptiles and the amounts compared	Bending towards Light No bending — Total auxin collected in each case is approximately the same	Whether light is present or not total auxin produced and collected is about the same. Light does not therefore inhibit auxin production nor does it subsequently destroy it.
	Thin glass plate to separate the two sides of the coleoptile — Light — Auxin collected either side of glass plate is measured	Amounts of auxin collected is approximately the same either side of the glass plate	While the glass plate does not affect the transmission of light, it does prevent auxin being transferred from one side to the other. Concentrations remain equal either side of the plate and no bending occurs.
	The glass plate is placed so that lateral transfer of auxin is possible at the tip — Auxin collected either side of glass plate is measured	Coleoptile bends towards light. Some auxin is redistributed to shaded area. 30% of total auxin collected on illuminated side. 70% of total auxin collected on shaded side.	Lateral transfer of auxin from illuminated to shaded side occurs near tip, above glass plate. 70% of total auxin passes down shaded side, increasing cell elongation so causing coleoptile to bend towards light.

Fig. 40.1 (cont.) Diagrammatic summary of the historical events leading to the discovery of auxin and an understanding of its mechanism of action

The rôle of auxin in producing a phototropic response is shown in Fig. 40.1. It appears that unilateral light causes a redistribution of auxin so that a greater amount travels down the shaded side. As one

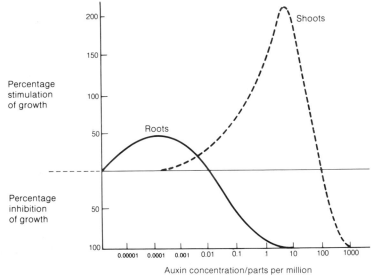

Fig. 40.2 Structure of indoleacetic acid (IAA)

effect of auxin is to cause cell elongation, the cells on the shaded side elongate more than those on the illuminated side. The shoot therefore bends towards the light. Exactly how unilateral light effects a redistribution is not clear.

The redistribution theory may seem plausible but does not immediately explain why many roots are negatively phototropic. If the same arguments are used, it follows that a root exposed to unilateral light should accumulate auxin on its shaded side. This side should elongate more rapidly and the root bend towards the light. Why then does it bend in the opposite direction? Experiments have revealed that roots are more sensitive to auxin than stems, i.e. they respond to lower concentrations of auxins. As the concentration of auxin increases, the growth of roots, far from being stimulated, becomes inhibited. At a concentration of auxin of one part per million, for example, the growth of roots is reduced while that of stems is considerably increased (see Fig. 40.3(a)). If the effect of unilateral light is to redistribute the auxin so that a concentration of one part per million passes down the shaded side, then in a stem, growth will increase on that side and it will bend towards the light, while in a root, growth will be inhibited and it will bend away from the light.

Auxin accumulates on shaded side

Light

Shoot – high auxin concentration *stimulates* growth. Causing a positive phototropic response

Root – high auxin concentration *inhibits* growth. Causing a negative phototropic response

Shoot – high auxin concentration *stimulates* growth. Causing a negative geotropic response

Gravity

Auxin accumulates on the underside

Root – high auxin concentration *inhibits* growth. Causing a positive geotropic response

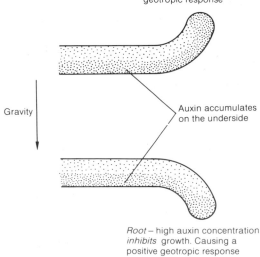

Fig. 40.3(b) Mechanism of auxin action in phototropic and geotropic responses of shoots and roots

Fig. 40.3(a) Relationship between growth and auxin concentration in roots and shoots

It is a similar redistribution of auxins which accounts for the responses of shoots and roots to gravity. It can be shown experimentally that a higher concentration of auxin occurs on the underside of horizontal roots and shoots. The concentrations found inhibit root growth, causing it to bend downwards, i.e. a positive geotropic response. The same concentrations stimulate stem growth, causing it to bend upwards, i.e. a negative geotropic response. Both phototropic and geotropic responses of stems and roots are summarized in Fig. 40.3(b). Phototropic and geotropic responses are due to auxins promoting cell elongation. Auxins, however, have other influences in plants. They stimulate cell division, help to maintain the structure of cell walls and, in high concentrations, inhibit growth. Their ability to stimulate cell division is most easily observed by treating the cut end of plant parts with auxins. Often, an area of disorganized and largely undifferentiated tissue results. The large swellings so produced are called **calluses**. By a similar means, auxins may stimulate fruit development without fertilization (**parthenocarpy**). The maintenance of the structure of cell walls by auxins inhibits **abscission**. This is the separation of leaves, flowers and fruits from a plant as the result of the middle lamellae between cell walls, at the base of a petiole, pedicel or peduncle, weakening to such an extent that small mechanical disturbances, e.g. wind, cause them to fall. Auxin prevents the formation of this abscission layer and so inhibits abscission.

The inhibition of growth by high concentrations of auxin results in **apical dominance**. This is where the bud at the apex of a shoot produces auxin in sufficient concentration to inhibit growth of the lateral buds further down the shoot. These lateral buds remain dormant unless the apical bud is removed, in which case one or more of them develops into side branches. This is the principle behind the pruning of many plants as a means of producing a more bushy form of growth. The effects of auxins are summarized at the end of this section, in Table 40.4.

40.2.2 Gibberellins

The name **gibberellins** was derived from the fungus *Gibberella* (since renamed *Fusarium*). This fungus was shown by Japanese scientists in the 1920s to be the cause of 'foolish seedling disease', a disorder which resulted in rice seedlings growing considerably taller than their healthy counterparts. An extract from this fungus produced an increase in growth when applied to other plants. A group of active substances was finally isolated from the extract; these were called gibberellins. The number of gibberellins now isolated exceeds 50, but all have a similar chemical make up.

The main influence of gibberellins on plants is to promote cell elongation and so increase growth. Unlike auxins, however, gibberellins can stimulate growth in dwarf varieties, thus restoring them to normal size. These dwarf varieties are thought to result from a genetic mutation which prevents them producing gibberellins naturally. They therefore require an external supply to make them grow to normal size. While the main effect of gibberellins is to cause elongation of the stem, they also influence cell division and differentiation to some extent. Their varied effects sometimes complement those of auxins, e.g. in promoting growth, but at other times they have antagonistic effects, e.g. while auxins promote the growth of adventitious roots, gibberellins inhibit their formation. Gibberellins have no rôle in phototropic or geotropic responses. Their varied effects are summarized in Table 40.4.

40.2.3 Cytokinins

Plant cells grown in synthetic nutrient media were found to remain alive but did not undergo cell division. Division only occurred if malt extract or coconut milk was added to the culture. It was later found that autoclaved samples of DNA also induced cell division and from these **kinetin**, a substance similar to adenine, was isolated. Partly as a consequence of their effects in promoting cell division (**cytokinesis**), substances like kinetin were termed **cytokinins**. A number of cytokinins are produced naturally by many plants; all are derivatives of adenine. They are found largely in actively dividing tissues, especially fruits and seeds, where they promote cell division in the presence of auxins.

Effect of gibberellins

One interesting effect of cytokinins is their ability to delay **senescence** (ageing) in leaves. In the presence of cytokinins, leaves removed from a plant remain green and active, rather than turning yellow and dying. Further effects are summarized in Table 40.4.

40.2.4 Abscissic acid

Abscissic acid inhibits growth and so works antagonistically to auxins, gibberellins and cytokinins. Its main effect, as its name suggests, is on abscission (leaf and fruit fall). The process results from a balance between the production of auxin and abscissic acid. As a fruit ripens, the level of auxin (which inhibits abscission) falls, while that of abscissic acid (which promotes abscission) increases. This leads to the formation of an abscission layer which causes the fruit to fall. Other influences of abscissic acid are listed in Table 40.4.

40.2.5 Ethene

Unlike the other hormones, **ethene** (ethylene) has a relatively simple chemical structure. It is produced as a metabolic by-product of most plant organs, especially fruits. Its main effect is in stimulating the ripening of fruits, but it also influences many auxin-induced responses. Other effects are summarized in Table 40.4.

40.2.6 Commercial applications of synthetic growth regulators

As the main function of plant hormones is to control growth, it is hardly surprising that they, or rather their synthetic derivatives, have been extensively used in crop production.

Synthetic auxins such as 2,4-dichlorophenoxyacetic acid (2,4-D) and 2,4,5-trichlorophenoxyacetic acid (2,4,5-T) are used as **selective weedkillers**. When sprayed on crops, they have a more significant effect on broad-leaved (dicotyledonous) plants than on narrow-leaved (monocotyledonous) ones. They so completely disrupt the growth of broad-leaved plants that they die, while narrow-leaved ones at most suffer a temporary reduction in growth. As cereal crops are narrow-leaved and most of their competing weeds are broad-leaved, the application of these hormone weedkillers is of much commercial value. They are also extensively used domestically for controlling weeds in lawns. Other details of these weedkillers are given in Section 29.3.3. Another synthetic auxin, naphthaleneacetic acid (NAA), is used to increase fruit yields. If sprayed on trees, it helps the fruit to set naturally, or in some species causes them to set without the initial stimulus of fertilization (parthenocarpy). This usually results in seedless fruits which may be a commercial advantage.

Gibberellins extracted from fungal cultures are used commercially in the same way. Auxins are the active constituent of rooting powders. The development of roots is initiated when the ends of cuttings are dipped in these compounds. As we have seen, cytokinins will delay leaf senescence. They are therefore sometimes used commercially to keep the leaves of crops, like lettuce, fresh and free from yellowing after they have been picked. Both gibberellins and cytokinins are sometimes applied to seeds to help break dormancy and so initiate rapid germination. The longer a seed remains ungerminated in the soil, the more vulnerable it is to being eaten, e.g. by birds.

TABLE 40.4 **A summary of the effects of plant hormones**

Hormone	Effects		Examples
Auxins	Promote cell elongation	Phototropic responses	Shoots bend towards light
		Geotropic responses	Roots grow down into the soil
	Stimulate cell division	Promotes development of roots	Hormone root powders are used to help strike cuttings
		Stimulates cambial activity	Callus development at the site of wounds
		Stimulates development of fruits	Natural fruit setting may be improved by spraying with synthetic auxins or they may develop parthenocarpically, e.g. in apples
	Maintain cell wall structure	Inhibits leaf abscission	If the supply of auxin from leaves exceeds that from the stem, the leaf remains intact
		Inhibits fruit abscission	If the supply of auxin from the fruit exceeds that from the stem, the fruit remains intact
	Inhibit growth in high concentrations	Apical dominance	Lateral buds remain dormant under the influence of auxin from the apical bud
		Disruption of growth	Synthetic auxins, e.g. 2,4-D and 2,4,5-T are used as selective weedkillers
Gibberellins	Reverse of genetic dwarfism		Dwarf varieties of peas and maize grow to normal size when gibberellin is applied
	Promote cell elongation		Increases the length of internodes
	Break dormancy of buds		Dormancy of many buds, e.g. birch, is broken by the addition of gibberellin
	Break dormancy of seeds		Ash and cereal seed dormancy may be broken, leading to germination
	Stimulate fruit development		Cherry and peach fruits develop more readily after application of gibberellins
	Remove need for cold period in vernalization		Carrots can be induced to flower without first being subjected to a period of cold
	Affect flowering		Promotes flowering in some long-day plants and inhibits it in some short-day ones
Cytokinins	Promote cell division		Increase growth rate in many plants, e.g. sunflower (*Helianthus*)
	Delay leaf senescence		Maintain leaf for some time once detached from plant
	Stimulate bud development		Promote development of buds on cutting of African violet (*Saintpaulia*) and protonemata of some mosses
	Break dormancy		Break dormancy in both seeds and buds
Abscissic acid	Inhibits growth		Retards growth in most plant parts
	Promotes abscission		Causes the formation of an abscission layer in the petioles and pedicels of leaves, flowers and fruits
	Induces dormancy		Promotes dormancy in the seeds and buds of many plants, e.g. birch (*Betula*) and sycamore (*Acer*)
	Closes stomata		Promotes stomatal closure under conditions of water stress
Ethene	Ripens fruit		Most citrus fruits ripen more rapidly in the presence of ethene
	Breaks dormancy		Ends dormancy of buds in some plants
	Induces flowering		Promotes flowering in pineapples

Abscissic acid may be sprayed on fruit crops to induce the fruits to fall so they can be harvested together. Ethene is applied to tomatoes and citrus fruits in order to stimulate ripening.

40.3 Control of flowering

Plants flower at different times of the year, daffodils and snowdrops appearing in early spring, roses in the summer and chrysanthemums in autumn. The seasonal differences in the time of flowering are related to two main climatic factors: daylength and temperature.

40.3.1 The phytochrome system

Many plant processes are influenced by light. Before a plant can respond to variations in light intensity, duration or wavelength it must first detect these changes. Some form of photoreceptor is necessary. In Section 23.2.2 we discussed the relationships between the absorption spectrum for chlorophyll and the action spectrum for photosynthesis. A similar relationship can be established between a pigment called **phytochrome** and a number of light-induced plant responses (See Fig. 40.4).

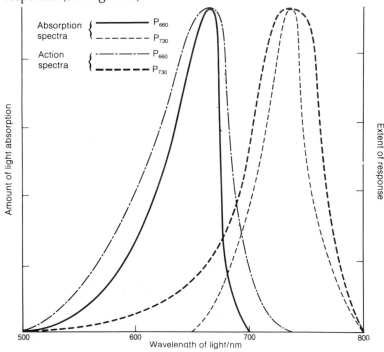

Fig. 40.4 Absorption and action spectra for phytochromes and the responses they control

Phytochrome, isolated in 1960, exists in two interconvertible forms:

1. Phytochrome 660 (P_{660}) – This absorbs red light (peak absorption at a wavelength of 660 nm).

2. Phytochrome 730 (P_{730}) – This absorbs light in the far-red region of the spectrum (peak absorption at 730 nm).

Even a short exposure to the appropriate light wavelength causes the conversion of one form into the other, as shown opposite: These conversions may also be brought about by daylight and darkness. During daylight, P_{660} is converted to P_{730}, while in the dark, a rather slower conversion of P_{730} to P_{660} occurs.

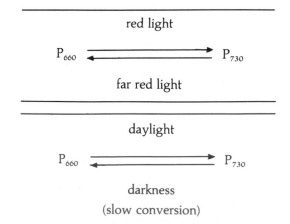

TABLE 40.5 **Summary of the effects of red light and far-red light**

Red light effects	Far-red light effects
Phytochrome 660 changes to phytochrome 730	Phytochrome 730 changes to phytochrome 660
Stimulates germination of some seeds, e.g. lettuce (*Lactuca*)	Inhibits germination of some seeds, e.g. lettuce (*Lactuca*)
Induces formation of anthocyanins (plant pigments)	Inhibits formation of anthocyanins
Stimulates flowering in long-day plants	Inhibits flowering in long-day plants
Inhibits flowering in short-day plants	Stimulates flowering in short-day plants
Elongation of internodes is inhibited	Elongation of internodes is promoted
Induces increase in leaf area	Prevents increase in leaf area
Causes epicotyl (plumule) hook to unbend	Maintains epicotyl (plumule) hook bent

Phytochrome comprises a protein and a pigment. It is distributed throughout the plant in minute quantities, being most concentrated in growing tips. The actions of the two forms are usually antagonistic, i.e. where P_{660} induces a response, P_{730} inhibits it. The various effects of the two forms are listed in Table 40.5.

40.3.2 Photoperiodism

One major influence on the timing of flowering is the length of the day or **photoperiod**. The effects of the photoperiod on flowering differ from species to species but plants fall into three basic categories:

1. Long-day plants (LDP) – These only flower when the period of daylight exceeds a critical minimum length. Examples of long-day plants include radish, clover, barley and petunia.

2. Short-day plants (SDP) – These only flower when the period of daylight is shorter than a critical maximum length. Examples of short-day plants include chrysanthemum, poinsettia, cocklebur and tobacco.

3. Day-neutral plants – These plants flower regardless of the length of daylight. Examples of day-neutral plants include cucumber, begonia, violet and carrot.

Intermediate varieties exist. For example, **short-long-day plants** only flower after a sequence of short days is followed by long ones. These plants will flower naturally in mid-summer when the days are long following the shorter ones of spring. **Long-short-day plants** flower after a sequence of long days is followed by short ones. These will flower naturally in the autumn after the long days of summer are followed by the shorter ones of early autumn.

It is rather unfortunate that historically plants were categorized as short-day or long-day, as it is the length of the dark period which is crucial in determining flowering. Short-day plants require a long dark period, whereas long-day plants require a short dark period. This fact was established by a series of experiments summarized in Fig. 40.5.

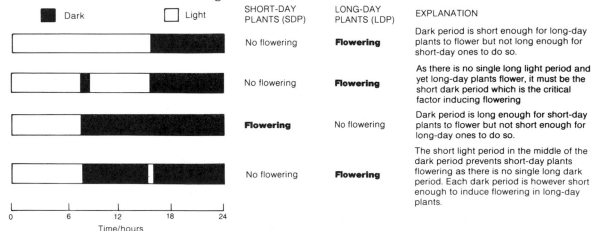

Fig. 40.5 *Flowering related to the length of dark period*

Interrupting a long dark period with red light is as effective as daylight in stopping short-day plants flowering. Far-red light, however, has no effect and short-day plants flower as if the dark period had been continuous. These and other experiments suggest that phytochrome is the photoreceptor detecting different light wavelengths and ultimately determining whether or not a plant flowers.

Although flowers are formed at the apex, experiments confirm that the light stimulus is detected by the leaves. In some cases only a single leaf needs to be subjected to the appropriate stimulus to induce flowering. A message must therefore pass from the leaves to the apex. As plants coordinate by chemical means, this message is assumed to be a hormone and has been called **florigen**, even though it has not yet been isolated. It is thought to be transported within phloem. Understanding of the mechanism by which phytochrome initiates flowering is poor. The process is clearly complex and, because they can have similar effects to red light, gibberellins may be involved. One possible mechanism is summarized in Fig. 40.6. As long-day plants are known to flower after a short exposure to red light, it is possible that red light (wavelength 660 nm) is absorbed by phytochrome 660 which is rapidly converted to phytochrome 730 which then induces flowering. Short-day plants by contrast flower in response to phytochrome 660. This is formed by absorption of far-red light (wavelength 730 nm) by phytochrome 730 which is then converted to phytochrome 660. This rapid conversion by far-red light can also take place, although far more slowly, in darkness. Hence a long dark period induces flowering in these short-day plants. One postulated mechanism by which florigen is produced suggests that there are two inactive forms of the hormone, one of which is converted to the hormone by P_{660}, the other by P_{730}. These events are summarized in Fig. 40.6.

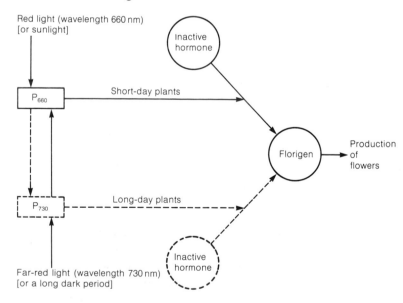

Fig. 40.6 Summary of one possible mechanism to explain the rôle of the phytochrome system in flowering

40.3.3 Vernalization

Many plants grow vegetatively but fail to flower if they are not exposed to a period of low temperature. This phenomenon is known as **vernalization**. Examples of plants exhibiting vernalization include long-day, short-day and day-neutral plants. There is evidence that it occurs in response to a hormone and this led to the name **vernalin** being adopted. It now appears that, rather than being a separate hormone, vernalin is in fact a gibberellin.

In temperate regions, the importance of vernalization and photoperiodism is in ensuring that flowering occurs at specific times of the year, for example at the time when the particular pollinating insects of the species are abundant. There would be little point in most insect-pollinating plants flowering in the middle of winter when few, if any, insects are active. It is likewise essential for successful cross pollination that a species of plant produces most of its flowers at the same time. As the photoperiod is consistent, regardless of climatic conditions, coordinating flower production to day length is clearly a most reliable means of achieving synchronized flowering.

The artificial control of temperature and the length of the light period has obvious commercial applications. By so doing it is possible for growers to produce many varieties of flower out of season, and so ensure a supply throughout the year.

40.4 Questions

1. Describe the part played by plant hormones in the life cycle of a flowering plant under the following headings.
    (i) dormancy
    (ii) growth
    (iii) response to stimuli
    (iv) fruit formation   *(20 marks)*
Joint Matriculation Board June 1989, Paper IIB, No. 3

2.   (a) What is *photoperiodism*?   *(3 marks)*
    (b) How does the photoperiod affect
        (i) flowering,   *(7 marks)*
        (ii) dormancy in plants,   *(4 marks)*
        (iii) breeding behaviour in animals?   *(6 marks)*
            (Total 20 marks)
Oxford and Cambridge Board June 1988, Paper II, No. 5

3.   (a) Explain what is meant by the term *growth*.
           (6 marks)
    (b) Describe one method of measuring the growth of a root or a shoot. Assess the validity of this method   *(6 marks)*
    (c) How would you investigate the hypothesis that a plant hormone is involved in the response of a coleoptile to unilateral illumination? *(10 marks)*
    (d) (i) What is a short-day plant?
        (ii) How would you investigate the relative importance of day length and night length for such a plant?
        (iii) How might a nurseryman make use of such information?   *(2,4,2 marks)*
            (Total 30 marks)

Oxford Local 1987, Specimen Paper II, No. 5

4.   (a) Describe, with practical details, how you would perform an experiment to show a phototropic response in the shoot of a plant.   *(8 marks)*
    (b) Explain how hormones (growth substances) are involved in photoperiodic behaviour in plants.
           (6 marks)
    (c) Describe *two* different kinds of photoperiodic behaviour in animals.   *(6 marks)*
            (Total 20 marks)

London Board June 1988, Paper II, No. 3

5. Plants differ in the amount of light required per day to induce flowering. They can be divided into three photoperiodic groups: long-day, short-day and day-neutral plants.
    (a) (i) How do day-neutral plants differ from the other two groups?
        (ii) The graph below shows the relationship between the number of hours of light per 24 hours and flowering in duckweed. Give **three** important features which would need to be considered when carrying out an investigation to produce results such as those shown below.

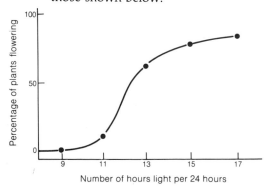

(iii) Which photoperiodic group does duckweed belong to? Give your reasoning. (*6 marks*)

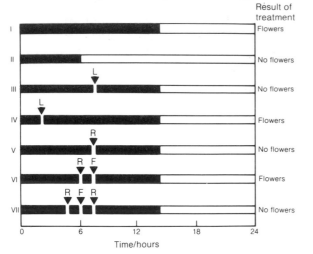

Time/hours

L white light R red light F far-red light

(b) Samples of plants from an unrelated species were subjected to a range of light and dark treatments (I to VII) as shown in the diagram. Shaded bars indicate darkness, unshaded bars indicate light periods. The letters refer to particular light flashes during the dark period. The effect of each treatment on flowering is also given.

(i) Using the results from I and II, state the photoperiodic group to which this plant belongs.

(ii) Using results from I to IV, explain whether the length of the day or the length of the night is the critical element in the light/dark cycle.

(iii) Considering V, VI and VII, what can you deduce about the effects of red and far-red light on this species?

(iv) How could these facts be used to produce flowers from this species out of season?
(*9 marks*)
(*Total 15 marks*)

Joint Matriculation Board June 1983, Paper IA, No. 5

6. The figure is a diagrammatic representation of various treatments designed to investigate the geotropic responses in the roots of maize. Any curvature resulting from the treatment is shown by arrows in the diagram.

(a) State the region where gravity is perceived in these roots. (*Give the reason for your answer.*)

(b) State the function of the glass barrier placed transversely across half the width of the root in **E** and **F**.

(c) Explain why a glass barrier was placed in the position indicated in **G**.

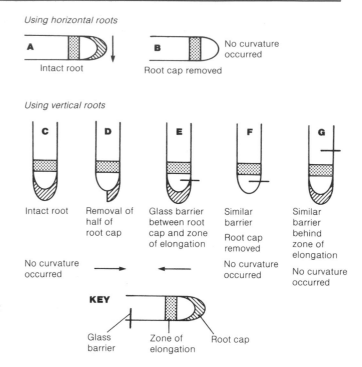

(d) The results of this series of experiments are claimed to demonstrate that an inhibitor of cell elongation is produced by the root cap. This diffuses back into the root away from the cap and under the influence of gravity becomes asymmetrically distributed, accumulating in the lower half of horizontal roots, and causing greater inhibition of cell elongation on that side.

Explain how this claim is supported by the results.

(*Total 10 marks*)

Cambridge Board June 1988, Paper III, No. 5

7. An investigation was carried out into the effect of gibberellin on the growth of leaves in dwarf bean plants. Equal amounts of the hormone were applied either to the stem or to the first leaves produced by the plants.

In one experiment the plants were left intact, but in a second experiment the growing point (apex) of each plant was removed when gibberellin was applied. In both experiments a control group of plants received no gibberellin.

Results are shown in the graphs on the next page.

By reference to the graphs, answer the following questions.

(a) Compare the effects of applying gibberellin to the stem and to the leaves of the intact bean plants. (*4 marks*)

(b) (i) Describe the effect on leaf growth of removal of the growing points from the bean plants when no gibberellin is applied.
(*2 marks*)

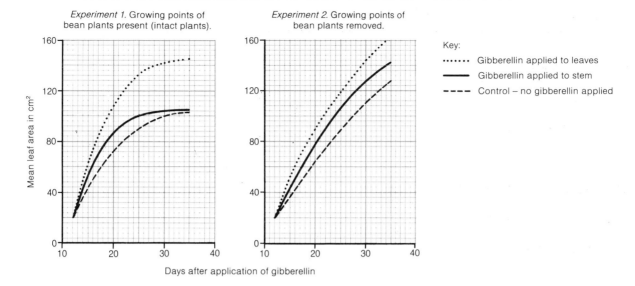

Experiment 1. Growing points of bean plants present (intact plants).

Experiment 2. Growing points of bean plants removed.

Mean leaf area in cm²

Days after application of gibberellin

Key:

······· Gibberellin applied to leaves

——— Gibberellin applied to stem

- - - - Control – no gibberellin applied

    (ii) Suggest *one* reason why removal of the growing point has this effect. (*1 mark*)

(c) (i) How did removal of the growing points from the bean plants affect the results of gibberellin application? (*3 marks*)

    (ii) Suggest *one* hypothesis which could explain this effect. (*2 marks*)

    (iii) Outline *one* additional experiment to test this hypothesis. (*3 marks*)

(d) (i) Suggest *two* cellular mechanisms which may lead to an increase in leaf area. (*2 marks*)

    (ii) Suggest a mechanism by which a hormone such as gibberellin may exert its effect on leaf growth. (*1 mark*)

(e) Give *one* similarity and *one* difference between a plant hormone such as gibberellin and a typical animal hormone. (*2 marks*)

(*Total 20 marks*)

London Board June 1989, Paper I, No. 17

41 Biotechnology

Biotechnology is the application of scientific and engineering principles to the production of materials by biological agents. Given its recent wide publicity, one could be forgiven for thinking it was a new branch of science. While there is no doubt that recent technological and biochemical advances have led to considerable developments in biotechnology, its origins go back a long way. Food for human consumption has always been vulnerable to spoilage by microorganisms. Ancient civilizations probably found that normally detrimental microbial contamination occasionally conferred some benefit: improved flavour or better preservation for instance. In this way beers would have been developed from 'spoilt' grain and wine from 'spoilt' fruit. Contamination of the alcohol by a different agent led to the production of vinegar which was then used to preserve food. Cheese, butter and yoghurt all resulted from various microbial contaminations of milk.

The modern biotechnology industry had its origins in the First World War. A naval blockade deprived Germany of the supply of vegetable fats necessary for the production of glycerol from which explosives were made. They turned to the fermentation of plant material by yeast as an alternative source. At the same time, the British were using *Clostridium acetobutylicum* to produce acetone and butanol as part of their war effort. In a similar way, the Second World War prompted the mass production of the antibiotic penicillin (discovered by Alexander Fleming in 1929) using *Penicillium notatum*.

Many other chemicals were produced thereafter by use of fermentation techniques, but it was in the 1980s that biotechnology underwent major expansion. This was almost entirely due to the development of **recombinant DNA technology** (Section 12.8).

41.1 Growth of microorganisms

Microorganisms (microbes) are found in every ecological niche – from deepest ocean to the limits of the stratosphere, from hot springs to the frozen poles. They are small, easily dispersed and quickly multiply given a suitable environment. They grow on a wide diversity of substrates making them ideal subjects for commercial application. Each species has its own optimum conditions within which it grows best.

41.1.1 Factors affecting growth

The factors which affect the growth of microorganisms are equally applicable to the growth of plant and animal cell cultures.

Nutrients
Growth depends upon both the types of nutrients available and their concentration. Cells are largely made up of the four elements: carbon,

hydrogen, oxygen and nitrogen with smaller, but significant, quantities of phosphorus and sulphur. Accounting as they do for 90% of the cell's dry mass, all six are essential for growth.

Needed in smaller quantities, but no less important, are the metallic elements: calcium, potassium, magnesium and iron, sometimes known as **macro-nutrients**. Required in smaller amounts still are the **micro-nutrients (trace elements)**: manganese, cobalt, zinc, copper and molybdenum – indeed not all may be essential to some species. A further group of chemicals, loosely termed **growth factors**, are also needed. These fall into three categories:

1. Vitamins

2. Amino acids

3. Purines and pyrimidines

Up to a point, the more concentrated a nutrient the greater the rate of growth, but as other factors become limiting the addition of further nutrients has no beneficial effect. More details of the nutrients used in culturing cells are given in Section 41.1.3.

Temperature
As all growth is governed by enzymes and these operate only within a relatively narrow range of temperature, cells are similarly affected by it. If the temperature falls too low the rate of enzyme catalysed reactions becomes too slow to sustain growth; if too high the denaturation of enzymes causes death. Most cells grow best within the range 20–45 °C although some species can grow at temperatures as low as − 5 °C, while others do so at 90 °C. Three groups are recognized according to their preferred temperature range:

1. **Psychrophiles** (e.g. *Bacillus globisporus*) – These have optimum growth temperatures below 20 °C, many continuing to grow down to 0 °C.

2. **Mesophiles** (e.g. *Escherichia coli*) – These have optimum growth temperatures in the range 20–40 °C.

3. **Thermophiles** (e.g. the alga *Cyanidium caldarium*) – These have optimum growth temperatures in excess of 45 °C, a few surviving in temperatures as high as 90 °C. These cells have enzymes which are unusual in not being denatured at high temperatures.

pH
Microorganisms are able to tolerate a wider range of pH than plant and animal cells, some species growing in an environment as acid as pH 2.5, others in one as alkaline as pH 9. Microorganisms preferring acid conditions, e.g. *Thiobacillus thio-oxidans* are termed **acidophiles**.

Oxygen
Many microorganisms are aerobic, requiring molecular oxygen for growth at all times: these are termed **obligate aerobes**. Some, while growing better in the presence of oxygen, can nevertheless survive in its absence; these are called **facultative anaerobes**. Others find oxygen toxic and do not grow well in its presence: these are the **obligate anaerobes**. Some of this group, while tolerating oxygen, nevertheless grow better when its concentration is very low. These are termed **microaerophiles**.

Osmotic factors

All microorganisms require water for growth. In most cases this is absorbed osmotically from the environment, although pinocytosis is used in certain protozoa and all groups produce a little water as a product of aerobic respiration. To ensure absorption, the water potential of the external environment must be less negative (higher) than the cell contents. For this reason most microorganisms cannot grow in environments with a high solute concentration — a fact made use of in preserving foods, e.g. salting of meat and fish, bottling of jam and fruit in sugar. A few, called **halophiles**, can survive, however, in conditions of high salt concentration.

Pressure

Although pressure is not a major factor affecting growth in most microorganisms, a few species inhabiting the ocean depths can grow under immense pressure. Some of these **barophiles** cannot grow in surface waters where the pressure is too low for their survival.

Light

Photosynthetic microorganisms require an adequate supply of light to sustain growth.

Water

In common with all organisms, microbes require water for a variety of functions. In addition, photosynthetic microbes use it as a source of hydrogen to reduce carbon dioxide. Chemoautotrophs may use alternative inorganic hydrogen sources, e.g. hydrogen sulphide, for this purpose.

41.1.2 Growth patterns

The growth of a culture of individual cells, e.g. of unicellular yeasts, protozoa or bacteria is, in effect, the growth of a population and follows the same pattern as that described in Section 10.1. The typical bacterial growth curve for a batch culture, illustrated in Figure 41.1, consists of four phases:

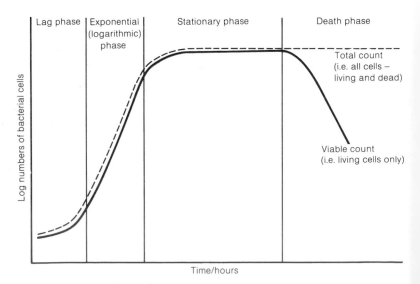

Fig. 41.1 Bacterial growth curve

1. The lag phase – This is the period after **inoculation** (the addition of cells to the nutrient medium) during which the growth rate increases towards its maximum. Growth during this time is slow initially as the bacteria adapt to produce the necessary enzymes needed to utilize the nutrient medium. The rate of cell division gradually increases during this phase.

2. The exponential (logarithmic) phase – During this period, with nutrients in good supply and few waste products being produced, the rate of cell division is at its maximum, cells sometimes dividing as frequently as every ten minutes. The culture is in a state of **balanced growth** with the doubling time, cell protein content and cell size all remaining constant.

3. The stationary phase – As nutrients are used up and toxic waste products accumulate, the rate of growth slows. The changed composition of the medium results in the production of cells of various sizes with a different chemical make-up. They are in a state of **unbalanced growth**. For the total count, this section of the graph is horizontal because the rate at which new cells are produced is equal to the rate at which dead ones are broken down.

4. The death phase – While the total number of cells remains constant, the number of living ones diminishes as an ever increasing number die from a lack of the nutrients necessary to produce cellular energy, or because of poisoning by their own toxic wastes.

41.1.3 Culture media

The correct balance of nutrients, an apppropriate pH and a suitable medium are essential to microbial growth.

The medium itself may be liquid (broth) or solid. Solid media are usually based upon **agar**, a seaweed extract which is metabolically inert and which dissolves in hot water but solidifies upon cooling. The more agar, the more solid is the resulting jelly. Any medium composed to satisfy the demands of a single species is called a **minimal medium**. One which provides the nutrients for a small group of microbes with similar requirements, e.g. acidophiles, is known as a **narrow spectrum medium**, whereas a medium for general purposes, designed to grow as wide a range of microorganisms as possible is called a **broad spectrum medium**. It is possible to select for the growth of specific types of organisms by use of **selective media** which permit growth of only a single species. The composition of a typical broad spectrum medium is given in Table 41.1.

41.1.4 Aseptic conditions

Both in the laboratory and on an industrial scale, pure cultures of a single type of microorganism need to be grown, free from contamination with others.

A number of techniques are used to sterilize equipment, instruments, media and other materials. Heat, either passing through a flame, dry heating in an oven, or using water in an autoclave (a type of pressure cooker) are effective in sterilizing instruments, small vessels and culture media. Where heating may affect the media, it may be filtered free of microorganisms using especially fine filters. Ultraviolet light can be used on equipment or even whole rooms.

TABLE 41.1 **A typical broad spectrum medium**

Ingredient	Quantity	Source of
Water	1.0 dm³	Metabolic and osmotic water, hydrogen and oxygen
Glucose	5 g	Carbohydrate energy source
Yeast extract	5 g	Organic nitrogen, organic growth factors
Dipotassium hydrogen phosphate	1 g	Potassium and phosphorus
Magnesium sulphate	250 mg	Magnesium and sulphur
Iron (II) sulphate	10 mg	Iron and sulphur
Calcium chloride	10 mg	Calcium and chloride
Cobalt, copper, manganese, molybdenum and zinc salts	trace	Metallic ion trace elements

Certain other equipment may be sterilized using a suitable disinfectant, e.g. hypochlorite.

41.2 Industrial fermenters and fermentation

Much of modern biotechnology involves the large scale production of substances by growing specific microorganisms in a large container known as a fermenter. Fermentation should strictly refer to a biological process which occurs in the absence of oxygen. However, the word is taken to include aerobic processes – indeed the supply of adequate oxygen is a major design feature of the modern fermenter.

41.2.1 Batch versus continuous cultivation

In **batch cultivation** the necessary nutrient medium and the appropriate microorganisms are added to the fermenter and the process allowed to proceed. During the fermentation air is added if it is needed and waste gases are removed. Growth is allowed to continue up to a specific point at which the fermenter is emptied and the product extracted. The fermenter is then cleaned and sterilized in readiness for the next batch.

With **continuous cultivation**, once the fermenter is set up, the used medium and products are continuously removed. The raw materials are also added throughout and the process can therefore continue, sometimes for many weeks.

The continuous process has the advantage of being quicker because it removes the need to empty, clean and refill the fermenter as regularly and hence ensures an almost continuous yield. In addition, by adjusting the nutrients added, the rate of growth can be maintained at the constant level which provides the maximum yield of product. Continuous cultivation is, however, only suited to the production of biomass or metabolites which are associated with growth. **Secondary metabolites**, like antibiotics, which are produced when growth is past its maximum, need to be manufactured by the batch process. The organisms used to produce antibiotics are in any case too unstable for growth by continuous fermentation. In addition, continuous fermentation requires sophisticated monitoring technology and highly trained staff to operate efficiently.

41.2.2 Fermenter design

The basic design of a **stirred-tank fermenter** is shown in Fig. 41.2. It consists of a large stainless steel vessel with a capacity of up to 500 000 dm³ around which is a jacket of circulating water used to control the temperature within the fermenter. An agitator, comprising a series of flat blades which can be rotated, is incorporated. This ensures that the contents are thoroughly mixed thus bringing nutrients into contact with the microorganisms and preventing the cells settling out at the bottom.

Where oxygen is required, air is forced in at the bottom of the tank through a ring containing many small holes – a process known as **sparging**. To assist aeration, increased turbulence may be achieved by adding baffles to the walls of the fermentation vessel. A series of openings, or **ports**, through which materials can be introduced or withdrawn, is provided. The **harvest line** is used to extract culture medium. An outlet to remove air and waste gases is needed, as well as one to allow small samples of the culture medium to be removed for analysis. Inlet tubes permit nutrients to be provided and, as the pH changes during fermentation, allow acid or base to be added to maintain the optimum pH. With air being forced into the medium, chemicals often need to be added to reduce foaming. Finally, it is essential to have an **inoculation port** through which the initial inoculum of cells can be introduced once the required conditions in the fermenter are achieved. **Probes**, which constantly register the temperature and pH within the vessel, are used to indicate when adjustments to these factors are necessary.

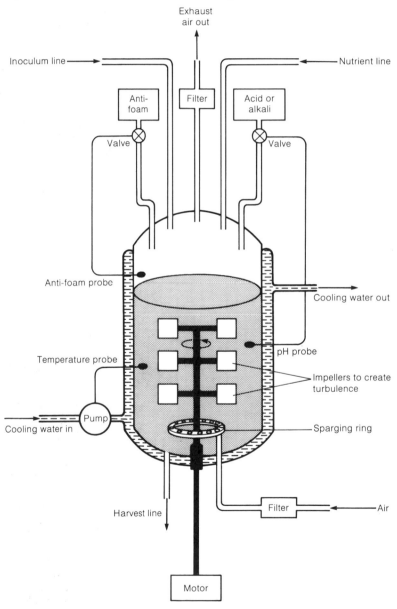

Fig. 41.2 A stirred-tank fermenter

The stirred-tank fermenter is a well-tried and tested design used extensively in the fermentation industry. It is, however, relatively costly to run largely as a consequence of the energy needed to drive the agitators and introduce the compressed air. Alternatives have therefore been designed where the air forced into the vessel to provide oxygen is used to circulate the contents, thus making an agitator unnecessary.

One such design, the **pressure-cycle fermenter**, is of two types. In the **air-lift type** the air is introduced centrally at the bottom making the medium less dense. It therefore rises through a central column in the vessel to the top where it escapes. The now more dense medium descends around the sides of the vessel to complete the cycle (Fig. 41.3a). Higher pressure at the bottom of the vessel increases solubility of oxygen, while lower pressure at the top decreases solubility of carbon dioxide, which as a consequence comes out of solution. In the **deep-shaft type**, the principle is similar but the air is introduced at the top (Fig. 41.3b). This has the advantage of giving a more even delivery of oxygen to the microorganisms.

The **tower-** or **bubble-column** (a variety of the air-lift fermenter) has horizontal rather than vertical divisions (Fig. 41.3c). This allows conditions in each section to be maintained at different levels if necessary. A microorganism being carried up from the bottom may therefore pass from a high pH and low temperature to a lower pH and higher temperature to suit each phase of its growth.

While all three types can be used continuously only the air-lift fermenter is used for batch processing. All types must be taller than the conventional stirred-tank vessel to operate effectively.

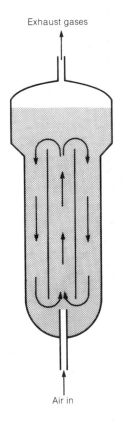

Fig. 41.3(a) Air-lift fermenter

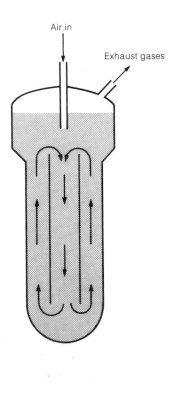

Fig. 41.3(b) Deep-shaft fermenter

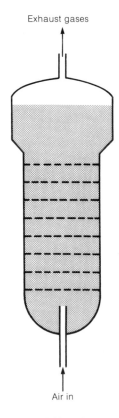

Fig. 41.3(c) Bubble-column fermenter

41.2.3 The operation and control of fermenters

There are two main problems associated with setting up a large scale fermentation process. Firstly, the inoculum containing the desired strain of microorganism has to be obtained in sufficient quantity; if too little is added to the fermenter, the lag phase is unacceptably long, making the process uneconomic. A small scale fermentation is set up in a vessel as small as 10 cm³, using frozen culture stock. This is then added to flasks, containing the appropriate nutrient medium, of increasing capacity, e.g. 300 cm³, 3000 cm³, 30 000 cm³, etc., until the final capacity of the end fermentation vessel is reached. This is known as the **fermenter train**. The problem is that the operational conditions that give the optimum yield in a 300 cm³ fermentation flask are often very different from those for a 300 000 cm³ vessel.

The other main problem arises from the fact that the microorganisms used in fermenters have been genetically selected for the properties (e.g. a high product yield) which make them suitable for use in a large-scale fermenter. They are often enfeebled mutant strains which have resulted from deliberately induced mutanogenesis. As efficient production depends upon rapid growth, i.e. many generations of the microorganism in a short period, there is a tendency for the strain to mutate naturally, often reverting to the parent type which has less desirable properties. One way around the problem is to prevent the microorganism producing the desired product until the final fermentation. This reduces the selection pressure which might alter the gene responsible for the product, but is not always feasible. The use of genetically engineered stock (Section 12.8) has the advantage that it is much easier to express ('switch on') the desired gene at the appropriate time by use of chemical triggers called **promoters**.

Once a sufficient quantity of the inoculum has been produced, usually between 1% and 10% of the total medium, it is added to the fermenter only after the medium within it is of the correct composition and at the desired temperature and pH. The problem now is to maintain all conditions throughout the fermentation. To achieve this the levels of various nutrients, oxygen, pH and temperature are constantly monitored using probes and the information fed to a computer for analysis, along with information on the composition of the exhaust gases. The necessary corrective changes can then be made. These may be complex – for example, an increase in oxygen uptake could be countered by reducing the air supply, the pressure within the vessel, the nutrient levels or the agitator speed. Which one is selected has implications for other factors and the choice therefore needs to be made advisedly. While the temperature of the medium may need to be increased initially by piping steam through it, once fermentation is under way the heat generated by the microorganisms necessitates continuous cooling by the water jacket around the vessel.

The processes involved in recovering the product are called **downstream processing**. This often involves separation of the cells from the medium which may be achieved in a number of ways:

1. Settlement – The cells may readily settle once agitation and sparging cease. The process can be accelerated by the addition of **flocculating agents**, many of which work by neutralizing the charges on the cells which otherwise keep them in suspension by electrostatic repulsion.

2. Centrifugation – The contents of the fermenter are spun at high speed in a centrifuge causing the cells to settle out. Continuous centrifugation is now possible.

3. Ultrafiltration – The fermenter contents are forced through filters with a pore size less than 0.5 μm which thereby traps cells allowing only liquid through. Some extracellular protein may also be retained.

Where the desired products are the entire cells themselves, these need only be washed, dried and compacted to complete the process. Where the product is contained within the cells, these must be disrupted by some means, the cell debris removed (e.g. by centrifugation or ultrafiltration), and the desired chemical recovered using precipitation, chromatographic or solvent extraction techniques. Where the product lies in the fermentation liquor rather than the cells, this is separated from unwanted enzymes and metabolites, again by precipitation, chromatography or solvent extraction as appropriate.

41.2.4 Sterilization during and after fermentation

The need for aseptic conditions and the basic mechanisms for achieving them were discussed in Section 41.1.4. In industrial fermentation this presents many practical problems considering that not only a very large vessel needs to be sterilized, but also all associated pipework and probes as well as the nutrients, air-supply, and other agents added during the process.

The equipment is designed so that any nooks and crannies which might harbour microorganisms are minimized. The vessel is highly polished, for example, and all components are designed to allow easy access by sterlizing agents. Having been thoroughly washed all equipment is steam sterilized.

The initial nutrient medium may be sterilized in the fermenter by heating it; any added later may be sterilized by heating *en route* to the vessel, as can all other liquid additions. Concentrated acids and alkalis used to adjust the pH may be so inhospitable to contaminating organisms as not to warrant sterilization.

Filtration is used to remove potential contaminants from the air supply although bacteriophages are small enough to pass through – with disastrous consequences. Heat treatment of incoming air can alleviate this problem. The exhaust air is also sterilized to prevent potentially harmful microorganisms being introduced into the atmosphere.

41.2.5 Immobilization of cells and enzymes

One problem with the fermentation processes described so far is that at some point the cell culture is removed and discarded. This is fine when the cells are the desired product, but if it is a metabolite they produce which is required, their removal takes away the manufacturing source which then needs to be replaced. Any mechanism for immobilizing the microorganism and/or the enzymes they produce, improves the economics of the process. The idea is not a new one – vinegar manufacture and some stages of sewage treatment have used the technique for a century or more.

There are three basic methods of immobilization:

1. Entrapment – Cells or enzyme molecules are trapped in a suitable meshwork of inert material, e.g. collagen, cellulose, carrageenan, agar, gelatin, polystyrene.

2. Binding – Cells or enzyme molecules become physically attached to the surface of a suitable material, e.g. sand or gravel.

3. Cross-linking – Cells or enzyme molecules are chemically bonded to a suitable chemical matrix, e.g. glutaraldehyde.

However immobilized, the cells or enzymes are made into small beads which are then either packed into columns, or kept in the nutrient medium. The nutrient can be continually added and the product removed without frequent removal of the microorganisms/enzymes. The process cannot be continued indefinitely. Impurities may accumulate preventing further enzyme action or contamination may occur.

41.3 Biotechnology and food production

Until Louis Pasteur showed in 1857 that wine fermentation was the consequence of microbial activity, no one had been aware of the rôle that microorganisms played in the manufacture of some foods. With Pasteur's discovery came further development of the use of microorganisms in food production, an expansion which continues today.

41.3.1 Baking

The use of yeasts in food production is the oldest, and most extensive contribution made by any group of microorganisms. In bread-making cereal grain is crushed to form flour thus exposing the stored starch. Water is added (making a dough) to activate the natural enzymes, e.g. amylases, in the flour which then hydrolyse the starch via maltose into glucose. The yeast, *Saccharomyces cerevisiae*, is added which uses the glucose as a respiratory substrate, producing carbon dioxide. This carbon dioxide forms small bubbles which become trapped in the dough; upon baking in an oven these expand giving the bread a light texture. Dough is often kneaded – a process which traps air within it. This not only helps to lighten the bread directly but also provides a source of oxygen so that the yeast can respire aerobically producing a greater quantity of carbon dioxide. Some anaerobic respiration nevertheless takes place and the alcohol produced is evaporated during baking.

41.3.2 Beer and wine production

Fermentation by yeasts produces alcohol according to the equation:

$$C_6H_{12}O_6 \longrightarrow C_2H_5OH + 2CO_2$$

hexose sugar — ethanol — carbon dioxide

Some alcohol produced in this way is for industrial use, but much goes to make beverages like beers and wines which may then be distilled to form spirits. The variety of such beverages is immense and depends largely on the source of the sugar and the type of yeast used to ferment it. Various additives further increase the diversity of alcoholic drinks.

To make wine, the sugar fermented is glucose obtained directly from grapes, whereas beers are made by fermenting glucose obtained from cereal grain (usually barley) which results from the breakdown of starch in the grain. The yeast used in wine production is often *Saccharomyces ellipsoideus* as this variety can tolerate the higher alcohol levels encountered in wines. Even more tolerant to high alcohol levels are *S. fermentati* and *S. beticus* and these are primarily used in making sherry. Beers are of two basic types – top fermenting

Brewing beer

varieties of yeast such as *S. cerevisiae* produce a typical British 'bitter' while *S. carlsbergensis*, a bottom fermenting variety, is used to make lager. In beer production, the barley grain is first malted by soaking it in water for two to three days. The grain is then spread on concrete floors and allowed to germinate (about ten days), during which time the natural amylases and maltases in the grain start to convert starch to glucose. This process is stopped by drying the grain and storing it – a process called **kilning** or **roasting**. The higher the temperature during this process, the darker the resulting beer. The germinated grain is often crushed during this stage. The dried, crushed germinated grain is now added to water and heated to the desired temperature – **mashing**. During mashing the remaining starch is converted to sugar to produce a liquid called **wort**. Yeast is added to the wort to convert it to alcohol, as well as hops and other additives, e.g. caramel, which are used to give each beer its characteristic flavour and colour.

The fermentation itself takes place in large deep tanks of around 500 000 dm^3 capacity. No air is introduced as anaerobic respiration is the aim. Although traditionally a batch process, beer production can also be carried out using continuous fermentation. This is both more economic and allows the carbon dioxide to be collected – a valuable by-product when converted to dry-ice.

The beer is finally separated from the yeast and clarified, and carbon dioxide is added. Sometimes the beer is pasteurized to extend its shelf-life. A good traditional beer, however, retains some yeast in the enclosed barrel which produces the carbon dioxide naturally. Such beers are, for obvious reasons, termed 'live' beers.

41.3.3 Dairy products

Microorganisms have long been exploited in the dairy industry as a means of preserving milk. From this a diverse number of different products have been manufactured which fall into three main categories: cheese, yoghurt and butter.

Cheese manufacture
An ancient process, cheese-making has altered little over the years. A group of bacteria known as **lactic acid bacteria** are used to ferment the lactose in milk to lactic acid according to the equation:

$$C_{12}H_{22}O_{11} \ + \ H_2O \longrightarrow 4CH_3CHOHCOOH$$

lactose   water   lactic acid

Most commercially used lactic acid bacteria are species of two genera – *Lactobacillus* and *Streptococcus*.

Cheese production begins with the pasteurization of raw milk which is then cooled to around 30 °C before a starter culture of the required lactic acid bacteria is added. The resultant fall in pH due to their activity causes the milk to separate into a solid **curd** and a liquid **whey** in a process called **curdling**. The addition of **rennet** at this stage encourages the casein in the milk to coagulate aiding curd formation. Originally extracted from the stomachs of calves slaughtered for food, rennet has now largely been replaced by **chymosin**, a similar enzyme produced by genetically engineered *Escherichia coli*. The whey is drained off and may be used to feed animals. The curd is heated in the range 32–42 °C and some salt added before being pressed into moulds for a period of time which varies according to cheese type.

Whey sampling during cheese manufacture

The ripening of the cheese allows flavour to develop as a result of the action of other milk enzymes or deliberately introduced microorganisms. In blue cheese, for example, *Penicillium* spp. are added. The duration of ripening varies with Caerphilly taking just a fortnight in contrast to a year required for mature Cheddar. Whereas hard cheeses ripen owing to the activity of lactic acid bacteria throughout the cheese, in soft cheeses it is fungi growing on the surface which are responsible.

Yoghurt manufacture

Made from pasteurized milk with much of the fat removed, yoghurt production also depends on lactic acid bacteria, in particular *Lactobacillus bulgaricus* and *Streptococcus thermophilus*. These are added to the milk in equal quantity and incubated at around 45 °C for five hours during which time the pH falls to around 4.0. Cooling prevents further fermentation and fruit or flavourings can then be added as required.

Butter manufacture

Not essentially a process requiring microorganisms, butter production is nevertheless frequently assisted by the addition to cream of *Streptococcus lactis* and *Leuconostoc cremoris* which help to sour it, give flavour and aid the separation of the butterfat. Churning of this butterfat produces the final product.

41.3.4 Single cell protein (SCP)

Single cell protein comprises the cells, or their products, of microorganisms which are grown for animal, including human, consumption. High in protein, the product also contains fats, carbohydrates, vitamins and minerals making it a useful food. The raw materials for SCP production have included petroleum chemicals, alcohols, sugars and a variety of agricultural and industrial wastes. The microorganisms used to ferment these have been equally diverse – bacteria, algae, yeasts and filamentous fungi. The success of various manufacturing processes has varied. The use of a waste product to produce food seems highly attractive and economical, but the demand it creates for the raw material ceases to make it a waste, its price rises and the process can become uneconomical. Excess food production in some parts of the world has meant the selling off of butter and grain 'mountains' and therefore reduced the need for alternative sources of food such as SCP. It has not therefore proved the success originally anticipated and many countries have ceased production altogether.

The world's largest continuously operating fermenter (600 tonnes) owned by ICI, produces a single cell protein called **Pruteen.** It comprises 80% protein and has a high vitamin content. The process uses methanol, a waste product of some of ICI's other activities, making the raw material relatively cheap. This is acted upon by the aerobic bacterium *Methylophilus methylotrophus* in a pressure-cycle fermenter with a capacity of 1500 m^3 to produce the odourless, tasteless, cream-coloured Pruteen which is used as an animal feed.

Similar projects are in use to produce a protein based on the fungus *Fusarium graminearum* which can be grown on flour waste. The product, **mycoprotein**, is intended for human consumption and, being high in protein and fibre, but low in cholesterol, is a healthy addition to the diet.

Primary fermentation of vinegar

41.3.5 Other foods produced with the aid of biotechnology

Vinegar
One of the oldest products of biotechnology is vinegar; originally produced from wine, much is now the product of fermenting malt. The alcohol produced by *Saccharomyces* spp. (Section 41.3.2) is converted to acetic acid by aerobic bacteria belonging to the genus *Acetobacter* according to the overall equation:

$$C_2H_5OH + O_2 \longrightarrow CH_3COOH + H_2O$$

| ethanol | oxygen | acetic acid (ethanoic acid) | water |

In the fermenter, the bacteria are immobilized on the surface of wood shavings or other inert material, while the ethanol is sprinkled over them. Air is forced upward through the shavings and the vinegar is drawn off at the bottom.

Sauerkraut
Cabbage leaves are shredded, mixed with salt and pressed in containers to exclude air. Lactic acid bacteria are added to ferment the leaves, reducing their pH to around 5.0. This prevents other microorganisms growing and hence acts as a form of preservation, as well as imparting flavour to the cabbage.

Olives and cucumber preservation
Both olives and cucumbers can be preserved by storing them in barrels containing brine. In these anaerobic conditions lactic acid bacteria thrive, reducing the pH to around 4.0, thus preventing other microbial activity.

TABLE 41.2 **Some enzymes produced by microorganisms used in the food industry**

Enzyme	Examples of microorganisms involved	Application
α-Amylases	*Aspergillus oryzae*	Breakdown of starch in beer production Improving of flour Preparation of glucose syrup Thickening of canned sauces
β-Glucanase	*Bacillus subtilis*	Beer production
Glucose isomerase	*Bacillus coagulans*	Sweetener for soft drinks Cake fillings
Lactase	*Kluyveromyces* sp.	Lactose removal from whey Sweetener for milk drinks
Lipase	*Candida* sp.	Flavour development in cheese
Pectinase	*Aspergillus* sp.	Clearing of wines and fruit juices
Protease	*Bacillus subtilis*	Meat tenderizers
Pullulanase	*Klebsiella aerogenes*	Soft ice cream manufacture
Sucrase	*Saccharomyces* sp.	Confectionery production

Preparation of coffee and cocoa beans
Both coffee and cocoa beans are surrounded by a mucilage coat which must be removed before further processing can take place. The beans are heaped together to allow naturally occurring microflora, including yeasts and bacteria to ferment the mucilage and also prevent the beans germinating by killing the embryo in the high temperatures, around 50 °C, which are created.

Soy sauce
The fermentation of soya beans and wheat by *Aspergillus oryzae* and other microorganisms is used in the manufacture of soy sauce.

41.3.6 Enzymes associated with the food industry

Many enzymes used in the food industry are produced by microorganisms. Some examples are given in Table 41.2.

41.4 Biotechnology and pharmaceuticals

Since the antibiotic penicillin was first produced on a large scale in the 1940s, there has been a considerable expansion in the use of biotechnology to produce a range of antibiotics, hormones and other pharmaceuticals.

41.4.1 Antibiotics

Antibiotics are chemical substances produced by microorganisms which are effective in dilute solution in preventing the spread of other microorganisms. Most inhibit growth rather than kill the microbes on which they act. Some, like **penicillin**, are only effective on relatively few pathogens – **narrow spectrum antibiotics**, while others, e.g. **chloramphenicol**, will inhibit the growth of a wide variety of pathogens – **broad spectrum antibiotics**. Although around 5000 antibiotics have been discovered only 100 have proved medically and commercially viable.

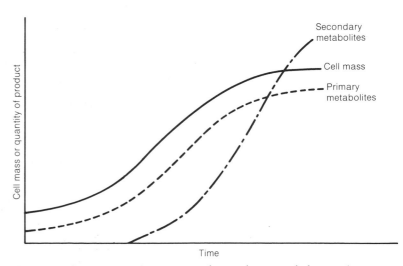

Fig. 41.4 *Comparison of primary and secondary metabolite production*

Penicillin fermentation

TABLE 41.3 **Some antibiotics and their producer organism**

Antibiotic	Producer organism	Type of organism
Penicillin	*Penicillium notatum*	Fungus
Griseofulvin	*Penicillium griseofulvum*	Fungus
Streptomycin	*Streptomyces griseus*	Actinomycete
Chloramphenicol	*Streptomyces venezuelae*	Actinomycete
Tetracycline	*Streptomyces aureofaciens*	Actinomycete
Colistin	*Bacillus colistinus*	Bacterium
Polymyxin B	*Bacillus polymyxa*	Bacterium

Antibiotics are made when growth of the producer organism is slowing down rather than when it is at its maximum. They are therefore secondary metabolites and their production takes longer than for primary metabolites. It also means that continuous fermentation techniques are unsuitable and only batch fermentation can be employed.

In penicillin manufacture, a stirred-tank fermenter is inoculated with a culture of *Penicillium notatum* or *Penicillium chrysogenum* and the fungus is grown under optimum conditions: 24 °C, good oxygen supply, slightly alkaline pH. Penicillin production typically commences after about 30 hours, reaching a maximum at around four days. Production ceases after about six days, at which point the contents of the fermenter are drained off. As the antibiotic is an extracellular product, the fungal mycelium is filtered off, washed and discarded. The liquid filtrate containing penicillin is chemically extracted and purified using solvents to leave a crystalline salt. After sterilization the fermenter is available for the next batch.

The development of resistance to antibiotics by pathogens means there is a continuing need to find new types. The emphasis has moved from searching for new natural antibiotics to the development of new strains using genetic engineering. However, this is not easy. Being secondary metabolites, antibiotics are the product of a long metabolic pathway involving numerous genes. Manipulation of these is complex and difficult. Programmes for enhancing the production rate of existing strains and using random mutation and selection methods to develop new ones, are continuously developed. Table 41.3 lists some other antibiotics and their producer organisms.

41.4.2 Hormones

With the advent of recombinant DNA technology (Section 12.8.1), it is now possible to use microorganisms to produce a wide range of hormones which previously had to be extracted from animal tissues. Hormone manufacture using fermentation techniques is relatively straightforward, but the high degree of purity of the final product which is essential, makes downstream processing a complex process. Ion-exchange, chromatography and protein engineering are some of the techniques employed to provide a high level of purity.

Hormones produced in this way include insulin, used in the treatment of diabetes, and human growth factor which prevents pituitary dwarfism. A number of steroids are also manufactured including cortisone and the sex hormones, testosterone and oestradiol. Others with possible commercial and medical value are relaxin which aids childbirth, and erythropoietin for the treatment of anaemia. Bovine somatotrophin (BST), a hormone administered to cows to increase milk yield, is also in current production.

41.5 Biotechnology and fuel production

The rise in oil prices in the early 1970s led to research into alternative means of producing fuel. With only a finite supply of oil available, work continues in this field, accelerated to some extent by the harmful consequences of burning traditional fossil fuels (Section 29.3.1). One method already tried is the fermentation of waste to yield **gasohol** (alcohol) or **biogas** (methane). It may be that in years to come, these fuels will be formed from crops specifically grown for the purpose.

41.5.1 Gasohol production

The 1970s rises in oil price hit third world, oil-importing countries, such as Brazil, especially hard, and prompted the initiation of the **Brazilian National Alcohol (or Gasohol) Programme**. The concept was simple – namely to use yeasts to ferment Brazil's plentiful supply of sugar cane into alcohol and so create a relatively cheap, renewable, home-produced fuel.

The programme began in 1975 and incorporated research into improving sugar cane production as well as fermenter technology. By 1985 sugar cane production had increased by a third, fermentation conversion by 10% and the fermentation time had been reduced by three quarters. Over 400 distilleries now yield more than $12 \times 10^9 \, dm^3$ of alcohol annually and all Brazilian cars have been converted to using the fuel, either entirely or mixed with petrol. (Some petrol is added to all alcohol fuels as a disincentive to people drinking them.) It is hoped by the year 2000 that all the country's energy needs will be supplied in this way. What makes the programme so successful in Brazil is that the sugar cane is not only a source of the fermentation substate, but also a fuel for the distilleries. Once the sugar is extracted from the sugar cane, the fibrous waste, called **bagasse**, can be dried and burnt as a power source for the distillery.

The actual process entails a number of stages:

1. Growing and cropping sugar cane.

2. Extraction of sugars by crushing and washing the cane.

3. The crystallizing out of the sucrose (for sale) leaving a syrup of glucose and fructose called **molasses**.

4. Fermentation of the molasses by *Saccharomyces cerevisiae* to yield dilute alcohol.

5. Distillation of the dilute alcohol to give pure ethanol, using the waste bagasse as a power source.

The special circumstances in Brazil have doubtlessly contributed to its success but schemes in some other countries, e.g. Kenya, have been abandoned as uneconomic. Nevertheless, the potential for solving both the problem of diminishing oil supplies and disposal of waste at the same time has its attractions. Waste straw, sawdust, vegetable matter, paper and its associated waste, and other carbohydrates are all possible respiratory substrates although many require enzyme treatment to convert them into glucose before yeast can act upon them.

41.5.2 Biogas production

The capacity of naturally occurring microorganisms to decompose wastes can be exploited to produce another useful fuel, methane (biogas). It has the advantage over alcohol (gasohol) production of not requiring complex distillation equipment – indeed, the process is very simple. A container known as a **digester** is filled with appropriate waste (domestic rubbish, sewage or agricultural waste can be used) to which is added a mixture of many bacterial species. The anaerobic fermentation of these yields methane which is collected ready for use for cooking, lighting or heating. Small domestic biogas fermenters are common in China and India.

Biogas generator, Nepal

41.6   Biotechnology and waste disposal

In the previous section we saw how microorganisms could be used not only to dispose of wastes but also yield a useful by-product at the same time. These are not the only ways in which microbes are used to dispose of unwanted material; the disposal of sewage, the decomposition of plastics and the breakdown of oil are further examples.

41.6.1 Sewage disposal

The basic details of sewage treatment have been discussed in Section 29.3.2 and therefore this section will confine itself to the microbial aspects of the process, in particular biological filtration and sludge digestion.

In a typical sewage filter bed a large variety of microflora are immobilized on layers of coke or stones. Openings above, below and in the sides of the filter, and the spaces between the 'clinker', maintain aerobic conditions. As the effluent trickles down the action of the microorganisms changes its composition. As a result, the species become stratified vertically in the filter bed. On the surface are numerous mobile protozoa as well as fungi of the genus *Fusarium*. In the upper layers occur *Zoogloea* spp., while *Nitrosomonas* spp., *Nitrobacter* spp. and stalked protozoa are found lower down. Between them, these and other species found in the filter oxidize the many organic substances in the sewage, to largely inorganic chemicals. Sludge digestion can be carried out either anaerobically or aerobically. In anaerobic digestion the sewage sludge is drained into large digester tanks where obligate anaerobes like *Clostridium* sp. ferment the protein, polysaccharides and lipids of the sludge into acetic acid, carbon dioxide and hydrogen. These products are then acted upon by *Methanobacterium* spp. which form methane (biogas) a useful fuel, which can be used to power the sewage treatment plant.

Aerobic digestion utilizes a technique known as **activated sludge**. Here the liquid from the primary settlement process is pumped to tanks where it is sparged with air by pumping it through diffusers at the base. Aerobic microflora, including *Zoogloea* spp., *Nitrobacter* spp., *Nitrosomonas* spp., *Pseudomonas* spp., *Beggiatoa* spp. and others, oxidize the organic components. The liquid is moved to other tanks where the remaining solids settle out in lumps called **flocs** − a process facilitated by bacteria, e.g. *Zoogloea ramigera*. This settled material can be returned to the aeration tanks for further treatment.

Any solid sludge not decomposed by either of these methods is led into storage lagoons where slow decomposition by anaerobic microorganisms over many years reduces it to a largely inorganic residue. It may otherwise be used for the production of fertilizer.

41.6.2 Biodegradable plastics

One of the commercially attractive features of plastics, polythene, polystyrene and related materials is their durability; they do not corrode or decay easily and are resistant to most forms of chemical attack or other forms of degradation apart from burning. It is this very feature, however, which creates the environmental problem of its disposal once it has ceased to be of use. Incineration produces

Sewage primary sedimentation

unpleasant, and sometimes dangerous, gases and hence there is a need to use a more acceptable and safer means of breaking it down. Polythene and polyester polyurethanes of low molecular mass have been developed which can be degraded by microorganisms such as the fungus *Cladosporium resinae*. In general, the more flexible plastics are broken down more easily than rigid ones. Such breakdown is desirable where the material has a short useful life, e.g. packaging, but not where its life-expectancy is longer, e.g. furniture and utensils. Research is therefore being carried out into finding new forms of biodegradable plastics for the packaging industry. These products are based on a storage chemical, **polyhydroxybutyrate**, of many microorganisms. Not only does this have the advantage of being biodegradable but it can also be manufactured by microorganisms using biotechnological methods. Another biodegradable packaging product is **Pullulan**, a commercially produced polysaccharide made by *Aureobasidium pullulans*.

41.6.3 Disposal of oil

Oil pollution of oceans, seas and rivers has unfortunately become all too common in recent years. The safe disposal, not only of oil spillages, but also the 'spent' oil from vehicles and machinery is highly desirable. Like the plastics made from it, oil is very resistant to microbial degradation largely because it contains little if any water. The same fungus, *Cladosporium resinae*, which degraded some plastics is equally effective in breaking down paraffin-based oils, and some forms of *Pseudomonas* spp., developed using recombinant DNA technology, are commercially used in cleaning up oil spills. These same organisms along with species of the fungi *Aspergillus* and *Cephalosporium* cause problems in fuel systems because their growth degrades the oil, and the fungal hyphae block filters and pipes. The problem is most acute where water, e.g. from condensation, contaminates the oil.

One means of dealing with oil pollution in water is to use emulsifiers to cause the oil to mix with the water and so both disperse it and speed up its microbial breakdown. One such emulsifier, a polysaccharide called **Emulsan**, is commercially produced by the bacterium *Acinetobacter calcoaceticus*.

41.6.4 Disposal of industrial wastes

The economics of industrial processes are such that waste disposal is often an unwanted expense and all too often in the past has led to materials simply being dumped at the nearest convenient point with little regard for the environmental consequences. An increased awareness of ecological issues and tighter legislation have prompted safer waste disposal. To offset the additional costs this entails, many methods incorporate an element of recycling which yields financial benefits.

In Section 41.5 we saw how certain wastes could be used to produce fuel. The use of domestic and agricultural wastes in producing biogas (methane) is equally applicable to many forms of organic industrial waste such as that from paper, cotton and wool mills. Brewery waste can be converted to citric acid, used extensively in the food industry, by the fungus *Aspergillus niger*. Animal feed is manufactured by another fungus *Paecilomyces* spp. from paper-mill waste, or by yeasts of the genera *Candida* and *Endomycopsis* from the waste of a potato-processing plant.

The viability of such schemes often depends on the ease with which the wastes can be converted into a form that microorganisms can utilize as a respiratory substrate. Genetically engineered microbes are making this task increasingly easy.

41.6.5 Recovery of valuable material from low-level sources including wastes

Few elements exist naturally in an uncontaminated form. Separating the desired material from its contaminants is frequently difficult and a point is reached where the level of the required material is so low that the cost of its extraction is no longer economic. The use of microorganisms in recovering such substances from low-level sources is not new (the Romans used the technique) but the range of materials obtained has expanded enormously, not least because of its value in removing environmentally harmful chemicals.

The bacterium *Thiobacillus thio-oxidans* is used to extract copper and uranium from waste ore which does not lend itself to metal extraction by conventional means. The ore is dumped and sprinkled with dilute acid to encourage the growth of *Thiobacillus* spp. (an acidophile) which oxidizes the copper sulphide according to the equation

$$CuS + 2Fe_2(SO_4)_3 \longrightarrow 2FeSO_4 + CuSO_4 + S$$

copper (II) sulphate   iron (III) sulphate   iron (II) sulphate   copper (II) sulphate   sulphur

The soluble metal sulphates so produced are washed out and collected – **metal leaching**. The pure metal can be extracted from the leachate by chemical means, e.g. by use of scrap iron.

$$CuSO_4 + Fe \longrightarrow Cu + FeSO_4$$

copper (II) sulphate   iron   copper   iron (II) sulphate

It is hoped that *Thiobacillus* spp. can also be used to remove the sulphur from high sulphur coal, thus making it more environmentally acceptable by reducing the sulphur dioxide it produces on combustion. The reverse of the above process by which insoluble metals are made soluble, can be used to remove toxic metal ions from waste. The bacterium *Desulfovibrio desulfuricans* will make metal sulphates into insoluble sulphides which then precipitate out. The alga *Scenedesmus* will absorb metal ions against a concentration gradient. Both species can therefore be used to remove harmful metal ions from industrial and mining waste before it is discharged. The use of denitrifying bacteria as a means of removing excess nitrates (the result of fertilizer use – see Section 29.3.2) from drinking water is being investigated.

The recovery of some oil also entails the use of microorganisms. With almost half of the world's underground oil being either trapped in rock or too viscous to recover by conventional means, microbes are used to reduce its viscosity. The process, known as **enhanced oil recovery**, allows the more fluid oil to drain out of the rock and be pumped to the surface. **Xanthan**, a polysaccharide gum produced by the bacterium *Xanthamonas campestris* is used to thicken water and improve its ability to drive the oil out of the rock.

41.7 Other products of the biotechnology industry

In addition to all the biotechnological applications of microorganisms already discussed, there is an assortment of other products made in this way. Some of these are given in Table 41.4.

TABLE 41.4 **A range of other commercial substances produced by microorganisms**

Product	Producer organism	Function of product
Protease	*Aspergillus oryzae* *Bacillus* spp.	Detergent additive Removal of hair from animal hides
Butanol and acetone	*Clostridium acetobutylicom*	Solvents
Indigo	*Escherichia coli*	Textile dye
Xanthan gum	*Xanthomonas campestris*	Thickener used in food, paints and cosmetics
Cellulases	*Trichoderma* spp.	Brightener in washing powders
Cyanocobalamin (vitamin B_{12})	*Propioni bacterium shermanii*	Food supplement
Gellan	*Pseudomonas* spp.	Food thickener
Glutamic acid	*Corynebacterium glutamicum*	Flavour enhancer
Streptokinase	*Streptomyces* spp.	Treatment of thrombosis
Interferon	*Escherichia coli*	Treatment of viral infections
Ergot alkaloids	*Claviceps purpurea*	Vasoconstricter used to treat migraine and in childbirth
Cyclosporin	*Cotypocladium inflatum*	Immunosuppressant drug

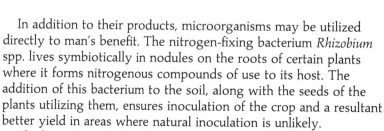

In addition to their products, microorganisms may be utilized directly to man's benefit. The nitrogen-fixing bacterium *Rhizobium* spp. lives symbiotically in nodules on the roots of certain plants where it forms nitrogenous compounds of use to its host. The addition of this bacterium to the soil, along with the seeds of the plants utilizing them, ensures inoculation of the crop and a resultant better yield in areas where natural inoculation is unlikely.

The bacterium *Bacillus thuringiensis* produces a protein which is highly toxic to insects. By contaminating the natural food of an insect pest with the bacterium, some control can be effected. The ecological consequences of using such 'natural' pesticides need further investigation but it seems likely that they will be less harmful than their artificial counterparts.

A recent, and controversial, development is the use of the bacterium *Pseudomonas syringae* to form artificial snow at winter holiday resorts. The bacterium, which is sprayed with water on to a fan, has surface properties which act as nuclei for the formation of ice crystals. The fact that this process could cause considerable frost-

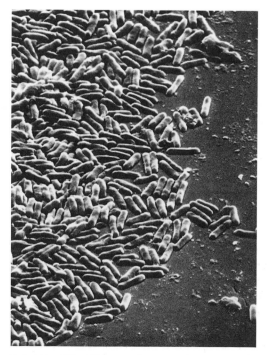

Escherichia coli (× 3000 approx.)

damage if used on food crops illustrates the risks that attend many biotechnological advances if used wrongly, either by accident or with malice.

41.8   Cell, tissue and organ culture

It is not only microorganisms that can be utilized to produce useful substances. Although more difficult to establish, plant and animal cells can also be grown in vitro in order to manufacture a variety of products. The culture requirements for eukaryotic cells are far more sophisticated than for prokaryotic cells.

41.8.1 Plant cell, tissue and organ culture

Section 20.3 dealt with the growth and development of plants and described how meristematic tissue was the basis of all plant growth. If a part of a plant containing a high proportion of meristematic tissue, e.g. bud, root tip or germinating seed, is removed and grown aseptically on a nutrient medium, an undifferentiated mass known as a **callus** frequently develops. Plant hormones – such as auxins, gibberellins and cytokinins (Section 40.2) – a nitrogen and carbohydrate source, as well as vitamins, trace elements and growth regulators need to be present in the nutrient medium. Undifferentiated callus produced in this way is an example of **plant tissue culture**. If the callus is suspended in a liquid nutrient medium and broken up mechanically into individual cells it forms a **plant cell culture**. These can be maintained indefinitely if sub-cultured giving rise to a **cell-line**. If pectinases and cellulases are added to these cultures, the plant cell walls can be removed and the resulting protoplasts can be fused with the protoplasts of different cells to give hybrid cells which can later be grown on into new plant varieties.

    Plant organ culture is achieved by taking apical shoot tips, sterilizing them and growing them on a nutrient medium containing cytokinin. A cluster of shoots develops, each of which may be grown on into new clusters. This form of micropropagation results in a large quantity of genetically identical individuals.

41.8.2 Applications of plant cell, tissue and organ culture

There are two main applications of these forms of culture. Firstly, the generation of plants for agricultural or horticultural use. Vast numbers of plants can be grown in sterile controlled conditions ensuring a much greater survival rate than would be the case where seeds were planted outdoors. These plants can all be identical if required thus ensuring a more uniform crop which incorporates desired characteristics. A particular advantage of this form of propagation is that the stock plants, raised as they are in sterile conditions, are completely pathogen free when planted out, by which time they have developed defence mechanisms against many diseases.

    The second application is the manufacture of useful chemicals by plant cultures. To date only one product, shikonin, a dye used in the silk industry and in the treatment of burns, has been produced commercially, but Table 41.5 lists some of the possible applications of plant cell culture once the processes become economically viable.

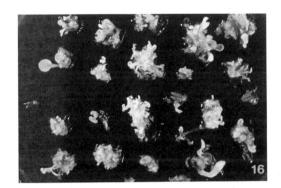

Anther calluses on agar

TABLE 41.5 **Some useful plant chemicals which might be produced using plant cell culture**

Product	Producer plant	Use of product
Atropine	Deadly nightshade (*Atropa belladonna*)	Ophthalmic use – dilation of pupil Treatment of certain heart conditions
Codeine	Opium poppy (*Papaver somniferum*)	Pain killer
Digoxin	Foxglove (*Digitalis* spp.)	Treatment of cardio-vascular complaints
Jasmine	Jasmine (*Jasminium* spp.)	Perfume
Menthol	Peppermint (*Mentha piperita*)	Food flavouring
Quinine	Chinchona tree (*Chinchona ledgeriana*)	Drug used to treat malaria Bitter flavouring in drinks
Pyrethrin	Chrysanthemum (*Chrysanthemum* spp.)	Insecticide

41.8.3 Animal cell culture

Only in recent years has it proved possible to culture vertebrate cells on any scale. The process begins by treating the appropriate tissue with a proteolytic enzyme like trypsin in order to separate the cells. These are transferred to an appropriate nutrient medium where they attach themselves to the bottom of the container and divide mitotically to give a monolayer of cells. This is referred to as the **primary culture**. Cells from these can be used to establish secondary cultures but their life span is limited, division often ceasing after 50–100 divisions.

It is possible to make these cultures continue to divide indefinitely by the addition of chemicals or viruses which induce the formation of cancer cells. These cell lines are said to be transformed and are **neoplastic**, i.e. they can induce cancer if transplanted into a related species.

41.8.4 Applications of animal cell cultures

Of the many applications of animal cell cultures, the production of viral vaccines is the oldest. Viruses are grown on culture cells, frequently monkey kidney cells or those from chick embryos, but increasingly, human cells. The culture medium is harvested and the viruses extracted by filtration. These are then treated to kill them, or, if of the attenuated type (see Section 32.2.2), stored at low temperature ready for use. Poliomyelitis, measles, German measles and rabies vaccines can be produced in this way.

Pharmaceutical products can be harvested from cell lines which over-produce particular products. These include interferon from certain human blood cells, human growth factor and the clotting factors used to treat haemophiliacs. This alternative supply has the added advantage of not transferring AIDS to the recipient, a risk, albeit very small, of the current practice of extracting these factors from donated blood.

Animal cell cultures are used in recombinant DNA technology where appropriate virus vectors can be cloned in them. Perhaps the most important use, however, is in the production of monoclonal antibodies — a topic dealt with in the following section.

41.8.5 Monoclonal antibodies

In Section 32.2 various aspects of the immune response were discussed including the production of antibodies from B-lymphocytes in the blood in response to a specific antigen. Any foreign material entering the body will possess more than one antigen and so induces a number of different B-lymphocytes to multiply, producing clones of themselves. The many clones then produce a range of antibodies known as **polyclonal antibodies**.

It is obviously of considerable therapeutic value to be able to produce antibodies outside the body, but until recently the inability to sustain the growth of B-lymphocytes prevented this. However, a cancer of B-lymphocytes produces myeloma cells which continue to divide indefinitely. These can be fused in the laboratory, using polyethylene glycol, with each specific B-lymphocyte to produce cells, called **hybridoma cells**, which produce a single antibody — **monoclonal antibody**.

41.8.6 Applications of monoclonal antibodies

The large scale production of antibodies using hybridoma cells is now underway and they are used medically to treat a range of infections. This is not their only application, however. Because they are specific to a single chemical (antigen), to which they become attached, they can be used to separate a particular chemical from a complex mixture. To do this, the monoclonal antibody for the required chemical is immobilized on resin beads which are then packed in a column. The mixture is passed over the beads and only the required chemical becomes attached to the antibodies. The chemical may then be obtained in a pure state by washing the beads with a solution which causes the antibodies to release it.

Monoclonal antibodies are also used in **immunoassays**. Here, the antibody is labelled in some way, e.g. radioactively or by a fluorescent dye, so that it can easily be detected. When added to a test sample they will attach to their specific antigen. Washing in solutions which remove only unattached antibodies leaves only those attached to the antigen. The amount of these in the sample is then apparent from the degree of radioactivity or fluorescence. For example, the presence of a particular pathogen in a blood sample can be detected by use of the appropriate monoclonal antibody tagged with a fluorescent dye.

Another technique is to immobilize the antibodies and pass the solution under test over them. Suppose we are testing for chemical X. If it is present in the solution it will attach to the antibody. A second type of antibody which has an enzyme attached is then added. It combines only with those original antibodies which are linked to chemical X. By adding a substrate which the enzyme causes to change colour, the amount of chemical X will be apparent by the extent of any colour change. This technique, called **Enzyme Linked Immunosorbant Assay (ELISA)**, has many uses including detecting drugs in athletes' urine, pregnancy testing kits and detecting the HIV virus (the AIDS test). It is also possible to link anti-cancer drugs to monoclonal antibodies which are attracted to cancer cells – the so-called '**magic bullets**'. An even more sophisticated technique is to tag monoclonal antibodies with an enzyme which converts an inactive form of the cytotoxic drug (**the prodrug**) into an active form. Once injected these antibodies link to the cancer cells. The prodrug is then administered in a relatively high dose as it is harmless in its inactive state. In the vicinity of normal cells the drug remains ineffective but in the presence of the cancer cells the enzyme on the attached antibody activates the drug which acts upon the cells, killing them. The technique is called **ADEPT (Antibody Direct Enzyme Prodrug Therapy)**.

41.9 Questions

1. Answer **BOTH** parts A **AND** B

PART A

A nutrient broth was inoculated with *Clostridium perfringens* (a bacterium) and grown under suitable conditions for 24 h. During this time samples were taken, diluted serially, plated onto an appropriate agar medium and the number of resulting colonies counted. The following table shows the number of bacteria present in the broth in terms of colony forming units (cfu) per cm³ of the original broth.

Time/h	cfu/cm³
0	5×10^3
2	4×10^3
4	2×10^4
6	1×10^5
8	2×10^6
12	9×10^8
16	6×10^8
20	5×10^7
24	5×10^6

(a) Present these data in a suitable graphical form, *either* using semilog graph paper, *or* using ordinary graph paper after the conversion of the bacterial counts to logarithmic values

(5 marks)

(b) On your graph indicate at least three phases of population change. Account for each phase in terms of factors which may slow or maintain the rate of reproduction. *(8 marks)*

PART B

A culture of yeast cells was initiated and after 3 days was inoculated with *Paramecium* (a protozoan which feeds on yeast cells).

(a) Sketch a graph indicating the population fluctuations you would expect of both the yeast and the *Paramecium* in such a mixed culture. (N.B. you are not expected to give realistic figures for cell counts but a reasonable time scale should be included). *(3 marks)*

(b) Give reasons for the precise form of your graph in part (*a*). *(4 marks)*

(Total 20 marks)

Northern Ireland Board June 1988, Paper 2, No. 5X

2. Write an account of the production of **one** of the following and indicate its practical applications. Define the **other two** terms.

(a) monoclonal antibodies

(b) protoplasts

(c) single cell protein *(12 marks, 4 marks)*

(Total 16 marks)

Oxford Local Board AS Specimen Paper, Option C, No. 4

Beer, Brewing and Biotechnology

3. Today's brewing industry is sophisticated and operates on a large scale. Most brewers use one of a number of strains of brewer's yeast *Saccharomyces cerevisiae*. The sort of yeast that brewers use is crucial,
5 because the microorganisms do more than just produce alcohol. The characteristic taste of a beverage depends on its 'flavour profile', which comes from the cocktail of organic chemicals made by the yeasts as they grow. All yeasts produce alcohol in the same way, by fermenting
10 sugars in the glycolytic pathway. But different strains vary in the extent to which they accumulate particular chemicals in other pathways.

Fermentation may be carried out either in batches or as a continuous process. In batch fermentation, the
15 reaction comes to an end, the products are removed, and the cycle is restarted with fresh reactants. Continuous fermentation involves a regular input of raw materials as products are tapped off and the process is on-going as long as nothing happens to upset it. Beer is produced in
20 batches rather than continuously, because the product must have a consistent flavour profile, and it is easier to reproduce the necessary conditions batch by batch, rather than risk the small differences that might arise during a continuous process. Continuous fermentation is also
25 more vulnerable to contamination by unwanted microorganisms. This is a particular problem in the brewing industry because it operates 'hygienically' rather than aseptically.

Ultimately, yeasts produce alcohol from the starch
30 stored in barley. Yeasts themselves cannot perform the first step in the breakdown of starch. Fortunately, a germinating barley grain can; it produces amylase that degrades starch. Hence brewers soak barley so that it germinates, and allow the amylase to break down the
35 starch to sugars, which the yeasts can exploit. This process is known as saccharification and the end product is called 'malt'. Brewers then boil the mixture to stop further action by the amylase. They filter it to produce the 'wort' and add hops to provide resins that contribute
40 to the flavour and also act as preservatives. The 'hopped wort' is then ready for fermentation and it is inoculated with the desired yeast.

During fermentation, yeasts produce bubbles of carbon dioxide which mix the beer and help to keep conditions
45 uniform inside the fermenter. Once the process is complete, manufacturers filter the beer to remove the yeast and impurities. They may then recycle a little of the yeast if they can, or use it elsewhere — as a food product, for instance.

50 So much is standard practice, yet there is room for improvement. Brewers would like new strains of yeast to reduce costs and improve the beer. For instance, a yeast that could use starch directly would make saccharification obsolete. So how can such improvements be created?
55 Plant and animal breeders have years of experience of

'improving' the characteristics of a population. They make an initial survey of the extent of variation in the population. Then, if necessary, mutations in existing strains are encouraged, e.g. by irradiation. Individuals
60 showing the desirable characteristics are selected and crossed with other strains to introduce the new features into the population. Unfortunately, brewer's yeast cannot be manipulated in this way. Yeasts will usually mate only with members of the same strain which makes it difficult
65 for geneticists to cross different yeasts together. The spores released by fungi which provide such accessible raw material for creating crosses are rarely produced by yeasts. Furthermore, industrially important yeasts tend to have unusual numbers of chromosomes, with multiple
70 or incomplete sets.

The advent of recombinant-DNA technology means that biologists can now bypass these difficulties. Over the past decade, researchers have produced plasmids that can introduce new DNA into the yeast. A plasmid in its
75 native state is simply a circular piece of DNA in the cytoplasm of a cell, distinct from the chromosome in the nucleus. Researchers have used such plasmids and placed them inside a virus to create a vector which can carry the foreign genes into a yeast cell. The plasmid DNA
80 must then be incorporated into the yeast's DNA so that it is transcribed along with the yeast's own genes. Genetic engineers make 'recombinant' plasmids that contain several segments of foreign DNA, each with a different function. First of all they insert the foreign gene of their
85 choice. This is flanked by signals for starting and stopping the transcription of RNA from the gene. Additional DNA sequences are then added which enable the yeast to make copies of the entire introduced plasmid. Finally, further lengths of DNA — the so-called 'selectable genes' are
90 inserted. These enable biologists to isolate the yeast cells into which the genetically-engineered plasmid becomes successfully integrated. These genes give the yeasts an ability to live on a particular culture medium, for instance, or resist an antibiotic. Using a recombinant plasmid like
95 this, genetic engineers can introduce all sorts of new genes into traditional strains of yeast.

One long-term goal is to make starch accessible to yeasts by giving them the genes that will produce amylase. One strain of yeast, *Saccharomyces diastaticus*,
100 already does contain a gene for amylase. The stumbling block is that *S. diastaticus* produces beers with an unpleasant taste! Recently, however, researchers have isolated the amylase gene from *S. diastaticus*. They have inserted the gene into other strains of yeast via a plasmid.
105 The genetically-engineered yeast metabolized starch efficiently. Unfortunately, the gene proved to be rather unstable in the plasmid vector, and the yeast rapidly lost the ability to degrade starch.

Brewers are also attempting to develop yeasts that are
110 more tolerant to alcohol. The fact that yeasts produce it as a waste product does not mean that they are immune to its toxic effects. Traditional fermenting techniques

achieve, at most, an alcohol concentration of around 6.5 per cent. At this concentration, and the temperature in
115 the fermentation vessel, most yeasts die. Economically speaking, the more alcohol a brewer can obtain from a fermentation the better. The sensitivity of the yeast to alcohol means that manufacturers now spend a lot of money and effort on removing alcohol from the vessel
120 before it kills too many yeast cells. An increase in the tolerance of yeast to alcohol would, therefore, save money.

Using the information in the passage and your own knowledge of biology, answer the following questions. Short, concise answers are required.

(a) Explain why brewers use different strains of yeast to produce different types of beer. (1 mark)

(b) Give **two** reasons why brewers use batch, rather than continuous, fermentation to produce beer. (2 marks)

(c) Explain what is meant by the statement that the brewing industry operates 'hygienically' rather than aseptically (line 27). (2 marks)

(d) List the steps and processes involved in the production of beer from barley. You may use a flow diagram in your answer if you wish. (3 marks)

(e) Explain the economic significance of the idea expressed in the statement 'yeast that could use starch directly would make saccharification obsolete' (lines 52–54). (1 mark)

(f) Explain how the unusual numbers of chromosomes in the yeasts used in industrial processes (lines 68–69) would make it difficult for geneticists to cross them successfully with different strains. (2 marks)

(g) Explain what is meant by 'recombinant-DNA technology' (line 71). (1 mark)

(h) From information in the passage, draw a labelled diagram of a plasmid to show the different sequences of DNA that would have to be inserted so that a new strain of yeast could be isolated after treatment with the recombinant plasmid. (4 marks)

(i) Explain how yeast cells in which an introduced plasmid is functioning normally can be distinguished from cells in which the plasmid has not been successfully inserted. (2 marks)

(j) Give **two** economic advantages of using genetically engineered yeasts which can withstand high concentrations of alcohol. (2 marks)
(Total 20 marks)

Joint Matriculation Board (Nuffield) June 1989, Paper IB,
No. 1

4. (a) What is metabolism? (1 mark)

(b) Primary metabolites of a microbial culture are products of metabolism during active growth and reproduction. Secondary metabolites are substances produced in significant quantities when the growth rate of the culture is slowing down. Little is known about why secondary metabolites are produced.
The table gives data for the industrial production of a metabolite by a culture of Bacillus ammoniagenes, supplied with a sugar substrate at the start of the process, over a period of eight days.

Time from start of production process /days	Residual sugar /mg dm^{-3}	Dry biomass of micro-organisms /mg dm^{-3}	Cumulative metabolite production /mg dm^{-3}
0	100	0.5	0.0
1	82	2.5	0.2
2	63	7.5	0.5
3	45	9.5	1.0
4	34	10.5	4.5
5	29	11.5	7.0
6	18	12.5	9.0
7	11	11.0	10.5
8	10	11.5	11.0

(adapted from Principles of Biotechnology – M. E. Bushell)

(i) Graph the data on a single set of axes. Use the same scale on one vertical axis for both dry biomass and cumulative metabolite production. Use a different scale on a second vertical axis for the residual sugar. (6 marks)

(ii) Describe and explain the growth stages shown by this culture over the eight days (6 marks)

(iii) Explain the relation between the residual sugar and the dry biomass. (2 marks)

(iv) Calculate the rates per day of dry biomass and metabolite production over the first three days and over the final five days. (3 marks)

(v) Give the ratio of metabolite production to dry biomass production over the first three days and over the final five days. Use this information to explain if this product is a primary or secondary metabolite. (3 marks)

(c) Give an example of the use of another micro-organism in the industrial production of a metabolite. Why is this metabolite of commercial importance? (3 marks)
(Total 24 marks)

Associated Examining Board June 1988, Paper II,
No. 4

5. (a) Explain how the nutrients of an appropriate culture medium satisfy the growth requirements of a typical mould or yeast.

(5 marks)

The graphs below show a fermentation process in which *Penicillium chrysogenum* is being grown using the batch culture method. From 40 hours, glucose was added slowly and continuously.

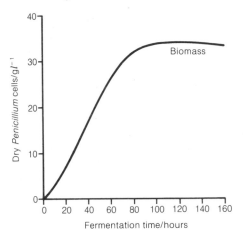

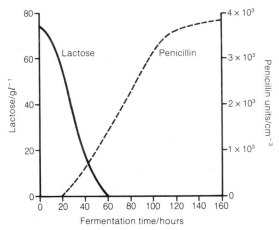

(b) (i) What is meant by a 'batch culture'?
  (ii) Explain the shape of the biomass curve.
  (iii) What evidence from the graph supports the conclusion that penicillin is a secondary metabolite?
  (iv) Glucose feeding starts at 40 hours. Why is this necessary at this time?
  (v) State, giving a reason, what is the most beneficial time to harvest the penicillin.
  (vi) Outline the procedures that might be required to obtain crystalline penicillin.

(10 marks)

(c) Outline various techniques which could be used to improve efficiency of antibiotic production by a particular species of fungus. How might this efficiency be tested?

(5 marks)

(Total 20 marks)

Joint Matriculation Board June 1989, Paper IB Option C, No. 4

6. In a study of yoghurt production, milk was incubated with *Lactobacillus bulgaricus* and *Streptococcus thermophilus* obtained from commercial natural yoghurt. Changes in the bacterial populations were recorded by estimating the numbers of organisms and by measuring the amount of lactic acid they produced. The graph shows the growth rate of the bacteria by numbers as determined by the serial dilution technique.

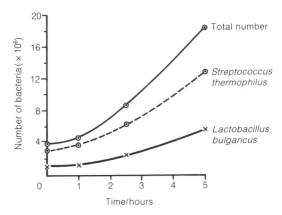

(a) (i) Sketch the shapes of the graphs you would expect if the results were plotted on the axes below.

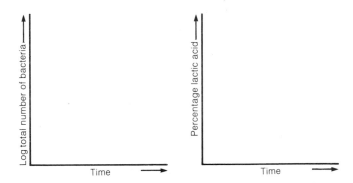

  (ii) Name the compound being used up to produce the lactic acid.
  (iii) Name the metabolic process taking place.

(4 marks)

(b) Outline the procedure you would use in order to obtain a sample of bacteria with a dilution factor of × 10 000, from the developing yoghurt.

(4 marks)

(c) (i) Describe how you would prepare a slide of the bacteria so that they are clearly visible under the microscope.
  (ii) How would you distinguish between *Lactobaccillus* sp. and *Streptococcus* sp?

(4 marks)

(d) Name **two** useful products, in addition to yoghurt, which can be manufactured by bacterial activity.

(2 marks)

(Total 14 marks)

Welsh Joint Education Committee June 1989, Paper A2, No. 11

7. The growth rate of a culture of the bacterium *Escherichia coli* in a standard nutrient solution and at the same constant temperature was measured by three different methods.

Method I

E. coli cultured in these conditions produces a gas which can be used as a measure of bacterial growth. The volume of gas produced was measured at regular intervals using the graduated scale on the pipette in the apparatus shown below.

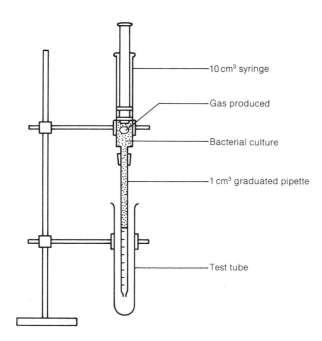

(a) Describe **one** precaution which should be taken to prevent contamination of the culture while setting up the investigation. *(1 mark)*

(b) *Outline* how you would use the apparatus to measure the rate of gas production. *(2 marks)*

Method II

At regular intervals a 1 cm³ sample of the culture was removed from a conical flask and transferred to a tube containing 9 cm³ of sterile saline solution. A series of increasing dilutions was prepared from this mixture. 0.1 cm³ volumes of each dilution were spread over the surfaces of separate nutrient agar plates. After incubation, the number of bacteria on each plate was determined by counting the colonies, assuming that each colony developed from a single bacterium.

(c) *Outline* how you would determine the number of bacteria present in the original 1 cm³ sample by this method. *(2 marks)*

Method III

At regular intervals a measured sample of the bacterial culture was transferred to a tube and its optical density determined with a colorimeter.

(d) Assuming that the appropriate filter was used, give **one** way of reducing the chances of experimental error when determining the optical density of the sample. *(1 mark)*

The results for each method are shown in the three graphs below.

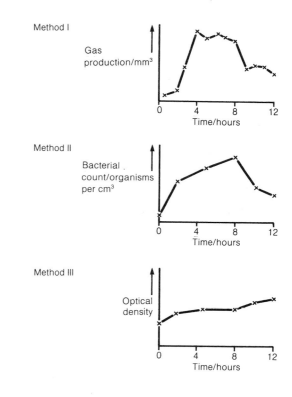

(e) Describe and explain the gas production results (Method I) during the first 4 hours. *(2 marks)*

(f) Suggest **one** explanation for the fall in number of bacteria after 8 hours in the results for Method II. *(1 mark)*

(g) How could you explain the continued increase in optical density from 8 to 12 hours (Method III) even though the bacterial counts and the gas production were decreasing? *(1 mark)*

(h) (i) Which **one** of the three methods would give the most reliable measure of the growth rate of the bacterial culture? *(1 mark)*

(ii) Explain your answer. *(1 mark)*

(Total 12 marks)

Joint Matriculation Board (Nuffield) June 1988, Paper IA, No. 5

8. A bacterial culture was established in liquid medium at an initial concentration of 1 million bacteria per mm³ of solution. The growth of the culture was monitored by sampling every ten hours and counting the number of bacteria in the samples. The results are shown on the graph at the bottom of the next page.

After 200 hours the culture was split into two equal volumes, labelled A and B in the figure. A protozoan that feeds on the bacteria was introduced into culture B and the subsequent changes in the protozoan numbers were plotted in curve C. The predatory protozoan was ciliated and used its cilia to swim and to create a feeding current which carried the floating bacteria down into its cytosome, or mouth.

(a) State the time range during which the bacterial culture is growing at its maximum rate in the absence of the protozoan.   (*2 marks*)

(b) State how long it takes the bacterial colony to double its numbers in the absence of the protozoan
   (i) after the colony has been growing for 40 hours,
   (ii) when the number present is 35 million per mm³.   (*2 marks*)

(c) Suggest an explanation for the difference between your answers to (*b*) (i) and (ii). (*2 marks*)

(d) Which of the following **best** explains the fact that little change in the number of bacteria takes place in culture A after 160 hours?
   (i) Food shortage has caused reproduction to fall to a low level.
   (ii) The death rate has risen to a point where it equals the reproductive rate.

(iii) Overcrowding is causing a steady decrease in reproductive rate.
(iv) Food shortage and/or overcrowding are causing a steady increase in mortality.
   (*1 mark*)

(e) Suggest a reason for
   (i) the fall in protozoa numbers after 230 hours;
   (ii) the rise in bacteria numbers after 240 hours;
   (iii) the rise in protozoa numbers after 290 hours.   (*3 marks*)

(f) After 420 hours the numbers of bacteria and protozoa settled down to more or less constant levels. The investigator explained this in the following words.

'After 340 hours the development of a thick growth of bacteria was noted on the inner walls of the vessels containing the culture B. Bacteria isolated from this vessel after the experiment showed a marked tendency to grow on solid surfaces or in large aggregates. In the presence of the protozoa, bacterial variants which adhere to the walls have a significant selective advantage. The system provides an example of niche diversification, the creation and utilization of a separate niche in what was a homogeneous experimental system.'

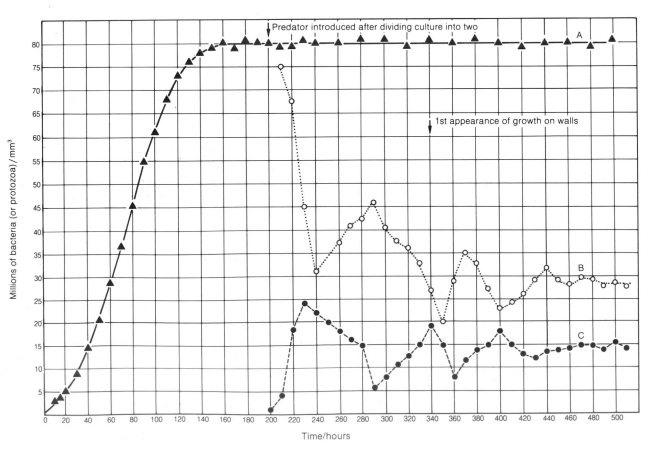

Explain what the investigator means by the following:
  (i) large aggregates;
  (ii) niche;
  (iii) homogeneous experimental system;
  (iv) selective advantage. (*4 marks*)
(*g*) (i) Explain fully why growth on the walls might be an advantage to the bacteria in this experiment.
  (ii) If the investigator's explanation of wall growth is correct, would you expect as much to occur in culture A? Explain your answer. (*4 marks*)
(*h*) If wall growth had not appeared in culture B, suggest what may have eventually happened to the bacterial and protozoan populations.
(*2 marks*)
(*Total 20 marks*)

Welsh Joint Education Committee June 1984, Paper AII, No. 1

9. Cells isolated from living plants can be grown in a nutrient solution under controlled conditions. This technique is known as tissue culture and it can be used to produce clones of plants in a relatively short time for commercial use.
Cells isolated from a rose plant were grown in cultures, and the changes in fresh mass, dry mass and DNA content were measured during a period of 12 days.
The results are shown graphically below.

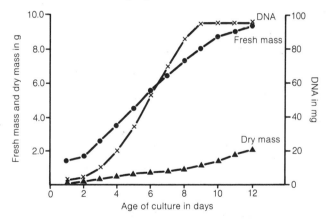

(*a*) Describe the changes in fresh mass and dry mass during the experimental period. Suggest an explanation for the differences between the changes in fresh and dry mass.
(*4 marks*)
(*b*) (i) What is the significance of the increase in DNA content during the experimental period? (*2 marks*)
  (ii) Comment on the observation that no increase in DNA content of the cultures occurred between days 9 and 12, while during the same period fresh mass continued to increase. (*3 marks*)

(*c*) (i) Name *two* minerals that you would expect to be present in the culture solution.
(*2 marks*)
  (ii) Name *one* condition that should be controlled in the cultures, and give *one* reason for your answer. (*2 marks*)
(*d*) (i) Cells in a five day old culture averaged 34 μm in diameter. At 12 days the cells in the same culture averaged 60 μm in diameter.
  Assuming the cells to be spherical, calculate the increase in average cell volume between days 5 and 12 in this culture.
  Show your working.
  Volume of sphere = $4/3 \pi r^3$,
  where $\pi = 3.14$ and r = cell radius
(*3 marks*)
  (ii) Cells from the 12 day old cultures in the above experiment were used to start new cultures. After a further 5 days, cells in the new cultures averaged 30 μm in diameter. What conclusion can be drawn concerning their relative rates of cell division and cell expansion?
  Explain your answer. (*2 marks*)
(*e*) Suggest *two* possible advantages of cloning plants using tissue culture techniques. (*2 marks*)
(*Total 20 marks*)

London Board January 1990, Paper I, No. 15

10. Read the passage below and then answer the questions which follow.

Insulin from pigs and horses has been widely used for treating diabetes mellitus, but it is not identical to human insulin and can have undesirable side effects. Human insulin can now be produced commercially using the techniques of genetic engineering. Reverse transcriptase may be used to generate cDNA from intron-free insulin mRNA. 'Start and stop' signals are added to the cDNA, together with sticky ends. The gene is then inserted into a suitable plasmid, such as pBR322, and the hybrid vectors are mixed with bacteria in conditions which encourage their uptake. The latter are then cloned, using suitable enrichment techniques, and screened for insulin production using DNA probes or antibodies. Finally the required clone is cultured, and the insulin is extracted and purified. One problem often found is that recombinant bacteria may lose their desired hybrid plasmids. In the absence of selective media the resulting daughter cells grow at a faster rate than their hybrid parents, so the rate of insulin production in a culture may decrease.

(a) In what way is reverse transcriptase an unusual enzyme? *(1 mark)*

(b) What is an intron? *(2 marks)*

(c) Human insulin genes, which contain two small introns, can be isolated from human chromosomes, but even if correctly incorporated into bacteria they cannot produce insulin. Why not? *(3 marks)*

(d) What is a clone? *(1 mark)*

(e) What is meant by the term hybrid vector? *(2 marks)*

(f) A plasmid vector often used for genetic engineering is pBR322, shown below:

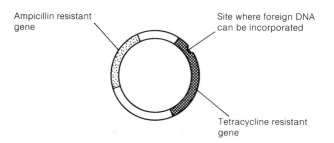

pBR322 was mixed with insulin cDNA and subsequently with bacteria. The bacteria were afterwards grown on agar. The colonies were then replicated onto the media shown below. Three types of colonies were found (A, B and C), some growing on only one medium, some on neither, and some on both.

Growth medium	+ Ampicillin added	+ Tetracycline added
Type A	Grows	Grows
Type B	Grows	No growth
Type C	No growth	No growth

Give an explanation of these results and show how this procedure can be used to identify those colonies which contain hybrid vectors. *(4 marks)*

(g) Suggest why bacteria which have lost their hybrid plasmids grow faster than those which still possess them. *(2 marks)*

(Total 15 marks)

Oxford Local Board AS Specimen Paper, Option C, No. 3

42 *Studying A-level Biology*

Most students embarking on an A-level Biology course will have studied Biology previously and so have some idea of what the subject entails. Some may have studied related subjects such as Human Biology, Environmental Biology or Social Biology. A few may be tackling the subject for the first time. For all students, whether or not they have previously studied the subject, the first problem will be to master the skills required at A-level. These skills differ markedly from those encountered at lower levels. Many students have no conception of the additional demands of A-level work, seeing it simply as a continuation of GCSE work.

The most fundamental difference encountered at A-level is the greater emphasis on understanding, interpreting and applying knowledge rather than the learning and recall of factual information which is common at lower levels. A-level candidates need to appreciate the underlying principles of a subject and understand the concepts and ideas behind them in addition to the factual content associated with them. Students may be presented with information on topics which they have never previously encountered and asked to apply their general biological knowledge in answering questions. Application of knowledge is dealt with in Section 42.3. The analysis and interpretation of data, presented in its many forms, is common at A-level.

A-level questions are more open-ended, requiring students to argue a case, justifying their points with supporting evidence. To do so effectively demands not only a body of factual knowledge but also the ability to communicate it effectively. Advice on communicating information in a variety of ways is given in Section 42.2.

A-level success depends largely upon the ability to acquire a wide range of knowledge. This is most effectively achieved by reading around the subject as variously as possible; class notes and a textbook alone being inadequate. Advice on reading is given in Section 42.1.3.

42.1 Study skills

Just as the techniques required at A-level differ from those required at lower levels so too do the methods of study. To a large degree, the method of study depends upon the subject involved and the character and temperament of the student. It therefore differs widely from individual to individual. The purpose of this chapter is *not* to dictate to students how to study, or to suggest that one method is better than another. It is simply to provide ideas which the student can experiment with. At the end of the day it is up to individuals to decide which method suits them best and brings most success. The main thing is to adopt a different approach from that used at

GCSE. Failure to do so is likely to reduce one's chances of success. The following outline concerns study skills which are relevant to A-level Biology.

42.1.1 Note taking

Notes may be made from books (see Section 42.1.3), periodicals, television, films and radio. All can make a valuable contribution to a student's knowledge and understanding. For most students, however, notes will be largely taken from lessons or lectures given by teachers. Whatever the source, there are a number of skills required for efficient note taking. They include:

1. Distinguishing between the main points and the peripheral material.

This is not quite the same as distinguishing between what is relevant and what is irrelevant. In most cases all the material will be relevant, but time won't allow everything to be written down. It is important to try to understand the main idea which is being put across and note this down along with any supporting evidence. There may well not be time to include background or additional information but this may be included at some future time. If at all possible, notes should be read again later the same day. The content will be fresh in your mind and you may well recall additional detail which can then be added at this time. You may not be able to understand all that you have written, in which case you should seek clarification from the subject teacher as soon as possible.

To distinguish the main points is not always easy. At first it may be necessary to ask whoever has given the lesson to check what you have written, or at least to tell them what you think were the main points and see if they agree.

2. Dividing notes into sections under clear headings

Little is more daunting than page after page of unbroken prose. Imagine this book, or even this section, written as one long essay with no sub-sections or headings. Remember that when you come to revise you will probably not feel much like the task anyway. If you are then faced with a seemingly endless collection of continuous notes you may well give up after a few minutes or even be put off starting altogether.

If possible, try to devise a scheme of heading values, where there are three or four levels or sub-sections. A typical scheme, with some examples, is given in Table 42.1.

TABLE 42.1 **How notes should be divided into sections under headings**

Heading value	Written as	Example
A. Major section	Upper case letters underlined in the centre of the page	**BLOOD AND CIRCULATION**
B. Major topic (within section)	Upper case letters underlined at the edge of the page	**THE HEART**
C. Sub-divisions of topic	Lower case letters underlined at the edge of the page	Heart beat (Cardiac rhythm)
D. Minor heading	Lower case letters not underlined	Factors modifying heart beat

Not always will all heading values be required and occasionally more may be needed, but for most purposes four will suffice. Such formality in note-taking may appear unnecessary and time-consuming. It will, however, make access to information easier at some future date and make revision easier to divide into manageable sections. In any case, the heading values will quickly become automatic and it will hardly matter if they are not one hundred per cent accurate. A word of warning — don't use a ruler for underlining during actual note-taking. It is unnecessarily time-consuming, so underline freehand. If you are especially fastidious about the appearance of your notes, do the underlining later, but remember, it is the content of the notes, not their appearance, which will bring success.

3. Writing quickly and concisely

Not everyone can write rapidly, but even slow writers can develop techniques which allow them to get down a considerable amount of information in a short time. These include:

(a) Writing in note form as grammatical prose is too time-consuming. *e.g.* Insulin, from pancreas (islets of Langerhans β-cells) glucose $\rightarrow$ glycogen — stored in liver.

(b) Using standard abbreviations of biological words, e.g. DNA, ATP, NADP.

(c) Using abbreviated forms of words, e.g. temp., sec., $2°$ (secondary).

(d) Developing your own shorthand, especially for commonly occurring biological words, e.g. P/S (photosynthesis), R/P (respiration). These must be restricted to your notes and *not* used in essays or examinations.

(e) Avoiding the use of rulers, coloured pencils, etc., unless essential for diagrams. Don't waste time underlining with a ruler or giving headings a different colour; it is more important to get down as much information as possible.

(f) Avoiding the use of correction fluid — if you make an error cross it out. Don't waste precious time using correction fluid and waiting for it to dry.

4. Continually adding to your notes

As you gain more information from reading or watching films and documentaries, add this to the relevant section of your notes. Loose-leaf files are more or less essential for this purpose. It also helps to leave a number of lines at the end of each sub-section so relevant information can be added at a later date. Always keep notes accessible and add new information immediately it comes to hand. It is good practice to keep the relevant notes at your side when reading a book or article on a particular topic. Adding material over the length of your course helps to keep you up-to-date. Remember that some notes will be nearly two years old by the time you sit the examination. In a subject which changes as rapidly as biology an awareness of recent developments is a distinct advantage.

42.1.2 Organizing notes

Having written reasonable notes it is essential to organize them efficiently. The use of appropriate headings is one means of achieving this. Another way is to ensure the notes are organized within a file, preferably using cardboard dividers between the major sections. This makes access much easier and saves time wasted thumbing through a pile of papers for notes on a particular topic.

An index at the start of each file is often useful, although page numbering is not essential as the addition of new material quickly makes this obsolete and necessitates renumbering. Essays on a topic can be added at the end of the relevant section of notes. Don't be afraid to reorganize notes from time to time. It is not always desirable to keep them in chronological order. The order in which topics are taught may depend on many factors such as the seasonal availability of materials, the availability of apparatus and the preferences of the teacher. Be prepared to arrange your notes in the order which suits you.

42.1.3 Reading skills

Wide and varied reading around the subject is an essential adjunct to A-level success. Dependence on class notes and a textbook alone may leave the student with a rather superficial knowledge of biology. Other sources of information should be used. These include books, periodicals, magazines and newspapers. All may contain material of relevance to an A-level course.

The problem is exactly what to read. Newspapers and magazines often carry articles of biological interest, especially where these relate to human activities. Medical discoveries, problems of pollution, advances in agriculture, discussions on diseases such as AIDS are just a few examples of relevant articles. Oil pollution from the Torrey Canyon (1967), the release of dioxin in Seveso (1976) and the explosion at the nuclear power plant at Chernobyl (1986) are examples of biologically important events which were first reported in newspapers. It is a good idea to browse through as many newspapers and magazines as possible and, provided they are not needed by others, articles can be removed and filed in the relevant section of your notes. A word of caution, however; newspaper and magazine articles may not give a balanced or scientific view. Read critically and use them as a basis for ideas and arguments rather than a source of factual information.

Periodicals relevant to A-level Biology include *Scientific American*, *New Scientist* and *Nature*, but perhaps the most relevant is *Biological Sciences Review*, a topical and interesting magazine written specifically for students of A-Level biological subjects. All these contain a wealth of information on the latest discoveries and ideas. Many of the articles go beyond A-level and may be difficult to understand. It will not matter if the fine detail is incomprehensible provided the general principles are clear. Reading these periodicals is the best way of keeping up-to-date, and some notes, or at least a reference, should be added to the relevant section of your class notes. Check the syllabus first to avoid writing copious notes on a topic which has no relevance to your course. In addition, there are many other journals on the market which contain useful and interesting information. Look around, sample a few, and perhaps subscribe to those you consider most useful. Books remain the major source of biological information and are used for a number of purposes:

1. As an aid to learning, helping the student to recall factual information.

2. For reference as a source of information for the writing of essays or for answering problems.

3. As a means of improving understanding and providing a greater depth of knowledge.

4. As a way of stimulating an interest in the subject.

5. As a means of generating new ideas.

6. For enjoyment.

The traditional large textbook is not intended to be read from beginning to end in sequence. Rather it should be used as a reference book, where specific pages or chapters are read in conjunction with the current topic being taught or this week's essay question. There is now a welcome trend towards the publication of shorter books on more specific topics. These may be read in much the same way as a chapter in a textbook. In some cases the entire book can be completed in an evening. The excellent *New Studies in Biology* series sponsored by the Institute of Biology and published by Edward Arnold is a good example. Each monograph deals with individual topics. All are relatively inexpensive and make excellent reading. Being designed primarily for A-level, the content is relevant, the language accessible and detail more than adequate. It would be an extremely dull student who could not find a number of these books which interested him. Indeed, one would question such a student's suitability for an A-level Biology course. A second series of even shorter monographs, typically 16–32 pages, is the *Oxford/Carolina Biology Readers.* The detail often goes beyond A-level, but they nevertheless can make a useful contribution to A-level reading.

One is often asked by students to recommend books to read. This is almost impossible to do as much depends on an individual's taste. Obviously it is easy to make recommendations on sources of information which may be used for specific essays, but 'general reading' presents a greater problem. The best answer is to read, initially at least, any book on a topic in which you have a major interest. It may be genetics or pollution or something more specific like DNA or blood clotting. You may have a passion for learning about how animals adapt to life in deserts, the territorial behaviour of certain fish or migration in birds. The nature of the topic hardly matters, provided you are interested. After all, reading is not intended as a form of punishment, but something to be enjoyed. The enforced reading of particular books is unlikely to engender a love of biology; it is far more likely to have you applying for a change of course. In any case, you are unlikely to complete a book you don't enjoy and so will have gained little if anything from the exercise.

Don't feel that all reading has to be directly relevant to your syllabus. Any reading has some value. It will improve your English through the constant use of sentence construction and grammar. It will improve both your biological and non-biological vocabulary as you meet new words (provided you use an appropriate dictionary to discover their meaning). In this way reading will help you to become a more efficient communicator and so put across your ideas more coherently. As you meet different styles and approaches you will be able to adapt your own writing to include those you find effective. At the same time, you will doubtlessly learn new biological information as well as reinforcing what you already know. You may also be stimulated to do further reading, possibly on topics you had not previously considered. Make brief notes during your reading and add these to your class notes. Jot down page references as you go; you may need to refer back to these when writing some future essay.

It is important to distinguish between reading notes during revision and general reading. The former requires careful attention to each word because, if the notes have been taken properly, they will contain little or no superfluous information. In addition, as you are revising, it may be helpful to read parts a number of times to commit the information to memory. General reading must be much more rapid. Too many students read out loud. That is, they effectively say each word, if not audibly, at least to themselves. You should try to perfect a technique whereby the eyes skim rapidly across the lines, much faster than if the words were spoken. The brain is well capable of taking in the subject content at a speed faster than that necessary to speak the words. This speed reading is essential if enough books are to be completed to have a significant influence on your performance. Reading speed can normally be increased without any loss of comprehension. Try it and attempt to perfect it through practice. You may even consider taking one of the many speed-reading courses available.

42.1.4 Use of teacher comments and criticism

If A-level success could be gained solely from reading books, students could be issued with a textbook and a reading list and asked to return nearly two years later to sit their examinations.

Students must see their teachers as more than providers of oral information and printed sheets. They should seek their advice and guidance throughout the course. Essays and other questions set during a course will normally be marked. When these are discussed it is vital to analyse your own shortcomings. Isolate where you have weaknesses. Is it that you don't read the questions carefully? Having read the question accurately did you misinterpret what was required? Was the subject content inaccurate or inadequate? Was detail lacking? Was there an absence of supporting evidence? Did you include too few examples? Did you fail to appreciate the underlying principles or did you get bogged down with unnecessary detail? Did you use the right references? Having determined where you went wrong, think of methods by which future work can avoid these problems. If necessary seek your teacher's advice.

Where individual comments are made on essays, take heed of them and determine not to make the same errors in future. A number of students ignore criticisms and simply repeat their failings time and time again. Apart from exasperating their subject staff, this approach is hardly likely to effect an improvement. Learn to see all criticisms as positive; a means of isolating errors and providing an opportunity of putting them right. After all, if your work is beyond criticism you may as well take the A-level without further delay.

Ask for clarification of any comments if you are not clear to what they refer. Seek guidance on the best way to improve. Don't expect sudden or dramatic changes as this is unlikely. Aim to make slow, steady progress in partnership with your subject teacher.

42.1.5 Organizing free time

Possibly for the first time, students studying A-levels find themselves with free lessons, or study periods as most teachers prefer to call them. The latter term is much more indicative of how they should be used. Whether supervised or not, these lessons should be organized in advance if you are not to find yourself with nothing to do. It is incomprehensible for any student studying two or three A-levels not to have enough work to occupy him or herself full-time. The absence of anything to do in study periods reflects inadequate planning rather than a lack of work needing to be done.

Writing of essays, where many books may be involved, is often inappropriate during study periods, especially where desk space is limited. Taking notes from a book in preparation for an essay is perfectly feasible. Study periods are ideal opportunities for general reading, especially of journals and periodicals which may be available in the school or college library. Planning essays or reading over completed ones can also be undertaken at this time. Answering questions which do not involve a large number of reference books is another way to occupy yourself. Always plan what you intend to do in advance and have a reserve assignment available in case it is needed. Do not rely on being able to do work given the lesson before; it may not be set, or may prove impractical to do.

On the whole, study periods are subject to fewer distractions than time spent at home. The attitude 'I work better at home' is often a cover-up for inadequate planning or an absence of self-discipline. The more work completed during study periods, the more flexibility you will have in organizing your evenings and weekends.

Time at home should be equally well organized. Draw up a timetable of your total 'free time'. Fill in long-standing commitments, e.g. part-time jobs, clubs, societies. Add times when you know work has to be done. For example, if experiments have to be written up the same day as they are carried out, earmark an appropriate length of time that day for this purpose. Include time for reading over notes done that day plus an allowance for general reading, essay preparation etc. Only when all this has been added should casual commitments be fitted in. If there is inadequate free time to complete all you wish to do it may be necessary to cut down on some social activities or even your part-time job. Above all, make the A-level work your priority. It should be in conjunction with leisure, however. It is inadvisable to spend your free time studying to the exclusion of all else.

Try to develop a regular pattern of work whereby certain evenings each week are given over to a particular activity. The pattern should not be too rigid, but permit enough flexibility for sudden changes of plan. Remember, however, that if a study evening is given up to attend a concert, a compensatory change from social to study use should be made on another night.

It is difficult to specify how much time should be spent in study, outside the prescribed lessons. Revision apart, a *minimum* of 5 hours per week per A-level subject is probably appropriate. For those aspiring to higher grades, this will need to be considerably increased to allow for the additional reading necessary.

42.2   Communicating information

There is little point in developing study skills, acquiring knowledge and learning how to apply it, but then being unable to communicate it effectively to others. The vast majority of assessment at A-level is through written work in one form or another, although not entirely in the form of essays.

42.2.1 Tables and graphs

As a scientific subject Biology involves data in many forms. The student may be required to present data collected during an experiment in an appropriate way or may be asked to analyse and interpret data which has been provided. It may be necessary at times to convert one form of data into another.

The various methods by which data may be presented are dealt with in Section 15.1. The skill lies in choosing the most appropriate method for a given set of data and guidance on this is also provided. Use a table for numerical or brief written information. If you are writing long accounts in each section of the table, it is probably not the most appropriate way of presenting it. Remember to name each column and/or row clearly and where applicable give units, e.g. time/sec, temperature/°C), rate/mm^{3}O$_2$ sec^{-1}.

When drawing graphs always:

1. Give the graph a clear title, e.g. 'Graph to show photosynthesis in *Elodea*, measured as mm^{3}O$_2$ released per second'.

2. Choose scales so they are easy to use and make maximum use of the graph paper.

3. Label axes clearly — state units and scale used.

4. Plot points carefully (in pencil initially) using a dot surrounded by a circle or a small cross.

5. Draw the best straight line or smooth curve through the points where there is good reason to think that the intermediate values fall on that curve. Do not attempt to join each adjacent point. The line need not pass through every point provided it is reasonably close to them. Where you do not know how the values vary between each plotted point, then join adjacent points with a straight line.

6. If two or more lines are to be drawn on the same axes, distinguish each by clearly labelling them.

If using a histogram, bar chart or pie chart avoid excessive time-consuming shading. If it is necessary to distinguish one block from another, use cross-hatching rather than shading the entire area.

42.2.2 Diagrams and drawings

Diagrams and drawings are an essential part of practical work, but they also play their part in theory. Some questions specifically ask for drawings, e.g. 'Give an illustrated account...', 'By means of labelled drawings...', 'With the aid of diagrams...' etc. In these cases drawings must be included. Where drawings are not specifically requested it is more difficult to advise on their use. If drawings are included they should be drawn well; rough sketches rarely bring credit. To do well, drawings take time, so ask yourself whether the information can be provided as clearly but more quickly by other means, e.g. in prose. If it can, avoid diagrams. If you do decide a drawing is appropriate, label and annotate it well. Do not make drawings for the sake of it, only include them where they are entirely relevant to the main theme of a question and they increase the understanding of the reader. Some students instinctively draw a diagram whenever they use a particular word, e.g. villus. In a question specifically on absorption of digested food in mammals, a drawing of a villus may be appropriate. In an essay on different mechanisms of feeding in animals it comprises such a tiny proportion of the total answer that it barely warrants a mention, let alone a diagram. If you decide a diagram is appropriate use the following guidelines.

1. Use a good quality, sharp pencil. The hardness will depend upon the individual but HB is best for most people.

2. Consider the size of what you intend to draw in relation to the size of your paper and leave room for labels and annotations.

3. Make large, clear line drawings without the use of ink, or coloured pencils except possibly when distinguishing oxygenated and deoxygenated blood.

4. Do not shade unless it is essential as a means of distinguishing various parts.

5. Give each drawing a suitable title.

6. State the magnification, scale or actual size of the object drawn.

7. Fully label all drawings and add appropriate annotations (notes attached to each label).

8. Do not label too close to the drawing and never on it.

9. Do not intersect (cross) label lines and if possible arrange labels vertically one beneath the other.

42.2.3 Essay writing

All examining boards include at least one essay paper in their A-level Biology examinations. The advantage of an essay is its flexibility. There is no set answer and many different responses can be equally valid. There is scope in essays to include detail and examples which have been acquired through additional reading. All essays are individual and their style and content will vary markedly from one student to the next. The very nature of essay questions encourages the individualistic approach and it would be wrong for anyone to try to stereotype essay writing. There are, however, a few fundamental skills which once mastered can improve essay technique. A single essay often covers a broad area of biology. Essays are often set on topics where there is conjecture and controversy, where views and opinions are important. It is therefore necessary to consider carefully what you intend to write in order to ensure a balanced and unbiased scientific view. Any arguments must be arranged in sequence and the content logically organized. All this requires careful planning.

Having carefully read the essay title two or three times, jot down the ideas which immediately come to mind. Do not worry at this stage about organization, simply note the key points as you think of them. Next read the parts of your notes, textbook and other books which relate to the essay, adding new ideas which they generate to your plan. Make separate, more detailed notes from these sources and aspects which you think important. Alternatively add page references to your plan. It is important to use a range of sources when researching essays. Essays are a valuable resource when it comes to revision. If they have been compiled solely from your own notes and the textbook they will contribute nothing original from which to revise. Think of essays not so much as a test of how much you have learnt, but rather as a means of expanding and varying your knowledge.

Having researched the essay over a period of time (anything from a few days to a few weeks), take the plan and identify the important points which are central to answering the question. Mark them clearly. Try to arrange them in a logical sequence, so that the essay has a theme running through it. This is not always practical and, depending on the essay, it may be necessary to develop a number of separate lines of argument rather than follow a central theme. Where an essay involves two or more separate viewpoints it is important to establish a fair balance between each one.

Having determined the key points and arranged them logically, write them out in order, leaving a space between each. Now consider the peripheral material which you have amassed and try to associate it with one of the key areas listed. It may take the form of examples to illustrate a point, substantiating evidence to support an argument or simply additional detail to clarify an issue. Add concise notes to each of the key areas. If one or other area has no additional material added to it, it may be that further research is needed, or possibly the point should be omitted altogether. A typical plan of this type is shown below.

ESSAY QUESTION:
How might a carbon atom in a molecule of carbon dioxide expired by a herbivore become part of a glycogen molecule in another herbivore?

PLAN

CO$_2$ FROM HERBIVORE

Result of respiration(?)
Expired from lungs $\rightarrow$ atmosphere(?)

ENTRY OF CO$_2$ INTO PLANT
Diffusion through stomata (open in daylight)
Intercellular spaces – dissolves in moist cell wall
In solution through cytoplasm $\rightarrow$ chloroplast

PHOTOSYNTHESIS
RBP + CO$_2$ $\rightarrow$ GP
NADPH + H$^+$$_2$ + ATP from light reaction
Triose $\rightarrow$ Hexose (glucose)
Glucose polymerized to starch (stored)
Calvin cycle(?) C$_4$ plants(?)

INGESTION BY HERBIVORE (e.g. cow)
Tongue helps cropping
Incisors/horny pad for tearing grass
Molars/premolars – grinding
Swallowed $\rightarrow$ 4-chambered stomach

DIGESTION
Rumen – symbiotic microorganisms – cellulases to release starch
Regurgitation and reswallowing
Stomach $\rightarrow$ duodenum
Pancreatic amylase – starch $\rightarrow$ maltose
Maltase – maltose $\rightarrow$ glucose
Active absorption – villi

GLYCOGEN FORMATION
Transport along HPV by blood
Condensation reaction $\rightarrow$ glycogen
Control by insulin

Not all plans will fit this pattern and this one is shown merely by way of illustration. Each essay demands its own style of plan. It is sometimes difficult to know exactly what to include. In the example given, some points are marked(?). These are considered more peripheral to the question and could be omitted if time were short. When writing the essay from the plan, follow these guidelines:

1. Keep the answer strictly relevant to the question asked. It pays to re-read the question from time to time to ensure this is being done.

2. Make points clearly and positively.

3. Support arguments with collaborative evidence.

4. Use specific examples to illustrate points, giving the names of specific organisms where appropriate.

5. Marshall points in a logical sequence.

6. Write legibly and fluently, avoiding long, complex sentences which are hard to follow. Write grammatically correct sentences, not notes.

7. Take care with spelling, especially over biological words, e.g. ileum (small intestine) and ilium (hip bone), carpal (wrist bone) and carpel (female part of flower), thymine (organic base in DNA) and thiamine (vitamin B).

8. Draw diagrams only if appropriate and where they make a useful contribution to the quality of the answer. Draw neatly – label and annotate.

9. Avoid repeating points and do not fill out essays with superfluous or irrelevant material.

10. Where an essay has a number of parts, allocate time in strict proportion to the mark distribution of each.

42.3   Application of knowledge

At A-level the emphasis is on the application of knowledge rather than its recall. This does not imply that students can get away without learning factual information. On the contrary, a large body of biological facts is essential before they can be applied to a problem. Sometimes students will be provided with familiar information and asked to apply their biological knowledge in answering questions about it. Increasingly, however, examination questions deliberately provide unusual and obscure material. This means that the student will not have met the information before. The answers cannot therefore have previously been prepared by the student and the quality of the response will depend entirely upon the student's ability to apply the facts he knows.

The most important point to bear in mind is that the information provided in the question must form the basis of the answer. If a graph is drawn, relate your answer to the information on the graph. If a table of data is provided, use the figures in the table to support your arguments. If a written passage is given, use words and sentences from it to substantiate your views. All too often students simply write answers from memory, without reference to the information in the question. To obtain most credit from these types of questions you must apply biological principles to the information provided.

That after all is what this type of question is designed to test. Reciting memorized answers or recalling lots of factual information will not, on its own, bring high marks.

There is a wide range of question types involving the application of knowledge and some of these are dealt with in Section 43.3.

42.4 The biology syllabus

In order to bring consistency to the syllabuses of the various examination boards in the United Kingdom, a common core of topics in most subjects has been established. In Biology this common core has been agreed by the various boards and will make up around 60% of each of their Biology A-level syllabuses. The aims of these common core syllabuses are:

1. To develop an understanding of biological facts and principles and an appreciation of their significance.

2. To be complete in themselves and perform a useful educational function for students not intending to study biology at a higher level.

3. To be suitable preparation for university and polytechnic courses in biology, for biological studies in other educational establishments and for professional courses which require students to have a knowledge of biology when admitted.

The major topics common to all syllabuses and the chapter where they are covered in this book are:

TOPIC	CHAPTER NUMBERS
1. Diversity of organisms	6, 7, 8, 9
2. Cell biology	3, 4, 5
3. Photosynthesis	23
4. Heterotrophic nutrition	24
5. Respiration	26
6. Homeostasis	36
7. Gaseous exchange	31
8. Chemical coordination in plants	40
9. Coordination in animals	37, 38
10. Transport in plants	33
11. Transport in animals	32
12. Reproduction in animals	19
13. Development in animals	20
14. Reproduction in flowering plants	18
15. Support, movement and locomotion	39
16. Genetics	12, 13, 14, 15

17. Evolution   16

18. Ecology   27, 28, 29

Within each of these topics the syllabuses vary slightly in the detail required and the examples used. There is nevertheless considerable conformity between the different boards despite the wide and varied nature of the subject. The terminology used by examining boards is consistent with that published in *Biological Nomenclature: Recommendations on Terms, Units and Symbols* (Institute of Biology and Association for Science Education − 1989).

Biology cannot be studied in complete isolation and any A-level syllabus inevitably requires a certain basic knowledge of other subjects, especially mathematics, physics and chemistry. The following list covers most of those required by the various examining boards. It is comprehensive and most individual syllabuses will not require knowledge of all the items. Check your own board's syllabus for their precise requirements. The physical and chemical knowledge generally required is:

The electromagnetic spectrum
Reflection, transmission and absorption of radiation
Isotopes, including radioactive isotopes
Laws of thermodynamics
Potential energy
Activation energy
Chemical bond energy
Covalent, electrovalent, hydrogen bonding
Latent heat
Atoms, molecules, ions and electrons
Acids, bases, pH and buffers
Solubility of gases, liquids and solids in water
Relative humidity, vapour pressure, partial pressures
Relationship between pressure, volume and temperature of a gas
Surface tension and cohesion
Colloidal state
Hydrolysis and condensation
Dynamic equilibrium
Oxidation, reduction, electron/hydrogen transport

The mathematical knowledge generally required is:

Line graphs − linear and logarithmic scales
Bar graphs, histograms and pie charts
Decimals, ratios, proportions, reciprocals and percentages
Arithmetic mean, mode and median
Variance and standard deviation
Chi squared test
Calculation of rates, e.g. from graphs

43 *How to approach A-level Biology examinations*

The culmination of around two years of studying biology will be the A-level examinations. Despite its many shortcomings the examination system is still with us and likely to be so for the foreseeable future. There has been a trend to increasing diversity in the papers. At one time almost all assessment was made on the basis of essay-style questions. These tended to favour those with the best command of English and a fluent essay style. While essay-style papers are still used by all examination boards, the proportion of the total assessment made by this means is now far less, often around one third of the total marks.

The two types of theory paper which have replaced much of the essay work are multiple choice and structured or short-answer papers. The former requires the response to be a letter or number representing the correct response chosen from a list of four or five. In other words, alternative answers are provided and it is up to the student to select the correct or best one. In structured or short answer questions the candidate must compose his own responses but these normally comprise a word or short statement rather than a long response. Further details of all forms of questions are given in Section 43.3.

43.1 Revision

The surest way to examination success is to prepare yourself adequately, through the practising of examination questions and careful revision of the relevant material. Without this, no amount of advice on examination technique will bring success.

Revision is very much a personal affair and each student must use the methods which suit him or her. Perhaps the best advice which one can give is to try as many variations as possible and then select those which bring the best results. Use tests and examinations throughout the course to try out different methods. It would be foolish to test a new technique during your final A-level revision; by that time you should have perfected your revision style.

At A-level, success is achieved by the gradual accumulation of knowledge and understanding throughout the course. The volume of information is too vast for it all to be absorbed during a few weeks' revision prior to the examinations. Try to read over notes daily, and if possible the whole week's work during the weekend. Prepare for tests and periodic examinations thoroughly – only by so doing can you effectively test the efficiency of the revision methods you are using. Students who are disappointed by their final grade often admit to not having revised adequately for earlier examinations. How then could they have expected to know how much revision was needed for their final, and all-important test? Always revise thoroughly for all examinations and analyse your results. If they were poor, change your revision methods.

To discover other methods of revising, it is often necessary to discuss the matter with others. Do not assume that the most successful student has the best method. They may simply be brighter than you and in any case their technique may not suit you. Try to experiment with as many styles as possible.

Whatever the detailed methods you employ, there are a few basic principles which should be adhered to:

1. Work in a place with the least likelihood of distraction. Some students prefer to work with background noise, e.g. music. This, however, may itself be distracting. Firstly, there is the inevitable tendency to sing along with the music — something which requires a degree of concentration. Secondly, there are periodic interruptions while you change the programme, disc or cassette. If at all possible, work in silence to allow your complete concentration to be focused on the revision.

2. Do not work for too long at one session. The power of the brain to concentrate, and so absorb material, diminishes rapidly after a while. The actual time varies between individuals but an hour at one continuous stretch is typical.

3. Take a short break of 10–15 minutes between each session of revision. Have a complete change during this time. Leave the room you are working in, even take a walk outside. Make a drink, listen to music, watch television — anything provided it gives you a complete break from the monotony of revision.

4. During the revision session do *not* allow yourself to be distracted. Let someone else answer the telephone or the door. Make it clear to others around you that you wish to be left alone.

5. Be aware of 'displacement activities'. These are the various activities which you carry out in order to bring relief from work. Revision is often boring and tedious and you will sub-consciously be looking for a means of avoiding it. The sudden craving for coffee or something to eat is no more than an excuse to stop work. You will hardly starve or die of thirst before your next break. There is no need to get up and look out of the window every time there is a noise, or to see whether or not it is raining. The cleaning of your shoes is not urgent and the dog can wait until later for its walk. Always be conscious of the dozens of trivial matters which suddenly assume great importance — and ignore them.

6. Vary the revision by changing topics or subjects from time to time. Variety may reduce the boredom.

7. Test yourself periodically or get others to do so. The testing is best done about a day after the revision. Closing the book and immediately writing down what you remember has little value as it tests only short-term memory. Success may depend on remembering information revised days or even weeks earlier. One useful technique is to write short questions as you revise and then try to answer them a day or so later. Even single words or dates can be jotted down, the significance of which you can later try to remember. Go back immediately and re-revise those questions you got wrong or could not answer.

8. Practise problems or essays, preferably from past papers. Do them within the allocated time so you can practise working within time limits. Study past papers to make yourself conversant with the styles of question. Be aware of any recent changes in the format or style of papers and questions. Predicting questions is a risky business and not worth the gamble.

9. Use spare minutes for revision. It should be possible to read notes, or test yourself during the many spare periods in a day. The ten minutes spent on a bus or train can be better occupied than reading the advertisements around you. The five minutes waiting for dinner or for someone to call can all be put to good effect. In themselves they may not be much but cumulatively they make a significant contribution to the total revision. They could be the difference between a particular grade and the one above – in some cases the difference between pass and fail.

10. Organize revision by making a *written* timetable well in advance. Be realistic. Do not make it so arduous that you fall behind schedule within the first week. Choose times to revise when you are less likely to be distracted. If you have an 'unfavourite' aunt who visits on Wednesday evenings, put down at least three hours for this time – you have a better chance of achieving it than the evening of the local disco or your favourite television programme. Leave yourself at least one day a week with no revision, and leave one week in four completely free. This means that if unexpected events, e.g. illness, arise you can use the 'free' days to compensate for lost time. If it does not prove necessary, then either use the time for additional revision or for a day off. The break may prove more valuable than revising. It will allow you to become refreshed, making future revision more effective. Do not make a timetable for all your revision time. It would be an extremely lucky student who did not have some misfortune which prevented him revising for at least a few days in the weeks prior to the examination.

43.2   Examination technique

Poor examination technique is often put forward as the reason for a disappointing result. It is certainly true that, without the necessary skills, students may not perform as well as they should. It would, however, be a foolish candidate who had not worked to perfect his technique by the time of his final examination. It is important to isolate weaknesses in technique at each examination and take measures to correct them. It is not an inborn fault to misread or misinterpret questions and even the slowest writer can learn means of conveying a lot of information in a short time.

One of the most likely reasons for a candidate performing badly in examinations while producing excellent term work is an overdependence on books and/or others. One skill which all students need is the ability to think for themselves. While books are essential as aids to success they should not be allowed to dominate a student's work. There is little value in copying wholesale from books, or with minor alterations in the hope of disguising the fact. Mimicking other students' ideas or always working together with others on essays and problems can be equally pointless in the long run. These activities mean that students never learn to think out things for themselves because they rely too heavily on books or others. The result is that, in the absence of such assistance in the examination, they perform much worse than the term work had suggested they would. The failure has little to do with poor examination technique.

Examination technique commences early during a course. It begins with learning to think for yourself and carrying out adequate revision before the day. That being so there are a number of rules to observe when sitting the actual papers. Most are simple common-sense points.

1. Read all instructions carefully. Do not assume them to be exactly the same as those you have seen on past papers.

2. Note the number of questions to be answered and keep strictly within the limits.

3. Act on any guidance given in the general instructions about the use of English, necessity for diagrams, need for orderly presentation etc.

4. Read all questions with scrupulous care. Do not be in a hurry to get started. Be sure you understand what the question requires before answering.

5. Where there is a choice of questions read *all* the questions first before making any selection. Read the paper a second time, making your choices, and finally read the selected questions a third time to ensure you have chosen ones you are competent to answer. Answer questions in any order but number them clearly.

6. Often the total marks for the whole paper are stated. Divide this number into the time provided in order to find approximately how many minutes are available for each mark, e.g. 100 marks on a three hour paper gives

$$\frac{180 \text{ minutes}}{100 \text{ marks}} = 1.8 \text{ minutes/mark}$$

Allowing time for reading this gives about 1.5 minutes/mark. You should therefore spend 15 minutes on a question worth 10 marks. Allocate your time for every question or part of a question in proportion to the marks available.

7. Refer back to the question a number of times during the writing of an answer to ensure you have not strayed from the point. Re-read the whole question once you have completed your answer in case you have omitted any part.

8. Try to isolate the key word or words in a question and answer precisely in accordance with them (see list on next page).

9. Try to be completely relevant, clear and concise in your answers. Do not ramble aimlessly.

10. Check during the last quarter of a paper that you have followed all instructions carefully and have answered (or are about to) the requisite number of questions. Do not leave this until the last five minutes — it will be too late to put right should you have made an error.

The above are only general guidelines, so more specific instructions relating to different types of examinations are included in Sections 43.3 and 43.4.

It is important for candidates to appreciate that the wording of questions is chosen very carefully, often after lengthy discussions and argument. Each word has a very precise meaning which must be understood and complied with if success is to be achieved. Many students seem to think that words like 'describe', 'explain', 'compare', 'distinguish' etc. all have the same meaning. They therefore write a similar answer regardless of the word used. Below is a list of words which frequently arise in questions, and their approximate meanings.

Describe	– give an account of, using factual descriptive detail
Explain / **Account for**	– show how and why, give the reasons for
Compare	– point out the similarities and differences
Distinguish / **Contrast**	– make distinctions between, point out differences
Discuss	– debate, giving the various viewpoints and arguments
Criticize	– point out faults and shortcomings
Survey	– give a comprehensive and extensive review
Comment on	– make remarks and observations on
Illustrated	– use figures, drawings, diagrams
Annotate	– add notes of explanation
Briefly / **Concisely**	– give a short statement of only the main points
Outline	– give the main points
List	– catalogue, often as a sequence of words one beneath the other
Significance	– importance of
State	– establish; set down
Suggest	– put forward ideas, thoughts, hypotheses
Devise	– construct, compose, make up

43.3 The theory examinations

The division between theory and practical work is an artificial one. Practical work requires a sound theoretical knowledge and the understanding of theory work is greatly enhanced from having carried out practical exercises. Theory papers reflect these facts and frequently require practical knowledge. This is especially true of practical experiments which have unpredictable results, are too time-consuming to be carried out in an examination, or require complex apparatus. Photographs of organisms, tissues and cell ultra-structure are becoming increasingly common on theory papers. It is therefore essential to include all practical work when undertaking revision for theory papers.

43.3.1 Multiple choice papers

Multiple choice papers are set by a number of examining boards in A-level biology. The style of questions varies slightly but in general comprises a statement or question followed by four or five alternative answers from which the candidate is required to select the correct one.

A typical example is:

Which of the following components of an ecosystem has the greatest biomass?
  A. decomposers
  B. primary producers
  C. primary consumers
  D. secondary consumers
  E. tertiary consumers

The correct answer is B.

In some cases all answers may be accurate to a greater or lesser extent, in which case the candidate must choose the best available response. It is particularly important therefore that you consider all options. Mistakes could arise if a candidate decides upon one of the answers and does not bother to read those which follow; one of these may be an even better response. Consider the following example:

The nerve impulse may be best described as:
  A. the sudden reversal of charges on a neurone membrane
  B. the flow of electrons along a nerve axon
  C. a self-propagating change of polarity along the neurone membrane
  D. the movement of sodium ions across the neurone membrane

Option A is accurate and having selected it, some candidates may ignore the remaining options. Options C and D are, however, also correct. Option C is the best of the three because it gives a much more comprehensive definition. Option B is inaccurate because nerve impulses arise from ionic changes and not electron flow.

One alternative style of multiple choice question is to provide three or four statements, any number of which may be correct. The candidate is required to select those he thinks are accurate and then respond by giving the correct letter according to a key which is provided. An example of a typical question and instructions is given below:

For the following question, determine which of the responses that follow are correct answers to the question. Give the answer A, B, C, D or E according to the key below:

  A. 1, 2 and 3 are correct
  B. 1 and 3 only are correct
  C. 2 and 4 only are correct
  D. 4 alone is correct
  E. 1 and 4 only are correct

The direct effects of follicle stimulating hormone (FSH) include:
  1. development of the corpus luteum
  2. development of the Graafian follicle
  3. ovulation
  4. stimulation of sperm production

The key word is 'direct' as only FSH *directly* causes the development of the Graafian follicle (option 2) and stimulates sperm production (option 4). Options 2 and 4 only are correct and reference to the key gives the response C.

The best approach to multiple choice questions is to read through the whole question and alternative responses first. On reading a second time, reject responses you think are incorrect. Always reject on a sound biological basis and not because they do not seem to fit or because the correct response is unlikely to be C for the sixth consecutive time. Should you be left with two answers, and are unable to decide between them, at least guess rather than leave the answer blank. Unless the instructions clearly indicate there is a penalty for wrong answers, you have nothing to lose and a one in four or five chance of gaining a mark.

43.3.2 Structured/short-answer papers

The style of structured or short-answer papers is immensely varied. In their simplest form they comprise a single word response to a straightforward question, e.g. 'What organic base absent in DNA is found in RNA?'

At the other extreme, an answer may involve the analysis and interpretation of complex data and an essay-style explanation. All stages between these extremes occur. Data in its many different forms is very popular. A candidate may be required to display any or all of the following skills:

1. Show knowledge and understanding of biological terms, concepts, principles and relationships.

2. Construct hypotheses.

3. Design experiments.

4. Interpret the results of experiments.

5. Draw conclusions and make inferences.

6. Assess and evaluate numerical and non-numerical information.

7. Explain observations and solve problems.

8. Present data in its many varied forms.

9. Comprehend, interpret and translate data.

10. Criticize material and exercise biological judgement.

11. Construct or label diagrams of biological importance.

12. Interpret or comment on photographs.

13. Collect, collate and summarize biological information in an appropriate form.

14. Appreciate the social, environmental, economic and technological applications of biology.

All points made under 'Examination techniques' in the previous section are applicable to this style of paper. Answers should, however, be especially concise. It is the content of the answer, rather than its style, which is being tested. Responses must of course be comprehensible but too much attention need not be paid to the niceties of grammar, and note-style answers are normally acceptable.

Use the space allocated in the answer book as a guide to the length of answer required.

A variety of examples of typical questions are given throughout the book, at the end of each chapter. These have been chosen to illustrate the many different styles of questions set in examination papers.

43.3.3 Essay papers

Despite the reduced emphasis on them, essay-style questions still form part of every examining board's A-level Biology papers. Unlike other forms of questioning, they tend to be open-ended and give much scope for a candidate to demonstrate his knowledge. They are not, however, without some constraints, and students must ensure they do not digress from the main theme of the question.

Essay questions may be structured, in which case they are divided into sections which give the candidate some guidance as to the lines along which they may be answered. A typical example might read:

> **(a)** Compare wind-pollinated flowers with insect-pollinated flowers. (6)
>
> **(b)** Describe briefly three mechanisms in flowers which favour inbreeding and three which favour outbreeding. (8)
>
> **(c)** Discuss the genetical consequences of inbreeding and outbreeding. (6)

These structured essays normally include a mark distribution and the time spent on each part should be allocated accordingly. In the example above, marginally more time should be given to part **(b)** than to **(a)** or **(c)**, to which equal time should be devoted. As the question itself guides the type of response, planning is less important. Candidates should nevertheless give some forethought to each part and note down the key points to be made before attempting to answer. The different parts should be answered separately, each being clearly labelled. Do not use the same information in two different parts of the answer; you will not be given credit twice.

The alternative form of essay is the unstructured one, e.g. 'Describe the ways in which green plants depend upon the activities of animals.'

This form of essay is much broader and more open-ended. In the above example there are numerous points which can be made, supported by an almost limitless number of examples. No candidate could be expected to cover everything in 30–40 minutes. The skill lies in making certain that one or two examples are quoted for each of the major key areas. These areas include:

1. Respiratory carbon dioxide for photosynthesis.

2. Supply of nitrogen from excretory activities.

3. Breakdown of organic material by flies, nematodes, earthworms to help recycle elements.

4. Soil improvement by earthworms etc.

5. Pollination.

6. Dispersal of seeds.

7. Control of plant parasites.

Any essay which concentrates on just one or two of these areas will not bring high marks regardless of the detail included. The best answers will include some detail and examples from all seven areas. The key to success therefore lies in identifying the key areas. Planning for this form of essay is therefore essential and 5–10 minutes spent doing so will almost certainly pay dividends. Some boards give marks specifically for the manner in which the essay is written in addition to those for biological content. Proper planning is the best means of securing these marks.

The general advice given in Section 42.2.3 on the planning and writing of essays is relevant to writing essays in examinations. It must of course be suitably modified to take account of the time limits imposed.

43.4   Practical assessment

The most obvious difference in the methods of examination used at A-level is the presence of practical assessment. This will comprise anything between 20% and 35% of the total A-level percentage. It may take the form of a practical examination, a project or continuous assessment or any combination of the three. For some students this emphasis on practical work is daunting, but most quickly overcome their apprehension. They soon appreciate the necessity of such work and its value in understanding the theory. It is a requirement of some boards that a minimum standard for practical work be achieved before an overall pass grade can be given.

43.4.1 Continual assessment and projects

There is unanimity among the examining boards that biology is best taught as a practical subject. It is, however, appreciated that much practical work is time-consuming and some experiments can only be performed over a long period, often many weeks. For this and other reasons most practical assessment now includes some form of teacher-marked work which is performed during the A-level course rather than in an examination at the end. These techniques provide an excellent opportunity for a student to demonstrate some of his skills without the pressure and time-limitations imposed by an examination. It is well to remember, however, that quantity is no substitute for quality. Just because work can be completed at relative leisure does not mean it needs to be long and complex. Even in writing up a project, work that is concise, detailed and to the point is likely to bring more credit than that which is long and rambling.

Make sure early on that you fully understand what, if any, practical work will be assessed during the course and what percentage of the final mark it will account for. Start all work as early as possible, bearing in mind that it may take many months for you to master some of the basic skills. For most students the majority of assessment will be done during the second half of a course, by which time you should have acquired the basic practical skills and mastered the necessary techniques.

Fieldwork is one aspect of practical work which is frequently assessed in this way. It might involve a course, possibly a week at a Field Centre, occasional visits to local habitats, or work within the grounds of the college or school. In most cases work will be done over a period of time. It will be written up by the student and presented for assessment by a teacher or examiner. The work done may, in addition, be tested in theory or written practical examinations. There is a trend towards testing dissection skills as part of continuous assessment rather than in an examination. This allows a longer period to be devoted to the dissection of each organism. More time can therefore be spent refining dissection techniques and learning the spatial relationships of internal organs.

43.4.2 The practical examination

This takes many forms and, as practical work is beyond the scope of this book, only some general comments will be made here. Depending on the particular paper, candidates may be expected to perform some, or all, of the following activities:

1. Microscope work – The student will need to demonstrate that he or she can use a microscope unaided and use low and high power magnifications as appropriate. The ability to make temporary mounts, sometimes involving simple staining techniques, is required by some boards. Suitable drawings of what is seen will need to be made (see Section 43.4.3.)

2. Identification and drawing of biological specimens – Candidates will be expected to identify the main groups to which plant and animal specimens belong, even though they may never have seen the particular specimen before. It is therefore essential to learn the major features of all groups on your syllabus. It is expected that candidates will be able to apply this knowledge to any specimen rather than learn to recognize by memory particular examples. The approach should be 'the specimen has jointed appendages and an exoskeleton, therefore it is an arthropod' rather than 'I recognize this as a locust, and I remember that locusts belong to the group Arthropoda'. The sub-groups to which a specimen belongs may be expected for specified examples.

Drawing of whole specimens or parts of them may be required. The student should be able to make annotated labels for each of the biologically significant features on a specimen. The ability to relate structure to function is important. Some boards expect candidates to be able to construct simple dichotomous keys (see Section 6.1.2).

Questions often require a candidate to make comparative observations of two or more specimens, or parts of them. The aim is to test the student's powers of observation and ability to apply biological knowledge rather than remember previously learned facts. For this reason the material provided may not be familiar to the candidate.

Specimens provided in practical examinations may be living or preserved. Photographs and photoslides may also be used. These allow questions to be set on rare or obscure specimens or ones which are otherwise too large for inclusion in an examination. As a result candidates must expect anything to turn up.

3. Dissections – While examining boards generally expect students to have done some dissection work, e.g. of organs, no examining board will set a dissection of a vertebrate as part of an examination for school and college candidates. Photographs of dissected material may be substituted, however.

Remember when doing dissections to display clearly all relevant parts so they can be seen without the observer having to move any tissues. Keep your dissection tidy. Remove all debris completely, do not leave it lying in the dish. Dissection of small organisms is often better carried out with the specimen covered by water as this helps to support the tissues. A dissecting lens is particularly helpful when dissecting small structures.

4. Experimental work – This is a component of almost all practical examinations. Details of the techniques and skills involved are given in Section 43.4.4.

43.4.3 Drawings for practical assessment

The making of accurate drawings is an essential skill for the A-level biologist. Both as part of the continual assessment and the examination, the student will be required to make drawings of whole specimens, parts of speciments, dissections, microscope slides and even photographs and photoslides. In constructing these drawings the candidate should:

1. Use a good quality, sharp pencil of a hardness that suits the individual (HB is normally the best).

2. Draw on good quality plain paper which is capable of withstanding erasure of pencil lines.

3. Ensure the diagram, along with all labels and annotations, will fit comfortably on the page.

4. Make large, clear, line drawings without the use of ink or coloured pencils.

5. Make single pencil lines without sketching or shading.

6. Keep the drawing simple by providing only an outline of all the basic structures.

7. Draw accurately and faithfully what can be seen. Never draw anything you cannot see, even if it is expected to be present. Never copy from books.

8. Draw individual parts of a specimen in strict proportion to each other.

9. Provide suitable headings which clearly indicate the nature of the drawing. For microscope drawings, the section (TS/LS etc.) should be stated.

10. State the magnification, scale or actual size of each specimen.

11. Label fully all biological features keeping labels away from the diagram, and never label on the actual drawing.

12. Avoid crossing label lines and if possible arrange labels vertically one beneath the other.

13. Use annotations (notes added to labels) if at all possible. In particular, try to relate structure to function.

14. With microscope drawings it may be necessary to include two drawings described as follows:

(a) a low power map indicating the main regions of each cell type, but *without drawing individual cells.*

(b) A high power drawing of a section or wedge passing through all the major cell areas. Draw a few cells of each type and in at least one cell include cellular detail, e.g. nucleus, cytoplasm, storage grains etc.

15. Keep all drawings for assessment carefully, e.g. in a hardback loose-leaf file. Not only are they a permanent record of your work, they are also invaluable for future reference.

43.4.4 Experimental work

Experimental work usually forms an important component of both continual assessment and the practical examination. Experimental work tests a number of skills:

1. Ability to follow instructions.

2. Construction of suitable hypotheses.

3. Planning of experiments.

4. Design and manipulation of apparatus.

5. Making accurate observations and recordings.

6. Making precise measurements.

7. Presenting results in a suitable form.

8. Interpreting results accurately.

9. Making logical deductions from results by applying biological knowledge.

10. Discussing critically the methods used and results obtained.

The writing up of experiments must be in accordance with any instructions given by a teacher or the rubric of an examination paper. In general the account will include:

1. A title – This should indicate the broad purpose of the experiment, e.g. 'Investigation of the effect of temperature on enzyme activity.'

2. The aim of the experiment (hypothesis) – This should give the precise aim of the experiment, e.g. 'to determine the rate of starch breakdown by amylase at temperatures in the range 0–100°C'.

3. Method – This is a precise, step-by-step account of the procedures carried out. It should be written in the past tense and impersonally, i.e. 'A test tube was taken ...' If properly written, the experiment should be capable of being perfectly imitated by another biologist following your account.

4. Results – These are a complete account of your recordings and observations. They should be presented in some appropriate form, e.g. descriptive prose, table, graph, histogram etc.

5. Conclusion – This is a *brief* statement of the single main fact determined by the experiment, e.g. 'The optimum temperature for the breakdown of starch by amylase was found to be 45°C'.

6. Discussion – This should be an attempt to relate known biological knowledge to the results in trying to explain them. It might also include:

(a) Criticisms of the method employed.

(b) Possible sources of error in the results.

(c) Suggested improvements to the experiment.

For continual assessment purposes, experiments may be spread over several weeks. Commonly these experiments are investigations on genetics, growth or ecological changes, but almost any investigation can be designed to be performed over a long period.

For examination purposes, the choice of experiment is restricted by the need for it to be carried out within about two hours. Examination investigations therefore often involve food tests, enzymes or osmotic experiments. Even within these topics there is a vast number of possible investigations. Provided appropriate instructions are given, unfamiliar materials may be provided.

When sitting a practical examination the following general points should be observed:

1. Read carefully all instructions. All questions are compulsory as a rule.

2. Read the *whole* paper through before starting, and read each question both before and while carrying it out.

3. Allocate time carefully according to the mark distribution. Speed and accuracy are required throughout.

4. Answer questions in any order, but bear the following in mind:

(a) Some specimens may be shared between candidates and you may only have them available to you for a limited period. Be sure you know when it is your turn and arrange your question order accordingly. Use the specimen as soon as possible to ensure you finish the work in your allocated time.

(b) Many physiological experiments require a sequence of readings over a period of time. These must be started early to ensure they are completed in time. Even where the experiment requires a shorter period, an early start will enable all or part of it to be repeated should something go badly wrong.

Further Examination Questions

1. Plants and animals of temperate climates are adapted to survive seasonal changes. Discuss the adaptations they show in response to such changes in terms of
    (i) feeding, and
    (ii) reproduction and dispersal.
Refer to **named** examples.

Joint Matriculation Board June 1983, Paper IIB, No. 1

2. Amplify and discuss **three** of the following statements:

The activities of an organism may be influenced by the lengths of day and night.

In many animals, smell is a most important factor for reproductive success.

Without the capacity to undergo dormancy, many plants would not survive.

The absorption of dietary fats is enhanced by the presence of bile salts and lacteals. *(20 marks)*

Oxford and Cambridge Board June 1984, Paper II, No. 5

3. Give details of the practical techniques that you would use for **two** of the following:
    (a) a comparison of the rates of respiration of two types of germinating seed;
    (b) investigation of the soil characteristics in an ecological study;
    (c) a study of learning behaviour in a small mammal;
    (d) variations in the rate of amylase production by germinating barley seeds over the first five days.
                              (2 × 10 marks)
Evaluate one of the methods and suggest improvements you could make on repeating the experiment. *(5 marks)*
                              (Total 25 marks)

Southern Universities Joint Board June 1986, Paper II, No. 10

4. Write an essay on the dependence of green plants on animals other than humans. *(20 marks)*

Associated Examining Board June 1985, Paper III, No. 1

5. The following practices are based on certain biological principles. Select **five** of the practices listed and, for each, state what you think these principles are.

    (a) Addition of enzymes (amylases) to flour in modern factory bread-making processes.
                              (3 marks)
    (b) Spraying apple orchards with auxins at blossom time. *(3 marks)*
    (c) Eating wholemeal, rather than white bread, and eating potatoes with their skins. *(3 marks)*

    (d) Shading greenhouses in summer and enriching the atmosphere with carbon dioxide. *(3 marks)*
    (e) Removing all lower leaves of cuttings (of plants such as fuchsia, chrysanthemum, geranium, etc.), but leaving the few at the apex. *(3 marks)*
    (f) Malting (partially germinating) barley before extracting with water to make into beer.
                              (3 marks)
    (g) Storing fruit in atmosphere of carbon dioxide to prevent deterioration. *(3 marks)*
                             (Total 15 marks)

Northern Ireland Board June 1983, Paper I, No. 4

6. (a) Describe the biological principles underlying any **four** of the following (all parts carry equal marks):
    (i) The draining by farmers of marshy fields.
    (ii) Avoidance of the discharge of organic effluent (e.g. raw sewage, silage effluent) into rivers.
    (iii) The annual cull of the herd of deer in Richmond Park, London. (A cull is the killing of a proportion of the animals in a herd.)
    (iv) Scratching the surface of sweet pea, and other difficult-to-germinate seeds; soaking parsley seeds **briefly** in boiling water; exposing some seeds (e.g. tree seeds) to a period of cold (at about 0°C) before sowing.
    (v) Daily measurement of body temperature and avoidance of intercourse during the three days before and after ovulation in women practising the rhythm method of contraception. *(12 marks)*
    (b) Similarly describe the biological principles underlying **four** of the following (all parts carry equal marks):
    (i) The advice to farmers to grow the same crops year after year in the one field.
    (ii) The argument by conservationists that farmers should use manure instead of artificial fertilisers.
    (iii) The use of ultrafiltration membranes in haemodialysis of patients with kidney disease. (Ultrafiltration membranes will allow free passage of small molecules in solution, but not larger molecules, such as proteins).
    (iv) 'Biological' washing powders contain enzymes with alkaline pH optima, and are recommended for low temperature washing (below 50°C).

(v) Use of *in vitro* fertilization with women who have blocked fallopian tubes as a result of pelvic inflammatory disease. (*8 marks*)

(*Total 20 marks*)

Northern Ireland Board June 1985, Paper II, No. 2

7. Select any **one** of the following topics (i–x) and write a balanced logical biological essay in clear continuous prose on the subject chosen. Credit will be given both for factual content and good style. Avoid the use of diagrams unless these are essential.

(i) The uptake and utilization of minerals by plants

(ii) Structure/function relationships in biology

(iii) The methods of study of animal behaviour

(iv) Excretion in animals and plants

(v) The functions of blood

(vi) Domestication of plants and animals

(vii) Acid rain

(viii) In vitro fertilization

(ix) The elucidation of the structure and function of DNA

(x) Mechanisms of osmoregulation in plants and animals. (*25 marks*)

Northern Ireland Board June 1985, Paper I, No. 1

8. What is meant by each of the following biological terms:

(*a*) polychaete; (*d*) polypeptide;

(*b*) polygamy; (*e*) polytene?

(*c*) polymerase; (*5 marks*)

Oxford Local June 1984, Paper I, No. 2

9. State where you would expect to find the following:

(*a*) Bowman's capsule; (*d*) Peyer's patches;

(*b*) Brunner's glands; (*e*) Purkinje's fibres.

(*c*) Glisson's capsule; (*5 marks*)

Oxford Local June 1984, Paper I, No. 15

10. Give, as precisely as possible, one site where the following may be found and concisely state one function for each.

	Site	Function
(*a*) Acetylcholine (ACh)		
(*b*) Endolymph		
(*c*) Fibrinogen		
(*d*) Synovial fluid		
(*e*) Vitreous humour		

(*10 marks*)

Oxford Local 1987, Specimen Paper I, No. 3

11. Give an explanation for each of the following statements.

(*a*) Relatively small volumes of concentrated urine are produced by a human in hot weather. (*3 marks*)

(*b*) The flowers of some plants have highly-branched stigmas projecting well beyond the rest of the flower. (*3 marks*)

(*c*) Clay soil becomes easily water-logged after rain. (*3 marks*)

(*d*) The leaves of green plants living in dry, exposed conditions often have a reduced surface area (*3 marks*)

(*e*) Protozoan blood parasites do not possess contractile vacuoles. (*3 marks*)

(*Total 15 marks*)

London Board June 1985, Paper I, No. 11

12. A predator fly, *Euxesta* sp., lays eggs (oviposits) beneath the bracts on immature fruits of *Costus woodsonii*, a plant of the ginger family which grows in Central Panama. On hatching from eggs the larvae destroy the seeds.

The plant secretes extrafloral nectar from bracts on the inflorescence and this is harvested by ants. Two species of ants are dominant, *Camponotus planatus* in the dry season (January to mid-May) and *Wasmannia auropunctata* in the wet season (mid-May to the end of December). The seasonal course of flowering by *Costus woodsonii* and ant occupation is shown in Fig. 1.

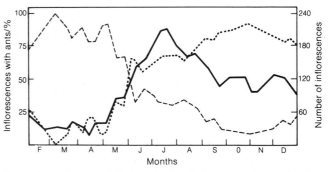

Fig. 1. Seasonal course of Costus woodsonii *flowering and ant occupation. The number of flowering inflorescences is indicated by the solid line. The percentage of inflorescences occupied by the ant* Camponotus planatus *is indicated by the dashed line, that of* Wasmannia auropunctata *by the dotted line.*

Individuals of *Camponotus* aggressively chase off the fly from the inflorescences and individuals of *Wasmannia*, because of their smaller size, can forage beneath the bracts and consume the larvae of the fly.

Experiments were devised to test the hypothesis that the ants lowered the probability of inflorescence utilization and egg laying by *Euxesta*. Some results are given in Tables 1 and 2, on the next page.

TABLE 1 **Differential response by the predator fly *Euxesta* sp. to inflorescences of *Costus woodsonii* with ants present (*controls*) or excluded (*treatments*)**

Category	Controls	Treatments
Number of observations of *Euxesta* sp. on *Costus woodsonii* inflorescences:		
Dry season	477	891
Wet season	128	872
Number of ovipositions by *Euxesta* sp.:		
Dry season	33	104
Wet season	4	69

TABLE 2 **Seed production by inflorescences of *Costus woodsonii* with ants present (*controls*) or excluded (*treatments*)**

Season	Mean (range)
Dry season:	
Controls	159 (0–823)
Treatments	55 (0–261)
Wet season:	
Controls	612 (23–1058)
Treatments	183 (0–398)

(a) Show how the figures presented in Tables 1 and 2 can be used as evidence which supports the hypothesis.

(b) Which of the two ants can be considered as a superior plant protector? Give reasons for your answer.

(c) What may the marked seasonal flowering behaviour of *Costus* and ant abundance indicate?

(*9 marks*)

Cambridge Board June 1983, Paper II, No. 8

13. Indicate, by writing the letter A, B, C, or D, the **one** correct alternative in each of parts (a) to (j).

(a) The mode of nutrition which best describes all producer organisms is

A.	holozoic	C.	autotrophic
B.	heterotrophic	D.	holophytic.

(b) The statement which **best** describes predators in a grassland community is:

A. In a food chain, they occupy a higher trophic level than their prey, and are therefore smaller organisms than the prey.

B. They do not have enemies therefore, in contrast with grazers, they are not usually camouflaged.

C. They are adapted to kill their prey and therefore have prominent canine teeth.

D. They devour grazers, other predators, or animals which feed in some other way, and are therefore usually less numerous than their prey.

(c) In the pyramid of numbers represented by the following diagram,

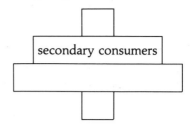

A. the primary producers are small plants

B. the primary producers are large plants

C. the biomass of the primary consumers is greater than that of the primary producers

D. the biomass of the tertiary consumers is equal to that of the primary producers.

(d) The diagram shows four groups of organisms in a food web. The arrows show the directions in which food passes between the groups.

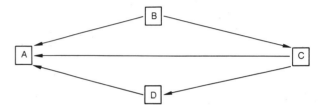

Indicate which group consists of decomposers.

(e) The structure(s) absent from the cells of higher plants is (are)

A. mitochondria

B. endoplasmic reticulum

C. centrioles

D. cell membrane.

(f) Ribosomes are dense, slightly angular structures about 15 nm in diameter and found in every type of cell. They can be clearly seen

A. with the aid of a × 20 hand lens

B. with a good light microscope

C. with a phase-contrast microscope

D. only with an electron microscope.

(g) In mammals, mitochondria are likely to be found in the greatest abundance in

A. an ovum

B. a spermatozoon

C. an epithelial cell

D. a leucocyte.

(h) One of the following statements is **not** true of cellulose and a protein:

A. both are found in the cell membrane;

B. both can be used as energy sources;

C. both are compounds of high molecular mass;

D. both are synthesized by the condensation of a number of simple units.

(i) The surface membrane of cells (the plasmalemma):
   A. Permits the free diffusion of any solute into and out of the cytoplasm.
   B. Is concerned with the active transport of water into and out of the cell.
   C. Is concerned with the active transport of some solutes against a concentration gradient.
   D. Permits the passage of large particles, by phagocytosis, into but not out of the cell.

(j) In investigations using the unicellular green alga *Acetabularia*, parts of cells from two species with different head structures were grafted together as shown in the diagrams.

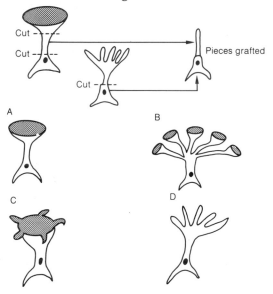

State which of the cells would be produced.

(*Total 10 marks*)

Welsh Joint Education Committee June 1984, Paper AI, No. 1

14. An investigation was made into the distribution of a particular species of aphid on mature sycamore trees. These trees occur both in exposed and sheltered conditions. The distribution of aphids on several trees is given in the table below.

| Height of leaves above ground/m | Mean number of aphids per unit area of leaf surface | | | |
| | Exposed trees | | Sheltered trees | |
	Upper surface	Lower surface	Upper surface	Lower surface
1	0	19	14	50
6	0	1	4	13
10	0	4	3	10

(a) (i) Give **three** different explanations which could account for the varied distribution of the aphids.

(ii) Give **two** precautions which must be taken to avoid errors when collecting data of this kind. (*5 marks*)

(b) Suggest **two** possible explanations for the fact that aphids were found most frequently near the veins which in sycamore leaves protrude from the leaf surface. (*2 marks*)

(*Total 7 marks*)

Joint Matriculation Board June 1982, Paper IA, No. 1

15. The diagram below shows the plan of a simple choice-chamber for investigating the responses of small invertebrates. It is composed of four plastic Petri dishes joined together and connecting by means of an aperture in each of their side walls. Chambers P and Q contain damp filter paper; chambers Q and R have been painted black to exclude light; chambers S and P are transparent.

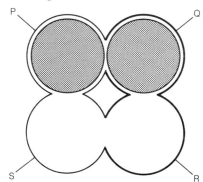

An experiment is started by placing five live woodlice in each of the four chambers and then replacing the lids. The choice-chamber is left in suitable temperature and light conditions (e.g. near a laboratory window but not in direct sunlight) for about 30 minutes.

In the table below (i)–(viii) are some hypothetical experimental results of woodlice distribution which might have been observed after half an hour. For each of these results, select one deduction from the second list A–J which could legitimately be made in order to explain the result. Write the letter of your choice of conclusion in the relevant space alongside the given result in the table. (Each letter may be used once, more than once or not at all.)

Results	P	Q	R	S	Conclusion
(i)	1	19	0	0	
(ii)	9	11	0	0	
(iii)	0	0	11	9	
(iv)	7	3	2	8	
(v)	9	0	2	9	
(vi)	0	0	4	16	
(vii)	6	5	4	5	
(viii)	0	8	11	1	

Conclusions:

Woodlice:
  A prefer to move away from darkness and towards a drier area.
  B are insensitive to humidity and show no preference for light.
  C are insensitive to humidity and show a preference for light.
  D are intolerant of high humidity and show a preference for light.
  E are insensitive to humidity and intolerant of light.
  F are insensitive to light and show a preference for high humidity.
  G are intolerant of humid conditions and show a preference for darkness.
  H are intolerant of light and show a preference for high humidity.
  I respond to dark conditions and prefer to hide there if it is damp.
  J are intolerant of high humidity and insensitive to light.

(*8 marks*)

Oxford Local June 1989, Paper I, No. 6

16. The **A, B, O** blood groups in humans are controlled by multiple alleles of a single autosomal gene. The gene locus is usually represented by the symbol **I**. There are three alleles represented by the symbols I^A, I^B and I^O. Alleles I^A and I^B are equally dominant and I^O is recessive to both.

(*a*) State all the possible genotypes of blood groups **A** and **O**.

(*b*) If a group **O** man married a group **AB** woman, state the possible blood groups that their children could have.

(*c*) (i) Explain, using the above symbols, the possible blood groups of the children whose parents are both heterozygous, the father for blood group **A** and the mother for blood group **B**.

Genotypes of parents
Genotypes of offspring
Blood groups

(ii) If two of these children are non-identical twins, what is the probability that both will have blood group **AB**?

(*Total 10 marks*)

Cambridge Board June 1988, Paper III, No. 3

17. (*a*) (i) Name a metabolic process in which ammonia is liberated from excess amino acids in mammals. (*1 mark*)

(ii) In which mammalian organ does this occur? (*1 mark*)

(iii) State *one* source, other than diet, of the amino acids involved in this process. (*1 mark*)

(iv) State *one* possible fate of the non-nitrogenous products of this process. (*1 mark*)

(*b*) (i) Name the compound produced in mammals when ammonia derived from excess amino acids is combined with carbon dioxide. (*1 mark*)

(ii) Name the metabolic pathway in which this compound is formed. (*1 mark*)

(iii) Name *two* other reactants involved in this pathway. (*2 marks*)

(*c*) (i) What is the advantage to mammals of combining ammonia with carbon dioxide? (*2 marks*)

(ii) Why is this pathway unnecessary in a freshwater teleost? (*2 marks*)

(*Total 12 marks*)

London Board June 1989, Paper I, No. 11

Index

Main entries are indicated by **bold type**